Sustainable Bioenergy Production

Edited by
Lijun Wang

CRC Press
Taylor & Francis Group
Boca Raton London New York

CRC Press is an imprint of the
Taylor & Francis Group, an **informa** business

CRC Press
Taylor & Francis Group
6000 Broken Sound Parkway NW, Suite 300
Boca Raton, FL 33487-2742

Printed on acid-free paper
Version Date: 20131217

International Standard Book Number-13: 978-1-4665-0552-0 (Hardback)

Library of Congress Cataloging-in-Publication Data

Sustainable bioenergy production / editor, Lijun Wang.
pages cm
Summary: "The energy, water, land and natural resources have been under great pressure to support increasing world population. The availability of fossil energy has been declining worldwide. It is essential to change the global fossil-dependent economy into a sustainable biobased economy. Environmental concern associated with fossil-based energy is another key driver for the emerging bioenergy industry. The bioenergy production from all renewable biological resources such as energy crops, crop residues, forest residues, animal wastes and food processing wastes influence all human activities and environmental sustainability"-- Provided by publisher.
Includes bibliographical references and index.
ISBN 978-1-4665-0552-0 (hardback)
1. Biomass energy. I. Wang, Lijun, 1971 March 20-

TP339.S865 2014
662'.88--dc23 2013044330

Visit the Taylor & Francis Web site at
http://www.taylorandfrancis.com

and the CRC Press Web site at
http://www.crcpress.com

Contents

SECTION III Sustainable Bioenergy Processing

SECTION IV Sustainable Biorefinery Systems

Preface

Energy, water, land, and natural resources are under great constraints to support increasing demands from the growing population worldwide. The availability of fossil energy has been declining worldwide. It is essential to change the global fossil fuel–dependent economy into a sustainable biobased economy. Environmental concern associated with fossil-based energy is another key driver for the emerging bioenergy industry. Bioenergy production from renewable biological resources such as energy crops, crop residues, forest residues, animal wastes, and food processing wastes influence human activities and environmental sustainability. Fertile land is a critical resource for food production. However, more and more cropland is being used to supply biomass for the production of bioenergy, thereby depleting and eroding such resources. The production and processing of energy crops may require large amounts of fossil energy–dependent fertilizers and pesticides as well as heat and power. Therefore, it is essential to assess the environmental impacts of biomass feedstock production and bioenergy processing toward the final development of sustainable bioenergy production systems. Sustainability is a key principle in biomass production, bioresource management, and bioenergy product development. The environmental sustainability of a bioenergy production system depends on the use of (1) the total energy from both fossil and renewable sources, (2) water, and (3) land. It involves operational efficiency, minimization of environmental impacts, economic viability, and social acceptability. This book, *Sustainable Bioenergy Production*, provides comprehensive knowledge and skills on the analysis and design of sustainable biomass production, bioenergy processing, and biorefinery systems to professionals in the field of bioenergy. It focuses on topics vital to the sustainability of the bioenergy industry, pinpoints the trends in future research and development, and assembles essential, complete references and data to address sustainability in the bioenergy industry.

This book will cover four key topics: (1) fundamentals of engineering analysis and design of bioenergy production systems, (2) sustainable biomass production and supply logistics, (3) sustainable bioenergy processing, and (4) sustainable biorefinery systems. Section I covers the fundamentals of genetic engineering, novel breeding, and cropping technologies applied in the development of energy crops. Modern computational tools have been developed to assist the design and analysis of bioenergy production systems. These are covered in this section. The life cycle assessment that is used to evaluate the environmental sustainability of biomass production and bioenergy processing technologies is addressed. Section II focuses on the analysis of the technical and economic feasibility and environmental sustainability of different biomass feedstocks. Emerging technologies to improve the sustainability of each biomass feedstock are discussed. Section III focuses on the analysis of the technical and economic feasibility and environmental sustainability of different bioenergy processing technologies. Emerging technologies to improve the sustainability of each bioenergy process are discussed. A biorefinery is a processing plant where biomass feedstocks are converted and extracted into a spectrum of valuable products. Section IV discusses the design and analysis of biorefineries and different biorefinery systems, including lignocellulosic feedstock, whole-crop, and green biorefinery.

This book has been written as an essential and comprehensive reference source for engineers and technologists in the bioenergy industry and for researchers in universities and research institutions.

It can also be used as a textbook for senior level undergraduate students and for graduate students in science and engineering who are interested in the development of bioenergy products and their corresponding processes.

MATLAB® is a registered trademark of The MathWorks, Inc. For product information, please contact:

The MathWorks, Inc.
3 Apple Hill Drive
Natick, MA 01760-2098 USA
Tel: 508-647-7000
Fax: 508-647-7001
E-mail: info@mathworks.com
Web: www.mathworks.com

Editor

Dr. Lijun Wang is an associate professor of biological engineering at North Carolina Agricultural and Technical State University, Greensboro, North Carolina. He has 20 years' experience as a process engineer working in the fields of bioenergy and energy efficiency technologies. His current research interests are in bioenergy processing technologies, waste-to-energy technologies, and computational process engineering. Dr. Wang regularly presents his research at academic meetings and publishes journal articles. He has published 2 books, 17 book chapters, and over 55 peer-reviewed journal articles and has given more than 60 conference presentations. He wrote the book entitled *Energy Efficiency and Management in Food Processing Facilities*, which was published by Taylor & Francis Group in 2008. Dr. Wang currently teaches undergraduate and graduate level courses in bioprocess engineering and advises graduate students in energy and environmental systems, as well as in computational engineering. He has served as a consultant in the field of process engineering to several corporations. He has been invited to attend 23 proposal review panels in the fields of bioenergy and environmental sustainability for the National Scientific Foundation, Department of Agriculture and Environmental Protection Agency in the United States. Dr. Wang has studied and worked in various universities in Asia, Europe, and North America. He earned his BS in chemical engineering from Zhengzhou University, China, MS in chemical engineering from South China University of Technology, and PhD in agricultural and food engineering from the National University of Ireland, Dublin. Dr. Wang is a licensed professional engineer in North Carolina and Nebraska and a certified energy manager with the Association of Energy Engineers. He is a member of the American Institute of Chemical Engineers and the American Society of Agricultural and Biological Engineers.

Contributors

Samuel Asomaning Agyemang
Department of Computational Science and Engineering
North Carolina Agricultural and Technical State University
Greensboro, North Carolina

Javaid Akhtar
Centre for Coal Technology
University of the Punjab
Lahore, Pakistan

Helena M. Amaro
Laboratory of Biodiversity of Aquatic Ecosystems
CIIMAR—Interdisciplinary Centre for Marine and Environmental Research
Porto, Portugal

Nor Aishah Saidina Amin
Faculty of Chemical Engineering
Universiti Teknologi Malaysia
Johor Bahru, Malaysia

Karin Arens
Department of Paper and Bioprocess Engineering
College of Environmental Science and Forestry
State University of New York
Syracuse, New York

Lorenzo Barbanti
Department of Agricultural Sciences (DipSA)
University of Bologna
Bologna, Italy

Pushpak Bhandari
Department of Biosystems and Agricultural Engineering
Oklahoma State University
Stillwater, Oklahoma

Rabah Boukhanouf
Faculty of Engineering
Department of Architecture and Built Environment
University of Nottingham
Nottingham, United Kingdom

Igathinathane Cannayen
Department of Agricultural and Biosystems Engineering
North Dakota State University
Fargo, North Dakota

Carlos A. Cardona
Department of Chemical Engineering
Institute of Biotechnology and Agribusiness
National University of Colombia at Manizales
Manizales, Colombia

Dan Cheng
Department of Energy and Environmental Systems
North Carolina Agricultural and Technical State University
Greensboro, North Carolina

Suani T. Coelho
Brazilian Reference Center on Biomass
Institute of Electrotechnics and Energy
University of Sao Paulo
Sao Paulo, Brazil

Chenyu Du
School of Biosciences
University of Nottingham
Loughborough, United Kingdom

Mario Richard Eden
Department of Chemical Engineering
Auburn University
Auburn, Alabama

Jose Goldemberg
Brazilian Reference Center on Biomass
Institute of Electrotechnics and Energy
University of Sao Paulo
Sao Paulo, Brazil

Patricia Guardabassi
Brazilian Reference Center on Biomass
Institute of Electrotechnics and Energy
University of Sao Paulo
Sao Paulo, Brazil

Reinout Heijungs
Institute of Environmental Sciences
Leiden University
Leiden, the Netherlands

and

Department of Econometrics and Operations Research
VU University Amsterdam
Amsterdam, the Netherlands

Yaohuan Huang
State Key Laboratory of Resources and Environmental Information Systems
Institute of Geographical Sciences and Natural Resources Research
Chinese Academy of Sciences
Beijing, People's Republic of China

Dong Jiang
State Key Laboratory of Resources and Environmental Information Systems
Institute of Geographical Sciences and Natural Resources Research
Chinese Academy of Sciences
Beijing, People's Republic of China

Selhan Karagoz
Department of Chemistry
Karabük University
Karabük, Turkey

Richa Kothari
Department of Environmental Science
Babasaheb Bhimrao Ambedkar University
Lucknow, India

Apostolis A. Koutinas
Department of Food Science and Human Nutrition
Agricultural University of Athens
Athens, Greece

Ajay Kumar
Department of Biosystems and Agricultural Engineering
Oklahoma State University
Stillwater, Oklahoma

Virendra Kumar
Department of Environmental Science
Babasaheb Bhimrao Ambedkar University
Lucknow, India

Dennis Y.C. Leung
Department of Mechanical Engineering
The University of Hong Kong
Pokfulam, Hong Kong, People's Republic of China

Carol Lin
School of Energy and Environment
City University of Hong Kong
Kowloon, Hong Kong, People's Republic of China

Hao Liu
Faculty of Engineering
Department of Architecture and Built Environment
University of Nottingham
Nottingham, United Kingdom

Shijie Liu
Department of Paper and Bioprocess Engineering
College of Environmental Science and Forestry
State University of New York
Syracuse, New York

Zheng Liu
Department of Chemical Engineering
Auburn University
Auburn, Alabama

F. Xavier Malcata
LEPABE—Laboratory for Engineering of Processes, Environment, Biotechnology and Energy
Department of Chemical Engineering
University of Porto
Porto, Portugal

Michael Mandl
Institute for Water, Energy and Sustainability Research
Joanneum Research
Graz, Austria

Jonathan Moncada
Department of Chemical Engineering
Institute of Biotechnology and Agribusiness
National University of Colombia at Manizales
Manizales, Colombia

Plinio M. Nastari
Brazilian Reference Center on Biomass
Institute of Electrotechnics and Energy
University of Sao Paulo
Sao Paulo, Brazil

N.L. Panwar
Department of Renewable Energy Sources
College of Engineering and Technology
Maharana Pratap University of Agriculture and Technology
Rajasthan, India

Liangcai Peng
National Key Laboratory of Crop Genetic Improvement
and
Biomass and Bioenergy Research Centre
and
College of Plant Science and Technology
and
College of Life Science and Technology
Huazhong Agricultural University
Wuhan, People's Republic of China

Lucas Reijnders
Institute for Biodiversity and Ecosystem Dynamics
University of Amsterdam
Amsterdam, the Netherlands

William L. Rooney
Plant Release Committee
Texas A&M University
College Station, Texas

Marty R. Schmer
Agricultural Research Service
United States Department of Agriculture
Agroecosystem Management Research Unit
University of Nebraska-Lincoln
Lincoln, Nebraska

Sergio O. Serna-Saldivar
Department of Biotechnology and Food Engineering
Monterrey Institute of Technology
Monterrey, Mexico

Abolghasem Shahbazi
Department of Natural Resources and Environmental Design
and
Department of Chemical, Biological and Bioengineering
North Carolina Agricultural and Technical State University
Greensboro, North Carolina

Ashokkumar Sharma
Department of Biosystems and Agricultural Engineering
Oklahoma State University
Stillwater, Oklahoma

H.S. Shekhar Sharma
Plant Health and Environmental Protection Branch
Agri-Food and Biosciences Institute
Belfast, Northern Ireland

Birger Solberg
Department of Ecology and Natural Resource Management
Norwegian University of Life Sciences
Ås, Norway

I. Sousa-Pinto
Laboratory of Biodiversity of Aquatic Ecosystems
CIIMAR—Interdisciplinary Centre for Marine and Environmental Research
and
Faculty of Sciences
Department of Biology
University of Porto
Porto, Portugal

Luis A. Toro
Department of Mathematics and Statistics
Institute of Biotechnology and Agribusiness
National University of Colombia at Manizales
and
Department of Physics and Applied Mathematics
Autonomous University of Manizales
Manizales, Colombia

V.V. Tyagi
Department of Physics
Manav Rachna College of Engineering
Haryana, India

Hui Wang
Department of Natural Resources and Environmental Design
North Carolina Agricultural and Technical State University
Greensboro, North Carolina

Lijun Wang
Department of Natural Resources and Environmental Design
and
Department of Chemical, Biological and Bioengineering
North Carolina Agricultural and Technical State University
Greensboro, North Carolina

Colin Webb
School of Chemical Engineering and Analytical Science
University of Manchester
Manchester, United Kingdom

Edi Iswanto Wiloso
Institute of Environmental Sciences
Leiden University
Leiden, the Netherlands

and

Research Center for Chemistry
Indonesian Institute of Sciences (LIPI)
Puspiptek, Indonesia

Xuan Wu
Department of Mechanical Engineering
The University of Hong Kong
Hong Kong, People's Republic of China

Guosheng Xie
National Key Laboratory of Crop Genetic Improvement
and
Biomass and Bioenergy Research Centre
and
College of Plant Science and Technology
Huazhong Agricultural University
Wuhan, People's Republic of China

Bo Zhang
Department of Natural Resources and Environmental Design
North Carolina Agricultural and Technical State University
Greensboro, North Carolina

and

School of Chemical Engineering and Pharmacy
Wuhan Institute of Technology
Wuhan, Hubei, People's Republic of China

Dafang Zhuang
State Key Laboratory of Resources and Environmental Information Systems
Institute of Geographical Sciences and Natural Resources Research
Chinese Academy of Sciences
Beijing, People's Republic of China

Section I

Fundamentals of Engineering Analysis and Design of Bioenergy Production Systems

1 Genetic Engineering of Bioenergy Crops toward High Biofuel Production

Guosheng Xie and Liangcai Peng

CONTENTS

1.1 INTRODUCTION

Fossil fuels have been widely used as a major energy source, which represents more than 85% of the total energy consumption for economic and social development around the world (Carroll and Somerville 2009). However, global consumption of fossil fuels and their impact on climate change due to greenhouse gas emissions have inspired the development of renewable and sustainable bioenergy for transportation fuels and industrial chemicals (Himmel et al. 2007; Solomon 2010). Bioenergy is considered as a renewable energy derived from biological sources that can be used for heat, electricity, fuel, and chemical products (Yuan et al. 2008; Himmel and Bayer 2009). Globally, land plants can produce approximately 200 billion tons of biomass per year, 70% of which is hypothesized to derive from plant cell walls, an unprecedented resource for biofuel production (Pauly and Keegstra 2008). Over the past years, the first generation of bioenergy systems, including starch- or sugar-based ethanol or plant oil–derived biodiesel, has already made a relatively small but significant contribution to global energy supplies. Technologies for the production of bioethanol from starch and biodiesel from lipid have become matured worldwide (Demirbas 2008; Soccol et al. 2010). However, the limited availability of resources and nonsustainability of food-based bioethanol, as well as their competitions with food supply, have triggered a paradigm shift toward the production of lignocellulosic bioenergy (Campbell et al. 2008). The second generation of biofuels such as the ethanol produced from lignocellulosic biomass has received increasing research and development interest (Goh et al. 2010; Mabee and Saddler 2010).

The production of lignocellulosic ethanol involves three major processes: biomass pretreatment, lignocellulose enzymatic digestibility, and sugar fermentation. Biomass recalcitrance, which is resistant to the pretreatment and enzymatic deconstruction, makes the current biomass process unacceptably expensive (Akin 2007; Himmel et al. 2007). As such, considerable efforts have been made recently to increase the lignocellulosic conversion rate (Li et al. 2013a; Zhang et al. 2013). In spite of these efforts, difficulty remains with biomass pretreatment and enzymatic degradation. Although the extreme pretreatment conditions, such as strong acid/base or extreme temperature/pressure, have been widely used for biomass pretreatment, they are costly and may cause environmental pollution. As biomass recalcitrance is essentially determined by the composition of plant cell wall and the interlinking of wall polymers, the selection of energy crops by genetic modification of plant cell wall is hence considered to be a promising solution to this problem. Although plant cell walls are mainly composed of cellulose, hemicelluloses, and lignin, their structural diversity and functional complexity make genetic engineering of energy crops difficult. This chapter describes major grass plants as desirable energy crop candidates and discusses various potential approaches for the genetic breeding thereof.

1.2 ENERGY CROPS AND BIOMASS RESOURCES

Lignocellulosic biomass from agricultural biological resources has been regarded as the primary source of biofuels and other value-added products (Carroll and Somerville 2009; Pauly and Keegstra 2010). Sustainable biofuel production depends on the selection of energy crops with high biomass production under various environmental conditions. Hence, energy crops maximize biomass yield per acre with minimum inputs (i.e., irrigation, fertilizer, and pesticides) and exhibit value-added traits that enhance their use as biofuel feedstock. In addition, the quality of energy crops is determined by their structural characteristics that are favorable for the harvest and processing for bioethanol and other chemical products.

Desirable energy crops can be grown on second-tier agricultural lands because the production of biofuels should not compete with the food supply on arable soils in the long run. Accordingly, energy crops can be divided into three groups based on their biomass composition: sugar- and starch-rich plants (e.g., cassava and sugarcane), lipid-rich plants (e.g., rapeseed, sunflower, and oil palm), and cellulose-rich plants (e.g., poplar, eucalyptus, and grasses). For instance, a number of C_4 grass species have been found to generate high yields of biomass with minimal inputs, such as giant *Miscanthus* (*Miscanthus giganteus* L.), switchgrass (*Panicum virgatum* L.), and sweet sorghum (*Sorghum bicolor* L.).

Perennial energy plants and annual food crops (e.g., rice, wheat, maize, and sweet sorghum) are the most common sources for bioenergy production worldwide (Ragauskas et al. 2006; Carpita and McCann 2008; Taylor 2008; Hodgson et al. 2011). *Miscanthus* (Dohleman and Long 2009) and switchgrass (Schmer et al. 2008) have been the focus for many research and development activities in Europe and the United States. *Miscanthus* and sweet sorghum species are currently considered as leading energy feedstock candidates in China (Xie and Peng 2011). On the other hand, rice, wheat, and maize are major food crops that also provide approximately 75% biomass resources of total agricultural residues in China and elsewhere. Potential energy food crops should maintain a high grain yield and good quality for the food supply with easy destruction of cell walls in their straws/stalks for biofuel production. Compared with food crops, energy plants should have their advantages for growth in different geographical regions and exhibit high use efficiency of resources such as light and water.

In order for large amounts of biomass supply, genetic breeding and biotechnology can be applied to increase the biomass yield in food crops and energy plants as shown in Table 1.1. The selection of energy crops from natural germplasm resources and genetic mutagenesis pools is considered as an initial approach. On the whole, genetic breeding strategies include traditional breeding and related molecular approaches such as the use of genetic markers and genome mapping. Hence, the genetic engineering of energy crops is essential to the discovery of key genes associated with plant growth traits for increasing biomass quantity and enhancing plant growth and resistance to various environmental stresses.

TABLE 1.1
List of the Potential Representative Energy Plants and Energy Crops

Species	Photosynthetic Type	Genome Size (Mbp)	Biomass Yield (Tons/ha Year)	Bioethanol Yield (Tons/ha Year)
Switchgrass	C_4	1,372–1,666	14–20	4–5
Miscanthus	C_4	4,300–6,800	60–80	6–8
Rice	C_3	430	15–30	1.5
Wheat	C_3	16,000	15–30	1.5
Maize	C_4	2,500	15–45	1.5
Sweet sorghum	C_4	740	60–80	4–6

Source: Data modified from Xie, G. and Peng, L., *J. Integr. Plant Biol.*, 53(2), 143, 2011.

1.3 PLANT CELL WALL AND BIOMASS RECALCITRANCE

Plants consist of many types of cells that form typical primary and secondary cell walls. Plant cell walls significantly differ in the chemical compositions and structural features of monocots and dicots as shown in Table 1.2 (Grabber et al. 2000; Vogel 2008; Lionetti et al. 2010; Scheller and Ulvskov 2010). There are two typical primary cell walls in plants: Type I comprises cellulose microfibrils cross-linked with xyloglucans and embedded in a complex matrix of pectins and proteins (Whitney et al. 1999), and type II is rich in glucuronoarabinoxylans (GAXs) and mixed-linkage glucans (MLGs) (Carpita 1984; Burton and Fincher 2012). In secondary cell walls, less branched or unbranched hemicelluloses are interlinked with the surface of cellulose fibrils via hydrogen bonds, whereas the side chains of hemicelluloses are covalently bonded with lignin to create enzyme-impenetrable cross-linking networks. Lignin–carbohydrate complexes are also thought to exclude water and prevent chemical or enzyme-catalyzed deconstruction of cell walls, predominantly determining the biomass recalcitrance (Chundawat et al. 2011). To date, the knowledge about the fine structures of plant cell walls is very limited, which hinders

TABLE 1.2
Components (% DW) of Cell Wall in Grass and Dicot Plants

	Grass Cell Wall		Dicot Cell Wall	
Components	Primary	Secondary	Primary	Secondary
Cellulose	20–30	35–45	15–30	45–50
Xylans	20–40	40–50	5	20–30
MLG	2–15	Minor	Absent	Absent
XyG	2–5	Minor	20–25	Minor
(Gluco)mannan	2	0–5	3–15	2–5
GAX	20–40	40–50	5	Absent
Pectins	5	0.1	20–35	0.1
Structural proteins	1	Minor	10	Minor
Phenolics	1–5	0.5–1.5	Minor	Minor
Lignin	Minor	20	Minor	7–10

Sources: Referred from Vogel, J., *Curr. Opin. Plant Biol.*, 11(3), 301, 2008; Scheller, H.V. and Ulvskov, P., *Annu. Rev. Plant Biol.*, 61, 263, 2010.

the identification of the crucial factors of biomass recalcitrance (Xie and Peng 2011; Huang et al. 2012; Li et al. 2013b; Zhang et al. 2013).

Genetic modification of plant cell walls in energy crops is considered a promising solution to recalcitrance. The discovery of appropriate genes is an initial step, and potential genetic engineering approaches should be considered in energy crops (Chen and Dixon 2007; Ambavaram et al. 2011). Such approaches involve altering wall polymer cross-linking, reducing lignocellulose crystallinity and lignin levels, increasing hemicellulose contents, and adding foreign cellulase enzymes and/or other components into cell walls (Weng et al. 2008; Xie and Peng 2011).

1.4 SELECTION OF NATURAL GERMPLASM RESOURCES

Perennial grasses, such as switchgrass and *Miscanthus*, are rich in natural germplasm resources. They have high efficiencies of using water and nutrient and are characterized by their wide geographic distribution and high biomass productivity. The selection of natural germplasm resources is considered as an initial and essential step in energy crop breeding. It not only can determine valuable genetic materials for energy crop breeding but also may directly select energy plants for biofuel production (Xie and Peng 2011).

Switchgrass (*Panicum virgatum* L.), which is a C_4 perennial forage grass, is currently being evaluated as a biofuel crop in North America and Europe. It has two distinct ecotypes: lowland tetraploid plants (2n=4x=36) with a tall coarse stature and exceptional biomass and upland octoploid plants (2n=8x=72) with a short stature and low biomass yield (Bouton 2007). Owing to the large variability within available switchgrass populations with respect to important agronomic traits, such as tillering, biomass digestibility, and biomass production, this important forage and biofuel crop holds great potential for germplasm improvement. As biomass yields of switchgrass vary greatly depending on cultivar, year, and locations, four switchgrass genotypes with similar cellulose and hemicellulose composition but significantly different lignin contents, especially in the ratio of lignin monomers and considerable levels of *p*-coumaric acid and ferulic acid, have been selected. In addition, the full genome sequence of switchgrass will allow us to use marker-assisted breeding to improve biomass yield and quality.

Miscanthus, which is another important C_4 perennial plant, has the highest biomass yield among grass plants. It was primarily originated from East Asia and the nearby Pacific islands, and 14 species of which have been identified to date. More than 1000 natural *Miscanthus* sp. accessions, including four major species (*Miscanthus sacchariflorus*, *Miscanthus lutarioriparius*, *Miscanthus sinensis*, and *Miscanthus floridulus*), have been collected in China (Xie and Peng 2011). Each species represents different ecological and geographical types, making it a diverse germplasm resource. Two hundred representative accessions with diverse cell wall composition and wide biomass digestibility have been previously determined to select possible *Miscanthus* species as energy plants (Xie and Peng 2011; Huang et al. 2012). Due to their distinct biomass degradation, several *Miscanthus* materials were selected as energy plants for biofuel production in this study as shown in Table 1.3. In addition, a genetic model based on the integrative analysis between cell wall composition and biomass digestibility is herein proposed for the potential genetic breeding of *Miscanthus* as biofuel feedstock.

1.5 SCREENING OF PLANT CELL WALL MUTANTS

During evolution, plants have to construct their typical cell walls to complete their life cycles rather than meet the requirements for biofuel production. Due to the diversity of plant cell wall structures and the complexity of their functions, the direct genetic modification of plant cell walls can unexpectedly lead to the alterations in growth and development of plant cell (Torney et al. 2007; Vega-Sánchez and Ronald 2010). Alternatively, large-scale screening of plant cell wall mutants is considered as practicable work. Mutagenesis of energy crops can be conducted by means of T-DNA

TABLE 1.3
Cell Wall Composition and Degradability of *Miscanthus* Ecotypes under Different Pretreatments[a]

	Cell Components (%)			Heat Degradation Efficiency (%)		Alkaline Degradation Efficiency (%)		Acid Degradation Efficiency (%)	
Ecotype	Cellulose	Hemicellulose	Lignin	C6-Sugar	C5-Sugar	C6-Sugar	C5-Sugar	C6-Sugar	C5-Sugar
MI10	28.25	38.74	33.01	19.07	3.87	47.07	24.54	31.98	31.31
MI108	33.33	38.11	28.56	8.59	2.60	45.86	26.06	30.61	33.96
MI1	44.50	27.95	27.56	3.29	1.24	21.78	14.08	14.56	24.63

[a] The degradability was subjective to hexoses/pentoses (percentage of total cell walls) released from both pretreatment and 0.4% cellulase digestion (Angel Comp. Limited, China). Heat degradation, the powders (40 mesh) of mature stem were heated at 121°C for 20 min, and the remaining residues incubated with cellulase for 48 h; alkaline/acid degradation, the powders treated either with 1% (w/w) NaOH for 2 h at 50°C or with 1% H_2SO_4 for 20 min at 121°C, and the remaining residues incubated with cellulase as just described.

insertion knockout, transposon insertion, as well as chemical (e.g., ethyl methanesulfonate-induced) and physical (e.g., cobalt-60 irradiation) treatments. The ideal cell wall–altered mutants should be able to maintain normal plant growth and similar grain and biomass yields but show enhanced biomass digestibility as compared with the wild types. Hence, these mutants can be directly used as energy crops for the production of biofuels. On the other hand, their identification would provide novel genes for genetic manipulation of such energy crops.

The selection of mutants is specifically helpful for the genetic breeding of energy food crops due to their inadequate natural germplasm resources. Rice, wheat, and maize provide major food sources worldwide. However, the enormous amounts of biomass residues from food crops have not been optimized for biofuel production. Their biomass is principally composed of cellulose (30%–45%), hemicelluloses (20%–30%), and lignin (25%–35%) (Pauly and Keegstra 2008, 2010). Through the genome sequencing of two rice cultivars, functional genomics and several microarray platforms have been established (Matsumura et al. 1999; Gowda et al. 2004; Nobuta et al. 2007). Rice transformation has been advanced using methods such as gene silencing and overexpression techniques (Miki and Shimamoto 2004; Nishimura et al. 2006). The variation-induced collections of tagged lines (Tos17, Ac/Ds, and T-DNA) and chemically or physically induced mutant lines (Hirochika et al. 2004; Zhang et al. 2006) have facilitated the discovery of new genes that determine cell wall properties. Recently, large-scale rice mutants have been investigated as an energy and food crop candidate (Xie and Peng 2011). Distinct from previously identified cell wall mutants that show abnormal phenotypes, such as dwarfism, irregular xylem (IRX), and even lethality (Goubet et al. 2003; Desprez et al. 2007), most of the rice mutants selected have exhibited agronomic traits and grain yields similar to those in the wild type but with a remarkable alteration in cell wall composition and structure. Under various mild physical and chemical pretreatments, several elite mutants demonstrated enhanced biomass enzymatic digestibility as shown in Table 1.4.

In maize, the uniform mutator (Mu) population has been developed by introgressing Robertson's Mu, an insertional mutation element, into established maize inbreds W22 and B73 (McCarty et al. 2005; Settles et al. 2007), and a database has been established to facilitate high-throughput molecular analysis of Mu-tagged mutants and gene knockouts (http://www. maizegdb.org/documentation/uniformmu/). The project of cell wall genomics has successfully identified a set of mutants in which cell wall biosynthesis–related genes are disrupted by a Mu element. On the other hand, four spontaneous brown midrib (*bm*) mutants in maize were recognized by the reddish-brown coloration of the vascular tissue in the leaf blade and sheath. The *bm1* mutation affects the lignin biosynthetic enzyme cinnamyl alcohol dehydrogenase (CAD) (Halpin et al. 1998), the *bm2* mutant contains

TABLE 1.4
Cell Wall Composition and Digestibility of Rice and Maize Mutants under Different Pretreatments[a]

	Cell Components (%)			Heat Degradation Efficiency (%)		Alkaline Degradation Efficiency (%)		Acid Degradation Efficiency (%)	
Mutants	**Cellulose**	**Hemicellulose**	**Lignin**	**C_6-Sugar**	**C_5-Sugar**	**C_6-Sugar**	**C_5-Sugar**	**C_6-Sugar**	**C_5-Sugar**
Rice									
RG53	34.13	33.73	32.14	77.0	7.20	75.40	23.50	68.30	30.00
RC2	38.81	37.44	23.76	21.80	4.80	55.00	29.60	42.80	32.50
RG37	46.67	29.28	24.05	22.00	4.70	37.10	21.90	35.10	25.90
WT	46.47	28.27	25.26	24.90	4.40	56.10	19.40	43.40	20.40
Maize									
CM10	31.12	34.64	34.24	37.40	7.40	57.10	29.10	47.80	29.80
CM9	41.30	29.01	29.69	23.10	6.40	47.50	25.60	39.60	26.70
WT	43.08	28.68	28.24	30.40	6.00	52.70	28.80	39.30	26.40

[a] The degradation efficiency was subjective to hexoses/pentoses (percentage of total cell walls) released from both pretreatment and 0.4% cellulase digestion (Angel Comp. Limited, China). Heat degradation, the powders (40 mesh) of mature stem were heated at 121°C for 20 min, and the remaining residues incubated with cellulase for 48 h; alkaline/acid degradation, the powders treated either with 1% (w/w) NaOH for 2 h at 50°C or with 1% H_2SO_4 for 20 min at 121°C, and the remaining residues incubated with cellulase as just described.

fewer guaiacyl and syringyl residues (Vermerris and Boon 2001), the *bm3* gene encodes the enzyme caffeic acid/5-hydroxyferulic acid *O*-methyltransferase (COMT) (Vignols et al. 1995), and the *bm4* mutant shares similar chemical composition with the *bm2* mutant (Barriere et al. 2004). Hydrolysis of stovers from *bm1* to *bm3* in an inbred A619 background resulted in enhanced saccharification, compared with the wild type. Several good maize mutant lines have been selected with increased biomass degradability in our laboratory as shown in Table 1.4.

Sweet sorghum (*Sorghum bicolor* L.) is a breeding line from the ordinary grain sorghum species. The stalk contains 17%–21% sugar content, and the biomass yield is 60–80 Mg ha^{-1} (Carpita and McCann 2008). The *bm* mutants were first isolated by chemical mutagenesis more than 30 years ago, and spontaneous mutants were identified since then. Most *bm* mutants were found to have an altered cell wall composition, particularly in lignin composition, and enhanced biomass saccharification (Bout and Vermerris 2003). Recently, dozens of ethyl methanesulfonate-induced mutants have been selected to show a pattern of plant growth and grain yield similar to that in the wild type. However, several mutants with high levels of soluble sugar and cell wall components in the stove and efficient biomass digestibility have been detected (Xie and Peng 2011).

1.6 GENETIC MODIFICATION OF PLANT CELL WALLS

1.6.1 Genes Involved in Cellulose Biosynthesis and Assembly

Cellulose biosynthesis has been extensively studied in higher plants. Generally, cellulose is synthesized at the plasma membrane by an enzymatic complex and deposited directly into the cell wall in a directional manner (Somerville 2006; Mutwil et al. 2008; Taylor 2008), undergoing a dynamic reorientation followed by the deposition that enables its anisotropic expansion (Anderson et al. 2010). Cellulose synthase (CESA) isoforms form rosette hexameric complexes, which presumably consist of 36 individual CESA proteins and some other accessory proteins. The CESA complexes are assembled in the Golgi apparatus and then exported to the plasma membrane via exocytosis, whereas the glucose residues come from UDP-glucose molecules that are present in the cytosol. Three major steps are associated with the polymerization of UDP-glucose and formation of glucan chains: chain initiation, chain elongation, and chain termination (Peng et al. 2002). The deposition of cellulose is oriented by an interaction between CESA protein and microtubules (Baskin 2001; Lane et al. 2001; Pagant et al. 2002). It is believed that the rosette CESA protein complexes move in the plasma membrane in a direction that is defined by cortical microtubules, thereby producing cellulose microfibrils that are deposited into the extracellular wall matrix at a location adjacent to the plasma membrane (Paredez et al. 2006; Li et al. 2013a). Many proteins may have a role in linking cortical microtubules to a membrane-associated scaffold in cellulose deposition. However, the precise roles of these proteins remain unclear (Somerville 2006; Bringmann et al. 2012; Li et al. 2012).

Genes from CESA family have been annotated in plants (Holland et al. 2000; Wang et al. 2010). In rice, OsCESA1, OsCESA3, and OsCESA8 form the CESA complex typically for primary cell wall biosynthesis, whereas OsCESA4, OsCESA7, and OsCESA9 are characterized for secondary cell wall formation (Tanaka et al. 2003; Wang et al. 2010). In maize, *ZmCESA1_9* is associated with cellulose synthesis in primary cell walls, whereas *ZmCESA10_12* is involved in secondary cell wall synthesis. In addition, several non-CESA proteins, such as KORRIGAN (KOR), COBRA, and KOBITO, have been reported to be involved in cellulose synthesis and modification (Pagant et al. 2002; Bhandari et al. 2006; Brady et al. 2007; Xie et al. 2013). The KOR proteins encode a membrane-bound β-1,4-glucanse (Nicol et al. 1998), but its precise role remains to be elucidated (Peng et al. 2002). The rice mutant brittle culm1 is a putative ortholog of AtCOBL4, which shows reduced level and an improper orientation of crystalline cellulose microfibrils (Li et al. 2003). COBRA families have 11 members in rice and 9 genes in maize (Brady et al. 2007). The spontaneous *Zea mays* brittle *stalk2* mutant is a putative glycosylphosphatidylinositol-anchored gene that plays a role in cellulose deposition into the cell wall (Ching et al. 2006; Sindhu et al. 2007). In addition, *KOBITO* gene mutation has been demonstrated to cause a randomized cellulose microfibril orientation in newly deposited cell wall layer (Pagant et al. 2002).

CESA-interactive protein 1 (CSI1) has been identified to interact with CESA isoforms for primary plant cell wall synthesis in *Arabidopsis*. As the CSI1 mutant affects the distribution and movement of CESA complexes in the plasma membrane, CSI1 may function as a scaffold protein for the assembly of the cellulose synthase complex (CSC) or in its interaction with microtubules (Gu et al. 2010). As UDP-glucose is the major substrate for cellulose biosynthesis in the cell wall, sucrose synthase (SuSy), sucrose phosphate synthase (SPS), invertase, and UDP-glucose pyrophosphorylase (UGPase) are important candidates involved in the regulation of carbon partitioning into cellulose and starch (Albrecht and Mustroph 2003).

Furthermore, the cell wall–related NAC, MYB, WRKY, and leucine zipper transcription factors have been partially identified to regulate cellulose biosynthesis in plants (Taylor 2008; Wang and Dixon 2012). In rice and maize, *OsSWNs* (1, 3, and 7) and *ZmSWNs* (1, 3, 6, and 7) were found to serve as the secondary wall–associated NACs (Zhong et al. 2011). Overexpression of an ATAP2 family transcription factor (*SHINE/WAX INDUCER1*) caused a 34% increase in cellulose and a 45% reduction in lignin content with an altered lignin composition, improved biomass digestibility, and no compromise in plant strength and performance (Ambavaram et al. 2011).

Enzymatic hydrolysis of biomass is synergistically catalyzed by cellulases, including endoglucanases, exoglucanases, and β-glucosidases (Mosier et al. 2005; Nguyen et al. 2010). Transgenic plants expressing exogenous microbial cellulase genes have been determined to exhibit enhanced hydrolysis activity and no side effects on plant growth and biomass yield (Himmel and Bayer 2009). Furthermore, the heterologous expression of the catalytic domain of endo-1,4-beta-D-glucanase E1 in *Acidothermus cellulolyticus* targeted to the apoplast in maize can enhance the efficiency of the conversion of the polysaccharides in corn stover into glucose (Ransom et al. 2007). In rice, the overexpression of a nonglycosylated mutant of a cellulose-bound module (CBM) from bacteria has led to a 30% increase in rumen digestibility of the rice straw compared with the wild type. In addition, expansins and expansin-like proteins are cell wall proteins involved in the extension and loosening of plant cell walls, and they have also been found to serve as endogenous CBMs in plants.

Lignocellulose crystallinity is a reportedly important parameter affecting biomass enzymatic digestibility in biofuel production (Somerville 2006; Xu et al. 2012). Cellulose crystallinity refers to the ratio of crystalline regions to noncrystalline regions in cellulose microfibrils, and it can thus be determined by the degree of polymerization (DP) of β-1,4-glucans and cross-linking style of cellulose microfibrils with hemicelluloses and lignin. Hence, it is rational to hypothesize that genetic manipulation of genes associated with cellulose biosynthesis and assembly can result in an alteration of lignocellulose crystallinity. On the other hand, the reduction in lignocellulose crystallinity may somewhat affect plant growth and fitness. To avoid such effect, inducible promoters and protein signals may be applied specifically for gene expression and protein tissue localization in energy crops.

1.6.2 Genes for Hemicellulose Biosynthesis

Hemicelluloses are synthesized in the Golgi apparatus by a plethora of glycosyltransferases (GTs) and subsequently modified by different hydrolases, esterases, and lyases (Leroxel et al. 2006). Cellulose synthase-like F (CSLF) genes encode Golgi-localized glycan synthases that are involved in hemicellulose biosynthesis. The ectopic expression of two rice *CSLF* genes (*OsCSLF2* and *OsCSLF4*) in *Arabidopsis* has confirmed the functions of CSLF proteins in MLG synthesis (Burton et al. 2006). In total, 25 xylem-specific GT genes from 7 GT families (GT2, GT8, GT14, GT31, GT43, GT47, and GT61) have been recommended for hemicellulose biosynthesis (Scheller and Ulvskov 2010). Several GT families, including GT43, GT47, and GT61, may be involved in xylan biosynthesis. The identification of *Arabidopsis* IRX mutants has implicated GT8, GT43, and GT47 families as potential glucuronoxylan (GX) biosynthetic genes (Zhong et al. 2005; Pena et al. 2007; Persson et al. 2007; Brown et al. 2009). Recent research has shown that the biosynthesis of GAXs would require at least three GTs: xylosyltransferase (XylT), arabinosyltransferase (AraT), and glucuronosyltransferase (GlcAT) (Zeng et al. 2008, 2010). Three wheat (*Triticum aestivum* L.) TaGT proteins from the GT43, GT47, and GT75 families have been identified as promising candidates involved in GAX synthesis termed as TaGT43-4, TaGT47-13, and TaGT75-4. In maize, UDP-glucose 6-dehydrogenase genes encode central enzymes of hemicellulose biosynthesis and appear to be essential for cell wall formation in young organs.

Biomass recalcitrance is collectively known as the natural resistance of plant cell walls to microbial and enzymatic deconstruction (Himmel and Bayer 2009). In nature, the types and abundance of hemicelluloses are considered to be factors of biomass recalcitrance (Gírio et al. 2010; Xu et al. 2012). As hemicelluloses are often cross-linked with lignin to form lignin–carbohydrate complexes, the degree of xylose residues substituted with arabinose residues has been considered an important parameter for the process efficiency of the cell wall (Li et al. 2013b; Zhang et al. 2013). As GAX other than MLG tightly links to lignin, MLG can be used to replace GAX by expressing CSLF and CSLH genes (Fry et al. 2008).

Feruloyl esterases play a key role in enhancing the accessibility of hydrolyzing enzymes to hemicellulose fibers by removing ferulic acid side chains and cross-linking bonds. The expression of a fungal ferulic acid esterase gene in transgenic ryegrass renders the cell walls more accessible to endoxylanases. Plant enzymes that remodel the cell wall architecture, such as expansins and xyloglucan endotransglycosylases/hydrolases (XTHs), can also be used to facilitate wall deconstruction. Expansins act to reduce hydrogen bonding by an unknown mechanism. For example, EXPB1, a β-expansin subfamily member, binds to maize cell walls, causes swelling of the maize cell wall, and facilitates the local movement and stress relaxation of arabinoxylan–cellulose networks (Yennawar et al. 2006). XTHs can generate additional modifications of the hemicellulose–cellulose networks that can potentially loosen cell wall structures. These enzymes cleave and ligate xyloglucans with new xyloglucan partners such that the wall architecture may be modified. Moreover, mannan transglycosylase/hydrolase can catalyze a transglycosylation reaction using mannan-derived oligosaccharide as a substrate. These transglycosylation activities may potentially generate novel heteropolymeric networks that impact biomass recalcitrance.

1.6.3 Genes Associated with Lignin Biosynthesis

Lignin is primarily composed of guaiacyl (G), syringyl (S), and hydroxycinnamates (H) in plant cell walls (Grabber et al. 2004). Three predominant *p*-hydroxycinnamyl alcohols (*p*-coumaryl, coniferyl, and sinapyl alcohols) form lignin monomers. Biosynthesis of lignin involves two major processes: the monolignol pathway (via the phenylpropanoid pathway) in the cytosol and polymerization of the monomers into the cell wall. In the first process, 10 enzymes are considered for monolignol biosynthesis: phenylalanine ammonia lyase (PAL), cinnamate 4-hydroxylase (C4H), 4-coumarate-CoA ligase (4CL), shikimate hydroxycinnamoyltransferase (HCT), coumarate 3-hydroxylase (C3H), caffeoyl-CoA 3-*O*-methyltransferase (CCoAOMT), cinnamoyl-CoA reductase (CCR), CAD, ferulate 5-hydroxylase (F5H), and COMT (Vogt 2010). Subsequent enzymatic steps that catalyze hydroxylations and methylations produce lignin monomers (Fry et al. 2000; Lindsay and Fry 2008). Members of the pfam PF02458 gene family have been recently found to potentially act as arabinoxylan feruloyl transferases in rice (Piston et al. 2010). The global transcriptome fingerprinting in maize has identified many gene families in lignin biosynthetic pathways (e.g., three PALs, one C4H, one 4CL, two CCoAOMTs, one CCR, one COMT, one F5H1, and one CAD) (Bosch et al. 2011).

Lignin appears to have two primary effects on biomass enzymatic hydrolysis: prohibiting cellulose fiber swelling to reduce surface area access to the cellulase enzyme and preventing cellulase action on the cellulose surface (Converse and Optekar 1993). A direct approach to reduce the side effects of lignin is to alter its content and composition by regulating lignin synthesis pathways (Vanhholme et al. 2010; Sun et al. 2013). Generally, downregulation of genes early in the monolignol biosynthetic pathway, such as PAL, C4H, HCT, and C3H, is the most effective means of reducing lignin contents. In contrast, downregulation of CCoAOMT, CCR, F5H, COMT, or CAD genes has been shown to affect their monomeric compositions (Simmons et al. 2010).

The *bm3* mutant, a well-characterized natural mutant, is defective in COMT gene and has less lignin and decreased syringyl units, resulting in improved biomass digestibility (Vignols et al. 1995). COMT maize antisense lines are also generated and exhibit a similar phenotype (Piquemal et al. 2002). The *bm1* mutant shows an altered lignin content and composition (Halpin et al. 1998). Allelic variations have also been evaluated for two CCoAOMT and COMT genes in 34 lines of maize selected for their varying digestibility (Guillet-Claude et al. 2004). In sweet sorghum, CCR catalyzes the last step of the phenylpropanoid pathway for monolignol biosynthesis. Plants contain multiple CCR-like genes, thereby complicating the rational selection of lignin-specific targets to maximize their impact. These candidate genes should be accessible to allow for genetic modifications of the content and composition of lignin in energy crops.

1.7 PERSPECTIVES

More than 1000 genes are associated with plant cell wall biosynthesis, degradation, and regulations (Yong et al. 2005; Torney et al. 2007; Vega-Sánchez and Ronald 2010). In total, 550 GT and 419 GH genes have been annotated in rice and maize (Penning et al. 2009). For the discovery of candidate genes, we can take advantage of the rice and maize mutants as described earlier and use genomic information available for the identification of the related orthologs in bioenergy crops. The close evolutionary relationship of the C_4 grasses from the Panicoid subfamily and the syntenic organization of grass genomes make it possible to determine the translation of genes that impact biomass characteristics into more genetically recalcitrant species. Notably, as major discrete transcription factors, such as NAC and MYB families, have been identified to regulate plant cell wall biosynthesis (Karpinska et al. 2004; Goicoechea et al. 2005), we may be able to increase biomass quantity and/or improve biomass quality by dynamically altering the gene expression time and level in energy grass crops. However, more genes are needed to be explored and determined in terms of the gene expression regulation during the biosynthesis of plant cell wall, which is affected by functional genotype, especially from a comparative genomics and proteomics perspective that evaluates all enzymes involved in biomass production. On the other hand, new metabolic engineering and genomics approaches hold great potential for understanding the molecular mechanism of lignocellulosic conversion into economically significant products.

1.8 CONCLUSIONS

Lignocellulose is regarded as a major biomass resource for biofuel production worldwide. Due to biomass recalcitrance, the current biofuel process is unacceptably costly. Genetic breeding of energy crops is considered a promising solution to reduce the production cost of biofuels. This chapter discussed major grass plants that can be used as potential bioenergy crops in terms of the selection of natural germplasm resources and screening cell wall mutants. Related genetic engineering strategies of cell wall were proposed to increase biomass quantity and enzymatic degradability of lignocellulosic energy crops.

REFERENCES

Akin DE. 2007. Grass lignocellulose: Strategies to overcome recalcitrance. *Applied Biochemistry and Biotechnology* 137–140(1–12):3–15.

Albrecht G and Mustroph A. 2003. Localization of sucrose synthase in wheat roots: Increased in situ activity of sucrose synthase correlates with cell wall thickening by cellulose deposition under hypoxia. *Planta* 217(2):252–260.

Ambavaram MM, Krishnan A, Trijatmiko KR, and Pereira A. 2011. Coordinated activation of cellulose and repression of lignin biosynthesis pathways in rice. *Plant Physiology* 155(2):916–931.

Anderson CT, Carroll A, Akhmetova L, and Somerville C. 2010. Real-time imaging of cellulose reorientation during cell wall expansion in *Arabidopsis* roots. *Plant Physiology* 152(2):787–796.

Barriere Y, Ralph J, Mechin V, Guillaumie S, Grabber JH, Argillier O, Chabbert B, and Lapierre C. 2004. Genetic and molecular basis of grass cell wall biosynthesis and degradability. II. Lessons from brown-midrib mutants. *Comptes Rendus Biologies* 327(9–10):847–860.

Baskin TI. 2001. On the alignment of cellulose microfibrils by cortical microtubules: A review and a model. *Protoplasma* 215(1–4):150–171.

Bhandari S, Fujino T, Thammanagowda S, Zhang D, Xu F, and Joshi CP. 2006. Xylem-specific and tension stress-responsive coexpression of KORRIGAN endoglucanase and three secondary wall-associated cellulose synthase genes in aspen trees. *Planta* 224(4):828–837.

Bosch M, Mayer CD, Cookson A, and Donnison IS. 2011. Identification of genes involved in cell wall biogenesis in grasses by differential gene expression profiling of elongating and non-elongating maize internodes. *Journal of Experimental Botany* 62(10):3545–3561.

Bout S and Vermerris W. 2003. A candidate-gene approach to clone the sorghum Brown midrib gene encoding caffeic acid O-methyltransferase. *Molecular Genetics and Genomics: MGG* 269(2):205–214.

Bouton JH. 2007. Molecular breeding of switchgrass for use as a biofuel crop. *Current Opinion in Genetics & Development* 17(6):553–558.

Brady SM, Song S, Dhugga KS, Rafalski JA, and Benfey PN. 2007. Combining expression and comparative evolutionary analysis. The COBRA gene family. *Plant Physiology* 143(1):172–187.

Bringmann M, Landrein B, Schudoma C, Hamant O, Hauser MT, and Persson S. 2012. Cracking the elusive alignment hypothesis: The microtubule-cellulose synthase nexus unraveled. *Trends in Plant Science* 17(11):666–674.

Brown DM, Zhang Z, Stephens E, Dupree P, and Turner SR. 2009. Characterization of IRX10 and IRX10-like reveals an essential role in glucuronoxylan biosynthesis in Arabidopsis. *The Plant Journal: For Cell and Molecular Biology* 57(4):732–746.

Burton RA and Fincher GB. 2012. Current challenges in cell wall biology in the cereals and grasses. *Frontiers in Plant Science* 3:130.

Burton RA, Wilson SM, Hrmova M, Harvey AJ, Shirley NJ, Medhurst A, Stone BA, Newbigin EJ, Bacic A, and Fincher GB. 2006. Cellulose synthase-like CslF genes mediate the synthesis of cell wall (1,3;1,4)-beta-D-glucans. *Science* 311(5769):1940–1942.

Campbell JE, Lobell DB, Genova RC, and Field CB. 2008. The global potential of bioenergy on abandoned agriculture lands. *Environmental Science & Technology* 42(15):5791–5794.

Carpita NC. 1984. Cell wall development in maize coleoptiles. *Plant Physiology* 76(1):205–212.

Carpita NC and McCann MC. 2008. Maize and sorghum: Genetic resources for bioenergy grasses. *Trends in Plant Science* 13(8):415–420.

Carroll A and Somerville C. 2009. Cellulosic biofuels. *Annual Review of Plant Biology* 60:165–182.

Chen F and Dixon RA. 2007. Lignin modification improves fermentable sugar yields for biofuel production. *Nature Biotechnology* 25(7):759–761.

Ching A, Dhugga KS, Appenzeller L, Meeley R, Bourett TM, Howard RJ, and Rafalski A. 2006. Brittle stalk 2 encodes a putative glycosylphosphatidylinositol-anchored protein that affects mechanical strength of maize tissues by altering the composition and structure of secondary cell walls. *Planta* 224(5):1174–1184.

Chundawat SP, Beckham GT, Himmel ME, and Dale BE. 2011. Deconstruction of lignocellulosic biomass to fuels and chemicals. *Annual Review of Chemical and Biomolecular Engineering* 2:121–145.

Converse AO and Optekar JD. 1993. A synergistic kinetics model for enzymatic cellulose hydrolysis compared to degree-of-synergism experimental results. *Biotechnology and Bioengineering* 42(1):145–148.

Demirbas A. 2008. Studies on cottonseed oil biodiesel prepared in non-catalytic SCF conditions. *Bioresource Technology* 99(5):1125–1130.

Desprez T, Juraniec M, Crowell EF, Jouy H, Pochylova Z, Parcy F, Hofte H, Gonneau M, and Vernhettes S. 2007. Organization of cellulose synthase complexes involved in primary cell wall synthesis in *Arabidopsis thaliana*. *Proceedings of the National Academy of Sciences of the United States of America* 104(39):15572–15577.

Dohleman FG and Long SP. 2009. More productive than maize in the Midwest: How does *Miscanthus* do it? *Plant Physiology* 150(4):2104–2115.

Fry SC, Nesselrode BH, Miller JG, and Mewburn BR. 2008. Mixed-linkage (1 → 3, 1 → 4)-beta-D-glucan is a major hemicellulose of *Equisetum* (horsetail) cell walls. *The New Phytologist* 179(1):104–115.

Fry SC, Willis SC, and Paterson AE. 2000. Intraprotoplasmic and wall-localised formation of arabinoxylan-bound diferulates and larger ferulate coupling-products in maize cell-suspension cultures. *Planta* 211(5):679–692.

Gírio FM, Fonseca C, Carvalheiro F, Duarte LC, Marques S, and Bogel-Lukasik R. 2010. Hemicelluloses for fuel ethanol: A review. *Bioresource Technology* 101(13):4775–4800.

Goh CS, Tan KT, Lee KT, and Bhatia S. 2010. Bio-ethanol from lignocellulose: Status, perspectives and challenges in Malaysia. *Bioresource Technology* 101(13):4834–4841.

Goicoechea M, Lacombe E, Legay S, Mihaljevic S, Rech P, Jauneau A, Lapierre C, Pollet B, Verhaegen D, Chaubet-Gigot N et al. 2005. EgMYB2, a new transcriptional activator from *Eucalyptus* xylem, regulates secondary cell wall formation and lignin biosynthesis. *The Plant Journal: For Cell and Molecular Biology* 43(4):553–567.

Goubet F, Misrahi A, Park SK, Zhang Z, Twell D, and Dupree P. 2003. AtCSLA7, a cellulose synthase-like putative glycosyltransferase, is important for pollen tube growth and embryogenesis in *Arabidopsis*. *Plant Physiology* 131(2):547–557.

Gowda M, Jantasuriyarat C, Dean RA, and Wang GL. 2004. Robust-LongSAGE (RL-SAGE): A substantially improved LongSAGE method for gene discovery and transcriptome analysis. *Plant Physiology* 134(3):890–897.

Grabber JH, Ralph J, and Hatfield RD. 2000. Cross-linking of maize walls by ferulate dimerization and incorporation into lignin. *Journal of Agricultural and Food Chemistry* 48(12):6106–6113.

Grabber JH, Ralph J, Lapierre C, and Barriere Y. 2004. Genetic and molecular basis of grass cell-wall degradability. I. Lignin-cell wall matrix interactions. *Comptes Rendus Biologies* 327(5):455–465.

Gu Y, Kaplinsky N, Bringmann M, Cobb A, Carroll A, Sampathkumar A, Baskin TI, Persson S, and Somerville CR. 2010. Identification of a cellulose synthase-associated protein required for cellulose biosynthesis. *Proceedings of the National Academy of Sciences of the United States of America* 107(29):12866–12871.

Guillet-Claude C, Birolleau-Touchard C, Manicacci D, Fourmann M, Barraud S, Carret V, Martinant JP, and Barriere Y. 2004. Genetic diversity associated with variation in silage corn digestibility for three O-methyltransferase genes involved in lignin biosynthesis. *Theoretical and Applied Genetics* 110(1): 126–135.

Halpin C, Holt K, Chojecki J, Oliver D, Chabbert B, Monties B, Edwards K, Barakate A, and Foxon GA. 1998. Brown-midrib maize (bm1)—A mutation affecting the cinnamyl alcohol dehydrogenase gene. *The Plant Journal: For Cell and Molecular Biology* 14(5):545–553.

Himmel ME and Bayer EA. 2009. Lignocellulose conversion to biofuels: Current challenges, global perspectives. *Current Opinion in Biotechnology* 20(3):316–317.

Himmel ME, Ding SY, Johnson DK, Adney WS, Nimlos MR, Brady JW, and Foust TD. 2007. Biomass recalcitrance: Engineering plants and enzymes for biofuels production. *Science* 15(5813):804–807.

Hirochika H, Guiderdoni E, An G, Hsing YI, Eun MY, Han CD, Upadhyaya N, Ramachandran S, Zhang Q, Pereira A et al. 2004. Rice mutant resources for gene discovery. *Plant Molecular Biology* 54(3):325–334.

Hodgson EM, Nowakowski DJ, Shield I, Riche A, Bridgwater AV, Clifton-Brown JC, and Donnison IS. 2011. Variation in *Miscanthus* chemical composition and implications for conversion by pyrolysis and thermo-chemical bio-refining for fuels and chemicals. *Bioresource Technology* 102(3):3411–3418.

Holland N, Holland D, Helentjaris T, Dhugga KS, Xoconostle-Cazares B, and Delmer DP. 2000. A comparative analysis of the plant cellulose synthase (CesA) gene family. *Plant Physiology* 123(4):1313–1324.

Huang J, Xia T, Li A, Yu B, Li Q, Tu Y, Zhang W, Yi Z, and Peng L. 2012. A rapid and consistent near infrared spectroscopic assay for biomass enzymatic digestibility upon various physical and chemical pretreatments in *Miscanthus*. *Bioresource Technology* 121C:274–281.

Karpinska B, Karlsson M, Srivastava M, Stenberg A, Schrader J, Sterky F, Bhalerao R, and Wingsle G. 2004. MYB transcription factors are differentially expressed and regulated during secondary vascular tissue development in hybrid aspen. *Plant Molecular Biology* 56(2):255–270.

Lane DR, Wiedemeier A, Peng L, Hofte H, Vernhettes S, Desprez T, Hocart CH, Birch RJ, Baskin TI, Burn JE et al. 2001. Temperature-sensitive alleles of RSW2 link the KORRIGAN endo-1,4-beta-glucanase to cellulose synthesis and cytokinesis in *Arabidopsis*. *Plant Physiology* 126(1):278–288.

Lerouxel O, Cavalier DM, Liepman AH, and Keegstra K. 2006. Biosynthesis of plant cell wall polysaccharides—A complex process. *Current Opinion in Plant Biology* 9(6):621–630.

Li A, Xia T, Xu W, Chen TT, Li XL, Fan J, Wang RY, Feng SQ, Wang YT, and Peng LC. 2013a. An integrative analysis of four CESA isoforms specific for fiber cellulose production between *Gossypium hirsutum* and *Gossypium barbadense*. *Planta* 3:1–13.

Li FC, Ren SF, Zhang W, Xu ZD, Xie GS, Chen Y, Tu YY, Li Q, Zhou SG, Li Y et al. 2013b. Arabinose substitution degree in xylan positively affects lignocellulose enzymatic digestibility after various NaOH/H_2SO_4 pretreatments in *Miscanthus*. *Bioresource Technology* 130:629–637.

Li S, Lei L, Somerville CR, and Gu Y. 2012. Cellulose synthase interactive protein 1 (CSI1) links microtubules and cellulose synthase complexes. *Proceedings of the National Academy of Sciences of the United States of America* 109(1):185–190.

Li Y, Qian Q, Zhou Y, Yan M, Sun L, Zhang M, Fu Z, Wang Y, Han B, Pang X et al. 2003. BRITTLE CULM1, which encodes a COBRA-like protein, affects the mechanical properties of rice plants. *The Plant Cell* 15(9):2020–2031.

Lindsay SE and Fry SC. 2008. Control of diferulate formation in dicotyledonous and gramineous cell-suspension cultures. *Planta* 227(2):439–452.

Lionetti V, Francocci F, Ferrari S, Volpi C, Bellincampi D, Galletti R, D'Ovidio R, De Lorenzo G, and Cervone F. 2010. Engineering the cell wall by reducing de-methyl-esterified homogalacturonan improves saccharification of plant tissues for bioconversion. *Proceedings of the National Academy of Sciences of the United States of America* 107(2):616–621.

Mabee WE and Saddler JN. 2010. Bioethanol from lignocellulosics: Status and perspectives in Canada. *Bioresource Technology* 101(13):4806–4813.

Matsumura H, Nirasawa S, and Terauchi R. 1999. Technical advance: Transcript profiling in rice (*Oryza sativa* L.) seedlings using serial analysis of gene expression (SAGE). *The Plant Journal: For Cell and Molecular Biology* 20(6):719–726.

McCarty DR, Settles AM, Suzuki M, Tan BC, Latshaw S, Porch T, Robin K, Baier J, Avigne W, Lai J et al. 2005. Steady-state transposon mutagenesis in inbred maize. *The Plant Journal: For Cell and Molecular Biology* 44(1):52–61.

Miki D and Shimamoto K. 2004. Simple RNAi vectors for stable and transient suppression of gene function in rice. *Plant and Cell Physiology* 45(4):490–495.

Mosier N, Hendrickson R, Ho N, Sedlak M, and Ladisch MR. 2005. Optimization of pH controlled liquid hot water pretreatment of corn stover. *Bioresource Technology* 96(18):1986–1993.

Mutwil M, Debolt S, and Persson S. 2008. Cellulose synthesis: A complex complex. *Current Opinion in Plant Biology* 11(3):252–257.

Nguyen TA, Kim KR, Han SJ, Cho HY, Kim JW, Park SM, Park JC, and Sim SJ. 2010. Pretreatment of rice straw with ammonia and ionic liquid for lignocellulose conversion to fermentable sugars. *Bioresource Technology* 101(19):7432–7438.

Nicol F, His I, Jauneau A, Vernhettes S, Canut H, and Hofte H. 1998. A plasma membrane-bound putative endo-1,4-beta-D-glucanase is required for normal wall assembly and cell elongation in *Arabidopsis*. *The EMBO Journal* 17(19):5563–5576.

Nishimura A, Aichi I, and Matsuoka M. 2006. A protocol for *Agrobacterium*-mediated transformation in rice. *Nature Protocols* 1(6):2796–2802.

Nobuta K, Vemaraju K, and Meyers BC. 2007. Methods for analysis of gene expression in plants using MPSS. *Methods in Molecular Biology* 406:387–408.

Pagant S, Bichet A, Sugimoto K, Lerouxel O, Desprez T, McCann M, Lerouge P, Vernhettes S, and Hofte H. 2002. KOBITO1 encodes a novel plasma membrane protein necessary for normal synthesis of cellulose during cell expansion in *Arabidopsis*. *The Plant Cell* 14(9):2001–2013.

Paredez AR, Somerville CR, and Ehrhardt DW. 2006. Visualization of cellulose synthase demonstrates functional association with microtubules. *Science* 312(5779):1491–1495.

Pauly M and Keegstra K. 2008. Cell-wall carbohydrates and their modification as a resource for biofuels. *The Plant Journal: For Cell and Molecular Biology* 54(4):559–568.

Pauly M and Keegstra K. 2010. Plant cell wall polymers as precursors for biofuels. *Current Opinion in Plant Biology* 13(3):305–312.

Pena MJ, Zhong R, Zhou GK, Richardson EA, O'Neill MA, Darvill AG, York WS, and Ye ZH. 2007. *Arabidopsis* irregular xylem8 and irregular xylem9: Implications for the complexity of glucuronoxylan biosynthesis. *The Plant Cell* 19(2):549–563.

Peng L, Kawagoe Y, Hogan P, and Delmer D. 2002. Sitosterol-beta-glucoside as primer for cellulose synthesis in plants. *Science* 295(5552):147–150.

Penning BW, Hunter CT, 3rd, Tayengwa R, Eveland AL, Dugard CK, Olek AT, Vermerris W, Koch KE, McCarty DR, Davis MF et al. 2009. Genetic resources for maize cell wall biology. *Plant Physiology* 151(4):1703–1728.

Persson S, Caffall KH, Freshour G, Hilley MT, Bauer S, Poindexter P, Hahn MG, Mohnen D, and Somerville C. 2007. The *Arabidopsis* irregular xylem8 mutant is deficient in glucuronoxylan and homogalacturonan, which are essential for secondary cell wall integrity. *The Plant Cell* 19(1):237–255.

Piquemal J, Chamayou S, Nadaud I, Beckert M, Barriere Y, Mila I, Lapierre C, Rigau J, Puigdomenech P, Jauneau A et al. 2002. Down-regulation of caffeic acid o-methyltransferase in maize revisited using a transgenic approach. *Plant Physiology* 130(4):1675–1685.

Piston F, Uauy C, Fu L, Langston J, Labavitch J, and Dubcovsky J. 2010. Down-regulation of four putative arabinoxylan feruloyl transferase genes from family PF02458 reduces ester-linked ferulate content in rice cell walls. *Planta* 231(3):677–691.

Ragauskas AJ, Williams CK, Davison BH, Britovsek G, Cairney J, Eckert CA, Frederick WJ, Jr., Hallett JP, Leak DJ, Liotta CL et al. 2006. The path forward for biofuels and biomaterials. *Science* 311(5760):484–489.

Ransom C, Balan V, Biswas G, Dale B, Crockett E, and Sticklen M. 2007. Heterologous *Acidothermus cellulolyticus* 1,4-beta-endoglucanase E1 produced within the corn biomass converts corn stover into glucose. *Applied Biochemistry and Biotechnology* 137–140(1–12):207–219.

Scheller HV and Ulvskov P. 2010. Hemicelluloses. *Annual Review of Plant Biology* 61:263–289.

Schmer MR, Vogel KP, Mitchell RB, and Perrin RK. 2008. Net energy of cellulosic ethanol from switchgrass. *Proceedings of the National Academy of Sciences of the United States of America* 105(2):464–469.

Settles AM, Holding DR, Tan BC, Latshaw SP, Liu J, Suzuki M, Li L, O'Brien BA, Fajardo DS, Wroclawska E et al. 2007. Sequence-indexed mutations in maize using the UniformMu transposon-tagging population. *BMC Genomics* 8:116.

Simmons BA, Loque D, and Ralph J. 2010. Advances in modifying lignin for enhanced biofuel production. *Current Opinion in Plant Biology* 13(3):313–320.

Sindhu A, Langewisch T, Olek A, Multani DS, McCann MC, Vermerris W, Carpita NC, and Johal G. 2007. Maize Brittle stalk2 encodes a COBRA-like protein expressed in early organ development but required for tissue flexibility at maturity. *Plant Physiology* 145(4):1444–1459.

Soccol CR, Vandenberghe LP, Medeiros AB, Karp SG, Buckeridge M, Ramos LP, Pitarelo AP, Ferreira-Leitao V, Gottschalk LM, Ferrara MA et al. 2010. Bioethanol from lignocelluloses: Status and perspectives in Brazil. *Bioresource Technology* 101(13):4820–4825.

Solomon BD. 2010. Biofuels and sustainability. *Annals of the New York Academy of Sciences* 1185:119–134.

Somerville C. 2006. Cellulose synthesis in higher plants. *Annual Review of Cell and Developmental Biology* 22:53–78.

Sun HY, Li Y, Feng SQ, Zou WH, Guo K, Fan CF, Si SL, and Peng LC. 2013. Analysis of five rice 4-coumarate: Coenzyme A ligase enzyme activity and stress response for potential roles in lignin and flavonoid biosynthesis in rice. *Biochemical and Biophysical Research Communications* 430(3):1151–1156.

Tanaka K, Murata K, Yamazaki M, Onosato K, Miyao A, and Hirochika H. 2003. Three distinct rice cellulose synthase catalytic subunit genes required for cellulose synthesis in the secondary wall. *Plant Physiology* 133(1):73–83.

Taylor NG. 2008. Cellulose biosynthesis and deposition in higher plants. *The New Phytologist* 178(2):239–252.

Torney F, Moeller L, Scarpa A, and Wang K. 2007. Genetic engineering approaches to improve bioethanol production from maize. *Current Opinion in Biotechnology* 18(3):193–199.

Vanholme R, Demedts B, Morreel K, Ralph J, and Boerjan W. 2010. Lignin biosynthesis and structure. *Plant Physiology* 153(3):895–905.

Vega-Sanchez ME and Ronald PC. 2010. Genetic and biotechnological approaches for biofuel crop improvement. *Current Opinion in Biotechnology* 21(2):218–224.

Vermerris W and Boon JJ. 2001. Tissue-specific patterns of lignification are disturbed in the brown midrib2 mutant of maize (*Zea mays* L.). *Journal of Agricultural and Food Chemistry* 49(2):721–728.

Vignols F, Rigau J, Torres MA, Capellades M, and Puigdomenech P. 1995. The brown midrib3 (bm3) mutation in maize occurs in the gene encoding caffeic acid O-methyltransferase. *The Plant Cell* 7(4):407–416.

Vogel J. 2008. Unique aspects of the grass cell wall. *Current Opinion in Plant Biology* 11(3):301–307.

Vogt T. 2010. Phenylpropanoid biosynthesis. *Molecular Plant* 3(1):2–20.

Wang HZ and Dixon RA. 2012. On-off switches for secondary cell wall biosynthesis. *Molecular Plant* 5(2):297–303.

Wang L, Guo K, Li Y, Tu Y, Hu H, Wang B, Cui X, and Peng L. 2010. Expression profiling and integrative analysis of the CESA/CSL superfamily in rice. *BMC Plant Biology* 10:282.

Weng JK, Li X, Bonawitz ND, and Chapple C. 2008. Emerging strategies of lignin engineering and degradation for cellulosic biofuel production. *Current Opinion in Biotechnology* 19(2):166–172.

Whitney SE, Gothard MG, Mitchell JT, and Gidley MJ. 1999. Roles of cellulose and xyloglucan in determining the mechanical properties of primary plant cell walls. *Plant Physiology* 121(2):657–664.

Xie G and Peng L. 2011. Genetic engineering of energy crops: A strategy for biofuel production in China. *Journal of Integrative Plant Biology* 53(2):143–150.

Xie GS, Yang B, Xu ZD, Li FC, Guo K, Zhang ML, Wang LQ, Zou WH, Wang YT, and Peng LC. 2013. Global identification of multiple OsGH9 family members and their involvement in cellulose crystallinity modification in rice. *PLoS One* 8(1): e50171.

Xu N, Zhang W, Ren S, Liu F, Zhao C, Liao H, Xu Z, Huang J, Li Q, Tu Y et al. 2012. Hemicelluloses negatively affect lignocellulose crystallinity for high biomass digestibility under NaOH and H_2SO_4 pretreatments in *Miscanthus*. *Biotechnology for Biofuels* 5(1):58.

Yennawar NH, Li LC, Dudzinski DM, Tabuchi A, and Cosgrove DJ. 2006. Crystal structure and activities of EXPB1 (*Zea m* 1), a beta-expansin and group-1 pollen allergen from maize. *Proceedings of the National Academy of Sciences of the United States of America* 103(40):14664–14671.

Yong W, Link B, O'Malley R, Tewari J, Hunter CT, Lu CA, Li X, Bleecker AB, Koch KE, McCann MC et al. 2005. Genomics of plant cell wall biogenesis. *Planta* 221(6):747–751.

Yuan JS, Tiller KH, Al-Ahmad H, Stewart NR, and Stewart CN, Jr. 2008. Plants to power: Bioenergy to fuel the future. *Trends in Plant Science* 13(8):421–429.

Zeng W, Chatterjee M, and Faik A. 2008. UDP-xylose-stimulated glucuronyltransferase activity in wheat microsomal membranes: Characterization and role in glucurono(arabino)xylan biosynthesis. *Plant Physiology* 147(1):78–91.

Zeng W, Jiang N, Nadella R, Killen TL, Nadella V, and Faik A. 2010. A glucurono(arabino)xylan synthase complex from wheat contains members of the GT43, GT47, and GT75 families and functions cooperatively. *Plant Physiology* 154(1):78–97.

Zhang J, Li C, Wu C, Xiong L, Chen G, Zhang Q, and Wang S. 2006. RMD: A rice mutant database for functional analysis of the rice genome. *Nucleic Acids Research* 34(Database issue):D745–D748.

Zhang W, Yi ZL, Huang JF, Li FC, Hao B, Li M, Hong SF, Lv YZ, Sun W, Ragauskas A, Hu F, Peng JH, and Peng LC. 2013. Three lignocellulose features that distinctively affect biomass enzymatic digestibility under NaOH and H_2SO_4 pretreatments in *Miscanthus*. *Bioresource Technology* 130:30–37.

Zhong R, Lee C, McCarthy RL, Reeves CK, Jones EG, and Ye ZH. 2011. Transcriptional activation of secondary wall biosynthesis by rice and maize NAC and MYB transcription factors. *Plant and Cell Physiology* 52(10):1856–1871.

Zhong R, Pena MJ, Zhou GK, Nairn CJ, Wood-Jones A, Richardson EA, Morrison WH, 3rd, Darvill AG, York WS, and Ye ZH. 2005. *Arabidopsis* fragile fiber8, which encodes a putative glucuronyltransferase, is essential for normal secondary wall synthesis. *The Plant Cell* 17(12):3390–3408.

2 Novel Cropping Technologies and Management Applied to Energy Crops

Lorenzo Barbanti

CONTENTS

2.1 INTRODUCTION

The targets for renewable energy production to alleviate the global energy crisis and greenhouse gas (GHG) reduction under national and international schemes involve all major renewable energy sources (RESs). The challenge lies in the fact that the energy sector is the most responsible for GHG emissions (Olivier et al. 1999), while the demand for energy is projected to increase by 55% between 2005 and 2030 (IEA 2007). Replacing fossil energy (about 80% of the present world supply) with renewable sources including dedicated crops is expected to significantly alleviate the double-faced problem in the energy sector. The general agreement on GHG reduction in the frame of the Kyoto

Protocol and the commitment to halve CO_2 emissions by 2050 (34th G8 summit of July 2008 in Japan) support the development of RESs.

Biomass has been the most important source of energy during the history of mankind. Now it represents only 11% of the world energy consumption as a main source for domestic cooking and heating and is foreseen to attain 16% by 2040 (Rosillo-Calle 2007). Biofuels still cover less than 1% of the total fuel supply, and most of them are produced in Brazil and the United States (85% of ethanol in the world market) and the European Union (EU) (75% of biodiesel in the world consumption) (IEA 2007).

Global biofuel use is expected to increase twofold between 2009 and 2015 (Klein et al. 2008): Brazil will remain the world's top exporter, while the United States are expected to perform the largest increase in biofuel use (more than 30% of the world biofuel consumption). In Europe, France and Germany are established producers of biofuels. The report also predicts that by the year 2015, other countries will significantly contribute to world biofuel production: Argentina, China, Colombia, Indonesia, Malaysia, the Philippines, and Thailand. More to this, another source (IEA 2012) predicts a 40% increase in global power generation from RES within the year 2017.

Dedicated crops represent one of the most promising options to replace fossil fuels in the short time: a variable estimate between 58 and 141 million hectares will be cultivated with energy crops in developed and developing countries in the next decades (Strengers et al. 2004), representing a significant component in a portfolio of climate mitigation measures (average savings, 1085 Mt CO_2-eq. year^{-1}; Sims et al. 2006). However, turning agricultural land to energy crops is a subject of controversy that is spurred by the threats for food security feared in many developing countries (FAO 2008). This is in contrast to the belief that a relevant expansion of energy crops would not significantly influence food prices (Smeets et al. 2007 and references therein). A second reason hampering energy crop expansion is that some aspects such as their environmental impact, economic sustainability and specific adaptation are still debated, requiring deeper insight and better understanding in view of large-scale diffusion.

Advanced and specific crop husbandry is of great relevance in growing plants for energy uses: enhancing the efficiency of limited or expensive crop inputs (water, energy, etc.), fitting in current agricultural systems, and complying with agroenvironmental legislations are important premises to minimize potential drawbacks undermining bioenergy development. Therefore, reviewing the state-of-the-art and the novel techniques potentially suited to these heterogeneous crops contributes to enhance their competitiveness, sustainability, and ultimate expansion.

Several interventions are carried out during crop cycle, addressing various aspects of crop management. A basic subdivision among five areas allows us better discuss specific issues: (1) crop planting, covering the operations from soil tillage to crop establishment; (2) plant nutrition, including chemical and organic fertilization; (3) water management, covering drainage, irrigation, and adaptation to drought; (4) crop protection from competing, feeding, and parasitizing organisms (weeds, pests and diseases); and (5) harvest and subsequent management (conditioning and storage), which represent the link between the agricultural and energy conversion phases.

Two crosscutting issues cover the complex of the five areas: the effects of crop management on the energy balance in the agricultural phase and the applications of precision farming. Energy balance addresses the principle that only crops ensuring a high energy return on the amount of energy invested (EROEI), in the two forms of net energy (NE) (output–input) gain and energy efficiency (EE) (output/input), may be retained among energy crops. Precision farming is seen as the last frontier in crop management, more realistically as a tool to improve the ways of tackling existing problems. Applications of precision farming lie on the combined technologies of global positioning system (GPS) and geographic information system (GIS). The GPS involves accuracy up to ~0.02 m in locating soil and crop records; thereafter, a GIS arranges site-referenced data in various layers of information. Reviewing the progress in crop management is the best way to acknowledge the areas where new approaches drive the current development or old constraints remain to be overcome.

2.2 CROP PLANTING

2.2.1 Soil Tillage

The planting phase is to a large extent responsible for plant growth and development, strongly influencing the final yield of most energy crops. Energy crops are intrinsically grown in the pursuit of high NE and EE, that is, high energy yield in exchange for a low use of subsidiary energy (fertilizers, fuels, etc.). Soil properties play a role in this frame, as a good soil status reduces the need for external inputs to fill the gap between actual and potential growth. The soil is the natural habitat of plant roots: a denser rooting means a more effective retrieval of available moisture and nutrients. Soil structure, the 3D pattern in which elementary particles aggregate, is the most important physical property whose expression summarizes other relevant traits such as bulk density and porosity. Soil strength, that is, the resistance to penetration, is the most expedient field measurement showing a remarkable correlation with soil structure and the aforementioned traits determined in laboratory conditions (Canarache 1990; Campbell and O'Sullivan 1991), as well as root growth (Ehlers et al. 1983). Assuring low soil mechanical impedance through low bulk density and high porosity is considered the best way to enhance root growth (Taylor 1983; Atwell 1993), in turn allowing good water and nutrient uptake from deep soil layers.

In soil tillage, different tools and techniques may lead to similar results, whereas at varying conditions, the same technique may determine contrasting results. The variety of approaches (instruments, soil depth, time of the year, etc.) adopted in each crop demonstrates the substantial heterogeneity among farmers operating in the same area. Therefore, a general strategy may not be envisaged.

A basic difference concerns the adoption of inversion versus noninversion tillage. The moldboard plow is the foremost tool for the former category, having received continuous improvements in the ca. 1000 years of life spent as a soil-inverting tool (Lal et al. 2007). Noninverting tools cover a much wider array of passive instruments, including disks, rippers, rigid and elastic cultivators and harrows, and power-operated instruments (rotary hoes at vertical or horizontal axis). Different kinds of passive tools are often arranged in combined machines, exerting their action at different soil depths, thanks to their weight and shape. Noninverting tools entail a modest mixing of the tilled layer and subsequent incorporation of plant residues, mineral and organic fertilizers, weed seeds, etc. However, in modern agriculture, the use of advanced means and techniques (chemical herbicides, fluid fertilizers, crop residue mulching at soil surface, etc.) reduces the need for soil contribution in the two areas of fertilization and weed control. Nevertheless, it is perceived that the progress in crop husbandry leads to a simplistic attitude in assigning each problem to a specific category and missing the wealth that only a comprehensive, integrated approach can provide. It is therefore sensed that the forthcoming development in the agroenergy sector may not be achieved in a sustainable way unless this concept is borne in mind.

The complexity of soil tillage effects originates the dilemma among conventional tillage, reduced/minimum tillage, and no tillage (no-till). Conventional tillage is normally represented by deep plowing (usually at a 0.3 m depth); minimum and no tillage are often grouped under the common category of conservation tillage. No tillage was envisaged as a viable option at the eve of energy crop development (Wolf et al. 1989), despite the fact that *Panicum virgatum* and *Eragrostis curvula*, the two investigated perennial grasses, only need tillage in the year of establishment. This first study specifically addressed the feasibility of no tillage and sod seeding at the light of the pesticides (herbicides and insecticides) and equipment (seeders) available to face the difficulties arising in the new system. The subsequent progress in these two fields has greatly reduced the problems responsible for the early failures, while the cumulated experience has indicated in which conditions no tillage is unadvisable because of structural (e.g., poor drainage under wet climate) or occasional constraints (e.g., soil compaction in previous crop harvest).

Overcoming these problems has prompted a shift in soil tillage research: the effort now lies in understanding at a deeper level why no till performs either better or worse than conventional tillage, producing an inconsistency that is often echoed in the scientific literature. In annual crop rotations including species at potential bioenergy use (maize and soybean), no tillage is responsible for lower soil temperature in the springtime (Chen et al. 2011), decreased surface runoff (Chen et al. 2011) and soil erosion (Zhang et al. 2007; Soane et al. 2012), increased water storage and use efficiency (He et al. 2007; Wang et al. 2007; Jin et al. 2009; Chen et al. 2011), thanks to better soil physical properties (Duiker and Beegle 2006; Zhang et al. 2007; Blanco-Canqui et al. 2011; He et al. 2011; Soane et al. 2012). The role of higher soil carbon content determined by no-till on improved structural properties was recently described by Blanco-Canqui et al. (2011) in three long-term (>20 years) experiments: especially tensile strength and mean weight diameter of air-dried aggregates outlined remarkable improvements as shown in Figure 2.1.

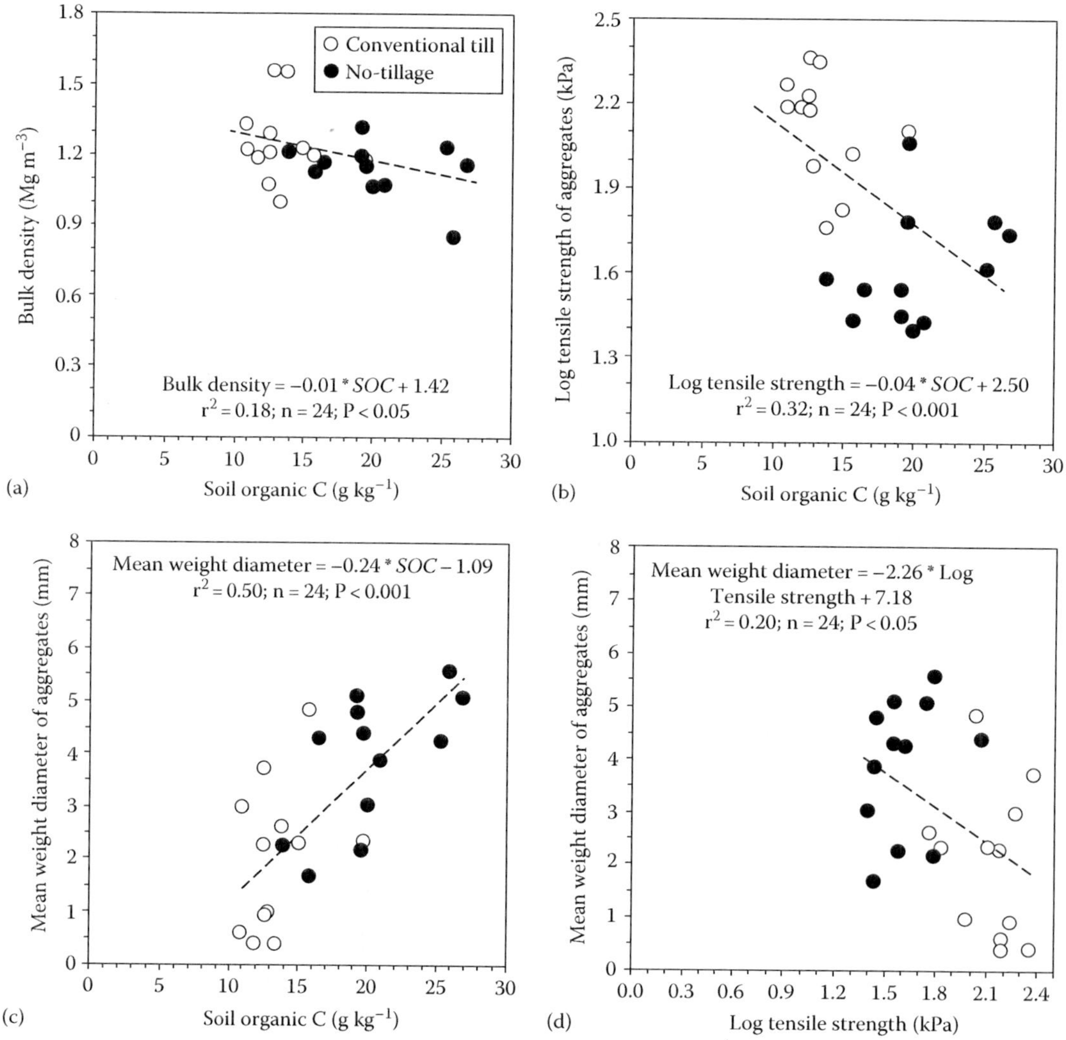

FIGURE 2.1 Relationship of soil organic carbon concentration with (a) bulk density, (b) tensile strength of aggregates, (c) mean weight diameter of aggregates, and (d) the relationship between tensile strength and mean weight diameter of aggregates in three long-term experiments. (Reproduced from *Agric., Ecosyst. Environ.*, 144, Blanco-Canqui, H., Schlegel, A.J., and Heer, W.F., Soil-profile distribution of carbon and associated properties in no-till along a precipitation gradient in the central Great Plains, 107–116, Copyright 2011, with permission from Elsevier.)

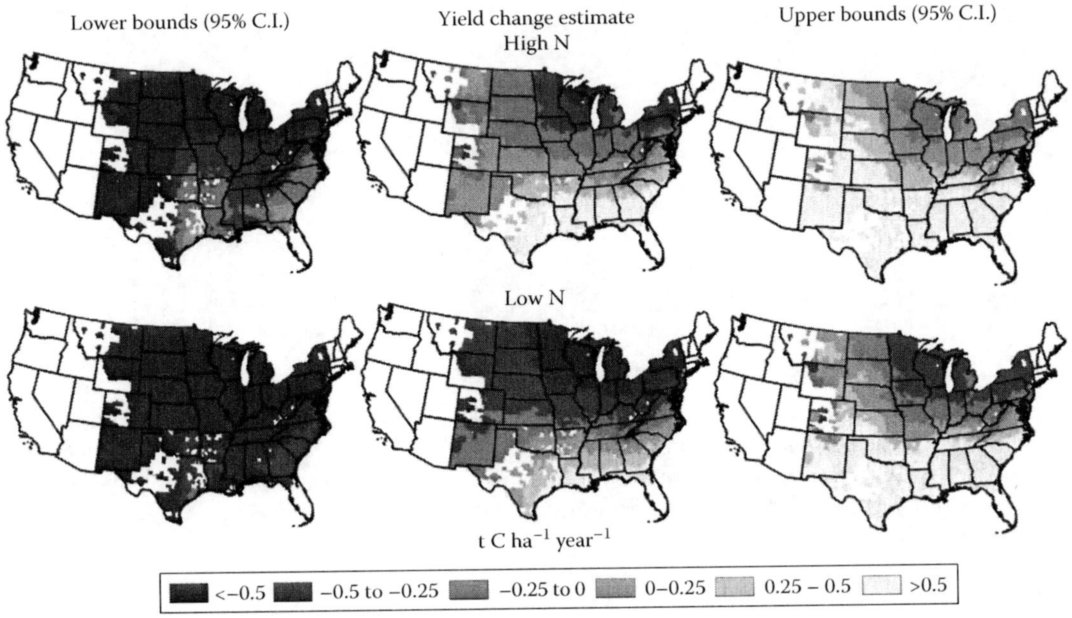

FIGURE 2.2 (Color figure at CRC website http://www.crcpress.com/product/isbn/9781466505520.) Predicted yield changes for maize expressed as tons of carbon per hectare and year (t C ha^{-1} $year^{-1}$) with the lower and upper bound for the 95% confidence interval. High N fertilization rates were estimated at 200 kg N ha^{-1}, and low N fertilization rates were estimated at 50 kg N ha^{-1}; the rates were based on the first and third quantile of N fertilization rate distribution from the experimental data. (Reproduced from *Agric., Ecosyst. Environ.*, 149, Ogle S.M., Swan, A., and Paustian, K., No-till management impacts on crop productivity, carbon input and soil carbon sequestration, 37–49, 2012, Copyright 2011, with permission from Elsevier.)

In dry regions, no-till yields tend to be higher than in conventional tillage (Zhang et al. 2007; Hansen et al. 2012); the opposite occurs in wet years (−10% to −15%; Zhang et al. 2007) and in cool, wet regions (Ogle et al. 2012). This is consistent with the overall yield reduction (−8.5%) observed for no tillage in a metaregression analysis over 47 studies across Europe, a vast area featuring severely dry conditions only in its southern part (Van den Putte et al. 2010). However, another review focusing on the same area reports comparable yields with the two techniques, and the modest reduction sometimes associated with no tillage is considered tolerable in view of the lower production costs (Soane et al. 2012). In the case of maize, predicted yield changes after no-till adoption vary depending on N fertilizer dose (Ogle et al. 2012): A higher nutrient supply is needed to support the crop under no-till management as shown in Figure 2.2. In addition, the yield gain of no tillage is enhanced in the case of double (Jin et al. 2009) and relay cropping (Johnson and Novak 2012), where it saves precious time and water with respect to conventional tillage. In recent times, the increase in fuel price and the concurrent decrease of nonselective herbicide cost have further contributed to the profitability of herbicide-intensive conservation tillage versus fuel-intensive traditional tillage (Nail et al. 2007).

Soil carbon sequestration is another subject of paramount importance associated with tillage intensity. No tillage is deemed responsible for a lower soil carbon mineralization resulting in higher C stocks, especially in the top (generally 0.1 m) layer (Sombrero and De Benito 2010). At higher depths, a sharp decline is often observed as the effect of carbon stratification near the soil surface (Duiker and Beegle 2006; Blanco-Canqui et al. 2011). However, averaged over the whole soil profile, no-till maintains a positive margin in soil carbon content over conventional (moldboard plow) and, to a lesser extent, reduced tillage (chisel), as shown by Duiker and Beegle (2006) after 25 years of continuous maize in Figure 2.3. Conversely, in case of poor no-till performance, a decline in C input is expected to reduce C storage with respect to conventional tillage (Ogle et al. 2012).

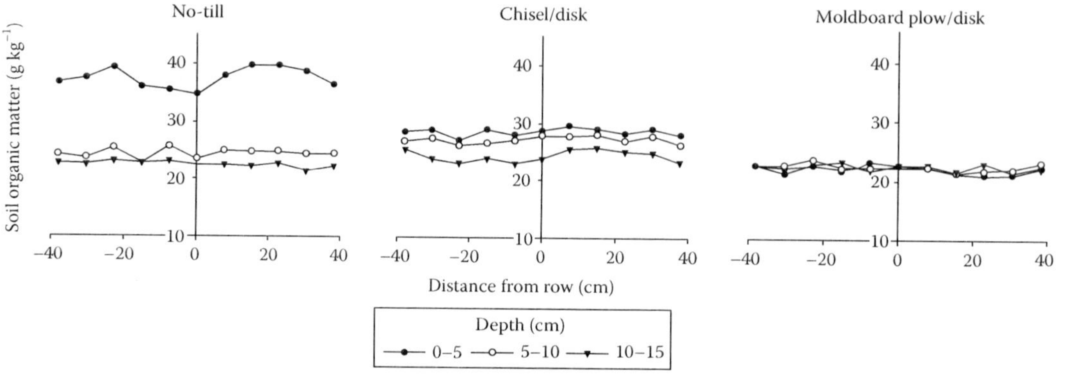

FIGURE 2.3 Horizontal and vertical distribution of soil organic matter contents with no-till, chisel/disking, and moldboard plow/disking after 25 years of continuous maize. (Reproduced from *Soil Till. Res.*, 88, Duiker, S.W. and Beegle, D.G., Soil fertility distributions in long-term no-till, chisel/disk and moldboard plow/disk systems, 30–41, 2006, Copyright 2005, with permission from Elsevier.)

In a recent survey of 16 experiments carried out in Europe, the gain in soil organic carbon (average 0–0.3 m depth) with no tillage versus plowing averaged a rough 400 kg C ha^{-1} year^{-1} (variation, 0–1300 kg C ha^{-1} year^{-1}) (Soane et al. 2012), representing a relevant contribution to soil organic matter status and GHG mitigation. Special attention should also be paid to the high volatility of soil carbon gains if the conservation practices are not pursued in the long time: this is proved by a 6% loss of soil organic matter observed after a single plowing following 7–9 years of conservation (noninversion) tillage (Koch and Stockfisch 2006).

A large difference in carbon sink potential exists between annual and perennial species, the latter benefitting from undisturbed soil after the year of establishment. This involves an easier approach to soil tillage in perennial species, based on the fact that the economic and environmental impact of plowing is diluted over a long (≥10 years) crop life (Fazio and Monti 2011). However, in temperate and cold areas, perennial species only include herbaceous (*Arundo*, *Miscanthus*, switchgrass, reed canary grass, etc.) and woody (poplar, willow, etc.) biomass plants, whereas in tropical areas they also include sugar- (sugarcane) and oil-rich plants (oil and coconut palms, *Jatropha*, and castor bean): This, in the former case, reduces the choice of energy conversion plants to those fed by biomass, if soil carbon storage is retained among top priorities of bioenergy production. Hopefully, second-generation processes for biofuel (biodiesel and ethanol) production from cellulosic materials are about to take off, relieving this limitation. Herbaceous perennial species have an additional advantage represented by their belowground storage organs (rhizomes) further contributing as C sinks, beside the soil carbon rhizodeposition operated by the fibrous root apparatus.

Despite the overall economic and environmental advantages, no tillage is slowly being adopted by farmers even in areas such as North China Plain, where large scientific evidence supports its diffusion. In major European countries, no tillage is estimated to span between 0.4% (Germany) and 5.2% (Spain) of the total arable area (Soane et al. 2012). Reduced tillage, which is generally carried out with noninverting passive tools at lower depths than plowing, represents a compromise between plowing and no-till, achieving limited gains in exchange for limited changes in farm management and equipment.

Compared to no tillage, the application of variable tillage in the frame of precision farming is still in a development phase, yet it might disclose valuable benefits. Operating with a combined disk-and-tine cultivator at variable tine depth as shown in Figure 2.4, Walther et al. (2007) recorded a decrease in tractor draft force with respect to tillage at a fixed depth while keeping a variable amount of crop residues at soil surface according to file slope gradient as a measure to prevent erosion.

FIGURE 2.4 Combined disk-and-tine cultivator modified for precision tillage at variable tine depth. (1) Tractor; (2) GPS system; (3) draft force measurement; (4) electronic hydraulic valve; (5) TTK GPS and antenna; (6) inclination sensor; (7) hydraulic cylinders with displacement transducer; (8) inclination sensor; and (9) cultivator. (Reproduced from *Bull. Univ. Agric. Sci. Vet. Med. Cluj-Napoca. Agric.*, 63, Walther, S., Schule, T., and Koller, K., Precision tillage: Online variation of tillage depth, 34–39, Copyright 2007, with permission from the authors.)

The general considerations on soil tillage, focusing on the differences between annual and perennial species, also apply to energy crops. At present, the main concern for soil fertility and productivity lies in the use of crop residues for complementary energy production or other uses, in view of a global valorization of produced biomass that goes under the concept of biorefinery (Kamm and Kamm 2004). In the case of maize, evidence shows that the more stover is removed from the soil surface, the higher is the risk of negative consequences for soil properties and subsequent crop yield. More specifically, removing more than 25% stover in continuous maize under no-till regime in erosion-prone soils significantly reduced productivity as shown in Figure 2.5, soil organic carbon, aggregate stability, and other soil properties at macro- and microscale (Blanco-Canqui and Lal 2007, 2008). In another experiment in a marginal area, 50% of stover removal in continuous no-till maize determined a slight but significant grain yield decrease even at optimum fertilizer supply (Varvel et al. 2008). Moreover, in this experiment the combined bioethanol (first plus second generation) produced from maize grain and harvested stover fraction did not equal the amount produced from switchgrass (*Panicum virgatum* L.), the perennial biomass species suited for marginal lands such as those under the US Conservation Reserve Program (Lee et al. 2007). Only in more favorable conditions as in central Iowa, stover removal up to 100% did not affect maize yield nor soil properties (Karlen et al. 2011).

2.2.2 Crop Establishment

Soil tillage influences relevant heterogeneous aspects (soil fertility, atmospheric GHG balance, etc.), yet only partially contributes to successful crop establishment, the result mainly depending on the actual planting phase. The operations carried out at planting exert a large influence on crop yield as well as crop relationship with the surrounding environment and the atmosphere. In fact, achieving a good establishment helps to contain land use change (LUC), since a good establishment is the premise to high yield per hectare, in turn reducing the crop surface to the minimum needed.

The scientific literature on crop planting is not abundant: the issues are normally dealt at a technical level, seeking practical solutions to the problems as they occur. A basic difference exists between species propagated by seed or by transplants of seedlings/vegetative organs. Tree crops, either for biomass (poplar, willow, etc.) or oil fruit production (palms, *Jatropha*, etc.), are normally transplanted, although they yield seed: a nursery is arranged to obtain established seedlings

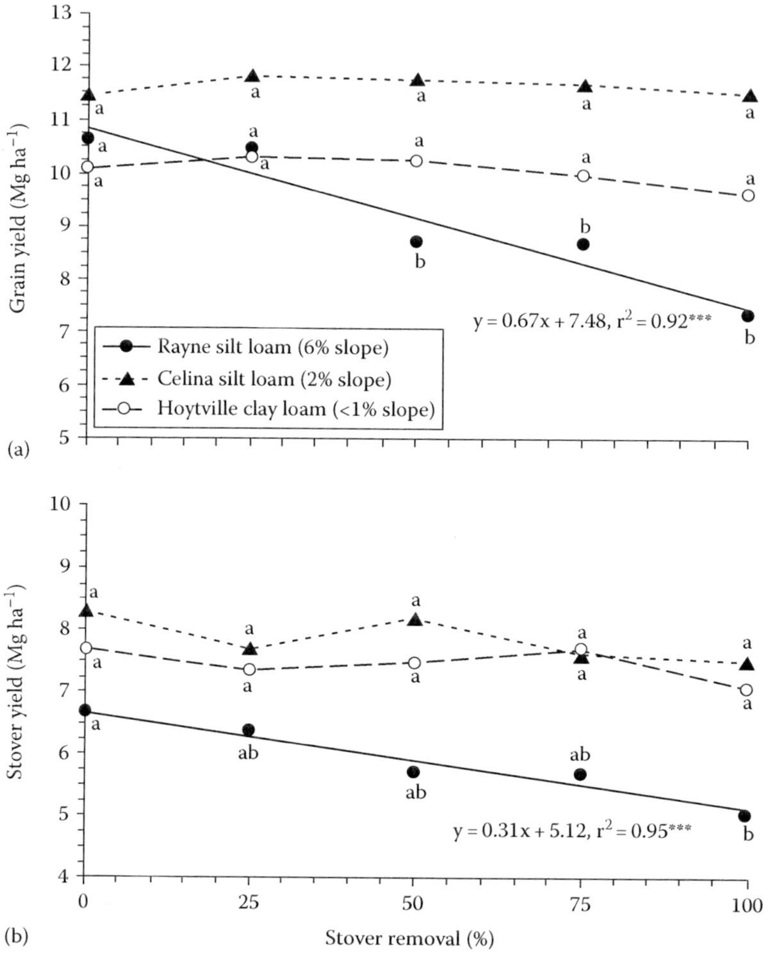

FIGURE 2.5 Yields of (a) maize grain and (b) stover in response to stover removal at 0%, 25%, 50%, 75%, and 100% from three no-till soils. Means followed by the same letters are not significantly different within the same soil. The correlations for the Rayne silt loam were all significant at $P \leq 0.001$ (***). (Reproduced from *Geoderma*, 141, Blanco-Canqui, H. and Lal, R., Soil and crop response to harvesting corn residues for biofuel production, 355–362, Copyright 2007, with permission from Elsevier.)

or cuttings, which are planted in the open field either by hand or with the help of a transplanter. Operating on species at relatively high density (e.g., poplar) in developed countries, a transplanter is normally used and accompanying measures (fertilizing and irrigations) are implemented to ensure rapid and complete establishment (Mitchell et al. 1999). Conversely, operating on species at low density in developing countries (e.g., oil palm; *Elaeis* spp.), the transplanting is done by hand, and the accompanying measures often consist of natural means such as organic fertilizers and the planting time corresponding to the rainy season (Barbanti and Fazio 2011).

Besides tree crops, some of the most promising herbaceous biomass species, namely, *Arundo* (also known as giant reed; *Arundo donax* L.) and *Miscanthus* (*Miscanthus* × *giganteus* Greef et Deuter), are sterile and need to be propagated by vegetative organs, most often rhizomes, stem cuttings, and, lately, seedlings obtained from tissue culture. Especially in *Arundo*, the research is very active at present: stem nodes with dormant buds were recently shown as the most effective fraction to initiate in vitro cultures achieving a good level (77%) of differentiated shoots (Cavallaro et al. 2011).

Most herbaceous energy crops are propagated by seed, considerably easing the planting operations. In the case of cereals, such as maize, sorghum, and the winter cereals (wheat, barley, etc.), seeding times, depth, and density are already acknowledged by farmers, who own or contract the appropriate equipment to carry out the operation. Conversely, the task is more critical in annual and perennial grasses or other species at small seed size (e.g., oilseed rape; *Brassica napus* L. var. *oleifera* D.C.), especially if the plant is intrinsically characterized by modest germination and low seed vigor (e.g., switchgrass). The lack of an adequate program of seed improvement in species at incipient diffusion is another cause for poor seed performance and failures in stand establishment.

The recent progress has mainly addressed sod seeding in no tillage cropping systems: cutting the mulch of plant residues and the soil surface to place the seed at the desired depth requires sturdy seeders and aptly-shaped organs. Good soil/seed contact and well-covered seed furrows are major requirements in order to ensure seed hydration and subsequent germination while avoiding seed preying by animals. In crops at wide row space (≥0.5 m), a row cleaner before the seed coulter may be useful to displace residues between the rows, clearing soil surface on the row for proper seed placement. In fact, the combination of a row cleaner plus rolling disk coulter attached before the furrow opener of a maize seeder as shown in Figure 2.6 reduced surface residue in a field covered by wheat straw and improved seedling emergence as shown in Table 2.1, compared to no seeder attachment (Fallahi and Raoufat 2008). Alternatively, seed coulters in maize may be shaped as wide-sweep openers to improve soil physical properties in the seed zone; this potentially benefited early root growth but at the expenses of a lower soil/seed contact (Vamerali et al. 2006). In small seeds and whenever using seeders devoid of press wheels as many narrow row-spaced machines (<0.25 m), the soil/seed contact may be enhanced by rolling the seedbed prior to seeding. In switchgrass, this improved seedling emergence from 56% to 70% (Monti et al. 2001). However, the boundary between beneficial soil compression and noxious compaction may easily be passed, leading to deleterious effects on seedling emergence and plant growth.

On conclusion, the variety of seeding elements and their outfits for no tillage are rapidly increasing to address all possible soil/crop combinations. However, nonunivocal results are obtained using the same tool on the same crop because they depend on many factors including soil moisture, texture, and seedbed firmness. Care and experience remain the first ingredients to perform good seeding as the premise to adequate plant stand.

2.3 PLANT NUTRITION

The focus on plant nutrition instead of fertilization reflects a priority in understanding the processes underpinning nutrient uptake and use by plants. The application of fertilizers is consequential to this. A single guideline in plant nutrition cannot be drawn in view of the extreme variation among energy crops. In addition to this, the effects of nutrient supply on yield and quality need to be appraised in view of the different energy sectors. Attention is also paid to the environmental implications of nutrient release outside the plant–soil system, especially in the case of nitrogen. The quest for a sustainable agroenergy development demands that nutrient supply be tailored to assure plant growth while restraining environmental impact. In this light, a special opportunity is also represented by the recycling of agroenergy residues as nutrient sources for energy crops to benefit financial and energy balances.

2.3.1 NITROGEN

Nitrogen is the foremost nutrient in agricultural systems, where a large share of N is exported from the system via the agricultural products. This is particularly true for the production of energy crops, as all of the recoverable aboveground biomass is removed from the field. The same also occurs when the residue of grain crops is turned to complementary production of energy and other

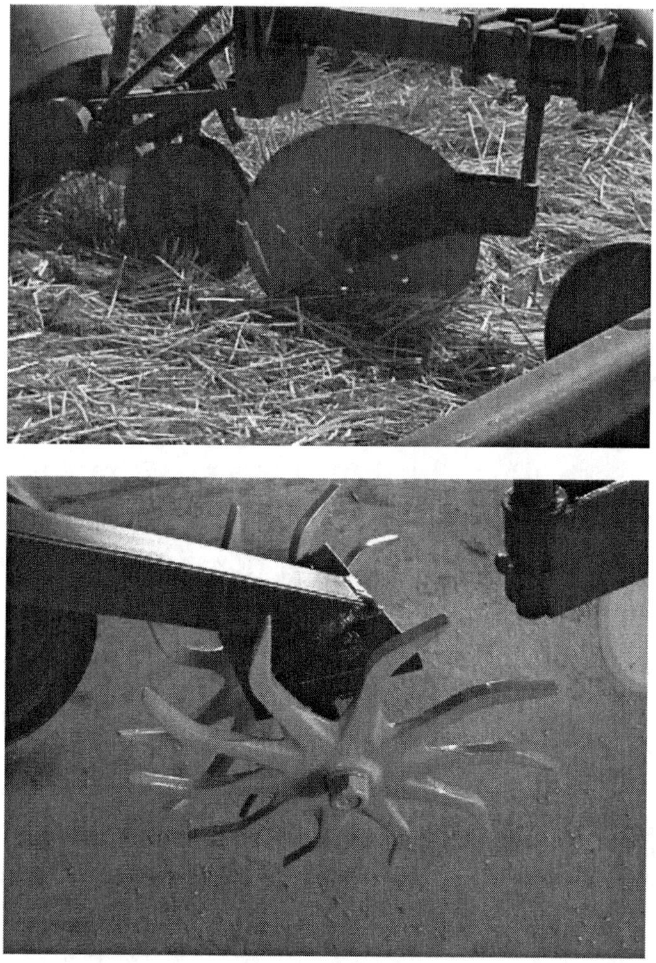

FIGURE 2.6 Plain rolling coulter (top) and row cleaner (bottom) attachments for a maize seeder operating in the presence of surface plant residue in conservation tillage. (Reproduced from *Soil Till. Res.*, 98, Fallahi, S. and Raoufat, M.H., Row-crop planter attachments in a conservation tillage system: A comparative study, 27–34, 2008, Copyright 2007, with permission from Elsevier.)

TABLE 2.1
Effect of Seeder Attachments on Surface and Subsurface Residues and Emergence Rate Index of Maize

Seeder Attachment	Surface Residue (kg m^{-2})	Subsurface Residue (kg m^{-2})	Emergence Rate Index (%, day^{-1})
None	0.38 a	3.06 c	15.1 d
Rolling coulter (C)	0.31 b	3.46 a	16.4 c
Row cleaner (R)	0.25 bc	3.15 c	18.5 a
R followed by C	0.19 c	3.30 b	17.5 b

Source: Adapted from *Soil Till. Res.*, 98, Fallahi, S. and Raoufat, M.H., Row-crop planter attachments in a conservation tillage system: A comparative study, 27–34, 2008, Copyright 2007, with permission from Elsevier.

a, b, etc., indicate significantly different means ($P \leq 0.01$).

TABLE 2.2
Average Increase in Plant Nutrient Removal (kg ha^{-1}) for Various Stover Harvest Scenarios in Central Iowa, United States

Harvest Scenario	Continuous Maize			Maize–Soybean Rotation		
	N	P	K	N	P	K
Whole plant	29	2.7	34	42	3.3	34
Upper half	22	2.1	27	34	3.5	28
Lower half	6	0.6	8	10	0.7	10
No removal	—	—	—	—	—	—

Source: Adapted from *Soil Till. Res.*, 115–116, Karlen, D.L., Birell, S.J., and Hess, J.R., A five-year assessment of corn stover harvest in central Iowa, USA, 47–55, Copyright 2011, with permission from Elsevier.

chemical products. In such a case, nutrient removal will require additional doses to be compensated as shown in Table 2.2 (Karlen et al. 2011).

Growing N-fixing leguminous species in energy crop rotations can reduce the amount of supplementary N required, although this may dilute the efforts for the supply of energy crop and require more land to produce the same amount of energy crops. The intercrops of cereal and leguminous species appear somewhat promising: they have been proved to achieve comparable yields at lower energy costs in the case of wheat–pea (Pelzer et al. 2012) and maize–bean (Dawo et al. 2007). However, they require a fine tuning of their management (varieties, respective plant densities, etc.) in specific crop conditions, in order to assure reliable results. Due to the higher nitrogen content, the intercropped biomass appears to be better suited for anaerobic digestion where N must be proportioned to carbon (optimum C/N range, 25–32) for efficient methane conversion (Angelidaki et al. 2003). Conversely, it appears to be less suitable for thermochemical conversions and especially combustion, since a high N content leads to high NO_x emissions (Lewandowski and Kauter 2003).

In the case of perennial herbaceous biomass crops, the use of a grass-leguminous mixture appears to be less feasible given the difference in plant height. In these rhizomatous species, recent studies on seasonal nutrient dynamics between above- and belowground organs offer a clue to significant nutrient savings: the accumulation of carbohydrates and nutrients in the rhizomes at the end of the growing season and the remobilization in the following year remarkably differed in *Miscanthus* subjected to autumn versus late-winter harvest (Strullu et al. 2011). The supply of nitrogen (120 kg ha^{-1} in spring application) enhanced dry biomass yield in the first but not in the second case (Figure 2.7), depending on the much higher autumn N flux from the above- to the belowground organs in late (winter) versus early (autumn) harvest (140 vs. 72 kg ha^{-1} in fertilized plots): The high belowground N stock cumulated in late harvest enhanced the regrowth and nutrient accumulation in the following year, whereas the depleted N stock associated with early harvest increased the need for fertilizer. Similar considerations on nitrogen storage are suggested by the trend of N content in the aboveground organs of *Miscanthus* and *Arundo* in the period between full summer and late winter (Nassi o Di Nasso et al. 2010).

In a similar example, when the harvest time of reed canary grass (*Phalaris arundinacea* L.), a perennial rhizomatous grass usually grown in northern and boreal environments, was shifted from the autumn to the following spring, the amount of nutrients removed could be reduced by an approximate 50%, which in turn could curb the unfavorable elements and enhance nutrient recirculation

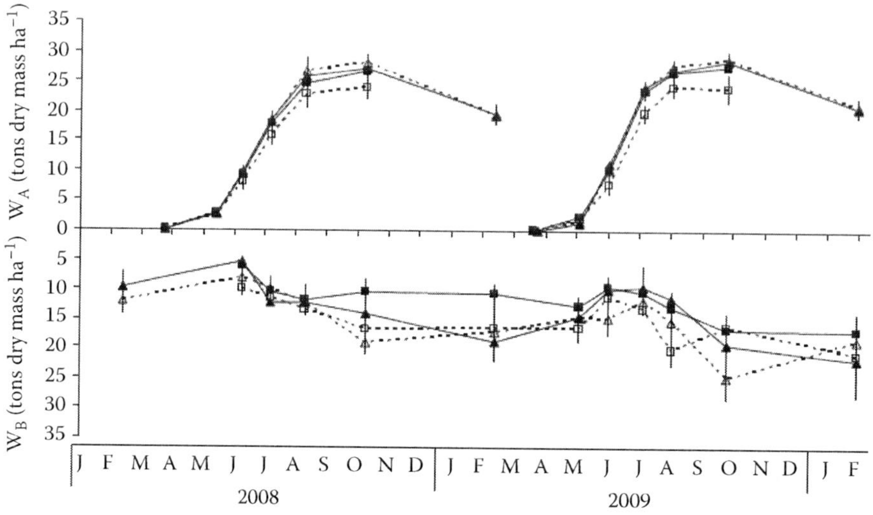

FIGURE 2.7 Effect of harvest date and nitrogen fertilization on aboveground (top) and belowground (bottom) biomass production of *Miscanthus* × *giganteus* in the third and fourth year of growth. Bars represent standard error. Nonfertilized treatments (N0) are represented by open symbols and broken lines and fertilized treatments (N1) by black symbols and continuous lines. Triangles correspond to late harvest (L) and squares to early harvest treatments (E). (Reproduced from *Field Crops Res.*, 121, Strullu, L., Cadoux, S., Preudhomme, M., Jeuffroy, M.H., and Beaudoin, N., Biomass production and nitrogen accumulation and remobilization by *Miscanthus* × *giganteus* as influenced by nitrogen stocks in belowground organs, 381–391, Copyright 2011, with permission from Elsevier.)

in the plant–soil system (Landström et al. 1996). A similar behavior is observed in switchgrass at single- versus double-cut harvest strategy. In the double-cut strategy (booting stage plus the autumn regrowth), the plant is always harvested unripe and its tissues are relatively rich in nitrogen (N content > 10 mg g^{-1}). Conversely, in the single-cut strategy, the plant is harvested at senescence when it is deprived of nitrogen (N content < 10 mg g^{-1}) and has completed nutrient mobilization to the belowground organs (Reynolds et al. 2000; Lemus et al. 2008a). The double-cut strategy involves a somewhat higher biomass yield on average, as reviewed by Monti et al. (2011) in a series of switchgrass experiments (Table 2.3). However, the double-cut strategy removes a quite higher amount of nutrient by multiplying the higher N content in the biomass and two cuts each year, which increases the need for fertilizers. On the other hand, the balance between applied and removed nitrogen tends to be positive in the single harvest system and negative in the double harvest system. A positive balance potentially leads to environmental release of the excess nitrogen as either nitrate (NO_3^-) leaching in soil profile, ammonia (NH_3) volatilization, or nitrous oxide (N_2O) emission to the atmosphere. All the three releases, but especially the last one, have serious environmental consequences: N_2O is a powerful GHG that has a global warming potential (GWP) relative to CO_2 of 298 (Forster et al. 2007). N_2O and methane (CH_4; relative GWP, 25) represent nearly all the GHG emission from agriculture, accounting for about 60% and 50% of global anthropogenic emissions of the two respective gases (Metz et al. 2007). Since the amount of N_2O emitted from the soil is related to N fertilization, reducing mineral and organic fertilizers while enhancing nutrient use efficiency appears to be a major strategy to curb potential emissions and improve crop sustainability.

Apart from environmental issues, the major agronomic task is to determine crop nitrogen demand from the mineral source. In the first instance, the amount of nutrient that will be removed from the soil could be assumed as the dose to be applied. This corresponds to the computation of nutrient budgets (Oenema et al. 2003), an old method that still retains useful information in assessing which crops/agricultural systems are prone to nutrient excesses or deficiencies. However,

TABLE 2.3
Biomass Yield with Single- versus Double-Cut Harvest in a Series of Experimental Cases in Switchgrass

Location	Cuts Year^{-1}	Biomass Yield (Mg ha^{-1})	Sources
Dallas, TX, United States	1	7.1	Sanderson et al. (1999)
	2	6.6	
Stephenville, TX, United States	1	14.8	Sanderson et al. (1999)
	2	8.7	
Knoxville, TX, United States	1	17.4	Reynolds et al. (2000)
	2	18.7	
Ames, IA, United States	1	13.5	Vogel et al. (2002)
	2	13.0	
Mead, NE, United States	1	11.0	Vogel et al. (2002)
	2	11.3	
Chickasha, OK, United States	1	16.0	Thomason et al. (2004)
	2	20.0	
Perkins, OK, United States	1	9.8	Thomason et al. (2004)
	2	10.4	
Princeton, KY, United States	1	12.7	Fike et al. (2006)
	2	14.5	
Raleigh, NC, United States	1	11.9	Fike et al. (2006)
	2	17.0	
Jackson, TN, United States	1	11.3	Fike et al. (2006)
	2	13.8	
Knoxville, TN, United States	1	18.7	Fike et al. (2006)
	2	21.3	
Blacksburg, VA, United States	1	11.0	Fike et al. (2006)
	2	13.6	
Blacksburg, VA, United States	1	10.7	Fike et al. (2006)
	2	16.2	
Orange, VA, United States	1	11.6	Fike et al. (2006)
	2	13.5	
Morgantown, WV, United States	1	14.5	Fike et al. (2006)
	2	13.4	
Ozzano, Bologna, Italy	1	14.9	Monti et al. (2008a)
	2	14.1	
St. Lambert, Québec, Canada	1	11.5	Massé et al. (2010)
	2	11.9	
Stillwater, OK, United States	1	15.9	Aravindhakshan et al. (2010)
	2	15.4	
Frederick and Burneyville, OK, United States	1	13.6	Guretzky et al. (2011)
	2	18.0	

Source: Monti, A., Barbanti, L., Zatta, A., and Zegada-Lizarazu, W.: The contribution of switchgrass in reducing GHG emissions. *GBC Bioenergy*. 2011. 4. 420–434. Copyright Wiley-VCH Verlag GmbH & Co. KGaA. Reproduced with permission.

the changes in N status occurring in the soil only marginally depend on the simple input–output accounting (Oenema et al. 2003), given the influence of biological and environmental factors on soil N availability. It may be wrong to assume that the surplus nitrogen will entirely be leached. A 9-year experiment on oilseed rape showed that only 13%–25% of the budget surplus of nitrogen was actually leached out of the soil (Sieling and Kage 2006).

To overcome such inconsistencies, decision support systems (DSSs) have been developed to associate nutrient budgets with the assessment of soil N status in specific conditions (time of the year, soil depth, analysis method) and other support elements. Several versions of these methods have been developed in different countries to be used on various crops. The principle is to assess the amount of available nitrogen in a given soil layer (generally, mineral N at 0.6 or 0.9 m depth), in order to forecast soil availability during the growing season. Other parameters are then assumed to adjust the calculation: expected mineralization from the soil, organic fertilizers, and previous crops' residues, nutrient requirement for a standard or personalized yield potential, and residual amount of soil nitrogen at the end of the season.

The first operational method based on these principles was the nitrogen balance sheet (Rémy and Hébert 1977; Rémy and Viaux 1982): the difference between the previously calculated natural supply and crop requirement was the amount of fertilizers to be applied. The method has been used since the early 1980s for large-scale crops such as sugar beet and winter wheat, which are two potential energy crops. Thereafter, computer software was developed for the use of this method (Machet et al. 1990). At present, the method has been implemented as a soil submodel in a dynamic soil–crop model for wheat (Figure 2.8) (Jeuffroy and Recous 1999). The crop submodel simulates the growth and nitrogen uptake that are affected by crop light interception, radiation use efficiency, and the assessment of plant N status based on nutrient dilution curve. The model appears very informative as it covers both soil and plant behavior. It could be extended to other types of crops such as summer cereals and oilseeds for energy uses, but the first attempts in oilseed rape have been abandoned in favor of a simpler plant-based DSS.

Another soil-based method is based on electro-ultrafiltration (EUF) to extract mineral and short-chain organic fractions of nitrogen (Németh 1982) in a relatively shallow soil layer (0.3 m depth) several months before the cropping season (preceding summer). The EUF method, which was first developed for sugar beet (Wiklicky 1982) and more recently adapted to wheat and other crops (Mengel et al. 2006), has been widely used in Central Europe where the weather pattern is relatively predictable. The organic fraction of EUF nitrogen adds consistency to the system, making it possible to better cope with climatic anomalies.

Both balance sheet and EUF have been conceived for species (sugar beet, wheat) where nitrogen plays a delicate role, that is, excesses are to be feared as much as deficiencies. So far, the two methods have little expanded to grain or biomass crops for energy, where simpler systems are preferred. For instance, in oilseed rape, which is the main source of oil for biodiesel in Europe, the *easy rule for nitrogen* has been developed in France as a web application (CETIOM 2012): At the end of winter, the fresh weight of plants is measured in small (1 m^2) field plots or visually estimated. Then a few soil and crop information are fed to the system (deep/shallow soil; organic amendments supplied or not; yield goal range; cropping region), and the software calculates the N dose to be supplied in one or two spring applications. In practice, the software adjusts the dose to the season's effects on plant growth: the higher the growth and the nutrient uptake in the autumn–winter, the lower the amount of complementary fertilizer needed in the springtime and vice versa. By this method, up to 100 kg ha^{-1} of N may be saved by enhancing oilseed rape growth in the autumn, which represents 50% of the amount required by a top-producing (5 Mg ha^{-1}) crop (Figure 2.9) (Rathke et al. 2005). Timely crop establishment can also improve NE yield, thanks to the savings in N fertilizers at high energy cost (66 MJ kg^{-1} of N as urea; Frischknecht and Jungbluth 2003). In the same study of Rathke et al. (2005), seed protein and oil content described opposite upward and downward trends, respectively, in response to nitrogen. Since oil is the major component for the production of biodiesel, it is perceived that bioenergy crops sometimes involve compromises in fertilizer supply between contrasting yield and quality goals.

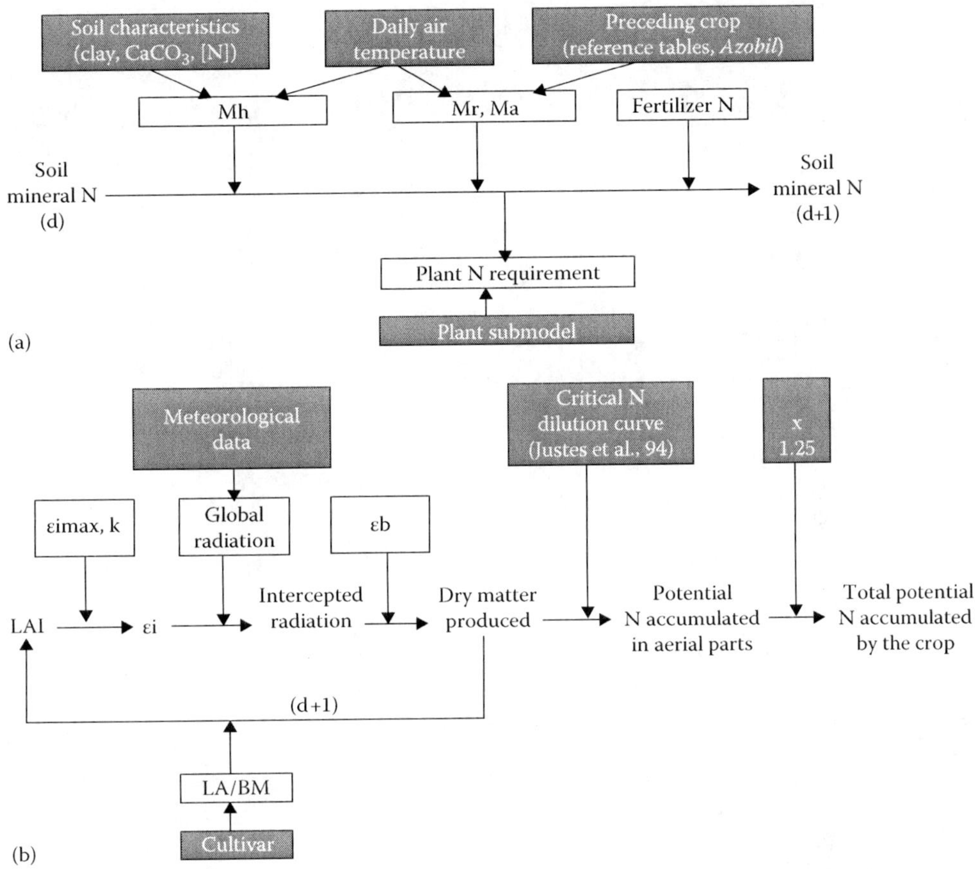

FIGURE 2.8 Flow charts of the simulation model of daily plant nitrogen requirements (a) and daily soil nitrogen availability (b). Mh, net mineralization of humus; Mr, net mineralization of crop residues; Ma, net mineralization of organic wastes; εi, fraction of incident radiation intercepted by the crop; εb, radiation use efficiency; εimax and k, maximum εi and extinction coefficient; LA/BM, ratio of leaf area over total aerial biomass; boxes in gray, submodel input. (Reproduced from *Eur. J. Agron.*, 10, Jeuffroy, M.H. and Recous, S., Azodyn: A simple model simulating the date of nitrogen deficiency for decision support in wheat fertilization, 129–144, Copyright 1999, with permission from Elsevier.)

Compared to cereals and oilseeds, less literature addresses the subject of nitrogen nutrition in biomass crops. The results of recent experiments may help to outline trends in use of N for these crops. Among annual species, biomass sorghum (*Sorghum bicolor* (L.) Moench) in its two forms of sweet and fiber sorghum is one of the most widely investigated. However, the research has addressed the effects of nitrogen on crop yield, quality, and associated traits such as nitrogen use efficiency (NUE) of the biomass sorghum, but has not tackled the issue of developing a DSS to assist in large-scale management. Generally speaking, biomass sorghum has showed a limited response to nitrogen fertilizer based on the two experiments on sweet sorghum in a Mediterranean environment (Cosentino et al. 1996, 2012): No significant yield increase was observed with an amount of nitrogen up to 120 and 180 kg ha^{-1} in the two respective cases, compared to a remarkable effect (>100% yield increase) determined by irrigation. Another experiment on sweet and fiber sorghum in a semi-continental environment (Barbanti et al. 2006) also showed that no yield response was observed with the use of nitrogen up to 120 kg ha^{-1} and the nutrient only increased the N content in plant tissues. However, since about 150 kg ha^{-1} nitrogen is removed with the biomass during harvesting (Cosentino et al. 1996, 2012; Barbanti et al. 2006), the supply of nitrogen is needed at least in the

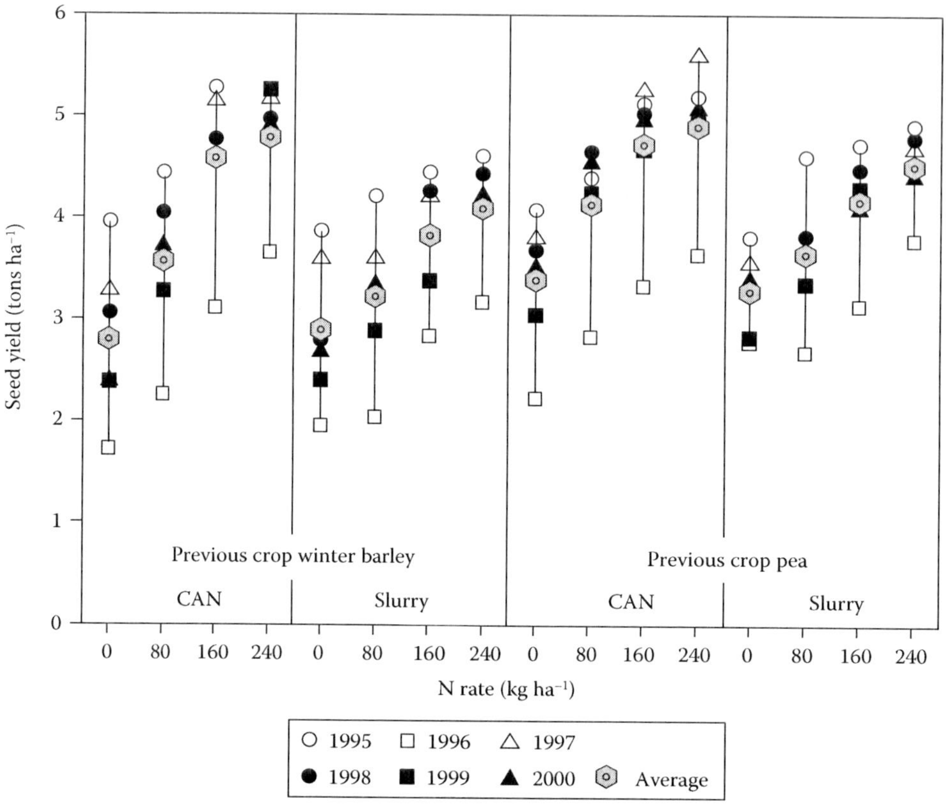

FIGURE 2.9 Seed yield of winter oilseed rape related to different rate and source of nitrogen supply. CAN, calcium ammonium nitrate; slurry, cattle manure slurry. (Reproduced from *Field Crops Res.*, 94, Rathke, G.W., Christen, O., and Diepenbrock, W., Effects of nitrogen source and rate on productivity and quality of winter oilseed rape (*Brassica napus* L.) grown in different crop rotations, 103–113, 2005, Copyright 2004, with permission from Elsevier.)

medium term. Considering the natural contribution from the soil, the applied N could be limited to an average 80–100 kg ha^{-1}, although a large variation may be envisaged between fields at good versus poor general fertility, variable use of organic residues, organic amendments (e.g., manure), etc. In all cases, the supply of N fosters a better utilization of soil-deriving N, as shown by the comparison between the low actual recovery of applied nitrogen (<20%) and the high apparent recovery (up to 100%) in fiber sorghum grown with a ^{15}N-labeled compost (Barbanti et al. 2011).

The response to applied N in perennial biomass crops is equally modest; the fact is consistent with the already discussed winter stock of N and subsequent remobilization in the springtime (Landström et al. 1996; Nassi o Di Nasso et al. 2010; Strullu et al. 2011). In *Arundo*, a positive response was often recorded with yield increases varying between 15% (100 vs. 50 kg ha^{-1} of N; Mantineo et al. 2009) and 30% (200 vs. 0 kg ha^{-1} of N; Angelini et al. 2005). In the latter case, fertilization (200–80–200 kg ha^{-1} of N–P–K) represented 78% of total energy inputs in the fertilized crop and determined only a 14% increase in NE yield compared to the unfertilized crop (Angelini et al. 2005). In cardoon (*Cynara cardunculus* L. var. *altilis* DC.), no response was observed beyond a basic supply of 50 kg ha^{-1} (Mantineo et al. 2009; Ierna et al. 2012). *Miscanthus* and switchgrass have featured a variable behavior: In a survey of 97 and 77 experimental crops of the two respective species, switchgrass showed a significantly stronger response to fertilizer N than *Miscanthus* (Figure 2.10) (Heaton et al. 2004). In specific field trials, *Miscanthus* responses to nitrogen ranged from insignificant (Christian et al. 2008; Mantineo et al. 2009) up to an approximate 50% and 100%

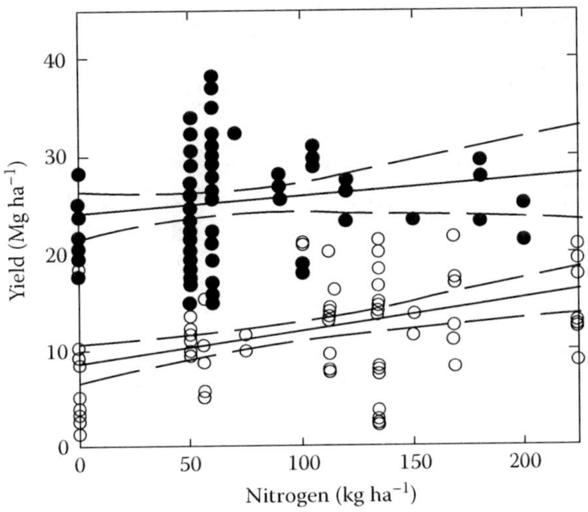

FIGURE 2.10 Response of the annual dry biomass yield of *M.* × *giganteus* (filled symbols) and *P. virgatum* (empty symbols) to nitrogen fertilizer in 97 and 77 cases of the two respective species. Solid lines indicate least square linear regression by crop; dashed lines represent 95% confidence limits. (Reproduced from *Biomass Bioenergy*, 27, Heaton, E., Voigt, T., and Long, S.P., A quantitative review comparing the yields of two candidate C_4 perennial biomass crops in relation to nitrogen, temperature and water, 21–30, 2004, Copyright 2003, with permission from Elsevier.)

under rainfed and irrigated conditions, respectively (Ercoli et al. 1999). The same large variation is perceived in switchgrass, where a survey on 13 nitrogen trials (Monti et al. 2011) showed an average 39% yield increase with a standard deviation almost as large (38%) (Table 2.4). As a general rule, 50–100 kg ha^{-1} year^{-1} sounds a reasonable amount of N to be supplied in perennial rhizomatous species based on the experimental results. As in sorghum, variations outside this range may be justified by local fertility, applications of manure, compost, etc.

2.3.2 Other Nutrients

Phosphorus (P) and potassium (K) are commonly addressed in crop fertilizer schemes. Other macro- (Ca, Mg, and S) and micronutrients (Cu, Fe, Mn, Zn, and B) are involved in specific soil/crop combinations. In energy crops, all these nutrients play a minor role, compared with nitrogen. However, a limited supply of the other nutrients will result in stunted growth and poor performance in sensitive species. This is the case of sulfur, exerting a special role in the nutrition of cruciferous species, thus showing a significant response in oilseed rape (Fismes et al. 2000).

In contrast to nitrogen, available P and K are normally determined in soil analysis and their contents do not substantially vary in the short term. Therefore, their assessment is a valuable guideline to support applications. Exchangeable K is generally well provided in fine-textured soils as in many South-European countries. Therefore, no K supply may be a common option even in crops where all the aboveground biomass is removed. Conversely, the P status may not easily be predicted and soil analysis remains a basic tool to support the decisions. Energy crops as a whole are neither profligate nor thrifty P and K consumers. When they are represented by highly demanding species such as maize, deficiency symptoms result in slow early growth (P) or stalk weakness (K), which will affect yield. On the other hand, when energy crops are represented by unpretentious species such as perennial rhizomatous grasses, deficiencies are seldom detected, although the lack of symptoms could hide the problem of a progressive loss in soil P and K status. The risk of incurring *soil mining*, as such circumstance is known (Van der Pol 1992), should therefore be avoided by checking P and K through soil analysis at regular intervals (3–5 years). Conversely, it makes little sense to deliberate

TABLE 2.4
Biomass Yield and Nutrient Balance with and without Nitrogen in a Series of Experimental Cases in Switchgrass

Location	Cuts Year^{-1}	Applied N (kg ha^{-1})	Biomass Yield (Mg ha^{-1})	N Balance (kg ha^{-1})	Sources
Beeville, TX, United States	1	0	3.8	—	Muir et al. (2001)
	1	168	14.3	—	
Stephenville, TX, United States	1	0	6.0	—	Muir et al. (2001)
	1	168	8.3	—	
Shorter, AL, United States	1	0	3.7	—	Ma et al. (2001)
	1	224	1.0	—	
Ames, IA, United States	2	0	8.0	−55	Vogel et al. (2002)
	2	120	12.0	0	
Mead, NE, United States	2	0	9.0	−90	Vogel et al. (2002)
	2	120	10.0	−10	
Chickasha, OK, United States	1	0	16.6	−149	Thomason et al. (2004)
	1	448	17.7	276	
Chickasha, OK, United States	3	0	20.9	−276	Thomason et al. (2004)
	3	448	22.5	129	
Perkins, OK, United States	1	0	9.2	−90	Thomason et al. (2004)
	1	448	10.3	322	
Perkins, OK, United States	3	0	11.5	−136	Thomason et al. (2004)
	3	448	11.9	264	
Blacksburg and Orange, VA, United States	2	0	11.0	−76	Lemus et al. (2008a)
	2	30	12.1	−71	
	2	90	13.3	−37	
Lucas and Wayne, IA, United States	1	0	3.9	−187	Lemus et al. (2008b)
	1	112	4.9	−113	
	1	224	5.2	−41	
Frederick and Burneyville, OK, United States	1	0	10.4	−41	Guretzki et al. (2011)
	1	135	15.0	32	
	1	180	15.5	67	
Frederick and Burneyville, OK, United States	2	0	12.2	−95	Guretzky et al. (2011)
	2	135	18.0	−48	
	2	180	22.8	−55	

Source: Monti, A., Barbanti, L., Zatta, A., and Zegada-Lizarazu, W.: The contribution of switchgrass in reducing GHG emissions. *GBC Bioenergy*. 2011. 4. 420–434. Copyright Wiley-VCH Verlag GmbH & Co. KGaA. Reproduced with permission.

fixed amounts of P and K to be supplied every year, although this approach is implicitly suggested by the structure of the experiments reported in the literature. In this respect, it should be noted that no research has so far addressed the use of P or K in specific field trials on biomass crops: The two nutrients have only been tested in association with N (Angelini et al. 2005), further proving the limited expectations associated with their use. It is therefore difficult to indicate reference P and K doses for energy crops. P and K have a modest cost in terms of energy (15 MJ kg^{-1} of P as triple superphosphate and 21 MJ kg^{-1} of K as potassium sulfate; Frischknecht and Jungbluth 2003); therefore, their supply does not significantly affect energy balance. More to this, the supply of P and K in no-till systems and in perennial herbaceous species raises the problem of a shallow stratification of these two nutrients. In no-till with annual species, the problem may be partially overcome by nonstandard application techniques such as soil injection with liquid fertilizers and placement aside or below the seed.

Both methods proved effective in maize growing on K-deficient soils after 12 years of no tillage (Vyn et al. 2002). In perennial herbaceous species, the problem could be prevented by massive applications before planting, followed by lighter doses during crop life. Associated with nutrient stratification is the problem of diagnosis: It is debated whether soil sampling should be maintained at a certain depth also in no tillage or carried out in a shallower layer; the latter case involves that specific calibration curves be available to account for higher nutrient concentration (Duiker and Beegle 2006).

2.3.3 Fertilizers

Fertilizer type does not represent a specific issue in energy crop production. Basically, the products to be used should deliver the nutrients required in the forms most suitable to comply with soil characteristics, in order to minimize processes limiting nutrient availability (immobilization, adsorption, etc.). The use of advanced products featuring various combinations of macro- and micronutrients, slow-release forms, etc., is often superfluous. Sometimes, also the use of complex fertilizers, that is, those featuring a N–P–K (or N–P, N–K) combination, is not convenient compared to the equivalent amounts of simple fertilizers at specific crop stages. Conversely, the use of organic fertilizers, biosolids, and slurries from agroindustrial and urban activities is gaining interest, as these products are a source of organic matter and nutrients for soils, which in areas at intensive farming are progressively being depleted. In addition, they foster a valorization of waste products outside the food chain, helping to fend off the fears for food security originated by wastes of various nature.

The application of biosolids and slurries from energy conversion plants to the soils cultivated at energy crops feeding the same plants allows to directly recycle nutrients. These products include ashes from combustion processes, slurries from anaerobic digestion (digestate), and a carbon residue of pyrolysis (biochar). They serve different uses, depending on their heterogeneous characteristics and suitability for specific aims. For instance, the liquid, solid, and solid composted fraction of a slurry from the anaerobic digestion of maize silage showed a different potential in terms of soil C storage (best treatment, composted slurry) and N supply to plants (best treatment, liquid slurry) (Grigatti et al. 2011). At present, further studies are needed to define the best application practices while complying with national and local legislations ruling their use. Finally, using organic fertilizers, by-products, and residues as nutrient sources entails significant savings in the amount of energy used for the production of mineral fertilizers, although such savings are partly offset by the higher energy consumption involved in the handling, transport, and application of organic products.

2.3.4 Site-Specific Management of Crop Nutrition

Variable rate application (VRA) of fertilizers is gaining consensus amid precision cropping techniques. The principle is to monitor the differences in soil characteristics through the field by sampling and analyzing according to a grid. Once acknowledged, the differences should lead to differential fertilizer doses in order to overcome constraints in areas at potentially high yields or to avoid excesses in areas at limited productivity. This approach is seldom followed in practice, in view of the high costs to analyze, for example, a 50 × 50 m grid, meaning four separate analyses per hectare. As an alternative, real-time monitoring of soil properties or crop behavior provides clues to the amount of nutrients, namely, nitrogen, to be applied in specific field spots. On-the-go soil sensors based on several chemical and physical principles determine various soil properties including texture, organic matter content, compaction/bulk density, pH, and nitrates (Adamchuk et al. 2004): among them, electrical and electromagnetic sensors are still the cheapest, most widely used and with a broad spectrum of analysis (Table 2.5). Electrochemical sensors addressing soil chemical fertility are on the rise, although the need for recurrent calibration and some data inconsistencies still hamper their use (Adamchuk et al. 2005).

On-the-go sensing of crop nutritional status encompasses a vast array of tools and techniques, spanning from the ground-based measurement of spectral reflectance through tractor-mounted

TABLE 2.5
Soil Properties Targeted with On-the-Go Sensing Methods

Soil Properties	Electrical, Electromagnetic	Optical, Radiometric	Mechanical	Acoustic, Pneumatic	Electrochemical
Soil texture	X	X		X	
Soil organic matter	X	X			
Soil moisture	X	X			
Soil salinity	X				X
Soil compaction, bulk density			X	X	
Depth variability[a]	X		X	X	
Soil pH		X			X
Residual NO_3, TKN[b]	X	X			X
K content					X
CEC[c]	X	X			

Source: Adapted from *Comput. Electron. Agric.*, 44, Adamchuk, V.I., Hummel, J.W., Morgan, M.T., and Upadhyaya, S.K., On-the-go soil sensors for precision agriculture, 71–91, Copyright 2004, with permission from Elsevier.

[a] Depth of topsoil or hard pan detection.

[b] Total Kjeldahl nitrogen.

[c] Cation exchange capacity (and other buffer indicators).

spectrometers to remote sensing, a term including both aerial (manned or drone aircrafts) and space (satellite) sensing. Ground-based sensing proved its effectiveness in assessing maize nitrogen status through reflectance measurements at different wavelengths of the visible and infrared spectra, discriminating plant nitrogen deficiencies during the vegetative stage (Graeff and Claupein 2003). Likewise, in winter wheat the simultaneous measurement of canopy reflectance and incident radiation by a tractor-mounted double spectrometer showed a close relationship ($R^2 = 0.95$) between canopy reflectance and nitrogen nutrition index (NNI), which is the trait dynamically indicating crop N status during growth (Mistele and Schmidhalter 2008). In both cases, ground-based reflectance successfully replaced time-consuming analysis of plant tissues, providing sensible information for real-time adjustments of fertilizer dose in a single-pass strategy (tractor rear-mounted fertilizer spreader).

Remote sensing has a greater potential on large crop surfaces than ground-based sensing: satellite sensing is a promising field of development, thanks to the continuous progress in resolution and wave band readings. However, so far aerial sensing has proved more reliable, representing a compromise between ground and satellite base. The advantages and disadvantages of the three sensing platforms are discussed and summarized by Scotford and Miller (2005) (Table 2.6); the picture is in rapid evolution, anyway.

The integration of different approaches is seen as the best strategy to enhance the benefits of a variable rate distribution of fertilizers while at the same time compensating for reciprocal weak points. From a conceptual point of view, yield mapping coupled with soil/crop sensing means passing from farming by balance (mapping approach) or by growth (real-time approach) to farming by sustainability (Auernhammer 2001) (Figure 2.11).

2.4 WATER MANAGEMENT

Water is the plant growth factor most often available in short supply. The natural sources, or *green water* according to the water footprint (WF) approach (Hoekstra et al. 2011), are insufficient to sustain the world output of food, feed, and nonfood products, including energy. Agriculture is thereby

TABLE 2.6
Summary of the Main Advantages and Disadvantages of Different Sensing Platforms

	Space	Aerial	Ground
Area covered per scan	Area scanned increases with platform height		
	Typically km^2	Typically m^2	Typically cm^2
Spatial resolution (pixel size)	Resolution coarseness increases with platform height		
	1–30 m^2	0.05–2 m^2	mm^2 to cm^2
Temporal resolution	Weeks	Days	Hours
Effect of cloud cover	Influence of cloud increases with platform height		
	Heavy influence	Moderate influence	Not affected
Effect of local illumination conditions	Influence of illumination conditions decreases with height		
	Not affected	Moderate influence	Heavy influence
Availability of data to end user	Long delays	Some delays	No delays
Control of end user	Limited control	Some control	Full control

Source: Reproduced from *Biosyst. Eng.*, 90, Scotford, I.M. and Miller, P.C.H., Applications of spectral reflectance techniques in Northern European cereal production: A review, 235–250, Copyright 2005, with permission from Silsoe Research Institute.

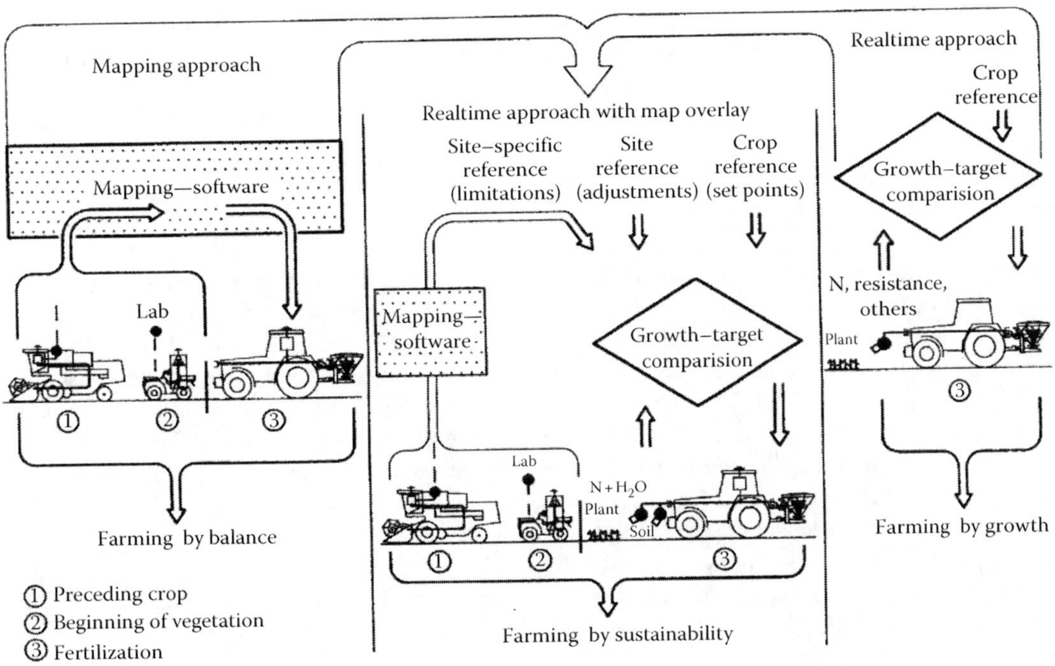

FIGURE 2.11 Systematic approaches for the implementation of site-specific fertilization. (Reproduced from *Comput. Electron. Agric.*, 30, Auernhammer, H., Precision farming—The environmental challenge, 31–43, Copyright 2001, with permission from Elsevier.)

the primary consumer of irrigation water (*blue water*), consuming up to 80% of available water resources at world level (Condon et al. 2004). In turn, irrigated agriculture supplies about 40% of world food production (Johnson et al. 2001). The growing population together with the increase in *per capita* consumptions and economic activities will increase the future competition for water resources. In this frame, agriculture shall be more efficient in coping with a decreasing availability while meeting the demand for products and services of agricultural origin. The commitment toward a more efficient use of water is further strengthened by the circumstance that the WF of agroenergy is 70–400 times higher than the WF of other RESs such as wind and thermal solar energy (Gerbens-Leenes et al. 2009); the wide range depends on the agricultural productivity in different areas of the globe.

It is generally acknowledged that providing food is the primary task of agricultural activities. Therefore, energy crops will only access a limited share of water resources, remaining more exposed to drought. The efforts to enhance water use efficiency (WUE), that is, the amount of dry biomass or energy obtained per unit water consumed (kg or MJ m^{-3}), are pursued not only through agronomic progress and breeding but also through the associated biotechnology that is used to improve plant water-related traits. For instance, they are credited with a 50% potential in enhancing photosynthetic efficiency at a leaf scale (Long et al. 2006), which is more than the potential to increase light interception and biomass partitioning efficiency exploited by breeding programs up to present.

2.4.1 Enhancing Green Water Efficiency

Enhancing the efficiency in the use of precipitation and soil moisture is the best strategy to cope with the limited resource. There are three basic ways in which a plant can adapt to drought stress (Ludlow 1989): (1) tolerance, involving the breeding of genotypes better performing under drought stress (active photosynthesis at lower leaf–water potentials, higher transpiration efficiency, etc.); (2) avoidance, indicating a plant's attitude to withstand a transient dry period through various mechanisms (osmotic adjustment, stomata closure, etc.); and (3) escape, involving a shift of plant cycle with respect to the dry period (e.g., autumn vs. spring seeding wherever feasible). The three strategies and the overall management of limited water resources should be combined in order to integrate improvement of crop performance with reduction of water losses. The task involves a major revision of the whole agricultural system, especially as it concerns the reduction of the amount of water lost to evaporation, runoff, and drainage (Passioura 2006). In the Mediterranean region, a water saving up to 52% at no yield cost is estimated if all the measures in improving water conveyance to the field, use of waste water for irrigation, better irrigation scheduling, and drip irrigation in row crops are implemented (Jacobsen et al. 2012).

A successful method to enhance green water efficiency is represented by rainfall harvesting: It consists of shaping soil surface in alternating ridges and furrows, in order to concentrate precipitation in the furrows where seeds are planted (Figure 2.12). In the semiarid Loess Plateau of Northwestern China, this practice has extensively been investigated in row crops such as maize and, lately, sweet sorghum. The best combination is represented by plastic-covered ridges and mulched furrows with gravel–sand or straw. Areas at decreasing yearly rainfall (down to 300 mm) necessitate higher ridge-to-furrow ratios (up to 1:1) to harvest more rainfall per unit soil actually planted (Li and Gong 2002). Rainfall harvesting was successfully associated with supplemental irrigation (SI) to better deal with a transient period of low moisture, enhancing yield and WUE of maize (Li and Gong 2002) and sweet sorghum (Wang et al. 2009). Doubts remain in the costs for the construction of the plastic-covered ridge and mulched furrow system: either the procedure is manually done in developing countries with cheap manpower, or a highly mechanized system must be developed in order to assure a consistent trade-off.

Beside rainfall harvesting, soil moisture accumulation during fallow may be enhanced by tillage practices: in sunflower growing on clayey soils in Morocco, plowing was seen as the best tool to increase soil moisture in wet years, whereas in years at dry fallow period, reduced tillage involved

FIGURE 2.12 A photograph showing ridge and furrow rainfall harvesting system. (Reproduced from *Agric. Water Manag.*, 99, Wang, Y., Xie, Z., Malhi, S.S., Vera, C.L., Zhang, Y., and Wang, J., Effects of rainfall harvesting and mulching technologies on water use efficiency and crop yield in the semi-arid Loess Plateau, China, 374–382, Copyright 2009, with permission from Elsevier.)

a lower exposure of soil layers to evaporation losses (Aboudrare et al. 2006). Moreover, a simulation study in the cited experiment showed a steadier yield and WUE at intermediate levels of soil moisture at seeding, because high moisture leads to an excess in leaf expansion responsible for more rapid moisture depletion during the reproductive stage.

It appears therefore that nonunivocal strategies should be followed in the quest for higher WUE in energy crops. Perennial species for biomass production are seldom investigated as it concerns this topic; thanks to their extended growth season and limited (e.g., switchgrass) or no biomass partitioning to seeds (e.g., *Arundo* and *Miscanthus*), these species should better comply with intermittent drought, exploiting periods of good moisture whenever they occur. This in turn should lead to more consistent behavior across years at different weather patterns.

2.4.2 Irrigation Scheduling

When water is available in full supply, the moral and legal obligations to save it for food crops cease, while the problem about how to efficiently use it arises. Irrigation science has developed several ways to enhance irrigation water use efficiency (IWUE), as the premise to use the lowest amount of water to obtain the highest result. Among them, deficit irrigation means supplying a lower amount of water with respect to potential consumption. When a plant is subjected to water shortage, it enacts mechanisms of water saving, for example, partial stomata closure, decreasing the amount of transpired water. Photosynthesis slowdown is less than proportional to transpiration reduction, and the plant resumes full activity as soon as moisture is restored. The amount of water used during crop evapotranspiration (ET_C) determines water consumption under nonlimiting conditions (Allen et al. 1998). In a field practice, only a fraction of the difference between ET_C and natural supply is added during deficit irrigation. Deficit irrigation has been frequently proved effective in restraining yield loss and enhancing WUE when applied at critical stages of potential energy crops including wheat (IWUE, +51%; Tavakkoly and Oweis 2004) (Figure 2.13), maize (WUE, +17%; Payero et al. 2009), and sunflower (WUE, +17%; Karam et al. 2007). Much work has been carried out to detect

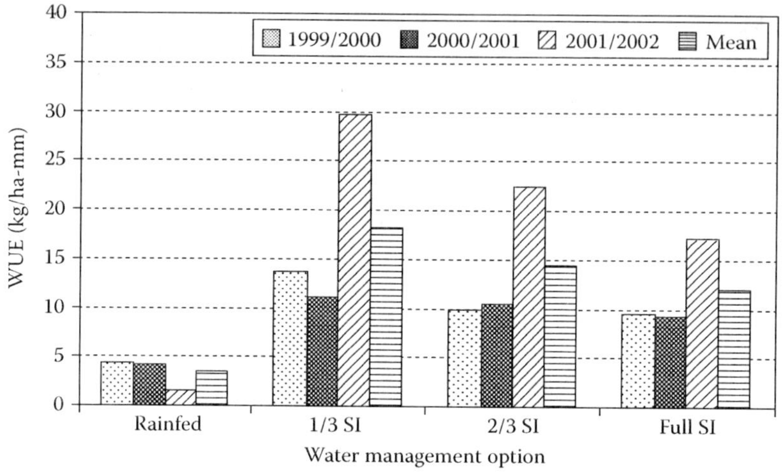

FIGURE 2.13 Precipitation and irrigation WUE in producing wheat grain with various management options under dry continental climate. IWUE was calculated as the ratio of the difference between irrigated yield and rainfed yield to the amount of irrigation water used. SI = supplemental irrigation. 1/3, 2/3, and full SI = 95, 151, and 207 mm, respectively (3-year averages). (Reproduced from *Agric. Water Manag.*, 65, Tavakkoli, A.R. and Oweis, T.Y., The role of supplemental irrigation and nitrogen in producing bread wheat in the highlands of Iran, 225–236, 2004, Copyright 2003, with permission from Elsevier.)

the critical stages for water supply: In grain crops such as cereals and oilseeds, they generally cover the early reproductive phase (flowering to seed set/milk dough); in biomass crops, they are more difficult to seize, as early vegetative stages may show a higher sensitivity than later ones (Mastrorilli et al. 1999). In perennial biomass crops, as the research on irrigation is still at an early stage, detecting critical stages does not appear a top priority.

A special form of regulated deficit irrigation is alternate partial root-zone drying (PRD): It consists in a frequent irrigation only on one side of the crop row, alternating sides at 10–15-day intervals. The progressive drying of half root system originates a signal (abscisic acid) regulating stomatal aperture, thus protecting the plant from excessive transpiration; shifting sides of the root apparatus assures the maintenance of a high hormonal signaling. The method that has been devised for vineyards and fruit tree crops at large row spaces involves two parallel drip irrigation lines. However, the potential benefits have been investigated also in herbaceous crops such as maize plants grown in split-root pots: PRD achieved a higher WUE (5.83 kg m^{-3}) than deficit irrigation (5.55 kg m^{-3}), compared to 5.21 kg m^{-3} for full irrigation (Wang et al. 2012). In the same experiment, PRD also determined a greater N accumulation in leaves, as the likely effect of enlarged root system. Due to the high costs, PRD could not be applied to energy crops with sprinkler or drip irrigation methods, whereas it is possible to use furrow irrigation, a method at intrinsic low water efficiency. In maize grown with this method, PRD exhibited contrasting results: higher grain yield (+24%; Kang et al. 2000) and also WUE (+8%; Du et al. 2010) than deficit irrigation, and analogous yield and IWUE in another experiment (Kirda et al. 2005).

As regards irrigation methods for energy crops, they are constrained by the cost of irrigation methods and aimed for a high efficiency in water application and use. Therefore, an irrigation method should have a reasonable cost and an acceptable efficiency in delivering water. A method at low application efficiency such as furrow irrigation should not be adopted in areas at limited water resources. Sprinkler irrigation in its various forms should play a key role in energy crops. There are several ways to improve the efficiency of irrigation. Among them, low-energy precision application (LEPA) is a method applied in center-pivot or lateral-move systems, which consist of small water sprayers hanging down from the spraying boom close to the soil, instead of spraying water from the

FIGURE 2.14 LEPA sprayers mounted on center-pivot irrigation system. (Reproduced from *Lo sviluppo delle colture energetiche in Italia—Il contributo dei progetti di ricerca SUSCACE e FAESI*, http://ing.entecra.it/Biomasse, Cosentino, S.L., Mastrorilli, M., Saita, A., and Scalici, G., The cultivation of biomass sorghum in southern Italy, 517–538, Copyright 2011, with permission from the authors.)

top (Figure 2.14). This method showed a remarkable efficiency in delivering water to the crop, in exchange for an affordable cost (Lyle and Bordovsky 1981).

Precision irrigation is promising although it is less popular than precision fertilization at this moment. Two ways are envisaged for a variable application of water: (1) In mobile irrigators (traveling guns), the travel speed over the field length may be varied in order to compensate for the uneven storage capacity of the soil (Al-Karadsheh et al. 2002) or a different occurrence of plant stress symptoms (Zhang et al. 2011); (2) in center-pivot and lateral-move irrigators, solenoid valves operate a differentiated control of each nozzle (Al-Karadsheh et al. 2002). In the latter case, water supply can better comply with the variation in soil moisture/plant stress, although a higher cost is involved. The VRA of water most often relies on remote sensing. Aerial sensing through airborne thermal and multispectral images was proved effective in assessing both acute and chronic water stress of wheat (Abuzar et al. 2009). For this purpose crop water stress index (CWSI) (Jackson et al. 1981), which is based on the relationship between leaf and air temperature, was modified to account for the chronic component that is significantly relevant in semiarid regions where full groundcover is rarely achieved. In a similar case, Landsat satellite images with a spatial resolution of 28.5 m were used to assess some traits that were implemented to determine variable rates of irrigation according to a water balance model (Figure 2.15) (El-Nahry et al. 2011). Among these traits, a prominent role was played by the normalized difference vegetation index (NDVI), based on the relationship between near infrared and visible radiation and correlated with plant size and vigor. A 23% potential saving of irrigation water (145 mm) was thereby possible on a 62 ha pivot surface cultivated with maize in Egypt.

2.4.3 Irrigating Energy Crops

Energy crops exhibit a variable suitability to irrigation. Among cereals, maize is intrinsically efficient in the use of water, thanks to a C_4 photosynthetic pathway, but is exposed to summer drought

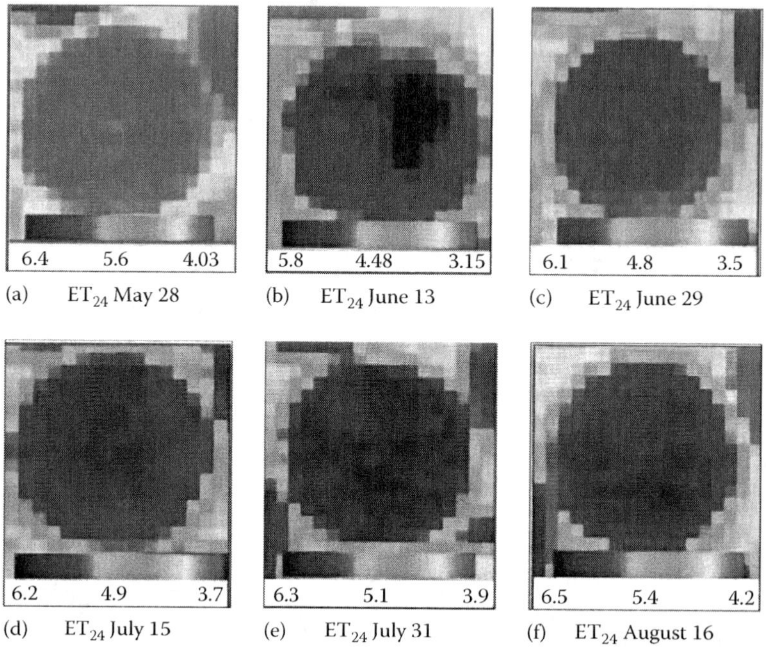

FIGURE 2.15 (Color figure at CRC website http://www.crcpress.com/product/isbn/9781466505520.) Instantaneous ET (mm day^{-1}) in different dates during maize growth in a warm dry environment. (Reproduced from *Agric. Water Manag.*, 98, El-Nahry, A.H., Ali, R.R., and El-Baroudy, A.A., An approach for precision farming under pivot irrigation system using remote sensing and GIS techniques, 517–531, 2011, Copyright 2010, with permission from Elsevier.)

in warm dry climates. Grain sorghum is another C_4 cereal showing a better adaptation to drought, thanks to higher WUE, IWUE, and harvest index and soil moisture depletion potential (Farré and Faci 2006). Grain sorghum can therefore substitute maize as a rainfed crop for energy uses (ethanol, biogas) in areas prone to dry spells where irrigation is not available.

Cold-season cereals for energy uses include winter wheat, barley, rye, triticale, and other minor species. They all have C_3 photosynthetic pathway, so are intrinsically less efficient in the use of water than C_4 cereals. However, as most of their growth occurs during the wet season in temperate–warm climates, they can take advantage of natural precipitation and close crop cycle at the beginning of summer drought. Winter wheat, the main representative of this group, is often grown as a rainfed crop, although the benefits from SI, that is, the irrigation applied in arid zones to winter crops normally grown in rainfed conditions (Oweis et al. 1998), have been reported in several trials in dry areas of China (Du et al. 2010) and in the West Asia–North Africa region (Tavakkoli and Oweis 2004), where irrigation between 1/3 and 2/3 of full supply proved effective in boosting wheat yield and WUE (Figure 2.13) (Tavakkoli and Oweis 2004; Karrou and Oweis 2012). Therefore, it may be argued that winter cereals should not be irrigated, although in regions where they may substitute summer cereals while saving considerable amounts of water, this option should not be excluded.

In oilseed crops, irrigation is a less common practice than in cereals. Oilseed rape is a cold-season species that has the same cycle as wheat and can take advantage of natural precipitation for most of its growth. However, in case of severe drought, irrigation at the flowering stage may be envisaged as the single intervention most contributing to sustain yield and WUE (Istanbulluoglu et al. 2010). Sunflower is another major oilseed that grows in the warm season, although its low thermal requirement and good resistance to cold foster a late-winter seeding in many temperate areas. This helps sunflower to better use soil moisture reserves and precipitation until the onset of summer drought. Nevertheless, irrigation may prove useful in the reproductive stage: in fact, the crop thrived when repeatedly irrigated between heading and milk stages (Göksoy et al. 2004), although WUE and the economic return were enhanced when irrigation was restrained to a single, large watering at flower bud stage (Rinaldi 2001) or at flowering (Göksoy et al. 2004).

In biomass crops irrigation has been so far less studied than in grain crops; only biomass sorghum (sweet and fiber genotypes) has been extensively investigated. In the subhumid environment of Northern Italy, irrigation did not enhance fiber sorghum growth in normally wet years (Amaducci et al. 2000). In a typical Mediterranean environment, the dry biomass of sweet sorghum increased almost fourfold from 7.5 Mg ha^{-1} (rainfed) to 27.5 Mg ha^{-1} (full irrigated), the largest single increase being obtained with deficit irrigation at 50% full irrigated volume (20.1 Mg ha^{-1}) (Cosentino et al. 2012). An even stronger increase was recorded in another Mediterranean environment, where sweet sorghum barely survived without irrigation (4.4 Mg ha^{-1} of dry biomass), attaining a peak yield of 32.4 Mg ha^{-1} at full irrigation (Sakellariou-Makrantonaki et al. 2007). Therefore, in the two cited experiments, the crop proved very responsive to irrigation, although the amounts of water applied at full irrigation (517 and 659 mm, respectively) appear almost prohibitive. Deficit irrigation applied in most sensitive stages appears the only strategy for sustainable sorghum cropping in Mediterranean environments.

Perennial rhizomatous grasses for biomass have been scarcely investigated in terms of irrigation. The interaction of irrigation and N fertilization could increase the yield of *Miscanthus*, which is another C_4 species. An approximate 40% increase in biomass and NE yield was observed at optimum N supply (Ercoli et al. 1999). In a typical Mediterranean environment, shifting irrigation from 25% to 75% of full supply increased the *Miscanthus* yield by 35% (Mantineo et al. 2009). Both results are consistent with the high responsiveness of *Miscanthus* to rising precipitation across 97 experimental cases (Figure 2.16) (Heaton et al. 2004).

Other perennial grasses have been less investigated than *Miscanthus*. Switchgrass was proved to be less responsive to natural precipitation across 77 experimental cases than *Miscanthus* (Figure 2.16) (Heaton et al. 2004), although an expansion outside its native range in the continental United States is predicted only if adequate moisture can be assured (Barney and DiTomaso 2010). *Arundo* is considered a

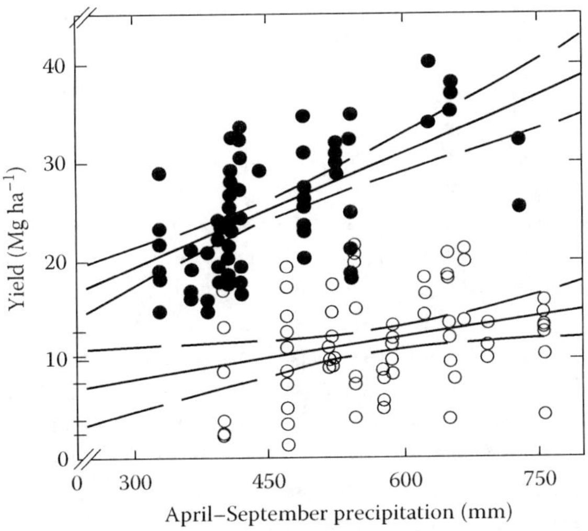

FIGURE 2.16 Response of the annual dry biomass yield of *M.* × *giganteus* (filled symbols) and *P. virgatum* (empty symbols) to seasonal precipitation in 97 and 77 cases of the two respective species. Solid lines indicate least square linear regression by crop; dashed lines represent 95% confidence limits. (Reproduced from *Biomass Bioenergy*, 27, Heaton, E., Voigt, T., and Long, S.P., A quantitative review comparing the yields of two candidate C_4 perennial biomass crops in relation to nitrogen, temperature and water, 21–30, 2004, Copyright 2003, with permission from Elsevier.)

thrifty species with respect to water: shifting irrigation from 25% to 75% of full supply in the aforementioned experiment with *Miscanthus* only increases the yield by 14% (Mantineo et al. 2009). At last, in the same experiment cardoon responded to the increase of irrigation with an intermediate 26% yield increase, but with a higher interannual variation (Mantineo et al. 2009).

Among tree crops for energy, interest has recently been raised by *Jatropha* (*Jatropha curcas* L.), the bush/tree of subtropical origin proposed as a source of nonedible oil for energy uses (Gübitz et al. 1999). *Jatropha* was observed to widely adapt evapotranspiration (ET) to actual moisture conditions, varying seed yield between 600 and 1560 kg ha^{-1} $year^{-1}$ (Kesava Rao et al. 2012). Therefore, suitable areas and techniques of cropping should be identified, in order to avoid competition with food crops for limited water resources. In tree species for short rotation forest (SRF), irrigation does not appear as a major topic and the focus is more on plant adaptation to drought. In poplar (*Populus* spp.), the wide interclonal variation in ecophysiological traits may be exploited to select genotypes suited for either moist (high photosynthetic rate) or dry (high belowground biomass and WUE) environments (Gong et al. 2011).

2.5 CROP PROTECTION

Crop protection from competing weeds, parasitizing microorganisms (fungi, bacteria, and viruses), and feeding pests (insects and other invertebrates) is a challenging task in energy crops. The comprehensive activity exerted by the three organisms goes under the definition of plant biotic stresses, mirroring the abiotic stresses determined by unfavorable ambient conditions such as drought, nutrient deficiency and frost. In food/feed crops, controlling biotic stresses involves a stronger commitment than in energy crops, in terms of the quantitative (growing world population) and qualitative (threats to food safety) implications. Nevertheless, in energy crops weeds, pests, and diseases curb yields, which in turn jeopardizes the supply of energy, increases the uncertainties of agroenergy initiatives, and reduces EE and sustainability to the detriment of the whole production chain. Moreover, curtailing the means of protection to save money contrasts the pursuit of economic efficiency whenever insufficient control becomes the limiting factor hampering a full expression of other crop inputs. Given this premise, energy crop protection will only be outlined as a vast subject for detailed discussion.

2.5.1 Potential and Actual Damage

Assessing potential damage is the cornerstone for a sustainable strategy of crop protection. Unfortunately, the sensitivity to weeds, pathogens, and pests greatly varies depending on plant group (e.g., perennial vs. annual species), natural versus bred species (e.g., *Arundo* vs. maize), species diffusion and crop intensification. This is the case of *Jatropha*, a promising oil species for tropical environments that is quite pest- and disease-free when growing in isolation, whereas in plantations it is exposed to attacks often requiring chemical sprayings (Franke and Nielsen 2010).

The overall damage due to pests and diseases in major arable crops and the efficacy of control measures were last assessed by Oerke and Dehne (2004): At a world level, weeds are considered the principal threat (32% loss potential) but also allowing the best control (71% efficacy), thanks to herbicides and also hand weeding in developing countries. Animal pests (namely, insects) and fungi plus bacteria own a respective 18% and 15% loss potential associated with a 42% and 34% control efficacy, respectively. Lastly, viruses are credited with the least potential damage (3%) but also the weakest control (13% efficacy). However, the general picture greatly varies when referring to different world regions: In developed areas, efficient means of control are deployed, namely, herbicides and pesticides, whereas in developing areas, less effective tools are available (tillage, crop rotation, mechanical/hand weeding, etc.). As a result, the potential damage due to pests and diseases is controlled up to 60% in the former areas and between 40% and 50% in the latter ones (Figure 2.17) (Oerke and Dehne 2004).

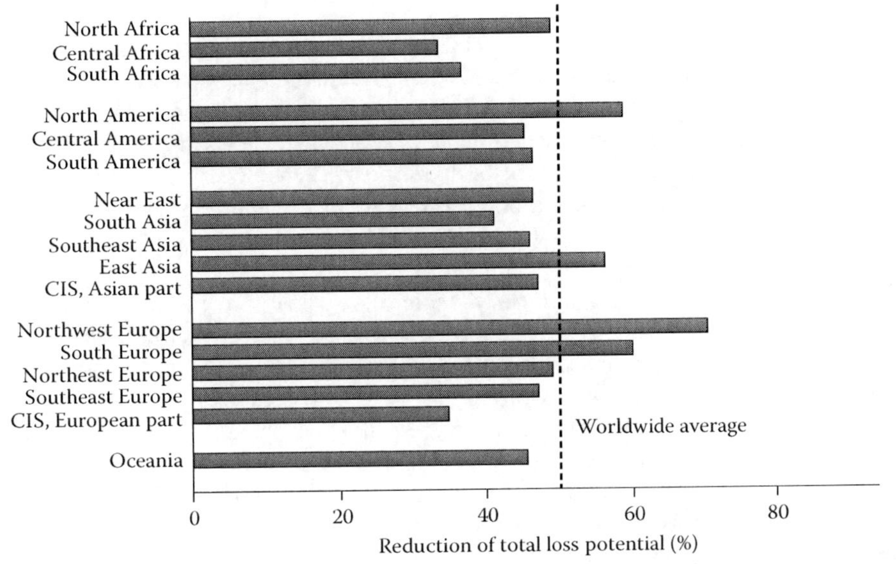

FIGURE 2.17 Regional differences in the overall efficacy of actual crop protection practices in wheat, rice, maize, barley, potatoes, soybean, sugar beet, and cotton. (Reproduced from *Crop Protect.*, 23, Oerke, E.C. and Dehne, H.W., Safeguarding production—Losses in major crops and the role of crop protection, 275–285, 2004, Copyright 2003, with permission from Elsevier.)

2.5.2 Integrated Pest Management

The financial constraints add to the policies of pesticide reduction that are pursued in many developed countries. In the EU, a review process addressing pesticide registration and usage is under way, resulting in the outplacement of several active ingredients: Under European Directive 2009/128/EC on the sustainable use of pesticides, national action plans are being implemented by member states, involving that pesticides be used in a framework of integrated pest management (IPM) in all farms by the year 2014. As a result of this process, it is feared that the reduction in the number of active ingredients will endanger the current level of agricultural production while increasing the risk of pesticide resistance in the targeted pests (Hillocks 2012). To avoid this, a deeper understanding of IPM practices and access to information and technical support will be needed, allowing farmers to adapt to the new scenario. Besides the limits set by legislation, the IPM approach to crop protection owns remarkable advantages, integrating all appropriate measures to control pest populations while keeping pesticides at economically justified levels and minimizing the risks to human health and the environment. The concept was originally developed for pests during the second half of the twentieth century (Dent 1995); thereafter, its principles were extended to weeds and diseases, although these latter organisms rely on fewer means of biological control compared to pests. IPM concepts are particularly well suited to energy crops, improving their energy and economic balances while coping with a limited number of active ingredients. The latter is a typical occurrence when growing crops of recent introduction (e.g., *Arundo* and *Miscanthus*), counting on very few active ingredients registered for specific use.

IPM practices fit in the larger category of integrated crop management (ICM) in view of enhancing environmental sustainability while supporting farm income. Integrated weed control is a typical example of IPM merging with broader ICM; the scarcity of registered herbicides in such crops as biomass sorghum has prompted a shrewd combination of chemical and mechanical weeding; false seeding, involving timely seedbed preparation and subsequent weed suppression (chemical or mechanical) just prior to seeding, greatly contributes to abate weed pressure before seeding. Thereafter, hoeing completes the effect of residual herbicides applied in row bands at seeding. In the case of diseases, the combination of direct means of control with sources of genetic resistance is another successful application of IPM: in sugar beet, a species at potential ethanol and biogas production, cultivars bred with significant levels of resistance to cercospora leaf spot delay disease progression and require a lower number of fungicide sprayings to keep a healthy canopy until harvest (Cioni et al. 2012).

The advent of genetically modified (GM) plant varieties provided with the resistance to nonselective herbicides or pests eases the task of protecting large-scale energy crops as maize, oilseed rape, and soybean. In the former case, a nonselective herbicide can be sprayed killing the whole weed flora while being detoxified by the GM plant. In the latter case, a pest of difficult control (e.g., maize stem borer) is thwarted by a toxin secreted by plant tissues, thanks to a DNA fragment from a pest parasite (e.g., the bacterium *Bacillus thuringiensis*). Despite such advantages, GM crops are still opposed by part of the public opinion especially in Europe. However, it is perceived that the persisting hostility should be mitigated when dealing with crops grown outside food/feed uses. This means that GM plants whose biomass is entirely consumed in the energy sector should be favored over plants whose by-products and residues are used in a food/feed chain. In this light, second-generation energy conversion processes exploiting the whole biomass should more easily gain the public approval than first-generation processes leaving relevant amounts of residues. Likewise, new uses should be sought for GM coproducts such as cereal dry distilled grains with solubles (DDGS) and oilseed meals in order to rule them out of the food/feed sector.

2.5.3 Precision Application of Pesticides

Precision spraying of pesticides is a practice of growing interest, avoiding off-target application errors and saving valuable amounts of active ingredients wherever superfluous; both effects contribute to the goals set by Directive 2009/128/EC and similar legislations around the world. In a recent study (Luck et al. 2010), map-based automatic boom section control technology applied to a 25 m boom sprayer with 30 control sections (application control resolution, 0.83 m) allowed an average 16% reduction in the sprayed area compared to the total area covered by the spray boom in fields at highly irregular shape. In the same experiment, a lower control resolution (5 m, corresponding to 5 sections in 25 m) determined an 11% reduction, indicating that a simpler arrangement could still achieve a significant saving. Compared to this, selective spraying on target weeds is a more complicated issue, although encouraging results were achieved in the detection and multicriteria approach to differential spraying of *Avena sterilis*, a competitive grass in cereal fields (Figure 2.18) (Tellaeche et al. 2008). Mixed weed populations will likely require additional efforts, but the way toward a smart application of active ingredients appears quite clearly traced.

2.6 HARVEST AND SUBSEQUENT MANAGEMENT

Harvest, conditioning, and storage are the link between the agricultural and industrial phases, that is, the connection between feedstock production and utilization. The variety of energy crop characteristics reflects on harvest equipment and techniques. The main difference is between plants where the organ of commercial interest is a grain as in cereals and oilseeds, a vegetative portion as in biomass crops, and a fresh fruit as in tropical oil crops (e.g., oil palm and *Jatropha*). In grain crops, combine harvesters are the most appropriate equipment, whereas in biomass crops, more complex machinery is needed, sometimes deriving from existing schemes for fodder crops and wood trees. Finally, in fresh fruit crops, the harvest is carried out either entirely by hand or through machines requiring a certain support from manpower.

2.6.1 Grain Crops

The recent progress in grain crop harvest benefits food as well as energy uses: combine harvesters of increasing size cruise large-scale crops, most often mapping real-time yield thanks to load

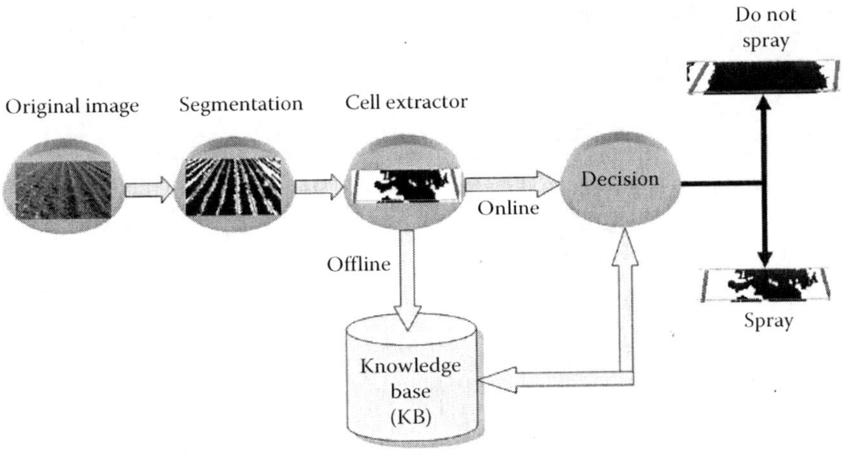

FIGURE 2.18 Vision-based decision process as the premise to differential weed spraying. (Reproduced from *Comput. Electron. Agric*, 60, Tellaeche, A., Burgos Artizzu, X.P., Pajares, G., Ribeiro, A., and Fernández-Quintanilla, C., A new vision-based approach to differential spraying in precision agriculture, 144–155, 2008, Copyright 2007, with permission from Elsevier.)

cells and GPS connections. This picture of technological smoothness implies a relevant work in crop husbandry and breeding, which is needed to mitigate crop differences affecting harvest quality, grain moisture, etc. In uniform field management, these differences arise from field variation in soil properties (texture, structure, nutrient and water status, etc.), topographic characteristics (slope, drainage, etc.), and biotic adversities (occurrence of weeds, pests, and diseases). The combined variation affects plant growth and development, in turn reflecting on traits such as resistance to lodging and time to achieve maturity; weeds also constitute a physical obstacle, occasionally. All these drawbacks depress harvest quality while boosting losses, impurities, inconsistent grain moisture, etc.

The techniques of precision farming discussed in this chapter are devised to smooth the variation and the consequent shortcomings through the application of differentiated amounts of inputs (fertilizers, water, pesticides, etc.). Breeding contributes to the process, by releasing cultivars more and more provided with multistress tolerance, resistance to lodging, and concentrated ripening. The progress in crop surface, which is pursued in modern agriculture through the merging of adjacent fields, adds to the inherent problem of yield variation: This is proved by a literature survey on 19 maize crops monitored through the GPS, showing an increase in yield variation at rising crop surface (Figure 2.19).

The progress in combine harvesters hardly needs to be discussed; it is only worth mentioning that the substitution of traditional combines at drum thresher and straw walkers with axial-flow machines determining a higher pulverization of crop residues and a more difficult recovery has come to an equilibrium, as the two types have reciprocal advantages and disadvantages also depending on specific models. A relevant progress in the harvest of oilseed rape has been made by the introduction of varieties, often hybrids, bred for lower plant size and more concentrated flowering and ripening. This reduces pod shattering and seed loss with respect to traditional, tall varieties at long bottom-to-top flowering/ripening, involving a seed loss up to 30% (Weiss 1983). To limit yield loss in traditional varieties, direct combining used to be replaced by swathing and windrowing, that is, cutting the plants at incipient maturity (yellowing pods at midheight) and leaving them to dry on the stubble, then threshing at full maturity with a combine harvester equipped with a pickup header. Apart from the higher costs, this operation is constrained by a narrow seed moisture interval: Above 35%, oil and protein content may be affected, while below 20%, some pod shattering occurs.

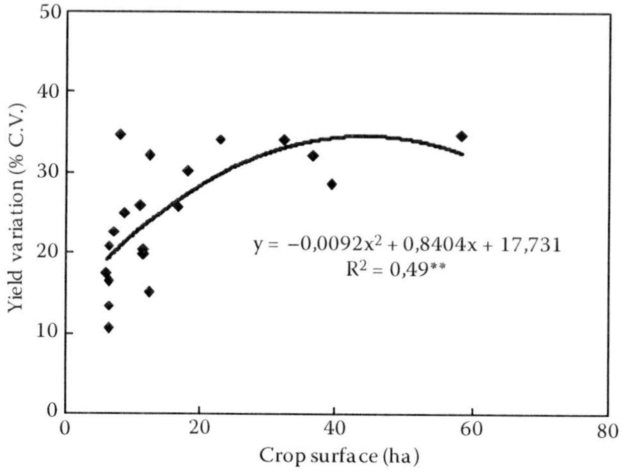

FIGURE 2.19 Relationship between grain yield variation (% CV) and crop surface in 19 maize cases monitored through the GPS and reported in the scientific literature. (From Ortega, R.A. and Santibañez, O.A., *Comput. Electron. Agric.*, 58, 49, 2007; Basso, B. et al., *Eur. J. Agron.*, 26, 82, 2007; Parent, A.C. et al., *Biosyst. Eng.*, 93, 134, 2008, Copyright 2007; Bocchi, S. and Castrignanò, A., *Field Crop Res.*, 102, 185, 2007; Marques da Silva, J.R. and Silva, L.L., *Biosyst. Eng.*, 101, 183, 2008; Reproduced from Matteo, A., Matteo Antonello's Thesis, Applicazioni dell'Agricoltura di Precisione nel "Corn Belt" Americano, University of Bologna, Bologna, Italy, (March 25, 2010), with permission from the author.)

The level of seed moisture allowing long-term storage while avoiding oxidation processes varies between cereals and oilseeds. About 13%–14% of moisture is a typical threshold in cereals, whereas 9% is the limit in many oilseeds. This means that especially warm-season crops maturing at the end of summer need air-drying prior to storage. Grain crops for energy are also subjected to the rule in order to prevent losses of energy potential due to seed rancidity and rotting.

Finally, the recovery of postharvest residues (straw, stover, etc.) to be used as a complementary energy source should be subjected to careful evaluation as discussed in the soil tillage section. Once planned, it is normally carried out by windrowing and baling the air-dried biomass. A particular case is represented by a prototype combine header designed to harvest cardoon, which is an oilseed as well as a biomass crop; the prototype performs capitula cutting and threshing while at the same time cutting at the base and windrowing stems and leaves (Figure 2.20) (Pari et al. 2009); both components are recovered in a satisfactory way, compared to normal two-pass harvest hampering a good recovery of residual biomass.

2.6.2 Herbaceous Biomass Crops

Herbaceous biomass crops have a variable height, stiffness, and moisture, necessitating a specific harvest organization. Three cases of increasing complexity may be outlined: (1) The equipment for harvest and conditioning of forage crops is also suited for biomass crops, (2) adaptations of existing machinery may solve minor problems of compliance with particular crop features, and (3) dedicated machinery is needed as those already available cannot assure good and consistent harvest quality. In the last case, the new equipment often consists of some appliance to be mounted on existing machines, for example, a new header for a chopper harvester: By exploiting the engine units of expensive machinery, this approach contributes to a higher utilization around the year, reducing fixed costs. Harvest and conditioning are also influenced by the maximum level of storage moisture: the wider the difference from harvest moisture, the greater the difficulty and the cost to bridge this gap, especially in the cold season.

FIGURE 2.20 Combine harvester equipped with a dedicated header for cardoon (*C. cardunculus*) simultaneous seed harvest and residual biomass windrowing. (Reproduced from *Proceedings of the 17th European Biomass Conference and Exhibition*, Pari, L., Civitarese, V., Assirelli, A., and Del Giudice, A., Prototype for *Cynara cardunculus* capitula threshing and biomass windrowing, 262–267, Hamburg, Germany, June 29–July 3, 2009, with permission from the authors.)

For a soft-stemmed plant like switchgrass, the typical procedures of haymaking may successfully be applied, in accordance with the species' original use as a forage crop: cutting and conditioning with a one-pass machine, then windrowing to air-dry the biomass prior to baling. Switchgrass for energy uses is generally cut once a year at the end of the growing season; the two-cut system (heading plus autumn regrowth) has generally been abandoned, as the regrowth significantly contributes to total biomass only when soil moisture is not limiting in the summertime (Reynolds et al. 2000; Thomason et al. 2004; Guretzki et al. 2011). This has been proved by the already-cited survey on 19 experimental cases comparing the two strategies in rainfed conditions: The double cut allowed a modest gain at 9% on average over the single cut, although 6 cases out of 19 passed the 20% gain threshold (Table 2.3). In addition to this, the double harvest repeated in the years leads to progressively weaker plant growth (Sanderson et al. 1999; Monti et al. 2008a), likely due to insufficient accumulation of belowground reserves. However, the double harvest retains a certain interest when switchgrass is set for anaerobic digestion (Massé et al. 2010; Frigon et al. 2012): The higher N content and lower lignification of the plant repeatedly harvested at a juvenile stage represent an advantage for the anaerobic digestion process (Angelidaki et al. 2003), while the high moisture is not disadvantageous in view of biomass storage as wet silage.

The single harvest per year remains the standard practice when switchgrass is set for thermochemical conversions. The choice remains between cutting at maturity, in the autumn, or in full-late winter. In the latter case, freeze-drying and weathering of standing plants involve a lower moisture and some loss of biomass, which is counterbalanced by better quality, thanks to a decrease of N, Si, Cl, and other undesired elements (Baxter et al. 1998). This effect, which is generally associated with leaf loss in biomass crops (Monti et al. 2008b), was specifically observed in *Miscanthus* (Lewandowski et al. 2003; Nassi o Di Nasso et al. 2011; Strullu et al. 2011) and *Arundo* (Nassi o Di Nasso et al. 2011). Biomass loss and moisture decrease between autumn and winter have been more precisely assessed in *Miscanthus* than in other perennial crops: An average 18% loss in dry biomass yield was observed in exchange for an almost 50% decrease in moisture (from 56% to 29% on a fresh weight basis) in five sites encompassing Northern to Southern Europe (Lewandowski et al. 2003), whereas a stronger biomass loss (average, 30%) was recorded in the complex of three other sites in Central Europe (Zub et al. 2011).

For switchgrass, biomass losses have been reported at several other steps during harvest and subsequent storage; in a hilly field, Monti et al. (2009) recorded a loss of 35%–45% of potentially harvestable biomass with the mowing–baling harvest system; uncut biomass accounted for a higher share of the losses (average, 29%) depending on field slope, but also baling losses associated with repeated windrowing were noteworthy (average, 11%). For switchgrass, large losses at harvest are also associated with drier biomass as a presumable effect of shattering (Sanderson et al. 1997). Storage losses of the baled biomass were monitored during storage. After 6 months and 1 year in an open environment, round bales lost only 13% and 16% of their dry weight, respectively (Sanderson et al. 1997). When sheltered from precipitation, dry weight losses were negligible over a variable number of months (Sanderson et al. 1997; Monti et al. 2009). It appears therefore that in contrast to harvest losses, storage losses of properly managed switchgrass biomass may be restrained to a minimum.

Compared to switchgrass, harvest involves a more complicate management for hard-stemmed, tall, and moist species such as biomass sorghum, depending on final energy uses. The two genotypes of sweet and fiber sorghum should actually have different end uses, but the onset of second-generation biofuels and the fading differences between commercial hybrids of the two types suggest to address the harvest requirements in view of specific energy sectors (biogas, first- and second-generation ethanol, and thermochemical processes), instead of genotype. With this premise, biomass sorghum is harvested at the end of the growing season when still moist (dry matter content, 25%–40%). Winter harvest in temperate climates is not advisable, being more prone to lodging.

First-generation ethanol is produced by fermenting the soluble sugars (glucose, fructose, and sucrose) that represent 12%–18% of the juice contained in stem pith. Processing fresh feedstock is of great importance, as wet stalk storage soon leads to decaying and consequent losses. Whole-stalk deterioration is quite rapid: only 24 h, according to Dalvi et al. (2012), or more than 4 days, according to Lingle et al. (2012), although in the latter paper, chopped stalks began to lose significant amounts of sugars after only 1–2 days (Figure 2.21). In contrast to this, a concentrated syrup extracted from stem juice (>35% degrees Brix [°Bx]) can be stored at ambient temperature up to 1 year, allowing ethanol production around the year in both temperate and warm areas (Kumar et al. 2010). In practice, the ideal harvesting of sweet sorghum for first-generation ethanol production consists of stripping leaves and panicles, which are subsequently used as animal feed (Kumar et al. 2010) and then storing either the whole stalks or 0.2–0.4 m long billets for a few days at ambient temperature. This labor-intensive procedure suits developing countries, where the recovery of separate plant parts fosters a remarkable multifunctionality, thus providing several food/feed and nonfood products beside ethanol (Ratnavathi et al. 2011). In developed countries, sweet sorghum for ethanol can be harvested by means of sugarcane harvesters. The large self-propelled machines cut off sorghum tops, blow away leaves, and cut stems into billets (typically, 0.3 m long) loaded on trailers. In such conditions, the recovery of residual plant parts is hardly feasible and they are normally plowed in, contributing to soil carbon balance and nutrient recycling. The alternative is to use forage choppers to shred the whole plant into small pieces to be processed within 24 h in order to prevent degradation.

Given the constraints posed by the short time of storage, extending the harvest window is one of the main goals in sweet sorghum research. Staggered seeding in subtropical areas at nonlimiting temperatures appears a solution, although stalk, juice, and sugar yield achieves a top level then decreases during the harvest window (Dalvi et al. 2012). Adopting genotypes at different maturities may help in the task without significant yield loss, until the differences between early- and late-maturing genotypes do not pass 15 days (Kumar et al. 2010). In fact, 1-month differences in physiological maturity date involved a lower ethanol yield in early genotypes (Figure 2.22) (Zhao et al. 2012). Harvesting at different plant stages is another way to extend the harvest window,

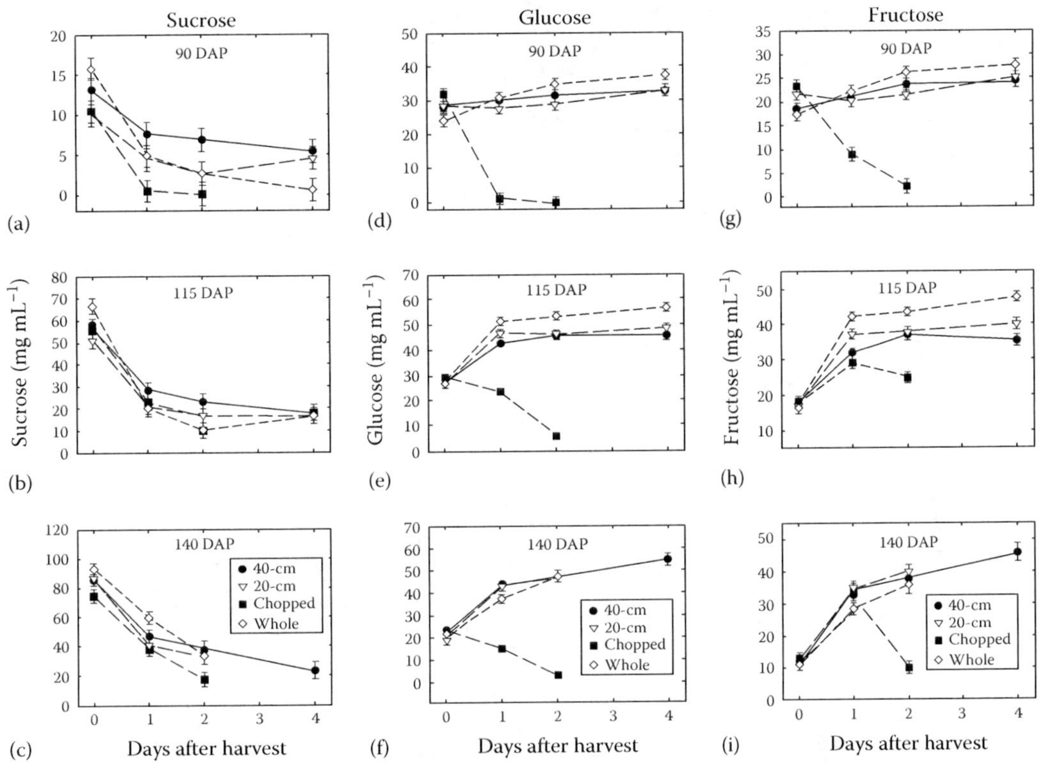

FIGURE 2.21 Sucrose, glucose, and fructose contents of sweet sorghum juice during storage after harvest at 90, 115, and 140 days after planting (DAP). Sweet sorghum stalks were cut into 40 or 20 cm billets, chopped, or left whole. Vertical bars represent ±SE. (Reproduced from *Bioenergy Res.*, Post-harvest changes in sweet sorghum I: Brix and sugars, 5, 2012, 158–167, Lingle, S.E., Tew, T.L., Rukavina, H., and Boykin, D.L., Figure 2.4, Copyright Springer Science and Business Media 2011, with kind permission from Springer Science and Business Media.)

although early stages such as flowering entail a yield gap ranging from 30% (Almorades et al. 2007) to 80% (Bhairappanavar and Sharanappa 2011) with respect to physiological maturity. Conversely, delaying harvest from physiological maturity (September–October) to December in a temperate–cold climate featuring first killing frosts in October–November determined only a 2%–12% decrease in ethanol yield in top-performing middle to late cultivars (Figure 2.22) (Zhao et al. 2012).

It appears therefore that the three strategies of varying seeding and harvest time and using genotypes at different maturity should be integrated in order to achieve a reasonable extension of the harvest window while limiting losses of yield potential. However, it should be noted that supplying sugar mill operations for a longer time involves additional costs in terms of yield gaps, as proved by the combined effect of optimum seeding and harvest time, determining a 30%–210% increase in fermentable sugar yield with respect to a vast range of alternative combinations (Han et al. 2012).

Harvesting biomass sorghum outside the ethanol production chain is less complicate. The high moisture is not a serious drawback in the case of sweet or fiber sorghum to be ensiled for biogas production: The standing crop is harvested with a forage chopper and directly transported to the silage platform. Conversely, when biomass sorghum is to be stored dry, mowing, conditioning, and subsequent windrowing must be performed as in haymaking in order to ease the natural drying in the open. However, the size and thickness of sorghum stem necessitate sturdier machines than mower–conditioners designed for grass fields; such machines are not easily available, as the dry use

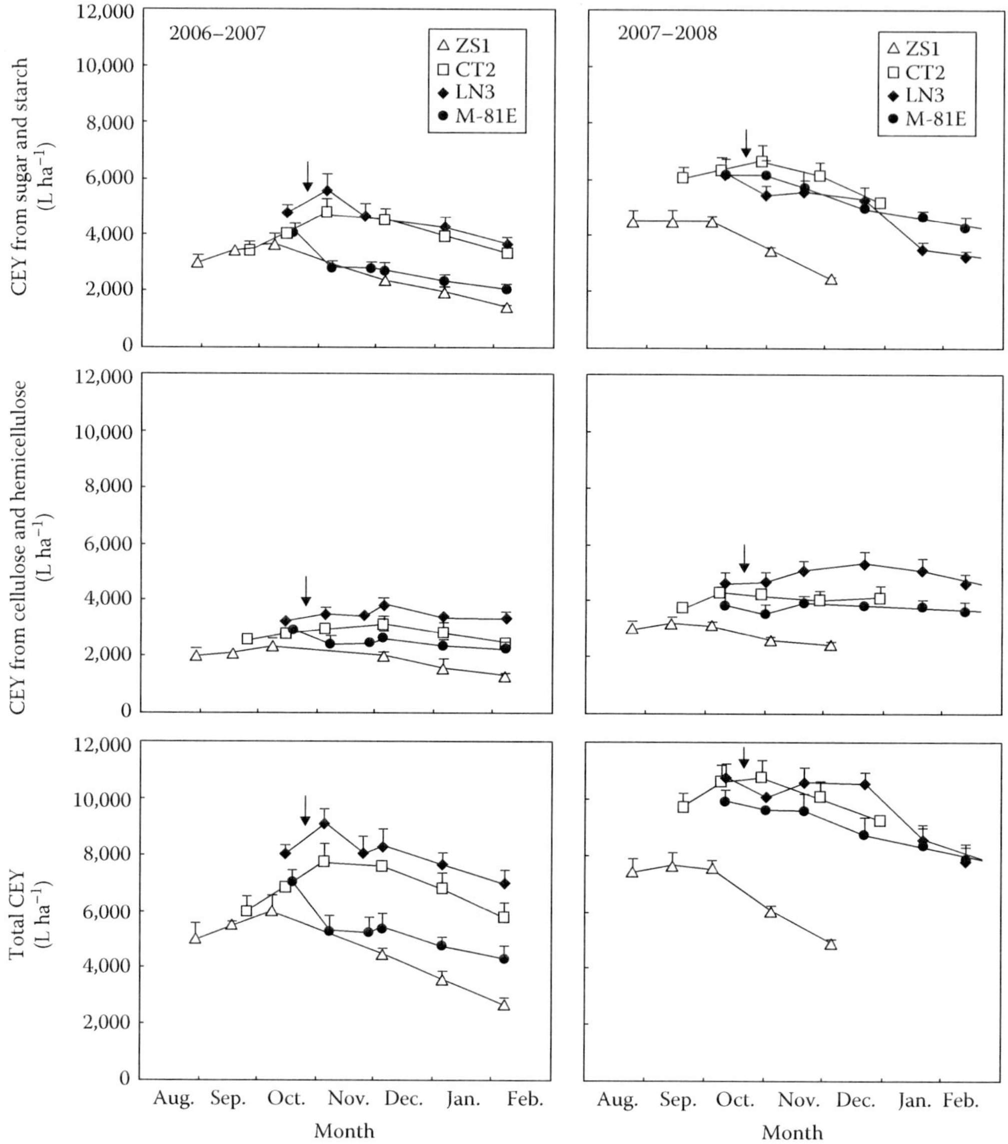

FIGURE 2.22 Effects of delayed harvest on calculated ethanol yield (CEY) from sugar, starch, cellulose, and hemicellulose in stems of sweet sorghum early (Zaoshu-1 [ZS1]), middle (Chuntian-2 [CT2]), and late cultivars (Lvneng-3 [LN3] and M-81E) in two harvest seasons. The vertical arrowheads indicate the first killing frost dates; vertical bars represent the standard deviation. (Reproduced from *Biomass Bioenergy*, 39, Zhao, Y.L., Steinberger, Y., Shi, M., Han, L.P., and Xie, G.H., Changes in stem composition and harvested produce of sweet sorghum during the period from maturity to a sequence of delayed harvest dates, 261–273, Copyright 2012, with permission from Elsevier.)

of sorghum biomass is only recently growing in thermochemical conversions. A commercial prototype was recently built consisting of a tractor-mounted header provided with cutting blades and coupled spiral conditioning screws for each crop row (Figure 2.23) (Pari et al. 2010). The equipment performs a successful mowing and conditioning of sorghum plants without significant choking/loss of leaves and tops. Once sufficiently dry to ensure adequate storage (moisture, 25%–30%), the biomass may be baled and stored for a long time.

FIGURE 2.23 Prototype of tractor-mounted mowing–conditioning header developed for fiber sorghum harvest. (Reproduced from *Proceedings of the 18th European Biomass Conference and Exhibition*, Pari, L., Assirelli, A., and Suardi, A., Developing and evaluating the fiber sorghum mower–conditioner prototype CRA-Ing/Cressoni, 463–467, Lyon, France, May 3–7, 2010, with permission from the authors.)

2.6.3 Arboreal Biomass Crops

Tree biomass is used as wood for thermochemical conversions and, lately, second-generation biofuels. Therefore, hardness and moisture are key challenges to tackle in tree biomass harvest, conditioning, and storage. Poplar is the main species in SRFs, also known as short rotation coppices (SRCs), for energy uses in temperate regions. These plantations entail a higher density (up to 10,000 plants ha^{-1}) and a more frequent coppicing (1–4 years) than plantations for paper pulp production. The harvest is carried out in the dormant season to avoid damage to new shoots. However, there is a high risk of soil compaction and stool mechanical damage in this frequently wet season. In a willow experiment, heavy compaction determined a significant yield reduction only in the first year after the treatment, and soil loosening treatments carried out between rows to alleviate compaction effects did not improve plant growth (Souch et al. 2004). Conversely, mechanical damage to stools could decrease the biomass yield up to 21%.

In plantations at frequent coppicing (every 1–2 years), the stem of poplars is soft enough to be harvested by sturdy forage choppers that produce fresh chips transported by tractor-towed trailers (Figure 2.24). In this case, the fresh chips should be used as soon as possible and are difficult to be used in combustion due to the high moisture (50%–60%). Coppicing at longer intervals requires special machines to fell older plants. The felled trees may be aligned between the rows or gathered in heaps on the headlands for weeks or months to lose moisture before being chipped for utilization. The recent trend is toward longer harvest frequencies. In a 12-year experiment on poplar, the annual yield of dry biomass for a 3-year harvest interval was 66% and 19% higher than that of a 1- and 2-year harvest interval, respectively (Figure 2.25) (Nassi o Di Nasso et al. 2010). This in turn led to a higher trade-off in the use of energy: NE yield increased from 172 to 299 GJ ha^{-1} $year^{-1}$ and EE from 19 to 61 GJ GJ^{-1} between 1- and 3-year interval. In the cited experiment, the wide gap in performances between 1- and 3-year interval was partly due to the shorter SRF life in the annual harvest, showing a stump survival of <35% after the seventh year. This is a further drawback of poplar SRF at frequent harvest also observed by Labrecque and Teodorescu (2005).

As in other biomass types, a lower chip moisture reduces microbial degradation, thus storage losses. Chip storage may be carried out in the open air or in sheltered environments, depending

FIGURE 2.24 SRC poplar harvest by means of a self-propelled mower–chopper delivering fresh chips onto a tractor-pulled trailer. (Reproduced from *Lo sviluppo delle colture energetiche in Italia—Il contributo dei progetti di ricerca SUSCACE e FAESI*, http://ing.entecra.it/Biomasse, Pari, L. and Civitarese, V., Harvesting yards for woody crops, 165–181, 2011, with permission from the authors.)

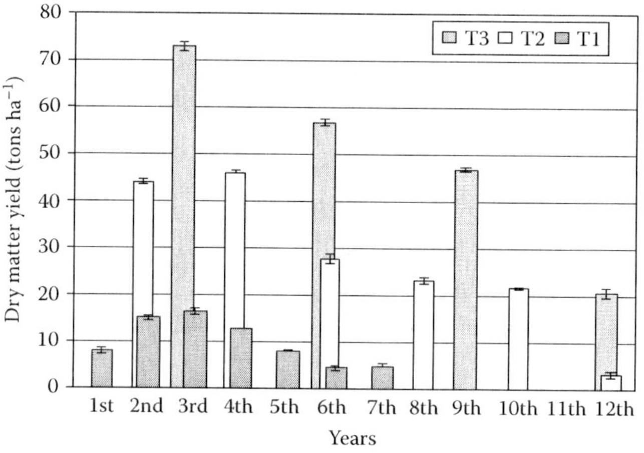

FIGURE 2.25 Dry matter yields of SRC poplar under different harvest frequencies: 1(T1)-, 2(T2)-, and 3(T3)-year rotations. Standard deviations are shown by vertical bars. (From Nassi o Di Nasso, N., Roncucci, N., Triana, F., Tozzini, C., and Bonari, E.: Biomass production and energy balance of a 12-year-old short-rotation coppice poplar stand under different cutting cycles. *GBC Bioenergy*. 2010. 2, 89–97, Copyright Wiley-VCH Verlag GmbH & Co. KGaA. Reproduced with permission.)

on mass moisture and availability of suitable facilities. In the open air, additional options to improve storage include plastic-covered versus uncovered heaps and heap mixing or not: Heap moisture, temperature, and season of storage (summer vs. winter) are the drivers determining the best factor combination.

2.7 CONCLUSIONS

Dedicated energy crops are undergoing a rapid development. Progress is not only sought in increasing crop surface and biomass utilization but also in advanced techniques for increasing the efficiency in the agricultural phase while restraining the drawbacks in energy cropping. The use of traditional food crops as an energy source only needs minor changes, whereas novel energy crops often necessitate starting from scratch in laying out their husbandry. Therefore, continuous improvement in large-scale food crops such as cereals and oilseeds also benefits their energy uses, whereas a variable gap remains to be filled in novel crops such as perennial rhizomatous grasses to keep abreast of current agricultural progress. Energy crops are aimed for a high efficiency in converting subsidiary energy and are devised to originate positive externalities in environmental sustainability and global warming abatement. The ongoing progress in crop techniques is therefore confronted with these two delicate topics. WF and land use are two other foremost issues related to energy crops at present. The fears for the depletions of specific resources such as fertilizer phosphates and for negative externalities such as land grabbing will likely be new issues to be discussed as regards energy crops. All these issues increase the pressure to improve the efficiency per unit land cultivated and per unit input of energy, seed, fertilizer, etc., in energy crop production. The agricultural practices reviewed in the five categories of planting, nutrition, water management, protection, and harvest are equally involved in the process. The potential echoed in several scientific sources is the best premise for optimistic prospects; the commitment in pursuing the current efforts is the best assurance to achieve them.

NOMENCLATURE

°Bx	Degree Brix
CSWI	Crop water stress index
CV	Coefficient of variation
DDGS	Dry distilled grains with solubles
DSS	Decision support system
EE	Energy efficiency
EROEI	Energy return on energy invested
ET	Evapotranspiration
ET_C	Crop evapotranspiration
EU	European Union
EUF	Electro-ultrafiltration
FAO	Food and Agriculture Organization of the United Nations
GHG	Greenhouse gas
GIS	Geographic information system
GM	Genetically modified (organisms)
GPS	Global positioning system
GWP	Global warming potential
ha	Hectare
ICM	Integrated crop management
IEA	International Energy Agency
IPM	Integrated pest management

IWUE Irrigation water use efficiency
LEPA Low-energy precision application (irrigation system)
LUC Land use change
N_2O Nitrous oxide
NDVI Normalized difference vegetation index
NE Net energy
NH_3 Ammonia
NNI Nitrogen nutrition index
NO_3^- Nitrate
NO_x Mononitrogen oxide
NUE Nitrogen use efficiency
PRD Partial root-zone drying
R^2 Regression coefficient
RES Renewable energy source
SRC Short rotation coppice
SRF Short rotation forest
VRA Variable rate application
WF Water footprint
WUE Water use efficiency

REFERENCES

Aboudrare, A., Debaeke, P., Bouaziz, A., and H. Chekli. 2006. Effects of soil tillage and fallow management on soil water storage and sunflower production in a semi-arid Mediterranean climate. *Agricultural Water Management* 83: 183–196.

Abuzar, M., O'Leary, G., and G. Fitzgerald. 2009. Measuring water stress in a wheat crop on a spatial scale using airborne thermal and multispectral imagery. *Field Crops Research* 112: 55–65.

Adamchuk, V.I., Hummel, J.W., Morgan, M.T., and S.K. Upadhyaya. 2004. On-the-go soil sensors for precision agriculture. *Computers and Electronics in Agriculture* 44: 71–91.

Adamchuk, V.I., Lund, E.D., Sethuramasamyraja, B., Morgan, M.T., Dobermann, A., and D.B. Marx. 2005. Direct measurement of soil chemical properties on-the-go using ion-selective electrodes. *Computers and Electronics in Agriculture* 4(48): 272–294.

Al-Karadsheh, E., Sourell, H., and R. Krause. 2002. Precision irrigation: New strategy irrigation water management. *Deutscher Tropentag 2002—Conference on International Agricultural Research for Development*, Berlin, Germany. www.tropentag.de/2002 (last consulted, September 6, 2012).

Allen, R.G., Pereira, L.S., Raes, D., and M. Smith. 1998. FAO irrigation and drainage paper no. 56—Crop evapotranspiration. Rome, Italy: Food and Agriculture Organization.

Almodares, A., Hadi, M.R., Ranjbar, M., and R. Taheri. 2007. The effects of nitrogen treatments, cultivars and harvest stages on stalk yield and sugar content in sweet sorghum. *Asian Journal of Plant Sciences* 6: 423–426.

Amaducci, S., Amaducci, M.T., Benati, R., and G. Venturi. 2000. Crop yield and quality parameters of four annual fibre crops (hemp, kenaf, maize and sorghum) in the North of Italy. *Industrial Crops and Products* 11: 179–186.

Angelidaki, I., Ellegaard, L., and B.K. Ahring. 2003. Applications of the anaerobic digestion process. In *Biomethanation II* (*Advances in Biochemical Engineering Biotechnology*), ed. B.K. Ahring, pp. 1–33. Berlin, Germany: Springer.

Angelini, L.G., Ceccarini, L., and E. Bonari. 2005. Biomass yield and energy balance of giant reed (*Arundo donax* L.) cropped in central Italy as related to different management practices. *European Journal of Agronomy* 22: 375–389.

Aravindhakshan, S.C., Epplin, F.M., and C.M. Taliaferro. 2010. Economics of switchgrass and *Miscanthus* relative to coal as feedstock for generating electricity. *Biomass and Bioenergy* 34: 1375–1383.

Atwell, B.J. 1993. Response of roots to mechanical impedance. *Environmental and Experimental Botany* 33: 27–40.

Auernhammer, H. 2001. Precision farming—The environmental challenge. *Computers and Electronics Agriculture* 30: 31–43.

Barbanti, L. and S. Fazio. 2011. Oil crops. In *Handbook on Biofuels and Family Agriculture in Developing Countries*, eds. A. Gevé, L. Barbanti, and S. Fazio, pp. 37–114. Bologna, Italy: Pàtron Editore. http://edeng.gvc-italia.org (last consulted, July 20, 2012).

Barbanti, L., Grandi, S., Vecchi, A., and G. Venturi. 2006. Sweet and fibre sorghum (*Sorghum bicolor* (L.) Moench), energy crops in the frame of environmental protection from excessive nitrogen loads. *European Journal of Agronomy* 25: 30–39.

Barbanti, L., Grigatti, M., and C. Ciavatta. 2011. Nitrogen release from a ^{15}N-labeled compost in a sorghum growth experiment. *Journal of Plant Nutrition and Soil Science* 174: 240–248.

Barney, J.N. and J.M. DiTomaso. 2010. Bioclimatic predictions of habitat suitability for the biofuel switchgrass in North America under current and future climate scenarios. *Biomass and Bioenergy* 34: 124–133.

Basso, B., Bertocco, M., Sartori, L., and E.C. Martin. 2007. Analyzing the effects of climate variability on spatial pattern of yield in a maize–wheat–soybean rotation. *European Journal of Agronomy* 26: 82–91.

Baxter, L.L., Miles, T.R., Miles, T.R. Jr. et al. 1998. The behavior of inorganic material in biomass-fired power boilers: Field and laboratory experiences. *Fuel Processing Technology* 54: 47–78.

Bhairappanavar, S.T. and Sharanappa. 2011. Bio-ethanol production potential of sweet sorghum (*Sorghum bicolor*) cultivars under different spacing and harvest stages. *Indian Journal of Agronomy* 56: 52–56.

Blanco-Canqui, H. and R. Lal. 2007. Soil and crop response to harvesting corn residues for biofuel production. *Geoderma* 141: 355–362.

Blanco-Canqui, H. and R. Lal. 2008. Corn stover removal impacts on micro-scale soil physical properties. *Geoderma* 145: 335–346.

Blanco-Canqui, H., Schlegel, A.J., and W.F. Heer. 2011. Soil-profile distribution of carbon and associated properties in no-till along a precipitation gradient in the central Great Plains. *Agriculture, Ecosystems and Environment* 144: 107–116.

Bocchi, S. and A. Castrignanò. 2007. Identification of different potential production areas for corn in Italy through multitemporal yield map analysis. *Field Crop Research* 102: 185–197.

Campbell, D.J. and M.F. O'Sullivan. 1991. The cone penetrometer in relation to trafficability, compaction, and tillage. In *Soil Analysis. Physical Methods*, eds. K.A. Smith and C.E. Mullins, pp. 399–430. New York: Marcel Dekker.

Canarache, A. 1990. PENETR—A generalized semi-empirical model estimating soil resistance to penetration. *Field Crop Research* 16: 51–60.

Cavallaro, V., Tringali, S., and C. Patanè. 2011. Large-scale in vitro propagation of giant reed (*Arundo donax* L.), a promising biomass species. *Journal of Horticultural Science and Biotechnology* 86: 452–456.

CETIOM. 2012. La Réglette azote colza. www.cetiom.fr (last consulted, September 6, 2012).

Chen, Y., Liu, S., Li, H. et al. 2011. Effects of conservation tillage on corn and soybean yield in the humid continental climate region of Northeast China. *Soil and Tillage Research* 115–116: 56–61.

Christian, D.G., Riche, N.E., and A.B. Yates. 2008. Growth, yield and mineral content of *Miscanthus × giganteus* grown as a biofuel for 14 successive harvests. *Industrial Crops and Products* 28: 320–327.

Cioni, F., Maines, G., Collina, M., Secor, G., and V. Rivera. 2012. A new integrated pest management (IPM) model for cercospora leaf spot of sugar beets in the Po river valley. *Proceedings of the 73rd Congress of the International Institute for Beet Research*, Brussels, Belgium, February 14–15, pp. 54–64. www.iirb.org (last consulted, September 6, 2012).

Condon, A.G., Richards, R.A., Rebetzke, G.J., and G.D. Farquhar. 2004. Breeding for high water-use efficiency. *Journal of Experimental Botany* 55: 2447–2460.

Cosentino, S.L., Mantineo, M., and G. Testa. 2012. Water and nitrogen balance of sweet sorghum (*Sorghum bicolor* (L. Moench)) cv. Keller under semi-arid conditions. *Industrial Crops and Products* 36: 329–342.

Cosentino, S.L., Mastrorilli, M., Saita, A., and G. Scalici. 2011. The cultivation of biomass sorghum in southern Italy. In *Lo sviluppo delle colture energetiche in Italia—Il contributo dei progetti di ricerca SUSCACE e FAESI*, ed. L. Pari, Rome, Italy, pp. 517–538. http://ing.entecra.it/Biomasse (last consulted, September 6, 2012).

Cosentino, S.L., Patanè, C., and P. Guarnaccia. 1996. Biomass, leaf area index and water use efficiency of sweet sorghum in limited water availability in Mediterranean environment. In *Proceedings of First European Seminar on Sorghum for Energy and Industry*, Toulouse, France, April 1–3, pp. 228–235.

Dalvi, U.S., Chavan, U.D., Shinde, M.S., and S.R. Gadakh. 2012. Effect of staggered planting on stalk yield, sugar content and ethanol yield of sweet sorghum for increasing harvest window. *Sugar Tech* 14: 144–147.

Dawo, M.I., Wilkinson, J.M., Sanders, F.E.T., and D.J. Pilbeam. 2007. The yield and quality of fresh and ensiled plant material from intercropped maize (*Zea mays*) and beans (*Phaseolus vulgaris*). *Journal of the Science of Food and Agriculture* 87: 1391–1399.

Dent, D.R. 1995. Principles of integrated pest management. In *Integrated Pest Management*, ed. D. Dent, pp. 8–46. London, U.K.: Chapman & Hall.

Du, T., Kang, S., Sun, J., Zhang, X., and J. Zhang. 2010. An improved water use efficiency of cereals under temporal and spatial deficit irrigation in north China. *Agricultural Water Management* 97: 66–74.

Duiker, S.W. and D.B. Beegle. 2006. Soil fertility distributions in long-term no-till, chisel/disk and moldboard plow/disk systems. *Soil and Tillage Research* 88: 30–41.

Ehlers, W., Köpke, U., Hesse, F., and W. Böhm, 1983. Penetration resistance and root growth of oats in tilled and untilled loess soil. *Soil and Tillage Research* 3: 261–275.

El-Nahry, A.H., Ali, R.R., and A.A. El-Baroudy. 2011. An approach for precision farming under pivot irrigation system using remote sensing and GIS techniques. *Agricultural Water Management* 98: 517–531.

Ercoli, L., Mariotti, M., Masoni, A., and E. Bonari. 1999. Effect of irrigation and nitrogen fertilization on biomass yield and efficiency of energy use in crop production of *Miscanthus*. *Field Crops Research* 63: 3–11.

Fallahi, S. and M.H. Raoufat. 2008. Row-crop planter attachments in a conservation tillage system: A comparative study. *Soil and Tillage Research* 98: 27–34.

FAO. 2008. Bioenergy, food security and sustainability towards an international framework. *High-Level Conference on World Food Security: The Challenges of Climate Change and Bioenergy*. FAO Report HLC/08/INF/3. Rome, Italy: FAO.

Farré, I. and J.M. Faci. 2006. Comparative response of maize (*Zea mays* L.) and sorghum (*Sorghum bicolor* L. Moench) to deficit irrigation in a Mediterranean environment. *Agricultural Water Management* 83: 135–143.

Fazio, S. and A. Monti. 2011. Life cycle assessment of different bioenergy production systems including perennial and annual crops. *Biomass and Bioenergy* 35: 4868–4878.

Fike, J.H., Parrish, D.J., Wolf, D.D. et al. 2006. Switchgrass production for the upper southeastern USA: Influence of cultivar and cutting frequency on biomass yields. *Biomass and Bioenergy* 30: 207–213.

Fismes, J., Vong, P.C., Guckert, A., and E. Frossard. 2000. Influence of sulphur on apparent N-use efficiency, yield and quality of oilseed rape (*Brassica napus* L.) grown on a calcareous soil. *European Journal of Agronomy* 12: 127–141.

Forster, P., Ramaswamy, V., Artaxo, P. et al. 2007. Changes in atmospheric constituents and in radiative forcing. In *Climate Change 2007: The Physical Science Basis. Contribution of Working Group I to the Fourth Assessment Report of the Intergovernmental Panel on Climate Change*, eds. S. Solomon, D. Qin, M. Manning, Z. Chen, M. Marquis, K.B. Averyt, M. Tignor, and H.L. Miller, pp. 129–234. Cambridge, U.K.: Cambridge University Press.

Franke, Y.J. and F. Nielsen. 2010. Plantation establishment and management. In *The Jatropha Handbook: From Cultivation to Application*, ed. J. de Jongh, pp. 9–26. Heindhoven, the Netherlands: Fact Foundation.

Frigon, J.C., Mehta, P., and S.R. Guiot. 2012. Impact of mechanical, chemical and enzymatic pre-treatments on the methane yield from the anaerobic digestion of switchgrass. *Biomass and Bioenergy* 36: 1–11.

Frischknecht, R. and N. Jungbluth. 2003. Implementation of life cycle impact assessment methods. Final report ecoinvent 2000. Dubendorf (CH), Zurich, Switzerland: Swiss Centre for LCI. www.ecoinvent.ch (last consulted, September 6, 2012).

Gerbens-Leenes, P.W., Hoekstra, A.Y., and T. van der Meer. 2009. The water footprint of energy from biomass: A quantitative assessment and consequences of an increasing share of bio-energy in energy supply. *Ecological Economics* 68: 1052–1060.

Göksoy, A.T., Demir, A.O., Turan, Z.M., and N. Dağüstü. 2004. Responses of sunflower (*Helianthus annuus* L.) to full and limited irrigation at different growth stages. *Field Crops Research* 87: 167–178.

Gong, J.R., Zhang, X.S., and Y.M. Huang. 2011. Comparison of the performance of several hybrid poplar clones and their potential suitability for use in northern China. *Biomass and Bioenergy* 7: 2755–2764.

Graeff, S. and W. Claupein. 2003. Quantifying nitrogen status of corn (*Zea mays* L.) in the field by reflectance measurements. *European Journal of Agronomy* 19: 611–618.

Grigatti, M., Di Girolamo, G., Chincarini, R., Ciavatta, C., and L. Barbanti. 2011. Potential nitrogen mineralization, plant utilization efficiency and soil CO_2 emissions following the addition of anaerobic digested slurries. *Biomass and Bioenergy* 35: 4619–4629.

Gübitz, G.M., Mittelbach, M., and M. Trabi. 1999. Exploitation of the tropical oil seed plant *Jatropha curcas* L. *Bioresource Technology* 67: 73–82.

Guretzky, J.A., Biermacher, J.T., Cook, B.J., Kering, M.K., and J. Mosali. 2011. Switchgrass for forage and bioenergy: Harvest and nitrogen rate effects on biomass yields and nutrient composition. *Plant and Soil* 339: 69–81.

Han, K.J., Alison, M.W., Pitman, W.D., Day, D.F., Kim, M., and L. Madsen. 2012. Planting date and harvest maturity impact on biofuel feedstock productivity and quality of sweet sorghum grown under temperate Louisiana conditions. *Agronomy Journal* 104: 1618–1624.

Hansen, N.C., Allen, B.L., Baumhardt, R.L., and D.J. Lyon. 2012. Research achievements and adoption of no-till, dryland cropping in the semi-arid U.S. Great Plains. *Field Crops Research* 132: 196–203.

He, J., Hongwen, L., Rasaily, R.G. et al. 2011. Soil properties and crop yields after 11 years of no tillage farming in wheat–maize cropping system in North China Plain. *Soil and Tillage Research* 113: 48–54.

He, J., Hongwen, L., Xiaoyan, W. et al. 2007. The adoption of annual subsoiling as conservation tillage in dryland maize and wheat cultivation in northern China. *Soil and Tillage Research* 94: 493–502.

Heaton, E., Voigt, T., and S.P. Long. 2004. A quantitative review comparing the yields of two candidate C_4 perennial biomass crops in relation to nitrogen, temperature and water. *Biomass and Bioenergy* 27: 21–30.

Hillocks, R.J. 2012. Farming with fewer pesticides: EU pesticide review and resulting challenges for UK agriculture. *Crop Protection* 31: 85–93.

Hoekstra, A.Y., Chapagain, A.K., Aldaya, M.M., and M.M. Mekonnen. 2011. *The Water Footprint Assessment Manual*. London, U.K.: Earthscan.

IEA. 2007. *World Energy Outlook 2007*. Paris, France: IEA. www.iea.org (last consulted, July 20, 2012).

IEA. 2012. *Medium-Term Renewable Energy Market Report 2012—Market Trends and Projections to 2017*. Paris, France: IEA. www.iea.org (last consulted, July 20, 2012).

Ierna, A., Rosario, P., Mauro, R.P., and G. Mauromicale. 2012. Biomass, grain and energy yield in *Cynara cardunculus* L. as affected by fertilization, genotype and harvest time. *Biomass and Bioenergy* 36: 404–410.

Istanbulluoglu, A., Arslan, B., Gocmen, E., Gezer, E., and C. Pasa. 2010. Effects of deficit irrigation regimes on the yield and growth of oilseed rape (*Brassica napus* L.). *Biosystems Engineering* 105: 388–394.

Jackson, R.D., Idso, S.B., Reginato, R.J., and P.J. Pinter. 1981. Canopy temperature as a crop water stress indicator. *Water Resources Research* 17: 1133–1138.

Jacobsen, S.E., Jensen, C.R., and F. Liu. 2012. Improving crop production in the arid Mediterranean climate. *Field Crops Research* 128: 34–47.

Jeuffroy, M.H. and S. Recous. 1999. Azodyn: A simple model simulating the date of nitrogen deficiency for decision support in wheat fertilization. *European Journal of Agronomy* 10: 129–144.

Jin, H., Qingjie, W., Hongwen, L., Lijin, L., and G. Huanwen. 2009. Effect of alternative tillage and residue cover on yield and water use efficiency in annual double cropping system in North China Plain. *Soil and Tillage Research* 104: 198–205.

Johnson, J.M.F. and J.M. Novak. 2012. Chapter 8—Sustainable bioenergy feedstock production systems: Integrating carbon dynamics, erosion, water quality, and greenhouse gas production. In *Managing Agricultural Greenhouse Gases*, eds. M.A. Liebig, A.J. Franzluebbers, and R. Follett, pp. 111–126. London, U.K.: Academic Press.

Johnson, N., Revenga, C., and J. Echeverría. 2001. Managing water for people and nature. *Science* 292: 1071–1072.

Kamm, B. and M. Kamm. 2004. Principles of biorefineries. *Applied Microbiology and Biotechnology* 64: 137–145.

Kang, S., Liang, Z., Panb, Y., Shic, P., and J. Zhang. 2000. Alternate furrow irrigation for maize production in an arid area. *Agricultural Water Management* 45: 267–274.

Karam, F., Lahoud, R., Masaad, R. et al. 2007. Evapotranspiration, seed yield and water use efficiency of drip irrigated sunflower under full and deficit irrigation conditions. *Agricultural Water Management* 90: 213–223.

Karlen, D.L., Birell, S.J., and J.R. Hess. 2011. A five-year assessment of corn stover harvest in central Iowa, USA. *Soil and Tillage Research* 115–116: 47–55.

Karrou, M. and T. Oweis. 2012. Water and land productivities of wheat and food legumes with deficit supplemental irrigation in a Mediterranean environment. *Agricultural Water Management* 107: 94–103.

Kesava Rao, A.V.R., Wani, S.P., Singh, P., Srinivas, K., and C. Srinivasa Rao. 2012. Water requirement and use by *Jatropha curcas* in a semi-arid tropical location. *Biomass and Bioenergy* 39: 175–181.

Kirda, C., Topcu, S., Kaman, H. et al. 2005. Grain yield response and N-fertiliser recovery of maize under deficit irrigation. *Field Crop Research* 93: 132–141.

Klein, T.W., Kosaka, L., Poirier, L. et al. 2008. *Global Ethanol and Biodiesel Outlook, 2009–2015*. Houston, TX: Hart Energy Publishing.

Koch, H.J. and N. Stockfisch. 2006. Loss of soil organic matter upon ploughing under a loess soil after several years of conservation tillage. *Soil and Tillage Research* 86: 73–83.

Kumar, A.A., Reddy, B.V.S., Reddy, C.R. et al. 2010. Enhancing the harvest window for supply chain management of sweet sorghum for ethanol production. *Journal of SAT Agricultural Research* 8: 1–5. http://eprints.icrisat.ac.in/907 (last consulted, September 20, 2012).

Labrecque, M. and T.I. Teodorescu. 2005. Field performances and biomass production of 12 willow and poplar clones in short-rotation coppice in southern Quebec (Canada). *Biomass and Bioenergy* 29: 1–9.

Lal, R., Recoisky, D.C., and J.D. Hanson. 2007. Evolution of the plow over 10,000 years and the rationale for no-till farming. *Soil and Tillage Research* 93: 1–12.

Landström, S., Lomakka, L., and S. Andersson. 1996. Harvest in spring improves yield and quality of reed canary grass as a bioenergy crop. *Biomass and Bioenergy* 11: 333–341.

Lee, D.K., Owens, V.N., and J.J. Doolittle. 2007. Switchgrass and soil carbon sequestration response to ammonium nitrate, manure and harvest frequency on conservation reserve program land. *Agronomy Journal* 99: 462–468.

Lemus, R., Brummer, E.C., Burras, C.L., Moore, K.J., Barker, M.F., and N.E. Molstad. 2008b. Effects of nitrogen fertilization on biomass yield and quality in large fields of established switchgrass in southern Iowa, USA. *Biomass and Bioenergy* 32: 1187–1194.

Lemus, R., Parrish, D.J., and O. Abaye. 2008a. Nitrogen-use dynamics in switchgrass grown for biomass. *BioEnergy Research* 1: 153–162.

Lewandowski, I., Clifton-Brown, J.C., Andersson, B. et al. 2003. Environment and harvest time affects the combustion qualities of *Miscanthus* genotypes. *Agronomy Journal* 95: 1274–1280.

Lewandowski, I. and D. Kauter. 2003. The influence of nitrogen fertilizer on the yield and combustion quality of whole grain crops for solid fuel use. *Industrial Crops and Products* 17: 103–117.

Li, X.Y. and J.D. Gong. 2002. Effects of different ridge: Furrow ratios and supplemental irrigation on crop production in ridge and furrow rainfall harvesting system with mulches. *Agricultural Water Management* 54: 243–254.

Lingle, S.E., Tew, T.L., Rukavina, H., and D.L. Boykin. 2012. Post-harvest changes in sweet sorghum I: Brix and sugars. *BioEnergy Research* 5: 158–167.

Long, S.P., Zhu, X.G., Naidu, S.L., and D.R. Ort. 2006. Can improvement in photosynthesis increase crop yield? *Plant, Cell and Environment* 29: 315–330.

Luck, J.D., Pitla, S.K., Shearer, S.A. et al. 2010. Potential for pesticide and nutrient savings via map-based automatic boom section control of spray nozzles. *Computers and Electronics in Agriculture* 70: 19–26.

Ludlow, M.M. 1989. Strategies of response to water stress. In *Structural and Functional Responses to Environmental Stresses*, eds. K.H. Kreeb, H. Ritcher, and T.M. Hinckley, pp. 269–281. The Hague, the Netherlands: SPB Academic Publishing.

Lyle, W.M. and J.P. Bordovsky. 1981. Low energy precision application (LEPA) irrigation system. *Transactions of the ASAE* 24: 1241–1245.

Ma, Z., Wood, C.W., and D.I. Bransby. 2001. Impact of row spacing, nitrogen rate, and time on carbon partitioning of switchgrass. *Biomass and Bioenergy* 20: 413–419.

Machet, J.M., Dubrulle, P., and P. Louis. 1990. AZOBIL: A computer program for fertilizer N recommendations based on a predictive balance sheet method. In *Proceedings of the First Congress of the European Society for Agronomy*, ed. A. Scaife, pp. 20–21. Paris, France: European Society of Agronomy, December 5–7.

Mantineo, M., D'Agosta, G.M., Copani, V., Patanè, C., and S.L. Cosentino. 2009. Biomass yield and energy balance of three perennial crops for energy use in the semi-arid Mediterranean environment. *Field Crops Research* 114: 204–213.

Marques da Silva, J.R. and L.L. Silva. 2008. Evaluation of the relationship between maize yield spatial and temporal variability and different topographic attributes. *Biosystems Engineering* 101: 183–190.

Massé, D., Gilbert, Y., Savoie, P., Bélanger, G., Parent, G., and D. Babineau. 2010. Methane yield from switchgrass harvested at different stages of development in Eastern Canada. *Bioresource Technology* 101: 9536–9541.

Mastrorilli, M., Katerji, N., and G. Rana. 1999. Productivity and water use efficiency of sweet sorghum as affected by soil water deficit occurring at different vegetative growth stages. *European Journal of Agronomy* 11: 207–215.

Mengel, K., Hütsch, B., and Y. Kane. 2006. Nitrogen fertilizer application rates on cereal crops according to available mineral and organic soil nitrogen. *European Journal of Agronomy* 24: 343–348.

Metz, B., Davidson, O.R., Bosch, P.R., Dave, R., and L.A. Meyer. 2007. *Climate Change 2007: Mitigation of Climate Change*. Cambridge, U.K.: Cambridge University Press.

Mistele, B. and U. Schmidhalter. 2008. Estimating the nitrogen nutrition index using spectral canopy reflectance measurements. *European Journal of Agronomy* 29: 184–190.

Mitchell, C.P., Stevens, E.A., and M.P. Watters. 1999. Short-rotation forestry—Operations, productivity and costs based on experience gained in the UK. *Forest Ecology and Management* 121: 123–136.

Monti, A., Barbanti, L., Zatta, A., and W. Zegada-Lizarazu. 2011. The contribution of switchgrass in reducing GHG emissions. *Global Change Biology Bioenergy* 4: 420–434.

Monti, A., Bezzi, G., Pritoni, G., and G. Venturi. 2008a. Long-term productivity of lowland and upland switchgrass cytotypes as affected by cutting frequency. *Bioresource Technology* 99: 7425–7432.

Monti, A., Di Virgilio, N., and G. Venturi. 2008b. Mineral composition and ash content of six major energy crops. *Biomass and Bioenergy* 32: 216–223.

Monti, A., Fazio, S., and G. Venturi. 2009. The discrepancy between plot and field yields: Harvest and storage losses of switchgrass. *Biomass and Bioenergy* 33: 841–847.

Monti, A., Venturi, P., and H.W. Elbersen. 2001. Evaluation of the establishment of lowland and upland switchgrass (*Panicum virgatum* L.) varieties under different tillage and seedbed conditions in northern Italy. *Soil and Tillage Research* 63: 75–83.

Muir, J.P., Sanderson, M.A., Ocumpaugh, W.R., Jones, R.M., and R.L. Reed. 2001. Biomass production of "Alamo" switchgrass in response to nitrogen, phosphorus, and row spacing. *Agronomy Journal* 93: 896–901.

Nail, E.L., Young, D.L., and W.F. Schillinger. 2007. Diesel and glyphosate price changes benefit the economics of conservation tillage versus traditional tillage. *Soil and Tillage Research* 94: 321–327.

Nassi o Di Nasso, N., Guidi, W., Ragaglini, G., Tozzini, C., and E. Bonari. 2010. Biomass production and energy balance of a 12-year-old short-rotation coppice poplar stand under different cutting cycles. *Global Change Biology Bioenergy* 2: 89–97.

Nassi o Di Nasso, N., Roncucci, N., Triana, F., Tozzini, C., and E. Bonari. 2011. Seasonal nutrient dynamics and biomass quality of giant reed (*Arundo donax* L.) and *Miscanthus* (*Miscanthus* × *giganteus* Greef et Deuter) as energy crops. *Italian Journal of Agronomy* 6: 152–158.

Németh, K. 1982. Electro-ultrafiltration of aqueous soil suspension with simultaneously varying temperature and voltage. *Plant and Soil* 64: 7–23.

Oenema, O., Kroes, H., and W. de Vries. 2003. Approaches and uncertainties in nutrient budgets: Implications for nutrient management and environmental policies. *European Journal of Agronomy* 20: 3–16.

Oerke, E.C. and H.W. Dehne. 2004. Safeguarding production—Losses in major crops and the role of crop protection. *Crop Protection* 23: 275–285.

Ogle, S.M., Swan, A., and K. Paustian. 2012. No-till management impacts on crop productivity, carbon input and soil carbon sequestration. *Agriculture, Ecosystems and Environment* 149: 37–49.

Olivier, J.G.J., A.F. Bouwman, Berdowski, J.J.M. et al. 1999. Sectoral emission inventories of greenhouse gases for 1990 on a per country basis as well as on 1° × 1° grid. *Environmental Science Policy* 2: 241–263.

Ortega, R.A. and O.A. Santìbañez. 2007. Determination of management zones in corn (*Zea mays* L.) based on soil fertility. *Computers and Electronics in Agriculture* 58: 49–59.

Oweis, T., Pala, M., and J. Ryan. 1998. Stabilizing rainfed wheat yields with supplemental irrigation and nitrogen in a Mediterranean-type climate. *Agronomy Journal* 90: 672–681.

Parent, A.C., Bélanger, M.C., Parent, L.C. et al. 2008. Soil properties and landscape factors affecting maize yield under wet spring conditions in eastern Canada. *Biosystems Engineering* 99: 134–144.

Pari, L., Assirelli, A., and A. Suardi. 2010. Developing and evaluating the fibre sorghum mower conditioner prototype CRA-Ing/Cressoni. *Proceedings of the 18th European Biomass Conference and Exhibition*, Lyon, France, May 3–7, pp. 463–467.

Pari, L. and V. Civitarese. 2011. Harvesting yards for woody crops. In *Lo sviluppo delle colture energetiche in Italia—Il contributo dei progetti di ricerca SUSCACE e FAESI*, ed. L. Pari, Rome, Italy, pp. 165–181. http://ing.entecra.it/Biomasse (last consulted, September 6, 2012).

Pari, L., Civitarese, V., Assirelli, A., and A. Del Giudice. 2009. Prototype for *Cynara cardunculus* capitula threshing and biomass windrowing. *Proceedings of the 17th European Biomass Conference and Exhibition*, Hamburg, Germany, June 29–July 3, pp. 262–267.

Passioura, J. 2006. Increasing crop productivity when water is scarce—From breeding to field management. *Agricultural Water Management* 80: 176–196.

Payero, J.O., Tarkalson, D.D., Irmak, S., Davison, D., and J.L. Petersen. 2009. Effect of timing of a deficit-irrigation allocation on corn evapotranspiration, yield, water use efficiency and dry mass. *Agricultural Water Management* 96: 1387–1397.

Pelzer, E., Bazot, M., Makowski, D. et al. 2012. Pea–wheat intercrops in low-input conditions combine high economic performances and low environmental impacts. *European Journal of Agronomy* 40: 39–53.

Rathke, G.W., Christen, O., and W. Diepenbrock. 2005. Effects of nitrogen source and rate on productivity and quality of winter oilseed rape (*Brassica napus* L.) grown in different crop rotations. *Field Crops Research* 94: 103–113.

Ratnavathi, C.V., Kalyana Chakravarthy, S., Komala, V.V., Chavan, U.D., and J.V. Patil. 2011. Sweet sorghum as feedstock for biofuel production: A review. *Sugar Tech* 13: 399–407.

Rémy, J.C. and J. Hébert. 1977. Le devenir des engrais dans le sol. *Comptes Rendus de l'Académie d'Agriculture de France* 63: 700–710.

Rémy, J.C. and P. Viaux. 1982. The use of nitrogen fertilizers in intensive wheat growing in France. In *Proceeding of the Symposium on Fertilizers and Intensive Wheat Production in EEC*, eds. P.F.J. van Burg, W.H. Prins, D.J. den Boer, and W.J. Sluiman, Vol. 211, pp. 67–92. London, U.K.: The International Fertiliser Society.

Reynolds, J.H., Walker, C.L., and M.J. Kirchner. 2000. Nitrogen removal in switchgrass biomass under two harvest systems. *Biomass and Bioenergy* 19: 281–286.

Rinaldi, M. 2001. Application of EPIC model for irrigation scheduling of sunflower in Southern Italy. *Agricultural Water Management* 49: 185–196.

Rosillo-Calle, F. 2007. Overview of biomass energy. In *The Biomass Assessment Handbook: Bioenergy for a Sustainable Environment*, eds. F. Rosillo-Calle, P. de Groot, S.L. Hemstock, and J. Woods, pp. 1–26. London, U.K.: Earthscan.

Sakellariou-Makrantonaki, M., Papalexis, D., Nakos, N., and I.K. Kalavrouziotis. 2007. Effect of modern irrigation methods on growth and energy production of sweet sorghum (var. Keller) on a dry year in Central Greece. *Agricultural Water Management* 90: 181–189.

Sanderson, M.A., Egg, R.P., and A.E. Wiselogel. 1997. Biomass losses during harvest and storage of switchgrass. *Biomass and Bioenergy* 12: 107–114.

Sanderson, M.A., Read, J.C., and R.L. Reed. 1999. Harvest management of switchgrass for biomass feedstock and forage production. *Agronomy Journal* 91: 5–10.

Scotford, I.M. and P.C.H. Miller. 2005. Applications of spectral reflectance techniques in Northern European cereal production: A review. *Biosystems Engineering* 90: 235–250.

Sieling, K. and H. Kage. 2006. N balance as an indicator of N leaching in an oilseed rape—Winter wheat—Winter barley. *Agriculture, Ecosystems and Environment* 115: 261–269.

Sims, R.E.H., Hastings, A., Schlamadinger, B., Taylor, G., and P. Smith. 2006. Energy crops: Current status and future prospects. *Global Change Biology* 12: 2054–2076.

Smeets, E., A. Faaij, Lewandowski, I., and W. Turkenburg. 2007. A bottom-up assessment and review of global bio-energy potentials to 2050. *Progress in Energy and Combustion* 33: 56–106.

Soane, B.D., Ball, B.C., Arvidsson, J. et al. 2012. No-till in northern, western and south-western Europe: A review of problems and opportunities for crop production and the environment. *Soil and Tillage Research* 118: 66–87.

Sombrero, A. and A. de Benito. 2010. Carbon accumulation in soil. Ten-year study of conservation tillage and crop rotation in a semi-arid area of Castile-Leon, Spain. *Soil and Tillage Research* 107: 64–70.

Souch, C.A., Martin, P.J., Stephens, W., and G. Spoor. 2004. Effects of soil compaction and mechanical damage at harvest on growth and biomass production of short rotation coppice willow. *Plant and Soil* 263: 173–182.

Strengers, B., Leemans, R., Eickhout, B. et al. 2004. The land-use projections and resulting emissions in the IPCC SRES scenarios as simulated by the IMAGE 2.2 model. *GeoJournal* 61: 381–393.

Strullu, L.S., Cadoux, S., Preudhomme, M., Jeuffroy, M.H., and N. Beaudoin. 2011. Biomass production and nitrogen accumulation and remobilisation by *Miscanthus* × *giganteus* as influenced by nitrogen stocks in belowground organs. *Field Crops Research* 121: 381–391.

Tavakkoli, A.R. and T.Y. Oweis. 2004. The role of supplemental irrigation and nitrogen in producing bread wheat in the highlands of Iran. *Agricultural Water Management* 65: 225–236.

Taylor, H.M. 1983. Managing root systems for efficient water use: An overview. In *Limitations to Efficient Water Use in Crop Production*, eds. H.M. Taylor, W.R. Jordan, and T.R. Sinclair, pp. 87–113. Madison, WI: ASA-CSSA-SSSA.

Tellaeche, A., BurgosArtizzu, X.P., Pajares, G., Ribeiro, A., and C Fernández-Quintanilla. 2008. A new vision-based approach to differential spraying in precision agriculture. *Computers and Electronics in Agriculture* 60: 144–155.

Thomason, W.E., Raun, W.R., Johnson, G.V. et al. 2004. Switchgrass response to harvest frequency and time and rate of applied nitrogen. *Journal of Plant Nutrition* 27: 1199–1226.

Vamerali, T., Bertocco, M., and L. Sartori. 2006. Effects of a new wide-sweep opener for no-till planter on seed zone properties and root establishment in maize (*Zea mays* L.): A comparison with double-disk opener. *Soil and Tillage Research* 89: 196–209.

Van den Putte, A., Govers, G., Diels, J., Gillijns, K., and M. Demuzere. 2010. Assessing the effect of soil tillage on crop growth: A meta-regression analysis on European crop yields under conservation agriculture. *European Journal of Agronomy* 33: 231–241.

Van der Pol, F. 1992. *Soil Mining: An Unseen Contributor to Farm Income in Southern Mali*, Bulletin 325. Amsterdam, the Netherlands: Royal Tropical Institute.

Varvel, G.E., Vogel, K.P., Mitchell, R.B., Follett, R.F., and J.M. Kimble. 2008. Comparison of corn and switchgrass on marginal soils for bioenergy. *Biomass and Bioenergy* 32: 18–21.

Vogel, K.P., Brejda, J.J., Walters, D.T., and D.R. Buxton. 2002. Switchgrass biomass production in the Midwest USA: Harvest and nitrogen management. *Agronomy Journal* 94: 413–420.

Vyn, T.J., Galic, D.M., and K.J. Janovicek. 2002. Corn response to potassium placement in conservation tillage. *Soil and Tillage Research* 67: 159–169.

Walther, S., Schule, T., and K. Koller. 2007. Precision tillage: Online variation of tillage depth. Bulletin of University of Agricultural Sciences and Veterinary Medicine Cluj-Napoca. *Agriculture* 63: 34–39.

Wang, X.B., Cai, D.X., Hoogmoed, W.B., Oenema, O., and U.D. Perdok. 2007. Developments in conservation tillage in rainfed regions of North China. *Soil and Tillage Research* 93: 239–250.

Wang, Y., Xie, Z., Malhi, S.S., Vera, C.L., Zhang, Y., and J. Wang. 2009. Effects of rainfall harvesting and mulching technologies on water use efficiency and crop yield in the semi-arid Loess Plateau, China. *Agricultural Water Management* 96: 374–382.

Wang, Z., Liu, F., Kang, S., and C.R. Jensen. 2012. Alternate partial root-zone drying irrigation improves nitrogen nutrition in maize (*Zea mays* L.) leaves. *Environmental and Experimental Botany* 75: 36–40.

Weiss, E.A. 1983. *Oilseed Crops*. London, U.K.: Longman Group.

Wiklicky, L. 1982. Application of the EUF procedure in sugar beet cultivation. *Plant and Soil* 64: 115–127.

Wolf, D.D., Parrish, D.J., Daniels, W.L., and J.R. McKenna. 1989. No-till establishment of perennial, warm-season grasses for biomass production. *Biomass* 20: 209–217.

Zhang, G.S., Chan, K.Y., Oates, A., Heenan, D.P., and G.B. Huang. 2007. Relationship between soil structure and runoff/soil loss after 24 years of conservation tillage. *Soil and Tillage Research* 92: 122–128.

Zhang, L., Clarke, M.L., Steven, M.D., and K.W. Jaggard. 2011. Spatial patterns of wilting in sugar beet as an indicator for precision irrigation. *Precision Agriculture* 12: 296–316.

Zhao, Y.L., Steinberger, Y., Shi, M., Han, L.P., and G.H. Xie. 2012. Changes in stem composition and harvested produce of sweet sorghum during the period from maturity to a sequence of delayed harvest dates. *Biomass and Bioenergy* 39: 261–273.

Zub, H.W., Arnoult, S., and M. Brancourt-Hulmel. 2011. Key traits for biomass production identified in different *Miscanthus* species at two harvest dates. *Biomass and Bioenergy* 35: 637–651.

3 Mathematical Modeling in Biomass and Bioenergy Systems

Lijun Wang, Samuel Asomaning Agyemang, and Abolghasem Shahbazi

CONTENTS

3.1 INTRODUCTION

Lignocellulosic biomass is an abundant renewable energy source. Energy crops such as switchgrass and sweet sorghum have been widely investigated. As biomass is characterized by a low energy density, conversion of biomass into different energy products is essential. The conversion of biomass into bioenergy products such as ethanol and syngas involves the interaction between chemical and physical processes. Experiments at a large scale are often expensive and complicated. Mathematical models can provide economical and powerful tools for the design and optimization of biomass production and conversion. Mathematical models have been widely used to simulate all aspects of bioenergy production systems including the growth kinetics of energy crops (van den Broek et al., 2001), conversion processes (Gomez-Barea and Leckner, 2010), production economics (Zhang et al., 2013b), supply logistics (Gan and Smith, 2011), and environmental impacts (Cherubini et al., 2009; Clarens et al., 2010).

Biomass resources and conversion techniques more often depend on local conditions for biomass feedstock supply and energy uses than fossil fuels. Biomass fuels are frequently by-products from other main industrial activities such as forestry, agriculture, and waste management. Bioenergy supply chains often share costs with other complementary activities. However, mathematical models are always developed by simplifying and generalizing real physical systems. It requires different strategies to predict many unique features of a real bioenergy system including the cost structure of energy production, information asymmetry, socioeconomic factors, household economics, strategic considerations, and policy uncertainties (Roos and Rakos, 2000). Once a mathematical model is developed, the values of a number of parameters have to be obtained as the inputs of the model. The experimental techniques should be available for the determination of the model inputs and the comparison with the predicted results (Grace and Taghipour, 2004). This chapter is to review the different mathematical models that have been developed to simulate the production of energy crops and the conversion of biomass into different energy products.

3.2 COMPUTER MODELS FOR BIOMASS PRODUCTION

3.2.1 Energy Crop Growth Models

Energy crop yield is an important parameter to assess the cost of biomass production. However, the yield of an energy crop is usually an uncertain factor in the supply chain of bioenergy systems because there is limited commercial experience to produce energy crops at a large scale around the world (van den Broek et al., 2001). A crop growth model is a predictive tool that can be used to estimate yields of energy crops in a region under different situations such as potential growth with sufficient supply of water and nutrients and free of weeds, pests, and diseases, water-limited growth, and actual growth under all limitations in practice. Van den Broek et al. (2001) developed an energy crop growth model to estimate potential and water-limited yields for a certain region on the basis of solar radiation and daily data for precipitation and pan evaporation. The potential growth rate of an energy crop at a specific region where temperature, fertilizers, and water are not limiting growth factors can be calculated by

$$G_p = \varepsilon \varphi f_{PAR} \left(1 - e^{-kLAI}\right) HI \tag{3.1}$$

where

G_p is the potential rate (g_{dm}/m^2 day)
ε is the crop radiation utilization coefficient (g_{dm}/MJ_{PAR})
φ is the solar radiation (MJ/m^2 day)
f_{PAR} is the amount of photosynthetically active radiation (—)
LAI is the leaf area index $\left(m^2_{leaf}/m^2_{ground}\right)$
k is the canopy extinction coefficient $\left(m^2_{ground}/m^2_{leaf}\right)$
HI is the harvest index (—)

The input data for the growth of *Eucalyptus camaldulensis* in Nicaragua are given in Table 3.1 (van den Broek et al., 2001).

The temperature, nitrogen, and water-limited growth rate can be calculated as a function of the potential growth rate by taking into account individual limiting factors, which is given by

$$G_l = f_T f_N f_W G_p \tag{3.2}$$

TABLE 3.1
Input Data for the Growth of *E. camaldulensis* in Nicaragua

Parameter	Value	Unit
ε	2.2	g_{dm}/MJ_{PAR}
k	0.5	m^2_{ground}/m^2_{leaf}
φ	57	TJ/(ha year)
HI	0.85	—
LAI	0.8–3.2[a]	m^2_{leaf}/m^2_{ground}

Source: van den Broek, R., *Biomass Bioenergy*, 21, 335, 2001.

[a] The leaf area index depends on the growth year and rotation.

In the previous equation, f_T, f_N, and f_W are limiting factors for temperature, nitrogen availability, and water in the soil (Katterer et al., 1997; van den Broek et al., 2001). The limiting factors can be determined by (Katterer et al., 1997)

$$f_T = \frac{T - T_{min}}{T_{max} - T_{min}} \tag{3.3}$$

$$f_N = \frac{N_{leaf} - N_{leaf,min}}{N_{leaf,max} - N_{leaf,min}} \tag{3.4}$$

$$f_W = \frac{E_t}{E_{tp}} \tag{3.5}$$

where
- T is the temperature
- N is the nitrogen
- E_t is the actual transpiration in the soil
- E_{tp} is the potential transpiration in the soil

The use of a crop growth model requires input data such as meteorological data, soil data, and crop characteristics. Some coefficients in the previous model are determined by the data on the actual yield at a reference location. The interactions between site-specific properties such as climate, available soil water and nutrients, and crop growth are complex. Therefore, the previous model assumes that the nutrient conditions and yield reduction by weeds, pests, and diseases at a specific location and reference location are the same (van den Broek et al., 2001). The predictions showed that the potential and water-limited yields for the growth of *E. camaldulensis* in Nicaragua, the Netherlands, were 35 and 23 tons dry mass/ha/year, respectively. The actual yield was 13 tons dry mass/ha/year, which was about 58% of its water-limited yield (van den Broek et al., 2001).

Algae are an attractive source of biomass energy because they do not compete with food crops and have higher yields per area than terrestrial energy crops such as switchgrass. Small algae can achieve higher rates of photosynthesis and have higher specific growth rate and faster uptake of nutrients per unit mass of biomass (Hein et al., 1995). Therefore, since algae at different sizes present in the cultivation system, it cannot be assumed that all cells have the same growth rate in a mathematical model for the simulation of algal growth (Concas et al., 2010). Concas et al. (2010) developed a mass-dependent growth kinetics model, which also considered the effects of light intensity, nutrients concentration, and inhibitors of dissolved oxygen on the microalgae growth kinetics. The changing rate of cell mass was given by

$$\frac{dm}{dt} = \left[\mu_{max} \frac{I^n}{I_K^n + I^n} \prod_{j=1}^{2} \frac{C_j}{K_j + C_j} \left(1 - \frac{C_{O_2}}{C_{O_2,max}} \right) - \mu_c \right] m \tag{3.6}$$

where

m is the single cell mass (g)
μ_{max} is the maximum specific rate of cell growth (s^{-1})
I is the photosynthetically active radiation within the culture ($\mu E\ m^{-2}\ s^{-1}$)
I_K is the half saturation constant for photosynthetically active radiation ($\mu E\ m^{-2}\ s^{-1}$)
n is the coefficient appearing in the rate of cell growth
K_j is the half saturation constant of jth nutrient ($g\ m^{-3}$)
C_j is the concentration of jth nutrient in culture medium ($g\ m^{-3}$)
C_{O_2} is the dissolved oxygen concentration in culture medium ($g\ m^{-3}$)
$C_{O_2,max}$ is the dissolved oxygen concentration at which algal growth is inhibited ($g\ m^{-3}$)
μ_c is the catabolic growth rate (s^{-1})

The death of the mother cell and the cell birth due to mitosis can be determined by probability partition function (Concas et al., 2010). Every microalgal species has optimal requirements of light intensity, nutrients and CO_2, temperature, and pH value. Concas et al. (2010) collected the model parameters for *Spirulina platensis* from literature, which are given in Table 3.2.

One of the main research efforts in the area of microalgal cultivation is to identify algal strains with high lipid content for the production of biodiesel and jet fuels. Many green microalgae can increase their cellular neutral lipid content when cultured in nitrogen-limited or high light conditions. A mathematical model can be used to predict the lipid production of microalgae with respect to nutrient and light conditions (Packer et al., 2011). Four ordinary differential equations as given below were developed to describe (1) the algal growth or the time-dependent algal biomass concentration excluding neutral lipids, A(t) (g/m^3, dry mass); (2) the neutral lipid concentration, L(t) (g/m^3); (3) the *chl a* content in biomass, H(t) (g chl/g biomass, dry mass); and (4) the extracellular

TABLE 3.2
Model Parameters Used for Simulation of *S. platensis* Growth

Parameter	Value	Unit
I_K	160	$\mu E\ m^{-2}\ s^{-1}$
N	1.49	—
μ_{max}	2.06×10^{-5}	s^{-1}
K_{NO_3}	5.314	$g\ m^{-3}$
$K_{H_2PO_4}$	0.028	$g\ m^{-3}$
$C_{O_2,max}$	47.9	$g\ m^{-3}$
μ_c	0	s^{-1}

nitrogen concentration, N(t) (g/m^3). The algal mass consists of nonneutral lipid biomass A and neutral lipids L. The assumption of the model was that when the algal growth of A(t) ceases, excess carbon and energy are used to produce neutral lipids, L(t):

$$\frac{dA(t)}{dt} = \mu(A,L,H,N)A(t) \tag{3.7}$$

$$\frac{dL(t)}{dt} = \left[p(A,L,H,N) - c\mu(A,L,H,N)\right]A(t) \tag{3.8}$$

$$\frac{dH(t)}{dt} = c\frac{\mu(A,L,H,N)}{p(A,L,H,N)}\rho v(A,N) - H(t)\mu(A,L,H,N) \tag{3.9}$$

$$\frac{dN(t)}{dt} = -v(A,N)A(t) \tag{3.10}$$

In the previous equations, the coefficients μ, p, and v were determined as functions of nitrogen and light conditions. The coefficients of c and ρ are C subsistence quota and maximum chl/N, respectively. The comparison between predicted and experimental data for *Pseudochlorococcum* sp. showed that the model could predict the algal growth and neutral lipid synthesis very well for growth media with low nitrogen concentration. However, the model overestimated the biomass for the growth media with high nitrogen concentration (Packer et al., 2011).

3.2.2 GIS Model for Cropping Management

A geographical information system (GIS) is a spatial modeling tool, which includes digital maps and tables of attributes linked to map identities. GIS can be used to combine various digital agronomic data such as the irradiation and the slope of land and to mark the areas with the most feasible energy crop production in each layer. Digital layers are subsequently combined according to hierarchical factors of contribution to the feasible growth of the energy crops. Figure 3.1 shows the agroclimate variables and their contribution for the growth of energy crop of *Brassica* sp. (Gasol et al., 2011). The total potential crop area (TPCA) can be determined by summing the scores of individual agronomic factors, AF_i, multiplied by their weighting factors, W_i, which is given by

$$TPCA = \sum AF_i \times W_i \tag{3.11}$$

A GIS model can be used to maximize the energy production of energy crops by identifying suitable land to grow them based on their specific characteristics and the current use of the land (Fiorese and Guariso, 2010). The integration of GIS and life cycle assessment (LCA) can provide an effective method to determine the implementation strategy to reduce energy consumption and CO_2 emissions for growing energy crops at a local and decentralized scale (Gasol et al., 2011).

3.3 COMPUTATIONAL PROCESS ENGINEERING FOR BIOMASS CONVERSION

Biomass can be converted into energy products in various biological processes such as fermentation and anaerobic digestion and thermochemical processes such as combustion, pyrolysis, and gasification. Combined process models and reaction kinetics provide various advanced computational tools for the design and optimization of both biological and thermochemical processes (Di Blasi, 2008). Process simulation can be conducted at a steady state where the variation of process properties over time is ignored and dynamic state where the process properties are time dependent. Equilibrium

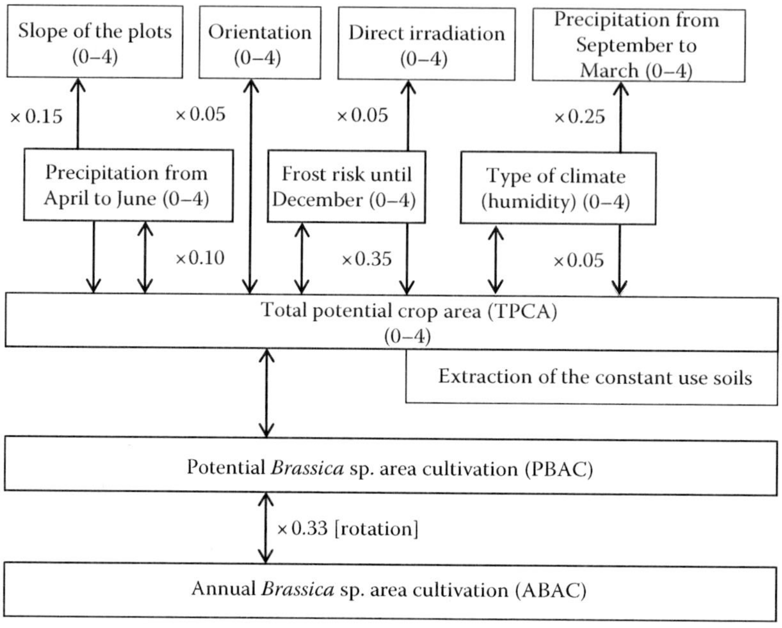

FIGURE 3.1 Agroclimate variables and their contribution for the growth of energy crop of *Brassica* sp. (Adapted from Gasol, C.M. et al., *Biomass Bioenergy*, 35, 2975, 2011.)

models assume that chemical equilibrium is reached for process reactions, which can be used to predict the highest conversion or thermal efficiency that can be possibly obtained for a given process at a steady state (Loha et al., 2011). Aspen Plus is one of the common commercial steady-state process simulators. ANSYS Fluent that is used to predict fluid flow, heat transfer, and mass transfer by solving a set of conservation equations is one of the most popular commercial dynamic-state process simulators (Ahmed et al., 2012). Reaction kinetic models can be used to estimate product formation rates and composition under varying operating conditions, which is essential knowledge for designing and operating a process.

3.3.1 Process Simulation Using Computational Fluid Dynamics (CFD)

Computational fluid dynamics (CFD) is a powerful tool used to study fluid flows and associated mass and heat transfer. In CFD calculation, the continuity equation, momentum conservation equation (also known as the Navier–Stokes transport equations), and energy conservation equation are numerically solved to give predictions of velocity, temperature, shear, pressure profiles, and other parameters in a fluid flow system. The balance of the conserved variables φ such as density, momentum, and enthalpy for each phase k, over a fixed domain or element of reactor volume, can be expressed by an Eulerian formula as (Gomez-Barea and Leckner, 2010)

$$\frac{\partial \varphi_k}{\partial t} + \mathrm{div}(\vec{u}_k \varphi_k) = \mathrm{div}(D_{\varphi_k} \mathrm{grad}\varphi_k) + S_{\varphi_k} \tag{3.12}$$

where

φ_k is the variable of velocity or temperature to be calculated
ρ is the density of materials
$\vec{u}$ is the velocity vector of flow
D is the property of materials such as apparent viscosity and thermal conductivity
S is the inner generation source such as reaction heat in the energy conservation equation

The specific conversion equations for the balances of density, momentum, and enthalpy can be given by the following equations:

Continuity equation:

$$\frac{\partial \rho_k}{\partial t} + \mathrm{div}(\vec{u}_k \rho_k) = S_{C,k} \tag{3.13}$$

Momentum equation:

$$\frac{\partial \rho_k \bar{u}_k}{\partial t} + \mathrm{div}(\rho_k \vec{u}_k \vec{u}_k) = -\mathrm{div}(p_k) + \mathrm{div}(\mu \cdot \mathrm{grad}\vec{u}_k) + S_{M,k} \tag{3.14}$$

Energy equation:

$$\frac{\partial \rho_k H_k}{\partial t} + \mathrm{div}(\rho_k \vec{u}_k H_k) = \mathrm{div}(\lambda_k \cdot \mathrm{grad}T_k) + S_H \tag{3.15}$$

where

ρ is the density of a fluid
μ is the viscosity
λ is the thermal conductivity
p is the pressure
H is the enthalpy
T is the temperature

3.3.1.1 CFD Modeling of a Multiphase Flow

The local accumulation rate of φ in each phase is caused by the rates of convection, $\mathrm{div}(\vec{u}_k \varphi_k)$, dispersion, $\mathrm{div}(D_{\varphi_k} \mathrm{grad}\varphi_k)$, and generation or consumption, S_k. The Eulerian–Eulerian, Lagrangian–Eulerian, and the volume of fluid multiphase models have been widely used in CFD simulations. The Eulerian–Eulerian multiphase model considers the different phases as interpenetrating continua with different phasic volume fractions. The Eulerian–Eulerian multiphase model solves the continuity, mass, and momentum equations for multiple phases such as fluid and solid phases, tracks the volume fraction of each phase, and then averages the Navier–Stokes equations over the volume fractions. For the Eulerian–Eulerian approach, the volume fraction of a phase should be introduced into Equation 3.12, which can be rewritten as

$$\frac{\partial \varepsilon_k \varphi_k}{\partial t} + \mathrm{div}(\varepsilon_k u_k \varphi_k) = \mathrm{div}(\varepsilon_k D_{\varphi_k} \mathrm{grad}\varphi_k) + S_{\varphi_k} \tag{3.16}$$

where ε_k is the volume faction of phase k.

The total phase volume fractions in each computational cell are equal to one (Singh et al., 2013). That is,

$$\sum \varepsilon_k = 1 \tag{3.17}$$

The Eulerian–Eulerian method requires process parameters such as drag, lift, collision, friction forces, and heterogeneous chemistry to be included as phase interaction terms in the governing equations (3.12) and (3.16) (Singh et al., 2013). The CFD simulation of a multiphase flow is strongly dependent on the proper modeling of the momentum exchange and drag coefficients among phases (Michele and Hempel, 2002).

The Lagrangian–Eulerian multiphase model treats a fluid phase as a continuum and the particulate phase as single dispersed particles. The basic assumption in this multiphase flow model is that the volume fraction of the dispersed or secondary phase (e.g., solid particle phase) is below 10%–12%, although the mass of the secondary phase can be greater than that of the primary phase (Singh et al., 2013). The primary phase is treated as a continuum by solving the time-averaged Navier–Stokes equation (3.12). The behaviors of particles or the particle trajectories in the particulate phase are obtained by calculating the forces of the primary phase flow field acting on a large number of the particles. The Lagrangian equation used to track particle trajectory is given by (Luo and Al-Dahhan, 2011)

$$m_p \frac{du_p}{dt} = \frac{1}{8}\pi\rho_p d^2 C_D \left|u_f - u_p\right|\left(u_f - u_p\right) + \frac{\pi d^3 \rho_f}{6}\frac{du_f}{dt} + \frac{\pi d^3 \rho_f}{12}\left(\frac{du_f}{dt} - \frac{du_p}{dt}\right) + \frac{\pi d^3}{6}\frac{\left(\rho_p - \rho_f\right)}{\rho_p} g \quad (3.18)$$

where
m_p is the particle mass
u is the velocity
d is the diameter of the particle
ρ is the density
g is the gravity
C_D is drag coefficient and the substitutes of p and f represent particle and fluid

The terms on the right side of the previous equation represent the drag, pressure, virtual mass, and buoyancy forces, respectively.

Particle trajectories are calculated in the given intervals during the primary flow phase calculations. The volume of fluid is to track the position of the interface between two or more immiscible fluids (ANSYS Inc., 2011). Dispersed and primary phases can exchange mass, energy, and momentum (Singh et al., 2013).

3.3.1.2 CFD Modeling of a Turbulent Flow

The transport equations of CFD can be applied to both laminar and turbulent flow conditions. The Navier–Stokes equation (3.14) can be solved directly for laminar flows. Turbulent flows are characterized by fluctuating velocity. Turbulent flow is characterized by fluctuating velocity of fluids. Turbulence affects the heat and mass transfer. However, it is impossible to use the Navier–Stokes equations to simulate the small fluctuations and high frequency of turbulent flows due to the limit of computational capacity. Turbulence modeling is a key issue in CFD simulations. The increase in turbulence can be represented by an increase in effective fluid viscosity.

The Reynolds-averaged Navier–Stokes (RANS) model is widely used to determine the effective viscosity by the average and fluctuating components. The large eddy simulation (LES) is typically used to simulate transient effects in turbulent flows. The RANS model is given by (Versteeg and Malalsekera, 1995)

$$\mu_{eff} = \mu + \mu_t \quad (3.19)$$

The eddy viscosity models such as κ-ε approach and second-order closure models are used to describe the flow turbulence if the effects of turbulence on the effective viscosity need to be considered (Versteeg and Malalsekera, 1995). There are three widely used turbulence models for CFD modeling of multiphase flows: mixture κ-ε, dispersed κ-ε, and two-phase κ-ε models. All three turbulence models use the same model constants but have different equations to determine the turbulence viscosity (ANSYS Inc., 2011). The most general turbulence model for multiphase flows solves a set of κ and ε transport equations for each phase, which is an appropriate choice when the turbulence transfer among the phases plays a dominant role.

The kinetic energy of turbulence, κ, can be determined by

$$\frac{\partial(\rho\kappa)}{\partial t}+\mathrm{div}(\rho u\kappa)=\mathrm{div}\left(\frac{\mu_{eff}}{\sigma_k}\,\mathrm{grad}\kappa\right)+G_k-\rho\varepsilon \tag{3.20}$$

The rate of dissipation of the kinetic energy of turbulence, ε, can be determined by

$$\frac{\partial(\rho\varepsilon)}{\partial t}+\mathrm{div}(\rho u\varepsilon)=\mathrm{div}\left(\frac{\mu_{eff}}{\sigma_\varepsilon}\,\mathrm{grad}\varepsilon\right)+C_{\varepsilon1}G_k\frac{\varepsilon}{\kappa}-C_{\varepsilon3}\rho\frac{\varepsilon^2}{\kappa} \tag{3.21}$$

The turbulent viscosity can be determined by

$$\mu_T=\rho C_\mu\frac{\kappa^2}{\varepsilon} \tag{3.22}$$

where

C_μ, $C_{\varepsilon1}$, $C_{\varepsilon2}$, and $C_{\varepsilon3}$ $\left(C_{\varepsilon3}=C_{\varepsilon2}+\left(C_\mu\rho\eta^3(1-\eta)/\eta_0\right)/\left(1+\beta\eta^3\right)\right)$ are constants of the turbulence model

μ_{eff} is the effective viscosity

G is the generation of the turbulent kinetic energy

σ_κ and σ_ε are the effective Prandtl numbers for κ and ε, respectively

Typical values of these constants are $C_\mu = 0.09$, $C_{1\varepsilon} = 1.44$, $C_{2\varepsilon} = 1.92$, $\sigma_k = 1$, and $\sigma_\varepsilon = 1.3$.

3.3.1.3 CFD Simulation Software

In the last few years, there has been continuous progress in the development of CFD codes. Some of the common commercial codes include CFX (http://www.software.aeat.com/cfx/), Fluent (http://www.fluent.com/), Phoenics (http://www.cham.co.uk/), and Star-CD (http://www.cd.co.uk). The computational procedure of most of commercial CFD packages is based on finite volume (FV) numerical method. In fact, the FV method was derived from the finite difference method. In the FV method, the domain is divided into discrete control volumes. The key step of the FV scheme is the integration of the transport equations over a control volume to yield a discretized equation at its nodal points (Versteeg and Malalsekera, 1995). Although CFD has been applied to industries such as aerospace, automotive, and nuclear for several decades, it has only recently been applied to the bioenergy industry due to the rapid development in computer and commercial software packages (Wu, 2013). Example applications of CFD in the bioenergy industry are CFD modeling of photobioreactors for microalgae production and CFD modeling of fluidized bed gasification of biomass. A review of CFD modeling of the fluidized bed gasification of biomass has been given by Gomez-Barea and Leckner (2010). Bitog et al. (2011) reviewed the applications of CFD in designing photobioreactors for microalgae production. Wu (2013) provided a review in the use of CFD to characterize, design, and optimize bioenergy systems with a focus on anaerobic digesters and photobioreactors.

3.3.1.4 CFD Modeling of Fluidized Bed Gasification of Biomass

Gasification has been considered as a promising technology to convert low-value solid biomass materials with a large variation in properties to a uniform gaseous mixture mainly consisting of CO, H_2, CO_2, and traces of CH_4. This gaseous mixture or syngas can be further used for heat and power generation, hydrogen production, and synthesis of liquid fuels (Wang et al., 2008). Compared to fossil resources such as coal, biomass resources have a large variation in properties such as moisture content, particle size, bulk density, chemical composition, and heating value. The variation in biomass properties increases the difficulty in developing a well-characterized biomass gasification process. A gasifier is designed to promote the interaction between biomass and gasifying agent to enhance the endothermic and heterogeneous reactions. Fluidized bed gasifiers have some advantages over fixed bed gasifiers including the uniform and controllable temperature distribution in the gasification zone. In a fluidized bed gasifier, a gasifying agent such as air or steam is blown upward through the solid bed to fluidize it, enhancing heat and mass transfer and thus gasification kinetics (Basu, 2006).

Fluidized bed gasification is a multiphase reactive flow phenomenon. The mechanical and thermal behaviors of reactive gas–solid flows in biomass fluidized bed gasifiers are not well understood due to complex physical phenomena that include particle–particle and particle–bubble interaction. Turbulent fluidization is particularly poorly understood due to the challenges of experimental and theoretical work related to this flow regime. As a consequence, the design and operation of gasifiers have been primarily based on empirical knowledge obtained through expensive and time-consuming laboratory experiments. The amount of experimental work could be substantially reduced if more detailed knowledge of the flow dynamics is gained from the analysis of these systems via mathematical models. A mathematical model can provide a fast and powerful prototyping tool for the design, control, optimization, and scale-up of a biomass gasification process (Gomez-Barea and Leckner, 2010). Although an equilibrium model can predict the thermodynamic limits of a gasification system, the yield and composition of syngas are dependent of the transport phenomena and reaction kinetics in a fluidized bed gasifier (Wang et al., 2011). The mechanical and thermal behaviors of reactive gas–solid flows in biomass fluidized bed gasifiers are not well understood due to complex physical and chemical phenomena (Di Blasi, 2008). The presence of solid particles in the turbulent gas flow in a gravitational field during fluidized bed gasification can lead to a significant reduction of turbulent mixing intensity, which affects the flow mechanical and thermal properties substantially (Barenblatt, 1996). Accurate simulation of microphysical phenomena like particle–particle interaction that causes solid-phase pressure gradient and collisional dissipation is another very important part of the mathematical model (Patil et al., 2005). CFD modeling of fluidized bed gasification is a relatively new and promising approach, but it requires significant improvements (Gomez-Barea and Leckner, 2010).

The biomass gasification occurs at two levels: the particle and the reactor levels (Gomez-Barea and Leckner, 2010). Modeling on the particle level requires the particle size and biomass properties such as density, specific heat, and thermal conductivity, which affect the devolatilization rate and dynamic composition of volatiles (Di Blasi, 2008). Depending on the size of biomass particles, internal or external transport process or both may significantly affect the rate of chemical reactions inside the particles. During the devolatilization of biomass, the biomass particle size is reduced by shrinkage, primary fragmentation, secondary and percolative fragmentation of char, and fines generation by attrition as shown in Figure 3.2.

Modeling of the comminution of biomass particles during gasification is required to determine the particle size distribution and the average size. The primary fragmentation probability, P_f, can be simply expressed as

$$P_f = 1 - \exp\left(-k_f t_v d_p\right) \tag{3.23}$$

where

$k_f = 0.11\ s^{-1}$ is the fragmentation constant

d_p is the particle size in meter

t_v is the time t, which primary fragmentation occurs at the beginning of the end of the devolatilization (Nemtsov and Zabaniotou, 2008)

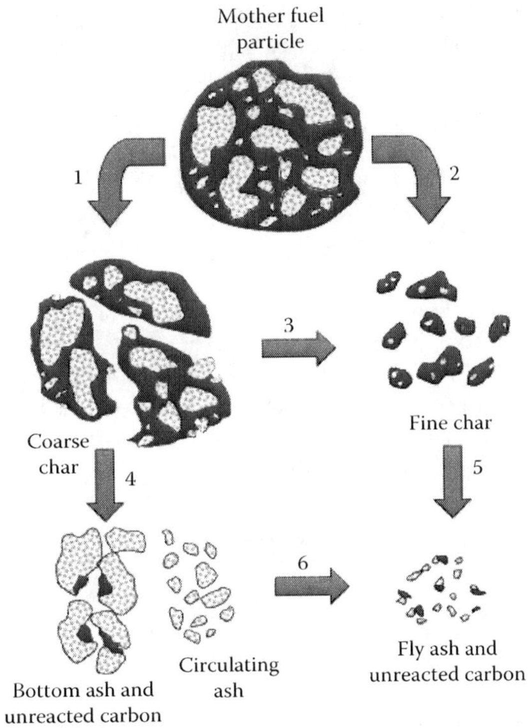

FIGURE 3.2 Comminution of biomass particles during gasification (1 and 2 are primary fragmentations, 3 is secondary fragmentation of coarse char, 4 is conversion and secondary fragmentation of coarse char, 5 is attrition of fine char, and 6 is attrition of fine char and ash). (Adapted from Gómez-Barea, A. and Leckner, B., *Prog. Energy Combust. Sci.*, 36, 444, 2010.)

Modeling on the reactor level requires the reactor design and operating parameters such as biomass feed points and rate. The movement of particles in a gasifier depends on the fluid dynamics of the bed. The gas–solid flow of a fluidized bed can be described by mass continuity, momentum, and energy conservation equations coupled with the chemical species transport equations presented in a variety of papers (Lathouwers and Bellan, 2001; Wang et al., 2005). A fluidized bed gasifier can be divided into three zones from the bottom to the top: dense bed, splash zone, and freeboard. The Eulerian–Eulerian two-fluid model is widely used to study the gas–solid particle flow in the dense bed. In the dense bed, there are three phases: gaseous phase, solid inert bed particulate phase, and solid biomass particulate phase. The solid biomass fuel generally constitutes 0.5%–5% by mass of the total solids in the bed, and the remaining 95%–99.5% solids are a bed material such as sand (Singh et al., 2013). Instead of using Eulerian equations to study the three phases, Papadikis et al. (2008, 2010) used a combination of Eulerian–Eulerian model and Eulerian–Lagrangian model to study the dense bed. In this study, the Lagrangian equation was used to study the movement of biomass particles, which were introduced to an Eulerian–Eulerian bubbling bed of inert sand bed. The Eulerian–Lagrangian method by tracking single particle or a particle parcel in a discrete phase can be used to study the gas–solid particle flow in the freeboard zone (Singh et al., 2013). However, there are few studies on the application of Eulerian–Eulerian with the consideration of thermochemical reactions in dense bed. No study was reported to simultaneously study both dense bed and freeboard using a CFD model. Also, it is challenging to determine the heat transfer coefficients between biomass and inert bed materials in the dense bed of fluidized bed gasifiers (Singh et al., 2013).

The reaction mechanism of biomass gasification includes biomass, tar, char, and several gaseous species such as H_2, CO, CH_4, CO_2, and H_2O and a series of reactions among them. The transport equations for the mass fraction of the jth species could be written (Lathouwers and Bellan, 2001; Wang et al., 2005) in a simplified form as follows:

$$\frac{\partial(\alpha\rho X_j)}{\partial t}+\nabla\cdot(\vec{u}\alpha\rho X_j)=\nabla\cdot(\alpha\rho D\nabla X_j)+R_j \tag{3.24}$$

where ρ, α, $\vec{u}$, D, and R_j are the density, volume fraction of a gas phase, flow velocity, diffusion coefficient, and production rate of the jth species, respectively.

The reactive gas–particle flow will be modeled at both the particle and the bed levels. The particle model includes the description of the release of gases and gasification of char. The gas–particle fluid motion in a biomass fluidized bed gasifier is governed by mass, momentum, and energy conservation equations, which can generally be expressed as Equation 3.12 (Lathouwers and Bellan, 2001; Wang et al., 2005).

Both laminar and turbulent flow regimes have been observed in industrial gasifiers. A high-order turbulence closure should be constructed specifically for the turbulent gas flow with solid sediment in the gravitational field. The developed closure can be integrated into the mathematical model and the numerical code. Accurate simulation of microphysical phenomena like particle–particle interaction that cause a solid-phase pressure gradient and particle collisional dissipation is another very important part of the mathematical model. Both effects caused by particle–particle interaction seriously affect dynamics of dense gas-fluidized beds. The solid-phase pressure gradients can be described by either a constant viscosity model or a kinetic theory of granular flow. The output data from the CFD simulation usually include the syngas composition, yield, and temperature.

3.3.1.5 CFD Modeling of Microalgae Cultivation Systems

The success of mass production of microalgae for biodiesel depends greatly on the design and performance of photobioreactors. The light intensity and distribution, gas injection and mixing, and hydrodynamics/flow pattern are the key factors to be considered for design and scale-up of a photobioreactor for the microalgal growth. Air enriched with 5% CO_2 at rates of 0.025–1 volume of air/s to volume of medium was found to be cost effective for mass cultivation of microalgae in a photobioreactor (Zhang et al., 2002). CFD technique has been used to analyze and visualize the flow characteristics in designing and operating photobioreactors with different design parameters such as diameter, height to diameter ratio, gas sparger type, and physical properties of the liquid (Bitog et al., 2011).

In the design of photobioreactors, mixing plays a key role in ensuring light intensity distribution, sufficient CO_2 transfer, and nutrients distribution and maintaining uniform pH and temperature. Mixing is also important to prevent algae sedimentation and avoid the algal cell attachment to the reactor wall. Airlift photobioreactors are a common design for the growth of microalgae. CFD model can be used to track algal cells and generate their trajectories in the reactor. There are two basic methods to model the multiphase flow in a photobioreactor. The Eulerian–Eulerian approach treats both the dispersed phase of bubbles and continuous phase of water as interpenetrating continuum. The Eulerian–Lagrangian approach treats the continuous phase using Eulerian equations and tracks the dispersed phase by solving the motion equations that are based on the balance of all relevant forces on the bubbles or particles such as pressure, gravity, drag, lift, and added mass. Luo and Al-Dahhan (2011) developed an Eulerian–Eulerian model to simulate the flow dynamics in a draft tube airlift photobioreactor. The interphase momentum transfer between different phases due to interfacial forces in Equation 3.14 can be expressed as

$$S_k = \sum S_{k,i} = S_{k,D} + S_{k,L} + \cdots \tag{3.25}$$

where $S_{k,D}$ and $S_{k,L}$ represent drag force and lift force. Other minor forces may be also considered.

Drag force is caused by the movement of bubbles at different velocities to the surrounded liquid phase, which can be determined by the local slip velocity between the continuous and dispersed phases as (Luo and Al-Dahhan, 2011)

$$S_{k,D} = \frac{3}{4}\frac{C_D}{d_b}\varepsilon_g\rho\left(u_g - u_l\right)\left|u_g - u_l\right| \tag{3.26}$$

The drag coefficient can be determined by different correlations such as the following:

$$C_D = 0.44\,(\text{constant}) \tag{3.27}$$

$$C_D = \frac{2}{3}\sqrt{\frac{(\rho_l - \rho_g)gd_b^2}{\sigma}}\quad(\text{Ishii}-\text{Zuber}) \tag{3.28}$$

$$C_D = \frac{24}{Re_b}\left(1 + 0.15Re_b^{0.678}\right)\quad(\text{Schiller}-\text{Naumann}) \tag{3.29}$$

$$C_D = \frac{4}{3}\frac{gd_b}{U_T^2}\frac{\nabla\rho}{\rho}\quad(\text{Grace}) \tag{3.30}$$

where

d_b is the bubble diameter
ε_g is the volume fraction of gas phase
ρ is the density
u is the velocity
g is the gravity and the subscripts of l and g represent liquid and gas phases
σ is the interfacial tension between the gas and liquid
$Re_b = \left(\rho_l\left|u_g - u_l\right|d_b\right)/\mu_l$ is the bubble Reynolds number
U_T is the terminal velocity for the bubble

The lift force is the interfacial forces acting on the direction perpendicular to the major flow direction, which can be determined by

$$S_{k,L} = \varepsilon_g\rho_l C_L\left(u_g - u_l\right)\times\left(\nabla u_l\right) \tag{3.31}$$

where C_L is nondimensional life force coefficient. Other forces such as virtual mass force and turbulence dispersion force may also be considered in a CFD model. Luo and Al-Dahhan (2011) used the Lagrangian equation (3.18) to track the trajectory of algal particles.

Light penetration is another important parameter that has to be considered carefully to scale up a photobioreactor. Bubble column and airlift photobioreactors are widely used for the growth of microalgae. Local light availability in bubble columns and airlift reactors is affected by aeration rate, gas holdup, and the mixing and turbulence of the liquid medium (Miron et al., 2000). The light intensity can be calculated as a function of culture depth, which is given by (Wu and Merchuk, 2002)

$$I = I_0 e^{-(k_x x + k_w)z} \quad (3.32)$$

where

I_0 is the light intensity at the surface, μE m^{-2} s^{-1}
k_x is the extinction coefficient for biomass, mL m^{-1} $(10^6 \text{ cell})^{-1}$
x is the biomass concentration, (10^6 cell) mL^{-1}
k_w is the extinction coefficient for water, m^{-1}
z is the depth, m

3.3.2 Process Simulation Using Aspen Plus

The commercial software, Aspen Plus from AspenTech, Inc., is a widely used simulation platform to analyze the mass and energy balance in a chemical process. Aspen Plus can be used to develop equilibrium process models. The equilibrium models are important to predict the highest conversion or thermal efficiency that can be possibly obtained by a given process. Aspen Plus has abundant library models for different unit operations such as reactions, separation, and heat exchange. It is also possible for users to develop their own models using *Fortran* codes nested with the Aspen Plus input file. Another advantage of Aspen Plus is that it has a large database for the properties of different common chemicals such as water and ethanol. Many key components such as biomass, cellulose, xylan, and lignin in a biorefinery are specified as nonconventional components in Aspen Plus. National Renewable Energy Lab (NREL), United States, has defined the properties of those biomass-related components in simulation models. An example is the ultimate and proximate analysis of organic wastes, which is given in Table 3.3 (Ramzan et al., 2011).

Aspen Plus simulation can be used to optimize design and operating variables of unit operations of a process such as lignocellulosic ethanol biorefinery (Aden et al., 2002; Sassner et al., 2008; Huang et al., 2009; Moncada et al., 2013; Quintero et al., 2013), biomass gasification (Nikoo and Mahinpey, 2008; Ramzan et al., 2011), biomass pyrolysis-based refinery (Ng and Sadhukhan, 2011), and organosolv biorefinery (Garcia et al., 2011) to analyze and improve the overall efficiency and economics.

TABLE 3.3
Feedstock Condition and Their Ultimate and Proximate Analysis

	Food Waste	Municipal Solid Waste	Poultry Waste
Source	Supermarket in the United Kingdom	Municipalities in the United Kingdom	Local chicken shops
Proximate analysis (mass percent)			
FC	14.6	7.6	8.4
VM	51.1	18.8	40.3
MC	29.3	50.9	7.5
Ash	4.9	22.7	43.9
Ultimate analysis (mass percent)			
Ash	6.9	46.3	43.4
C	56.65	36.35	22.4
H	8.76	4.96	3.8
O	23.54	10.13	27.1
N	3.95	1.43	2.6
S	0.19	0.83	0.7

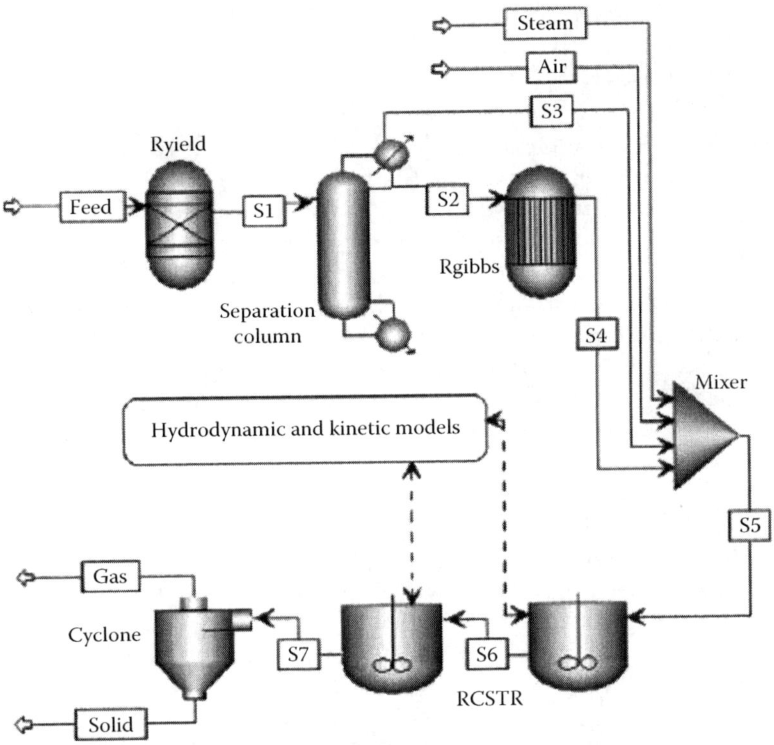

FIGURE 3.3 The diagram of the Aspen Plus simulation of fluidized bed biomass gasification. (Adapted from Nikoo, M.B. and Mahinpey, N., *Biomass Bioenergy*, 32, 1245, 2008.)

Aspen Plus has been used to simulate the biomass gasification process (Nikoo and Mahinpey, 2008; Ramzan et al., 2011; Zheng et al., 2013). The overall gasification process was assumed to consist of four unit operations including decomposition of the feed, volatile reactions, char gasification, and gas–solid separation. The library models including (1) the yield reactor, *RYield*; (2) Gibbs reactor, *RGibbs*; (3) CSTR reactor, RCSTR; and (4) *Cyclone* in the Aspen Plus were then selected to simulate the four major unit operations, respectively, as shown in Figure 3.3.

In the yield reactor, RYield, the biomass was converted into its constituting components of carbon, hydrogen, oxygen, sulfur, nitrogen, and ash. The yields of the decomposed components were determined according to the biomass ultimate analysis. A *separation column* model was then used to separate the char and volatiles exited from the yield reactor. The homogeneous reactions among the volatiles were simulated by the Gibbs equilibrium using the RGibbs model. The char gasification was simulated by the reaction rate kinetics using the *CSTR* model. The separation of syngas and char was finally simulated using the *Cyclone* model. The simulation results for the product gas composition and carbon conversion efficiency versus temperature, equivalence ratio (ER), steam-to-biomass ratio, and biomass average particle size were well agreed with the experimental results (Nikoo and Mahinpey, 2008).

The Aspen Plus models could not directly predict the yields of char and tar and the composition of tar during the biomass gasification, which should be determined by experimental data (Nikoo and Mahinpey, 2008; Ramzan et al., 2011; Zheng et al., 2013). Zheng et al. (2013) developed an Aspen Plus model to simulate the biomass integrated gasification combined cycle. In their model, the amount of char produced during biomass gasification was set at 19.7% of the total carbon entering the gasifier, which was the measured yield of char generated in a similar biomass gasification process. The general chemical formula of tar can be expressed as $C_xH_yO_z$, and the x, y, and z values are affected by the properties of biomass feedstock and the gasification conditions, which can be analyzed by experiments. In the model developed by Zheng et al. (2013), the tar was represented by C_6H_6O, while the N, S, and Cl in biomass were assumed to be converted to NH_3, H_2S, and HCl during the gasification, respectively.

The Aspen Plus software may not have proper library models to simulate some process units such as ethanol fermentation and adsorption. However, user subroutines can be developed using Microsoft Excel and MatLab to simulate those process units, which can be integrated into the simulation platform of Aspen Plus (Quintero et al., 2013). Aspen Plus-based process models have been widely integrated with economic models and life cycle models to assess the process economics and environmental impacts (Huang et al., 2009; Quintero et al., 2013; Zheng et al., 2013). The mass and energy balance through a bioenergy production system can be calculated by the Aspen Plus software. The process simulation results can be used as the inputs of the software of Icarus Process Evaluator developed by AspenTech Inc. to estimate the sizes and the costs of the unit operations (Huang et al., 2009). The process simulation results can also be used to supply inventory data for the life cycle assessment of the environmental impacts of a bioenergy system (Corti and Lombardi, 2004; Carpentieri et al., 2005).

3.3.3 Metabolic Flux Analysis of Biological Conversion Processes

Bioprocesses have been widely applied for the production of biofuels from starch and lignocellulosic biomass. Since biochemical reactions are catalyzed by the enzymes from growing microorganisms, it is necessary to know how the cell capability of producing the target products is realized and enhanced in a process. Cells uptake and convert substrates such as carbon and nitrogen sources from the environment for their growth and maintenance and excrete metabolite products such as CO_2 and ethanol to the environment. The intracellular flux distribution is useful when identifying targets for genetic engineering. However, intracellular fluxes are difficult to be measured. Metabolic flux analysis is the calculation and analysis of the flux distribution of the entire biochemical reaction network in the system. By using a metabolic flux analysis, the intracellular metabolic fluxes can be quantified by the measurement of extracellular metabolite concentrations in combination with the stoichiometry of intracellular reactions (Yang et al., 1998).

A metabolic network is an abstract representation of cellular metabolism. Each compound in a biological system may be involved in several biochemical reactions in the cell. For each relevant metabolite, a mass balance can be derived in which both the transport rates over the cell membrane and the intracellular reaction rates are included. The conversion rate of a metabolite in a biological system is a function of the stoichiometry and the metabolic flux through the respective reactions. The net conversion rate of each compound can be generally expressed as

$$r_i = \frac{dm_i}{dt} = \left[N_{j,i}\right]^T v_j \tag{3.33}$$

In the previous equation, $N_{j,i}$ is the stoichiometric matrix containing all coefficients of reactions with one row per reaction; v_j is the flux vector through reaction j; r is the net formation rates of metabolite i; i is the total number of compounds including substrates, products, and intracellular metabolites; and j is the number of fluxes. A set of linear equation (3.33) gives the relationship between the mass balances of each metabolite and a set of unknown of fluxes. To find the solution to Equation 3.33, the number of mass balances (equations) should equal to the number of fluxes to be calculated. However, the number of mass balances is usually smaller than the number of fluxes, and the set of linear equations is thus underdetermined. To solve an underdetermined network, either some fluxes have to be experimentally determined or extra constraints have to be added (van Winden et al., 2005; Cai et al., 2011). The cells can be grown on ^{13}C-labeled substrate, and the labeling patterns of the specific metabolites can be measured using advanced analytical instruments such as gas chromatography–mass spectrometry (GC–MS) and nuclear magnetic resonance (NMR) spectroscopy (Yang et al., 2002).

The metabolic pathway of *Saccharomyces cerevisiae* for ethanol production is well characterized as shown in Figure 3.4 (Franzen, 2003). By quantifying the intracellular fluxes, it might be possible to (1) understand the biochemistry of the cell at a quantitative level, (2) analyze the nutrient requirements of cells and thus redesign the culture medium, and (3) trace metabolic pathways that limit growth and production for precise genetic modification. The metabolic flux analysis can be used to select better genetic traits for heterologous expression, to express multiple genes at optimal levels, to combine various beneficial traits into single strains, and to ferment mixed sugars for targeted fuels and chemicals.

3.3.4 Modeling of Biomass Conversion Kinetics

The biological or thermochemical conversion of biomass involves reactions. A reaction kinetic model describes how the reaction rate depends on the key local environmental variables such as temperature and pressure. Both empirical and mechanistic equations were developed to describe the kinetics of reactions.

3.3.4.1 Biological Conversion Kinetics

The microbial growth profiles include linear, exponential, logistic, and fast acceleration/slow deceleration as shown in Figure 3.5. Empirical equations can be used to simply fit the experimental data of microbial growth by nonlinear regression. The following equations have been developed to describe these curves for solid-state fermentation as shown in Figure 3.5 (Mitchell et al., 2004):

Linear:

$$\frac{dX}{dt} = K \quad \text{or} \quad X = Kt + X_0 \tag{3.34}$$

Exponential:

$$\frac{dX}{dt} = \mu X \quad \text{or} \quad X = X_0 e^{\mu t} \tag{3.35}$$

Logistic:

$$\frac{dX}{dt} = \mu X\left(1 - \frac{X}{X_m}\right) \quad \text{or} \quad X = \frac{X_m}{1 + \left[(X_m/X_0) - 1\right]} e^{\mu t} \tag{3.36}$$

Two phase:

$$\begin{gathered} \frac{dX}{dt} = \mu X, (t < t_a) \quad \text{or} \quad X = X_0 e^{\mu t} \\ \frac{dX}{dt} = \left[\mu L e^{-k(t-t_a)}\right] X, (t \geq t_a) \quad \text{or} \quad X = X_A \exp\left[\frac{\mu L}{k}\left(1 - e^{-k(t-t_a)}\right)\right] \end{gathered} \tag{3.37}$$

where

- K is the reaction constant
- μ is the specific growth rate
- X is the biomass concentration
- X_m is the maximum possible biomass concentration
- t_a is the time that the exponential is switched to deceleration phase
- L is the ratio of the specific growth rate at the start of the deceleration phase to the specific growth rate during the exponential phase

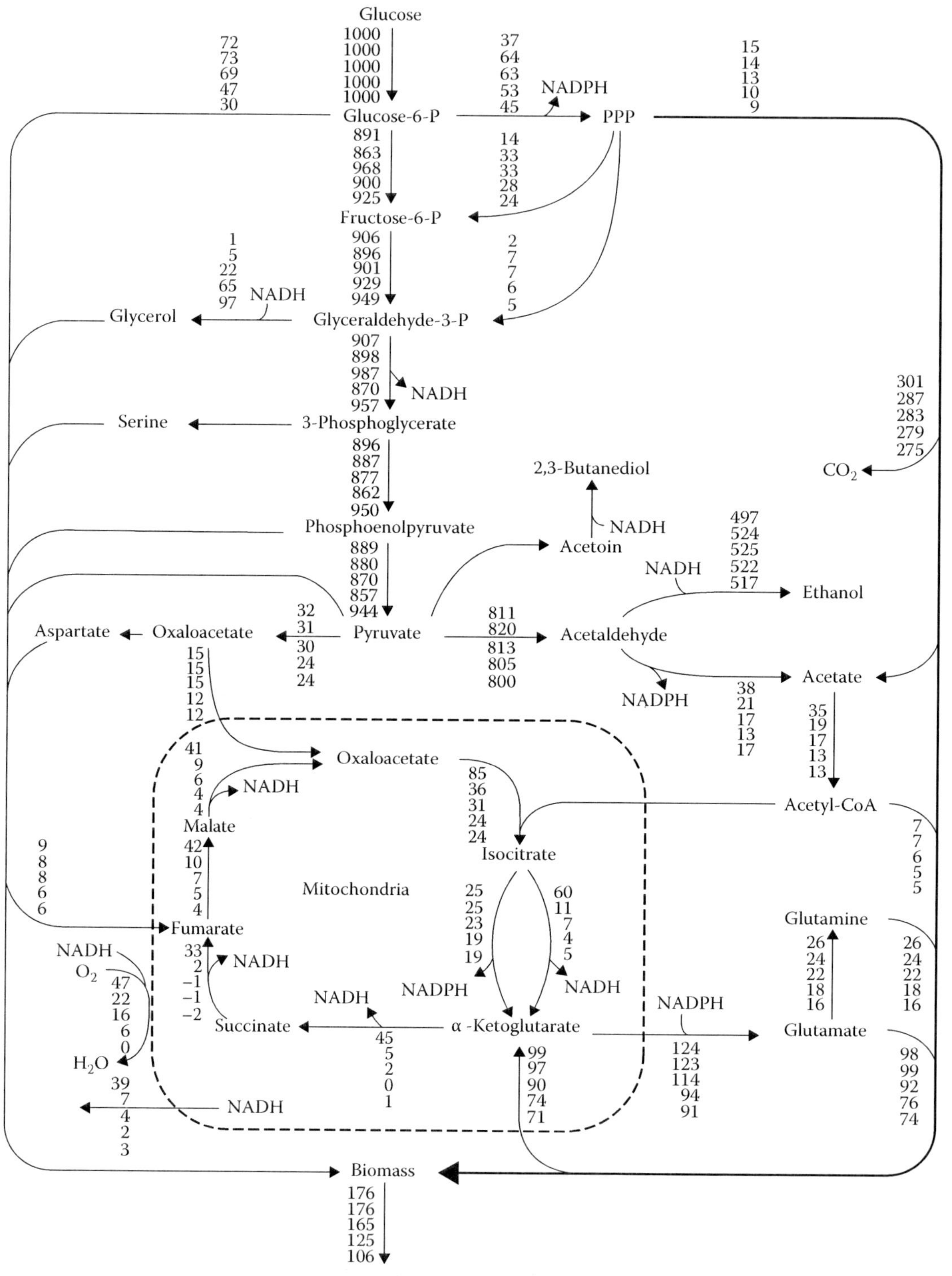

FIGURE 3.4 Estimated carbon flux distribution in the unit of mmol carbon/C-mol of glucose uptake at D = 0.25 h^{-1} and various RQ (from top to bottom RQ = 6, 12, 17, 50 and anaerobic). (Adapted from Franzén, C.J., *Yeast*, 20, 117, 2003.)

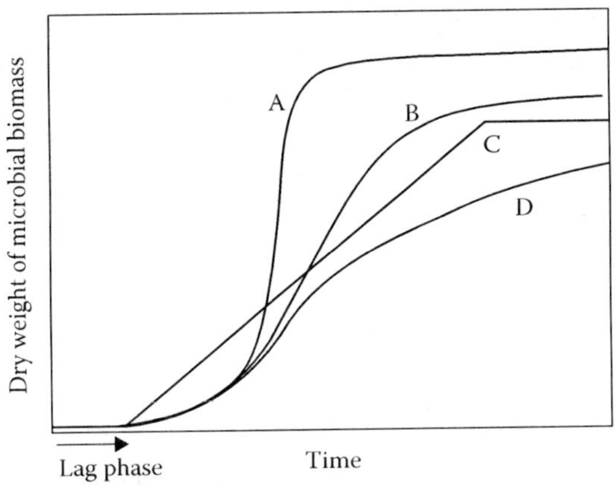

FIGURE 3.5 Kinetic profiles of solid-state fermentation (A) exponential, (B) logistic, (C) linear, and (D) fast acceleration/slow deceleration. (Adapted from Mitchell, D.A. et al., *Biochem. Eng. J.*, 17, 15, 2004.)

The modified Gompertz equation is another kinetic model to fit the experimental data of cell growth and product formation curves in order to estimate the maximum specific growth/production rates, lag phase, and maximum cell/product concentration. The modified Gompertz equation is given by (Wang et al., 2006; Zajsek and Gorsek, 2010)

$$\log\left(\frac{X}{X_0}\right) = A \cdot \exp\left\{-\exp\left[\frac{\mu_{max} \cdot e}{A}\left(t_{lag} - t\right) + 1\right]\right\} \tag{3.38}$$

where

μ_{max} is the maximum specific growth/production rate
A is the potential maximum cell or production concentration
t_{lag} is the delayed time
X and X_0 are the cell concentration at any time t and at the beginning

During biological conversion, the substrate and product may have significant inhibition on the maximum specific growth rate and maximum ethanol production rate (Jin et al., 2012). The modified Monod kinetic model is the most widely used empirical model to describe the microbial growth under the inhibition of substrate and product, which is given by (Song et al., 2010)

$$\frac{dX}{dt} = \mu X = \mu_{max}\left(\frac{S}{S + K_S + S^2/K_I}\right)\left(1 - \frac{P}{P_d}\right)^n X \tag{3.39}$$

In the previous equation, K_S and K_I are the substrate saturation and inhibition constants (g/L), respectively, P is the product concentration (g/L), P_d is the product concentration at which no cell growth occurs (g/L), and n is the degree of product inhibition.

The effect of temperature on the microbial growth kinetics can be determined by the modified Ratkowsky model, which is expressed as (Zwietering et al., 1991; Wang et al., 2006)

$$\mu_{max} = \left[\alpha\left(T - T_{min}\right)\right]^2 \cdot \left\{1 - \exp\left[\beta\left(T - T_{max}\right)\right]\right\} \tag{3.40}$$

where

α and β are constants
T_{min} and T_{max} are the minimum and maximum temperatures

The production rate may or may not be associated with the growth rate of the microorganism used in a biological process. The relationship between the production rate and the cellular growth rate can be generally expressed as (Luedeking and Piret, 2000; Song et al., 2010)

$$\frac{dP}{dt} = \alpha\frac{dX}{dt} + \beta X \quad (3.41)$$

where

α, g/g and β, and g/(g h) are the growth-associated and non-growth-associated constants, respectively

P and X are the product and cell concentrations, respectively

The coefficients in the kinetic models discussed previously can be obtained by fitting experimental data of cell concentration-time profiles and product concentration-time profiles obtained at different process conditions such as temperature and nutrients using nonlinear regression analysis, which can be done using the Microsoft Excel solver. The predictions using both models showed good agreement with the experimental data (Song et al., 2010; Zajsek and Gorsek, 2010).

3.3.4.2 Thermochemical Conversion Kinetics

A review on mathematical modeling the biomass pyrolysis was given in the literature (Di Blasi, 2008). A single global reaction is the simplest way to represent the whole pyrolysis reaction, which is usually treated as first-order reaction by (Di Blasi, 2008)

$$\text{Biomass} \xrightarrow{k} \alpha \cdot \text{volatiles} + (1-\alpha) \cdot \text{char} \quad (3.42)$$

where α is the coefficient of distribution between volatiles and char to be determined by experiments and k is the reaction kinetic constant:

$$\frac{dY}{dt} = -kY \quad (3.43)$$

The effect of temperature on the kinetic constant is usually assumed to follow the Arrhenius law, which is given by (Ahmed et al., 2012)

$$k = A\exp\left(\frac{-E}{RT}\right) \quad (3.44)$$

Experiments can be conducted with fine biomass particles to determine the constants in a kinetic model because of uniform temperature and negligible mass and heat transfer resistance through the particle. Measurements can be taken in thermogravimetric analyzer (TGA) with low heating rates or in a flash pyrolysis device where the heat rates are very high (e.g., 500°C/s) (Lu et al., 2010).

The improvement on the previous pyrolysis kinetic model is to use a combination of series and parallel reactions as shown in Figure 3.6.

The tar cracking is considered to follow the overall reaction as (Wang and Yan, 2008)

$$Tar_1 \xrightarrow{k} a_1CO + a_2CO_2 + a_3H_2 + a_4CH_4 + a_5Tar_2 \quad (3.45)$$

The stoichiometric coefficients, a_{1-5}, and kinetic constant, k, can be found in the literature (Wurzenberger et al., 2002; Corella and Sanz, 2005).

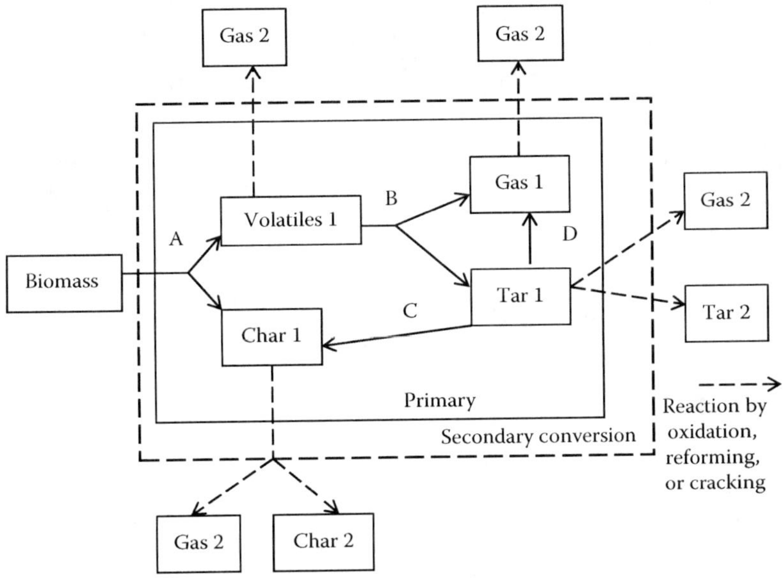

FIGURE 3.6 Simplified pyrolysis model. (Adapted from Di Blasi, C., *Prog. Energy Combust. Sci.*, 34, 47, 2008.)

The simple pyrolysis models as discussed previously can only predict the amounts of char, tar, and gas released during pyrolysis. However, they cannot predict the composition of the gas. Structural models such as bio-FLASHCHAIN model can be used to predict the yields of individual gas (Niksa, 2000). In the bio-FLASHCHAIN model, pyrolysis consists of three independent mechanisms: competitive chemical reactions, fragmentation statistics, and flash distillation. Fragmentation statistics determines the probability that a specific chemical reaction will release a product with a certain molecular mass during the transformation. Flash distillation determines whether a particular fragment will enter a vapor phase or remain in char in a condensed phase. All three mechanisms determine the release rates of a gaseous species (Niksa, 2000). Ranzi et al. (2008) proposed a mechanistic kinetic model that treated the overall devolatilization of biomass as the combination of the decomposition of the three constituent components of cellulose, hemicelluloses, and lignin. All the released species in the gas phase are then described with their primary propagation and decomposition reactions in the gas phase to predict the final products of H_2, CO, CO_2, CH_4, C_2, and H_2O. The lumped kinetic models for the pyrolysis of cellulose, hemicelluloses, and lignin were given in Table 3.4 (Ranzi et al., 2008). The main assumption for the mechanistic kinetic model is that there are no interactions in the thermal decomposition of the three individual constituent components in a biomass feedstock. This assumption may not be right for the pyrolysis of woody biomass. The mechanistic kinetic model can be improved by accurately measuring the volatile species using FTIR and TGA-MS analyses.

During gasification, the char generated continues to react with gases of O_2, H_2O, CO_2, and H_2. The reactivity of char can be determined as a function of the concentration of the reactant gas, C_g, which is given by

$$r_m = -\frac{1}{m_c}\frac{dm_c}{dt} = k_m C_g^n \ \left(\text{kg/kg s}\right) \tag{3.46}$$

The kinetic constant, k_m, and the order of the reaction, n, can be determined by experiments. The effect of temperature on the kinetic constant k_m can be determined by the Arrhenius law as shown

TABLE 3.4
Lumped Kinetic Models for the Pyrolysis of Cellulose, Hemicelluloses, and Lignin

Reaction	Rate Constant (1/s)
Cellulose	
Cell → cella	$k = 8 \times 10^{13} \exp(-46{,}000/RT)$
Cella $\rightarrow 0.95$ HAA $+ 0.25$ glyoxal $+ 0.20\ CH_3CHO + 0.20\ C_3H_6O$ $+0.25$ HMFU $+ 0.20\ CO_2 + 0.15\ CO + 0.1\ CH_4 + 0.90\ H_2O + 0.65$ char	$k = 1 \times 10^{9} \exp(-30{,}000/RT)$
Cella → LVG	$k = 4T \exp(-10{,}000/RT)$
Cell $\rightarrow 5\ H_2O + 6$ char	$k = 8 \times 10^{7} \exp(-32{,}000/RT)$
Hemicellulose	
HCE → 0.4 HCE1+ 0.6 HCE2	$k = 1 \times 10^{10} \exp(-31{,}000/RT)$
HCE1 $\rightarrow 2.5\ H_2 + 0.125\ H_2O + CO + CO_2 + 0.5CH_2O$ $+0.25\ CH_3OH + 0.125\ C_2H_5OH + 2$ char	$k = 3 \times 10^{9} \exp(-27{,}000/RT)$
HCE1 →xylose	$k = 3T \exp(-11{,}000/RT)$
HCE2 $\rightarrow 1.5\ H_2 + 0.125\ H_2O + 0.2CO_2 + 0.7CH_2O + 0.25\ CH_3OH$ $+0.125\ C_2H_5OH + 0.8G\{CO_2\} + 0.8G\{COH_2\} + 2$ char	$k = 1 \times 10^{10} \exp(-33{,}000/RT)$
Lignin	
$LIG-C \rightarrow 0.35\ LIG_{CC} + 0.1$ p-coumaryl $+ 0.08$ phenol $+ 1.49\ H_2 + H_2O$ $+1.32\ G\{COH_2\} + 7.05$ char	$k = 4 \times 10^{15} \exp(-48{,}500/RT)$
$LIG-H \rightarrow LIG_{OH} + C_3H_6O$	$k = 2 \times 10^{13} \exp(-37{,}500/RT)$
$LIG-O \rightarrow LIG_{OH} + CO_2$	$k = 1 \times 10^{9} \exp(-25{,}500/RT)$
$LIG_{CC} \rightarrow 0.3$ p-coumaryl $+ 0.2$ phenol $+ 0.3C_3H_4O_2 + 1.2\ H_2 + 0.7H_2O + 0.25CH_4$ $+0.25C_2H_4 + 0.5G\{CO\} + 1.3\ G\{COH_2\} + 7.5$ char	$k = 5 \times 10^{6} \exp(-31{,}500/RT)$
$LIG_{OH} \rightarrow LIG + 0.5\ H_2 + H_2O + CH_3O + G\{CO\} + 1.5\ G\{COH_2\} + 5$ char	$k = 1 \times 10^{13} \exp(-49{,}500/RT)$
$LIG \rightarrow C_{11}H_{12}O_4$	$k = 8T \exp(-12{,}000/RT)$
$LIG \rightarrow 0.7\ H_2 + H_2O + 0.2CH_2O + 0.5CO + 0.2CH_2O + 0.4CH_3OH + 0.2CH_3CHO$ $+0.2C_3H_6O_2 + 0.4CH_4 + 0.5C_2H_4 + G\{CO\} + 0.5G\{COH_2\} + 6$ char	$k = 1.2 \times 10^{9} \exp(-30{,}000/RT)$

Source: Ranzi, E., *Energy Fuels*, 22, 4292, 2008.
Notes: Cell, cellulose; cella, active cellulose; HAA ($C_2H_4O_2$), hydroxyl acetaldehyde, glyoxal $C_2H_2O_2$; LVG, levoglucosan; HMFU, 5-hydroxymethylfurfural; LIG-C, $C_{15}H_{14}O_4$; LIG-O, $C_{16}H_{10}O_6(OCH_3)_4$, and LIG-H, $C_{17}H_{13}O_4(OCH_3)_5$.

in Equation 3.36 (Di Blasi, 2009). The heterogeneous char–gas-phase reactions and their kinetics are given in Table 3.5.

The homogeneous reactions among the gases generated during biomass gasification also take place. The homogeneous gas-phase reactions and kinetics during biomass gasification are given in Table 3.6.

3.4 COMPUTER-AIDED BIOMASS SUPPLY LOGISTICS

The logistics of biomass feedstock supply including harvest, storage, and delivery are an integral part of the overall bioenergy supply system. The biomass supply chain consists of multiple harvesting, storage, preprocessing, and transport options. Mathematical models have been developed to analyze and optimize complex biomass supply systems.

TABLE 3.5
Heterogonous Char–Gas Reactions and Kinetics during Biomass Gasification

Reaction	Rate	Reference
$2\frac{\eta+1}{\eta+2}C+O_2 \rightarrow \frac{2\eta}{\eta+2}CO+\frac{2}{\eta+2}CO_2$	$r=1.5\times10^6\exp(-13{,}078/T)\cdot p_{O_2}\cdot(1-X_C)^{1.2}$ (1/s) $\eta = 3\times10^8\exp(-30{,}178/T)$	Di Blasi et al. (2009); Monson et al. (1995)
$C + CO_2 \rightarrow 2\,CO - 172$ kJ/mol	$r = 4{,}364\exp(-29{,}844/T)\cdot c_{CO_2}$ $(\mathrm{mol/m^2/s})$	Biggs and Agarwal (1997)
$C + H_2O \rightarrow CO + H_2 - 131$ kJ/mol	$r=\frac{4.93\times10^3\exp(-18{,}522/T)\cdot p_{H_2O}}{1+1.11\times10^1\exp(-3{,}548/T)\cdot p_{H_2O}+1.53\times10^{-9}\exp(25{,}161/T)\cdot p_{H_2}}$ (1/s)	Muhlen et al. (1985)

TABLE 3.6
Homogeneous Gas-Phase Reactions and Kinetics during Biomass Gasification

Homogenous Reactions	Rate (mol/m³/s)	References
$CO + H_2O \rightarrow CO_2 + H_2 + 41.1$ kJ/mol	$r = 2.78 \times 10^{-3} \exp\left(\frac{-1510}{T}\right)\left(c_{CO}c_{H_2O} - \frac{c_{CO_2}c_{H_2}}{K}\right)$	De Souza-Santos (1989)
$2CO + O_2 \rightarrow 2CO_2 + 566$ kJ/mol	$r = 3.98 \times 10^{14} \exp\left(\frac{-20,119}{T}\right)\left(c_{CO}c_{O_2}^{0.25}c_{H_2O}^{0.5}\right)$	Groppi et al. (2000)
$2H_2 + O_2 \rightarrow 2H_2O + 484$ kJ/mol	$r = 2.19 \times 10^{12} \exp\left(\frac{-13,127}{T}\right)\left(c_{H_2}c_{O_2}\right)$	Groppi et al. (2000)
$CH_4 + 2O_2 \rightarrow CO_2 + 2H_2O + 35.7$ kJ/mol	$r = 1.58 \times 10^{13} \exp\left(\frac{-24,343}{T}\right)\left(c_{CH_4}^{0.7}c_{O_2}^{0.8}\right)$	Groppi et al. (2000)

Note: T is the temperature in K and C is the concentration in mol/m³.

A switchgrass-based bioethanol supply chain is shown in Figure 3.7. Ten items along the switchgrass-based ethanol supply chain that include land, cultivation, harvest, bale storage, bale transportation, loose chop preprocessing, preprocessed biomass transportation, biomass refinery, and ethanol transportation contribute to the total cost of the switchgrass to ethanol supply chain. A linear model was developed to determine the total cost of the switchgrass-based bioethanol supply chain, which was given by (Zhang et al., 2013a)

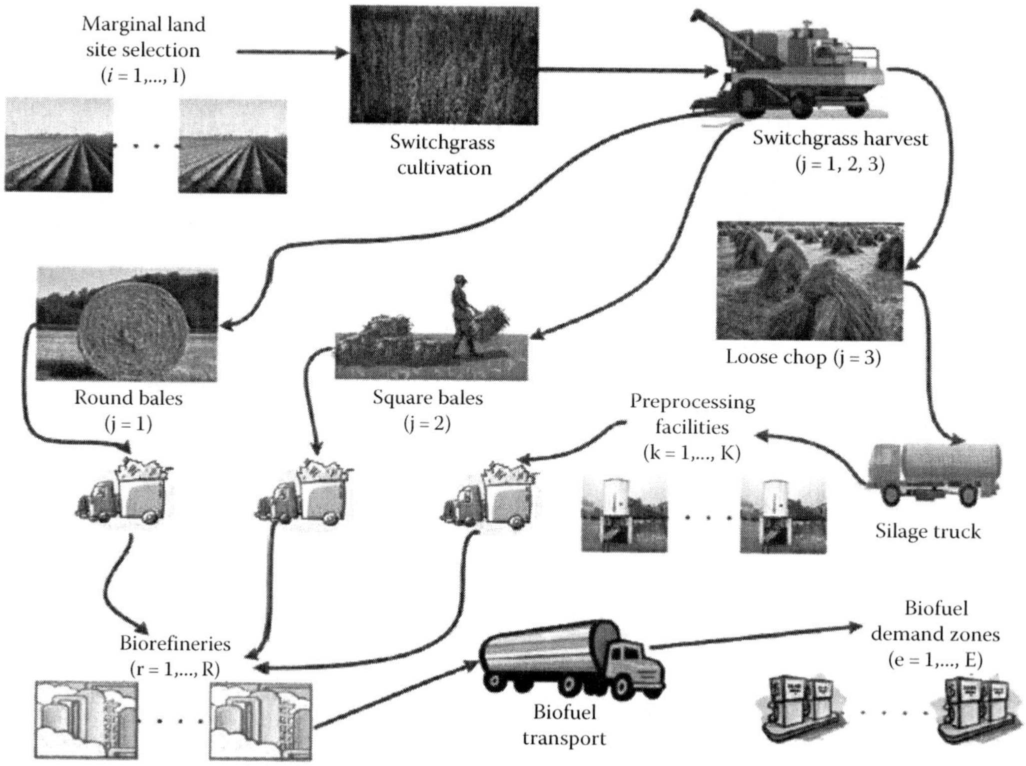

FIGURE 3.7 A switchgrass-based bioethanol supply chain. (Adapted from Zhang, J. et al., *Appl. Energy*, 102, 1205, 2013a.)

$$\theta = \sum_{i=1}^{I}\sum_{j=1}^{3} C_i X_{ij} + \sum_{i=1}^{I}\sum_{j=1}^{3} P_i X_{ij} + \sum_{i=1}^{I}\sum_{j=1}^{3} H_i X_{ij}$$

$$+\sum_{i=1}^{I}\sum_{j=1}^{2}\sum_{t=1}^{T} \xi_j S_{ijt} + \sum_{i=1}^{I}\sum_{j=1}^{2}\sum_{r=1}^{R}\sum_{t=1}^{T} F_{jir} D_{ir} V_{ijrt}$$

$$+\sum_{i=1}^{I}\sum_{k=1}^{K} \alpha_{ik} D_{ik} V_{ik} + \sum_{k=1}^{K}\sum_{r=1}^{R}\sum_{t=1}^{T} \gamma_{kr} D_{kr} V_{krt} + \sum_{i=1}^{I}\sum_{k=1}^{K} U_i V_{ik} + \sum_{k=1}^{K} W_k Y_k$$

$$+\sum_{r=1}^{R}\sum_{t=1}^{T} N_r Z_{rt} + \sum_{r=1}^{R}\sum_{q=1}^{Q} G_{rq} Y_{rq}$$

$$+\sum_{e=1}^{E}\sum_{r=1}^{R}\sum_{t=1}^{T} \psi_{re} D_{re} Z_{ret} \tag{3.47}$$

where

ψ_{re} is the bioethanol transportation cost from biorefinery r to demand zone e
ξ_j is the storage cost per unit mass for harvest method j
γ_{kr} is the transportation cost per unit volume of densified biomass and unit distance from preprocessing facility k to biorefinery r
α_{ik} is the transportation cost per unit volume of loose biomass and unit distance from supply i to preprocessing facility k
C_i is the rental cost per unit area in supply zone i
D_{ik} is the distance between supply i and preprocessing facility k
D_{ir} is the distance between supply i and biorefinery r
D_{kr} is the distance between k and r
D_{re} is the distance between biorefinery r and demand zone e
F_{jir} is the traditional bale transportation cost from i to r for bale method j
G_{rq} is the annualized capital cost of biorefinery with capacity level q at location r
H_i is the harvest cost per unit area in supply zone i
N_r is the bioethanol production cost per unit volume of ethanol in refinery r
P_i is the cultivation cost per unit area in supply zone i
S_{ijt} is the amount of biomass stored at supply zone i by harvest method j during time period t
U_k is the preprocessing cost per unit volume in facility k
V_{ik} is the amount of loose chop biomass transported from supply i to preprocessing facility k
V_{jirt} is the amount of traditional bales from i to r during time period t for bale method j
V_{kr} is the amount of densified biomass transported from k to r
W_k is the annualized capital cost of preprocessing facility in location k
X_{ij} is the marginal land area in supply zone i harvested by method j
Y_k is the number of preprocessing facilities in location k
Y_{rq} is the number of biorefineries with capacity level q in r location
Z_{ret} is the volume of ethanol from r to e during time period t
Z_{rt} is the volume of bioethanol produced in refinery r during time period t

The analysis and optimization of a biomass supply chain starts with the flow chart of the supply chain like the one as shown in Figure 3.7. The inventory data for the model parameters should be carefully collected and evaluated to ensure the reliability of the analysis and optimization using the model. The simulation using the previous model showed that if the biorefineries are not optimally located, the total cost of the switchgrass-based bioethanol supply chain in North Dakota in the United States can increase by 17% and the resulting transportation cost can increase by up to 400% (Zhang et al., 2013a).

A bioenergy supply chain can be optimized in terms of total cost, energy use, and environmental impact. The integrated biomass supply analysis and logistics (IBSAL) model, which was developed at the Oak Ridge National Laboratory, United States, can be used to estimate the costs, energy use, and greenhouse gas emissions for different biomass supply chains (Kumar and Sokhansanj, 2007). The model has different submodules for harvesting, processing, storage, and transportation. The input data of the model include local weather data, average yield of biomass, proportion of land for the biomass supply, crop harvesting information, capacity of the biorefinery, dry matter loss in storage, biomass moisture content at the time of harvest, and capital and operating costs of agricultural machinery. The output data of the model include the cost to deliver biomass to a biorefinery, GHG emission, and energy consumption (Kumar and Sokhansanj, 2007). Kumar and Sokhansanj (2007) found that for the supply of switchgrass at 2000 dry tons/day, the delivered cost for the different supply systems was \$37–\$48/dry ton. Energy consumption varied from 4.8% to 6.3% of the energy content of switchgrass. The emissions varied from 75 to 100 kg of CO_2/dry ton of switchgrass delivered. Dry matter loss in the delivery of switchgrass to a biorefinery varied from 3% to 4%.

3.5 LIFE CYCLE ASSESSMENT

Life cycle assessment is an emerging tool to measure and compare the environmental impacts of human activities. The details of LCA are discussed in Chapter 4. The LCA has four steps: goal definition and scoping, inventory, impact assessment, and interpretation (International Standard Organization, 1997). The impact indicators used in the LCA may include global warming potential (GWP), acidification, eutrophication, photochemical oxidation, and energy use (Lundie and Peters, 2005). GWP is a widely used impact indicator. Carbon dioxide, methane, and N_2O are the three gases that contribute to global warming. Methane has a GWP 25 times of CO_2. N_2O is another gas that causes global warming. The GWP of N_2O is 310 times of CO_2. The global warming score of a system can be determined by

$$GW = \sum_{s} GWP_s \times m_s \tag{3.48}$$

where

GW is the global warming score
s are the substances (the different greenhouse gases such as CO_2 and CH_4)
GWP_s is the GWP of the sth substance
m_s is the emitted amount of substance s in kg

Assumptions may be made to specify the scope of the LCA. It is very important to carefully identify and describe the inputs and outputs for each step and each process involved in the system to be assessed at the inventory step. The data quality strongly affects the reliability of the LCA (Aresta et al., 2005). LCA has been used to evaluate the environmental impacts of biomass production, bioenergy conversion, or the whole bioenergy system (Clarens et al., 2010; Collet et al., 2011; Buonocore et al., 2012).

The use of LCA for the biomass production requires the quantification of the regional climate, water availability, nutrient supply, and harvesting technology. A life cycle model was used to compare the environmental impacts of microalgae, switchgrass, canola, and corn farming (Clarens et al., 2010). The analysis showed that conventional crops of switchgrass, canola, and corn had lower environmental impacts than algae in energy use, greenhouse gas emissions, and water consumption although algae had better efficiency of land use and eutrophication potential.

The use of flue gas as the source of CO_2 and wastewater as a source of fertilizer could significantly reduce the environmental impacts of algal production to make the algae to be competitive as an energy source (Clarens et al., 2010).

The results of LCA may vary significantly even for similar bioenergy systems due to the differences in the type and management of raw materials, conversion technologies, end-use technologies, system boundaries, and reference energy system for comparison. Biofuels can contribute to the mitigation of greenhouse gas emissions in the transport sector only if significant emissions from land use change such as soil carbon pools and nitrous oxide emission from soils are avoided and appropriate production technologies are used (Cherubini et al., 2009). It is impossible to give exact values for the amount of greenhouse gas emissions and fossil energy consumption saved by a certain product because many key issues associated with a bioenergy system are site-specific and many factors affect the outcome (Cherubini et al., 2009). However, the LCA can be used to compare the environmental impacts of different biomass production and conversion technology.

3.6 CONCLUSIONS

Lignocellulosic biomass is an abundant renewable energy source. Various mathematical models have been developed to predict the growth rate of energy crops affected by the environment, the transport phenomena during biomass conversion, the energy and material balances in biomass processing facilities and cells, the conversion reaction kinetics, the biomass supply logistics, and the environmental impacts of bioenergy supply. Energy crops such as switchgrass and sweet sorghum have been widely investigated. However, there is limited commercial experience to produce energy crops at a large scale around the world. A crop growth model can be used to estimate yields of energy crops in a region under different growth conditions. A GIS model can be used to maximize the energy production of energy crops by indentifying suitable land to grow them based on their specific characteristics and the current use of the land. Biomass can be converted into energy products in various biological and thermochemical processes. Combined process models and reaction kinetics provide various advanced computational tools for the design and optimization of the biomass conversion processes. Process simulation can be conducted at a steady state where the variation of process properties over time is ignored and dynamic state where the process properties are time dependent. Equilibrium models such as an Aspen Plus-based model assume that chemical equilibrium is reached for process reactions, which can be used to predict the highest conversion or thermal efficiency that can be possibly obtained for a given process at a steady state. A CFD model can be used to predict transient fluid flow, heat transfer, and mass transfer by solving a set of conservation equations. Reaction kinetic models can be used to estimate product formation rates and composition under varying operating conditions, which is an essential knowledge for designing and operating a process. During the biological conversion of biomass, cells uptake and convert substrates such as carbon and nitrogen sources from the environment for their growth and maintenance and excrete metabolite products such as CO_2 and ethanol to the environment. Metabolic flux analysis can be used to select better genetic traits for heterologous expression, to express multiple genes at optimal levels, to combine various beneficial traits into single strains, and to ferment mixed sugars for targeted fuels and chemicals. An LCA model can be used to compare the environmental impacts of different biomass production and conversion technologies.

ACKNOWLEDGMENTS

This chapter was partially supported by funds provided by the U.S. Department of Energy (grant number: DE-EE0003138) and by the NSF (grant number: HRD-1242152). Its contents are solely the responsibility of the authors and do not necessarily represent the official views of the two funding agencies. Mention of a trade name, a company name, or proprietary products is for presentation clarity only and does not imply endorsement by the authors or the university.

REFERENCES

Aden A., Ruth M., Ibsen K., Jechura J., Neeves K., Sheehan J. et al. 2002. Lignocellulosic biomass to ethanol process design and economics utilizing co-current dilute acid prehydrolysis and enzymatic hydrolysis for corn stover. National Renewable Energy Laboratory Report No. NREL/TP-510-32438. Online: www.nrel.gov/docs/fy02osti/32438.PDF

Ahmed T.Y., Ahmad M.M., Yusup S., Inayat A., and Khan Z. 2012. Mathematical and computational approaches for design of biomass gasification for hydrogen production: A review. *Renewable and Sustainable Energy Reviews* 16: 2304–2315.

ANSYS Inc. 2011. ANSYS Fluent Software Package User's Manual (Version 14). Canonsburg, USA: ANSYS Inc.

Aresta M., Dibenedetto A., and Barberio G. 2005. Utilization of macro-algae for enhanced CO2 fixation and biofuels production: Development of a computing software for and LCA study. *Fuel Processing Technology* 86: 1679–1693.

Barenblatt G. 1996. *Scaling, Self-Similarity, and Intermediate Asymptotics*. Cambridge, U.K.: Cambridge University Press.

Basu P. 2006. *Combustion and Gasification in Fluidized Bed*. Boca Raton, FL: CRC Press.

Biggs M.J. and Agarwal P.K. 1997. The CO/CO_2 product ratio for porous char particle within an incipiently fluidized bed: A numerical study. *Chemical Engineering Science* 52: 941–952.

Bitog J.P., Lee I.B., Lee G.G., Kim K.S., Hwang H.S., Hong S.W., Seo I.H., Kwon K.S., and Mostafa E. 2011. Application of computational fluid dynamics for modeling and designing photobioreactors for microalgae production: A review. *Computers and Electronics in Agriculture* 76: 131–147.

Buonocore E., Franzese P.P., and Ulgiati S. 2012. Assessing the environmental performance and sustainability of bioenergy production in Sweden: A life cycle assessment perspective. *Energy* 37: 69–78.

Cai G., Jin B., Monis P., and Saint C. 2011. Metabolic flux network and analysis of fermentative hydrogen production. *Biotechnology Advances* 29: 375–387.

Carpentieri M., Corti A., and Lombardi L. 2005. Life cycle assessment (LCA) of an integrated biomass gasification combined cycle (IBGCC) with CO_2 removal. *Energy Conversion and Management* 46: 1790–1808.

Cherubini F, Bird N.D., Cowie A., Jungmeier G., Schlamadinger B., and Woess-Gallasch S. 2009. Energy- and greenhouse gas-based LCA of biofuel and bioenergy systems: Key issues, ranges and recommendations. *Resources, Conservation and Recycling* 53: 434–447.

Clarens A., Resurreccion E.P., White M.A., and Colosi L.M. 2010. Environmental life cycle comparison of algae to other bioenergy feedstocks. *Environmental Science & Technology* 44: 1813–1819.

Collet P., Helias A., Lardon L., Ras M., Goy R.A., and Steyer J.P. 2011. Life-cycle assessment of microalgae culture coupled to biogas production. *Bioresource Technology* 102: 207–214.

Concas A., Pisu M., and Cao G. 2010. Novel simulation model of the solar collector of BIOCOIL photobioreactors for CO_2 sequestration with microalgae. *Chemical Engineering Journal* 157: 297–303.

Corella J. and Sanz A. 2005. Modeling circulating fluidized bed biomass gasifiers. A pseudo-rigorous model for stationary state. *Fuel Processing Technology* 86: 1021–1053.

Corti A and Lombardi L. 2004. Biomass integrated gasification combined cycle with reduced CO_2 emissions: Performance analysis and life cycle assessment (LCA). *Energy* 29: 2109–2124.

De Souza-Santos M.L. 1989. Comprehensive modeling and simulation of fluidized bed boilers and gasifier. *Fuel* 68: 1507–1521.

Di Blasi C. 2008. Modeling chemical and physical processes of wood and biomass pyrolysis. *Progress in Energy Combustion Science* 34: 47–90.

Di Blasi C. 2009. Combustion and gasification rates of lignocellulosic chars. *Progress in Energy Combust Science* 35: 121–140.

Fiorese G. and Guariso G. 2010. A GIS-based approach to evaluate biomass potential from energy crops at regional scale. *Environmental Modeling & Software* 25: 702–711.

Franzén C.J. 2003. Metabolic flux analysis of RQ-controlled microaerobic ethanol production by *Saccharomyces cerevisiae*. *Yeast* 20: 117–132.

Gan J. and Smith C.T. 2011. Optimal plant size and feedstock supply radius: A modeling approach to minimize bioenergy production costs. *Biomass and Bioenergy* 35: 3350–3359.

Garcia A., Alriols M.G., Llano-Ponte R., and Labidi J. 2011. Energy and economic assessment of soda and organosolv biorefinery processes. *Biomass and Bioenergy* 35: 516–525.

Gasol C.M., Gabarrell X., Rigola M., Gonzalez-Garcia S., and Rieradevall J. 2011. Environmental assessment: (LCA) and spatial modeling (GIS) of energy crop implementation on local scale. *Biomass and Bioenergy* 35: 2975–2985.

Gómez-Barea A. and Leckner B. 2010. Modeling of biomass gasification in fluidized bed. *Progress in Energy and Combustion Science* 36: 444–509.

Grace J.R. and Taghipour F. 2004. Verification and validation of CFD models and dynamic similarity for fluidized beds. *Powder Technology* 139: 99–110.

Groppi G., Tronconi E., Forzatti P., and Berg M. 2000. Mathematical modeling of catalytic combustors fuelled by gasified biomass. *Catalysis Today* 59: 151–161.

Hein M., Foldager M.P., and Sand-Jensen K. 1995. Size-dependent nitrogen uptake in micro- and macroalgae. *Marine Ecological Progress Series* 118: 247–253.

Huang H.J., Ramaswamy S., Al-Dajani W., Tschirner U., and Cairncross R.A. 2009. Effect of biomass species and plant size on cellulosic ethanol: A comparative process and economic analysis. *Biomass and Bioenergy* 33: 234–246.

International Standard Organization. *Environmental Management—Life Cycle Assessment—Principles and Framework*. International Standard ISO 14040; Geneva, Switzerland: ISO 1997.

Jin H., Liu R., and He Y. 2012. Kinetics of batch fermentations for ethanol production with immobilized *Saccharomyces cerevisiae* growing on sweet sorghum stalk juice. *Procedure Environmental Sciences* 12: 137–145.

Katterer T., Eckersten H., Andren O., and Pettersson R. 1997. Winter wheat biomass and nitrogen dynamics under different fertilization and water regimes: Application of a crop growth model. *Ecological Modeling* 102: 301–314.

Kumar A. and Sokhansanj S. 2007. Switchgrass (*Panicum vigratum*, L.) delivery to a biorefinery using integrated biomass supply analysis and logistics (IBSAL) model. *Bioresource Technology* 98: 1033–1044.

Lathouwers D. and Bellan J. 2001. Modeling of biomass pyrolysis for hydrogen production: The fluidized bed reactor. *Proceedings of the 2001 DOE Hydrogen Program Review*, NREL/CP-570-30535. Golden, CO: National Renewable Energy Laboratory.

Loha C., Chatterjee P.K., and Chattopadhyay H. 2011. Performance of fluidized bed steam gasification of biomass-modeling and experiment. *Energy Conversion and Management* 52: 1583–1588.

Lu Q., Zhang Y., Tang Z., Li W., and Zhu X. 2010. Catalytic upgrading of biomass fast pyrolysis vapors with titania and zirconia/titania based catalysts. *Fuel* 89: 2096–2103.

Luedeking R. and Piret E.I. 2000. A kinetic study of the lactic acid fermentation. Batch process at controlled pH. *Biotechnology and Bioengineering* 67: 636–644.

Lundie S. and Peters G.M. 2005. Life cycle assessment of food waste management options. *Journal of Cleaner Production* 13: 275–286.

Luo H.P. and Al-Dahhan M.H. 2011. Verification and validation of CFD simulations for local flow dynamics in a draft tube airlift bioreactor. *Chemical Engineering Science* 66: 907–923.

Michele V. and Hempel D.C. 2002. Liquid flow and phase holdup-measurement and CFD modeling for two- and three-phase bubble columns. *Chemical Engineering Science* 57: 1899–1908.

Miron S.A., Camacho G.F., Gomez C.A., Grima E.M., and Chisti Y. 2000. Bubble-column and airlift photobioreactors for algal culture. *AIChE Journal* 46: 1872–1887.

Mitchell D.A., von Meien O.F., Krieger N., and Dalsenter F.D.H. 2004. A review of recent developments in modeling of microbial growth kinetics and intraparticle phenomena in solid-state fermentation. *Biochemical Engineering Journal* 17: 15–26.

Moncada J., El-Halwagi M.M., and Cardona C.A. 2013. Techno-economic analysis for a sugarcane biorefinery: Colombian case. *Bioresource Technology* 135: 533–543.

Monson C.R., Germane G.J., Blackham A.U., and Smoot L.D. 1995. Char oxidation at elevated pressures. *Combust and Flame* 100: 669.

Muhlen H.J., van Heek K.H., and Jungen H. 1985. Kinetic studies of steam gasification of char in the presence of H_2, CO_2 and CO. *Fuel* 64:944–949.

Nemtsov D.A. and Zabaniotou A. 2008. Mathematical modeling and simulation approaches of agricultural residues air gasification in a bubbling fluidized bed reactor. *Chemical Engineering Journal* 143: 10–31.

Ng K.S. and Sadhukhan J. 2011. Process integration and economic analysis of bio-oil platform for the production of methanol and combined heat and power. *Biomass and Bioenergy* 35: 1153–1169.

Nikoo M.B and Mahinpey N. 2008. Simulation of biomass gasification in fluidized bed reactor using ASPEN PLUS. *Biomass and Bioenergy* 32: 1245–1254.

Niksa S. 2000. Predicting the rapid devolatilization of diverse forms of biomass with bio-flash chain. *Proceeding of 28th International Symposium on Combustion* 8: 2727–2733.

Packer A., Li Y., Andersen T., Hu Q., Kuang Y., and Sommerfeld M. 2011. Growth and neutral lipid synthesis in green microalgae: A mathematical model. *Bioresource Technology* 102: 111–117.

Papadikis K., Bridgwater A.W., and Gu S. 2008. CFD modeling of the fast pyrolysis of biomass in fluidized bed reactors, part A: Eulerian computation of momentum transport in bubbling fluidized beds. *Chemical Engineering Science* 63: 4218–4227.

Papadikis K., Gu S., and Bridgwater A.W. 2010. A CFD approach on the effect of particle size on char entrainment in bubbling fluidized bed reactors. *Biomass and Bioenergy* 34: 21–29.

Patil D.J., van Sint Annaland M., and Kuipers J.A.M. 2005. Critical comparison of hydrodynamic models for gas–solid fluidized beds—Part I: bubbling gas–solid fluidized beds operated with a jet. *Chemical Engineering Science* 60: 57–72.

Quintero J.A., Moncada J., and Cardona C.A. 2013. Techno-economic analysis of bioethanol production from lignocellulosic residues in Colombia: A process simulation approach. *Bioresource Technology* 139: 300–307.

Ramzan N., Ashraf A., Naveed S., and Malik A. 2011. Simulation of hybrid biomass gasification using ASPEN PLUS: A comparative performance analysis for food, municipal solid and poultry waste. *Biomass and Bioenergy* 35: 3962–3969.

Ranzi E., Cuoci A., Faravelli T., Frassoldati A., Migliavacca G., Pierucci S., and Sommariva S. 2008. Chemical kinetics of biomass pyrolysis. *Energy Fuels* 22: 4292–4300.

Roos A. and Rakos C. 2000. The limits of modeling. Experiences with bioenergy in practice-could models have predicted this outcome? *Biomass and Bioenergy* 18: 331–340.

Sassner P., Galbe M., and Zacchi G. 2008. Techno-economic evaluation of bioethanol production from three different lignocellulosic materials. *Biomass and Bioenergy* 32: 422–430.

Singh R.I., Brink A., and Hupa M. 2013. CFD modeling to study fluidized bed combustion and gasification. *Applied Thermal Engineering* 52: 585–614.

Song H., Eom M.H., Lee S., Lee J., Cho J.H., and Seung D. 2010. Modeling of batch experimental kinetics and application to fed-batch fermentation of *Clostridium tyrobutyricum* for enhanced butyric acid production. *Biochemical Engineering Journal* 53: 71–76.

van den Broek R., Vleeshouwers L., Hoogwijk M., van Wijk A., and Turkenburg W. 2001. The energy crop growth model SILVA: Description and application to *Eucalyptus* plantations in Nicaragua. *Biomass and Bioenergy* 21: 335–349.

Van Winden W.A., van Dam J.C., Ras C., Kleijn R.J., Vinke J.L., van Gulik W.M., and Heijnen J.J. 2005. Metabolic-flux analysis of *Saccharomyces cerevisiae* CEN.PK 113-7D based on mass isotopomer measurements of ^{13}C-labeled primary metabolites. *FEMS Yeast Research* 5: 559–568.

Versteeg H.K. and Malalsekera W. 1995. *An Introduction to Computational Fluid Dynamics: The Finite Volume Method*. New York: Wiley.

Wang L., Shahbazi A., and Hanna M.A. 2011. Characterization of corn stover, distiller grains and cattle manure for thermochemical conversion. *Biomass and Bioenergy* 35: 171–178.

Wang L., Weller C.L., Jones D.D., and Hanna M.A. 2008. Contemporary issues in thermal gasification of biomass and its application to electricity and fuel production. *Biomass and Bioenergy* 32: 573–581.

Wang L.J., Amezquita A., and Weller C.L. 2006. An integrated model of heat transfer and temperature-dependent bacterial growth in cooked hams during air blast chilling. *Transactions of the ASABE* 49(5): 1437–1446.

Wang L.J., Weller C.L., and Hanna M.A. 2005. An integrated mathematical model of fluid dynamics, heat transfer and reaction kinetics for fluidized bed gasification of biomass. *Proceedings of the 2005 AIChE Annual Meeting*, Cincinnati, OH.

Wang Y. and Yan L. 2008. Review: CFD studies on biomass thermochemical conversion. *International Journal of Molecular Sciences* 9: 1108–1130.

Wu B. 2013. Advances in the use of CFD to characterize, design and optimize bioenergy systems. *Computers and Electronics in Agriculture* 93: 195–208.

Wu X. and Merchuk J.C. 2002. Simulation of algae growth in a bench-scale bubble column reactor. *Biotechnology and Bioengineering* 80: 156–168.

Wurzenberger J.C., Wallner S., Raupenstrauch H., and Khinast J.G. 2002. Thermal conversion of biomass: Comprehensive reactor and particle modeling. *AIChE Journal* 48: 2398–2411.

Yang C., Hua Q., and Shimizu K. 2002. Quantitative analysis of intracellular metabolic fluxes using GC-MS and two-dimensional NMR spectroscopy. *Journal of Bioscience and Bioengineering* 93: 78–87.

Yang Y.T., Bennett G.N., and San K.Y. 1998. Metabolic flux analysis of *E. coli* deficient in the acetate production pathway and expressing the *B. subtilis* acetolactate synthase. *Metabolic Engineering* 1: 26–34.

Zajsek K. and Gorsek A. 2010. Modelling of batch kefir fermentation kinetics for ethanol production by mixed natural microflora. *Food and Bioproducts Processing* 88: 55–60.

Zhang J., Osmani A., Awudu I., and Gonela V. 2013a. An integrated optimization model for switchgrass-based bioethanol supply chain. *Applied Energy* 102: 1205–1217.

Zhang K., Kurano N., and Miyachi S. 2002. Optimization aeration by carbon dioxide gas for microalgal production and mass transfer characterization in a vertical flat-plate photobioreactor. *Bioprocess and Biosystems Engineering* 25: 97–101.

Zhang Y., Brown T.R., Hu G., and Brown R.C. 2013b. Techno-economic analysis of monosaccharide production via fast pyrolysis of lignocellulose. *Bioresource Technology* 127: 358–365.

Zheng H., Kaliyan N., and Morey R.V. 2013. Aspen plus simulation of biomass integrated gasification combined cycle systems at corn ethanol plants. *Biomass and Bioenergy* 56: 197–210.

Zwietering M.H., De Koos J.T., Hasenack B.E., De Wit J.C., and Van't Riet K. 1991. Modeling of bacterial growth as a function of temperature. *Applied and Environmental Microbiology* 57(4): 10940–1101.

4 Life Cycle Assessment of Bioenergy Systems

Reinout Heijungs and Edi Iswanto Wiloso

CONTENTS

4.1 INTRODUCTION

The concept of a bio-based economy has been gaining ground in the last few years. The four *f's*—food, feed, fiber, and fuel—are topics of continuous innovation. Governments are developing biofuel policies, and multinationals are spending large investments for green production.

In some sense, the current increasing interest in bioenergy is a return to the past. Until the industrial revolution, man largely relied on biomaterials such as wood for timber and wool for garments, bioenergy such as wood for heating and cooking, and biotic sources for nutrition. Although the materials of nonbiotic origin were in use for ages, as witnessed by the eras of Stone Age, Iron Age, and Bronze Age, the ratio between the uses of nonbiotic to biotic materials increased tremendously in the nineteenth and twentieth centuries. Although biotic resources are still used in houses, furniture, and garments, these items typically contain a much larger amount of iron, glass, and plastics.

There are several reasons underlying this change, which include the following:

- The finite reserves of fossil fuels and mineral ores, leading to their depletion and high prices
- Geopolitical concerns on dependence of other countries for the supply of resources

- The issue of carbon dioxide, a major cause of climate change, which is released by combusting fossil fuels
- Socioeconomic support of the agricultural sector

It is well known that the introduction of large-scale biomaterials and bioenergy has resulted in some important problems: environmental (deforestation), social (increasing food prices), and technical (less than expected energy efficiency). For sure, certain mistakes have been made in the introduction and upscaling of the so-called first-generation bioenergy crops. The second-generation bioenergy program is trying to do better by taking nonedible crops and agricultural residues as feedstocks. However, it seems wise to learn from the experiences with the first-generation bioenergy, not only in fixing some design errors but also more fundamentally in analyzing direct and indirect social, economic, and environmental effects before the second-generation bioenergy program reaches the point where there is no return.

Life cycle assessment (LCA) is one of the important tools to do this analysis. LCA is primarily a tool for environmental decision support, analyzing the impacts on the environment of products and technologies in a life cycle perspective. In this chapter, we present the LCA method at a general level as well as its relation to bioenergy systems. LCA is a rather general method for analyzing the environmental sustainability of products. There is no specific bias toward energy systems or bioenergy feedstocks. In contrast, such systems present specific problems in terms of data, method, and interpretation. Therefore, this chapter presents after the general introduction to LCA a survey of some points of attention and development in applying the LCA technique to bioenergy systems.

4.2 LCA ACCORDING TO ISO STANDARDS

LCA refers to the process of compiling and evaluating the inputs, outputs, and the potential environmental impacts of a product system throughout its life cycle (ISO, 2006). In this section, the main features of the LCA method are discussed.

4.2.1 Overview of the ISO Standards

LCA has come a long way and continues to be improved. Since a decade or so ago, there has been a broadly accepted set of principles that can be claimed as the present-day LCA framework. In this chapter, this LCA framework is reviewed based on a handbook of LCA recently published by Curran (2012). The International Organization for Standardization (ISO) has issued a series of standards and technical reports for LCA, referred to as the 14040 series. This series consists of the documents listed in Table 4.1.

The standards are organized into the different phases of an LCA study. These are

- Goal and scope definition
- Inventory analysis
- Life cycle impact assessment (LCIA)
- Life cycle interpretation

The relationships among these phases have been illustrated in Figure 4.1, which has become a type of logo of LCA.

LCA typically starts with the definition of goal and scope, then proceeds to the inventory analysis, then optionally continues to impact assessment, and ends with the interpretation. However, as indicated in Figure 4.1, a LCA study is a highly iterative process, so that a LCA practitioner may need to go back to goal and scope after the preliminary inventory work, to move back from impact assessment to inventory analysis, and to have a look at the interpretation in an early stage. The main

TABLE 4.1
ISO Documents on LCA

Number	Type	Title	Year
14040	International standard	Principles and framework	1996, 2006
14041	International standard	Goal and scope definition and inventory analysis	1998[a]
14042	International standard	Life cycle impact assessment	2000[a]
14043	International standard	Life cycle interpretation	2000[a]
14044	International standard	Requirements and guidelines	2006[b]
14047	Technical report	Examples of application of ISO 14042	2003
14048	Technical report	Data documentation format	2001
14049	Technical report	Examples of application of ISO 14041	2000

[a] Updated in 2006 and merged into 14044.
[b] Replaces 14041, 14042, and 14043.

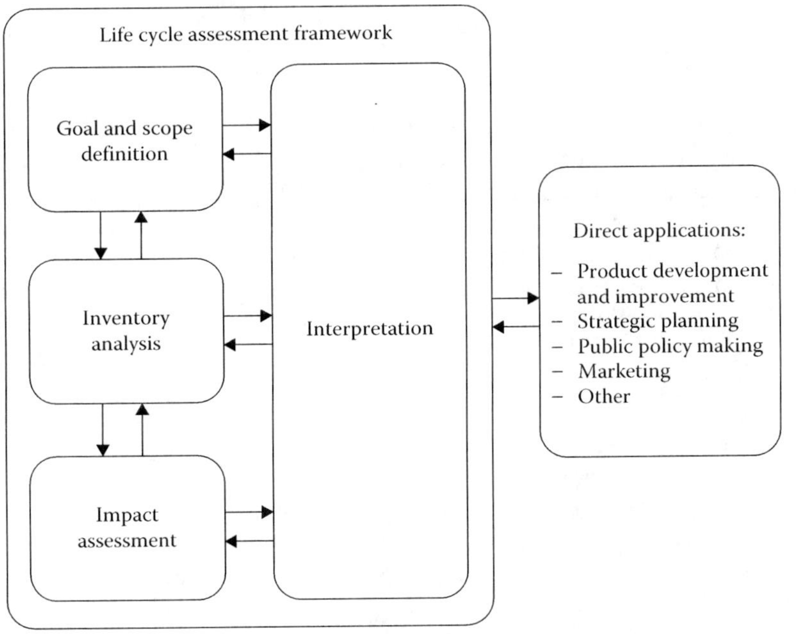

FIGURE 4.1 The general methodological framework for LCA. (From ISO, Environmental management—Life cycle assessment—Requirements and guidelines (ISO 14044), International Organization for Standardization, Geneva, Switzerland, 2006.)

idea and content of the four phases are discussed in the following separate subsections. All quotations are taken from the ISO documents, unless otherwise indicated.

4.2.2 Goal and Scope Definition

The first phase of an LCA centers on formulating the question and stating the context of answering this question. In the goal and scope definition, no data are collected and no results are calculated. Rather, it is a place where the plan of the LCA study is defined as clearly and unambiguously as possible. Likewise, in an LCA report, it should help readers to quickly find out the precise question addressed and main principles chosen. The goal of the LCA should deal with the following topics:

- The intended application
- The reasons for carrying out the study
- The intended audience
- Whether the results are to be used in comparative assertions disclosed to the public

The choices made here have an influence on the rest of the LCA procedure. For instance, depending on the intended audience, a critical review may be needed, and it may be important that an external expert takes this task. In the scope definition, a number of major choices are made: first of all, the product system or systems to be studied and the function that the system delivers (or in case of a comparative LCA, the functions that the systems deliver). For instance, one might be interested in the product systems of biodiesel versus fossil-based diesel, with the function of running a passenger car.

An important aspect of the scope definition is the functional unit. It is obviously suboptimal to compare 1 L of biodiesel with 1 L of fossil-based diesel, as the energy content differs. The ultimate function is not to have diesel but to have a transportation service. The functional unit expresses the function of the products and, thereby, offers a way to equalize differences in performance. A functional unit for analyzing fuels could thus better be phrased in terms of their function, for instance, "running a passenger car over 100 km." As LCA mathematically employs a linear calculation rule, the results can be scaled up by choosing a numerically different functional unit (say, "running a passenger car over 500 km"), but the alternatives considered will be scaled up or down consistently, so this will not affect the conclusions. A consequence is, however, that the LCA cannot tell if a product is *sustainable* or *environmentally friendly* but can only indicate if product X is *more sustainable* or *more environmentally friendly* than product Y or that the use phase is the *least sustainable* or *least environmentally friendly* part of the life cycle for product Z.

The scope definition further sets the main outline on a number of subjects that are discussed and further refined in more detail in the later phases. These include

- System boundaries
- Impact categories
- Treatment of uncertainty

The ISO standard and some other texts in places suggest that these topics are implemented in detail in the scope definition. This is wrong: As the goal and scope definition is not concerned with collecting data or calculating results, no concrete details on such topics can be specified at this phase.

4.2.3 Inventory Analysis

ISO defines life cycle inventory (LCI) analysis as the "phase of life cycle assessment involving the compilation and quantification of inputs and outputs for a product throughout its life cycle." It will be clear that quantification is an important aspect here, and numbers, in terms of data and calculations, are of central concern in the inventory analysis. The LCI is built on the basis of the unit process. A unit process is the "smallest element considered in the life cycle inventory analysis for which input and output data are quantified." Examples of unit processes are coal mining, steel production, refining of oil, production of furniture, use of a television set, recycling of waste paper, and transport by lorry. Each of these processes is described in quantitative terms as having inputs and outputs. In LCA, a unit process is treated as a black box that converts a bundle of inputs into a bundle of outputs as shown in Figure 4.2. Inputs come in several types: products (including components, materials, and services), waste for treatment, and natural resources (including fossils, ores,

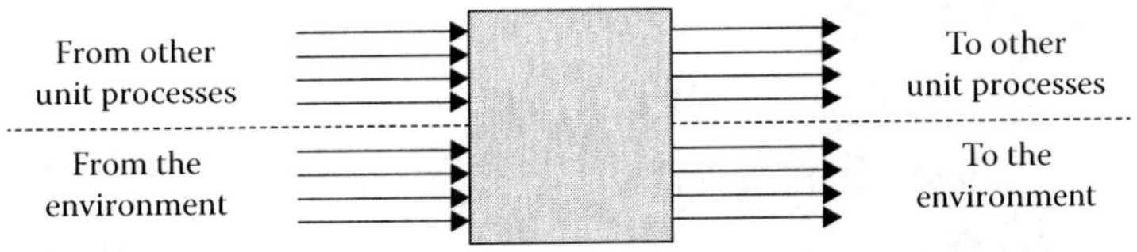

FIGURE 4.2 General template of a unit process.

biotic resources, and land). Outputs come in several types as well: products (including components, materials, and services), waste for treatment, and residuals to the environment (including pollutants to air, water, and soil, waste heat, and noise).

The process (gray rectangle) is considered as a black box, having inputs (left-hand side) and outputs (right-hand side) from and to other unit process (top arrows) and from and to the environment (bottom arrows). Unit processes form the building blocks of an LCA. This is because products are not harmful for the environment as such, except for the processes involved in products. Producing, using, and disposing products creates the burden to the environment. Therefore, these processes assume a central position in LCA. The essential feature of LCA in which it distinguishes itself from the analysis of an industrial or agricultural process is that it connects different unit processes into a system. A flow diagram is a graphical representation of the system comprised of connected unit processes. Figure 4.3 shows a fragment of such a flow diagram. Because the purpose of a flow diagram is to show how unit processes are connected, only the flows from and to other unit processes are displayed, and flows from and to the environment are hidden. All transport, packaging, and so on have been left out as well.

As we can see, some unit processes are connected with one another in simple upstream–downstream connections, for example, TV production is upstream connected to semiconductor production. But there are also more complicated connections, for example, electricity linking to different parts of the system and recycling feeding back to production. Flow diagrams are in fact huge webs of interconnected unit processes. In the present era of digital databases, LCA studies can easily comprise several thousands of unit processes. LCA is primarily a quantitative model. In the LCI, all unit processes included have to be quantified. This means that we have to specify the sizes of the inflows and outflows, per unit process. As an example, let us take the unit process of aluminum production. An aluminum plant may specify its technology in terms of inputs and outputs by stating input requirements (e.g., 2 kg aluminum oxide and 20 kWh electricity per kg aluminum produced) and emissions (e.g., 2 g dust per kg aluminum produced). We must translate these inputs and outputs into our template for unit processes as shown in Table 4.2.

For each of the unit processes included, quantitative data should be collected. Moreover, in order to be able to process the data and perform the calculations automatically, a clear and unambiguous representation is needed. This implies, among other things, harmonization of nomenclature

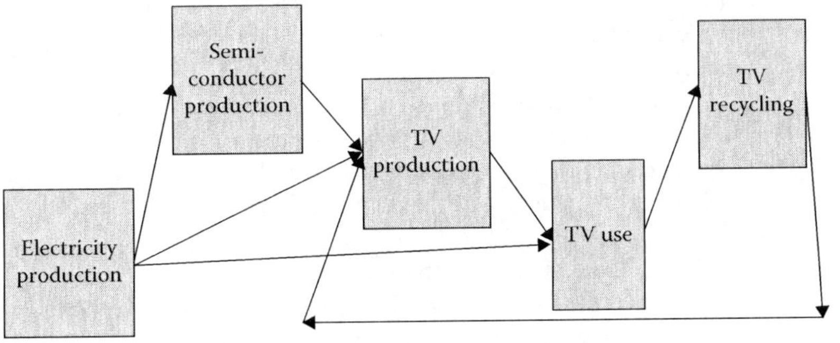

FIGURE 4.3 Fragment of a simplified flow diagram for an LCA on TV sets.

TABLE 4.2
Example of Unit Process Specification in an Aluminum Plant

Type of Flow	Name	Amount	Unit
Inputs from other unit processes	Aluminum oxide	2	kg
Inputs from other unit processes	Electricity	20	kWh
Outputs to other unit processes	Aluminum	1	kg
Outputs to the environment	Dust	0.002	kg

(e.g., not using *carbon dioxide* for one unit process and CO_2 for another) and harmonization of units (e.g., not mixing up kilograms [kg] and pounds [lb]). In Table 4.2, the unit process data are given per unit of output, here per kg of aluminum. In an LCA, we must find out how much of each specific output we need. For instance, the process may need 3 kg of aluminum, not 1 kg. The basic assumption of the LCA model is that technologies are linear. This means that we can scale up (or down) the data of a unit process by a simple multiplication factor. In the example, 3 kg of aluminum would require 6 kg of aluminum oxide and 60 kWh of electricity, while it would release 6 g of dust. The assumption of linear technology is an important restriction of LCA and yet it is an important step in making the calculation and data collection feasible.

In scaling up (or down) the unit processes, the weblike nature of the system quickly creates complication, as everything depends upon each other. The calculation of the scaling factors, and with that of the emissions to and extraction from the environment, is greatly simplified by considering the problem as a system of linear equations: one unknown (the scaling factor) for every unit process and one equation (a balance) for every flow. Thus, solutions may be obtained by matrix algebra. The details of this solution procedure are not discussed here but see the literature published by Heijungs and Suh (2002).

The approach mentioned earlier may fail in a number of cases. We mention two complications:

- For some products, upstream production processes or downstream disposal processes may be difficult to quantify.
- For some unit processes, the balance equations become impossible due to the fact that these processes produce not just one product but several coproducts.

The first issue can be solved by a procedure known as cutoff, and the second one by allocation.

Cutoff is a solution to the problem that the system is theoretically infinitely large. To refine a biofuel, we need machines, and these machines are produced by machines, and these machines in turn need machines, etc. But we have an intuitive idea that some very distant upstream processes will not be quite important to the study. This means that we will cut off certain inputs: Although we know that something is needed, and we sometimes even know how much is needed, we do not specify how these inputs are produced, for reasons of time and budget efficiency or because some of the data are unknown. It turns out to be difficult to specify reliable criteria when cutoff is allowed or to estimate how large the error is when a cutoff is made. Conducting a difference analysis facilitates matters considerably: In comparing air traffic on biofuels with air traffic on fossil fuels, we may leave out the airport and communication facilities, whereas they would in a noncomparative LCA in theory be included.

The second problem, coproduct allocation, has given rise to one of the biggest controversies in LCA theory. The problem can be demonstrated by this example: if a transportation process needs gasoline, the upstream unit process is a refinery that produces not only gasoline but also diesel, kerosene, heavy oils, and many other products. The direct impacts of the refinery are the pollutants such as CO_2. However, the flows to and from other processes that may also lead to impacts (e.g., from oil drilling) may be argued not be attributable to gasoline only but need to be distributed over gasoline, diesel, and all the other coproducts. This is hardly contested, but the debate now focuses on how to do this. To make the issue more concrete, the question can be stated as follows: How much of the CO_2 from a refinery is allocated to the gasoline? Different schools have provided different arguments, and none of which has been completely compelling so far. Some solutions lead to strange results, while other solutions may be very difficult to carry out (due to, e.g., the lack of data or appropriate software). Still others are rejected outright by many experts. To complicate the issue, the problem does not only occur in unit processes that produce several coproducts but also in unit processes that treat more than one type of waste, or where waste is recycled into a useable good, as so often happens in the case of second-generation biofuels. It is not even agreed upon if the multi-output case, the multi-input case, and the recycling case must be treated in the same way or not.

Within ISO, a preferred order for solving the multifunctionality problem has been designed. It distinguishes several solutions (dividing the unit process into two or more subprocesses, expanding the system to include the additional functions, partitioning on the basis of a physical parameter, partitioning on the basis of an economic parameter), separated by clauses like "wherever possible" and "where... cannot be established." This stepwise procedure is a clear comprise. In practice, it leaves so much freedom that LCA studies based on the ISO standard can give conflicting results.

After appropriate cutoff and allocation steps, the final inventory results can be calculated. Typically, this is a table with the quantified inputs from and outputs to the environment, for each of the alternative systems considered, expressed in relation to the functional unit. With the present-day software and databases, this inventory table may be 1000 lines or longer. It contains not only the familiar pollutants and resources, such as CO_2, NO_x, and crude oil, but also more exotic items, such as 1-pentanol, cyprodinil, and dolomite.

4.2.4 Impact Assessment

LCIA, or impact assessment for short, is the "phase of life cycle assessment aimed at understanding and evaluating the magnitude and significance of the potential environmental impacts for a product system throughout the life cycle of the product." Its motivation comes from two observations:

- The final result of the inventory analysis, or the inventory table, is too long (e.g., 1000 different items) to handle.
- The inventory table contains many items that require expert knowledge (such as 2-methyl-2-butene) to understand in terms of importance.

Impact assessment, and in particular the characterization step, solves both issues: It "involves the conversion of LCI results to common units and the aggregation of the converted results within the same impact category." While the unit process is the central element of the inventory analysis, the central element in impact assessment is the impact category. ISO defines it as a "class representing environmental issues of concern to which life cycle inventory analysis results may be assigned." Perhaps more helpful are some examples: climate change, toxicity, and depletion of fossil energy carriers.

As climate change (often used interchangeable with global warming) is a well-known issue, we will illustrate the main ideas of impact assessment with this case. The inventory table contains a number of greenhouse gases: CO_2, CH_4, N_2O, etc. All of these are known to contribute to the phenomenon of climate change. Climate change involves a long sequence of causal mechanisms: emissions of greenhouse gases lead to changes in the composition of the atmosphere, which lead to a change in the radiation balance, which in turn leads to a change in the temperature distribution, which leads to changes in climate, which leads to changes in ecosystems and human activities, etc. The further we proceed in this causal chain, the more uncertain and speculative our knowledge becomes. While quite some scientific evidence is available with respect to the composition of the atmosphere, the impacts on biodiversity are debated. Many of these later impacts are even conditional on our future activities, including future emission scenarios and mitigating actions. To be able to quantitatively model the emissions of different greenhouse gases into an impact indicator for climate change, we must do several things.

First, we must choose a certain point in the causal mechanism. This can be at the front end (a so-called midpoint indicator, like change in radiation balance), at the back end (change of biodiversity), or somewhere in between (a so-called endpoint indicator, like change in temperature). Regardless of the choice between midpoint and endpoint, the indicator chosen is referred to as the impact category indicator, or category indicator for short.

Second, a way must be found to convert the emission data into the chosen impact indicator. Scientists in chemistry, meteorology, ecology, etc., have developed model fragments to estimate the atmospheric lifetimes of greenhouse gases, their effect on the radiation balance and the formation of clouds, and the effects of temperature on the distribution of species. These fragments have been combined by workgroups from the United Nation-based International Panel on Climate Change (IPCC) into quantitative models of the impacts of greenhouse gas emissions. Part of this is the global warming potentials (GWPs), which are quantitative measures of the strength of different greenhouse gases. Many midpoint LCIA methods apply GWPs for climate change. We will illustrate their usage in the following. For now, it is sufficient to say that GWPs provide one example of a set of characterization factors and that the IPCC model from which they are derived is an example of a characterization model. Note, by the way, that IPCC has not developed this model as a characterization model for LCIA but that the LCA community has adopted this model as such and its derived GWPs as characterization factors. Also, it should be noted that the characterization model itself is not used by LCA practitioners and only the characterization factors that have been derived from it as a one-time exercise are used. Characterization factors are often tabulated in LCA guidebooks and are implemented in many LCA software packages, while the characterization models often require supercomputers and expert knowledge.

The selection of impact categories is needed before one can select a category indicator and a characterization model with the characterization factor. Some LCA studies concentrate on just one impact category. For instance, the carbon footprint is considered as a form of LCA that addresses climate change at the midpoint level through GWPs. At the other extreme, some LCA studies incorporate fifteen or more impact categories. For consistency reasons, the choice of impact categories is often made on the basis of a recommended impact assessment guidebook or its implementation in software. In the reports of LCA studies, we can often see the terms of "IMPACT2002+," "TRACI," "CML-IA," "ReCiPe," "ILCD," etc. All these methods comprise a recommended set of impact categories with a category indicator and set of characterization factors. ISO does not specify any choice in these matters. Table 4.3 gives an overview of some frequently used impact categories and category indicators. We can see that the column with endpoint indicators contains the same term (e.g., *loss of life years*) many times. This suggests that impact categories can be aggregated into fewer endpoint indicators than midpoint indicators.

As a concrete example of how characterization works, let us study a fragment of a hypothetical inventory table, containing the following information: emission of 100 kg CO_2, emission of 1 kg CH_4, and emission of 1 kg SO_2. Characterizing greenhouse gases with GWPs requires a table with GWPs. In such a table, one can find that the GWP of CO_2 is 1 (by definition) and that the GWP of CH_4 is 25 (kg CO_2 equivalent/kg CH_4). SO_2 has no GWP and it is assumed not to contribute to climate change. Characterization now proceeds in the case of climate change by calculating

$$1\times100+25\times10=350 \text{ kg } CO_2-\text{equivalent} \tag{4.1}$$

TABLE 4.3
Overview of Widely Used Impact Categories with Examples of Category Indicators at Midpoint and Endpoint Levels

Impact Category	Midpoint Category Indicator	Endpoint Category Indicator
Climate change	Infrared radiative forcing	Loss of life years, fraction of species disappeared
Ozone layer depletion	Change in tropospheric ozone concentration	Loss of life years
Acidification	H+ concentration	Fraction of species disappeared
Eutrophication	Biomass potential	Fraction of species disappeared
Human toxicity[a]	Time-integrated exposure, corrected for hazard	Loss of life years
Ecotoxicity[b]	Time-integrated exposure, corrected for hazard	Fraction of species disappeared
Depletion of energy carriers	Primary energy requirement	Decreased availability
Depletion of material resources	Amount of material used, corrected for availability and/or importance	Decreased availability
Land use impacts	Amount of land occupied or transformed	Fraction of species disappeared
Water use impacts	Amount of water used or displaced	Decreased availability

[a] Sometimes split into carcinogenics, noncarcinogenics, respiratory effects, etc.
[b] Sometimes split into aquatic toxicity, terrestrial toxicity, marine toxicity, etc.

For the more general case, this can be written as

$$GW = \sum_{s} GWP_s \times m_s \quad (4.2)$$

where
GW is the global warming score
s the substance (the different greenhouse gases)
GWP_s is the GWP of substance s
m_s is the emitted amount of substance s in kg

This may be further generalized as

$$I_c = \sum_{s} CF_{c,s} \times m_s \quad (4.3)$$

where
c stands for the impact category
I represents the indicator result for category c
$CF_{c,s}$ is the characterization that links substance s to impact category c

This formula is the operational formula for characterization. With a table of characterization factors specified, it makes clear that

- LCIA builds on the results of LCI (as is clear from the term m_s)
- Characterization converts the results of LCI into a common metric (as is clear from the multiplication by $CF_{c,s}$)
- Characterization aggregates the converted LCI results (as is clear from the summation symbol)

- The result from characterization is a list of numbers, for instance, a score for climate change and a score for toxicity. ISO refers to such numbers as *category indicator results*, but most LCA practitioners prefer names like *score*, sometimes expanded with the name of the impact (as in *toxicity score*). The complete list is known by names such as *LCIA profile* and *characterization table*

After this sketch of the principle of characterization, let us have a look at the more formal ISO point of view. Impact assessment in ISO is structured into a number of steps:

- Selection of impact categories, category indicators, and characterization models
- Classification
- Characterization
- Normalization
- Grouping
- Weighting
- Data quality analysis

Although the characterization is just one of the steps of impact assessment, the former term is often used as a pars pro toto for the latter term. Indeed, in discussing the principle of characterization earlier, we have touched upon the steps of selecting categories, indicators, and models and characterization. These are mandatory steps for ISO. The step of classification is mandatory as well, but few LCA studies report it. ISO defines it as the "assignment of LCI results to the selected impact categories." Its purpose is to clearly show which emissions and extractions are treated under which impact category, but it involves no numerical conversion into a common metric, as is the case for characterization.

Normalization refers to calculating "the magnitude of the category indicator results relative to some reference information." It is an optional step for ISO, and indeed, many LCIA studies stop at the characterization. The reference information is, in most cases, the total impact in a certain region in a certain time period, for example, in the country of decision in 1 year. Normalization helps *to better understand the relative magnitude for each indicator result*. Without normalization, the indicator results are in quite different units, for example, kg CO_2 equivalent for climate change and MJ primary energy for fossil energy depletion. To put these results in perspective, the normalization expresses them as a share of the total impact size in the region. Arbitrary differences due to the units disappear, and it becomes clear to which impact category a product contributes relatively much. The units of the normalization indicator results are equal; nevertheless, such numbers cannot meaningfully be added because the severity of the different impact categories has not yet been accounted for. This can be done in the weighting step as discussed later. Normalization fulfils several functions: It provides insight into the meaning of the impact indicator results, it helps to check for errors, and it prepares for a possible weighting step.

Grouping is another optional step, although it is seldom seen in LCA studies. Weighting, like characterization, converts and aggregates, but while characterization does so for the LCI results, weighting starts with the characterization (or normalization) results. Typically, weighting factors are applied, either to the characterization indicator results or to their normalized results. The weighting factors themselves are supposed to reflect value judgments, such as social and political priorities. Weighting typically produces one final number, by means of

$$W = \sum_{c} WF_c \times I_c \tag{4.4}$$

where

I_c is the impact score (or normalized impact score) for impact category c
WF_c the weighting factor for this impact category
W the weighted result

Well-known examples of such weighted results are the eco-indicator and the environmental load unit (ELU).

The data quality analysis, finally, involves an analysis of completeness, uncertainties, etc. We treat this more systematically in the following section of interpretation.

4.2.5 Interpretation

ISO defines interpretation as the "phase of life cycle assessment in which the findings of either the inventory analysis or the impact assessment, or both, are evaluated in relation to the defined goal and scope in order to reach conclusions and recommendations." Several elements of interpretation are listed by ISO:

- Identification of significant issues
- An evaluation that considers completeness, sensitivity, and consistency checks
- Conclusions, limitations, and recommendations
- Appropriateness of the definitions of the system functions, the functional unit, and system boundary
- Limitations identified by the data quality assessment and the sensitivity analysis

The text of ISO on interpretation is very concise, and no details are given on procedures and techniques to be employed. The same applies to most guidebooks on LCA. They mention carrying out an uncertainty analysis but give no clear guidance on how this should be done.

In another context (Heijungs and Kleijn, 2001), we have introduced the distinction between procedural and numerical approaches. Procedural approaches include all types of analyses that deal with the data and results in relation to other sources of information, like expert judgment, reports on similar products, intuition, and reputation of data suppliers. Numerical approaches include those approaches that somehow deal with the data that are used during the calculations, without reference to those other sources of information, but as algorithms that use and process the data in different ways, so as to produce different types of *smart* data reduction that provide an indication of reliability, key issues, discernibility, robustness, and so on.

This distinction helps us understand some important roles of interpretation. On the one hand, it is to compare the data and results with previous findings and put the results in the context of decision making and limitations. On the other hand, it is devoted to a systematic analysis with the help of statistical and other decision-analytic techniques. The latter type may be incorporated in software, and indeed, an increasing number of software packages contain options for running the Monte Carlo analysis, doing sensitivity analyses, carrying out statistical significance tests, etc. For instance, in the CMLCA software, we have implemented, among others

- Contribution analysis
- Comparative analysis
- Uncertainty analysis
- Perturbation analysis
- Key issue analysis
- Discernibility analysis

The iterative nature of the ISO framework (Figure 4.1) shows up in this context. Whenever the uncertainties are too high, we may go back to collect better data. Whenever sensitivity analysis shows that some decisions are crucial, we may go back and do a more refined analysis. In this way, the interpretation helps to prepare for a balanced decision while helping improve the LCA.

The practice of LCA is quite meager. We still see many LCA studies without uncertainty or sensitivity analysis, even though software increasingly facilitates this. There is a psychological argument that a contractor pays for finding out something, not for increasing the doubt. And as many LCA practitioners spend several months on collecting data, it is never a nice thing to waste this effort in a last-minute uncertainty analysis. But decision making obviously means also to take into account the limits of knowledge. Moreover, as discussed before, a proper analysis of uncertainties and sensitivities helps to prioritize the steps earlier on in the framework: collecting data, setting boundaries, and making choices.

4.2.6 LCA in Practice

In the earlier text, the emphasis has been on the generally accepted practice. This is a mix of the ISO standards and a set of guidebooks. These latter include the ILCD Handbook, the Dutch Handbook on LCA, the guidelines from EPA, and those from JEMAI. All these texts interpret, add, refine, or modify the ISO standards. As has been indicated at a few places, the practice in LCA is sometimes different from what the ISO standards prescribe.

There are also additional standards, dictated by the use of software and databases. Many software packages for LCA have built-in options for impact assessment and uncertainty analysis, but nearly always in a restricted form, allowing some variants and prohibiting other variants. LCI databases are often constructed with predefined allocation methods and cutoff rules, so the user cannot choose and carry out sensitivity analyses.

The important aspect is the incorporation of new insights in the practice of LCA. The ISO standards were dated from the period 1997–2000, and the 2006 update is not really an update (just a merging). Meanwhile, numerous developments have taken place. Some of these developments are already becoming the mainstream of LCA by now, while other developments may disappear. Anyhow, the practice of LCA is evolving, and it will continue to do so.

4.3 LCA AND BIOENERGY SYSTEMS: SOME ISSUES

LCA is a very general method that can be and actually is applied to a wide range of topics, from beverage packaging to solar panels and from cars to ammunition. Nevertheless, there are various types of products for which conducting an LCA is more difficult. Difficulties can be related to the problem of finding appropriate data, but methodological issues can also pop up. In particular, agricultural products, whether it is for food, feed, fiber, or fuel, turn out to be complicated. This has been recognized a long time ago (Wegener Sleeswijk et al., 1996), but with the revival of the bio-based economy, these problems deserve to be emphasized once more (Guinée et al., 2009). In this section, we will discuss some of these issues, concentrating on bioenergy. We build in this overview on our previous work (Heijungs and Wiloso, 2012; Wiloso and Heijungs, 2013).

4.3.1 Choice of Alternatives

Most LCA studies are comparative case studies, in which a number of alternative products are compared. For instance, one compares an electric car with a conventional one.

In the case of bioenergy, the choice seems easy, but it is not always so. For instance, consider the case of electricity produced from agricultural residues. In a comparative setting, one might hope to compare this to the *normal* way of producing electricity. But what is *normal*? Is it the average domestic electricity production? Or is it the average domestic electricity consumption? Or is it perhaps not an average but a marginal production technology instead?

Such choices can be highly influential in determining the comparative performance of new bio-based options.

4.3.2 Choice of Functional Unit

In an LCA, a unit of analysis must be specified. For instance, for an LCA of a TV, one might choose *1000 h watching TV* as a functional unit. There are many cases where the choice of an appropriate functional unit is difficult. Even for the TV example, this is not straightforward. Given that TVs have a standby mode, one might reformulate the functional unit as *1000 h using TV, of which 200 h watching and 800 h standby*. Obviously, the ratio of watching and standby may influence the relative ranking of different types of TV.

A similar problem exists for LCA of bio-based options. Take the example of bioethanol as a source of energy. The properties of this fuel differ from the more conventional forms. So, how will we define a functional unit here? In terms of liters, kilograms, megajoules, etc.?

4.3.3 Biogenic CO_2

One of the alleged advantages of biofuels over fossil fuels is that they are *carbon neutral*. There are at least two points to make in that respect. First, if there is anything neutral, it is LCA as an analytical tool. If biofuels are carbon neutral, this will follow from the LCA study, instead of being a starting point of the LCA study. Second, there are several situations where the carbon neutrality is challenged. One of these occurs in the situation that some of the CO_2 absorbed is not released as CO_2 but as CH_4, a greenhouse gas that is much stronger. This happens, for instance, when the biotic feedstock is subjected to a process of digestion with leakages occurring along the way. Another case is that of a cultivation process with two or more products (e.g., palm oil and palm oil cake), where part of the absorbed CO_2 is allocated to each of the coproducts (see the following). A final example is the fact that there is a time difference between CO_2 fixation and release, and dynamic LCA methods are in development to account for such situations.

4.3.4 System Boundaries

System boundaries reflect what is or isn't included in the analysis. Even though LCA studies should be *complete*, certain processes are excluded in practice. This is done for various reasons.

An important reason is the trade-off between the relevance and difficulty of obtaining the data. In doing an LCA for a bioethanol plant, the computers that control the process can be expected to be insignificant, while including them requires a lot of extra data. Another reason has already been mentioned earlier, when we spoke on cradle-to-gate studies that exclude the use and/or waste treatment stage. This may be attractive, but it is only admissible when there is no difference between these stages. The comparison between a plastic and a paper cup (e.g., for drinking tea) can in this way leave out the use stage (and thereby the growing of tea plants, the processing of tea leaves, the boiling of water, etc.). But the waste treatment of plastic cups is obviously different from that of paper cups.

4.3.5 Decisions on Goods and Waste

The term *waste* is used in daily life in an imprecise way. It may refer to waste to be processed (like waste that has been thrown in a waste bin), waste emissions released to the environment (such as carbon dioxide emissions), final waste after waste treatment (such as the bottom ash after incineration of municipal waste), and by-products (like iron scrap that can be upgraded). Distinguishing among these types of waste is essential in the systematic approach of LCA.

The logic in LCA is as follows:

- Any stream that goes directly to the environment is an environmental outflow (emission).
- Any stream that goes to another process is an intermediate outflow.

Within this second category, we further distinguish between goods and wastes:

- Any intermediate outflow that is sold to the next process is a good.
- Any intermediate outflow that is handed over to the next process along with compensation is a waste.

We can easily memorize this by stating the following:

- A product with a positive price is a good.
- A product with a negative price is a waste.

Typically, LCA follows the upstream of products. When diesel is produced from oil, we follow the oil upstream to the production site. But when the material is a waste, we do not follow upstream. If some waste stream from agriculture is used to produce a fuel, we do not include how the waste was produced. Its production is free of environmental burden. This, however, requires a clear distinction between goods and wastes. There are quite some cases where we do not know for sure if the price of an agricultural residue is positive or negative, especially when it remains within one company or farm. Further, due to technological developments, fluctuations in markets, and governmental intervention, goods may rapidly turn into a waste or the other way around.

4.3.6 Process Data

LCA relies on a lot of data. One needs data on the inflows and outflows of the processes that are included and moreover in a quantitative way. This includes, among others, emission data, energy use, and waste production. There are many cases where the data are not known. This can be due to various reasons:

- The process characteristics have never been measured.
- The data are there, but they are not available for reasons of confidentiality.
- The data are highly variable, as they depend on soil conditions, weather, agricultural practice, etc.
- The data are only available at a pilot plant level, not for the real-world situation.
- The data are known for some other situation, for example, for another country.

Doing LCA, by the way, is much easier than 10 years ago. There are now a number of online data repositories, in different continents: some for free and some for subscribers only. Some of these databases are quite extensive. The ecoinvent database, for example, contains thousands of processes, from electricity production to transport by truck and from palm oil production to pesticide production.

4.3.7 Allocation

Allocation refers to a computational trick to ensure that processes used to produce more than one product cause no computational problems. It can be explained as follows. Whenever a cow breeding farm produces milk and meat, it seems fair to allocate a part of the environmental burden to the milk and a part to the meat. But how exactly this should be done remains a topic of debate. Despite ISO's attempts to harmonize in these matters, the topic is not at all settled, and debates continue.

Allocation shows up in bio-based products in a temporal perspective as well. Crop rotation is a case in fact. Here, the farmer's long-term perspective determines what he is exactly doing, and it seems wise to include the full rotation cycle in the allocation decisions. Allocation matters in practice, and it is in fact an important source of variation in biofuel studies. Several national standards, such as the British PAS2050, have standardized the allocation. This will surely help to give more clarity, but from a scientific point of view, such *truth by legislation* is unsatisfactory (Wardenaar et al., 2012).

4.3.8 Impact Categories

LCA addresses a more or less complete set of environmental impacts. This is an important observation, because the key idea of LCA is to prevent misoptimization or biased comparisons. In fact, that is precisely the reason to take the life cycle into account. But with a limited set of impact categories, misoptimization and biased comparisons are still likely to occur.

Restricting the analysis of bio-based activities to global warming only (so-called carbon footprints) obviously misses the point. Standard life cycle impact (LCIA) methods are available, by names like ReCiPe, EDIP, TRACI, LIME, and CML-IA. These methods typically include a consistently selected set of 10 or more impact categories besides global warming such as acidification, eutrophication, toxicity, and resource depletion. While there is some agreement on some impact categories (for instance, how to address global warming), there is much less agreement on some other impact categories. Particularly notorious are the impacts related to land use, water use, biodiversity, and genetically modified organisms, while these are crucial in assessing the bio-based economy. Another open point is how energy itself is to be assessed. In a comparison of sources of energy, this is a critical question (Liao et al., 2011).

4.4 CONCLUSIONS

Applying LCA to bioenergy systems and other bio-based options is not easy. There are more open questions than the LCA of products like ceramic tiles, bicycles, or bridges. That is perhaps amazing: Bioenergy and bioproducts have been produced for centuries, and the most basic product of foods we use belongs to this class. Bioproducts are so elementary that it is difficult to separate economy from environment. One movement in the second-generation biofuels is the use of harvest residues as a feedstock. This can be a subject for LCA. But one of the issues for the use of the residues is the potential for depleting soil quality by removing the residues from the cultivated land after harvest. One way to include this is to develop a new impact indicator for the decline in net primary production (NPP) as a function of biomass removal. Another way is to model the future increased need for nutrient supplements to maintain the soil productivity at a sustainable level. Producing these extra nutrients then implies extra burdens for the environment. Between these two options, it is not clear which one is theoretically better, practically easier, or more acceptable to decision-makers. Similar considerations hold for the other topics mentioned in Section 4.3. LCA is not ready, there are undeniably problems, but there is a great promise, and the progress can be seen at many places.

REFERENCES

CMLCA. Scientific Software for LCA, IOA, EIOA, and more. http://www.cmlca.eu/. Accessed on 27 November, 2013.

Curran, M.A. (Ed.). 2012. *Life Cycle Assessment Handbook. A Guide for Environmentally Sustainable Products*. Beverly, CA: Scrivener Publishing.

Guinée, J.B., Heijungs, R., and van der Voet, E.. 2009. A greenhouse gas indicator for bioenergy: Some theoretical issues with practical implications. *International Journal of Life Cycle Assessment* 14: 328–339.

Heijungs, R. and Kleijn, R. 2001. Numerical approaches to life-cycle interpretation. Five examples. *International Journal of Life Cycle Assessment* 6: 141–148.

Heijungs, R. and Suh, S. 2002. *The Computational Structure of Life Cycle Assessment*. Dordrecht, the Netherlands: Kluwer Academic Publishers.

Heijungs, R. and Wiloso, E.I. 2012. LCA of second generation bioenergy and biomaterials: Progress and problems. In: Sudiyani, Y., Dwianto, W., Abimanyu, H., Haryono, A., and Lelono, A. (Eds.), *Proceedings of the Second Korea-Indonesia Workshop and International Symposium on Bioenergy from Biomass*. Puspiptek,Serpong, Indonesia: Research Center for Chemistry, Indonesian Institute of Sciences.

ISO. 2006. Environmental management—Life cycle assessment—Requirements and guidelines (ISO 14044). Geneva, Switzerland: International Organization for Standardization.

Liao, W., Heijungs, R., and Huppes, G. 2011. Is bioethanol a sustainable energy source? An energy-, exergy-, and emergy-based thermodynamic system analysis. *Renewable Energy* 36: 3479–3487.

Wardenaar, T., van Ruijven, T., Mendoza Beltran, A., Vad, K., Guinée, J., and Heijungs, R. 2012. Differences between LCA for analysis and LCA for policy: A case study on the consequences of allocation choices in bio-energy policies. *International Journal of Life Cycle Assessment* 17: 1059–1067.

Wegener Sleeswijk, A., Kleijn, R., Meeusen-van Onna, M.J.G., Leneman, H., Sengers, H.H.W.J.M., van Zeijts, H., and Reus, J.A.W.A. 1996. *Application of LCA to Agricultural Products; 1. Core Methodological Issues; 2. Supplement to the LCA Guide; 3. Methodological Background*. Leiden, the Netherlands: Centre of Environmental Science.

Wiloso, E.I. and Heijungs, R. 2013. Key issues in conducting life cycle assessment of bio-based renewable energy sources. In: Singh, A., Pant, D., and Olsen, S.I. (Eds.), *Life Cycle Assessment of Renewable Energy Sources*, Series: Green Energy and Technology. Springer-Verlag, 2013.

Wiloso, E.I., Heijungs, R., and de Snoo, G. 2012. LCA of second generation bioethanol: A review and some issues to be resolved for good LCA practice. *Renewable & Sustainable Energy Reviews* 16, 5295–5308.

5 Sustainability of Bioenergy Systems

Lucas Reijnders

CONTENTS

5.1 INTRODUCTION

Current bioenergy systems generate around 50 EJ year^{-1}, representing about 10% of the world's energy supply (Haberl et al. 2006; Sims et al. 2006; Beringer et al. 2011; Offerman et al. 2011). Bioenergy systems are understood here to encompass complete biofuel life cycles: from *seed* to emissions and the disposal of *wastes* originating in the production and consumption of biofuels. Bioenergy systems have been divided in *traditional* and *modern* categories (Goldemberg and Coelho 2004; Sims et al. 2006; Kuchler 2010; Gurung and Oh 2013). Although the definitions of traditional and modern bioenergy systems are not very precise, it would seem that liquid transport biofuels, such as bioethanol and biodiesel, and purified biogas originating in the anaerobic digestion of biomass and woodchips used in power plants are in the category of modern bioenergy systems. Modern bioenergy systems currently supply worldwide about 10 EJ year^{-1} (Sims et al. 2006; Beringer et al. 2011). It would seem that traditional bioenergy systems produce bioenergy products such as wood, charcoal, dung cake, and crop residues used for small-scale cooking and heating. Worldwide, traditional bioenergy systems currently supply about 40 EJ year^{-1}. Percentagewise, the share of traditional biomass in worldwide energy supply has been roughly constant since the 1970s (Grubler 2012). This chapter will deal with both traditional and modern bioenergy systems. Currently, biofuels are overwhelmingly terrestrial. However, there have also been proposals for algal biofuels. These will be dealt with in this chapter too.

Sustainability is a word that in practice may have many meanings (Babcicky 2013). In the context of bioenergy, there are mainly two different types of operational definitions of sustainability. The first type of definition refers to biofuels that are generated by practices that are environmentally and/or socially improved if compared with *business as usual* (e.g., Bucholz et al. 2009;

Palmujoki 2009; Bartle and Abadi 2010; Solomon 2010; Tay et al. 2011). In this category, for instance, biodiesel is made from palm oil produced by the plantations, which meet the standards of the Roundtable on Sustainable Palm Oil (Palmujoki 2009) and an "integrated biorefinery with maximum economic performance, while causing minimum environmental impact" (Tay et al. 2011). The second type of operational definition refers to bioenergy systems, which may continue indefinitely in line with the concept of a *steady-state economy* (Daly 1973). This type of definition aims at safeguarding for the future: natural resources, environmental quality, and ecosystem services. So-defined sustainability has been applied to forestry that is used for bioenergy and other purposes in Central Europe for centuries (Gamborg and Larsen 2003; Du Pisani 2009) and has also been proposed for the bioenergy systems that are not based on forestry (Springer 2012). In this chapter, the second type of operational definition will be used. Sustainable bioenergy systems are defined here as having no negative impacts on the availability and quality of natural resources, ecosystem services, and environmental quality for future generations (Goodland and Daly 1996; Reijnders 2006).

In view of the renewable character of bioenergy, sustainability as defined here implies that consumption of bioenergy feedstocks should not exceed the addition to the stocks thereof. This requirement is often not met in the case of *traditional* tree-derived bioenergy feedstocks (Chettri and Sharma 2009; Bailis et al. 2012; Ramos and de Albuquerque 2012). Also, for sustainability, there should be a good energetic return on energy investment (EROI) to maintain the quality of life when humanity runs out of readily available abundant fossil fuels (Hall et al. 2009). Hall et al. (2009) have suggested that this EROI should be >5. Furthermore, to be sustainable, bioenergy systems should be robust or resilient to variability in aspects of the physical environment that affect their functioning. Fourthly, sustainability, as defined here, implies conditions for bioenergy systems regarding their impact on other natural resources than biomass stock, on ecosystem services, and on pollution, which allow for *steady-state* bioenergy systems (cf. Daly 1973). In the following, the main aspects of sustainable bioenergy systems will be considered. Section 5.2 deals with the EROI and the robustness to variability and change in aspects of the physical environment. Section 5.3 will regard the sustainable use of natural resources needed for bioenergy systems. Section 5.4 will consider bioenergy systems and ecosystem services and Section 5.5 will deal with persistent pollution.

5.2 ENERGETIC RETURN ON ENERGY INVESTMENT AND ROBUSTNESS TO VARIABILITY AND CHANGE IN ASPECTS OF THE PHYSICAL ENVIRONMENT AFFECTING BIOENERGY SYSTEMS

The criterion that the EROI should be >5 can be met for forest-derived solid fuels such as firewood and woodchips used by power plants (Reijnders and Huijbregts 2009). In the case of current liquid biofuels, the EROI is often <5 (Reijnders and Huijbregts 2009; Clarens et al. 2011; Reijnders 2013). For instance, for ethanol from US corn, the EROI has been estimated to be on average 1.07, with a standard deviation of 0.2 (Murphy et al. 2011). For prospective liquid and gaseous algal biofuels, life cycle assessments currently also show an EROI<5 (Liu and Ma 2009; Luo et al. 2010; Chowdhury et al. 2012; Langlois et al. 2012; Reijnders 2012; Vasudevan et al. 2012).

The generation of feedstocks for bioenergy systems can be impacted by the aspects of the physical environment, which shows variability (Mu et al. 2011). Variations in precipitation resulting in drought or flooding may negatively impact feedstock availability (Mu et al. 2011). It has been argued that more extreme weather linked to current climate change may increase variability in the generation of feedstocks for bioenergy systems (Tirado et al. 2010; Stromberg et al. 2011). Also, climate change may be linked to negative effects on forest biomass by increasing the impact of pests. For instance, in British Columbia (Canada), forest stocks are negatively affected because the increases in temperature linked to climate change extend the range of the mountain pine beetle (Cullingham et al. 2011).

A slump in the local generation of feedstocks might lead to an increase in feedstock prices. It may be noted in this context that transport of harvested feedstocks over large distances to make up for local shortages or to counteract high local feedstock prices may be unattractive because feedstocks tend to have a relatively low energy content per unit of mass, and for this reason, long-distance transport is costly. High feedstock prices may negatively affect the viability of bioenergy systems. For instance, the combination of increased feedstock prices combined with falling oil prices has recently forced major producers of corn ethanol into bankruptcy in the United States (Mu et al. 2011).

It would seem impossible to increase robustness to the extent that bioenergy systems become invulnerable against any variability or change in aspects of the physical environment that affects the supply of bioenergy feedstocks. However, some adaptations of bioenergy systems may make them more robust against such variability. Temporary strong local reductions in feedstock harvest may be counteracted by multiannual feedstock storage. Robustness to variability in the availability of specific feedstocks may be increased by applying conversion technologies that are able to accommodate multiple feedstocks (Mu et al. 2011).

5.3 NATURAL RESOURCES

5.3.1 Natural Resources Used in Feedstock Production and Competition with Food Production

Bioenergy systems are dependent on natural resources. Two main categories of natural resources are often distinguished: resources that have been formed in slow geological processes (*nonrenewables*) and renewable resources. Nonrenewables are, for instance, phosphate ores, fossil water, and fossil fuels. Renewable resources are, for instance, surface water, mineral soils, and soil organic carbon. Apart from biomass stocks, several natural resources are important for and can be affected by bioenergy or biofuel systems. For microalgal biofuels grown in areas that are unsuitable for growing terrestrial biomass, the natural resources of water, nutrients, and fossil fuels tend to be important (Reijnders and Huijbregts 2009; Borowitzka and Moheimani 2013). Sea-based cultivation of macroalgae is associated with the exploitation of wild macroalgal stocks, providing *raw material* (Buschmann et al. 2001), whereas the conversion of macroalgae to biofuels is currently largely dependent on fossil fuel inputs (Langlois et al. 2012). For terrestrial biofuels, the important natural resources include soils (including aspects thereof affecting fertility such as soil organic matter), mineral nutrients, water, and fossil fuel stocks (Reijnders 2006; Gerbens-Leenes et al. 2009; Stone et al. 2010; Miao et al. 2011; Pfister et al. 2011).

Liao et al. (2012) have calculated that by 2008, modern bioenergy systems in the main producer countries required about 0.23% of the total anthropogenic natural resources supply. A major increase of modern biofuel production can lead to a much larger demand for natural resources. For instance, Pate et al. (2011) have estimated that meeting a goal of 10 billion barrels of algal oil per year in the United States might require about 44% of the present total US ammonia consumption and 20% of the present total US phosphorus (P) consumption. Supplying one billion tons of terrestrial agricultural biomass per year in the United States has been estimated to increase US synthetic fertilizer demand by a factor of 5.5 over the 1997 baseline (Han et al. 2011).

Crop-based bioenergy systems, either based on food crops or inedible feedstocks, and food production compete for the same scarce resources such as land, water, and nutrients (Reijnders and Huijbregts 2009; Finco and Doppler 2012; Stürmer et al. 2013). It has been argued that there is no such competition for scarce resources when crop residues are used (Champagne 2007; Archer et al. 2008; Damzartzis and Zabaniotou 2011; Goh and Lee 2011). However, crop residues are not wastes (contrast, e.g., Champagne 2007; Sahito et al. 2012). Crop residues contribute to soil nutrient and carbon stocks, which in turn serve as soil fertility or productivity (Lal 2005; Reijnders 2008b; Blanco-Canqui and Lal 2009). Recycling enough good-quality organic material and nutrients to soils to conserve fertility may strongly limit the availability of crop residues for sustainable bioenergy systems (Reijnders 2008b; Blanco-Canqui and Lal 2009).

The increase in demands on food production due to the growth of the world population and changes in eating habits may affect the sustainability of bioenergy systems (Harberl et al. 2011). A *food first approach* might shift bioenergy production to less productive land, which may negatively affect profitability and may also strongly limit future bioenergy supply (Delucchi 2010; Haberl et al. 2011). The natural resources of soil and soil organic matter, mineral nutrients, water, and fossil fuels and their sustainable use will be further discussed in the following.

5.3.2 Soil and Soil Organic Matter

Soil and soil organic matter (also called soil organic carbon) are important determinants of terrestrial biomass production. Soil organic matter serves the conservation of water and nutrients, improves soil structure, increases robustness against compaction by heavy machinery, and limits erosion (Blanco-Canqui and Lal 2009; Homma et al. 2012). The production and harvesting of feedstocks for bioenergy can give rise to a net loss of mineral soil by erosion and to the loss of soil organic matter (Blanco-Canqui and Lal 2009; VandenBygaart et al. 2012). Sustainability of the use of mineral soils would imply that the net loss of soil due to erosion does not exceed mineral soil formation due to processes such as natural weathering. Current practices for producing food crop-based feedstocks often do not meet this criterion (Riksen et al. 2003; Nordstrom and Hotta 2004; Reijnders 2006). Also, the stock of soil organic matter should not decrease in quantity or quality in sustainable bioenergy systems. Again, this condition is often not met in food crop-based biofuel production (Blanco-Canqui and Lal 2009; Bonin and Lal 2012). When perennial plants from mineral soils are used as bioenergy feedstocks, there is a scope for sustainability of soil organic matter stocks when considerable amounts of harvest residues are left (Bonin and Lal 2012).

It may also be noted that current climate change might, dependent on humidity and the nature and accessibility of soil organic matter, lead to increased soil respiration that negatively affects soil organic carbon stocks (Dungait et al. 2012; Selstedt et al. 2012). Soil organic carbon stocks might also be impacted by the sensitivity of photosynthesis to temperature, but the overall sign (positive or negative) and the size of this impact are uncertain and may show large regional variations (Booth et al. 2012; Smith 2012; Stavi and Lal 2012).

Especially soils under conventional tillage from which large amounts of harvest residues are withdrawn and soils of forests in the case of whole-tree harvesting are likely to be characterized by decreasing amounts of soil organic matter (Zinn et al. 2002; Lal 2006; Blanco-Canqui and Lal 2009; Eisenbies et al. 2009; Stavi and Lal 2012). The cultivation of bioenergy feedstocks on peaty soils is usually unsustainable because of the net loss of peat (mainly soil organic matter) linked to cultivation practices such as drainage (Danielsen et al. 2009).

The return of feedstock harvest residues to mineral soils may reduce soil loss due to erosion and conserve soil organic matter stocks. The latter is apparently linked to larger C inputs and the suppression of soil organic matter decomposition by returned residues (Kochsiek and Knops 2012). There are also other measures that can reduce erosion. These include planting windbreaks and terracing on slopes and buffer strips to catch runoff (Nordstrom and Hotta 2004; Reijnders 2006). In case of conservation tillage, there may be a scope for a limited withdrawal of harvest residues to serve as bioenergy feedstock without negatively affecting soil organic carbon stocks (Zinn et al. 2002; Blanco-Canqui and Lal 2007; Lafond et al. 2009; Lemke et al. 2010). It is likely that in sustainable bioenergy systems, there is very little scope for removal of harvest residues from soils under conventional tillage (Reijnders 2008b; Hamelin et al. 2012).

In principle, the scope for harvest residue removal might be increased when there is the return to arable soils of carbonaceous residues originating in the bioenergy life cycle beyond cultivation, for instance, from biofuel production plants. Apart from the digestate of biogas plants, such recycling is so far very limited (Reijnders and Huijbregts 2008; Eggleston 2010; Yuan et al. 2011;

Hermann 2012; Möller and Müller 2012; Nitayavardhana and Khanal 2012). A contributing factor to poor recycling is the presence in biofuel production residues of substances, which might negatively affect soil productivity (Nitayavardhana and Khanal 2001; Reijnders 2008b; Ernst et al. 2008; Menetrez 2010; Dong et al. 2011; Alburquerque et al. 2012).

The relative effectiveness as to the conservation of soil carbon stocks by harvest residues and manure on one hand and residues from biofuel production plants on the other hand is a matter to be considered. De la Fuente et al. (2013) noted that fresh anaerobically transformed wastes did not do well in conserving soil organic carbon stocks and recommended the use of composting. Schouten et al. (2012) found that the generation of bioenergy from manure by anaerobic digestion and pyrolysis of the solid residues might lead to declining C yields over the bioenergy chain.

5.3.3 Mineral Nutrients

There is a variety of minerals that are nutrients for microorganisms in bioenergy production. The production of bioenergy feedstocks is strongly dependent on the presence of adequate amounts of these minerals (Truax et al. 2012). Mineral nutrients include nitrogen (N), phosphorus (P), sulfur (S), calcium (Ca), potassium (K), and magnesium (Mg) (Miao et al. 2011; Wall 2012). Mineral nutrients are made available by natural processes. For sustainable bioenergy systems as defined here, in a first approximation, the losses of nutrients from bioenergy systems should not exceed the additions to nutrient stocks by natural processes. Mineral N may be supplied to feedstock generation by N-fixation, which utilizes N_2 present in the atmosphere, by deposition from the air and by the conversion of biomolecules containing N. Other minerals mentioned earlier may be added to soil by natural processes such as soil deepening, weathering, flooding, and deposition from air (Ranger and Turpault 1999; Reijnders 2006; Mead and Smith 2012). It is currently thought that in the long term, such natural processes allow for moderate biomass productivity (Miao et al. 2011). When biomass is harvested, minerals are taken away from biomass generation systems. The removal of minerals depends on harvesting practices and, for example, will be larger in case of whole-tree harvesting than in case of stem-only harvesting (Wall 2012). When the minerals are removed during the downstream processing of biomass feedstocks after harvesting and no recycling of minerals to the feedstock generation systems occurs, this may lead to nutrient losses from feedstock generation. There also may be nutrient losses from the field due to leaching and erosion (Tilman et al. 2002). Such losses tend to be much higher for terrestrial annual biofuel crops (Tilman et al. 2002) and lignocellulosic crops such as *Eucalyptus* (e.g., Hou et al. 2010) than those of the original (natural) ecosystems that they replaced. Replacement of arable land for annual crops by extensively used grassland and perennial grass-type crops may mitigate nutrient loss (Cherry et al. 2008). Nutrient loss from the generation of algae for bioenergy may occur due to the emission of N_2O and, more importantly, due to the discharge of wastewater (Clarens et al. 2011; Hou et al. 2011; Reijnders 2013). In the case of uncontrolled burning of biomass, large losses of N and S and substantial losses of K and P to air can occur (Heard et al. 2006).

To the extent that nutrient withdrawal and other losses exceed natural additions to nutrient stocks in feedstock generation systems, such as arable land, this raises the matter of sustainability of feedstock generation (Rytter 2002; Reijnders 2006). This matter can, in principle, be addressed by using the measures conducive to biological N-fixation and recycling minerals that have been withdrawn to reuse them as nutrients in bioenergy generation (Steenari and Lindquist 1997; Reijnders 2006; Wang et al. 2010; Mead and Smith 2012). In practice, such recycling of mineral nutrients is very limited (Mahmoudkhani et al. 2007). Moreover, recycling of post–bioenergy generation minerals might show deficiencies in minerals such as N and S. In practice, the recycling might give rise to unintended effects such as increased leaching of N and accumulation of Cd and plant growth inhibitors. This is the reason that several European countries have prohibited or restricted the recycling of post–bioenergy generation minerals (Reijnders and Huijbregts 2009; Wang et al. 2010; Bougnom et al. 2012; Mead and Smith 2012).

Current agricultural practices may result in nutrient deficits in bioenergy systems based on unfertilized forestry, the removal of nutrient-rich lignocellulosic materials, and low-input agriculture (Rytter 2002; Reijnders 2006; Eisenbies et al. 2009; Mead and Smith 2012; Wall 2012). Such deficits might in principle be compensated for by adding synthetic fertilizer to soils. High bioenergy biomass productivity currently tends to require large additions of synthetic fertilizer (Han et al. 2011; Miao et al. 2011). The use of synthetic fertilizer in bioenergy systems exploits natural resources that largely originate in slow geological processes. For some nutrients (e.g., Ca and Mg), natural resource stocks are very large (Reijnders 2006). For others, such as S and P, stocks are much more limited, which raises sustainability concerns (Reijnders 2006). The most critical natural nutrient resource is probably phosphate ore (Cordell et al. 2009). The stock thereof is relatively limited if compared with current withdrawals, and assuming business as usual, depletion thereof may well occur in this or the next century (Cordell et al. 2009; van Vuuren et al. 2010).

To aim at sustainability of bioenergy systems, efficient recycling of mineral nutrients that are present in bioenergy feedstocks, in a way which allows for bioavailability, would seem necessary. Also nutrient losses from the generation of feedstocks should be much reduced. For the purpose of the latter, the following interventions in terrestrial biofuel systems would seem useful: minimization of soil erosion (see Section 5.3.2), strip cutting instead of clear cutting in forestry, restriction of nutrient inputs, well-designed riparian plantings, and adaptation of cultivation practices to minimize losses linked to inefficient nutrient use by plants and to increase nutrient conservation in soils (Sharma et al. 2001; Tilman et al. 2002; Robertson et al. 2008; Cao et al. 2011; Cordell et al. 2011; Schröder et al. 2011; Mead and Smith 2012; Ra et al. 2012; Verbruggen et al. 2012). In the case of algal biofuels, recycling of treated wastewater may much reduce nutrient losses from biofuel life cycles. Such wastewater recycling is practiced in some types of horticulture (Newman 2004).

5.3.4 Water

Sustainability of good-quality freshwater consumption would imply that good-quality freshwater stocks are not reduced. Biofuel life cycle consumption of freshwater has been divided in three types, called blue, green, and grey water (Gerbens-Leenes et al. 2009; Yeh et al. 2011). Blue water is freshwater that is withdrawn from surface water or groundwater. Green water is rainwater available to plants. Grey water regards the volume of freshwater needed to dilute pollutant emissions to the extent that current water quality standards are met. As an overall indicator, a water footprint has been defined, which is the sum of blue, green, and grey water (Gerbens-Leenes et al. 2009; Mekkonen and Hoekstra 2011; Gleeson et al. 2012). It should be noted that this footprint needs additional interpretation to become an indicator of impact (Berger and Finkbeiner 2013) and estimates of grey water are uncertain as water quality standards differ and often assumptions are made to estimate pollutant emissions (Mekkonen and Hoekstra 2011; Chiu and Wu 2012). Some types of algal feedstocks may be served by saline water. However, microalgae, which have been suggested as sources for bioenergy, are often dependent on the supply of freshwater (Reijnders 2013). Per GJ of energy, the freshwater requirement of algal oil derived from freshwater algae is about two orders of magnitude larger than the freshwater requirement of crude fossil oil (Bonin and Lal 2012). Even if algae to be used as bioenergy feedstock are adapted to salt water, evaporative losses occurring in ponds on land may require inputs of freshwater to compensate such losses (Borowitzka end Moheimani 2013). In the case of microalgal oil derived from algae adapted to saline water, and when not including grey water, the water footprint may be similar to the water footprint of mineral oil (Vasudevan et al. 2012). Terrestrial bioenergy life cycles are associated with a large consumption of good-quality freshwater (Berndes 2002; Gerbens-Leenes et al. 2008, 2009, 2012; Dominguez-Faus et al. 2009; Mishra and Yeh 2011; Scown et al. 2011; Yeh et al. 2011). This water consumption regards both the production of current terrestrial bioenergy feedstocks and biomass-to-liquid-biofuel and biomass-to-biogas conversion technologies.

Water footprints GJ^{-1} for terrestrial biomass have been estimated to be in the range of 24–243 m^3, whereas estimated water footprints of primary fossil fuels (natural gas, coal, and crude oil) are in the range of 0.11–1.06 m^3 water GJ^{-1} and the estimated water footprint of hydropower is in the range of 1.55–22.3 m^3 water GJ^{-1} (Gerbens-Leenes et al. 2008; Herath et al. 2011; Bonin and Lal 2012). Water footprints for liquid biofuels (bioethanol, biodiesel) made from food crops exceed 10^3 L per liter biofuel and vary by more than a factor of 10^2 (Gerbens-Leenes et al. 2009; Mekkonen and Hoekstra 2011). Water footprints of terrestrial crop-based liquid biofuels tend to be at least one order of magnitude larger than the water footprint of fossil fuel-based competitors (Scown et al. 2011) and much larger than the water footprints of concentrating solar power systems (Burkhardt et al. 2011), photovoltaics (PVs) (Harto et al. 2010), and wind power (Li et al. 2012). The water footprints of ethanol from corn stover, wheat straw, and switch grass have been estimated to be of the same order of magnitude as the water footprints of grain- or sugar beet-based ethanol (Bonin and Lal 2012; Chiu and Wu 2012). Major expansions of terrestrial bioenergy systems are thus linked with large increases in freshwater footprint and consumption. For instance, Beringer et al. (2011) have estimated that an expansion of lignocellulosic bioenergy production to 270 EJ $year^{-1}$ might require doubling irrigation water demand.

Terrestrial bioenergy feedstocks may be grown under conditions characterized by high variability in weather and/or other factors impacting water availability, leading to a highly variable freshwater supply (Jewitt and Kunz 2011). And in semiarid areas, freshwater may well be the main limiting factor for the generation of bioenergy feedstocks (Ong and Leaky 1999). In such cases, there is a very good reason to consider the effects of bioenergy feedstock production on human food production and ecosystems. More in general, demands on freshwater should be viewed against the background of a decreasing trend of good-quality freshwater stocks in large areas of the world (Berndes 2002; Stone et al. 2010; Pfister et al. 2011; Gerbens-Leenes et al. 2012; Glazer and Likens 2012). Falling groundwater stocks and/or decreasing volumes of surface water during growing seasons are currently found at many locations (Stone et al. 2010; Glazer and Likens 2012; Hoekstra et al. 2012) and inter alia regard large aquifers that are critical to agriculture such as the aquifers of Western Mexico, the US High Plains, North Arabia, Iran, the Upper Ganges, and the North China Plain (Gleeson et al. 2012). Locally, the current decreasing trend in freshwater stocks already negatively affects the production of human food and ecosystems (Stone et al. 2010), and an additional pressure on freshwater stocks following from the cultivation of biofuel feedstocks may increase water stress on ecosystems and food production (Berndes 2002; Gerbens-Leenes et al. 2012). Constraints on biofuel production may mitigate freshwater stress that negatively impacts food production and ecosystems.

Current climate change tends to intensify the hydrological cycle by an estimated 8%°C^{-1} warming (Durack et al. 2012). The distribution of overall increased precipitation is however uneven. Land areas that are already wet tend to become wetter, and dry regions become dryer (Durack et al. 2012). Ongoing change of land cover may also impact moisture availability (Bagley et al. 2012). Net effects of climate change and land cover change on moisture and plant productivity may vary geographically and depend on bioenergy feedstock affected by changes in moisture availability (Cochrane and Barber 2009; Tirado et al. 2010; Bagley et al. 2012; Hember et al. 2012; Kirschbaum et al. 2012; Marlon et al. 2012). Bagley et al. (2012) estimate ongoing land cover change to have negative impacts of varying size on moisture and yield from growing currently important bioenergy feedstocks such as North American corn and South American soybeans. For southwestern British Columbia (Canada) and New Zealand, there is evidence for increased tree productivity under current climate change (Hember et al. 2012; Kirschbaum et al. 2012). For the western United States, there is evidence that warmer periods are associated with increased wildfire risk, which might negatively affect forest biomass stocks (Marlon et al. 2012). Similarly, ongoing climate change may increase forest fire risk in the Amazon (Cochrane and Barber 2009). Knowledge about the effects of climate change on freshwater demand by biofuel crops is as yet very limited. Marta et al. (2012) found that in the past 55 years, climate change has been associated with a decreasing overall water footprint for maize-based ethanol in Tuscany (Italy), presumably linked to a shorter growing cycle of maize, though blue water demand increased due to changes in precipitation and evaporation (Marta et al. 2012). On the other hand, for short-rotation lignocellulosic *Populus* plantations, an increase in water use linked to an increase of CO_2 concentration has been argued to be likely (Tricker et al. 2009).

The generation of grey water (water pollution) linked to bioenergy feedstock production depends on the type of feedstock production. Land use change from nature to feedstock production is likely to be linked to increased water pollution, whereas the replacement of annual feedstock crops by perennial feedstock crops might reduce water pollution (Costello et al. 2009; Diaz-Chavez et al. 2011; Love et al. 2011). It may be noted, though, that the reduction in water pollution by the replacement of annual crops with perennial feedstock crops may be relatively low when perennial feedstock production aims at maximizing biomass accumulation (Drewer et al. 2012; Schrotenboer et al. 2011). Feedstock production practices also matter. Judicious choice of pesticides, conservation tillage, precision agriculture, and low-intensity cropping might limit grey water production (Diaz-Chavez et al. 2011; Love et al. 2011). More efficient production processes and wastewater recycling and treatment may limit grey water production during converting biomass into liquid biofuels (Diaz-Chavez et al. 2011).

A reduced quantitative impact on water resources may be achieved by selecting biofuel feedstocks with relatively low water footprints (Wu and Liu 2012), by the development of crops that are more resistant to drought (Zhuang et al. 2010), and by changing production practices, for example, the substitution of full-field irrigation by drip irrigation and the substitution of once-through water use in biofuel production factories by water recycling (Diaz-Chavez et al. 2011).

5.3.5 Fossil Fuel Stocks

Current modern bioenergy systems are often heavily dependent on fossil fuels (Reijnders and Huijbregts 2009). In US corn ethanol production, the energetic return on fossil energy investment has, for instance, been estimated at about 1.07 (Murphy et al. 2011). Much of the fossil energy input in liquid biofuel life cycles is used for synthetic fertilizer production, processing biomass into biofuel and transport (Pimentel et al. 2002; Galloway et al. 2003). As fossil fuel resources are limited (Legget and Ball 2012), this practice cannot be continued indefinitely. The use of fossil methane for the Haber synthesis to generate N-fertilizers may be replaced by other hydrogen donors such as solar-powered hydrolysis of water and may be partly replaced by increased reliance on biogenic nitrogen fixation (Sharma et al. 2001; Service 2004). More efficient use of N-fertilizer may also contribute to more sustainable bioenergy systems (Tilman et al. 2002). Fossil fuels for transport and processing purposes may be replaced by power sources such as wind and solar radiation and by improved energy efficiency, which do not necessarily raise sustainability concerns regarding the consumption of natural resources (Turner 1999; Reijnders 2009; Leggett and Ball 2012), though they still might raise concerns regarding the impact on ecosystem services (Lovich and Ennen 2011).

5.4 ECOSYSTEM SERVICES

Bioenergy systems may affect ecosystem services. Ecosystem services are services to mankind provided by ecosystems, such as the provisioning and retention of freshwater, prevention or limitation of invasions by invasive species, pollination, soil formation and maintenance of soil fertility or productivity, regulation of climate, seed dispersal, improvement of water quality and air quality, and control of pests and erosion (Landis et al. 2008; Gardiner et al. 2010; Landis and Werling 2010; Zhang et al. 2010; Gasparatos et al. 2011; Polasky et al. 2011; Thompson et al. 2011). Biodiversity, reflecting diversity among and within species, is a main determinant of ecosystem services, though the contribution of species and varieties to ecosystem services may vary substantially (Thompson et al. 2011; Hooper et al. 2012; Pocock et al. 2012; Reich et al. 2012). This also holds for agroecosystems (Pocock et al. 2012).

In the biofuel life cycle, the generation of bioenergy feedstocks tends to be the most important determinant of the effect of bioenergy systems on ecosystem services. As demand for food and feed are highly inelastic (Searchinger et al. 2008), increasing production of agricultural biofuel feedstocks tends to add to the existing pressure on ecosystem services. This holds both for feedstocks based on food crops and lignocellulosic crops or inedible oils (Finco and Doppler 2012; Popp et al. 2012). This added pressure may come from the intensification of existing land use, the neglect of on-farm functional biodiversity, and land use change by converting areas under natural vegetation or set-aside land into arable land, which ultimately reduces land availability for living nature (Searchinger et al. 2008; Fargione et al. 2010; Beringer et al. 2011; Gasparatos et al. 2011; German et al. 2011; Gevers et al. 2011; Polasky et al. 2011; Erb et al. 2012; Gibbs 2012; Mehaffey et al. 2012; Tscharntke et al. 2012). Microalgal biofuels might negatively affect ecosystem services by land use change and wastewater emissions (containing, e.g., nutrients). Sea-based cultivation of macroalgae for bioenergy might negatively affect ecosystem services by impacting sedimentation processes (Bushmann et al. 1996) and sea grass beds (Eklöf et al. 2005), by overexploitation of wild stocks (Buschmann et al. 2001), by competition for nutrients (Wang et al. 2007), by allelopathy (Wang et al. 2007), by negative effects of pesticides used to safeguard macroalgae (Buschmann et al. 1996), and by the causation of algal blooms (Liu et al. 2009).

In the context of sustainability, the impact of bioenergy systems on ecosystem services for future generations has a center stage (Goodland and Daly 1996). Thus, the requirement of sustainability regarding ecosystem services leads to a focus on the prevention of poorly reversible or irreversible impacts on ecosystem services. Such impacts may follow from the loss of natural species and/or genetic diversity from forests or native grasslands (e.g., Danielsen et al. 2009; Edwards and Laurance 2012) and the introduction of bioinvasive species (e.g., DiTomasso et al. 2010).

The current predictability of the impacts of expanding bioenergy systems on ecosystem services is limited (Landis and Werling 2010). However, some impacts can be confidently predicted. Increased use of forestry residues as fuel is likely to negatively affect the species of animals, plants, and fungi, which depend on the presence of dead wood (Landis and Werling 2010; Lassauce et al. 2012; Toivanen et al. 2012). The sharing of characteristics between sought-after biofuel crops and invasive plant species may, in the absence of proper policies, increase biological invasions, which may negatively impact ecosystem services (Genovesi 2011; Barney et al. 2012; McCormick and Howard 2012). Species that are currently used (*Jatropha curcas*) or considered (e.g., *Arundo donax*, *Miscanthus*, *Panicum virgatum*, *Prosopis*) for bioenergy feedstock production have been found invasive in some places (Barney et al. 2012; McCormick and Howard 2012). Cultivation practices also matter. Monocultures and high inputs of pesticides and nutrients are likely to have relatively large negative impacts on ecosystem services (Gasparatos et al. 2011; Popp et al. 2011; Riffell et al. 2011; Mehaffey et al. 2012). It is likely that for bioenergy systems based on agricultural crops, ceteris paribus, the increase in land use will tend to be the most important determinant of negative impacts on (future) ecosystem services (Fargione et al. 2010; Frank et al. 2013). Fargione et al. (2010) have estimated that an additional production of 10^6 L year $^{-1}$ of food crop-based biofuel is associated with the change of land use on 65–365 ha. Frank et al. (2013) estimated that the European Union biofuel targets for 2020 might drive losses of 2.2 M ha of highly biodiverse areas. Beringer et al. (2011) have estimated that an expansion of lignocellulosic bioenergy systems to 270 EJ year^{-1} might require the conversion of up to 454 M ha of natural ecosystems and the doubling of current land use change. In this regard, it is important to note that land claims for the bioenergetic conversion of solar energy will be relatively large if compared with, for instance, PVs. This is illustrated in Table 5.1, which compares the net energy yields ha^{-1} of two relatively high-yielding bioenergy crops (sugar cane and oil palm) and PV modules in the tropics. As appears from Table 5.1, in converting solar energy to energy that is usable by mankind, Si-based PV systems appear to be much more efficient than bioenergy systems (Reijnders and Huijbregts 2009).

PV systems can be integrated in the built environment (e.g., roof-integrated PV modules), which does not lead to land use change. Grossmann et al. (2010) have calculated that to meet the present

TABLE 5.1
Net Energy Yields of Two Biofuels from Highly Productive Bioenergy Crops and Multicrystalline Si PV Modules in the Tropics

Method for the Conversion of Solar Radiation	Product	Net Energy Yield in GJ ha^{-1} $Year^{-1}$
Sugarcane (Brazil)	Ethanol	100–175
Oil palm (S.E. Asia)	Palm oil	142–180
Commercial PV modules (multicrystalline Si) (Brazil)	Electricity	7600–8700

Sources: Reijnders, L., *Trends Biotechnol.*, 26, 349, 2008a; Triana, C.A.R., *Energy Policy*, 39, 4605, 2011.

yearly demand for energy (about 500 EJ $year^{-1}$), 0.23%–0.51% of the global land that are covered with PV systems would be sufficient. Alternatively, 0.7%–1.5% of the global desert area covered by PV systems could generate about 500 EJ $year^{-1}$ (Grossmann et al. 2010).

For a specified yield of bioenergy feedstock, land use will increase when energy yields ha^{-1} are lower. This implies that relatively large negative impacts may be expected for the cultivation of bioenergy crops with a relatively low energy yield ha^{-1} (Havlik et al. 2011; Stoms et al. 2011), which would seem likely when the advice is taken to use degraded and abandoned lands for biofuel production (Fargione et al. 2010). Moreover, as much of degraded and abandoned land is currently used for subsistence of rural populations, including low-intensity food production, (Berndes et al. 2003) a land use change favoring bioenergy feedstock production might shift subsistence-related activities to areas that are currently in the category nature. It would also seem likely that the requirements regarding the sustainable use of natural resources such as nutrients and water, as defined here, would lead to relatively low yields (much lower than the yields of sugarcane and oil palm indicated in Table 5.1) (Reijnders 2006; Miao et al. 2011). All this would mean that the relative impact of palm oil and sugarcane ethanol on ecosystem services, if compared with PV modules, might well be larger than suggested by the data in Table 5.1. All in all, it would seem unlikely that a major expansion of biofuel production would be achievable without a detrimental effect on future ecosystem services.

5.5 FUTURE ENVIRONMENTAL QUALITY AS AFFECTED BY POLLUTANTS

Bioenergy systems give rise to a variety of emissions, which may contribute to climate change, deterioration of the ozone layer, acidification, eutrophication, human toxicity, and ecotoxicity (Lardon et al. 2009; Reijnders and Huijbregts 2009; Clarens et al. 2011; Uchida and Hayashi 2012). In the context of sustainability, the impact of biofuels on future environmental quality has center stage (Goodland and Daly 1996). This leads to a focus on pollutants that have long-lasting effects. These include the substances giving rise to acidification (e.g., Tomlinson 2003) and eutrophication (Babel et al. 2011; Bernal et al. 2012); persistent toxics; genotoxic compounds affecting germline DNA; the greenhouse gases CO_2, N_2O, and CH_4 (Nolt 2011); and persistent halocarbons, which are greenhouse gases and negatively affect the ozone layer (Velders et al. 2009).

It would seem that for food crop–derived and microalgal transport biofuels, the life cycle emissions of compounds giving rise to acidification and eutrophication tend to be higher than for fossil fuels such as diesel and gasoline (Reijnders and Huijbregts 2009; Delucchi 2010; Clarens et al. 2011). The emission of persistent halocarbons, which negatively affect the ozone layer by bioenergy systems, would seem to be very small (e.g., Uchida and Hayashi 2012). The emission of persistent toxics linked to bioenergy systems varies strongly, dependent on the type of biofuel (Reijnders and Huijbregts 2009; Alburquerque et al. 2012). The impact of biofuels on greenhouse gas emissions should consider the use of fossil fuels as inputs in bioenergy life cycles, the emission of N_2O and CH_4, and the emission of carbonaceous gases linked to changes in agroecosystem carbon stocks (Reijnders and Huijbregts 2009; Chen et al. 2011; Odlare et al. 2012; Popp et al. 2012). Reductions of the latter may be irreversible or subject to slow reversion. For bioenergy systems linked to sustainable forestry and in the case of Brazilian sugarcane ethanol life cycle, greenhouse gas emissions may well be smaller than those linked to their competitors based on fossil fuels, but for many terrestrial food crop-derived liquid biofuels, the opposite holds (Gibbs et al. 2008; Searchinger et al. 2008; Reijnders and Huijbregts 2009; Delucchi 2010; Achten and Verschot 2011; Repo et al. 2012).

A variety of approaches may reduce the future pollutant burden of bioenergy systems. These include reduced inputs of fossil fuels (see also Section 5.2.4), improved nutrient and energy efficiency, improved pest management, conservation tillage, and a switch of annual to perennial crops (Reijnders and Huijbregts 2009, 2011; Robertson et al. 2011).

5.6 CONCLUSION

Current bioenergy systems are often unsustainable. Much can be done to improve the sustainability, as defined here, of bioenergy systems. However, at levels of demand for food and feed expected in the next decades, a major expansion of bioenergy production would, for instance, seem ultimately incompatible with the conservation of present ecosystem services and water resources. Regarding the sustainability of water resources and ecosystem services, for example, Si-based solar power systems integrated in the built environment would seem to be superior to bioenergy systems.

REFERENCES

Achten, W.M.J. and L.V. Verschot. 2011. Implications of biodiesel-induced land use changes for CO_2 emissions: Case studies in tropical America, Africa and Southeast Asia. *Ecology and Society* 16(4): 14.

Alburquerque, J.A., C. de la Fuente, A. Ferrer-Costa et al. 2012. Assessment of fertilizer potential of digestates from farm and agroindustrial residues. *Biomass and Bioenergy* 40: 181–189.

Archer, A., J. Self, G. Guha et al. 2008. Cost and carbon savings from innovative conversion of agricultural residues. *Energy Sources Part B* 3: 103–108.

Babcicky, P. 2013. Rethinking the foundations of sustainability measurement: The limitations of the environmental sustainability. *Social Indicators Research* 113: 133–157.

Babel, M.S., B. Shrestha, and S.R. Perret. 2011. Hydrological impact of biofuels production: A case study of the Khlong Phlo watershed in Thailand. *Agricultural Water Management* 101: 8–20.

Bagley, J.E., A.R. Desai, P.A. Dirmeyer et al. 2012. Effects of land cover change on moisture availability and potential crop yield in the world's breadbaskets. *Environmental Research Letters* 7: 014009 (9pp.).

Bailis, R., J.L. Chattelier, and A. Ghilardi. 2012. Ecological sustainability of woodfuel as an energy source in rural communities. In: De Clerck F., J.C. Ingram, C. Rumbaitis del Rio (eds.), *Integrating Ecology and Poverty Reduction*. Springer, London, U.K., pp. 299–325. DOI: 10.1007/978-1-4419-0633-5_18.

Barney, J.N., J.M. Mann, G.B. Kyser et al. 2012. Assessing habitat susceptibility and resistance to invasion by the bioenergy crops switchgrass and *Miscanthus × giganteus* in California. *Biomass & Bioenergy* 40: 143–154.

Bartle, J.R. and A. Abadi. 2010. Toward sustainable production of second generation bioenergy feedstocks. *Energy & Fuels* 24: 2–9.

Berger, M. and M. Finkbeiner. 2013. Methodological challenges in volumetric and impact-oriented water footprints. *Journal of Industrial Ecology* 17: 79–89. DOI: 10.1111/j.1530-0290.2012.00495.x.

Beringer, T., W. Lucht, and S. Schaphoff. 2011. Bioenergy production potential of global biomass plantations under environmental and agricultural constraints. *Global Change Biology: Bioenergy* 3: 299–312.

Bernal, S., L.O. Hedin, G.E. Likens et al. 2012. Complex response of the forest nitrogen cycle to climate change. *Proceedings of the National Academy of Sciences USA* 109: 3406–3411.

Berndes, G. 2002. Bioenergy and water: The implications of large scale bioenergy production for water use and supply. *Global Environmental Change* 12: 253–271.

Berndes, G., Hoogwijk, M., and van den Broek, R. 2003 The contribution of biomass in the future global energy supply: A review of 17 studies. *Biomass and Bioenergy* 25: 1–28.

Blanco-Canqui, H. and R. Lal. 2007. Soil and crop response to harvesting corn residues for biofuel production. *Geoderma* 141: 355–362.

Blanco-Canqui, H. and R. Lal. 2009. Crop residue removal impacts in soil productivity and environmental quality. *Critical Reviews in Plant Science* 28: 139–163.

Bonin, C. and R. Lal. 2012. Bioethanol potentials and life-cycle assessments of biofuel feedstocks. *Critical Reviews in Plant Sciences* 31: 271–289.

Booth, B.B., C.D. Jones, M. Colins et al. 2012. High sensitivity of future global warming to land carbon cycle processes. *Environmental Research Letters* 7: 024002 (8pp.).

Bougnom, B.P., C. Niederkofler, B.A. Knapp et al. 2012. Residues from renewable energy production; their value for fertilizing pastures. *Biomass and Bioenergy* 40: 290–295.

Buchholz, T., V.A. Luzadis, and T.A. Volk. 2009. Sustainability criteria for bioenergy systems: Results from expert survey. *Energy Policy* 37: S86–S98.

Burkhardt, J.J., G.A. Heath, and C.S. Turchi. 2011. Life cycle assessment of a parabolic trough concentrating solar power plant and the impacts of key design alternatives. *Environmental Science & Technology* 45: 2457–2464.

Buschmann, A.H., J.A. Correa, R. Westermeier et al. 2001. Red algal farming in Chile: A review. *Aquaculture* 194: 203–220.

Buschmann, A.H., D.A. Lopez, and A. Medina. 1996. A review of the environmental effects and alternative production strategies of marine aquaculture in Chile. *Aquacultural Engineering* 6: 397–421.

Cao, Y., T. Lehto, T. Repo, R. Silvennoinen et al. 2011. Effects of planting orientation and density of willows on biomass production and nutrient leaching. *New Forests* 41: 361–377.

Champagne, P. 2007. Feasibility of producing bioethanol from waste residues in Canada. *Resources Conservation and Recycling* 50: 211–230.

Chen, R., M. Senbayram, X. Lin et al. 2011. Origin of positive delta ^{13}C of emitted CO_2 from soils after application of biogas residues. *Soil Biology and Biochemistry* 43: 2194–2199.

Cherry, K.A., M. Shepherd, P.J.A. Withers et al. 2008. Assessing the effectiveness of actions to mitigate nutrient loss from agriculture. *Science of the Total Environment* 406: 1–23.

Chettri, N. and E. Sharma. 2009. A scientific assessment of traditional knowledge on firewood and fodder values in Sikkim, India. *Forest Ecology and Management* 257: 2073–2078.

Chiu, Y. and M. Wu. 2012. Assessing county-level water footprints of different cellulosic-biofuel feedstock pathways. *Environmental Science & Technology* 46: 9155–9162.

Chowdhury, R., S. Viamajala, and R. Gerlach. 2012. Reduction of environmental and energy footprint of microalgal biodiesel production through material and energy integration. *Bioresource Technology* 108: 102–111.

Clarens, A.F., H. Nassau, E.P. Resurreccion et al. 2011. Environmental impacts of algae derived biodiesel and bioelectricity for transportation. *Environmental Science & Technology* 45: 7554–7560.

Cochrane, M.A. and C.F. Barber. 2009. Climate change, human land use and future fires in the Amazon. *Global Change Biology* 15: 601–612.

Cordell, D., J. Drangert, and S. White. 2009. The story of phosphorus: Global food security and food for thought. *Global Environmental Change* 19: 292–305.

Cordell, D., A. Rosemerin, J.J. Schröder et al. 2011. Towards global phosphorus security: A systems framework for phosphorus recovery and reuse options. *Chemosphere* 84: 747–758.

Costello, C., M.W. Griffin, A.E. Landis et al. 2009. Impact of biofuel crop production on the formation of hypoxia in the Gulf of Mexico. *Environmental Science & Technology* 43: 7985–7991.

Cullingham, C.I., J.E.K. Cooke, S. Dang et al. 2011. Mountain pine beetle host-range expansion threatens the boreal forest. *Molecular Ecology* 20: 2157–2171.

Daly, H. 1973. *Toward a Steady State Economy*. San Francisco, CA: Freeman.

Damzartis, T. and A. Zabaniotou. 2011. Thermochemical conversion of biomass to second generation biofuels through integrated process design: A review. *Renewable and Sustainable Energy Reviews* 25: 366–378.

Danielsen, F., H. Beukema, N.D. Burgess et al. 2009. Biofuel plantations on forested lands. Double Jeopardy for biodiversity and climate. *Conservation Biology* 23: 348–358.

De la Fuente, C., J.A. Alburquerque, R. Clemente et al. 2013. Soil C and N mineralisation and agricultural value of the products of an anaerobic digestion system. *Biology and Fertility of Soils* 49: 313–322.

Delucchi, M.A. 2010. Impacts of biofuels on climate change, water use and land use. *Annals of the New York Academy of Sciences* 1195: 28–45.

Diaz-Chavez, R., G. Berndes, D. Neary et al. 2011. Water quality assessment of bioenergy production. *Biofuels, Biproducts and Biorefining* 5: 445–463.

DiTomaso, J.M., J.K. Reaser, C.P. Dionigi et al. 2010. Biofuel vs. bioinvasion: Seeding policy priorities. *Environmental Science & Technology* 44: 6906–6910.

Dominguez-Faus, R., S.P. Powers, J.G. Burken et al. 2009. The water footprint of biofuels: A drink or drive issue? *Environmental Science & Technology* 43: 3005–3010.

Dong, X., M. Dong, Y. Lu et al. 2011. Antimicrobials and antioxidant activities of lignin from residue of corn stover to ethanol production. *Industrial Crops and Products* 34: 1629–1634.

Drewer, J, J.W. Finch, C.R. Lloyd et al. 2012. How do soil emissions of N_2O, CH_4 and CO_2 form perennial bioenergy crops differ from arable annual crops. *Global Change Biology: Bioenergy* 4: 408–419.

Du Pisani, J.A. 2009. Sustainable development—Historical roots of the concept. *Environmental Sciences* 3(2): 83–96.

Dungait, J.A.J., D.W. Hopkins, A.S. Gregory et al. 2012. Soil organic matter turnover is governed by accessibility not recalcitrance. *Global Change Biology* 18: 1781–1796.

Durack, P.J., S.E. Wijffels, and R.J. Matear. 2012. Ocean salinities reveal strong global water cycle intensification during 1950–2000. *Science* 336: 455–458.

Edwards, D.P. and S.G. Laurance. 2012. Green labeling, sustainability and the expansion of tropical agriculture: Critical issues for certification schemes. *Biological Conservation* 151: 60–64.

Eggleston G. 2010. Future sustainability of the sugar and sugar-ethanol industries. In: Eggleston G. (ed.), *Sustainability of the Sugar and Sugar-Ethanol Industries*. ACS Symposium Series. Washington, DC: American Chemical Society, pp. 1–19.

Eklöf, J.S., M. de la Torre Castro, L. Adelsköld et al. 2005. Differences in macrofaunal and seagrass assemblages in seagrass beds with and without seaweed farms. *Estuarine, Coastal and Shelf Science* 63: 385–396.

Elsenbies, M.H., E.D. Vance, W.M. Aust et al. 2009. Intensive utilization of harvest residues in Southern Pine plantations: Quantities available and implications for nutrient budgets and sustainable site productivity. *Bioenergy Research* 2: 90–98.

Erb, K., H. Harberl, and C. Plutzar. 2012. Dependency of global primary crop potentials in 2050 on food systems, yields, biodiversity conservation and political stability. *Energy Policy* 47: 260–269.

Ernst, G., A. Müller, H. Göhler et al. 2008. C and N turnover of fermented residues from biogas plants in soils in the presence of three different earthworm species (*Lumbricus terrestris, Aporrectodea longa, Aporrectodea caliginosa*). *Soil Biology and Biochemistry* 40: 1413–1420.

Fargione, J.E., R.J. Plevin, and J.D. Hill. 2010. The ecological impact of biofuels. *Annual Review of Ecology, Evolution, and Systematics* 41: 351–377.

Finco, M.V.A. and W. Doppler. 2012. Bioenergy and sustainable development: The dilemma of food security and climate change in the Brazilian savannah. *Energy for Sustainable Development* 14: 194–199.

Frank, S., H. Böttcher, P. Havlik et al. 2013. How effective are the sustainability criteria accompanying the European Union 2020 biofuel targets? *Global Change Biology: Bioenergy* 5: 306–314.

Galloway, J.N., J.D. Aber, J.W. Erisman et al. 2003. The nitrogen cascade. *BioScience* 53: 341–356.

Gamborg, C. and J.B. Larsen, 2003. Back to nature—A sustainable future for forestry? *Forest Ecology and Management* 179: 559–571.

Gardiner, M.A., J.K. Tuell, R. Isaacs et al. 2010. Implications of three biofuel crops for beneficial arthropods in agricultural landscapes. *Bioenergy Research* 3: 6–19.

Gasparatos, A., P. Stromberg, and K. Takeuchi. 2011. Biofuels, ecosystem services and human wellbeing: Putting biofuels in the ecosystem narrative. *Agriculture, Ecosystems and Environment* 142: 111–128.

Genovesi, P. 2011. European biofuel policies may increase biological invasions. *Current Opinion in Environmental Sustainability* 3: 66–70.

Gerbens-Leenes, W., A.Y. Hoekstra, and T.H. van der Meer. 2008. *Water Footprint of Bio-Energy and Other Primary Energy Carriers*. Delft, the Netherlands: UNESCO-IHE, pp. 1–44.

Gerbens-Leenes, W., A.Y. Hoekstra, and T.H. van der Meer. 2009. The water footprint of bioenergy. *Proceedings of the National Academy of Sciences USA* 196: 10219–10223.

Gerbens-Leenes, W., A.R. van Lienden, A.Y. Hoekstra et al. 2012. Biofuel scenarios in a water perspective: The global blue and green water footprint of road transport in 2030. *Global Environmental Change* 22: 764–776.

German, L., G.C. Schoneveld, and P. Pacheca. 2011. Local social and environmental impacts of biofuels; global comparative assessment and implications for governance. *Ecology and Society* 16(4): 29.

Gevers, J., T.K. Hoye, C.J. Topping et al. 2011. Biodiversity and the mitigation of climate change through bioenergy: Impacts of increased maize cultivation on farmland wildlife. *Global Change Biology: Bioenergy* 3: 472–482.

Gibbs, H. 2012. Trading forests for yields in the Peruvian Amazon. *Environmental Research Letters* 7: 011007 (2pp.).

Gibbs, H.K., M. Johnston, J.A. Foley et al. 2008. Carbon pay-back times for crop based biofuel expansion in the tropics: The effects of changing yield and technologies. *Environmental Research Letters* 3: 034001.

Glazer, A.M. and G.E. Likens. 2012. The water table: The shifting foundations of life on land. *AMBIO* 41: 657–669.

Gleeson, T., Y. Wada, M.F.P. Bierkens et al. 2012. Water balance of global aquifers revealed by groundwater footprint. *Nature* 488: 19–200.

Goh C.S. and K.T. Lee. 2011. Second-generation biofuel (SGB) in Southeast Asia via lignocellulosic biorefinery: Penny-foolish but pound-wise. *Renewable and Sustainable Energy Reviews* 25: 2714–2718.

Goldemberg, J. and S.T. Coelho. 2004. Renewable energy—Traditional biomass vs. modern biomass. *Energy Policy* 32: 711–714.

Goodland, R. and H. Daly. 1996. Environmental sustainability: Universal and non-negotiable. *Ecological Applications* 6: 1002–1017.

Grossmann, W.D., I. Grossmann, and K. Steininger. 2010. Indicators to determine winning renewable energy technologies with application to photovoltaics. *Environmental Science & Technology* 44: 4849–4855.

Grubler, A. 2012. Energy transitions research: Insights and cautionary tales. *Energy Policy* 50: 8–16.

Gurung, A. and S.E. Oh. 2013. Conversion of traditional biomass into modern bioenergy systems: A review in context to improve the energy situation in Nepal. *Renewable Energy* 50: 206–213.

Haberl, H, K. Erb, F. Krausman et al. 2011. Global bioenergy potentials from agricultural land in 2050: Sensitivity to climate change, diets and yields. *Biomass and Bioenergy* 35: 4753–4769.

Haberl, H, H. Weisz, C. Amann et al. 2006. The energetic metabolism of the European Union and the United States. *Journal of Industrial Ecology* 10(4): 151–171.

Hall, C., S. Balogh, and D. Murphy. 2009. What is the minimum EROI that a sustainable society must have. *Energies* 2: 25–47.

Hamelin, L., U. Jorgensen, B.M. Petersen et al. 2012. Modelling the carbon and nitrogen balances of direct land use changes from energy crops in Denmark: A consequential life cycle inventory. *Global Change Biology: Bioenergy* 41: 889–907.

Han, F.X, R.L. King, J.S. Lindner et al. 2011. Nutrient fertilizer requirements for sustainable biomass supply to meet U.S. bioenergy goal. *Biomass and Bioenergy* 35: 253–262.

Harto, C., R. Meyers, and E. Williams. 2010. Life cycle water use of low carbon transport fuels. *Energy Policy* 38: 4933–4944.

Havlik, P., U.A. Schneider, E. Schmid et al. 2011. Global land-use implications of first and second generation biofuel targets. *Energy Policy* 39: 5690–5702.

Heard, J., C. Cavers, and G. Adrian. 2006. Up in smoke—Nutrient loss with straw burning. *Better Crops* 90(3): 10–11.

Hember, R.A., W.A. Kurz, J.H. Metsaranta et al. 2012. Accelerating regrowth of temperate-maritime forests due to environmental change. *Global Change Biology* 18: 2026–2040.

Herath, I, M. Deurer, D. Horne et al. 2011. The water footprint of hydroelectricity: A methodological comparison from a case study in New Zealand. *Journal of Cleaner Production* 19: 1582–1589.

Hermann, A. 2012. Biogas production from maize: Current state, challenges and prospects. 2. Agronomic and environmental aspects. *Bioenergy Research* 5: 1027–1042.

Hoekstra, A.Y., M.M. Mekonnen, A.K. Chapagain et al. 2012. Global monthly water scarcity; blue water footprints versus blue water availability. *PloS One* 7(2): e32688 (9pp.).

Homma, S.K., H. Tokeshi, L.W. Mendes et al. 2012. Long-term application of biomass and reduced use of chemical alleviate soil compaction and improve soil quality. *Soil and Tillage Research* 120: 147–153.

Hooper, D.U, E.C. Adar, B.J. Cardinale et al. 2012. A global synthesis reveals biodiversity loss as a major driver of ecosystem change. *Nature* 486: 105–108.

Hou, J., P. Zhang, X. Yuan et al. 2011. Life cycle assessment of biodiesel from soybean, Jatropha and microalgae in China conditions. *Renewable and Sustainable Energy Reviews* 15: 5081–5091.

Hou, X., C. Duan, C.T. Tang et al. 2010. Nutrient relocation, hydrological functions, and soil chemistry in plantations as compared to natural forests in central Yunnan, China. *Ecological Research* 25: 139–148.

Jewitt, G. and R. Kunz. 2011. The impact of biofuel feedstock production on water resources: A developing country perspective. *Biofuels, Bioproducts and Biorefining* 5: 387–398.

Kirschbaum, M.U.F., M. Watt, A. Tait et al. 2012. Future wood productivity of *Pinus radiata* in New Zealand under expected climatic changes. *Global Change Biology* 18: 1342–1356.

Kochsiek, A.E. and J.M.H. Knops. 2012. Maize cellulosic biofuels: Carbon loss can be a hidden cost of residue removal. *Global Change Biology: Bioenergy* 4: 229–233.

Kuchler, M. 2010. Unraveling the argument for bioenergy production in developing countries: A world economy perspective. *Ecological Economics* 69: 1336–1343.

Lafond, G.P., M. Strumborg, R. Lemke et al. 2009. Quantifying straw removal through baling and measuring the long-term impact on soil quality and wheat production. *Agronomy Journal* 101: 529–537.

Lal, R. 2005. World crop residues production and implications of its use as a biofuel. *Environment International* 31: 575–584.

Lal, R. 2006. Enhancing crop yields in the developing counties through restoration of the soil organic carbon pool in agricultural lands. *Land Degradation and Development* 17: 197–209.

Landis, D.A., M.M. Gardiner, W. van der Werf et al. 2008. Increasing corn for biofuel production reduces biocontrol services in agricultural landscapes. *Proceedings of the National Academy of Sciences of the USA* 105: 20552–29557.

Landis, D.A. and B.P. Werling. 2010. Arthropods and biofuel production systems in North America. *Insect Science* 17: 220–236.

Langlois, J., J. Sassi, G. Jard et al. 2012. Life cycle assessment of biomethane from offshore cultivated seaweed. *Biofuels, Bioproducts and Biorefining* 6: 387–504.

Lardon, L., A. Helia, B. Sialve et al. 2009. Life cycle assessment of biodiesel production from microalgae. *Environmental Science & Technology* 43: 6475–6481.

Lassauce, A., F. Leutier, and C. Bouget. 2012. Woodfuel harvesting and biodiversity conservation in temperate forests: Effects of logging residue characteristics on saprophytic beetle assemblages. *Biological Conservation* 147: 204–212.

Leggett, L.M.W. and D.A. Ball. 2012. The implication for climate change and peak fossil fuel of the continuation of the current trend in wind and solar energy production. *Energy Policy* 41: 610–617.

Lemke, R.L., A.J. VandenBygaart, C.A. Campbell et al. 2010. Crop residue removal and fertilizer N: Effects on soil organic carbon in a long-term crop rotation on Udic Boroll. *Agriculture, Ecosystems and Environment* 135: 42–51.

Li, X., K. Feng, Y.L. Siu et al. 2012. Energy-water nexus of wind power in China: The balancing act between CO_2 emissions and water consumption. *Energy Policy* 45: 440–446.

Liao, W., R. Heijungs, and G. Huppes. 2012. Natural resource demand of global biofuels in the Anthropocene. *Renewable and Sustainable Energy Reviews*10: 996–1003.

Liu, D., J.K. Keesing, Q. Xing et al. 2009. World's largest macroalgal bloom caused by expansion of seaweed aquaculture in China. *Marine Pollution Bulletin* 58: 888–895.

Liu, J. and X. Ma. 2009. The analysis on energy and environmental impacts of microalgae-based fuel methanol. *Energy Policy* 37: 1479–1488.

Love, B.J., M.D. Einheuser, and A.P. Nejadhashemi. 2011. Effects on aquatic and human health due to large scale bioenergy crop expansion. *Science of the Total Environment* 409: 3215–3229.

Lovich, J.E. and J.R. Ennen. 2011. Wildlife conservation and solar energy development in the desert Southwest, United States. *BioScience* 61: 982–992.

Luo, S., Z. Hu, D.G. Choi et al. 2010. Life cycle energy and greenhouse gas emissions for an ethanol production process based on blue green algae. *Environmental Science & Technology* 44: 8670–8677.

Mahmoudkhani, M., R. Richards, and H. Theliander. 2007. Sustainable use of biofuel by recycling ash to forests: Treatment of biofuel ash. *Environmental Science & Technology* 41: 4118–4128.

Marlon, J.R., P.J. Barlein, D.G. Gavin et al. 2012. Long-term perspective on wildfires in the western USA. *Proceedings of the National Academy of Sciences USA* 109: E535–E543.

Marta, A.D., M. Mancini, F. Natali et al. 2012. From water to bioethanol: The impact of climate variability on the water footprint. *Journal of Hydrology* 444–445: 180–186.

Mead, D. and C. Smith. 2012. Principles of nutrient management for sustainable forest bioenergy production. *WIRE's Energy and Environment* 1: 152–164.

Mehaffey, M., E. Smith, and R. van Remortel. 2012. Midwest U.S. landscape change to 2020 driven by biofuel mandates. *Ecological Applications* 22: 8–19.

Mekkonen, M. and A. Hoekstra. 2011. The green, blue and green water footprint of crops and derived crop products. *Hydrology and Earth System Sciences* 15: 2261–2273.

Menetrez, M.Y. 2010. The potential impact of waste from cellulosic ethanol production. *Journal of the Air and Waste Management Association* 60: 245–250.

Miao, Y., B.A. Stewart, and F. Zhang. 2011. Long-term experiments for sustainable nutrient management. *Agronomy for Sustainable Development* 31: 397–414.

Mishra, G.S. and S. Yeh. 2011. Life cycle water consumption and withdrawal requirements of ethanol from corn grain residues. *Environmental Science & Technology* 45: 4563–4569.

Möller, A. and T. Müller.2012. Effects of anaerobic digestion on digestate nutrient availability and crop growth: A review. *Engineering in Life Sciences* 12: 242–257.

Mu, D., T.P. Seager, P.S.C. Rao, J. Park, and F. Zhao. 2011. A resilience perspective on biofuel production. *Integrated Environmental Assessment and Management* 7: 348–359.

Murphy, D.J., C.A.S. Hall, and B. Powers. 2011. New perspectives on the energy return on (energy) investment (EROI) of corn ethanol. *Environment Development and Sustainability* 13: 179–202.

Newman, S.F. 2004. Disinfecting irrigation water for disease management. *Proceedings of the 20th Annual Conference on Pest Management in Ornamentals*, Society of American Florists, San Jose, CA.

Nitayavardhana, S. and S.K. Khanal. 2012. Biofuels residues/wastes: Ban or boon? *Critical Reviews in Environmental Science and Technology* 42: 1–43.

Nolt, J. 2011. How harmful are the average American greenhouse gas emissions? *Ethics, Policy and Environment* 24: 3–10.

Nordstrom, K.F. and S. Hotta. 2004. Wind erosion from cropland in the USA: A review of problems, solution and prospects. *Geoderma* 121: 157–167.

Odlare, M., J. Abubaker, J. Lindmark et al. 2012. Emission of N_2O and CH_4 from agricultural soils amended with two types of biogas residues. *Biomass and Bioenergy* 44: 112–116.

Offermann, R., T. Seidenberger, D. Thrän et al. 2011. Assessment of global bioenergy potentials. *Mitigation and Adaptation Strategies to Global Change* 16: 103–115.

Ong, C.K. and R.R.B. Leaky. 1999. Why tree crop interactions in agroforestry appear at odds with tree-grass interactions in tropical savannahs. *Agroforestry Systems* 45: 109–129.

Palmujoki, E. 2009. Global principles for sustainable biofuel production and trade. *International Environmental Agreements* 9: 135–151.

Pate, R., G. Klise, and B. Wu. 2011. Resource demand implications for US algae biofuels production scale-up. *Applied Energy* 88: 3377–3388.

Pfister, S., P. Bayer, A. Koehler, and S. Hellweg. 2011. Environmental impacts of water use in global crop production: Hotspots and trade-offs with land use. *Environmental Science and Technology* 45: 5761–5768.

Pimentel, D., M. Hertz, M. Glickstein et al. 2002. Renewable energy: Current and potential issues. *BioScience* 52: 111–1120.

Pocock, M.J.O., D.M. Evans, and J. Memmott. 2012. The robustness and restoration of a network of ecological networks. *Science* 335: 973–977.

Polasky, S., E. Nelson, D. Pennington, and K.A. Johnson. 2011. The impact of land-use change on ecosystem services, biodiversity and returns to landowners: A case study in the state of Minnesota. *Environmental and Resource Economics* 48: 218–242.

Popp, A, M. Krause, J.P. Dietrich et al. 2012. Additional emissions from land use change –forest conservation as a precondition for sustainable production of second generation bioenergy. *Ecological Economics* 74: 64–70.

Popp, A., H. Lotze-Campen, M. Leimbach et al. 2011. On sustainability of bioenergy production; integrating co-emissions from agricultural intensification. *Biomass and Bioenergy* 35: 4770–4780.

Ra, K., F. Shiotsu, J. Abe, and S. Morita. 2012. Biomass yield and nitrogen use efficiency of cellulosic energy crops for ethanol production. *Biomass and Bioenergy* 37: 330–334.

Ramos, M.A. and I.P. de Albuquerque. 2012. The domestic use of firewood in rural communities of the Caatinga: How seasonality interferes with patterns of firewood collection. *Biomass and Bioenergy* 39: 147–158.

Ranger, J. and M.P. Turpault. 1999. Input-output nutrient budgets as a diagnostic tool for sustainable forest management. *Forest Ecology and Management* 122: 139–154.

Reich, P.B., D. Tilman, F. Isbell et al. 2012. Impacts of biodiversity loss escalate through time as redundancy fades. *Science* 336: 589–592.

Reijnders, L. 2006. Conditions for the sustainability of biomass based fuel use. *Energy Policy* 34: 863–876.

Reijnders, L. 2008a. Do biofuels from microalgae beat biofuels from terrestrial plants? *Trends in Biotechnology* 26: 349–350.

Reijnders, L. 2008b. Ethanol production from crop residues and soil organic carbon. *Resources, Conservation and Recycling* 52: 653–658.

Reijnders, L. 2009. Fuels for the future. *Journal of Integrative Environmental Sciences* 6: 279–299.

Reijnders, L. 2013. Lipid-based liquid biofuels from autotrophic microalgae: Energetic and environmental performance. *Wiley Interdisciplinary Reviews: Energy and Environment* 2: 73–85.

Reijnders L. and M.A.J. Huijbregts. 2008. Palm oil and the emission of carbon-based greenhouse gases. *Journal of Cleaner Production* 18: 477–482.

Reijnders, L. and M.A.J. Huijbregts. 2009. *Biofuels for Road Transport*. London, U.K.: Springer.

Reijnders, L. and M.A.J. Huijbregts. 2011. Nitrous oxide emissions from liquid biofuel production in life cycle assessment. *Current Opinion in Environmental Sustainability* 3: 432–437.

Repo, A., R. Känkänen, J. Tuovinen et al. 2012. Forest bioenergy climate impact can be improved by allocating forest residue removal. *Global Change Biology: Bioenergy* 4: 202–212.

Riffell, S., J. Verschuyl, D. Miller et al. 2011. A meta-analysis of bird and mammal response to short-rotation woody crops. *Global Change Biology: Bioenergy* 3: 313–321.

Riksen, M., F. Brouwer, and J. de Graaf. 2003. Soil conservation measures to control wind erosion in northwestern Europe. *Catena* 52: 309–326.

Robertson, G.P., V.H. Dale, O.C. Doering et al. 2008. Sustainable biofuels redux. *Science* 322: 49–50.

Robertson, G.P., S.K. Hamilton, S.J. Del Grosso et al. 2011. The biogeochemistry of bioenergy landscapes: Carbon, nitrogen and water considerations. *Ecological Applications* 21: 1055–1067.

Rytter, L. 2002. Nutrient content in stems of hybrid aspen as affected by tree age and tree size and nutrient removal with harvest. *Biomass and Bioenergy* 23: 13–25.

Sahito, A.R., R.B. Mahar, A.A. Menon et al. 2012. Assessment of waste agricultural biomass for prevailing management, quantification and energy potential at Sanghar Pakistan. *Waste and Biomass Valorization* 3: 275–284.

Schouten, A., J.W. van Groeningen, O. Oenema et al. 2012. Bioenergy from cattle manure? Implications of anaerobic digestion and subsequent pyrolysis for carbon and nitrogen dynamics in soil. *Global Change Biology: Bioenergy*. DOI: 10.1111/j.1757-1707.2012.01163x.

Schröder, J.J., A.J. Smit, D. Cordell et al. 2011. Improved phosphorus use efficiency in agriculture: A key requirement for its sustainable use. *Chemosphere* 84: 822–831.

Schrotenboer, A.C., M.S. Allen, and C.M. Malmstrom. 2011. Modification of native grasses for biofuel production may increase virus susceptibility. *Global Change Biology: Bioenergy* 3: 360–374.

Scown, C.D., A. Horvath, and T.E. McKone. 2011. Water footprint of US transportation fuels. *Environmental Science & Technology* 45: 2541–2553.

Searchinger, T., R. Heimlich, R.A. Houghton et al. 2008. Use of U.S. croplands for biofuels increases greenhouse gases through emissions from land-use change. *Science* 319: 1238–1240.

Selsted, M.B., L. van der Linden, A. Ibrom et al. 2012. Soil respiration is stimulated by elevated CO_2 and reduced by summer drought: Three years of measurements in a multifactor ecosystem manipulation experiment in a temperate heathland (CLIMAITE). *Global Change Biology* 18: 1216–1230.

Service, R.F. 2004. The hydrogen backlash. *Science* 305: 958–961.

Sharma, E., S.C. Rai, and R. Sharma. 2001. Soil, water and nutrient conservation in mountain farming systems: Case study from the Sikkim Himalaya. *Journal of Environmental Management* 61: 123–135.

Sims, R.E.H, A. Hastings, B. Schlamadinger et al. 2006. Energy crops: Current status and future prospects. *Global Change Biology* 12: 2054–2076.

Smith, P. 2012. Soils and climate change. *Current Opinion in Environmental Sustainability*. DOI: 10.1016/j.cosust.2012.06.005.

Solomon, B.D. 2010. Biofuels and sustainability. *Annals of the New York Academy of Sciences* 1185: 119–134.

Springer, T.L. 2012. Biomass yield from an urban landscape. *Biomass and Bioenergy* 37: 82–87.

Stavi, I. and R. Lal. 2012. Agriculture and greenhouse gases, a common tragedy. A review. *Agronomy and Sustainable Development*. DOI: 10.1007/s13593-012-0110-0.

Steenari, B.M. and O. Lindquist. 1997. Stabilisation of biofuel ashes for recycling to forest soil. *Biomass and Bioenergy* 13: 39–50.

Stoms, D.W., F.W. Davis, M.W. Jenner et al. 2011. Modeling wildlife and other trade-offs with biofuel crop production. *Global Change Biology: Bioenergy* 4: 330–341.

Stone, K.C., P.G. Hunt, K.B. Cantrell et al. 2010. The potential impacts of biomass feedstock production on water resource availability. *Bioresource Technology* 101: 2014–2025.

Stromberg P.M., M. Estban, and A. Gasparatos. 2011. Climate change effects on mitigation measures: The case of extreme wind events and Philippines' biofuel plan. *Environmental Science & Policy* 14: 1079–1090.

Stürmer, B., J. Schmidt, E. Schmid et al. 2013. Implications of agricultural bioenergy crop production in a land constrained economy—The example of Austria. *Land Use Policy* 30: 570–581.

Tay, D.H.S., D.K.S. Ng, N.E. Sammons Jr. et al. 2011. Fuzzy optimization approach for the synthesis of a sustainable integrated biorefinery. *Industrial Engineering and Chemistry Research* 30: 1652–1665.

Thompson, I.D., K. Okabe, J.M. Tylianakis et al. 2011. Forest biodiversity and the delivery of ecosystem goods and services: Translating science into policy. *BioScience* 61: 972–981.

Tilman, D., K.C. Cassman, P.A. Matson et al. 2002. Agricultural sustainability and intensive production practices. *Nature* 418: 671–677.

Tirado, M.C., M.J. Cohen, N. Aberman et al. 2010. Addressing the challenges of climate change and biofuel production for food and nutrition security. *Food Research International* 43: 1729–1744.

Toivanen, T., A. Markkanen, J.S. Kothiatho et al. 2012. The effect of fuel harvesting on the fungal diversity of clear-cuts. *Biomass and Bioenergy* 39: 84–95.

Tomlinson, G.H. 2003. Acid deposition, nutrient leaching and forest growth. *Biogeochemistry* 65: 51–81.

Triana, C.A.R. 2011. Energetics of Brazilian ethanol: Comparison between assessment approaches. *Energy Policy* 39: 4605–4613.

Tricker, P.J., M. Pecchiari, S.M. Bunn et al. 2009. Water use of a bioenergy plantation increase in a future high CO_2 world. *Biomass and Bioenergy* 33: 200–208.

Truax, B., D. Gagnon, J. Fortier et al. 2012. Yield in 8 year-old hybrid poplar plantations on abandoned farmland along climatic and soil fertility gradients. *Forest Ecology and Management* 267: 228–239.

Tscharntke, T., Y. Clough, T.C. Wanger et al. 2012. Global food security, biodiversity conservation and the future of agricultural intensification. *Biological Conservation* 151: 53–59.

Turner, J.A. 1999. A realizable renewable energy future. *Science* 285: 687–690.

Uchida, S. and K. Hayashi. 2012. Comparative life cycle assessment of improved and conventional cultivation practices for energy crops in Japan. *Biomass and Bioenergy* 36: 302–315.

van Vuuren, D.P., A.F. Bouwman, and A.H.W. Beusen. 2010. Phosphorus demand for the 1970–2100 period: A scenario analysis of resource depletion. *Global Environmental Change* 20: 428–439.

VandenBygaart, A.J., D. Kroetsch, E.G. Gregorich et al. 2012. Soil C erosion and burial in cropland. *Global Change Biology* 18: 1441–1452.

Vasudevan, V., R.W. Straton, M.N. Pearlson et al. 2012. Environmental performance of algal biofuel technology options. *Environmental Science and Technology* 46: 2451–2459.

Velders, G.J.M., D.W. Fahey, J.S. Daniel et al. 2009. Large contribution of projected HDC emissions to future climate forcing. *Proceedings of the National Academy of Sciences USA* 106: 10949–10954.

Verbruggen, E., E.T. Kiers, P.N.C. Bakelaar et al. 2012. Provision of contrasting ecosystem services by soil communities from different agricultural fields. *Plant and Soil* 350: 43–55.

Wall, A. 2012. Risk analysis of effects of whole-tree harvesting on site productivity. *Forest Ecology and Management* 282: 175–184.

Wang, P., B.A. Olson, H. Arvidsson et al. 2010. Short-term effects of nutrient compensation following whole-tree harvesting on soil and water chemistry in a young Norway spruce stand. *Plant and Soil* 336: 323–336.

Wang, Y., Z. Yu, X. Song et al. 2007. Effects of the macroalgae *Ulva pertusa* (Chlorophyta) and *Gracilaria lemaneiformis* (Rhodophyta) on growth of four species of bloom-forming dinoflagellates. *Aquatic Botany* 86: 139–147.

Wu, Y. and S. Liu. 2012. Impact of biofuels production alternatives on water quantity and quality in the Iowa river basin. *Biomass and Bioenergy* 36: 182–191.

Yeh, S., G. Berndes, G.S. Mishra et al. 2011. Evaluation of water use for bioenergy at different scales. *Biofuels, Bioproducts and Biorefining* 5: 361–374.

Yuan, X., X. Shi, S. Zeng et al. 2011. Activated carbons prepared from biogas residue: Characterization and methylene blue adsorption capacity. *Journal of Chemical Technology and Biotechnology* 86: 361–366.

Zhang, X., R.C. Izauralde, D. Manowitz et al. 2010. An integrative modeling framework to evaluate the productivity and sustainability of biofuel crop production. *Global Change Biology: Bioenergy* 2: 258–277.

Zhuang, J., R.W. Gdentry, G. Yu et al. 2010. Bioenergy sustainability in China: Potential and impacts. *Environmental Management* 46: 525–530.

Zinn, Y.L., D.V.S. Resck, and J.E. da Silva. 2002. Soil organic carbon as affected by afforestation with *Eucalyptus* and *Pinus* in the Cerrado region of Brazil. *Forest Ecology and Management* 166: 285–294.

Section II

Sustainable Biomass Production and Supply Logistics

6 Production and Supply Logistics of Microalgae as an Energy Feedstock

Helena M. Amaro, F. Xavier Malcata, and I. Sousa-Pinto

CONTENTS

6.1 INTRODUCTION

Recent decades have witnessed a worldwide energy crisis, derived from irreversible depletion of fossil fuels; their use as a major form of energy supply—namely, for internal combustion engines—is sustainable, besides their net contribution to atmospheric greenhouse gases (GHGs) that bring about global warming. Carbon-neutral renewable liquid biofuels are indeed needed to displace petroleum-derived fuels in the near future.

The most widely available form of biodiesel at present comes from oil crops such as palm, oilseed rape, and soybean. However, a few major concerns have been raised on the sustainability of this mode of production: these include competition for arable land, as well as excess of fertilizer runoff, high freshwater usage, and unfavorable life cycle assessments (Amaro et al., 2011). The development of advanced biofuels from more sustainable feedstocks such as microalgae is thus in order—in attempts to overcome food security issues while reducing the ecological footprint and energy consumption.

Biodiesel and biohydrogen may be obtained from plant oils and electrolysis of water, respectively, but microalgae-mediated routes are an effective alternative. Microalgae possess a remarkable metabolic plasticity. They exhibit much higher productivities than land plants (doubling times may be as short as 3.5 h), can accumulate up to 20%–50% (w/w_{DW}) triacylglycerols (TAGs), and do not require high-quality agricultural land (or even no land at all, if offshore farming is considered) (Scott et al., 2010; Singh and Gu, 2010). Despite their growing in aqueous media, microalgae need lower rates of water renewal than terrestrial crops need irrigation water; hence, depletion of freshwater sources is alleviated, as even brackish water and seawater may actually be used for cultivation. Furthermore, there is no need for herbicides or pesticides, and only sunlight is required, as well as a few simple and nonexpensive nutrients (Aslan and Kapdan, 2006); environmental impacts are thus kept to a minimum while permitting waste upgrade—in terms of NH_4^+, NO_3^-, and PO_4^{3+} that contaminate the effluents from agrofood processing but may be used as nutrients (Mata et al., 2010).

On the other hand, the advance in metabolic engineering has been able to release molecular hydrogen by microalgae via photosynthetic routes; this may easily be converted to electricity in fuel cells, with only water vapor as exhaust product (Amaro et al., 2012). A critical factor when assessing economic viability of microalgae is the possibility of operating in large photobioreactors—able to handle biomass and metabolites to sufficiently high levels and in extractable forms (Sánchez et al., 2008; Amaro et al., 2012). However, large-scale systems based on microalgae toward the manufacture of biodiesel and biohydrogen have been economically hampered by their still poor volumetric efficiencies; excessively high costs are indeed observed relatively to current petrofuel prices. This means that technological improvements are in order—both of biocatalyst and bioreactor.

In the coming years, as the conventional petroleum oil resources decrease, the production of advanced biofuels from microalgae will be further sought—provided that the market responds favorably and the associated supply logistics are improved. The current bottlenecks that preclude full industrial exploitation of microalgae will be briefly reviewed and critically discussed in this chapter, with an emphasis on biodiesel and biohydrogen production and supply as alternative energy feedstock.

6.2 BIODIESEL PRODUCTION BY MICROALGAE

Biodiesel consists of a mixture of fatty acid alkyl esters, obtained by the transesterification of oils or fats—composed of at least 90% triglycerides, with the remainder being mono- and diglycerides and free fatty acids—and traces of phospholipids, phosphatides, carotenes, tocopherols, sulfur compounds, and water (Bozbas, 2008).

The value chain of microalgal biodiesel is schematically depicted in Figure 6.1. It starts with the selection of the most appropriate species—usually dependent on local environmental conditions and intended configuration of the cultivation system. Then it goes through harvesting and processing of biomass, including extraction of oil, and ends with the transesterification of oil into biodiesel. The by-products generated may meanwhile be recycled or used for energy generation related with each of these steps—as detailed in the following sections.

6.2.1 Microalgal Species Selection

Several microalgal species can be induced to accumulate a considerable content of lipids. Although the lipid content of microalgae averages between 1% and 70%, some species may

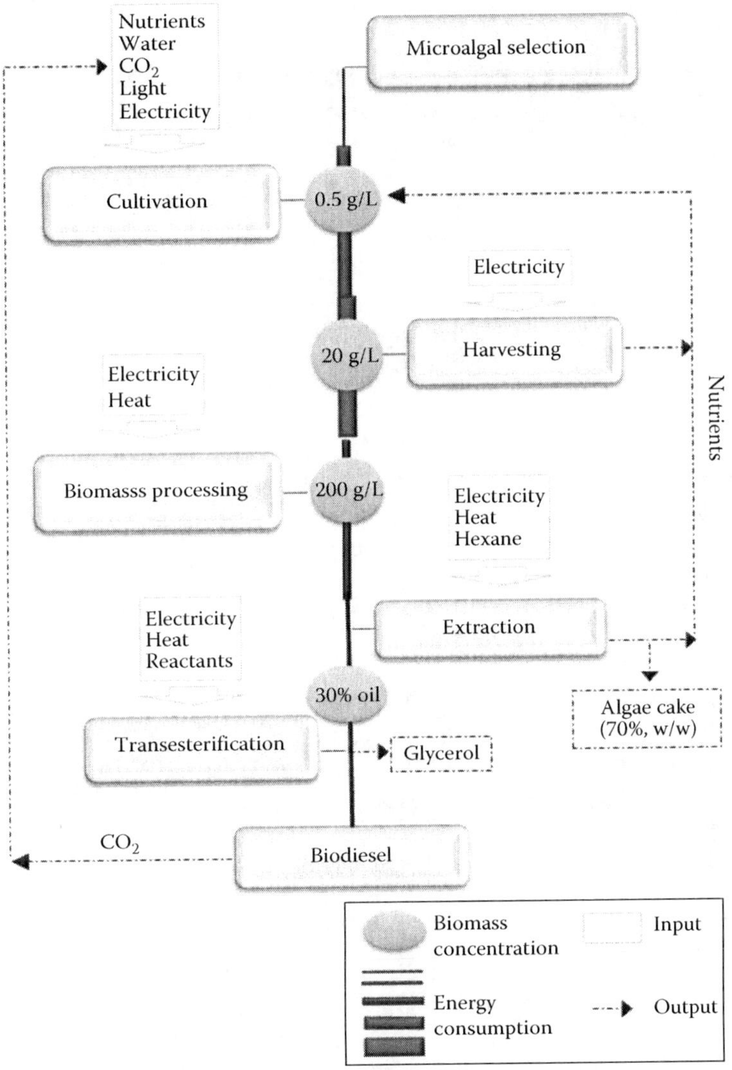

FIGURE 6.1 Schematic representation of the various stages of manufacture of microalgal biodiesel—including strain selection, biomass cultivation and harvesting, and processing, leading to oil extraction and transesterification to biodiesel—and of percent energy consumption at each stage (proportional to area of flow chart arrow), coupled with indication of major inputs and outputs.

reach 90% (w/w_{DW}) as shown in Table 6.1 (Chisti, 2007). It should be noted that a few differences are apparent even within the same genus and a high level of lipids does imply a high productivity. Some popular microalgae (e.g., *Chlorella*, *Dunaliella*, *Isochrysis*, *Nannochloris*, *Nannochloropsis*, *Neochloris*, *Nitzschia*, *Phaeodactylum*, and *Porphyridium* spp.) possess oil levels between 20% and 50%, as well as high productivities. In particular, *Chlorella* appears to have a good potential for biodiesel production. Conversely, the oil content of *Botryococcus braunii* reaches 75% (w/w_{DW}), but a low productivity is observed and besides the oil is secreted into the cell wall (Metzger and Largeau, 2005). Furthermore, larger lipid productivities are observed in marine microalgae. Since high salinity in seawater prevents extensive contamination while allowing seawater to be directly used instead of freshwater, marine microalgae appear more suited for mass production (Amaro et al., 2011).

TABLE 6.1
Lipid Content and Productivity of Selected Microalgal Species

Marine Microalgal Species	Lipid Content (w/w_{DW})	Lipid Productivity (mg L^{-1} day^{-1})
Botryococcus sp.	25.0–75.0	—
Chaetoceros calcitrans	14.6–16.4/39.8	17.6
Chaetoceros muelleri	33.6	21.8
Chlorella emersonii	25.0–63.0	10.3–50.0
Chlorella protothecoides	14.6–57.8	1214
Chlorella pyrenoidosa	2.0	—
Chlorella sorokiniana	19.0–22.0	44.7
Chlorella sp.	10.0–48.0	42.1
Chlorella sp.	18.0–57.0	18.7
Chlorella vulgaris	5.0–58.0	11.2–40.0
Chlorococcum sp.	19.3	53.7
Dunaliella primolecta	23.1	—
Dunaliella salina	6.0–25.0	116.0
Dunaliella sp.	17.5–67.0	33.5
Dunaliella tertiolecta	16.7–71.0	—
Ellipsoidion sp.	27.4	47.3
Haematococcus pluvialis	25.0	—
Isochrysis galbana	7.0–40.0	—
Isochrysis sp.	7.1–33	37.8
Nannochloris sp.	20.0–56.0	60.9–76.5
Nannochloropsis oculata	22.7–29.7	84.0–142.0
Nannochloropsis sp.	12.0–53.0	60.9–76.5
Neochloris oleoabundans	29.0–65.0	90.0–134.0
Pavlova lutheri	35.5	40.2
Pavlova salina	30.9	49.4
Phaeodactylum tricornutum	18.0–57.0	44.8
Scenedesmus obliquus	11.0–55.0	—
Scenedesmus quadricauda	1.9–18.4	35.1
Scenedesmus sp.	19.6–21.1	40.8–53.9
Spirulina platensis	4.0–16.6	—

Besides high productivity and lipid content, the selection of suitable strains should consider the ability to uptake nutrients or grow under specific environmental conditions. The fatty acid profile of the microalgal cell is also relevant because it significantly affects the heating value of the biodiesel produced. Most of the fatty acids in microalgae are saturated and unsaturated fatty acids, containing 12–22 carbon atoms often of the ω3 and ω6 types. Microalgae exhibit the fatty acid profiles relatively similar to each other. In general, they can synthesize C14:0, C16:0, C18:1, C18:2, and C18:3 fatty acids (Guedes et al., 2011), whereas the contents of other fatty acid residues are affected directly by nutrient and processing factors, cultivation conditions, and growth phases, for example, cultures under nitrogen deficiency and salt stress undergo accumulation of C18:1 in most species and C20:5 to a lesser extent in *B. braunii* (Mata et al., 2010).

Therefore, the selection of a specific strain should take into account at least the following factors: (1) growth rate; (2) lipid quantity and quality, especially in terms of fatty acid residue profile; (3) response to temperature, nutrient input, and light and ability to compete with other microalgal and/or bacterial species; (4) nutrient requirements and rate of uptake thereof—in particular CO_2 but also nitrogen and phosphorus (especially when carbon sequestration and upgrade of brackish waters and agricultural effluents are the target); (5) ease of biomass harvesting, oil extraction, and further processing; and (6) possibility of obtaining high added-value chemicals, coupled with application to the formulation of foods, cosmetics, and pharmaceuticals (Hill et al., 2006; Gressel, 2008).

A few microalgae species have been reported to be potential sources for producing biodiesel. As most wild species currently available in culture collections are not particularly promising, hence, genetic engineering may be considered in attempts to improve productivity and yield (Gressel, 2008). Metabolic engineering is also a possibility toward increasing the yields of acylglycerols, or even toward different molecules holding a better performance as biodiesel. In higher plants, several studies have already explored for the overexpression of enzymes involved in lipid synthesis, yet little change in oil content was achieved in species containing higher levels of acetyl-CoA carboxylase—the rate-limiting step in fatty acid biosynthesis—possibly because of the complex regulation of this enzyme (Amaro et al., 2011).

Despite its potential, a few barriers to implementation of the aforementioned methodology exist. A very few strains of microalgae have been genetically modified so far; the mechanisms underlying regulation of gene expression are still poorly understood; and no specific molecular biology tools are readily available (e.g., efficient nuclear transformation, availability of promoter and selectable marker genes, and stable expression of transgenes). Moreover, legislation is restrictive about biological contamination, which has also precluded the broader utilization of genetically engineered microalgae (Amaro et al., 2011).

Furthermore, the elucidation of molecular pathways and forms of regulation of lipid metabolism is still incipient in the case of microalgae. Nevertheless, bioinformatics applied to the microalgal genomes sequenced so far has unfolded essentially similar biochemical routes. On the other hand, experimental validation of putative enzyme activities is limited (Hu et al., 2008)—with the notable exception of recent work by Moellering et al. (2010) and Wang et al. (2009).

Metabolic evidence of lipid accumulation, for example, induction in microalgae by nitrogen deprivation, provides a useful experimental basis to monitor changes in gene transcription, protein synthesis, and metabolic activities prevailing during lipid accumulation in microalgae (Scott et al., 2010). Moreover, N-depletion coupled with RNAi suppression causes changes in both lipid and protein qualitative and quantitative profiles in *Chlamydomonas reinhardtii* during lipid droplet formation. The total content of fatty acids, on a per cell basis, in cultures transferred to N-depleted media increased by 2.4-fold within 72 h, and 65% of fatty acids were esterified to TAGs. It thus became apparent that de novo synthesis of fatty acids takes place in oil bodies. Proteomic analysis provided further evidence of a *major lipid droplet protein* (MLDP) that was rather abundant in those lipid bodies; the mRNA transcript abundance of this protein followed the observed increase in lipid droplets after N-depletion, together with the realization that RNAi lines of *C. reinhardtii* with a 55%–60% reduction of MLDP transcription produced lipid droplets characterized by diameters 40% larger than the control, suggesting the involvement of that protein in the regulation of the lipid droplet size.

Another efficient way of increasing lipid yield is to delete *redundant* pathways in the targeted microorganism. Wang et al. (2009) reported an increase in TAG contents in a mutant of *C. reinhardtii* with less starch productivity due to deficient adenosine diphosphate (ADP)-glucose pyrophosphorylase (an essential enzyme in starch production). After 48 h of N-depletion, wild-type cells increased their lipid droplet content by 15-fold, but the mutant cells increased up to 30-fold, thus leading to a TAG concentration of ca. 17 ng cell^{-1} versus only 10 ng cell^{-1} in the control.

6.2.2 Microalgal Biomass Production

Microalgae have been designed by nature to screen their environment for suitable nutrients and energy and to optimize the efficiency of their utilization including storage of energy in chemical form. Nutritional requirements for microalgal biomass growth lie mainly upon inorganic carbon substrates, vitamins, and salts. When considering microalgae specifically for biodiesel production to meet given specifications, it is also important to quantitatively assess the influence of operational parameters (pH, temperature, and light intensity) and their interactions, in order to master their manipulation to eventually obtain high-quality biomass and metabolites for the production of biodiesel (Amaro et al., 2011).

In what pertains to biomass production, microalgae may undergo distinct types of metabolism, as they can shift it in response to changes in environmental conditions. *Chlorella vulgaris*, *Haematococcus pluvialis*, and *Arthrospira platensis* (Spirulina) are examples of microalgal species that can grow under photoautotrophic, heterotrophic, or mixotrophic conditions (Mata et al., 2010). Other strains such as *Selenastrum capricornutum* and *Scenedesmus acutus* can grow either photoautotrophically, heterotrophically, or photoheterotrophically (Chojnacka and Marquez-Rocha, 2004).

Under phototrophic cultivation, there is a large variation in biomass lipid content—ranging from 5% to 68%, depending on the species. The highest lipid productivity reported up to date is ca. 180 mg L^{-1} day^{-1} by *Chlorella* sp., under bubbling of 2% (v/v) CO_2 in air and a renewal rate of 0.25 vvm. When compared with other modes of cultivation, contamination is typically less severe.

Limitation of available light may be circumvent by using heterotrophic cultivation, and thus allowing higher biomass production and productivity—with no relevant decrease in growth rate caused by opacity associated with increasing cell density (Huang et al., 2010). This cultivation mode has generated considerable interest for enhanced oil productivities. It has been report that the lipid content of *Chlorella protothecoides* increased by 40% when it changed from phototrophic to heterotrophic cultivation (Liang et al., 2009).

Under heterotrophic metabolism, microalgae can assimilate several organic carbon sources, for example, glucose, acetate, glycerol, fructose, sucrose, lactose, galactose, and mannose (Xiong et al., 2008). To improve economic feasibility, less expensive organic carbon sources have also been tested instead of sugars. For instance, corn powder hydrolyzate produced a good result in terms of biomass productivity (2 g L^{-1} day^{-1}) and lipid content (932 mg L^{-1} day^{-1}) (Cheng and Timilsina, 2011). The highest lipid productivity (3700 mg L^{-1} day^{-1}) was reported by Xiong et al. (2008) using a 5-L fermentor operated with an improved fed-batch culture strategy: there was a 20-fold increase in productivity compared to phototrophic cultivation.

Unfortunately, sugar-based heterotrophic systems are proved to be vulnerable to contamination (Carvalho et al., 2008), especially in open systems at a large scale. The cost of organic carbon sources is also a concern from a commercial standpoint. On the other hand, scaling up is easier when compared with phototrophic cultivation—which makes this method technologically attractive.

6.2.2.1 Design of Bioreactors

A major decision to be made is whether closed photobioreactors or open ponds should be used for microalgal cultivation. The latter can attain much larger areas and are relatively cheap to build and easy to operate. However, in open pond systems, contamination is more likely; stable environmental conditions (particularly temperature) are difficult to maintain; cell density is relatively low because of mutual shading; and there are substantial costs of harvesting afterward (Scott et al., 2010).

Conversely, contamination can be avoided in closed photobioreactors if operated under sterility, so one may reach higher biomass productivities. These reactors can take several configurations, for example, vertical reactors, flat-plate reactors, annular reactors, arrangements of plastic bags, and various forms of tubular reactors—which are stirred by rotating shafts or airlifting (El-Shishtawy et al., 1997). Photobioreactors are more useful for the optimization of external parameters that directly affect the rate of biomass production—for example, light path. Some photobioreactors combine a light-receiving face with a reflective sheet to enhance light supply; others are composed of three concentric glass cylinders, with incandescent lamps placed directly on the photobioreactor (Issarapayup et al., 2009). Zhu et al. (2008) proposed, in turn, a flat-plate airlift photobioreactor and found the optimum downcomer-to-riser area ratio for *H. pluvialis*

NIES-144. Most reactors of this type entail a high power consumption, with capital costs being widely variable (Scott et al., 2010).

6.2.2.2 Optimization of Processing Parameters

In the case of biomass produced under phototrophic conditions, both the light spectrum and intensity are critical factors that impact upon growth rate (either indoors or outdoors). It is estimated that 8 photons of photosynthetically active radiation (ca. 48% of incident solar flux) are needed to fix one molecule of CO_2 into carbohydrate, which corresponds to a maximum photosynthetic efficiency of only 12%. Under favorable conditions (i.e., low to moderate light levels), the efficiencies reported so far range in 4.5%–7% (Ma and Hanna, 1999; Scragg et al., 2002), which translates to a yield of 30–40 $g_{dry\ biomass}$ m^{-2} day^{-1} (Scott et al., 2010).

Several efforts have meanwhile been developed to take full advantage of incident light toward microalgal growth. Sourcing (i.e., external to the cell) and sinking (i.e., by the cell itself) are to be considered to further length. In an ideal scenario, light production and conduction should be efficient and cost-effective. However, this implies that installing and operating artificial light sources have relatively high costs. A number of alternative illumination strategies have been proposed (Melis et al., 1998). In a system using artificial light, the oil yield may reach 172 m^3 ha^{-1}. Such high yield is primarily a consequence of the stability in time of the light intensity, with a clear advantage of fluorescent lamps over incandescent ones owing to their relatively low power consumption per unit PAR. Since their capital costs are higher for a given total light intensity, this will be alleviated only by light-emitting diodes (LED). The use of optical fibers (OFs) to illuminate the inner portions of a microalgal bioreactor may increase the efficiency of light use and reduce the electricity. For example, OF that is exposed to solar energy can be combined with solar panels and wind power generators for a more efficient cultivation (Chen et al., 2011).

Commercial microalgal cultivation is carried out in open ponds, using solar energy as sole light source that is available at essentially no cost. However, the performance of outdoor systems is typically poor and the oil yield is between 100 and 130 m^3 ha^{-1}. The open ponds require large land surface areas (Chisti, 2007).

A different strategy aimed at optimizing light utilization focuses on the cell itself, particularly on its photosynthetic apparatus. If the pigments forming the so-called antenna are reduced, each cell will become less opaque, thus facilitating light transmission (Chen et al., 2011). This may be achieved by growing the microalga at high light intensity for a few hours or by manipulation of nutrient levels. Deprivation of some nutrients leading to increased productivity of TAGs may also cause a reduction in the chlorophyll level per cell (Stephenson et al., 2010). Another (invasive) method is mutating the genes responsible either for components of the antenna itself or for regulation of the antenna biogenesis. Strains modified under this perspective have exhibited enhanced photosynthetic efficiency also under high light levels (Scott et al., 2010). Although the available results so far look promising, there is a pending risk that reduced antenna sizes may also imply reduced fitness, thus compromising the overall strain productivity.

Biodiesel yield by microalgae depends on not only the concentration of biomass attained but also the oil content of the individual cells. In general, lipid productivity and content are inversely correlated to each other. Stress conditions like deprivation of nitrogen (or phosphate, to a lesser extent) limit cell growth and increase lipid content (Khozin-Goldberg and Cohen, 2006; Rodolfi et al., 2009). The general principle is that excess carbon from photosynthesis is channeled into storage molecules, for example, TAGs or starch, when there is insufficient nitrogen for protein synthesis as required by growth (Scott et al., 2010). Neutral lipid contents are enhanced in microalgae upon shift to lipid storage following exposure to nitrogen deprivation—and unlike straight-chain saturated hydrocarbons, multibranched and/or polyunsaturated components are predominate (Amaro et al., 2011). However, this type of limitation does not trigger a similar behavior in all microalgae. For example, *Dunaliella bardawil* and *Dunaliella salina* underwent a lipid fraction decrease of ca. 10% (w/w_{DW}), but a concomitant metabolic shift was observed toward carbohydrate synthesis and storage

(Mata et al., 2010). *C. vulgaris* grown under nutrient-sufficient conditions had lipid contents of 14–30% (w/w_{DW}) (Illman et al., 2000), but increases of up to 70% have been reported under nutrient-deficient conditions (Scott et al., 2010). *Chlorella emersonii* and *Chlorella minutissima* increased in lipid content by 63% and 56%, respectively (Santos et al., 2001).

A two-stage cultivation process has been reported by Rodolfi et al. (2009) to produce 0.2 kg_{oil} m^{-3} day^{-1} with photosynthetic microalgae. In their setup, cells were first grown under nutrient-sufficient conditions to enhance biomass accumulation, and only subjected to the nutrient deprivation afterward for lipid synthesis (Rodolfi et al., 2009). However, Stephenson et al. (2010) claimed that the most effective strategy to achieve high lipid contents in *C. vulgaris* is to allow cells to naturally deplete their original nitrogen source rather than transfer them into a medium devoid of that element. In the case of *Botryococcus* spp., neutral lipids (that comprise a major fraction of total lipids) reached the highest production rate in the stationary stage, whereas the greatest yield was observed upon turning from the green to the brown growth phase (Metzger and Largeau, 2005). Likewise, *Isochrysis* sp. accumulates higher fractions of lipids (ca. 25% of microalgal cell weight) for 10 days following nitrogen starvation.

In photoautotrophic microalgal cultures, CO_2 provided by aeration showed to play a major role in biomass and lipid contents of *Nannochloropsis oculata* (Richmond, 2004) as well as *Scenedesmus obliquus* and *Chlorella kessleri* strains isolated from a treatment pond of a coal-fired thermoelectric power plant in Brazil. These strains exhibit a high potential for biofixation of CO_2 (Chiu et al., 2008). However, CO_2 supply in shallow suspensions at near-neutral pH is difficult to achieve and control, since the bubble residence time is insufficient for adequate mass transfer—thus leading to most CO_2 being lost to the open atmosphere (Amaro et al., 2011). Under alkaline conditions, absorption of CO_2 is accelerated by hydration of CO_2 and subsequent acid–base reaction to form HCO^{3-}, as well as direct reaction of CO_2 with OH^- to form HCO^{3-}. The rate of the former is faster at pH values below 10, whereas the latter dominates above; between pH 8 and 10, their relative magnitude is around unity (Raja et al., 2008).

Finally, high iron concentrations were shown to induce lipid accumulation in *C. vulgaris*, which suggests that some metabolic pathways may be modified upon exposure to high levels of that oligo-element in the medium (Liu et al., 2008).

6.2.3 Microalgal Cell Harvesting

After metabolic synthesis, lipids accumulate in the cells—so biomass has to be harvested to extract them, which includes exposure to conditions that help release those lipids prior to eventual trans-esterification to produce biodiesel. Hence, efficient harvesting of biomass is essential for mass production of biodiesel.

Harvesting basically consists of concentrating a diluted microalgal suspension to a thick paste to obtain a slurry with at least 2%–7% (w/w_{DW}). This process may account for 20%–30% of the total production cost (Mata et al., 2010). Since it involves large biomass and medium volumes, suitable harvesting usually takes place through more than one step—of physical, chemical, or biological nature. This regularly encompasses bulk harvesting—with the purpose of separating the microalgal biomass from the bulk suspension, followed by thickening—which is aimed at concentrating the slurry. The latter usually resorts to filtration or centrifugation that often demands more energy than bulk harvesting (Golueke and Oswald, 1965). After comparative evaluation of filtration, flotation, centrifugation, precipitation, ion exchange, electrodialysis, and ultrasonic vibration, Amaro et al. (2011) concluded that only centrifugation and chemical precipitation are economically feasible options (Amaro et al., 2011). Selection of the most appropriate harvesting technique depends on the properties of the microalgae, for example, its density and size—as well as the specifications of the desired product (Golueke and Oswald, 1965). In particular, an optimal harvesting method for biofuel production should be species-independent, demand as low chemicals and energy as possible, and preferentially release intracellular materials for posterior collection. Unfortunately, a universal harvesting method does not exist, so this raises indeed great opportunities for research and development (R&D).

The major techniques available for microalgae harvesting include centrifugation, flocculation, filtration, sedimentation, flotation, and electrophoresis (Danquah et al., 2009). All of them are relatively expensive, because the initial mass fractions are typically low, and cells normally carry negative charge and excess algogenic organic material that enhance stability in the dispersed state (Gong and Jiang, 2011).

In mass cultivation of microalgae, a high overflow rate upward favors flotation of cells—instead of sedimentation—so gravity settling is suitable to harvest chiefly large-sized microalgal cells, for example, *Spirulina* spp. (Xiong et al., 2008). Gravity settling on its own can concentrate microalgal suspensions to 1.5% (w/w) solids at a reasonable cost. Conversely, flotation may be assisted via addition of an oxidant to destabilize suspended microalgal cells (Amaro et al., 2011). Recovery of microalgae via sedimentation may be enhanced via addition a flocculant instead. Use of microstrainers has also been suggested for large-sized microalgae. However, incomplete solid removal and difficulty in handling fluctuations in solid contents have been pointed at as shortcomings (Rossignol et al., 1999). Another relevant issue is the increase in bacterial and microalgal slime building on the microfabric (Rossignol et al., 1999)—which calls for regular cleaning; the mesh surface may be modified, thus reducing the affinity between dissolved (or colloidal algogenic organic) materials and collector surface.

The main costs of applying tangential flow filtration arise from membrane replacement and pumping so that large-scale harvesting based on this method will eventually be constrained. For instance, Rossignol et al. (1999) were successful in the use of polymer membranes for continuous harvesting of two marine microalgal species, *Haslea ostrearia* and *Skeletonema costatum*. However, the interactions between microalgal cells and membrane surfaces manufactured from different materials remain essentially unclear (Xiong et al., 2008). Deliberate changes in hydrophobicity of the membrane surface appear promising for reduction of fouling throughout harvesting.

Electrolytic methods can separate cells without the need to add chemicals. However, the high power required produces a temperature increase that was proved to damage the system and cathode fouling remains a major disadvantage.

6.2.4 Microalgal Biomass Processing

Cell disruption is the next step after harvesting and it should be applied when microalgal lipids are located intracellularly. This step is the most important because the overall yield of biodiesel depends strongly on the disruption method. This step should be as energy-efficient and economical as possible—namely, by avoiding the use of large amounts of organic solvents and maximizing the pool of liquid biofuel without significant recovery of other products, for example, DNA and chlorophyll (Liu et al., 2008).

Autoclaving, bead-beating, microwave application, sonication, and 10% (w/v) NaCl addition have been tested toward cell disruption. Sonication found its place therein, which basically cracks the cell wall and membrane via cavitation (Lee et al., 1998; Geciova et al., 2002). However, none of these methods is particularly efficient—and is far from being universal. The efficiency of cell disruption for lipid extraction differs, depending on the microalgal species and the extraction method employed. Microwaving appears to be the simplest, easiest, and most effective one in this regard. Lee et al. (1998) tested those methods in attempts to disrupt *Botryococcus* sp., *C. vulgaris*, and *Scenedesmus* sp. The highest lipid yield was obtained with a microwave oven, that is, shattering cells via high-frequency waves. Bead-beating and microwaving have been proven the most efficient, specifically for *Botryococcus* sp. ($P < 0.05$)—with 28.1% and 28.6% yields, respectively—whereas sonication produced the lowest efficiency (8.8%). Furthermore, bead-beating, based on direct mechanical damage to cells arising from high-speed spinning of fine beads, had previously been shown to be better in extraction of lipids in *B. braunii* than sonication, homogenization, French press, or lyophilization

(Geciova et al., 2002). Despite the high efficiency of the bead-beating method, it is not easy to scale up. On the other hand, autoclave and microwave oven methods turned out to be the most efficient in the case of *C. vulgaris*, whereas bead-beating was the worst (7.9%). For *Scenedesmus* sp., the microwave oven method led to the best results, yet the efficiencies of bead-beating, sonication, and osmotic shock were statistically similar to each other ($P < 0.05$). Although osmotic shock is simple and has results similar to bead-beating in the case of *C. vulgaris* and *Scenedesmus* sp., it requires longer treatments (for at least 48 h).

6.2.5 Microalgal Oil Extraction

Microalgal lipid extraction entails several processing factors—and the exact effect of each should be fully understood. Hence, scale-up to commercial level is still somewhat difficult. As microalgae are cultivated in aqueous media, the removal of water in the paste that typically has a moisture content of 10%–30% (w/w_{DW}) is cost-intensive mainly because of the high latent heat of vaporization of water. This means that the lipid extraction technology of choice should be directly applied to wet feedstocks.

The ideal lipid extraction method should not only be lipid-specific—to minimize coextraction of nonlipid materials—but also selective, so that only a few lipid fractions, for example, neutral lipids containing mono-, di-, and trienoic fatty acid residues, are extracted (Macías-Sánchez et al., 2007). The classic Folch chloroform-based lipid extraction protocol is particularly suitable for most microalgal lipids. As hexane is less efficient and also less toxic and has a marginal affinity for nonlipid contaminants, it appears more appropriate (Ma and Hanna, 1999; Geciova et al., 2002).

Supercritical carbon dioxide extraction is a promising green technology that may eventually override traditional organic solvents in lipid extraction. Its main advantages include tunable solvating power, low solvent toxicity, and favorable mass transfer rates—due to diffusion/viscosity and density/solvation properties of the solvent being between those of a gas or of a liquid, respectively (Mendes et al., 2005; Thana et al., 2008). Typical yields with this method go up to 25.0% (w/w_{DW}) in the case of *Nannochloropsis* sp. and 28.6% (w/w_{DW}) for *B. braunii* (Halim et al., 2011). Lipid yields tend to decrease with temperature and increase with pressure. The differences are hardly found in lipid profile under different processing conditions. Soxhlet extraction with hexane was found to be significantly less efficient than supercritical extraction: for a comparable lipid yield (ca. 0.058 g $g^{-1}_{dried\ microalgae}$), a period almost sixfold longer was necessary (Halim et al., 2011).

6.2.6 Microalgal Oil Transesterification

Biodiesel typically consists of a mixture of fatty acid alkyl esters—obtained by transesterification of oils or fats originally composed of 90%–98% triglycerides (Bozbas, 2008). This chemical reaction includes multiple steps in the presence of a catalyst—including reversible hydrolysis in series, as triglycerides are converted to diglycerides, then diglycerides to monoglycerides, and finally monoglycerides to fatty acids and glycerol—followed by reesterification with a short-chain alcohol (usually methanol) to assure a high volatility (Amaro et al., 2011). As processing costs may be considerably high, in situ transesterification is a new approach. This process facilitates the conversion of fatty acids to their alkyl esters directly inside the biomass, thereby eliminating the solvent extraction step and alleviating the need for biomass drying in harvesting. Alcoholysis of oil in the biomass matrix leads to biodiesel yields higher than that if the conventional route were followed and wastes are also considerably reduced (Ehimen et al., 2010).

6.2.7 Sustainability of Microalgal Biodiesel

It is expected that microalga-based biofuel may eventually overcome the use of arable land to produce biodiesel that would otherwise be used for food crops. Up to date, however, no such system has met with economic feasibility. Hence, biodiesel production from microalgae should be accompanied by the production and recovery of other products to help improve cost-effectiveness (Stephens et al., 2010). This biorefinery approach is particularly relevant in the case of biohydrogen, because its volatile nature leaves the biomass essentially intact.

The oil cake generated during oil extraction contains an amount of energy almost equivalent to the amount of energy stored in the biodiesel produced. In addition, glycerol will also be available as coproduct from transesterification (Figure 6.1), which may be refined and sold to the pharmaceutical industry or be used as livestock feed (Shirvani et al., 2011). Of the energy inventory of the microalgal cake, 35%–73% is accounted for by carbohydrates and proteins (Hu et al., 2008) so that at least three different coproduct utilization options may be (and have been) considered: (1) adjacent coal-fired power system that cofires biomass residues (Xu et al., 2006), (2) direct combustion of cake in integrated biomass-heating system, or (3) biomass-based combined heat and power unit (Huang et al., 2010). On the other hand, the biomass remaining after oil extraction may be fermented into ethanol or methane, which are biofuels as well; incorporated in livestock feed; used as organic fertilizer owing to its high N/P ratio; or simply burned for energy cogeneration (Bozbas, 2008; Mata et al., 2010).

Furthermore, a wide range of chemicals that are suitable for functional food formulation may be extracted from algal biomass, depending on the species (Raja et al., 2008). These products include bulk commodities such as sugars for the production of bioethanol and biomethane via fermentation, intermediate-value products such as proteins for animal feedstock, and high-value products such as active principles with antimicrobial, antioxidant, antitumoral, and anti-inflammatory activities for pharmaceutical purposes. Algal biomass may finally be pyrolyzed to produce sequestered carbon in the form of biochar, which holds a value as soil enhancer (Kruse and Hankamer, 2010).

6.3 HYDROGEN PRODUCTION BY MICROALGAE

Molecular hydrogen is one of the most promising biofuels, in terms of its large enthalpy and the absence of polluting products after combustion. Hydrogen is released by many phototrophic organisms (Kruse and Hankamer, 2010), including several hundred species from different groups of microalgae, cyanobacteria, and anaerobic photosynthetic bacteria (Dasgupta et al., 2010). At present, *C. reinhardtii* remains the best photosynthetic eukaryotic hydrogen producer, with *Nostoc* and *Synechocystis* spp. as the bacterium with the highest potential (El-Shishtawy et al., 1997).

6.3.1 Microalgal Hydrogen Production Mechanism

Photoproduction of H_2 by microalgae follows two processes: one direct and the other indirect. Both depend on reduced ferredoxin (Fd) as electron donor, coupled to (and required by) the action of hydrogenases (Kruse and Hankamer, 2010). The former includes photooxidation of water, in which both photosystem I (PSI) and photosystem II (PSII) play a role in supplying reductants (or electrons) to Fd via the photosynthetic electron transfer chain. The indirect pathway involves, in turn, oxidative carbon metabolism (e.g., starch degradation), but NADP-plastoquinone oxidoreductase and PSI activities are required to supply reductants. Either electron source (i.e., water or starch) can be used, but the actual contribution depends on the type of strain, culture conditions, extent of damage of PSII, and specific metabolic constraints (Amaro et al., 2012).

Light absorption by the photosynthetic apparatus is crucial for the generation of molecular hydrogen, as it causes oxidation of water that releases electrons and protons while facilitating endergonic transport of electrons to Fd and serves as physiological electron donor to Fe-hydrogenase, as it links the said enzyme to the electron transport chain in the chloroplasts (Florin et al., 2000). However, hydrogenase works only transiently under these conditions: it lasts from a mere several seconds to a few minutes, because the (light-dependent) oxidation of water also causes the release of molecular oxygen that is a powerful inhibitor of Fe-hydrogenase.

Besides the aforementioned role of PSII-dependent H_2 photoevolution that involves water as electron source and produces H_2/O_2 in 2:1 stoichiometric amounts, an alternative mechanism has been described by Gfeller and Gibbs (1984): upon dark anaerobic incubation and consequent induction of hydrogenase, electrons become available from catabolism of endogenous substrate and corresponding oxidative carbon metabolism.

6.3.2 Genetic Engineering of Species

Genetic transformation has a potential to improve H_2 production, especially in the case of *C. reinhardtii*, since its genome has already been fully elucidated (besides being a quite good producer). Several mutants have been generated with mutations at various levels, for example, hydrogenase, sulfate permease, and ribulose-1,5-bisphosphate carboxylase (RuBisCO) enzyme, as well as at the PSI and PSII (Torzillo et al., 2009). One illustrative example is *C. reinhardtii* strain *Stm6*. It is able to produce five-fold more H_2 than its wild counterpart (Kruse et al., 2005). The associated strains of Stm6Glc4 and Stm6Glc4T7 represent an extra improvement over Stm6, in terms of yield of biomass and hydrogen. A hexose symporter system from *Ch. kessleri* was specifically inserted into strain Stm6glc4, thus enabling the efficient production of hydrogen via water photolysis and from external sugars in parallel with feeding of the associated H^+ ions and electrons through the plastoquinone pool (Doebbe et al., 2007). Stm6Glc4 was further engineered to reduce its antenna size, thus yielding Stm6Glc4T7. It is characterized by 1.2–1.3-fold light capture efficiencies, higher light saturation threshold (800 μE), and ability to release up to 50% more H_2. The overall average production rates by these strains varied between 400 and 600 mL H_2 L^{-1} within 3–4 days, which corresponds to 8 mL H_2 L^{-1} h^{-1}.

The aforementioned poor efficiency of light utilization may be minimized by reducing the cross-sectional area of the light-harvesting antenna of PSII via genetic engineering (Mussgnug et al., 2007; Melis, 2009), which prevents the oversaturation of the photochemically active reaction centers. Furthermore, this strategy reduces photodamage (Melis, 2009), while allowing light (which would otherwise be dissipated) to penetrate into the culture deeper. *In vitro* testing has shown that the biomass production efficiency was doubled (Geciova et al., 2002). Outdoor trials have meanwhile been conducted to assess the efficiency.

Optimization of technical and physiological parameters of cultivation has also been attempted to increase hydrogen production efficiency. For instance, Torzillo et al. (2009) reported on the interplay between light intensity, chlorophyll concentration, and culture mixing on H_2 production in *C. reinhardtii*. They found that the best rates were obtained at 140 μE, with a chlorophyll concentration of 24 mg L^{-1} (ca. 2.4×10^8 cells L^{-1}).

6.3.3 Design of Bioreactors

Closed photobioreactors are the only configuration that is suitable for hydrogen production. However, they are limited by physical parameters associated with the reactor including light penetration, area to volume ratio, temperature, transparency and durability of construction material, rate of gas exchange, and level of stirring. They are also limited by physicochemical parameters that affect the biochemical pathways pertaining to hydrogen production. These physicochemical parameters include pH value, temperature, light intensity, dissolved oxygen and CO_2, shear due to agitation, and nature and ratio of carbon and nitrogen sources.

Based on the mode of operation, closed photobioreactors reported with regard to hydrogen production can be classified as batch, continuous, and fed-batch (Dasgupta et al., 2010). Continuous operation will, in principle, be preferred, for both suspended and immobilized systems. One advantage is that sulfur depletion and repletion can be achieved rapidly. For example, the productivity of hydrogen with immobilized *C. reinhardtii* was 5.5 mL L^{-1} day^{-1} (Laurinavichene et al., 2006).

One of the most cost-effective closed bioreactors is the Biocoil system, which consists of flexible tubing wrapped around a cylindrical support structure through which the microalgal culture is pumped (Borowitzka, 1999). It is relatively cheap, and its modular design makes the scale-up easy. Current R&D efforts focus on the design of higher-efficiency and lower-cost photobioreactors, which require low energy input and are able to generate biomass under aerobic condition and hydrogen under anaerobic condition (Lehr and Posten, 2009).

The highest rates of H_2 release have so far been obtained under mixotrophic conditions with the addition of acetate as carbon source. The acetate is rapidly utilized when the culture switches from aerobic into anaerobic and it is not used as substrate source for hydrogenase (Kruse and Hankamer, 2010).

In terms of light capture, Fresnel lenses have been employed with success to focus light onto the bioreactor (Masojídek et al., 2003) and OFs have been used to conduct light into the bioreactor (Ogbonna and Tanaka, 2000; Janssen et al., 2003). Optimization of culture mixing and cell immobilization has also been studied (Fedorov et al., 2005), as well as compact bioreactor designs. Optimal light capture may be facilitated via light dilution, which spreads strong incident solar radiation over extended surface areas (Amaro et al., 2012).

Microalgal hydrogen production should be conducted in regions characterized by high light intensity so as to maximize incident energy, while minimizing the area required for light capture. However, it is known that microalgae typically become saturated at light levels of ca. 100 μE, and solar radiation levels in subtropical regions are 1000–1500 μE (Torzillo et al., 2009); up to 90% of the incident radiation will actually be dissipated through nonphotochemical quenching under such operating conditions (Torzillo et al., 2009), which is a photoprotective mechanism, thus hampering the yield of H_2 on light.

6.3.4 Optimization of Processing Parameters

Since aerobic environments are detrimental for H_2 production, chemical and mechanical methods have been developed to remove O_2 released by photosynthesis, which includes the addition of oxygen scavengers or reductants or purging with an inert gas. Unfortunately, these methods are relatively expensive and so inconvenient for scale-up (Ghirardi et al., 2000).

In the absence of sulfur but in the presence of light, *C. reinhardtii* decreases its PSII activity to the rate of O_2 uptake by respiration. This implies that its cells will deplete O_2 internally to a sufficiently fast process that can generate an anaerobic microenvironment. Therefore, cells will induce (reversible) hydrogenase and produce H_2 for up to 4 days (Ghirardi et al., 2000). If sulfate is subsequently readded to the spent cultures at high concentration, further cycles of cell growth and H_2 production will occur (Ghirardi et al., 2000). Likewise, in the presence of O_2, a PSII inhibitor, such a process generates 2:1 stoichiometric amounts of H_2/CO_2. Hence, following a sufficiently long dark anaerobic incubation, release of H_2 will take place at high rates upon illumination of the microalgae in the presence of DCMU (Happe and Naber, 1993).

6.4 BARRIERS TO MICROALGAE AS ENERGY FEEDSTOCK

Despite the expected technological feasibility of producing biofuels from microalgae, there are still a number of technical and economic challenges. The development of cost-competitive algal biofuels will require further long-term research, development, and demonstration (Oltra, 2011). If advanced biofuels are to effectively contribute to reduce dependence on fossil fuels, numerous hurdles in such areas as technology, economics, environment, and social politics are still to be scrutinized individually as outlined in the following.

6.4.1 Technological Barriers

There are several processes involved in the production of biodiesel from microalgae, which include culturing, harvesting, drying, extraction, and oil transesterification. As a consistent set, they pose significant challenges in fundamental microalgal physiology and mass culture together with process and overall system engineering (Lundquist, 2010). Substantial improvements in microalgae productivity have meanwhile been achieved through both process and strain improvements (Wijffels and Barbosa, 2010). At this stage, microalgal biofuels will eventually constitute a technically viable alternative to first-generation biofuels (Oltra, 2011).

The technological viability of biohydrogen is strongly dependent on the development of cost-effective, large-scale, sustainable H_2 production systems that can replace the classical processes of steam reforming of natural gas, petroleum refining, and coal gasification (Rupprecht et al., 2006). Furthermore, contributions from material engineering will be a must, which can provide efficient containers and absorbers/adsorbers for liquid hydrogen that minimize leakage and risk of explosion.

6.4.2 Economic Barriers

Large-scale deployment of biofuels from microalgae requires a clear-cut economic feasibility in order to be able to compete with first-generation biofuels and oil diesel. These will not turn out successfully unless production costs and energy requirements are reduced while maximizing lipid productivity (Wijffels and Barbosa, 2010). Production costs vary significantly with the production scale and system of production (open or closed). In particular, harvesting of microalgae remains expensive due to its energy requirements and capital costs (Shirvani et al., 2011).

The use of microalgal biomass that is rich in lipids holds the potential to significantly reduce the use of land compared with other oil crops. In fact, the cost of production of microalgal biomass with an oil content of 30% is 1.11 € kg^{-1} using an open pond and 1.80 \$ kg^{-1} using a photobioreactor (and assuming that CO_2 is available for free). Oil extraction at an approximate yield of 1.14 L kg^{-1} costs at least threefold. This compares unfavorably with crude palm oil, which costs a mere 0.41 € L^{-1}, whereas petrodiesel sells at retail for 0.52–0.77 € L^{-1} (Cheng and Timilsina, 2011).

Overall economic analysis suggests that microalgal biofuel feasibility depends critically on the possibility of production of coproducts with a high market value (Stephens et al., 2010). Stand-alone microalgal systems in a mature biofuel market are not expected until the efficiency of conversion of solar energy to fuel increases substantially: The oil content should increase 2.5-fold, from the current productivity of 20 g m^{-2} day^{-1} with an oil content of 25% dry weight to 50 g m^{-2} day^{-1} and 50%, respectively (Cordell et al., 2009; Stephens et al., 2010). Raising biomass yield by 2.5-fold and oil content from 25% (w/w) (i.e., 23 MJ kg^{-1}) to 50% (i.e., 28 MJ kg^{-1}) would require a 1.2-fold increase in energy conversion under an average irradiation of 22 MJ m^{-2} day^{-1}. In other words, the current photosynthetic conversion efficiency of ca. 2.1% should rise to at least 6.4% (Janssen et al., 2003). Finally, the expensive steps of harvesting and extraction may be overridden by resorting to the secretion of oil by microalgae (or *milking*) (Ramachandra et al., 2009), which simplifies oil collection via solvent interfaces.

It is estimated that the cost of releasing hydrogen from water electrolysis lies somewhere within 2.40–9.40 € kg^{-1} depending on the price of electricity, while hydrogen production by microalgae is estimated to cost ca. 3.8 € kg^{-1} (Lemus and Martínez-Duart, 2010). In what pertains to costs of production of biodiesel, they are likely between 2.35 € and 3 € L^{-1} of oil for photobioreactors and raceway ponds, respectively, assuming that biomass contains 30% (w/w) oil and CO_2 is available for free (e.g., from flue gas) (Demirbas, 2010). Microalgal fuels have a much higher yield than terrestrial oil crops, and microalgae produce 30–100 times more energy per hectare than the terrestrial oil crops.

If consumer demand cannot turn the production of microalgal biofuels into profitable business, governmental incentives may play a critical role (Morrone et al., 2009). In countries such as the United States, the market of first-generation biofuels depends indeed on public subsidies and bio-products are *policy pushed* rather than *market pulled* (Oltra, 2011).

6.4.3 Environmental Barriers

Environmental sustainability of microalgal biofuel production is still a subject to be debated as very few life cycle analyses have been carried out in full with the scarcity of available data (Mendes et al., 2005). Furthermore, the production of advanced biofuels from algal biomass is heavily dependent on direct and indirect energy inputs along the life cycle (Shirvani et al., 2011).

The overall energy balance and the environmental impacts are accordingly the main issues to be considered in this regard. The ratio between the energy that biofuels generate upon combustion and energy required for their production should be carefully calculated as a key element of process sustainability. Achieving a positive energy balance is mandatory for any ecoinnovation—yet production, harvesting, and oil extraction have high energy inputs associated (Lardon et al., 2009). Environmental sustainability of the process will also depend on other impacts such as abiotic and ozone depletion, radiation with a global warming potential, impacts on existing biodiversity, and toxicity to humans (Lardon et al., 2009).

6.4.4 Sociopolitical Barriers

Several social, political, and regulatory challenges are faced by microalgal-derived biofuels, which have not yet received sufficient attention. The level of political acceptance, the type (and extent of use) of various political instruments to promote bio- and ecoinnovation, and the interests of economic agents—further to public acceptance and expectations—are likely to further influence this emerging technology.

Societal impacts of advanced biofuels should be addressed, as long as microalgal biofuels have the potential to supply a significant fraction of the fuel needs of any country. Another significant question is whether advanced biofuels may face the same vulnerabilities of first-generation biofuels, such as those related to land use, food security, equity issues, and loss of family farms (Mol, 2007). Biofuels from microalgae are supposed to have a positive impact on energy security and GHG emissions in advanced societies. However, it is still questionable whether they will be the solution to energy and climate change–elated problems (Oltra, 2011).

6.5 CONCLUSIONS

Response to environmental issues such as global warming cannot be delayed much longer. Otherwise, mankind may be at risk. Biofuels produced from microalgae are one piece of the puzzle in what concerns energy security, since they may eventually replace traditional fossil fuels. Combined with measures on energy efficiency and savings, logistics along the whole value chain, and educated changes in consumer behavior, such biofuels could be helpful to meet energy demands in the near future in a sustainable fashion.

Microalgae offer novel bioenergy systems characterized by much higher oil yields and much lower water demand than terrestrial biomass, as well as much lower costs to produce hydrogen than electrolysis of water. However, the microalgal processes available to date are somewhat marginal in terms of net energy balance and global warming mitigation. It is still a long way before microalgae will entail a profitable contribution as a source of energy. R&D efforts aimed at improving reactor designs and integrating processes are definitely required, which address both reaction engineering and product separation schemes.

Major breakthroughs are accordingly necessary to reduce processing costs, while increasing product yields. Integrated studies in the form of well-funded R&D programs may eventually aid at selecting (and improving) microalgal strains specifically adapted to regional conditions and in optimizing the corresponding processes. In particular, the biorefinery issue will be critical as it will allow upgrading of spent biomass via the production of alternative bulk or fine chemicals that contribute positively to the overall economic framework.

Careful life cycle assessments are also *a sine qua non*, since some biofuels may fail to provide a favorable contribution from the point of view of global sustainability. Critical analysis is required along all such steps, so as to provide an impartial picture of the scenarios available and to eventually back up the correct choices.

NOMENCLATURE

UNITS

% (v/v)	percentage volume per volume
% (w/v)	percentage weight per volume
DW	dry weight
$g\ L^{-1}\ day^{-1}$	biomass productivity
(ca. 0.058 $g\ g^{-1}_{dried\ microalgae}$)	concentration
$g_{dry\ biomass}\ m^{-2}\ day^{-1}$	biomass productivity
$kg_{oil}\ m^{-3}\ day^{-1}$	oil productivity
$m^3\ ha^{-1}$	yield
$mg\ L^{-1}\ day^{-1}$	productivity
$MJ\ kg^{-1}$	energy production
$MJ\ m^{-2}\ day^{-1}$	energy productivity
$mL\ H_2\ L^{-1}\ h^{-1}$	hydrogen productivity
$mL\ H_2\ L^{-1}$	hydrogen production rate
$ng\ cell^{-1}$	concentration
vvm	aeration rate volume per volume per minute
w	weight

ABBREVIATIONS

ADP	adenosine diphosphate
CO_2	carbon dioxide
DCMU	3-(3,4-dichlorophenyl)-1,1-dimethylurea
Fd	ferredoxin
GHG	greenhouse gas
H^+	hydrogen ion
H_2	hydrogen
HCO_3	bicarbonate
LED	light-emitting diodes
MLDP	major lipid droplet protein
mRNA	messenger ribonucleic acid
N	nitrogen
NaCl	sodium chloride
NADP	nicotinamide adenine dinucleotide phosphate
NH_4^+	ammonium
NO_3^-	nitrate
O_2	oxygen
OF	optical fiber
OH^-	hydroxide
PO_4^{3-}	phosphate
PSI	photosystem I
PSII	photosystem II
R&D	research and development
RNAi	interference ribonucleic acid
RuBisCO	ribulose-1,5-bisphosphate carboxylase
Stm6Glc4T7	Stm6Glc4, mutant strains of *C. reinhardtii*
TAGs	triacylglycerols
μE	light concentration (microeinstein)

ACKNOWLEDGMENTS

A PhD fellowship (ref. SFRH/BD/62121/2009, supervised by author F.X.M. and I.S.P.) was granted to author Helena M. Amaro, under the auspices of ESF (III Quadro Comunitário de Apoio) and the Portuguese State. This work received partial financial support by project MICROPHYTE (PTDC/EBB-EBI/102728/2008), coordinated also by author F.X.M. and under the auspices of ESF and the Portuguese State.

REFERENCES

Amaro, H. M., Guedes, A. C., and Malcata, F. X. 2011. Advances and perspectives in using microalgae to produce biodiesel. *Applied Energy*, 88, 3402–3410.

Amaro, H. M., Macedo, A. C., and Malcata, F. X. 2012. Microalgae: An alternative as sustainable source of biofuels? *Energy*, 44, 158–166.

Aslan, S. and Kapdan, I. K. 2006. Batch kinetics of nitrogen and phosphorus removal from synthetic wastewater by algae. *Ecological Engineering*, 28, 64–70.

Borowitzka, M. A. 1999. Commercial production of microalgae: Ponds, tanks, tubes and fermenters. *Journal of Biotechnology*, 70, 313–321.

Bozbas, K. 2008. Biodiesel as an alternative motor fuel: Production and policies in the European Union. *Renewable and Sustainable Energy Reviews*, 12, 542–552.

Carvalho, A. P., Meireles, L. A., and Malcata, F. X. 2008. Microalgal reactors: A review of enclosed system designs and performances. *Biotechnology Progress*, 22, 1490–506.

Chen, C. Y., Yeh, K. L., Aisyah, R., Lee, D. J., and Chang, J. S. 2011. Cultivation, photobioreactor design and harvesting of microalgae for biodiesel production: A critical review. *Bioresource Technology*, 102, 71–81.

Cheng, J. J., Timilsina, G. R. 2011. Status and barriers of advanced biofuel technologies: A review. *Renewable Energy*; 36, 3541–3549.

Chisti, Y. 2007. Biodiesel from microalgae. *Biotechnology Advances*, 25, 294–306.

Chiu, S.-Y., Kao, C.-Y., Chen, C.-H., Kuan, T.-C., Ong, S.-C., and Lin, C.-S. 2008. Reduction of CO_2 by a high-density culture of *Chlorella* sp. in a semicontinuous photobioreactor. *Bioresource Technology*, 99, 3389–3396.

Chojnacka, K. and Marquez-Rocha, F.-J. 2004. Kinetic, stoichiometric relationships of the energy and carbon metabolism in the culture of microalgae. *Biotechnology Advances*, 3, 21–34.

Cordell, D., Drangert, J.-O., White, S. 2009. The story of phosphorus: global food security and food for thought. *Global Environmental Change*, 19, 292–305.

Danquah, M. K., Ang, L., Uduman, N., Moheimani, N., and Forde, G. M. 2009. Dewatering of microalgal culture for biodiesel production: Exploring polymer flocculation and tangential flow filtration. *Journal of Chemical Technology & Biotechnology*, 84, 1078–1083.

Dasgupta, C. N., Jose Gilbert, J., Lindblad, P., Heidorn, T., Borgvang, S. A., Skjanes, K., and Das, D. 2010. Recent trends on the development of photobiological processes and photobioreactors for the improvement of hydrogen production. *International Journal of Hydrogen Energy*, 35, 10218–10238.

Demirbas, A. 2010. Use of algae as biofuel sources. *Energy Conversion and Management*, 51, 2738–2749.

Doebbe, A., Rupprecht, J., Beckmann, J., Mussgnug, J. H., Hallmann, A., Hankamer, B., and Kruse, O. 2007. Functional integration of the HUP1 hexose symporter gene into the genome of *C. reinhardtii*: Impacts on biological H2 production. *Journal of Biotechnology*, 131, 27–33.

Ehimen, E. A., Sun, Z. F., and Carrington, C. G. 2010. Variables affecting the in situ transesterification of microalgae lipids. *Fuel*, 89, 677–684.

El-Shishtawy, R. M. A., Kawasaki, S., and Morimoto, M. 1997. Biological H2 production using a novel light-induced and diffused photoreactor. *Biotechnology Techniques*, 11, 403–407.

Fedorov, A. S., Kosourov, S., Ghirardi, M. L., and Seibert, M. 2005. Continuous hydrogen photoproduction by *Chlamydomonas reinhardtii*. In: Davison, B. H., Evans, B. R., Finkelstein, M., and McMillan, J. D. (eds.) *Applied Biochemistry and Biotechnology. Proceedings of the 26th Symposium on Biotechnology for Fuels and Chemicals*. Chattanooga, TN: Humana Press.

Florin, L., Tsokoglou, A., and Happe, T. 2000. A novel type of Fe-hydrogenase in the green alga Scenedesmus obliquus is linked to the photosynthetic electron transport chain. *Journal of Biological Chemistry*, 276, M008470200.

Geciova, J., Bury, D., and Jelen, P. 2002. Methods for disruption of microbial cells for potential use in the dairy industry—A review. *International Dairy Journal*, 12, 541–553.

Gfeller, R. P., Gibbs, M. 1984. Fermentative metabolism of *Chlamydomonas reinhardtii*: I. Analysis of fermentative products from starch in dark-light. *Plant Physiology*, 75, 212–218.

Ghirardi, M. L., Zhang, L., Lee, J. W., Flynn, T., Seibert, M., Greenbaum, E., and Melis, A. 2000. Microalgae: A green source of renewable H2. *Trends in Biotechnology*, 18, 506–511.

Golueke, C. G. and Oswald, W. J. 1965. Harvesting and processing sewage-grown planktonic algae. *Journal Water Pollution Control Federation*, 37, 471–498.

Gong, Y. and Jiang, M. 2011. Biodiesel production with microalgae as feedstock: From strains to biodiesel. *Biotechnology Letters*, 33, 1269–1284.

Gressel, J. 2008. Transgenics are imperative for biofuel crops. *Plant Science*, 174, 246–263.

Guedes, A. C., Amaro, H. M., Barbosa, C. R., Pereira, R. D., and Malcata, F. X. 2011. Fatty acid composition of several species of microalgae and cyanobacteria, with a focus on eicosapentaenoic, docosahexaenoic and α-linolenic acids, for eventual dietary uses. *Food Research International*, 44, 2721–2729.

Halim, R., Gladman, B., Danquah, M. K., and Webley, P. A. 2011. Oil extraction from microalgae for biodiesel production. *Bioresource Technology*, 102, 178–185.

Happe, T. and Naber, J. D. 1993. Isolation, characterization and N-terminal amino acid sequence of hydrogenase from the green alga *Chlamydomonas reinhardtii*. *European Journal of Biochemistry*, 214, 475–481.

Hill, J., Nelson, E., Tilman, D., Polasky, S., and Tiffany, D. 2006. Environmental, economic, and energetic costs and benefits of biodiesel and ethanol biofuels. *Proceedings of the US National Academy of Sciences*, 103, 11206–11210.

Hu, Q., Sommerfeld, M., Jarvis, E., Ghirardi, M., Posewitz, M., Seibert, M., and Darzins, A. 2008. Microalgal triacylglycerols as feedstocks for biofuel production: Perspectives and advances. *Plant Journal*, 54, 621–639.

Huang, G., Chen, F., Wei, D., Zhang, X., and Chen, G. 2010. Biodiesel production by microalgal biotechnology. *Applied Energy*, 87, 38–46.

Illman, A. M., Scragg, A. H., and Shales, S. W. 2000. Increase in Chlorella strains calorific values when grown in low nitrogen medium. *Enzyme and Microbial Technology*, 27, 631–635.

Issarapayup, K., Powtongsook, S., and Pavasant, P. 2009. Flat panel airlift photobioreactors for cultivation of vegetative cells of microalga *Haematococcus pluvialis*. *Journal of Biotechnology*, 142, 227–232.

Janssen, M., Tramper, J., Mur, L. R., and Wijffels, R. H. 2003. Enclosed outdoor photobioreactors: Light regime, photosynthetic efficiency, scale-up, and future prospects. *Biotechnology and Bioengineering*, 81, 193–210.

Khozin-Goldberg, I. and Cohen, Z. 2006. The effect of phosphate starvation on the lipid and fatty acid composition of the fresh water eustigmatophyte *Monodus subterraneus*. *Phytochemistry*, 67, 696–701.

Kruse, O. and Hankamer, B. 2010. Microalgal hydrogen production. *Current Opinion in Biotechnology*, 21, 238–243.

Kruse, O., Rupprecht, J., Mussgnug, J. H., Dismukes, G. C., and Hankamer, B. 2005. Photosynthesis: A blueprint for solar energy capture and biohydrogen production technologies. *Photochemical & Photobiological Sciences*, 4, 957–970.

Lardon, L., Hélias, A., Sialve, B., Steyer, J. P., and Bernard, O. 2009. Life-cycle assessment of biodiesel production from microalgae. *Environmental Science & Technology*, 43, 6475–6481.

Laurinavichene, T. V., Fedorov, A. S., Ghirardi, M. L., Seibert, M., and Tsygankov, A. A. 2006. Demonstration of sustained hydrogen photoproduction by immobilized, sulfur-deprived *Chlamydomonas reinhardtii* cells. *International Journal of Hydrogen Energy*, 31, 659–667.

Lee, S., Yoon, B.-D., and OH, H.-M. 1998. Rapid method for the determination of lipid from the green alga *Botryococcus braunii*. *Biotechnology Techniques*, 12, 553–556.

Lehr, F. and Posten, C. 2009. Closed photo-bioreactors as tools for biofuel production. *Current Opinion in Biotechnology*, 20, 280–285.

Lemus, R. G. and Martínez-Duart, J. M. 2010. Updated hydrogen production costs and parities for conventional and renewable technologies. *International Journal of Hydrogen Energy*, 35, 3929–3936.

Liang, Y., Sarkany, N., and Cui, Y. 2009. Biomass and lipid productivities of *Chlorella vulgaris* under autotrophic, heterotrophic and mixotrophic growth conditions. *Biotechnology Letters*, 31, 1043–1049.

Liu, Z.-Y., Wang, G.-C., and Zhou, B.-C. 2008. Effect of iron on growth and lipid accumulation in *Chlorella vulgaris*. *Bioresource Technology*, 99, 4717–4722.

Lundquist, T. J., Woertz, I. C., Quinn, N. W. T., Benemann, J. 2010. A realistic technology and engineering assessment for algal biofuels production. Energy Biosciences Institute, Univ California, Berkley, CA.

Ma, F. and Hanna, M. A. 1999. Biodiesel production: A review. *Bioresource Technology*, 70, 1–15.

Macías-Sánchez, M. D., Mantell, C., Rodríguez, M., Martínez de la Ossa, E., Lubián, L. M., and Montero, O. 2007. Supercritical fluid extraction of carotenoids and chlorophyll a from *Synechococcus* sp. *The Journal of Supercritical Fluids*, 39, 323–329.

Masojídek, J., Papáček, Š., Sergejevová, M., Jirka, V., Červený, J., Kunc, J., Korečko, J. et al. 2003. A closed solar photobioreactor for cultivation of microalgae under supra-high irradiance: Basic design and performance. *Journal of Applied Phycology*, 15, 239–248.

Mata, T. M., Martins, A. A., and Caetano, N. S. 2010. Microalgae for biodiesel production and other applications: A review. *Renewable and Sustainable Energy Reviews*, 14, 217–232.

Melis, A. 2009. Solar energy conversion efficiencies in photosynthesis: Minimizing the chlorophyll antennae to maximize efficiency. *Plant Science*, 177, 272–280.

Melis, A., Neidhardt, J., and Benemann, J. 1998. Dunaliella salina (Chlorophyta) with small chlorophyll antenna sizes exhibit higher photosynthetic productivities and photon use efficiencies than normally pigmented cells. *Journal of Applied Phycology*, 10, 515–525.

Mendes, R. L., A. D. Reis, and A. F. Palavra. 2006. Supercritical CO_2 extraction of γ-linolenic acid and other lipids from *Arthrospira* (Spirulina) maxima: Comparison with organic solvent extraction. *Food Chemistry*, 99, 57–63.

Metzger, P. and Largeau, C. 2005. *Botryococcus braunii*: A rich source for hydrocarbons and related ether lipids. *Applied Microbiology and Biotechnology*, 66, 486–496.

Moellering, E. R., Benning, C. 2010. RNA interference97-106. Silencing of a major lipid droplet protein affects lipid droplets sizes in *Chlamydomonas reinhardtii*. *Eukaryot Cell*, 9, 97–106.

Mol, A. P. J. 2007 Boundless biofuels. *Sociologia Ruralis*, 47, 297–315.

Morrone, M., Stuart, B. J., McHenry, I., and Buckley, G. 2009. The challenges of biofuels from the prespectives of small-scale producers in Ohio. *Energy policy*, 37, 522–530.

Mussgnug, J. H., Thomas-Hall, S., Rupprecht, J., Foo, A., Klassen, V., Mcdowall, A., Schenk, P. M., Kruse, O., and Hankamer, B. 2007. Engineering photosynthetic light capture: Impacts on improved solar energy to biomass conversion. *Plant Biotechnology Journal*, 5, 802–814.

Ogbonna, J. C. and Tanaka, H. 2000. Light requirement and photosynthetic cell cultivation—Development of processes for efficient light utilization in photobioreactors. *Journal of Applied Phycology*, 12, 207–218.

Oltra, C. 2011. Stakeholder perceptions of biofuels from microalgae. *Energy Policy*, 39, 1774–1781.

Raja, R., Hemaiswarya, S., Kumar, N. A., Sridhar, S., and Rengasamy, R. 2008. A perspective on the biotechnological potential of microalgae. *Critical Reviews in Microbiology*, 34, 77–88.

Ramachandra, T. V., Mahapatra, D. M., and Karthick, B. 2009. Milking diatoms for sustainable energy: biochemical engineering versus gasoline-secreting diatom solar panels. *Industrial & Engineering Chemistry Research*, 48, 8769–8788.

Richmond, A. 2004. *Handbook of Microalgal Culture: Biotechnology and Applied Phycology*, Oxford, U.K.: Wiley-Blackwell

Rodolfi, L., Chini Zittelli, G., Bassi, N., Padovani, G., Biondi, N., Bonini, G., and Tredici, M. R. 2009. Microalgae for oil: Strain selection, induction of lipid synthesis and outdoor mass cultivation in a low-cost photobioreactor. *Biotechnology and Bioengineering*, 102, 100–112.

Rossignol, N., Vandanjon, L., Jaouen, P., and Quéméneur, F. 1999. Membrane technology for the continuous separation microalgae/culture medium: Compared performances of cross-flow microfiltration and ultrafiltration. *Aquacultural Engineering*, 20, 191–208.

Rupprecht, J., Hankamer, B., and Mussgnug, J. H. 2006. Perspectives and advances of biological H_2 production in microorganisms. *Applied Microbiology and Biotechnology*, 72, 442–449.

Sánchez, J. F., Fernández, J. M., Acién, F. G., Rueda, A., Pérez-Parra, J., and Molina, E. 2008. Influence of culture conditions on the productivity and lutein content of the new strain Scenedesmus almeriensis. *Process Biochemistry*, 43, 398–405.

Santos, G. M. D., Macedo, R. V. T., and Alegre, R. M. 2001. Influência do teor de nitrogênio no cultivo de Spirulina maxima em dois níveis de temperatura—Parte II. Produção de lipídios. *Ciência e Tecnologia Alimentar de Campinas* 21, 183–186.

Scott, S. A., Davey, M. P., Dennis, J. S., Horst, I., Howe, C. J., Lea-Smith, D. J., and Smith, A. G. 2010. Biodiesel from algae: Challenges and prospects. *Current Opinion in Biotechnology*, 21, 277–286.

Scragg, A. H., Illman, A. M., Carden, A., and Shales, S. W. 2002. Growth of microalgae with increased calorific values in a tubular bioreactor. *Biomass and Bioenergy*, 23, 67–73.

Shirvani, T., Yan, X., Inderwildi, O. R., Edwards, P. P., and King, D. A. 2011. Life cycle energy and greenhouse gas analysis for algae-derived biodiesel. *Energy & Environmental Science*, 4, 3773–3778.

Singh, J. and Gu, S. 2010. Commercialization potential of microalgae for biofuels production. *Renewable and Sustainable Energy Reviews*, 14, 2596–2610.

Stephens, E., Ross, I. L., King, Z., Mussgnug, J. H., Kruse, O., Posten, C., Borowitzka, M. A., and Hankamer, B. 2010. An economic and technical evaluation of microalgal biofuels. *Nature Biotechnology*, 28, 126–128.

Stephenson, A. I., Dennis, J. S., Howe, C. J., Scott, S. A., and Smith, A. G. 2010. Influence of nitrogen-limitation regime on the production by *Chlorella vulgaris* of lipids for biodiesel feedstocks. *Biofuels*, 1, 47–58.

Thana, P., Machmudah, S., Goto, M., Sasaki, M., Pavasant, P., and Shotipruk, A. 2008. Response surface methodology to supercritical carbon dioxide extraction of astaxanthin from *Haematococcus pluvialis*. *Bioresource Technology*, 99, 3110–3115.

Torzillo, G., Scoma, A., Faraloni, C., Ena, A., and Johanningmeier, U. 2009. Increased hydrogen photoproduction by means of a sulfur-deprived *Chlamydomonas reinhardtii* D1 protein mutant. *International Journal of Hydrogen Energy*, 34, 4529–4536.

Wang, Z. T,. Ullrich, N., Joo, S., Waffenschmidt, S., Goodenough, U. 2009. Algal lipid bodies: Stress induction, purification, and biochemical characterization in wild-type and starch-less *Chlamydomonas reinhardtii*. *Eukaryot Cell*, 8, 1856–1868.

Wijffels, R. H. and Barbosa, M. 2010. An outlook on microalgal biofuels. *Science* 329, 796–799.

Xiong, W., Li, X., Xiang, J., and Wu, Q. 2008. High-density fermentation of microalga *Chlorella protothecoides* in bioreactor for microbio-diesel production. *Applied Microbiology and Biotechnology*, 78, 29–36.

Xu, H., Miao, X., and Wu, Q. 2006. High quality biodiesel production from a microalga Chlorella protothecoides by heterotrophic growth in fermenters. *Journal of Biotechnology*, 126, 499–507.

Zhu, X. G., Long, S. P., Ort, D. R. 2008. What is the maximum efficiency with which photosynthesis can convert solar energy into biomass? *Current Opinion in Biotechnology*. 19, 153–159.

7 Production and Supply Logistics of Switchgrass as an Energy Feedstock

Marty R. Schmer and Igathinathane Cannayen

CONTENTS

7.1 INTRODUCTION

Switchgrass (*Panicum virgatum* L.) is a warm-season (C_4), perennial grass that is native to the tall grass ecoregion of North America (Figure 7.1). Historically, switchgrass has been used for summer forage, hay, and ensiling or in conservation plantings. At the end of the twentieth century, switchgrass was designated as a bioenergy crop in the United States with the goal of converting the lignocellulosic portion to a renewable transportation fuel. Switchgrass has high biomass production potential, wide adaptability, and low fossil fuel energy requirements and is compatible with modern agriculture practices making it an ideal herbaceous energy crop. At present, significant research has been conducted on switchgrass to better understand switchgrass genetics and agronomic management and develop reliable supply chains for a mature bioenergy industry. Although maximizing biomass production is important, a holistic approach needs to be acknowledged to meet numerous ecosystem services for long-term sustainability and profitability. In this chapter, we identify best management practices to ensure switchgrass productivity and efficient logistics for an emerging bioenergy industry.

7.2 BIOLOGY AND GENETICS

Switchgrass originated approximately two million years ago (Huang et al. 2003). North American glacial and interglacial periods had a major influence on switchgrass habitat and diversity with the majority of diversity originating from the Atlantic Seaboard region following the last glacial episode (Zhang et al. 2011). Switchgrass has several ploidy levels that are found in nature (Moser and Vogel 1995) but primarily tetraploid or octaploid with a base chromosome number of nine ($x=9$)

FIGURE 7.1 Switchgrass field in Nebraska, United States. Switchgrass has been used for forage and conservation plantings in the Great Plains and is being developed as a bioenergy crop on marginal lands. (Photo courtesy of Marty Schmer [USDA-ARS].)

(Taliaferro et al. 1999). Switchgrass is primarily self-incompatible and postfertilization incompatibility occurs when plants of two different ploidy levels are crossed (Martinez-Reyna and Vogel 2002). Switchgrass has two distinct ecotypes but intermediates or natural *hybrids* have been found in nature (Zhang et al. 2011). Understanding different switchgrass ecotypes and their respective biogeographic adaptations ensures optimal production and survivability.

7.2.1 Morphological and Physiological Description

Switchgrass leaf blades are flat with a patch of hair on the blade near the collar. The stems of switchgrass are round, smooth, and erect. Most genotypes are caespitose in appearance with some genotypes being rhizotamus (Vogel 2004). Switchgrass propagates vegetatively and by seeds through open pollination (Moser and Vogel 1995). The inflorescence of switchgrass is a diffuse panicle with spikelets at the end of long branches. Spikelets are two flowered with the first floret being sterile and the second floret being fertile. Switchgrass is photoperiod sensitive, requiring shortened day length to induce flowering (Benedict 1941). Switchgrass morphology is strongly correlated to day of year (DOY) and growing degree days (GDDs) (Mitchell et al. 1997). Switchgrass has a determinate growth habit where most vegetative growth terminates with inflorescence development (Dahl and Hyder 1977; Mitchell et al. 1997). Switchgrass contains a fibrous root system that extends 2.7–3.3 m below the soil surface (Weaver and Darland 1949).

7.2.2 Ecotypes

Switchgrass is morphologically divided into upland and lowland ecotypes. Lowland ecotypes are taller and have longer, more bluish-green leaves, longer ligules, higher yielding, and a coarser stem than upland ecotypes. Upland ecotypes are shorter growing, have a finer stem, and are more tolerant of dry climatic conditions than lowland ecotypes (Moser and Vogel 1995). Upland ecotypes, under drought stress, have higher CO_2 exchange rates and greater leaf water potential and recover from stress faster than lowland ecotypes (Nickell 1973). Lowland ecotypes are mainly tetraploid, while upland ecotypes are tetraploid, hexaploid, and octaploid (Taliaferro et al. 1999). Examples of upland ecotype cultivars are *Cave-in-Rock*, *Trailblazer*, and *Sunburst* (Figure 7.2a). Examples of popular lowland ecotype cultivars are *Alamo* and *Kanlow* (Figure 7.2a). In general, lowland ecotypes are adapted to the southern United States, while upland ecotypes are adapted to the northern United States (Figure 7.2b). Further subdividing of ecotypes into northern upland, southern upland, northern lowland, and southern lowland has been used to quantify switchgrass responses to latitudinal effects (Casler et al. 2004). Proper cultivar selection for bioenergy purposes is important for long-term crop viability and profitability. The use of plant adaptation regions (Figure 7.2a), a combination of ecoregions and hardiness zones, has been found to be compatible with empirical agronomics and population stability (Vogel et al. 2005; Casler et al. 2007).

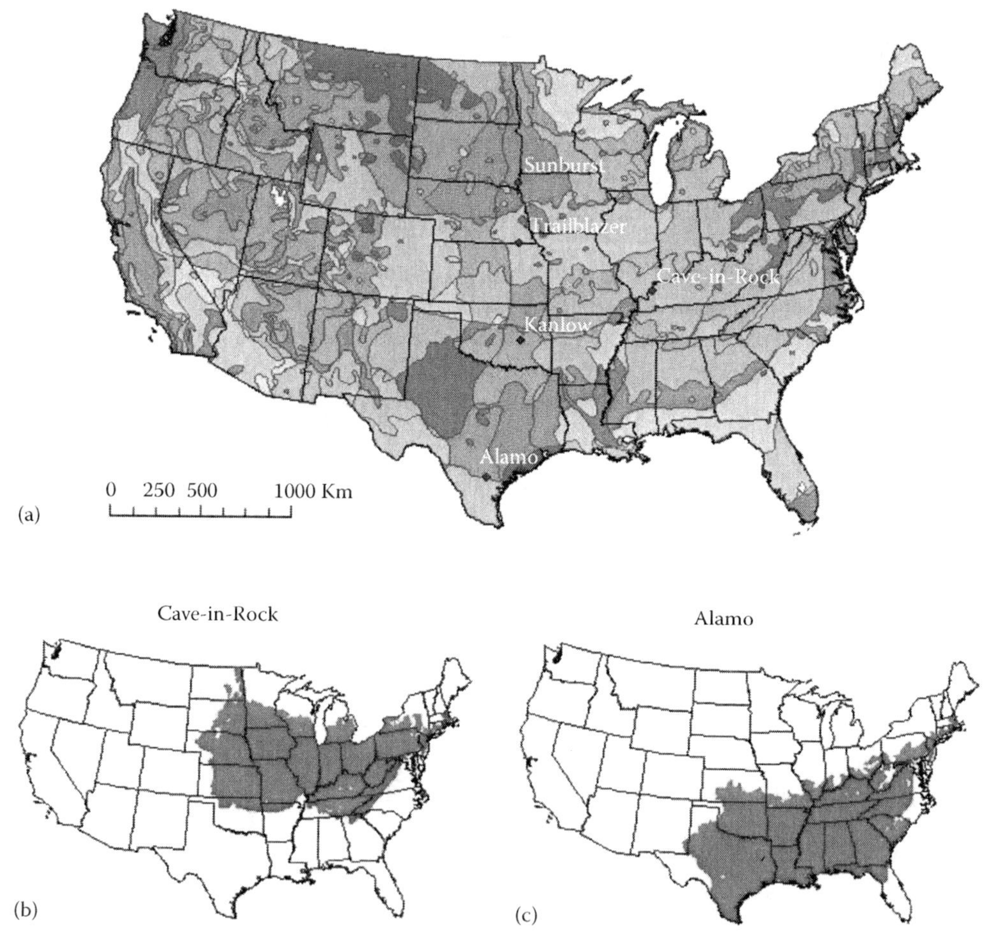

FIGURE 7.2 Plant adaptation region map for the conterminous United States showing the origins of popular switchgrass upland and lowland cultivars (a). Plant adaptation regions are the combination of ecoregions and USDA plant hardiness zones. (From Vogel, K.P. et al., *Rangeland Ecol. Manage.*, 58, 315, 2005.) Plant adaptation regions for upland cultivar *Cave-in-Rock* (b) and lowland cultivar *Alamo* (c).

7.2.3 Breeding

The majority of released switchgrass cultivars were developed for conservation or traits desirable for pasture grazing. Current switchgrass varieties tend to be from (1) seed increases from random plants originating from a remnant prairie or (2) improved cultivars based on selection of defined traits such as yield, digestibility, seed dormancy, or disease resistance (Casler et al. 2009). Switchgrass cultivars have largely been developed in the Central Plains region of the United States (Figure 7.2). Traditional breeding techniques have increased yield performance of switchgrass by 20%–30% from existing parent types (McLaughlin and Kszos 2005). Extensive genetic variability exists among and within switchgrass cultivars for biomass and nutrient efficiency (Yang et al. 2009). The potential for increasing biomass yield in switchgrass is significant due to large genetic variation from existing populations. Development of F1 hybrids derived from upland and lowland ecotypes has shown increased biomass yields compared with the parental lines (Vogel and Mitchell 2008). Breeding for specific bioenergy traits will be dependent on the conversion process used within a given region (Sanderson et al. 2006). It is expected that further improvements in both genetics (hybrid cultivars, molecular markers) and agronomics (production system management practices and inputs) will further improve switchgrass biomass yields.

7.3 AGRONOMICS

The Green Revolution greatly enhanced the capacity of agriculture to increase food supplies throughout the world by the use of improved genetics and agronomic management (Khush 1999). Maize (*Zea mays* L.) genetic gains in the United States have contributed 50%–60% to the overall grain yield gains with the remaining improvement from agronomic management (Duvick and Cassman 1999). Management research on switchgrass has a shorter history than maize, but potential gains in biomass yields are likely similar. Switchgrass for bioenergy purposes will require rapid establishment and effective weed control to maximize yield potential and reduce economic costs (Perrin et al. 2008).

7.3.1 Establishment

Establishment success is dependent on proper planning. Poor stands can cause economic loss and reduce biomass yield potential for one or more years (Schmer et al. 2006; Perrin et al. 2008). Key factors to consider prior to establishment are measuring soil fertility status, proper field preparation, proper cultivar selection, effective weed control, and using high-quality seed (Sanderson et al. 2012). Soil sampling before seeding is recommended to determine soil nutrient status and pH. Management practices to transition from conservation land to switchgrass may require autumn or spring hay harvest, prescribed burning in autumn or spring, pre- and postherbicide treatments, tillage, and the possible use of companion crops. Planting conservation land or pastures with herbicide-resistant soybeans or spring-planted cover crop mixtures 1–2 years before switchgrass establishment is an effective management tool for seedbed preparation (Sanderson et al. 2012).

7.3.1.1 Seed Quality

Seed quality parameters include viability, purity, cleanliness, and vigor (Vogel 2002). Seed vigor relates to the ability of a seed to germinate and establish a viable seedling under field conditions. Switchgrass seedlots with the same germination percentage but different seedling vigor have resulted in emergence differences of more than 40% (Vogel 2002). Average seed weight for switchgrass is 850 seeds g^{-1} but variation exists among and within cultivars for seed size (Vogel 2004). High germination rates and greater emergence have been reported for heavy switchgrass seed than lighter switchgrass seed (Aiken and Springer 1995; Green and Bransby 1995). Switchgrass seedlings from heavy seed had greater germination and earlier growth than lighter seed, but no growth differences were detected 8–10 weeks after emergence (Smart and Moser 1999).

Seed dormancy is common in perennial grasses and negatively affects seedling vigor and establishment. Dormancy is likely caused by genetics and environmental effects during seed production, harvesting, and processing. Primary dormancy of switchgrass seed can generally be broken by an after-ripening period or by cold stratification (Moser and Vogel 1995). Genetic selection for low-dormancy seed has been proven to lower overall primary dormancy in lowland ecotypes (Sanderson et al. 1996). Chemical treatments such as gibberellic acid, sodium nitroprusside, potassium ferrocyanide, potassium nitrate, polyethylene glycol, and hydrogen peroxide have been used to reduce residual dormancy in switchgrass, but overall success of each treatment is cultivar-specific (Madakadze et al. 2000; Sarath and Mitchell 2008).

7.3.1.2 Field Preparation

Switchgrass establishes best on well-drained, fertile soils, but it can establish and persist under highly variable soil conditions ranging from sand to clay loam (Vogel 2004). Optimal seed germination occurs between pH values of 6 and 8 (Hanson and Johnson 2005), but seedlings can tolerate pH values of 3.7–7.6 under field conditions (Vogel 2004; McLaughlin and Kszos 2005; Parrish and Fike 2005). Switchgrass establishment through direct seeding using a cultipacker drill, grain drill, or no-till drill is recommended. A firm seedbed is required for proper seed placement regardless of planting method since switchgrass is planted at a shallow depth (1–2 cm). Seedbed preparation is based on equipment accessibility, soil erosion concerns, previous crop, and initial soil moisture conditions.

Various tillage systems can be used to establish switchgrass (Parrish and Fike 2005). Conventional tillage reduces cool-season weed populations prior to seeding as well as reduces residue from previous cropping systems that interfere with shallow seed placement. Soil temperatures are generally higher under conventional tillage than no-tillage practices during spring seeding. Following a tillage practice, soil firming is required for proper seed placement and to provide good seed-to-soil contact at planting. Conventional tillage practices are not recommended if soil erosion is a concern. Soil carbon loss by tillage practices can be significant, especially on land that was previously in set aside programs or other perennial grassland system (Fargione et al. 2008).

No-tillage seeding of switchgrass has been effective in establishing switchgrass under variable climate conditions and previous cropping systems, especially soybeans. No-tillage seeding provides greater water conservation benefits than conventional tillage especially near the soil surface. Establishing switchgrass on land formerly in conservation or pastures has been done using either conventional tillage or no-tillage methods (Parrish and Fike 2005). For bioenergy purposes, both preemergent and postemergent herbicides are critical under no-tillage practices to control or reduce weed populations during the establishment year (Sanderson et al. 2012).

7.3.1.3 Planting Date

Switchgrass planting dates are similar to planting times for maize. Successful switchgrass establishment is determined by seeding depth, soil texture, soil moisture, and soil temperature (Parrish and Fike 2005). Switchgrass has a panicoid type of seedling root development that places the crown node (from which adventitious roots develop) near the soil surface by elongation of the subcoleoptile internode or mesocotyl (Hoshikawa 1969). The subcoleoptile internode has a relatively small xylem cross-sectional area in switchgrass that limits the amount of water and nutrients that can be transported (Moser 2000). The most important factor for switchgrass to develop adventitious roots is adequate soil surface moisture over a period of time (Newman and Moser 1988). Warm-season grasses will initiate adventitious roots at 2–4 weeks after emergence under favorable environmental conditions (Moser 2000). Adventitious roots have a larger xylem cross-sectional area than the subcoleoptile internode resulting in higher water uptake and nutrient absorption than the primary root system (Moser 2000). Switchgrass will typically have one adventitious root by the three-leaf emergence stage (Newman and Moser 1988). Soil temperatures above 20°C are required for switchgrass germination and seedling growth (Hsu and Nelson 1986). Optimal germination of several switchgrass cultivars is between 27°C and 30°C (Dierberger 1991), but germination tolerance to temperature varies by cultivar (Seepaul et al. 2011).

7.3.1.4 Seeding Rate

Seeding rates for switchgrass are traditionally based on pure live seed (PLS). PLS is calculated as purity (%) multiplied by germination (%) divided by 100. Switchgrass seed purity is determined on a mass balance (pure seed weight/seedlot weight). Recommended seeding rates for switchgrass are 200–400 PLS m^{-2}. Seeding rates less than 200 PLS m^{-2} will give sufficient stands for conservation plantings with adequate weed control (Vogel 1987). Low to high seeding rates have resulted in similar biomass yields during postestablishment years in the southeastern United States (West and Kincer 2011). Seed vigor tests that evaluate emergence from depth or accelerated aging tests have been recommended as alternative methods in determining field seeding rates (Asay and Johnson 1980; Hall and Wiesner 1990). Switchgrass seeding rates based on seedling emergence tests were better indicators of field establishment than the standard PLS method (Mitchell and Vogel 2012).

7.3.1.5 Weed Control

Weed competition is a major cause for switchgrass establishment stand failure. Acceptable switchgrass production can be delayed by one or more years by weed competition and poor stand establishment (Schmer et al. 2006). Weeds that are the most common in establishing warm-season grasses are annual grasses like crabgrass [*Digitaria sanguinalis* (L.) Scop.], green foxtail [*Setaria viridis* (L.) Beauv.], yellow foxtail [*Setaria glauca* (L.) Beauv.], fall panicum (*Panicum dichotomiflorum* Michx.), and barnyardgrass [*Echinochloa crusgalli* (L.) Beauv.] (Mitchell and Britton 2000). Recommended practices for controlling weeds in switchgrass fields are the use of preemergent herbicides particularly for annual grass control. Nonselective herbicides, such as glyphosate [*N*-(phosphonomethyl) glycine], are effective in weed control prior to switchgrass emergence. Switchgrass treated with preemergent and postemergent combinations of herbicides on upland and lowland ecotypes has been found to be effective to control annual weeds during the establishment year (Boydston et al. 2010; Mitchell et al. 2010a). Broadleaf weed control using a mechanical treatment (mowing) is successful when broadleaf weeds are taller than switchgrass and the mowing application can be done to minimize switchgrass leaf loss (Elbersen et al. 2004). If adequate weed control occurs during the establishment year, limited herbicide use is required postestablishment.

7.3.2 Postestablishment Management

A sustainable perennial bioenergy system will require maximizing energy output per unit of land, maximizing nutrient use efficiency, maintaining or enhancing ecosystem services, and minimizing greenhouse gas (GHG) emissions. Switchgrass harvests that maximize biomass yield potential while efficiently utilizing nutrient and soil resources are dependent on proper management. Nutrient management of switchgrass for bioenergy is required for optimizing biomass yields, quality composition characteristics, and costs. Key management practices for switchgrass production are efficient nutrient management, pest control, and harvest strategies to ensure long-term sward viability.

7.3.2.1 Nutrient Management

Nitrogen (N) fertilizer is not recommended during the planting year since N will encourage weed growth, increase competition for establishing seedlings, increase establishment cost, and increase economic risk associated with establishment if stands should fail (Mitchell et al. 2008). Available N found in the soil during the establishment year is generally sufficient for switchgrass development. Nitrogen is the primary limiting nutrient for switchgrass requiring fertilizer for optimal yield production during the postestablishment phase. Nitrogen management will be dependent on genotype, soil type, harvest management, and climate conditions. Switchgrass follows a quadratic yield response to increased N in aboveground and belowground biomass with the amount of N required dependent on climate and harvest date (Haas 1958; Heggenstaller et al. 2009). For example, Cave-in-Rock switchgrass had optimal N application of 120 kg ha^{-1} in the Central Plains, while N application of 50 kg ha^{-1} was optimal in Virginia (Parrish and Wolf 1992; Vogel et al. 2002). In Europe, switchgrass is generally harvested after senescence and yield response to N is minimal (Christian et al. 2002; Lasorella et al. 2011). Fertilizer timing is important in increasing nutrient use efficiency. Applying N too early in the spring or too late in the summer encourages cool-season grass invasion because of the inability of warm-season grasses to efficiently take up N at these times of the year.

Applying 10–12 kg N ha^{-1} for each Mg ha^{-1} biomass removed under a single late-summer harvest or autumn (postkilling frost) harvest is recommended in the Central Plains (Vogel et al. 2002). Increasing the N rate in switchgrass will increase N removal rates (Stout et al. 1991; Vogel et al. 2002). Aboveground switchgrass tiller N concentrations are lower after a killing frost harvest than from a growing season harvest (Vogel et al. 2002). Nitrogen rates can be reduced on switchgrass harvested after a killing frost because N is translocated down to the crown, roots, and rhizomes (see Section 7.3.3.3).

Available soil phosphorus (P) levels exceeding 25 mg kg^{-1} are recommended when establishing warm-season grasses (Rehm et al. 1976; Rehm 1984). Lemus et al. (2009) recommended available soil P levels ranging from 2 to 15 mg kg^{-1} to be sufficient for established switchgrass stands. In general, P requirements are low because of a symbiotic relationship between arbuscular mycorrhizal fungi (AMF) and switchgrass (Morris et al. 1982; Brejda 2000). Colonization of switchgrass roots by AMF increases plant uptake of P even at low soil P levels (Brejda et al. 1993). Mycorrhizal fungi enhance drought tolerance, water uptake, and nutrient uptake. The presence of AMF in sandy soils is especially important for warm-season grasses to become established. Brejda et al. (1993) found that switchgrass had a higher root weight, a higher shoot weight, and higher phosphorous uptake when inoculated with mycorrhizae in sandy soils. Phosphorus application rates of 10–39 kg ha^{-1} did not increase switchgrass dry matter (DM) production with the control (0 kg ha^{-1}) in Texas (Muir et al. 2001) nor did P application rates of 20 or 40 kg ha^{-1} affect switchgrass DM production in Pennsylvania (Jung et al. 1988). Applications of P fertilizer at 45 kg ha^{-1} were recommended to increase biomass yields on low-fertility soils in Oklahoma (Kering et al. 2012). Manure is also effective in supplying nutrients to switchgrass and results in slow N release (Lee et al. 2007).

7.3.2.2 Pests and Control

Research on insect and disease control on switchgrass grown for bioenergy has been limited. Few insects have been identified as potential pests. Switchgrass grows in functional monoculture clones within a diverse prairie ecosystem (Mitchell et al. 2010b) and has coevolved with insects and disease in North America. Large-scale, monoculture switchgrass plantings may be impacted by certain diseases and insects, however. Rust (*Puccinia emaculata*) symptoms have been more severe on cultivars from northern origins but heritability exists for rust resistance (Gustafson et al. 2003). Biomass yield losses from smut (*Tilletia maclaganii*) in switchgrass fields can be significant indicating that additional research on disease management is needed to maintain optimal production (Thomsen et al. 2008). Establishment year insecticide applications like carbofuran (2,3-dihydro-2,2-dimethly-7-bensofuranyl) have had variable success on switchgrass yield and stands (Bryan et al. 1984; Parrish and Fike 2005). Insect injury on switchgrass seedlings, in general, is dependent on climatic conditions, seeding dates, and weed populations (Sanderson et al. 2012).

7.3.3 Harvest

Switchgrass conversion to a biofuel will involve either a biochemical or thermochemical process (Laser et al. 2009). Biochemical conversion efficiency is dependent on feedstock cell wall composition, pretreatment methods, and fermentation technology (Dien et al. 2006). Switchgrass with low mineral (ash) concentrations is advantageous for thermochemical conversion (Sanderson and Wolf 1995; El-Nashaar et al. 2009). Mineral macronutrients from switchgrass can impact conversion efficiency through the formation of slag or alkali components and contribute to air pollution at the conversion stage. Cellulosic feedstocks with high N and ash concentrations will increase slagging of boilers and reduce boiler efficiency (McLaughlin et al. 2002). High potassium (K) levels also have the potential to damage turbines used in electrical generation (Sanderson and Wolf 1995). Productive stands can be maintained for more than 10 years with proper harvest management (Mitchell et al. 2010b). Harvest management practices mentioned in the following can optimize switchgrass characteristics for desired conversion processes.

7.3.3.1 Harvest Timing

Harvesting switchgrass as a forage crop requires evaluating both forage quality and yield (Moser and Vogel 1995). For bioenergy purposes, the conversion process is an important consideration when selecting the optimum harvest date. Optimum biomass yields for switchgrass in the Central Plains are with a single harvest at the R3–R5 (inflorescence emerging to postanthesis) maturity stage (Moore et al. 1991; Vogel et al. 2002). In the Southern Plains, a mid-September single harvest recorded the highest yields (Sanderson et al. 1999). An early September harvest had higher DM yields than an early November harvest of switchgrass in Virginia (Parrish and Wolf 1992). Another alternative harvest time is to leave switchgrass standing in the field over winter and harvest the following spring (Adler et al. 2006). A killing frost harvest (October 15–November 15) was found to have higher biomass yields than a spring harvest (April 6–May 4) in Pennsylvania (Adler et al. 2006). Yield losses associated with delaying harvest until spring may be acceptable if wildlife cover during winter is important (Adler et al. 2006). In the establishment year, switchgrass harvest after a killing frost is recommended in order to maximize translocation of carbon to belowground components.

7.3.3.2 Harvest Frequency

Switchgrass biomass can be maximized with a one-cut or multiharvest system depending on the region and climate conditions (Sanderson et al. 1999; Vogel et al. 2002; Monti et al. 2008). A single-cut, late-season harvest system is expected for perennial systems managed for bioenergy under rain-fed conditions (Vogel et al. 2002; Parrish and Fike 2005). In Alabama, a one- or two-cut system yielded approximately the same amount of biomass yield (McLaughlin et al. 1999). A two-cut harvesting system for the Central and Southern Plains had a lower yield than a single one-cut system (Sanderson et al. 1999; Vogel et al. 2002). Mitchell et al. (2010b) estimated sufficient regrowth to warrant a second harvest after a killing frost occurs about 1 year out of 4 in eastern Nebraska. A multiharvest system has maximized biomass potential under irrigated conditions for perennial grasses managed as a bioenergy crop (Robins 2010). However, harvesting two or more times within a growing season has resulted in stand reductions limiting long-term persistence (Newell and Keim 1947). Any regrowth following a harvest after inflorescence development will occur only from retillering because of switchgrass's determinate growth habitat (Mitchell and Schmer 2012). Meta-analysis from switchgrass biomass production studies conducted at 39 field sites in 17 states supports a single harvest for bioenergy (Wullschleger et al. 2010). Switchgrass used both as cattle forage and as biomass for energy may be viable in regions where livestock production is an important enterprise (Guretzky et al. 2011).

7.3.3.3 Nutrient Removal

Removing biomass from a production system also removes nutrients that need to be replaced to optimize yield potential and maintain long-term stand persistence. Nitrogen is the largest nutrient removed from the system and is also the most expensive to replace. Nitrogen removal in switchgrass production systems is a function of biomass yield and N concentration, with biomass N concentration increasing as N fertilization rates increase (Vogel et al. 2002; Jung and Lal 2011). Reducing N removal has a positive effect on the economic and environmental sustainability of the switchgrass production system. Multiharvest systems in a single growing season require higher N application rates than does an annual harvest system as greater N removal rates are associated with multiharvest systems.

Mineral concentrations will vary by plant maturity indicating that harvest management will influence nutrient removal rates. Nitrogen and K concentrations result in curvilinear declines with increased plant maturity (Sanderson and Wolf 1995). Harvesting after a killing frost minimizes nutrient removal, especially N, and improves thermochemical conversion or biopower efficiency. Calcium (Ca) concentrations of switchgrass by harvest date have been mixed, with studies showing little changes by harvest date (Sanderson and Wolf 1995; Lemus et al. 2009) to Ca differences by harvest date (Adler et al. 2006). Increased Ca availability does not affect aboveground P, K, sulfur (S), or magnesium (Mg) concentrations for switchgrass (Schmer et al. 2012a). Adler et al. (2006) reported that for fall and spring (overwinter) switchgrass harvests, concentrations of Cl, K, P, and Mg declined with spring harvests, while Ca, N, and S concentrations increased with spring harvests. Switchgrass under a one-cut harvest system had nutrient removal rates of 71 kg N ha^{-1}, 14 kg P ha^{-1}, 96 kg K ha^{-1}, and 48 kg Ca ha^{-1} (Lemus et al. 2009).

7.4 PRODUCTION

Solar radiation, temperature, water, and nutrient availability all contribute to switchgrass yield production. Unlike annual cropping systems and established grassland systems, temporal effects for recently established switchgrass will be affected by both weather conditions and stand development. Switchgrass, in general, does not reach full yield potential until 1–2 years after establishment. Proper agronomic practices with normal climatic conditions can result in establishment year biomass yields of 50% of full yield potential. Switchgrass, in long-term evaluations (>10 years), has been shown to have consistent biomass yields over time when stands are mature (Fike et al. 2006).

7.4.1 Yield Potential

Yield potential is defined in this chapter as maximum biomass yield that has been attained at any time for a specific genotype of switchgrass under field conditions (McLaughlin et al. 2006). Switchgrass mean biomass yield across 17 states was 8.7 ± 4.2 Mg ha^{-1} for upland cultivars and 12.9 ± 5.9 Mg ha^{-1} for lowland cultivars with the yield differences between ecotypes being significant (Wullschleger et al. 2010). A similar yield result was found by Tulbure et al. (2012) using an empirical model with mean biomass yields of 9 Mg ha^{-1} for upland ecotypes and 13 Mg ha^{-1} for lowland ecotypes across 15 states. Temperature, precipitation, N application, latitude, and ecotype contributed to regional differences in biomass yield (Wullschleger et al. 2010). An ecotype-specific empirical model showed both ecotypes having a positive correlation between yield and precipitation with upland ecotypes having a significant positive yield response to increased soil moisture (Figure 7.3a) (Jager et al. 2010). Small-plot yield evaluations across the United States have shown variation in productivity based on regional climatic conditions and germplasm used (McLaughlin et al. 2006). The biomass yield of switchgrass from plot studies was as high as 34.6 Mg ha^{-1} in southeastern United States (McLaughlin and Kszos 2005). Meta-analysis has indicated that switchgrass yields are similar between small plots and field-scale sites (Wullschleger et al. 2010). Field-scale information on switchgrass suggests that spatial patterns affect biomass yields to some degree (Di Virgilio et al. 2007; Lee et al. 2009; Schmer et al. 2010; Thelemann et al. 2010). Collins et al. (2008) reported average yield irrigated switchgrass in the Pacific Northwest ranged from 14.5 to 20.4 Mg ha^{-1}.

Biomass evaluations in Europe have indicated that lowland ecotypes yield more than upland ecotypes (Sanderson et al. 2012). Averaged over 4 years and 2 systems of cutting, lowland ecotypes

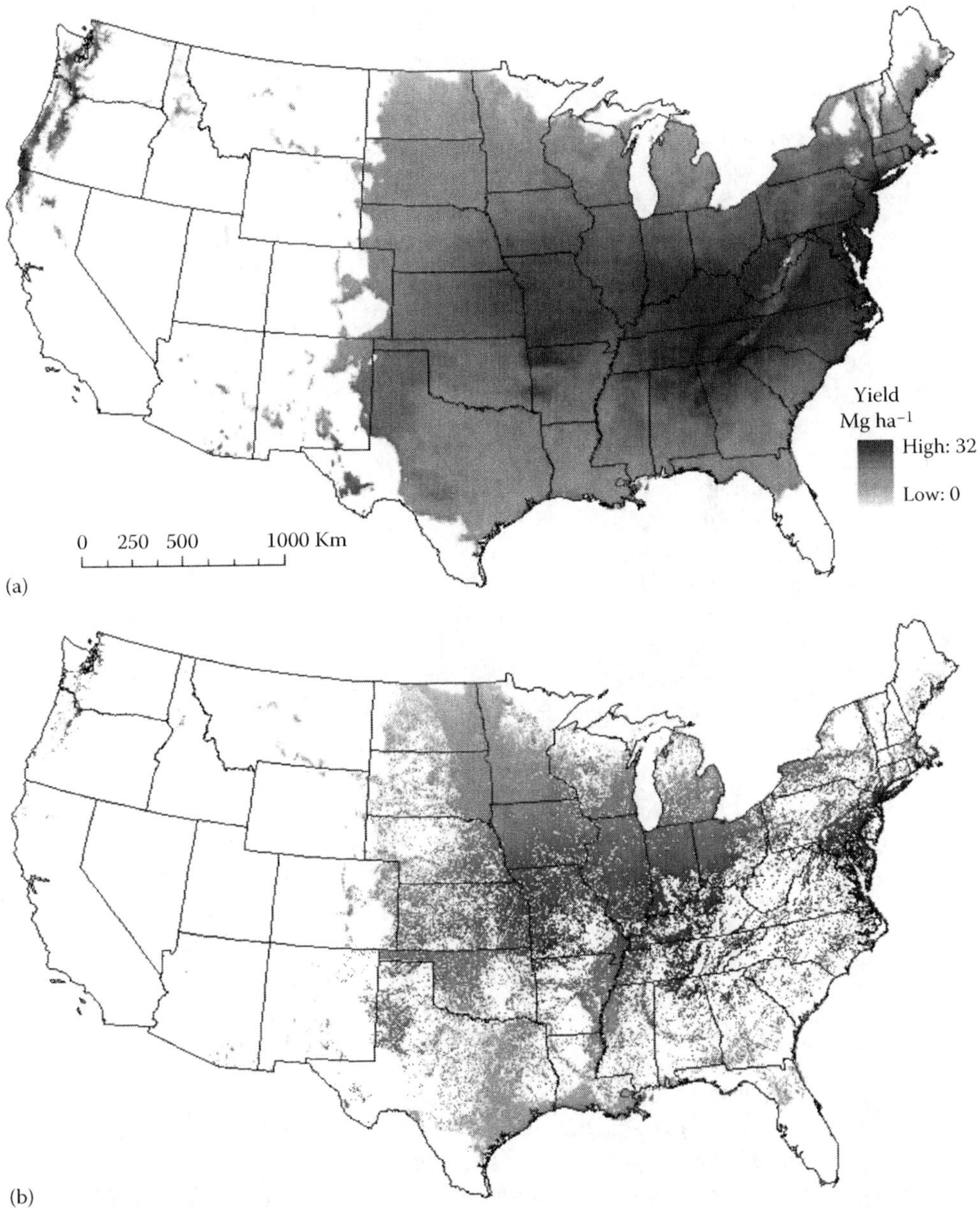

FIGURE 7.3 Switchgrass yield map (a) of the conterminous United States representing maximum upland and lowland ecotype biomass yields using an ecotype-specific empirical model. (From Jager, H.I. et al., *GCB Bioenergy*, 2, 248, 2010.) Switchgrass yield map is available at https://bioenergykdf.net/. Switchgrass land availability map (b) omitting private and public forest land cover (US National Forest's forest distribution map available at http://fsgeodata.fs.fed.us/rastergateway/forest_type/) and grassland/shrubland land cover west of the 100th Meridian. (USDA-NASS 2011 cropland data layer available at http://nassgeodata.gmu.edu/CropScape/.)

averaged 14.9 Mg ha^{-1} and upland ecotypes averaged 11.7 Mg ha^{-1} in Italy (Monti et al. 2008). Biomass yields for several switchgrass varieties ranged from 7.1 to 21.3 Mg ha^{-1} in Greece and 0.9 to 26.1 Mg ha^{-1} in Italy (Alexopoulou et al. 2008). Biomass yields ranged from 15.4 to 24 Mg ha^{-1} under irrigation in Greece for a 2-year period (Vamvuka et al. 2010).

7.4.2 Geographical Distribution and Land Availability

In North America, switchgrass is adapted to habitats east of the Rocky Mountains and south of 55°N latitude (Stubbendieck et al. 1997). Switchgrass is likely to be grown on land that is marginal for crop production based on either erosion potential or low crop productivity (Sanderson et al. 2012). An estimated 5.1×10^6 to 11.8×10^6 ha could potentially be allocated to switchgrass production in the United States by 2030 (USDOE 2011). Greatest biomass yields (Figure 7.3a) are predicted to occur in a region from the mid-Atlantic region to eastern Kansas and Oklahoma (Wullschleger et al. 2010). These regions, however, are largely composed of forested areas making large-scale conversion to switchgrass unlikely (Figure 7.3b). Areas suitable for switchgrass production in the Pacific Northwest are also largely composed of existing forests (Figure 7.3b). Switchgrass yields with the highest temporal stability were estimated to be in the US Corn Belt region (Tulbure et al. 2012). The Corn Belt region will likely have switchgrass planted on land classified as marginal or highly erodible. In the United States, the majority of land for perennial grass production is projected to come from the reallocation of existing cropland with land currently enrolled in the Conservation Reserve Program (CRP) and pastures being second and third, respectively (Walsh et al. 2003). Grasslands or pastures west of the 100th meridian will likely not be converted to switchgrass because of limited precipitation, and current land base is allotted for beef cattle production (Figure 7.3b).

Biomass refineries are expected to have four to five times higher capital costs than similar-sized grain ethanol plants based on first-generation biomass refining technology (Wright and Brown 2007). Reliable feedstock supplies will be critical in maintaining stable operational costs for biorefineries. Land availability and land type for conversion to switchgrass will vary by region. For example, switchgrass was found to be a suitable candidate on the corners of center-pivot irrigation systems in eastern Nebraska to support multiple cellulosic refineries without greatly affecting land dedicated to food crops (Mitchell et al. 2012). Another spatial analysis in Nebraska showed that riparian and roadway buffers, marginal cropland, and brownfield sites could meet up to 22% of the energy requirements of the state (Gopalakrishnan et al. 2009). A multifeedstock supply chain may be required for certain regions to meet overall bioenergy demands.

7.4.3 Feedstock Quality

Switchgrass cell walls consist of cellulose, hemicellulose, and lignin polymers with small amounts of pectin, proteins, and minerals (Vogel and Jung 2001). Cellulose is a hexose glucose polymer, linked by β-1, 4 glycosidic bonds (McKendry 2002). Hemicellulose in switchgrass is primarily xylan and forms hydrogen bonds with the cellulose microfibril surface (McKendry 2002). Lignin, an abundant phenolic polymer in cell walls, maintains plant structural support (Vogel and Jung 2001). Cellulose, hemicellulose, and lignin concentrations increase in cell walls with increasing plant maturity. Concentrations of ash, proteins, and organic acids decline with switchgrass maturity (Dien et al. 2006). Sucrose is the main soluble carbohydrate in switchgrass cells, starch is the main storage polysaccharide, and glucose is the main cell wall carbohydrate (Adler et al. 2006; Dien et al. 2006). In general, total C varies less in switchgrass than for a specific carbohydrate indicating plasticity in C portioning between carbohydrate structures (Vogel et al. 2011). Total hexose (cell wall polysaccharides and soluble sugars) concentration ranged from 342 to 398 g kg^{-1}, while pentose (arabinose and xylose) concentration ranged from 216 to 245 g kg^{-1} across switchgrass fields in the Great Plains (Schmer et al. 2012b). Cell wall glucose, found primarily in cellulose,

TABLE 7.1
Total Carbon and Ash Contents across Five Harvest Years from a Regional On-Farm Switchgrass Study in the Central and Northern Plains, United States

	Harvest Years									
Location	**Est.**	**2**	**3**	**4**	**5**	**Est.**	**2**	**3**	**4**	**5**
	Total carbon (g kg^{-1})					Ash (g kg^{-1})				
Munich, ND	434	441	444	441	443	165	91	66	78	63
Streeter, ND	—	440	443	446	441	—	90	82	64	72
Bristol, SD	446	450	448	445	444	80	70	62	70	59
Highmore, SD	445	444	446	448	445	125	93	68	67	66
Huron, SD	448	449	—	447	448	80	86	—	63	57
Ethan, SD	442	446	446	442	448	91	78	71	96	52
Atkinson, NE	447	454	451	453	—	81	73	65	49	—
Crofton, NE	427	445	451	443	445	104	78	65	71	61
Douglas, NE	435	446	448	444	450	86	77	66	70	57
Lawrence, NE	427	441	449	446	448	84	88	67	67	57

comprised 66%–82% of the total switchgrass hexose concentration, while xylose comprised 77%–90% of the total pentose concentration (Schmer et al. 2012b). They found that annual biomass yields and annual glucose concentrations were positively correlated ($r=0.54$, $P < 0.01$) across all sites. Selective breeding can alter cell wall polymer concentrations as well as variation within switchgrass populations can be exploited for improved conversion properties. Screening large populations for cell wall characteristics has been effectively done using near-infrared reflectance spectroscopy (Vogel et al. 2011).

Optimal cell wall concentrations will vary by conversion technology processes. Biochemical processes (e.g., simultaneous saccharification and fermentation, consolidated bioprocessing) will likely have increased conversion efficiency for feedstocks with high cellulose content and low lignin concentrations. Thermochemical conversion (e.g., gasification or pyrolysis) efficiency is less dependent on cell wall composition than biochemical processes. Pyrolysis of switchgrass showed similar gas yields by different harvest dates (Boateng et al. 2006). Total C concentration in switchgrass is consistent over sward age and regions (Table 7.1). In contrast, ash concentration shows temporal variability with establishment year switchgrass having higher ash concentrations than postestablishment switchgrass (Table 7.1).

7.4.4 Economics

Assuming a farm gate price of \$44 to \$66 Mg^{-1}, an estimated 5.1×10^6 to 11.8×10^6 ha, respectively, could potentially be allocated to switchgrass production in the United States by 2030 (USDOE 2011). The majority of acres converted into switchgrass production would be from existing cropland that is on marginal or highly erodible land (Vogel 1996). Results from a regional farm trial in the Northern Plains, United States, showed an average farm gate cost of \$66 Mg^{-1} (Perrin et al. 2008). Monti et al. (2007), based on a 4-year study in northern Italy, reported average yield at 6 Mg ha^{-1} with annualized costs at \$150 Mg^{-1}. Other cost estimates range from \$29 Mg^{-1} (Epplin 1996) for the Oklahoma plains to \$73 Mg^{-1} (Hallam et al. 2001) and \$96 Mg^{-1} (Duffy and Nanhou 2002) for Iowa cropland and \$98 Mg^{-1} in Illinois (Khanna et al. 2008). Delivered costs to a biorefinery (1814 Mg d^{-1}) excluding land costs were estimated to be \$37 to \$48 Mg^{-1} (Kumar and Sokhansanj 2007). Delivered costs in Indiana ranged from \$80 to \$96 Mg^{-1} (Brechbill et al. 2011). Differences between farm gate prices from previous studies are largely dependent on land type being converted, respective land rental cost variation by region, assumptions on yield potential, and rotation time length. For example, Perrin et al. (2012) using recent land rental costs in Nebraska showed a farm gate price of \$75 Mg^{-1} for switchgrass yield at 6.7 Mg^{-1} ha and \$60 Mg^{-1} at 13.5 Mg ha^{-1}. In Tennessee, breakeven prices ranged from \$46 to \$69 Mg^{-1} depending on land type and yield potential (Mooney et al. 2009). A breakeven price of \$115 Mg^{-1} was found for the Great Lakes region for estimated yields of 9 Mg ha^{-1} (James et al. 2010). Highest economic gains for producers were found to be a single harvest after senescence with a maximum N fertilizer rate of 65 kg ha^{-1} (Haque et al. 2009). Most costs in producing switchgrass are variable with respect to yield, with approximately 25% of total costs being fixed (Perrin et al. 2012). Switchgrass farm gate costs tend to decline over time with highest costs on a per mass basis occurring during the establishment year, a result of high input costs and low biomass yields (Perrin et al. 2008).

7.5 SUSTAINABILITY

Bioenergy sustainability must meet interconnected environmental, economic, and societal needs. For an alternative transportation fuel to be a viable substitute for conventional gasoline, the alternative fuel should (1) have superior environmental benefits, (2) be economically competitive, (3) have meaningful supplies to meet energy demands, and (4) have a positive energy return on fossil fuel energy investment (Hill et al. 2006). Environmental, societal, and economic trade-offs need to be considered when transitioning back to the use of perennial feedstocks to meet rural and urban energy demands. Switchgrass enhances ecosystem services more than annual row crop systems including reduced soil erosion, improved water quality, soil carbon storage, increased wildlife habitat, and reduced farm-level fossil fuel requirements, (McLaughlin et al. 2002; Farrell et al. 2006; Fargione et al. 2009). Converting idle, pasture, or conservation land to perennial energy crops may lead to wildlife habitat loss, decline in plant community diversity, and a shift in livestock production from pastures to confinement feeding depending on the region, feedstock used, and management practices implemented (Sanderson et al. 2012). In sum, bioenergy sustainability must satisfy renewable energy needs, avoid competition with food demand, and enhance environmental quality. The value of switchgrass in terms of ecosystem services will be dependent on management, land use changes, and regional conditions.

7.5.1 Carbon Storage

Soil organic carbon (SOC) typically increases rapidly when annual cropland is converted to switchgrass (Schmer et al. 2011; Mitchell et al. 2012). Soil carbon sequestration can decrease the amount of carbon dioxide being released into the atmosphere by accumulating carbon in the form of SOC (McLaughlin et al. 2002). Fields that were once cultivated, but are now in CRP, stored an average of 1.1 Mg C ha^{-1} $year^{-1}$ during the first 5 years in the Great Plains, United States (Gebhart et al. 1994).

The amount of carbon sequestered depends on the climate, soil type, SOC content, time, and placement depth of carbon (Conant et al. 2001). Switchgrass grown and managed for bioenergy on three marginally productive cropland sites in Nebraska resulted in an average SOC increase of 2.9 Mg C ha^{-1} $year^{-1}$ in the top 1.2 m of soil in a 5-year period (Liebig et al. 2008). Over a 9-year period in eastern Nebraska, increases in SOC exceeded 2 Mg C ha^{-1} $year^{-1}$ in the top 1.5 m of soil for switchgrass managed as a bioenergy crop with over 50% of the SOC increase occurring below 30 cm (Follett et al. 2012). McLaughlin et al. (2002) reported an average of 1.7 Mg C ha^{-1} $year^{-1}$ sequestered in the southeast region on switchgrass experimental plots. Soil C levels on low-input switchgrass fields have been shown to increase over time and across soil depths and are higher than adjacent cropland fields in the Northern Plains (Liebig et al. 2005). A similar result was found between switchgrass and a maize–soybean (*Glycine max* (L.) Merr.)–alfalfa

(*Medicago sativa* L.) rotational system in Iowa (Al-Kaisi et al. 2005). In the southeast, an estimated 0.17–0.21 Mg C ha^{-1} $year^{-1}$ can be sequestered based on SOC that was near steady state (Garten and Wullschleger 2000). Soil organic C increased after establishment of switchgrass throughout North America at rates ranging from 1.7 to 10.1 Mg C ha^{-1} $year^{-1}$ (Garten and Wullschleger 2000; Zan et al. 2001; Frank et al. 2004; Lee et al. 2007).

7.5.2 Life Cycle Assessment

Life cycle assessment (LCA) is a method of estimating the various energy requirements and demands to determine the environmental and societal risks/benefits for a given material or process (see Chapter 4). Lignocellulosic feedstocks such as switchgrass, woody plants, and man-made prairie mixtures of grasses and forbs have been proposed to offer energy, environmental, and economic advantages over current biofuel sources (grain or oilseed crops) because these feedstocks require low agricultural inputs and can be grown on agriculturally marginal lands (Hill et al. 2006). Typically, studies have used net energy values (NEVs) and net energy ratios (NERs) to measure biofuel sustainability. The higher the values of NEV and NER, the greater the renewable energy return on initial energy investment (Hammerschlag 2006). Net energy gain or yield (NEY) per hectare of biofuels is affected by both crop yield, conversion rate, and energy inputs required to produce, deliver, and process the feedstock [NEV (MJ L^{-1}) × biofuel yield (L ha^{-1})]. Estimated mean NEY on established switchgrass fields is approximately 60 GJ ha^{-1} $year^{-1}$ (Schmer et al. 2008). The measure of NEV, NEY, and NER are not consistent from study to study since these comparisons are largely dependent on the assumptions of agricultural inputs, the feedstock source for bioenergy, the time period in which the agricultural inputs are referenced, and whether coproducts are included (Shapouri et al. 2003; Farrell et al. 2006).

7.5.3 Greenhouse Gas Emissions

Switchgrass has lower GHG emissions than current grain or oilseed biofuels when evaluated by LCA. GHGs evaluated for bioenergy systems are primarily CO_2, CH_4, and N_2O. Limited studies have been published that have quantified CO_2, CH_4, and N_2O fluxes from switchgrass fields, but research on this topic is ongoing (Schmer et al. 2012c). For switchgrass, studies have estimated the amount of GHG displaced by the amount of harvested material that is converted to ethanol (Wang 2001; Kim and Dale 2004; Farrell et al. 2006), while others have determined the amount of GHG displaced by the amount of harvested material and by the amount of C sequestered into the soil profile (Spatari et al. 2005; Wu et al. 2006; Schmer et al. 2008). The use of lignin from biomass residue for combustion energy at a cellulosic biorefinery is the main reason why switchgrass has theoretically lower GHG emissions than biofuels from annual crops where processing energy currently is derived from fossil fuels (Hill et al. 2006). Cellulosic ethanol from switchgrass on average had 94% lower GHG emissions than from conventional gasoline (Schmer et al. 2008). Displacement of fossil fuels was found to be the largest C and GHG sink for switchgrass conversion to bioenergy (Adler et al. 2007).

Nitrous oxide has the highest global warming potential of the three GHGs, and agriculture is a major producer of direct and indirect N_2O emissions (Crutzen et al. 2008; Millar et al. 2010). Switchgrass with its well-developed root system and fertilizer application under appropriate N levels will have lower direct and indirect N_2O emissions than annual cropping systems. Switchgrass grown for bioenergy will affect regional N_2O emissions depending on the use and area of agricultural land being displaced. Agricultural N_2O emissions vary regionally depending on weather conditions, N application rates, fertilizer type, soil characteristics, and production practices (Del Grosso et al. 2006; Millar et al. 2010). As stated earlier, switchgrass will likely be grown on conservation land or marginal land currently in crop production (McLaughlin et al. 2002; Walsh et al. 2003; Wu et al. 2006; Fargione et al. 2008). Increased switchgrass planting on cropland dedicated to maize, for example, would result in lower regional N_2O emissions since recommended switchgrass N application rates are 25%–50% lower than current maize N rates (Vogel et al. 2002; Farrell et al. 2006; Schmer et al. 2008; Robertson et al. 2010). However, if switchgrass replaces formerly idle or conservation land, regional N_2O emissions would increase from baseline emissions. Based on C sequestration potential, switchgrass tends to be a CO_2 sink but can also be a CO_2 source for a given time scale based on climatic conditions (Skinner and Adler 2010). Most grasslands are net sinks of CH_4 or are CH_4 neutral (Liebig et al. 2006). Soil-derived CH_4 emissions for switchgrass (Schmer et al. 2012c) are similar to previous grassland studies, while CH_4 emissions during conversion will likely be small (Qin et al. 2006). In the future, better understanding of local and regional GHG emissions from switchgrass under different management practices will become available.

7.6 HARVEST TECHNOLOGY

Switchgrass has the advantage that it can be harvested and handled with conventional haymaking equipment (Cundiff and Marsh 1996). Harvest equipment selection should be based on size of operation, field condition, availability, and economic considerations. Biomass harvest technology requirements are to efficiently cut switchgrass tillers from the basal portion of the plant, consolidate switchgrass biomass, and package (compact) switchgrass for further handling and storage.

Current harvesting and collection equipment cannot avoid field collection losses arising from equipment usage, weather, and crop conditions during these operations. Switchgrass mechanical harvesting and collection losses can range from 35% to 45% compared to hand harvest methods (Sanderson et al. 1997; Monti et al. 2009). These harvesting losses occurred either due to biomass not picked up by the baler (up to 17%) or biomass left uncut by the harvesting machine (up to 29%). Reduction of harvest losses provides an opportunity to increase switchgrass production and calls for advancement in harvesting techniques. Logistically, future bioenergy demand can only be met through large-scale machinery and custom harvesting and delivery operations. Recent developments in farm machinery make it possible that high-capacity and large-scale machines with several electronic controls, ergonomic features, global positioning systems, yield monitoring, agricultural field logistics, and a host of various custom features are being developed and made available to users to assist with cellulosic biomass production and logistics.

7.6.1 HARVESTERS

Desirable features of biomass harvesters are uniform cutting height, high forward speed, wide swath width, adjustable height of cut, biomass collected with low soil contamination, processing features (e.g., chopping, crimping), low fuel consumption, and being economically viable. Major categories of harvesters are simple mowers, advanced mower–conditioners, and forage choppers. Common harvesting widths range from 1.5 to 5.5 m.

When switchgrass is considered as energy feedstock, one of the significant factors to be considered is the height of cut from the ground. While more biomass is at the bottom portion of the plant, the lower the cut, the higher the biomass production. Lignin content in switchgrass is highest at the basal portion of tillers and decreases toward the top of a mature tiller indicating that harvest height can influence composition to the biorefinery (Sarath et al. 2007). Biomass with less soil contamination (lower ash content) is valued by processing industries, as it will be more efficient to process and less damaging to industrial equipment. One recommendation is switchgrass should be cut at a 0.15 m height to reduce equipment tires puncturing by tillers and better stand longevity (Garland 2010). Based on modeling results, optimization of the switchgrass harvesting schedule (area harvested in a given time span) led to about 9% reduction in the delivered cost and about 30% increase in the total profit (Shastri et al. 2011). Thus, harvesting height and proper harvest scheduling management will influence switchgrass supply and overall economics.

7.6.1.1 Simple Mower

For nonwoody crops such as switchgrass and other grasses, harvesting can be performed with or without chopping or conditioning. A basic mechanical harvester, such as a mower, cuts the plants at a desired height and collects and conveys the crop onto wide swaths or windrows to be picked up by other equipment (e.g., baler). Sickle bar or rotary disc cutting elements are two common harvesting mechanisms. These types of harvesters can be operated by tractors that are mounted or pulled by tractors or can be self-propelled units. Most of the tractor-operated units are offset so that the tractor runs clear off the standing crop, while the harvesting unit works on crop rows. Self-propelled units have the cutting elements mounted in front and convey the cut crops behind the unit to make windrows. Some units have weighing arrangements or yield monitors. Tractor-mounted or tractor-pulled harvesters are usually of small capacity, while self-propelled units are of large capacity suitable for custom harvesting or large-scale farm operations.

7.6.1.2 Mower–Conditioner

Mower–conditioner either has the sickle bar or rotary disc cutterhead as the cutting mechanism. The conditioner component on a mower–conditioner accelerates switchgrass drying by crushing plant stems but not altering plant structure and consolidates the switchgrass into a windrow (Hess et al. 2009). Conditioning of crops is achieved by a set of crimping rollers rotating in opposite directions that compress and convey the harvested plants. Altering the working pressure between the rolls, selecting the roll material (steel or rubber) or speed of rolls, based on the requirement of crops, adjusts the severity of conditioning. Quicker drying of the harvested crop means more product quality retention and a better window of harvest flexibility to the producers. Self-propelled harvesters with rotary heads are preferred for harvesting high-yielding (>15 Mg ha^{-1}) switchgrass fields (USDOE 2011).

7.6.1.3 Forage Chopper

Choppers can be exclusive units that pick up harvested crops from the swath or windrow and make chops or can be a self-propelled unit that both harvests and chops the crop. In either case, the harvested crops are positively fed by toothed feed rollers and continuously chopped by a multiknife cutterhead rotor. Knife spacing on the cutterhead rotor can be adjusted to alter chop length. Chopped material is then moved by a blower mechanism (paddles on shaft) and transferred through the spout into a collecting truck following the chopper in parallel. Chopping methodology, at least at present, may not be suitable as a preprocessing logistics option because the chopped material cannot be easily transported to processing facilities unlike bales. However, bulk consolidation of chopped material is currently under research.

7.6.2 Balers

Balers mechanically pick up harvested materials that were left in the field as swaths or windrows and consolidate them into bales. Bales are compact units of biomass and can be readily transported to a stacking location for storage or delivered directly to a bioenergy facility. Features of bales and baling include shape, size, staking ease, compactness, density, and wrapping material. Two major types of balers are round and large rectangular. There are advantages and disadvantages for the round baling or large rectangular baling methods, but both are capable of processing switchgrass and are commercially available to producers.

As baling is a form of loose densification process, any increased bale density is always desired and is a function of the packing mechanism employed by the baler. Modern balers now utilize a chopping operation of the material before packing as well as increasing the number of packing piston strokes, both of which leads to increased bale density. Large round and square baled switchgrass densities at 15% moisture levels vary from 143 to 200 kg m^{-3} (Cundiff and Marsh 1996). The density of a round bale or a large rectangular bale will vary depending on harvest period with anthesis harvest bales having a greater density than postkilling frost harvest bales. Increased bale density results in reduced transport trips as well as less storage space requirements. Increased bale densities preserve biomass quality as well as minimize moisture-related damage when stored outside. Securing the compressed material is done either by high-strength strings that tie the bales (both round and square) at specified spacing or by nets that wrap the entire bale (round). Bales can be collected using an automatic bale picker or using a truck or trailer in combination with a loader. Automatic bale picker is a more efficient collection method especially for large-scale operations.

7.6.2.1 Round Bales

Round bales are common and popular as the size of equipment serves well for farm-scale operations and the equipment is less expensive. Smaller bioenergy feedstock producers may be opted for the round baler methods because of the lower capital costs or may outsource harvest and baling to custom harvesting enterprises that are equipped with large rectangular balers. However, for the supply of biomass-to-bioenergy facilities, big square bales that weigh about a ton (short, US) or more are preferred. The dimensions of round bales from various equipment manufacturers range from 0.6 to 1.8 m diameters with widths from 1.2 to 1.6 m.

7.6.2.2 Square or Rectangular Bales

Square bales can be easily stacked without void spaces in field stacks, storage, and transportation trucks. The ranges of bale dimensions for rectangular bales are 0.36–1.2 m width, 0.46–1.2 m height, 0.3–2.7 m length. Rectangular bales tend to be easier to handle and load for transport without road width restrictions. Loading requirements time onto semitrailers is double for round bales than it is for rectangular bales (Hess et al. 2009). Unless cellulosic biorefineries stipulate a certain baling method or alternative harvest method, both baling methods will likely occur for a given region.

7.6.2.3 Other Collection Methods

The Biomass Research and Development Board (BRDB) report indicates that although existing machines can be utilized to perform the operations, supply systems using existing technologies may not meet cost or capacity targets because these machines were not designed as a component of a biomass supply chain (BRDB 2010). Consolidation methods other than baling may be implemented in regions where weather conditions or existing infrastructure enterprises allow for alternative harvesting scenarios (Bransby et al. 2005; Kumar and Sokhansanj 2007; Mitchell et al. 2008; Cundiff et al. 2009; Digman et al. 2010). Wet storage methods have been proposed for switchgrass in regions where drying conditions for baling operations are not possible because of high relative humidity and increased chance of a postharvest precipitation event (Digman et al. 2010). Alternate methods that would eliminate baling and bale grinding (necessary before utilization) procedures in the supply system are delivering the biomass in a loose form using a forage harvester or advanced wagon (chopping and self-loading combined). Southeastern US researchers have proposed increased densification of chopped switchgrass by using modulizing technology developed for the cotton industry (Bransby et al. 2005; Cundiff et al. 2009).

Self-loading wagons handling chopped biomass use less energy and labor than the traditional forage harvesting system (Tremblay et al. 1991). The wagon system is more efficient than forage harvesters and was considered optimum for a bioenergy plant with 8 km radius supply area (Brownell et al. 2009). Dry switchgrass can be chopped to 25–50 mm particle size, blown into forage wagon, and piled into large cones or can be wet chopped, moved by wagon, unloaded into bunker or silo, and compacted to enhance fermentation to produce silage (Sokhansanj et al. 2009). Blowing the chopped biomass as well as storing the bales in the silage bags or tunnels is another collection method for which equipment is commercially available. Synthetic bags or tunnels are in situ biomass farm storage methods and provide moisture protection of material stored from rain that also allows for biochemical processes of silage making.

Loafing is another option for collecting biomass, wherein dry biomass (<15% wet basis [w.b.]) is picked up from windrows and made into a large package of about 2.4 m wide by 3.6 m high and up to 6 m long, using a loafer or stacker (Sokhansanj et al. 2009). Compacted loaf stack roof assumes a doom shape and diverts water protecting the biomass inside. Field capacity of a loafer stacker is lower than either conventional baling system or a forage chopper. Once the loafer stacker is full, the operator needs to immediately transport the biomass stack to the edge of the field or use specialized trailers to transport the biomass stack after harvest. Loaf stacks are susceptible to biomass loss in regions with significant wind velocities if placed perpendicular to the prominent wind direction.

A uniform-stacking feedstock supply design that can preprocess switchgrass and other cellulosic materials regardless of collection method for use in a large-scale cellulosic biorefinery (≥4535 Mg d^{-1}) has been proposed that would increase regional and producer flexibility to harvest and collect switchgrass (Hess et al. 2009). New technology and equipment with large capacities should be developed to meet the target requirements. Furthermore, the development of new technologies creates the opportunity to reduce environmental and social impacts.

7.6.3 Infield Bale Aggregation

Bales need to be efficiently aggregated (collected into groups and transported) to a field-edge stack (Figure 7.4) or a temporary storage before transport to a bioenergy facility. Bales can be collected using a self-loading bale picker or using a truck or trailer in combination with a loader. Igathinathane et al. (2012) simulated various aggregation scenarios and determined total distances involved and evaluated the effects of various field operating parameters (field size, number of bales handled, field shape, biomass yield, swath width, bale layout, and collection path on bale aggregation).

Igathinathane et al. (2012) found significant differences among aggregation methods while considering the basic single-bale loader as control method. Field shape, swath width, biomass yield, and bale layout randomness did not affect aggregation performance. Although the aggregation efficiency increases with the number of bales handled per trip, it is not significant beyond 12 bales. The most efficient strategy to collect bales is the application of the self-loading bale picker, followed by parallel run of loader and truck, diagonal grouping, and bale accumulator, while the least efficient is central grouping of bales.

FIGURE 7.4 Biomass bales scattered on a field waiting to be aggregated into a field stack. (Photo courtesy of Igathinathane Cannayen [ABEN, NDSU].)

7.7 FEEDSTOCK STORAGE

Proper feedstock storage is essential as biomass demand is year-round, while switchgrass harvest is seasonal. Switchgrass being a biodegradable feedstock can result in quantitative and qualitative changes during storage (McLaughlin et al. 1996). After harvest, poor switchgrass storage conditions can result in storage losses amounting to 25% in a single year as well as reduced biomass quality (USDOE 2011). The main objective of storage is to minimize mass and quality losses without incurring excessive cost. Understanding of various technologies and factors that influence the quality characteristics of stored material is central to manage the feedstock storage.

7.7.1 Storage Management

Substantial amounts of switchgrass biomass will need to be safely stored on a year-round basis to supply a cellulosic biorefinery. As current biomass consolidation method has low bulk densities, field storage takes area otherwise available for cultivation or other utilization. Therefore, field storage would need to be planned so as not to interfere with other cultural practices (Jenkins et al. 1981).

The fundamental strategy for preservation of biomass as hay is prohibiting microbial (yeast, mold, and bacteria), enzymatic (proteases), and chemical (browning) processes by drying the forage to below 20% w.b. (Pitt 1990). Efforts should be taken to ensure increased drying rate and provide required time so that harvested material quickly reaches the baling moisture content (18%–20% w.b.). Similarly, the fundamental strategy for preservation of biomass as silage is to exclude oxygen from the biomass and to reduce the pH rapidly through bacterial (lactic acid bacteria) fermentation, and the method is suitable for moistures above 40% w.b. (Pitt 1990).

Baling can be considered as a storage-related process as the wrapping of biomass protects the inner portion of the bale compared to leaving the material in loose form on windrows. The type of bale wrapping, such as twine, net, breathable film, and plastic film, will influence moisture and gas exchange, hence affecting the storage characteristics. Bales stored outdoors unprotected can be affected by rain or snow. Shape (square or round) and orientation (laid along edge or ends) of bales on stack will affect moisture gathering from rain or snow. Unlike rainwater that drains quickly, collected snow can stay on the bales (Figure 7.5) for extended periods of time until snowmelt occurs and releases water slowly that can penetrate inside the bale. Bale orientation can also influence DM loss. Bale oriented north/south had a small but significant effect on DM loss than bales with an east/west orientation (Huhnke 1993).

FIGURE 7.5 Snow getting collected on bales stored outdoors without protection. (Photo courtesy of Igathinathane Cannayen [ABEN, NDSU].)

Square bales hold more moisture than round bales during a rainfall event and thus result in higher DM losses than round bales (Khanchi et al. 2010). Round bales when stacked on their edges tend to shed rainwater because of the curvature, but square bales and round bales stacked on their flat ends provide a horizontal surface that result in water infiltration. Thus, with square bales, a protective cover is required for outdoor storage over an extended period of time.

Expected losses include DM (weight loss), leaching of solubles, and changes in compositional quality. Air and material temperature, nature of substrate, air relative humidity, material moisture content, density of packing, porosity of the packaged biomass, air diffusion rates, and microbial activity influence DM loss of stored biomass.

Chemical application such as an acid treatment helps in preservation and extends the feedstock storage life. Freshly harvested switchgrass, when blended in the first stage with sulfuric acid (100 g kg^{-1}), stored anaerobically, and a second-stage thermal treatment (1 h, 120°C), resulted in significantly higher conversion of cellulose to ethanol (63% maximum) as well as higher hydrolysate levels of xylose (Digman et al. 2012). On-farm acidification has proven to be valuable as crop preservation technique as well as found to lower the required pretreatment severity at the biorefinery. Storage in big bales and in the larger storage capacities was found to be cost-effective (Huisman 2003).

7.7.2 Storage Platforms

Usually, feedstocks are stored dry, while silage processing is an option for fresh wet biomass. Feedstock dry storage platforms can be broadly classified as (1) outdoor and (2) indoor methods. Other advanced methods that are being experimented are application of plastic tube to store bales or chopped materials and storage of chopped material with or without protective covering. For storing fresh biomass, silos of various designs such as aboveground upright cylindrical or underground horizontal are available. Proper management of ensiling process reduces avoidable losses and the measures depend largely on the moisture content of the crop, silo type, proper sealing and management of the silo, and the rate and method loading and unloading (McDonald et al. 1991). The ideal storage platform is for switchgrass to enter and leave the storage phase in an unaltered state (Mitchell and Schmer 2012).

7.7.2.1 Outdoor Storage

Unprotected storage can be simply stacking the biomass units on various substrate platforms. As the biomass moisture content is the most influential factor of spoilage, it is necessary to protect the biomass from various forms of precipitation. In outdoor storage, protection from precipitation can be performed by covering the stacked biomass by waste materials, plastic sheeting, or tarpaulin. These covering materials can be used on stacked biomass bales or chopped biomass piles. Plastic sheeting and tarps are less expensive but involve laborious methods, which require much inspection and repair after storms (Huisman 2003). Placing bales on well-drained surfaces including bare soil, crushed rock, or raised platforms (e.g., wooden poles or pallets) are near-term storage strategies. Switchgrass bale exposed to moist soil loses more mass because of an increase in the microbial degradation of cellulose upon contact with moist and biologically active soil than bales kept on a dry surface (Sanderson et al. 1997). Visible discoloration of the bales (stored unprotected) on the area outside of the core comprising 22%–29% of bale DM was observed in the first 4 months, but no further discoloration changes occurred through 12 months (Cundiff and Marsh 1996).

7.7.2.2 Indoor Storage

Storage in regular structures requires more capital investment, but simple roof structures are sufficient and even recommended for natural ventilation. Possible structures for indoor storage are pole barns with no walls, metal buildings with walls, fabric buildings, and truss arches (Huisman 2003). In southern Europe, round and rectangular switchgrass bales showed minimal storage loss and no visible microbial activity when stored under a sheltered roof (Monti et al. 2009). Application of cheaper storage options (plastic-covered ambient storage versus pole frame metal roof with plastic cover versus closed warehouse with drying) can lead to significant cost reduction in the whole biomass logistic system (Rentizelas et al. 2009). This reduction exceeds the extra costs of biomass material losses and increased handling in the simple storage option. A multifeedstock approach (storing several biomass feedstocks) appears to be attractive for systems where expensive storage options are used that had reduced storage space requirements.

7.7.3 Storage Losses

Significant differences in DM bale losses have been reported between indoor and outdoor storages of switchgrass (Sanderson et al. 1997), as well as between untarped and polyethylene-tarped outdoor storages (Khanchi et al. 2010). Round bales lose less DM than square bales in outdoor storage. Tarped treatments are equally effective to resist DM loss as compared with inside storage treatments. Khanchi et al. (2010) found bales stored outdoors on pallets and gravel lost less DM than bales stored on the ground, but Sanderson et al. (1997) reported no significant difference between sod and gravel pad platforms. Dry switchgrass bales (22.5% w.b.) stored outdoors for 9–11 months averaged 5.7%, 5.4%, 9.0%, 9.3%, and 15.4% DM loss for bales wrapped with plastic film, breathable film, net wrap, plastic twine, and sisal twine, respectively, while bales stored indoors had 4.9% DM loss (Shinners et al. 2010). Switchgrass round bales had DM losses of 13% in 6 months for 275 kg bales on grass sod and 5% loss in 12 months for 375 kg bales under outdoor storage (Sanderson et al. 1997). Other studies reported no significant change in DM content after 3 months in storage, but after 6 months, significant DM losses were observed in round and square bales stored outside without protection (Khanchi et al. 2010).

7.7.4 Compositional Changes during Storage

Compositional changes during storage will likely affect biochemical conversion more than either thermochemical conversion technology or direct combustion for electrical generation (Mitchell and Schmer 2012). The chemical composition of the bale can vary spatially (outer layer, core, and base) when removed from storage, and considerable loss of cellulose and hemicellulose during storage can occur (Shinners et al. 2010). Spatial distribution of moisture (Figure 7.6) on stored bales is affected by bale wrapping method that results in spatially varied chemical composition. Storing dry bales under cover results in the most uniform dry biomass feedstock.

Spoilage of bales in storage occurs mainly due to the biochemical reactions of respiration by microbes (Greenlees et al. 2000). Substantial moisture increase of about 204% in square untarped and 57% in round untarped bales from initial moisture content of 10.9% w.b. has been reported (Khanchi et al. 2010).

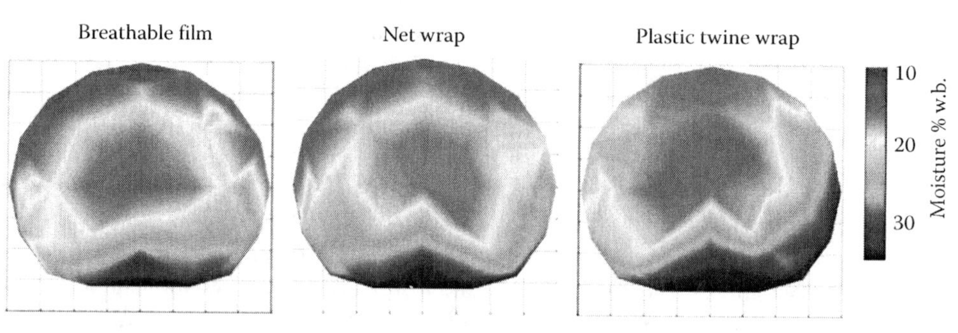

FIGURE 7.6 Example of relative spatial moisture distribution (% w.b.) in switchgrass bales stored outdoors on ground for 293 days. (From Shinners, K.J. et al., *Trans. ASABE*, 53, 359, 2010.)

Carbohydrates and soluble sugars such as sucrose, glucose, and fructose (water-soluble extractives) were found to be major components of switchgrass DM loss (18%–27%) during storage (Chen et al. 2010). The extractable sugars are easily fermented to ethanol using biochemical conversion processes (Wiselogel et al. 1996). Indirect increase in cellulose, lignin, and protein due to loss of extractives and sugars during storage was observed with switchgrass round and square bales stored uncovered, while ash content was unchanged (Khanchi et al. 2010).

7.8 PREPROCESSING TECHNOLOGY

7.8.1 Size Reduction

Size reduction is essentially a mechanical operation with little or no change in cell wall composition. Size reduction requires one-third of total energy inputs for biomass-to-ethanol conversion (Aden et al. 2002). Thus, size reduction of biomass is an energy-intensive process that necessitates energy efficiency in this process so biofuel competitiveness can be increased. Size reduction of biomass is a preprocessing operation that increases bulk density, improves flow properties, generates new surface area (Drzymala 1993), increases pore size, and increases rates of hydrolysis reactions (Schell and Harwood 1994). Biomass size reduction power requirements are related to biomass species, type of machine and size, moisture content, machine operating parameters (e.g., rotational speed, clearance), initial and final particle sizes, and feed rate of the material. Classifying screens fitted to size reduction devices having a fixed opening dimension will give only an indication of the product size as the ground product will have a range of sizes and a particle size distribution. In general, most downstream operations require particle sizes less than 6.4 mm. Particles of dimensions larger than sieve openings will pass through due to a *falling-through* effect (Igathinathane et al. 2009).

7.8.1.1 Types of Size Reduction Devices

Typical size reduction machines for biomass include hammer mill, knife mill, and disc mills, and various choppers, chippers, and shredders (Yu et al. 2003). Other possible devices applicable to biomass size reduction are ball mill, pin mill, rod mill, and chain mill. These devices are primarily used for brittle materials such as rocks and minerals but can be used for biomass size reduction with little or no modifications. A forage chopper with flywheel- or rotor-based cutting devices can also be considered as primary size reduction devices.

Hammer mills being a simple design, rugged, and versatile have wide applicability in biomass size reduction. Tub grinders that can handle large-scale grinding are essentially large-scale hammer mills. Hammer mills can finely grind a greater variety of materials than any other grinding machine (Scholten and McEllhiney 1985). Hammer mills use high-velocity rotating shafts to which fixed or swinging hammers are attached to impart kinetic energy to the processed material (Figure 7.7). The tip of the fast-moving hammers beats the material and gets the material disintegrated by the force of impact with the hammer as well as the material hitting the casing and the holding sieves, until the particles are small enough to pass through the sieves. Swinging hammer elements can be inverted and rotated such that each hammer can be used in two or four different positions, which contributes to the low maintenance requirement for the hammer mill machine (Yu et al. 2003). Hammer mill machines can process both fresh and dry materials.

Knife mill, also known cutter mill, has a rotor on which knife blades are attached forming the cutterhead, a stationary shear bar attached to the casing, and a classifying screen. The number of knife blades, their bevel angle, and the clearance between the blades and shear bar will vary based on the design. The mechanism of size reduction in knife-mill is shearing combined with impact, and the material is sheared between the rotating knives and the shear bar. Knife mill can accept input materials of large dimensions. The materials stay in the cutting chamber until they fall through the openings of the classifying screen. A knife mill can process both fresh and dry materials.

FIGURE 7.7 Inside view of a hammer mill showing the rotor, swinging hammers, and sieve. (Photo courtesy of Igathinathane Cannayen [ABEN, NDSU].)

Disc mill, also known plate mill, has two circular discs with rough surfaces, one of which is stationary and the other is rotary. Material to be ground is fed through the center, and the ground product emerges from the periphery along the gap between the two discs due to centrifugal force. The mechanism of size reduction is primarily attrition and is suitable for producing fine powders. The clearance between the discs controls the size of the final product. Discs will have different patterns of roughness. Disc mills accept input materials of smaller dimensions already ground by other machines. Disc mills are commonly used in grinding food grains and in the paper and pulp industries. The disc mill is mostly suitable for dry materials.

7.8.1.2 Energy Requirement in Size Reduction

Energy requirement in size reduction of switchgrass varies widely from 3.6 to 162 MJ Mg^{-1}, based on the type of grinding machine, operating parameters, input material condition, and product size desired. Samson et al. (2000) reported a specific energy consumption of 162 MJ Mg^{-1} for a hammer mill with a screen size of 5.6 mm with switchgrass. Energy requirement increased with increase in speed of the grinder. Other studies reported 53.6 MJ Mg^{-1} for switchgrass chopped to 25–100 mm particle size (Jannasch et al. 2001) and 84.6 MJ Mg^{-1} using a hammer mill to produce a 3.2 mm particle size (Mani et al. 2002). Switchgrass had the highest specific energy consumption of 99 MJ Mg^{-1} among wheat straw, barley straw, and corn stover for 3.2 mm screen size in hammer mill grinding (Mani et al. 2004). A 37% increase in grinding energy of switchgrass when hammer mill speed increased from 2000 to 3600 rpm was observed (Bitra et al. 2009a). Optimum total specific energies of 114.4 MJ Mg^{-1} for switchgrass were obtained for 2000 rpm for a mass feed rate of 2.5 kg min^{-1} and a 3.2 mm screen with 90° angled 6.4 mm thick hammers. The corresponding total specific energy per unit size reduction was 14.9 MJ Mg^{-1} mm^{-1} (Bitra et al. 2009a).

For size reduction with knife mill, total and effective specific energy (total less no load energy) decreased with an increase in screen size (12.7–50.8 mm) for switchgrass (Bitra et al. 2009b). A knife mill screen size of 25.4 mm, speed of 250 rpm, and feed rate of 7.6 kg min^{-1} was found optimum for size reduction of switchgrass requiring a total specific energy of 27.3 MJ Mg^{-1} and effective specific energy of 4.6 MJ Mg^{-1}. Thus, the energy utilization ratio for knife mill size reduction of switchgrass was 16.8% (Bitra et al. 2009b). A linear knife grid device offered an efficient first-stage size reduction of product dimensions of 50–100 mm and was greater for high-moisture than low-moisture switchgrass requiring 4.5 ± 4.4 and 3.6 ± 3.3 MJ dry Mg^{-1}, respectively (Igathinathane et al. 2008).

7.8.2 Densification

Bulk density of loose cut dry switchgrass (~9% w.b.) can be as low as 45 kg m^{-3} for a geometric mean dimension of 12.3 mm (Chevanan et al. 2010) making it economically inefficient to transport such low-density material. Further size reduction such as chopping will increase the bulk density of the biomass due to better arrangement of the smaller length particles. For example, the bulk density of 60–80 kg m^{-3} of 20–40 mm chopped switchgrass can be increased to 120 kg m^{-3} in loose fill and 200 kg m^{-3} in tapped fill if particle size is reduced to 1.5 mm (Sokhansanj et al. 2009). Thus, size reduction can be viewed as a densification operation. Size reduction is a preparatory step for other densification processes that further increase bulk density. Densification processes like briquetting (350 kg m^{-3}), cubing (400 kg m^{-3}), and pelleting (500–700 kg m^{-3}) all result in high switchgrass bulk densities (Sokhansanj et al. 2009). Preprocessing steps such as pelletizing or briquetting switchgrass provide decreased storage losses and transportation costs (Tumuluru et al. 2011).

7.8.2.1 Pelleting

Pelleting wood from sawdust while making lumber is a well-established technique with most wood pellet industries colocated in existing lumber industries. In the case of switchgrass, the material should be moved from the point of production to the densification centers. The technology options of making switchgrass pellets from raw biomass are illustrated in Figure 7.8.

Before pelleting, switchgrass should be dry and ground to a particle size of about 1 mm (Sokhansanj et al. 2009). When properly baled, the moisture content will be suitable for grinding. Successful pelleting requires an optimum moisture content, particle size distribution, pellet die size, chemical composition (presence of binding material), heat, and pressure. Moisture requirement for pelleting can be controlled in the conditioning stage using steam conditioners that adjust the moisture of input material before feeding into the pellet mill. Heat is generated during the pelleting process and pellets need to be cooled immediately to impart strength and durability. Cooled pellets should be stored in ventilated storages before shipping. An example of technical specification that has proven suitable for switchgrass pellet production was (1) first-pass grinding with a tub grinder, (2) second-pass grinding through a 112 kW champion hammer mill with a 2.78 mm screen, and (3) pelleting with a 149 kW Sprout-Waldron pelleter using a 6.35 mm die with a 57.15 mm thickness to provide a 9:1 length/diameter ratio (Samson et al. 2008). Samson et al. (2008) found that the pellets produced from the fall-harvested switchgrass were slightly inferior to the spring-harvested switchgrass.

7.8.2.2 Briquetting

Grover and Mishra (1996) describes various aspects of biomass briquetting including available technologies, preheating requirements, material and energy balances, setting up a briquetting plant, economic analysis, and applications of briquettes. Even though reference to specifically switchgrass was not included in this report, the technology of briquetting of other biomass can be applied to switchgrass. Kaliyan and Morey (2007a) studied various operating parameters of almond-shaped

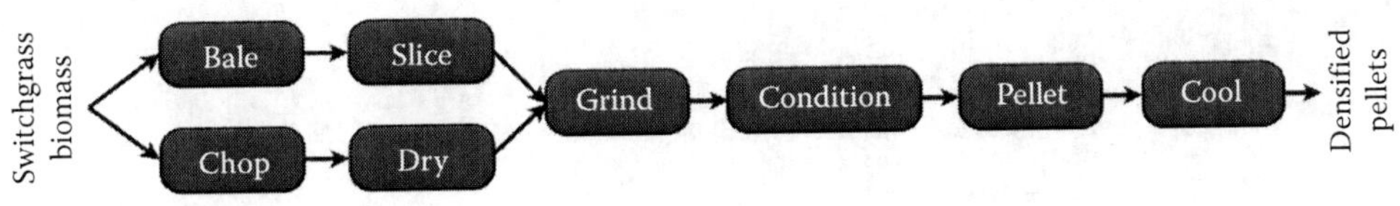

FIGURE 7.8 Production technology options of densified switchgrass pellets from raw switchgrass biomass. (Adapted from Sokhansanj, S. et al., *Biofuels Bioprod. Biorefin.*, 3, 124, 2009.)

briquettes having 28.7–31.3 mm length using pilot-scale roll press briquetting machine. They found that for switchgrass, temperatures around 75°C obtained by steam conditioning were necessary to produce good-quality briquettes. Switchgrass pellets had a bulk density of 420 kg m^{-3}, durability of 70%, and crushing strength of 171 N, and these values were somewhat less than but comparable to the pellets (9.6–9.8 mm diameter). They concluded that roll press briquetting is a promising low-cost, low-energy, and high-capacity approach for commercial production of densified switchgrass.

A densification concept of corn stover briquetting that outputs a flowable product but does not require the addition of heat or a reduced particle size was studied (Thoreson et al. 2010). A uniaxial piston cylinder pressed stover into a die (similar to an extruder) with a tapered conical outlet and produced briquettes of 640 kg m^{-3} density at a compression pressure of 10.3 MPa. Such experimental extrusion processes can also be applied to switchgrass, and successful densified products can be generated by controlling the various machines (e.g., pressure, die size, heat) and material parameters (e.g., moisture content, steam application, particle size, binding agent).

In a piston barrel single briquette (18.8 mm diameter) apparatus, a forming pressure of 150 MPa, preheating of switchgrass to 100°C without additional binding agents, produces a briquette of 67% durability compared to 0% durability without preheating (Kaliyan and Morey 2007b). In addition to preheating mixing of either 20% (mass) of corn stover or 2% (mass) of lignin-sulfonate-based binders or 5% (mass) lime produced switchgrass briquettes with durability near 80%. Kaliyan and Morey (2007b) concluded that preheating was essential to activate the natural binding agents in switchgrass with optimized particle size of 0.49–0.89 mm, moisture content ≥10% w.b., and preheating temperature ≥75°C and produce switchgrass pellets/briquettes with durability of ≥80%.

7.8.3 Another Densification Technology

Another technology being researched is bulk-format compaction of chopped switchgrass. A commercial-grade compactor compressed chopped switchgrass into a 57 m^3 reinforced, ejector-style transfer trailer and a hydraulic ejector ram offloaded the compressed material (Groothuis et al. 2011). Loose-filled bulk density of 89.4 kg m^{-3} could be increased to 186.4 kg m^{-3} at 15.4% w.b. moisture level by the bulk compaction method. Results indicate that in-bulk compaction is a promising method to improve transportation of chopped switchgrass as it doubled the load per semitrailer compared to a noncompacted load.

7.9 SUPPLY LOGISTICS

The four major components of switchgrass biomass feedstock logistics include (1) harvest and collection, (2) storage (input material and product), (3) preprocessing, and (4) transportation with a view to supply the biomass from *field to factory* (Figure 7.9).

Gold and Seuring (2011) reviewed supply chain and logistic issues of bioenergy production and recommended relationship-focused management (strategic management and social network theory)

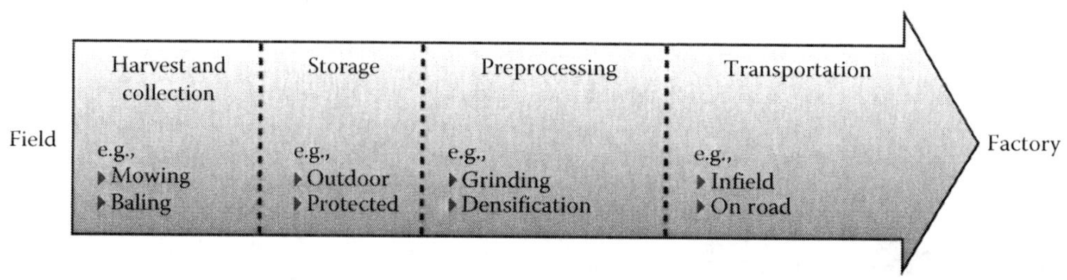

FIGURE 7.9 Major components of switchgrass biomass feedstock logistics showing *field to factory* movement of materials.

aspects of conducting bioenergy chains should be strengthened. Economies of scale can be achieved as the transport-vehicle capacities rise, and the biomass price tends to asymptotically decrease as the processing facility capacity increased (Tatsiopoulos and Tolis 2003). Scheduling the emptying of satellite storage locations (SSLs) based on travel time gave a lower average cost than assigning loaders to a given sector of the production region using discrete event simulation model (Ravula et al. 2008). A truck fleet ranging from 32 to 36 trucks was required to achieve full capacity for a 50 Mg h^{-1} bioprocessing plant. Truck operation (hauling) was about 75% of the total cost (truck ownership, fuel, loader moving, loader operating, unloader operating). Transportation cost was minimized when the loads hauled per day were maximized or loading and unloading times were reduced (Ravula et al. 2008).

7.9.1 Logistic Models and Simulations

Sokhansanj et al. (2006) developed a dynamic Integrated Biomass Supply Analysis and Logistics (IBSAL) model to simulate the collection, storage, and transport operations for supplying agricultural biomass to a biorefinery. The model considers time-dependent availability of biomass and effects of weather conditions on harvest progress. The model predicts the number and size of equipment to meet the rate of harvest and biorefinery demand schedule for feedstock and calculated the biomass delivery cost. Geographic information systems and network analysis were utilized to perform detailed analysis of the agricultural and forestry biomass resources locations and determine optimal sites for bioenergy plants using existing transport networks (Perpiñá et al. 2009).

Ebadian et al. (2011) developed the Integrated Biomass Supply Analysis and Logistics Multi Crop (IBSAL-MC) model based on the framework of IBSAL (Sokhansanj et al. 2006). IBSAL-MC is a hybrid push–pull logistic model. This model handles a mixture of agricultural feedstocks that push the field operations to harvest and collect as much biomass as possible within the harvest season while pulling downstream operations to meet the daily demand of the ethanol plant. A mathematical programming-based approach was utilized to determine SSLs for temporary storage and loading of round biomass bales and equipment routes for total cost minimization (Judd et al. 2012). The use of a side-loading rack system results in average savings of 21.3% over a briquetting densification system, while a rear-loading rack system is the most expensive. Judd et al. (2012) indicated that utilization of mobile equipment results in average savings of 14.8% over permanently placed equipment and a densification system is not justifiable for transportation distances less than 81 km.

Zhu et al. (2011) analyzed the challenges in designing the logistic system for dedicated switchgrass biomass-to-bioenergy industry. Unique features of switchgrass feedstock included low bulk density, restrictions on harvesting season and frequency, quality variations, weather effects, and crop scattered distribution over a wide geographical area. The model provided strategic decisions on the supply chain design and tactical decisions on annual operation schedules. The study showed that the logistic operations were significantly different for harvesting and nonharvesting seasons, and mass production with a steady and sufficient supply of biomass can increase the unit profit of bioenergy using a well-designed biomass logistic system.

Marvin et al. (2012) conducted an optimization study of the net present value of a biomass-to-ethanol supply chain in a nine-state region in the United States. They concluded that after technology stabilization, there is an incentive for 17.8 billion liter $year^{-1}$ of cellulosic ethanol to be developed in the Midwestern region. However, it was also estimated that with the current estimated parameter variability of the system, there was a 21.5% chance that this industry will not develop and if developed it could become uneconomical approximately 15% of the time.

7.9.2 Logistic Supply Costs

Sokhansanj et al. (2009) outlined four potential transport options: truck, train, barge, and pipeline. Truck transport is well developed and usually the most cost-effective but becomes expensive as travel distance increases. In case of rail and barge, a major component of fixed cost is the investment in the loading and unloading terminals as compared to the operating cost of loading and unloading. Transporting chopped biomass in the form of slurry with water through a pipeline is an attractive transport option, but its feasibility and cost are less certain. Combination of several different transport methods should be considered. For example, transport via rail, barge, or pipeline requires integrated operations with truck transport. Truck transport is economically desirable for distances less than 160 km, while rail transport is cost-effective above 160 km. The cost of truck transport ranged from \$15.42 Mg^{-1} for baled biomass at a yield of 10 Mg ha^{-1} to \$11.65 Mg^{-1} for pelletized biomass at a yield of 30 Mg ha^{-1}. Total cost of delivered switchgrass was \$80.64 Mg^{-1} for 10 Mg ha^{-1} and \$39.33 Mg^{-1} for 30 Mg ha^{-1} (Sokhansanj et al. 2009). Delivered costs of a switchgrass logistic system using a chopper with a cutterheader for harvesting and stretch-wrap baling technology may reduce costs by 25% when compared to various combinations of conventional logistic systems using hay harvest equipment for a 189 million liter $year^{-1}$ biorefinery located in East Tennessee (Yu et al. 2011). Savings using this stretch-wrap baling technology could result in a reduction of ethanol production costs by \$0.05 L^{-1}. On the storage logistic component, application of an affordable storage option (plastic-covered ambient storage versus pole frame metal roof with plastic cover versus closed warehouse with drying) leads to significant cost reductions in the whole biomass logistic system (Rentizelas et al. 2009). This cost reduction exceeds by far the extra costs of biomass material losses and increased handling in the simpler storage option.

7.9.3 BRDB View of Logistic Issues and Recommendations

Biomass logistic research issues are mainly focused on the various activities of a biomass supply assembly system so that biomass quantity requirements are delivered in a cost-effective and sustainable manner (BRDB 2010). The BRDB (2010) report identifies various barriers and recommendations (presented after ⇒) and is outlined as follows:

1. Biomass feedstocks from available resources have low mass and energy density with current harvest and collection technologies ⇒ develop densification technologies.
2. Moisture content of biomass at the time of harvest or collection is higher than desired and leads to degradation and decreased system efficiency ⇒ develop high-moisture biomass collection technology.
3. Currently available equipment for biomass feedstock logistic systems is inefficient ⇒ develop innovative equipment with higher throughput and increased efficiency through system integration.
4. Quality characteristics of biomass feedstocks are variable and inconsistent ⇒ develop quality standards and quality determination instrumentation development.
5. Transportation of biomass is costly and can strain transportation networks ⇒ develop understanding impacts of payload regulations due to increased traffic on rural roads and new transportation technology including improved containers and lighter vehicles to reduce impacts on existing road networks. A clear understanding of the issues and recommendations leads to research and development that will help in achieving the cellulosic biofuel targets from homegrown biomass.

7.10 CONCLUSIONS

Switchgrass will become an important feedstock for a mature cellulosic bioenergy industry. Continued research on the effects of harvest and storage management, determining compositional changes during storage on biofuel production, long-term research evaluating, harvesting management on nutrient removal, and developing strategies for maintaining sustainable soil nutrient status will be important as landscape-scale deployment of switchgrass for bioenergy moves forward. Further improvements in preprocessing, feedstock logistic systems, and specialized harvest and handling equipment will increase efficiency and reduce operational costs.

REFERENCES

Aden, A., M. Ruth, K. Ibsen, J. Jechura, K. Neeves, J. Sheehan, B. Wallace, L. Montague, A. Slayton, and J. Lukas. 2002. *Lignocellulosic Biomass to Ethanol Process Design and Economics Utilizing Co-Current Dilute Acid Prehydrolysis and Enzymatic Hydrolysis for Corn Stover*. NREL/TP-510-32438. Golden, CO: National Renewable Energy Laboratory.

Adler, P.R., S.J.D. Grosso, and W.J. Parton. 2007. Life-cycle assessment of net greenhouse-gas flux for bioenergy cropping systems. *Ecological Applications* 17: 675–691.

Adler, P.R., M.A. Sanderson, A.A. Boateng, P.J. Weimer, and H.-J.G. Jung. 2006. Biomass yield and biofuel quality of switchgrass harvested in fall or spring. *Agronomy Journal* 98: 1518–1525.

Aiken, G.E. and T.L. Springer. 1995. Seed size distribution, germination, and emergence of 6 switchgrass cultivars. *Journal of Range Management* 48: 455–458.

Alexopoulou, E., N. Sharma, Y. Papatheohari, M. Christou, I. Piscioneri, C. Panoutsou, and V. Pignatelli. 2008. Biomass yields for upland and lowland switchgrass varieties grown in the Mediterranean region. *Biomass and Bioenergy* 32: 926–933.

Al-Kaisi, M.M., X. Yin, and M.A. Licht. 2005. Soil carbon and nitrogen changes as influenced by tillage and cropping systems in some Iowa soils. *Agriculture, Ecosystems and Environment* 105: 635–647.

Asay, K.H. and D.A. Johnson. 1980. Screening for improved stand establishment in Russian Wild Ryegrass. *Canadian Journal of Plant Science* 60: 1171–1177.

Benedict, H.M. 1941. Effect of day length and temperature on the flowering and growth of four species of grasses. *Journal of Agricultural Research* 61: 661–672.

Bitra, V.S.P., A.R. Womac, N. Chevanan, P.I. Miu, C. Igathinathane, S. Sokhansanj, and D.R. Smith. 2009a. Direct mechanical energy measures of hammer mill comminution of switchgrass, wheat straw, and corn stover and analysis of their particle size distributions. *Powder Technology* 193: 32–45.

Bitra, V.S.P., A.R. Womac, C. Igathinathane, P.I. Miu, Y.T. Yang, D.R. Smith, N. Chevanan, and S. Sokhansanj. 2009b. Direct measures of mechanical energy for knife mill size reduction of switchgrass, wheat straw, and corn stover. *Bioresource Technology* 100: 6568–6585.

Boateng, A., K. Hicks, and K. Vogel. 2006. Pyrolysis of switchgrass (*Panicum virgatum*) harvested at several stages of maturity. *Journal of Analytical and Applied Pyrolysis* 75: 55–64.

Boydston, R.A., H.P. Collins, and S.C. Fransen. 2010. Response of three switchgrass (*Panicum virgatum*) cultivars to mesotrione, quinclorac, and pendimethalin. *Weed Technology* 24: 336–341.

Bransby, D.I., H.A. Smith, C.R. Taylor, and P.A. Duffy. 2005. Switchgrass budget model: An interactive budget model for producing & delivering switchgrass to a bioprocessing plant. *Industrial Biotechnology* 2: 122–125.

BRDB. 2010. Biomass Research and Development Board (BRDB). Biofuel feedstock logistics: Recommendations for research and commercialization. A report by Feedstock Logistics Interagency Working Group. Washington, DC: BRDB.

Brechbill, S., W. Tyner, and K. Ileleji. 2011. The economics of biomass collection and transportation and its supply to Indiana cellulosic and electric utility facilities. *BioEnergy Research* 4: 141–152.

Brejda, J.J. 2000. Fertilization of native warm-season grasses. In *Native Warm-Season Grasses: Research Trends and Issues*. K.J. Moore and B.E. Anderson (eds.), pp. 177–200. Madison, WI: CSSA and ASA.

Brejda, J.J., D.H. Yocom, L.E. Moser, and S.S Waller. 1993. Dependence on three Nebraska sandhills warm-season grasses on vesicular-arbuscular mycorrhizae. *Journal of Range Management* 46: 14–20.

Brownell, D.K., J. Liu, J.W. Hilton, T.L. Richard, G.R. Cauffman, and B.R. Macafee. 2009. Evaluation of two forage harvesting systems for herbaceous biomass harvesting. ASABE Paper Number 097390. St. Joseph, MI: ASABE.

Bryan, W.B., T.A. Mills, and J.A. Cronauer. 1984. Sod-seeding switchgrass and tall fescue into hill land pasture. *Journal of Soil and Water Conservation* 39: 70–72.

Casler, M.D., E.A. Heaton, K.J. Shinners, H.G. Jung, P.J. Weimer, M.A. Liebig, R.B. Mitchell, and M.F. Digman. 2009. Grasses and legumes for cellulosic bioenergy. In *Grassland: Quietness and Strength for a New American Agriculture*. W.F. Wedin and S.L. Fales (eds.), pp. 205–219. Madison, WI: ASA-CSSA-SSSA.

Casler, M.D., K.P. Vogel, C.M. Taliaferro, N.J. Ehlke, J.D. Berdahl, E.C. Brummer, R.L. Kallenbach, C.P. West, and R.B. Mitchell. 2007. Latitudinal and longitudinal adaptation of switchgrass populations. *Crop Science* 47: 2249–2260.

Casler, M.D., K.P. Vogel, C.M. Taliaferro, and R.L. Wynia. 2004. Latitudinal adaptation of switchgrass populations. *Crop Science* 44: 293–303.

Chen, S.F., R.A. Mowery, R.S. Sevcik, C.J. Scarlata, and C.K. Chambliss. 2010. Compositional analysis of water-soluble materials in switchgrass. *Journal of Agricultural and Food Chemistry* 58: 3251–3258.

Chevanan, N., A.R. Womac, V.S.P. Bitra, C. Igathinathane, Y.T. Yang, P.I. Miu, and S. Sokhansanj. 2010. Bulk density and compaction behavior of knife mill chopped switchgrass, wheat straw, and corn stover. *Bioresource Technology* 101: 207–214.

Christian, D., A.B. Riche, and N.E. Yates. 2002. The yield and composition of switchgrass and coastal panic grass grown as a biofuel in Southern England. *Bioresource Technology* 83: 115–124.

Collins, H.P., S. Fransen, A. Hang, R.A. Boydston, and C.E. Kruger. 2008. Biomass production and nutrient removal by switchgrass (*Panicum virgatum*) under irrigation. *ASA-CSSA-SSSA International Annual Meetings*, Houston, TX.

Conant, R.T., K. Paustian, and E.T. Elliott. 2001. Grassland management and conversion into grassland: Effects on soil carbon. *Ecological Applications* 11: 343–355.

Crutzen, P.J., A.R. Mosier, K.A. Smith, and W. Winiwarter. 2008. N_2O release from agro-biofuel production negates global warming reduction by replacing fossil fuels. *Atmospheric Chemistry and Physics* 8: 389–395.

Cundiff, J.S., J.H. Fike, D.J. Parrish, and J. Alwang. 2009. Logistic constraints in developing dedicated large-scale bioenergy systems in the Southeastern United States. *Journal of Environmental Engineering* 135: 1086–1096.

Cundiff, J.S. and L.S. Marsh. 1996. Harvest and storage costs for bales of switchgrass in the southeastern united states. *Bioresource Technology* 56: 95–101.

Dahl, B.E. and D.N. Hyder. 1977. Developmental morphology and management implications. In *Rangeland Plant Physiology*. R.E. Sosebee (ed.), pp. 257–290. Denver, CO: Society for Range Management.

Del Grosso, S.J., W.J. Parton, A.R. Mosier, M.K. Walsh, D.S. Ojima, and P.E. Thornton. 2006. DAYCENT national-scale simulations of nitrous oxide emissions from cropped soils in the United States. *Journal of Environment Quality* 35: 1451–1460.

Dien, B., H. Jung, K. Vogel, M. Casler, J. Lamb, L. Iten, R. Mitchell, and G., Sarath. 2006. Chemical composition and response to dilute-acid pretreatment and enzymatic saccharification of alfalfa, reed canarygrass, and switchgrass. *Biomass and Bioenergy* 30: 880–891.

Dierberger, B.L. 1991. Switchgrass germination as influenced by temperature, chilling, cultivar and seedlot. MS thesis, University of Nebraska-Lincoln, Lincoln, NE.

Digman, M., K. Shinners, R. Muck, and B. Dien. 2010. Full-scale on-farm pretreatment of perennial grasses with dilute acid for fuel ethanol production. *BioEnergy Research* 3: 335–341.

Digman, M.F., B.S. Dien, and R.D. Hatfield. 2012. On-farm acidification and anaerobic storage for preservation and improved conversion of switchgrass into ethanol. *Biological Engineering Transactions* 5: 47–58.

Di Virgilio, N., A. Monti, and G. Venturi. 2007. Spatial variability of switchgrass (*Panicum virgatum* L.) yield as related to soil parameters in a small field. *Field Crops Research* 101: 232–239.

Drzymala, Z. 1993. *Industrial Briquetting-Fundamentals and Methods*. Studies in mechanical engineering, Vol. 13. Warszawa, Poland: PWN-Polish Scientific Publishers.

Duffy, M. and V. Nanhou. 2002. Costs of producing switchgrass for biomass in southern Iowa. In *Trends in New Crops and New Uses*. J. Janick and A. Whipkey (eds.), pp. 267–274. Alexandria, VA: ASHS Press.

Duvick, D.N. and K.G. Cassman. 1999. Post-green revolution trends in yield potential of temperate maize in the North-Central United States. *Crop Science* 39: 1622–1630.

Ebadian, M., T. Sowlati, S. Sokhansanj, M. Stumborg, and L.T. Smith. 2011. A new simulation model for multi-agricultural biomass logistics system in bioenergy production. *Biosystems Engineering* 110: 280–290.

Elbersen, H.W., D.G. Christian, N.E. Bassam, G. Sauerbeck, E. Alexopoulou, N. Sharma, and I. Piscioneri. 2004. A management guide for planting and production of switchgrass as a biomass crop in Europe. *Second World Conference on Biomass for Energy, Industry and Climate Protection*, May 10–14, Rome, Italy.

El-Nashaar, H.M., G.M. Banowetz, S.M. Griffith, M.D. Casler, and K.P. Vogel. 2009. Genotypic variability in mineral composition of switchgrass. *Bioresource Technology* 100: 1809–1814.

Epplin, F.M. 1996. Cost to produce and deliver switchgrass to an ethanol conversion facility in the Southern Plains of the United States. *Biomass and Bioenergy* 11: 459–467.

Fargione, J., J. Hill, D. Tilman, S. Polasky, and P. Hawthorne. 2008. Land clearing and the biofuel carbon debt. *Science* 319: 1235–1238.

Fargione, J.E., T.R. Cooper, D.J. Flaspohler et al. 2009. Bioenergy and wildlife: Threats and opportunities for grassland conservation. *BioScience* 59: 767–777.

Farrell, A.E., R.J. Plevin, B.T. Turner, A.D. Jones, M. O'hare, and D.M. Kammen. 2006. Ethanol can contribute to energy and environmental goals. *Science* 311: 506–508.

Fike, J., D. Parrish, D. Wolf, J. Balasko, J. Green Jr, M. Rasnake, and J. Reynolds. 2006. Long-term yield potential of switchgrass-for-biofuel systems. *Biomass and Bioenergy* 30: 198–206.

Follett, R., K. Vogel, G. Varvel, R. Mitchell, and J. Kimble. 2012. Soil carbon sequestration by switchgrass and no-till maize grown for bioenergy. *BioEnergy Research*. DOI 10.1007/S12155-012-9198-y.

Frank, A.B., J.D. Berdahl, J.D. Hanson, M.A. Liebig, and H.A. Johnson. 2004. Biomass and carbon partitioning in switchgrass. *Crop Science* 44: 1391–1396.

Garland, C.D. 2010. Growing and harvesting switchgrass for ethanol production in Tennessee. UT Biofuels Initiative. UT Extension Publication Number SP701-A. Knoxville, TX: University of Tennessee.

Garten, C.T. and S.D. Wullschleger. 2000. Soil carbon dynamics beneath switchgrass as indicated by stable isotope analysis. *Journal of Environmental Quality* 29: 645–653.

Gebhart, D.L., H.B. Johnson, H.S. Mayeux, and H.W. Polley. 1994. The CRP increases soil organic carbon. *Journal of Soil and Water Conservation* 49: 488–492.

Gold, S. and S. Seuring. 2011. Supply chain and logistics issues of bio-energy production. *Journal of Cleaner Production* 19: 32–42.

Gopalakrishnan, G., M.C. Negri, M. Wang, M. Wu, S.W. Snyder, and L. LaFreniere. 2009. Biofuels, land, and water: A systems approach to sustainability. *Environmental Science and Technology* 43: 6094–6100.

Green, J.C. and D.I. Bransby. 1995. Effects of seed size on germination an seedling growth of 'Alamo' switchgrass. In *Proceedings of the Fifth International Rangeland Congress*. N.E. West (ed.), pp. 183–184. Salt Lake City, UT: Society for Range Management.

Greenlees, W.J., H.M. Hanna, K.J. Shinners, S.J. Marley, and T.B. Bailey. 2000. A comparison of four mower conditioners on drying rate and leaf loss in alfalfa and grass. *Applied Engineering in Agriculture* 16: 15–21.

Groothuis, M.D., A.R. Womac, G. Braswell, and V.S.P. Bitra. 2011. Compaction experiences with bulk-format switchgrass in commercial transfer systems. ASABE Paper Number 1110819. St. Joseph, MI: ASABE.

Grover, P.D. and S.K. Mishra. 1996. Biomass briquetting: Technology and practices. Field Document Number 46. FAO Regional Wood Energy Development Programme in Asia GCP/RAS/154/NET, Bangkok, Thailand.

Guretzky, J.A., J.T. Biermacher, B.J. Cook, M.K. Kering, and J. Mosali. 2011. Switchgrass for forage and bioenergy: Harvest and nitrogen rate effects on biomass yields and nutrient composition. *Plant Soil* 339: 69–81.

Gustafson, D.M., A. Boe, and Y. Jin. 2003. Genetic variation for *Puccinia emaculata* infection in switchgrass. *Crop Science* 43: 755–759.

Haas, H.J. 1958. Effect of fertilizers, age of stand, and decomposition on weight of grass roots and of grass and alfalfa on soil nitrogen and carbon. *Agronomy Journal* 50: 5–9.

Hall, R.D. and L.E. Wiesner. 1990. Relationship between seed vigor tests and field performance of 'Regar' meadow bromegrass. *Crop Science* 30: 967–970.

Hallam, A., I. Anderson, and D.R. Buxton. 2001. Comparative economic analysis of perennial, annual, and intercrops for biomass production. *Biomass and Bioenergy* 21: 407–424.

Hammerschlag, R. 2006. Ethanol's energy return on investment: A survey of the literature 1990—present. *Environmental Science and Technology* 40: 1744–1750.

Hanson, J.D. and H.A. Johnson. 2005. Germination under various temperature and pH regimes. *Seed Technology* 27: 203–210.

Haque, M., F.M. Epplin, and C.M. Taliaferro. 2009. Nitrogen and harvest frequency effect on yield and cost for four perennial grasses. *Agronomy Journal* 101: 1463–1469.

Heggenstaller, A.H., K.J. Moore, M. Liebman, and R.P. Anex. 2009. Nitrogen influences biomass and nutrient partitioning by perennial, warm-season grasses. *Agronomy Journal* 101: 1363–1371.

Hess, J.R., K.L. Kenney, L.P. Ovard, E.M. Searcy, and C.T. Wright. 2009. *Uniform-Format Bioenergy Feedstock Supply System: A Commodity-Scale Design to Produce and Infrastructure-Compatible Bulk Solid from Lignocellulosic Biomass*. Idaho Falls, ID: Idaho National Laboratory.

Hill, J., E. Nelson, D. Tilman, S. Polasky, and D. Tiffany. 2006. Environmental, economic, and energetic costs and benefits of biodiesel and ethanol biofuels. *Proceedings of the National Academy of Sciences* 103: 11206–11210.

Hoshikawa, K. 1969. Underground organs of the seedling and the systematics of gramineae. *Botanical Gazette* 130: 192–203.

Hsu, F.H. and C.J. Nelson. 1986. Planting date effects on seedling development of perennial warm-season forage grasses. II. Seedling growth. *Agronomy Journal* 78: 38–42.

Huang, S., X. Su, R. Haselkorn, and P. Gornicki. 2003. Evolution of switchgrass (*Panicum virgatum* L.) based on sequences of the nuclear gene encoding plastid acetyl-CoA carboxylase. *Plant Science* 164: 43–49.

Huhnke, R.L. 1993. Round bale orientation effects on alfalfa hay storage. *Applied Engineering in Agriculture* 9: 349–351.

Huisman, W. 2003. Optimising harvesting and storage systems for energy crops in The Netherlands. ASAE Publication Number 701P1103e. St. Joseph, MI: ASAE. In *Proceedings International Conference on Crop Harvesting and Processing*, February 9–11, 2003, Louisville, KY.

Igathinathane, C., C. Gustafson, D. Archer, J. Hendrickson, and S. Kronberg. 2012. Effects of field variables on infield biomass bales aggregation strategies. ASABE Paper Number 121337357. St. Joseph, MI: ASABE.

Igathinathane, C., L.O. Pordesimo, E.P. Columbus, W.D. Batchelor, and S. Sokhansanj. 2009. Sieveless particle size distribution analysis of particulate materials through computer vision. *Computers and Electronics in Agriculture* 66: 147–158.

Igathinathane, C., A.R. Womac, S. Sokhansanj, and S. Narayan. 2008. Knife grid size reduction to pre-process packed beds of high- and low-moisture switchgrass. *Bioresource Technology* 99: 2254–2264.

Jager, H.I., L.M. Baskaran, C.C. Brandt, E.B. Davis, C.A. Gunderson, and S.D. Wullschleger. 2010. Empirical geographic modeling of switchgrass yields in the United States. *GCB Bioenergy* 2: 248–257.

James, L.K., S.M. Swinton, and K.D. Thelen. 2010. Profitability analysis of cellulosic energy crops compared with corn. *Agronomy Journal* 102: 675–687.

Jannasch, R., Y. Quan, and R. Samson. 2001. A process and energy analysis of pelletizing switchgrass. Final Report, Natural Resources Canada. Sainte-Anne-de-Bellevue, QC, Canada: REAP-Canada.

Jenkins, B.M., B.C. Horsfield, J.B. Dobie, and G. Miller. 1981. Agricultural residues: Renewable energy for utility power companies. *Transactions of the ASAE* 24: 197–201, 207.

Judd, J.D., S.C. Sarin, and J.S. Cundiff. 2012. Design, modeling, and analysis of a feedstock logistics system. *Bioresource Technology* 103: 209–218.

Jung, G.A., J.A. Shaffer, and W.L. Stout. 1988. Switchgrass and big bluestem responses to amendments on strongly acid soil. *Agronomy Journal* 80: 669–676.

Jung, J.Y. and R. Lal. 2011. Impacts of nitrogen fertilization on biomass production of switchgrass (*Panicum Virgatum* L.) and changes in soil organic carbon in Ohio. *Geoderma* 166: 145–152.

Kaliyan, N. and R.V. Morey. 2007a. Roll press briquetting of corn stover and switchgrass: A pilot scale continuous briquetting study. ASABE Paper Number 076044. St. Joseph, MI: ASABE.

Kaliyan, N. and R.V. Morey. 2007b. Strategies to improve durability of switchgrass briquettes. ASABE Paper Number 076182. St. Joseph, MI: ASABE.

Kering, M., J. Biermacher, T. Butler, J. Mosali, and J. Guretzky. 2012. Biomass yield and nutrient responses of switchgrass to phosphorus application. *BioEnergy Research* 5: 71–78.

Khanchi, A., C. Jones, and B. Sharma. 2010. Characteristics and compositional variation in round and square switchgrass bales under different storage conditions. ASABE Paper Number 1009098. St. Joseph, MI: ASABE.

Khanna, M., B. Dhungana, and J. Clifton-Brown. 2008. Costs of producing miscanthus and switchgrass for bioenergy in Illinois. *Biomass and Bioenergy* 32: 482–493.

Khush, G.S. 1999. Green revolution: Preparing for the 21st century. *Genome* 42: 646–655.

Kim, S. and B.E. Dale. 2004. Cumulative energy and global warming impact from the production of biomass for biobased products. *Journal of Industrial Ecology* 7: 147–162.

Kumar, A. and S. Sokhansanj. 2007. Switchgrass (*Panicum vigratum* L.) delivery to a biorefinery using integrated biomass supply analysis and logistics (IBSAL) model. *Bioresource Technology* 98: 1033–1044.

Laser, M., H. Jin, K. Jayawardhana, B.E. Dale, and L.R. Lynd. 2009. Projected mature technology scenarios for conversion of cellulosic biomass to ethanol with coproduction thermochemical fuels, power, and/or animal feed protein. *Biofuels, Bioproducts and Biorefining* 3: 231–246.

Lasorella, M.V., A. Monti, E. Alexopoulou et al. 2011. Yield comparison between switchgrass and *Miscanthus* based on multi-year side by side comparison in Europe. *Proceeding of the 19th European Biomass Conference and Exhibition*, Berlin, Germany.

Lee, D.K., V.N. Owens, A. Boe, and B.-C. Koo. 2009. Biomass and seed yields of big bluestem, switchgrass, and intermediate wheatgrass in response to manure and harvest timing at two topographic positions. *Global Change Biology Bioenergy* 1: 171–179.

Lee, D.K., V.N. Owens, and J.J. Doolittle. 2007. Switchgrass and soil carbon sequestration response to ammonium nitrate, manure, and harvest frequency on Conservation Reserve Program land. *Agronomy Journal* 99: 462–468.

Lemus, R., D. Parrish, and D. Wolf. 2009. Nutrient uptake by 'Alamo' switchgrass used as an energy crop. *BioEnergy Research* 2: 37–50.

Liebig, M., H. Johnson, J. Hanson, and A. Frank. 2005. Soil carbon under switchgrass stands and cultivated cropland. *Biomass and Bioenergy* 28: 347–354.

Liebig, M.A., J.R. Gross, S.L. Kronberg, J.D. Hanson, A.B. Frank, and R.L. Phillips. 2006. Soil response to long-term grazing in the Northern Great Plains of North America. *Agriculture, Ecosystems and Environment* 115: 270–276.

Liebig, M.A., M.R. Schmer, K.P. Vogel, and R.B. Mitchell. 2008. Soil carbon storage by switchgrass grown for bioenergy. *BioEnergy Research* 1: 215–222.

Madakadze, I.C., B. Prithiviraj, R.M. Madakadze, K. Stewart, P. Peterson, B.E. Coulman, and D.L. Smith. 2000. Effect of preplant seed conditioning treatment on the germination of switchgrass (*Panicum virgatum* L.). *Seed Science and Technology* 28: 403–411.

Mani, S., L.G. Tabil, and S. Sokhansanj. 2002. Grinding performance and physical properties of selected biomass. ASAE Paper Number 026175. St. Joseph, MI: ASAE.

Mani, S., L.G. Tabil, and S. Sokhansanj. 2004. Grinding performance and physical properties of wheat and barley straws, corn stover and switchgrass. *Biomass and Bioenergy* 27: 339–352.

Martinez-Reyna, J.M. and K.P. Vogel. 2002. Incompatibility systems in switchgrass. *Crop Science* 42: 1800–1805.

Marvin, W.A., L.D. Schmidt, S. Benjaafar, D.G. Tiffany, and P. Daoutidis. 2012. Economic optimization of a lignocellulosic biomass-to-ethanol supply chain. *Chemical Engineering Science* 67: 68–79.

McDonald, P., A.R. Henderso, and S.J.E. Heron. 1991. *The Biochemistry of Silage*, 2nd edn. Buckinghamshire, U.K.: Chalcombe Publications.

McKendry, P. 2002. Energy production from biomass (part 2): Conversion technologies. *Bioresource Technology* 83: 47–54.

McLaughlin, S. and A.L. Kszos. 2005. Development of switchgrass (*Panicum virgatum* L.) as a bioenergy feedstock in the United States. *Biomass and Bioenergy* 28: 515–535.

McLaughlin, S.B., J. Bouton, D. Bransby, B.V. Conger, W. Ocumpaugh, D. Parrish, C.M. Taliaferro, K.P. Vogel, and S.D. Wullschleger. 1999. Developing switchgrass as a bioenergy crop. In *Perspective on New Crops and New Uses*. J. Janick, (ed.), pp. 282–299. Alexandria, VA: ASHS Press.

McLaughlin, S.B., D.G. De la Torre Daniel, C.T. Garten Jr., L.R. Lynd, M.A. Sanderson, V.R. Tolbert, and D.D. Wolf. 2002. High-value renewable energy from prairie grasses. *Environmental Science and Technology* 36: 2122–2129.

McLaughlin, S.B., J.R. Kiniry, C.M. Taliaferro, and D.G. De la Torre Daniel. 2006. Projecting yield and utilization potential of switchgrass as an energy crop. *Advances in Agronomy* 90: 267–297.

McLaughlin, S.B., R. Samson, D. Bransby, and A. Wiselogel. 1996. Evaluating physical, chemical, and energetic properties of perennial grasses as biofuels. In *Bioenergy'96—Proceedings of the Seventh National Bioenergy Conference*, Vol. 96, pp. 1–8. Nashville, TN.

Millar, N., G. Robertson, P. Grace, R. Gehl, and J. Hoben. 2010. Nitrogen fertilizer management for nitrous oxide (N_2O) mitigation in intensive corn (Maize) production: An emissions reduction protocol for US Midwest agriculture. *Mitigation and Adaptation Strategies for Global Change* 15: 185–204.

Mitchell, R.B. and C.M. Britton. 2000. Managing weeds to establish and maintain warm-season grasses. In *Native Warm-Season Grasses: Research Trends and Issues*. K.J. Moore and B.E. Anderson, (eds.), pp. 159–176. Madison, WI: CSSA and ASA.

Mitchell, R.B., K.J. Moore, L.E. Moser, J.O. Fritz, and D.D. Redfearn. 1997. Predicting developmental morphology in switchgrass and big bluestem. *Agronomy Journal* 89: 827–832.

Mitchell, R.B. and M.R. Schmer. 2012. Switchgrass harvest and storage. In *Switchgrass: A Valuable Biomass Crop for Energy*. A. Monti, (ed.), pp. 113–127. London, U.K.: Springer-Verlag.

Mitchell, R.B. and K.P. Vogel. 2012. Germination and emergence tests for predicting switchgrass field establishment. *Agronomy Journal* 104: 458–465.

Mitchell, R.B., K.P. Vogel, J. Berdahl, and R.A. Masters. 2010b. Herbicides for establishing switchgrass in the Central and Northern Great Plains. *BioEnergy Research* 3: 321–327.

Mitchell, R.B., K.P. Vogel, and G. Sarath. 2008. Managing and enhancing switchgrass as a bioenergy feedstock. *Biofuels, Bioproducts and Biorefining* 2: 530–539.

Mitchell, R.B., K.P. Vogel, M.R. Schmer, and D. Pennington. 2010a. Switchgrass for biofuel production *National eXtension*. Available at http://www.extension.org/pages/26635/switchgrass-panicum-virgatum-for-biofuel-production (accessed on November 26, 2013).

Mitchell, R.B., K.P. Vogel, and D.R. Uden. 2012. The feasibility of switchgrass for biofuel production. *Biofuels* 3: 47–59.

Monti, A., G. Bezzi, G. Pritoni, and G. Venturi. 2008. Long-term productivity of lowland and upland switchgrass cytotypes as affected by cutting frequency. *Bioresource Technology* 99: 7425–7432.

Monti, A., S. Fazio, V. Lychnaras, P. Soldatos, and G. Venturi. 2007. A full economic analysis of switchgrass under different scenarios in Italy estimated by BEE model. *Biomass and Bioenergy* 31: 177–185.

Monti, A., S. Fazio, and G. Venturi. 2009. The discrepancy between plot and field yields: Harvest and storage losses of switchgrass. *Biomass and Bioenergy* 33: 841–847.

Mooney, D.F., R.K. Roberts, B.C. English, D.D. Tyler, and J.A. Larson. 2009. Yield and breakeven price of 'Alamo' switchgrass for biofuels in Tennessee. *Agronomy Journal* 101: 1234–1242.

Moore, K.J., L.E. Moser, K.P. Vogel, S.S. Waller, B.E. Johnson, and J.F. Pedersen. 1991. Describing and quantifying growth stages of perennial forage grasses. *Agronomy Journal* 83: 1073–1077.

Morris, R.J., R.H. Fox, and G.A. Jung. 1982. Growth, P uptake, and quality of warm and cool-season grasses on a low available P soil. *Agronomy Journal* 74: 125–129.

Moser, L.E. 2000. Morphology of germinating and emerging warm-season grass seedlings. In *Native Warm-Season Grasses: Research Trends and Issues*. K.J. Moore and B.E. Anderson, (eds.), CSSA Special Publication 30 ed. pp. 35–47. Madison, WI: CSSA and ASA.

Moser, L.E. and K.P. Vogel. 1995. Switchgrass, big bluestem, and indiangrass. In *Forages: An Introduction to Grassland Agriculture*. R.F. Barnes (ed.). Ames, IA: Iowa State University Press.

Muir, J.P., Sanderson, M.A., Ocumpaugh, W.R., Jones, R.M., and Reed, R.L. 2001. Biomass production of 'Alamo' switchgrass in response to nitrogen, phosphorus, and row spacing. *Agronomy Journal* 93: 896–901.

Newell, L.C. and F.D. Keim. 1947. Effects of mowing frequency on the yield and protein content of several grasses grown in pure stands. Nebraska Agricultural Extension Station Research Bulletin 150: 36 pp. Lincoln, NE: University of Nebraska-Lincoln.

Newman, P.R. and L.E. Moser. 1988. Seedling root development and morphology of cool-season and warm-season forage grasses. *Crop Science* 28: 148–151.

Nickell, G.L. 1973. The physiological ecology of upland and lowland *Panicum virgatum*. PhD dissertation Abstract Int. B, University of Oklahoma, Norman, OK 33:3512.

Parrish, D. and J. Fike. 2005. The biology and agronomy of switchgrass for biofuels. *Critical Reviews in Plant Sciences* 24: 423–459.

Parrish, D.J. and D.D. Wolf. 1992. Managing switchgrass for sustainable biomass production. In *Liquid Fuels from Renewable Resources*. J. S. Cundiff, (ed.), pp. 34–39. St. Joseph, MI, Nashville, TN: ASAE.

Perpiñá, C., D. Alfonso, A.P. Navarro, E. Peñalvo, C. Vargas, and R. Cárdenas. 2009. Methodology based on geographic information systems for biomass logistics and transport optimization. *Renewable Energy* 34: 555–565.

Perrin, R., J. Sesmero, K. Wamisho, and D. Bacha. 2012. Biomass supply schedules for Great Plains delivery points. *Biomass and Bioenergy* 37: 213–220.

Perrin, R., K. Vogel, M. Schmer, and R. Mitchell. 2008. Farm-scale production cost of switchgrass for biomass. *BioEnergy Research* 1: 91–97.

Pitt, R. 1990. *Silage and Hay Preservation*. Ithaca, NY: Northeast Regional Agricultural Engineering Service.

Qin, X., T. Mohan, M. El-Halwagi, G. Cornforth, and B.A. McCarl. 2006. Switchgrass as an alternate feedstock for power generation: An integrated environmental, energy and economic life-cycle assessment. *Clean Technologies and Environmental Policy* 8: 233–249.

Ravula, P.P., R.D. Grisso, and J.S. Cundiff. 2008. Comparison between two policy strategies for scheduling trucks in a biomass logistic system. *Bioresource Technology* 99: 5710–5721.

Rehm, G.W. 1984. Yield and quality of a warm-season grass mixture treated with N, P, and atrazine. *Agronomy Journal* 76: 731–734.

Rehm, G.W., R.C. Sorenson, and W.J. Moline. 1976. Time and rate of fertilizer application for seeded warm-season and bluegrass pastures: I. Yield and botanical composition. *Agronomy Journal* 68: 559–564.

Rentizelas, A.A., A.J. Tolis, and I.P. Tatsiopoulos. 2009. Logistics issues of biomass: The storage problem and the multi-biomass supply chain. *Renewable and Sustainable Energy Reviews* 13: 887–894.

Robertson, G.P., S.K. Hamilton, S.J. Del Gross, and W.J. Parton. 2010. *Growing Plants for Fuel: Predicting Effects on Water, Soil, and the Atmosphere*. Washington, DC: Biofuels and Sustainability Reports. Ecological Society of America

Robins, J.G. 2010. Cool-season grasses produce more total biomass across the growing season than do warm-season grasses when managed with an applied irrigation gradient. *Biomass and Bioenergy* 34: 500–505.

Samson, R., P. Duxbury, M. Drisdelle, and C. Lapointe. 2000. Assessment of Pelletized Biofuels. Available at http://reap-canada.com/online_library/feedstock_biomass/15%20Assessment%20of.PDF (accessed on November 26, 2013). Sainte-Anne-de-Bellevue, QC, Canada: REAP-Canada.

Samson, R., S.B. Stamler, and C.H. Lem. 2008. Optimization of switchgrass management for commercial fuel pellet production. Final Report prepared by REAP-Canada for the Ontario Ministry of Food, Agriculture and Rural Affairs (OMAFRA) under the Alternative Renewable Fuels Fund. Sainte-Anne-de-Bellevue, QC, Canada: REAP-Canada.

Sanderson, M.A., P.R. Adler, A.A. Boateng, M.D. Casler, and G. Sarath. 2006. Switchgrass as a biofuels feedstock in the USA. *Canadian Journal of Plant Science* 86: 1315–1325.

Sanderson, M.A., R.P. Egg, and A.E. Wiselogel. 1997. Biomass losses during harvest and storage of switchgrass. *Biomass and Bioenergy* 2: 107–114.

Sanderson, M.A., J.C. Read, and R.L. Reed. 1999. Harvest management of switchgrass for biomass feedstock and forage production. *Agronomy Journal* 91: 5–10.

Sanderson, M.A., R.L. Reed, S.B. McLaughlin et al. 1996. Switchgrass as a sustainable bioenergy crop. *Bioresource Technology* 56: 83–93.

Sanderson, M.A., M.R. Schmer, V. Owens, P. Keyser, and W. Elberson. 2012. Crop management of switchgrass. In *Switchgrass: A Valuable Biomass Crop for Energy*. A. Monti ed. pp. 87–112. London, U.K.: Springer-Verlag.

Sanderson, M.A. and D.D. Wolf. 1995. Switchgrass biomass composition during morphological development in diverse environments. *Crop Science* 35: 1432–1438.

Sarath, G., L. Baird, K. Vogel, and R. Mitchell. 2007. Internode structure and cell wall composition in maturing tillers of switchgrass (*Panicum virgatum*. L). *Bioresource Technology* 98: 2985–2992.

Sarath, G. and R.B. Mitchell. 2008. Aged switchgrass seed lot's response to dormancy-breaking chemicals. *Seed Technology* 30: 7–16.

Schell, D.J. and C. Harwood. 1994. Milling of lignocellulosic biomass: Results of pilot-scale testing. *Applied Biochemistry and Biotechnology* 45/46: 68–159.

Schmer, M.R., J.D. Hanson, and H.A. Johnson. 2012a. Switchgrass and intermediate wheatgrass aboveground and belowground response to nitrogen and calcium. *Journal of Plant Nutrition* 35: 1065–1079.

Schmer, M.R., M.A. Liebig, J.R. Hendrickson, D.L. Tanaka, and R. Phillips. 2012c. Growing season greenhouse gas flux from switchgrass in the northern Great Plains. *Biomass and Bioenergy* 45: 315–319.

Schmer, M.R., M.A. Liebig, K.P. Vogel, and R.B. Mitchell. 2011. Field-scale soil property changes under switchgrass managed for bioenergy. *GCB Bioenergy* 3: 439–448.

Schmer, M.R., R.B. Mitchell, K.P. Vogel, W.H. Schacht, and D.B. Marx. 2010. Spatial and temporal effects on switchgrass stands and yield in the Great Plains. *BioEnergy Research* 3: 159–171.

Schmer, M.R., K.P. Vogel, R.B. Mitchell, B.S. Dien, H.G. Jung, and M.D. Casler. 2012b. Temporal and spatial variation in switchgrass biomass composition and theoretical ethanol yield. *Agronomy Journal* 104: 54–64.

Schmer, M.R., K.P. Vogel, R.B. Mitchell, L.E. Moser, K.M. Eskridge, and R.K. Perrin. 2006. Establishment stand thresholds for switchgrass grown as a bioenergy crop. *Crop Science* 46: 157–161.

Schmer, M.R., K.P. Vogel, R.B. Mitchell, and R.K. Perrin. 2008. Net energy of cellulosic ethanol from switchgrass. *Proceedings of the National Academy of Sciences* 105: 464–469.

Scholten, R.L. and R.R. McEllhiney. 1985. The effects of prebreaking in hammer mill particle size reduction. ASAE Paper Number 853542. St. Joseph, MI: ASAE.

Seepaul, R., B. Macoon, K.R. Reddy, and B. Baldwin. 2011. Switchgrass (*Panicum virgatum* L.) intraspecific variation and thermotolerance classification using in vitro seed germination assay. *American Journal of Plant Science* 2: 134–147.

Shapouri, H., J.A. Duffield, and M. Wang. 2003. The energy balance of corn ethanol revisited. *Transactions of the ASAE* 46: 959.

Shastri, Y., A. Hansen, L. Rodríguez, and K.C. Ting. 2011. Development and application of BioFeed model for optimization of herbaceous biomass feedstock production. *Biomass and Bioenergy* 35: 2961–2974.

Shinners, K.J., G.C. Boettcher, R.E. Muck, P.J. Weimer, and M.D. Casler. 2010. Harvest and storage of two perennial grasses as biomass feedstocks. *Transactions of the ASABE* 53: 359–370.

Skinner, R.H. and P.R. Adler. 2010. Carbon dioxide and water fluxes from switchgrass managed for bioenergy production. *Agriculture, Ecosystems and Environment* 138: 257–264.

Smart, A.J. and L.E. Moser. 1999. Switchgrass seedling development as affected by seed size. *Agronomy Journal* 91: 335–338.

Sokhansanj, S., A. Kumar, and A.F. Turhollow. 2006. Development and implementation of integrated biomass supply analysis and logistics model (IBSAL). *Biomass and Bioenergy* 30: 838–847.

Sokhansanj, S., S. Mani, A. Turhollow, A. Kumar, D. Bransby, L. Lynd, and M. Laser. 2009. Large-scale production, harvest and logistics of switchgrass (*Panicum virgatum L.*)—current technology and envisioning a mature technology. *Biofuels, Bioproducts and Biorefining* 3: 124–141.

Spatari, S., Y. Zhang, and H.L. MacLean. 2005. Life cycle assessment of switchgrass and corn stover-derived ethanol-fueled automobiles. *Environmental Science and Technology* 39: 9750–9758.

Stout, W.L., J.A. Staley, J.A. Shaffer, and G.A. Jung. 1991. Quantitative effects of soil depth and soil fertilizer nitrogen on nitrogen uptake by tall fescue and switchgrass. *Communications in Soil Science and Plant Analysis* 22: 1647–1660.

Stubbendieck, J.L., S.L. Hatch, and C.H. Butterfield. 1997. *North American Range Plants*. 5th edn. Lincoln, NE: University of Nebraska Press.

Taliaferro, C.M., K.P. Vogel, J.H. Bouton, S.B. McLaughlin, and G.A. Tuskan. 1999. Reproductive characteristics and breeding improvement potential of switchgrass. In *Biomass, a Growth Opportunity in Green Energy and Value Added Products*. R.P. Overand and E. Chornet, (eds.), pp. 147–152. Oxford, U.K.: Elsevier Science.

Tatsiopoulos, I.P. and A.J. Tolis. 2003. Economic aspects of the cotton-stalk biomass logistics and comparison of supply chain methods. *Biomass and Bioenergy* 24: 199–214.

Thelemann, R., G. Johnson, C. Sheaffer, S. Banerjee, H. Cai, and D. Wyse. 2010. The effect of landscape position on biomass crop yield. *Agronomy Journal* 102: 513–522.

Thomsen, P.M., E.C. Brummer, J.M. Shriver, and G.P. Munkvold. 2008. Biomass yield reductions in switchgrass due to smut caused by *Tilletia maclaganii*. *Plant Health Progress* DOI:10.1094/PHP-2008-0317-01-RS.

Thoreson, C., M. Darr, and K. Webster. 2010. Corn stover densification methods and their large-scale logistical impacts—Preliminary analysis. ASABE Paper Number 1009074. St. Joseph, MI: ASABE.

Tremblay, D., P. Savoie, and R. Theriault. 1991. Self-loading wagon power requirement for coarse chopping of forage. *Canadian Agricultural Engineering* 33: 31–38.

Tulbure, M.G., M.C. Wimberly, A. Boe, and V.N. Owens. 2012. Climatic and genetic controls of yields of switchgrass, a model bioenergy species. *Agriculture, Ecosystems and Environment* 146: 121–129.

Tumuluru, J.S., C.T. Wright, J.R. Hess, and K.L. Kenney. 2011. A review of biomass densification systems to develop uniform feedstock commodities for bioenergy application. *Biofuels, Bioproducts and Biorefining* 5: 683–707.

U.S. Department of Energy. 2011. *U.S. Billion-Ton Update: Biomass Supply for a Bioenergy and Bioproducts Industry*. R.D. Perlack and B.J. Stokes (Leads), 227p. ORNL/TM-2011/224. Oak Ridge, TN: Oak Ridge National Laboratory.

Vamvuka, D., V. Topouzi, and S. Sfakiotakis. 2010. Evaluation of production yield and thermal processing of switchgrass as a bio-energy crop for the Mediterranean region. *Fuel Processing Technology* 91: 988–996.

Vogel, K.P. 1987. Seeding rates for establishing big bluestem and switchgrass with preemergence atrazine applications. *Agronomy Journal* 79: 509–512.

Vogel, K.P. 1996. Energy production from forages (or American agriculture—back to the future. *Journal of Soil and Water Conservation* 21: 137–139.

Vogel, K.P. 2002. The challenge: High quality seed of native plants to ensure successful establishment. *Seed Technology* 24: 9–15.

Vogel, K.P. 2004. Switchgrass. In *Warm-Season (C_4) Grasses*. L.E. Moser, B.L. Burson, and L.E. Sollenberger, (eds.), pp. 561–580. Madison, WI: ASA-CSSA-SSSA.

Vogel, K.P., J.J. Brejda, D.T. Walters, and D.R. Buxton. 2002. Switchgrass biomass production in the Midwest USA: Harvest and nitrogen management. *Agronomy Journal* 94: 413–420.

Vogel, K.P., B. Dien, H. Jung, M. Casler, S. Masterson, and R.B. Mitchell. 2011. Quantifying actual and theoretical ethanol yields for switchgrass strains using NIRS analyses. *BioEnergy Research* 4: 96–110.

Vogel, K.P. and H.J. Jung. 2001. Genetic modification of herbaceous plants for feed and fuel. *Critical Reviews in Plant Sciences* 20: 15–49.

Vogel, K.P. and R.B. Mitchell. 2008. Heterosis in switchgrass: Biomass yield in swards. *Crop Science* 48: 2159–2164.

Vogel, K.P., M.R. Schmer, and R.B. Mitchell. 2005. Plant adaptation regions: Ecological and climatic classification of plant materials. *Rangeland Ecology and Management* 58: 315–319.

Walsh, M., D.G. De la Torre Daniel, H. Shapouri, and S. Slinsky. 2003. Bioenergy crop production in the United States. *Environmental Resource and Economics* 24: 313–333.

Wang, M. 2001. *Development and Use of GREET 1.6 Fuel-Cycle Model for Transportation Fuels and Vehicle Technologies, USDOE*. Lemont, IL: Argonne National Laboratory.

Weaver, J.E. and R.W. Darland. 1949. Soil-root relationships of certain native grasses in various soil types. *Ecological Monographs* 19: 303–338.

West, D.R. and D.R. Kincer. 2011. Yield of switchgrass as affected by seeding rates and dates. *Biomass and Bioenergy* 35: 4057–4059.

Wisclogel, A.E., F.A. Agblevor, D.K. Johnson, S. Deutch, J.A. Fennell, and M.A. Sanderson. 1996. Compositional changes during storage of large round switchgrass bales. *Bioresource Technology* 56: 103–109.

Wright, M.M. and R.C. Brown. 2007. Comparative economics of biorefineries based on the biochemical and thermochemical platforms. *Biofuels, Bioproducts and Biorefining* 1: 49–56.

Wu, M., Y. Wu, and M. Wang. 2006. Energy and emission benefits of alternative transportation liquid fuels derived from switchgrass: A fuel life cycle assessment. *Biotechnology Progress* 22: 1012–1024.

Wullschleger, S.D., E.B. Davis, M.E. Borsuk, C.A. Gunderson, and L.R. Lynd. 2010. Biomass production in switchgrass across the United States: Database description and determinants of yield. *Agronomy Journal* 102: 1158–1168.

Yang, J., E. Worley, M. Wang, B. Lahner, D. Salt, M. Saha, and M. Udvardi. 2009. Natural variation for nutrient use and remobilization efficiencies in switchgrass. *BioEnergy Research* 2: 257–266.

Yu, M., A.R. Womac, and L.O. Pordesimo. 2003. Review of biomass size reduction technology. ASAE Paper Number 036077. St. Joseph, MI: ASAE.

Yu, T.H., J.A. Larson, B.C. English, and Y. Gao. 2011. Evaluating the optimal logistics system of biomass feedstocks for a biorefinery with alternative harvest, storage and preprocessing options: A case study of East Tennessee. ASABE Paper Number 1110879. St. Joseph, MI: ASABE.

Zan, C.S., J.W. Fyles, P. Girouard, and R.A. Samson. 2001. Carbon sequestration in perennial bioenergy, annual corn and uncultivated systems in southern Quebec. *Agriculture, Ecosystems and Environment* 86: 135–144.

Zhang, Y., J. Zalapa, A.R. Jakubowski, D.L. Price, A. Acharya, Y. Wei, E.C. Brummer, S.M. Kaeppler, and M.D. Casler. 2011. Natural hybrids and gene flow between upland and lowland switchgrass. *Crop Science* 51: 2626–2641.

Zhu, X., X. Li, Q. Yao, and Y. Chen. 2011. Challenges and models in supporting logistics system design for dedicated-biomass-based bioenergy industry. *Bioresource Technology* 102: 1344–1351.

8 Production and Supply Logistics of Sweet Sorghum as an Energy Feedstock

Sergio O. Serna-Saldivar and William L. Rooney

CONTENTS

8.1 INTRODUCTION

The interest in alternative energy sources continues for several reasons including the nonrenewable supply of fossil fuels, climate change, and energy security. In the past 12 years, ethanol production and use as an automotive fuel has increased at least 12 times. According to the Renewable Fuels Association, the worldwide and US biorefineries generated nearly 110 and 53 billion liters of fuel ethanol in the year 2011. Current ethanol production is based primarily on maize and sugarcane as feedstocks (Berg 2004, Renewable Fuels Association 2012), and the use of these raw materials causes concerns related to food security especially as world population exceeds 7 billion (FAOSTAT 2012).

Further increases in liquid biofuel production are expected. In the United States, the Energy Independence Act mandates production of 136 billion liters of liquid biofuels by 2022. India intends to use a 10% ethanol blend or E10 gasoline to save an estimated 80 million liters year^{-1} to ease the country's growing need for petroleum and to reduce carbon emissions (Arbatti 2001). Even without a food–fuel argument, these increases in ethanol (or any other feedstock-based biofuel) cannot be based on starch and sugar sources. For example, if the ethanol produced from corn alone is used to meet the Energy Independence Act requirements previously mentioned, all of the US corn production would be required to produce ethanol. This is clearly not feasible. Therefore, US biomass research programs are directed at the future conversion of lignocellulosic biomass to energy using dedicated energy crops such as switchgrass (*Panicum virgatum*), sorghum (*Sorghum bicolor*), and energy cane (*Saccharum* spp.) as well as plant residues (McCutchen et al. 2008).

Energy crops provide potential biofuel production while minimizing the issues of food versus fuel. In addition, these crops can be developed to meet specific environmental constraints such as drought tolerance, fertilization requirements, pesticide use, and desirable composition for conversion. The use of maize (*Zea mays* L.) or sugarcane (*Saccharum officinarum*) for the ethanol production has caused strong criticism because these two crops are extensively used for direct food purposes. The new approach is the development of dedicated energy crops that could be planted in marginal areas or serve to protect ecosystems. An ideal dedicated energy crop must meet several requirements such as high biomass yield and growth rate, reduction in input necessities, full adaptability to different geographic regions, ease of manipulation via genetic improvement, noninvasiveness, high tolerance to stress, and a good carbon sequestration rate (McCutchen et al. 2008, Jessup 2009).

One of the species identified as a potential bioenergy crop is sorghum (*S. bicolor* L. Moench). Sorghum, which was originated in the Horn of Africa, has spread to other neighboring African regions, Southern Asia, Europe, Australia, and the Americas. The plant grows to a height of 1–6 m, depending on the varieties and growing conditions and can be an annual or short perennial crop. Sorghum is considered as a multifunctional crop for the production of fuel ethanol because the plant can generate starchy grains, sweet juice, and fibrous biomass that can be efficiently converted into fuel ethanol using either first- or second-generation processes. Sorghum can be classified in three broad groups: grain, sweet, and forage/high biomass. All belong basically to the same species and virtually there are no biological or taxonomic differences (Wang et al. 2009). The forage sorghum is characterized as a high-biomass crop, which has been boosted by intensive research programs worldwide, focused in the design of new varieties tailored for second-generation ethanol production (Rooney et al. 2007).

Of the types of sorghums, sweet sorghum offers the fastest and most efficient means of producing large quantities of biomass with good conversion efficiency using currently available technologies (Gnansounou et al. 2005, Chuck-Hernandez et al. 2011, Serna-Saldivar et al. 2012). Highly productive sweet sorghums planted with adequate agronomic practices are capable of yielding more than 7000 L anhydrous ethanol ha^{-1} (3500 from sweet juice, 3000 from spent bagasse, and 500 L from the starchy grains) after 4 months of sowing. Because of this potential, numerous regions of the world have identified sweet sorghum as a primary bioenergy crop (Lee 1997, Grassi 2001, FAOSTAT 2012).

Sorghum is particularly adapted to tropical, subtropical, and arid regions (Reddy et al. 2005, Zhang et al. 2010). Sorghum is a C4 plant and is resistant to biotic and abiotic factors such as insects, drought, salinity, and soil alkalinity. Furthermore, this crop has one of the best rates of carbon assimilation (50 g^{-1} m^2 day^{-1}), which in turn allows a fast growth and a better rate of net CO_2 use (Prasad et al. 2007). According to Woods (2000), Almodares and Hadi (2009), and Wu et al. (2010), sorghum is considered to be one of the most drought-resistant agricultural crops because it only requires 30%–40% of the water required by sugarcane and 75%–90% of the water required by maize. Furthermore, the use of fertilizers by energy sorghum is approximately one-third of the fertilizer required by sugarcane (Kim and Day 2010). Because the growth cycle is usually 3–5 months, there is the potential to produce up to two or three crops per year instead of one commonly obtained with sugarcane. Finally, given its rapid growth cycle, the procedure for sorghum breeding improvement is faster than that of perennial crops and there is a large and established set of genetic resources and germplasm. This facilitates plant breeding and development of new cultivars adapted to different regions around the globe (Zhang et al. 2010).

Sweet sorghums typically produce grain, biomass, and sugar. Sugar is the primary product obtained from sweet sorghums, which is in the juice produced and accumulated in the stalks in a similar fashion as sugarcane. Sugars extracted in the juice are primarily sucrose but also include measurable amounts of both glucose and fructose. These sugars are ideally suited for the production of first-generation ethanol with efficiencies around 90% (Wu et al. 2010, Chuck-Hernandez et al. 2011, Davila-Gomez et al. 2011, Serna-Saldivar et al. 2012). According to Almodares and Hadi (2009), sorghum yields a better energy output/input ratio compared to other feedstocks such as sugarcane, sugar beet (*Beta vulgaris*), maize, and wheat (*Triticum aestivum*). Besides the juice, the residue or bagasse can be also converted to ethanol or used for ruminant feed (Heredia-Olea et al. 2012, Serna-Saldivar et al. 2012).

Sorghum is a crop well adapted to adverse climatic conditions that at this time is one of the growing concerns in agronomic projections because of the change of rain patterns and climate, the greenhouse effect, and the steady rise of world temperature. Given all these advantages of sorghum as a potential source of biofuels, the objective of this chapter is to review the potential of sweet sorghum for first- and second-generation ethanol production. The chapter especially focuses on production and supply logistics and efficiencies of conversion of the sweet sorghum juice and spent bagasse into bioethanol.

8.2 SORGHUM BOTANICAL FEATURES AND AGRONOMIC CHARACTERISTICS

Sorghum is a member of Poaceae family, a high-efficient photosynthetic crop, well adapted to tropical and arid regions and climates. As a result, sorghum is extraordinarily efficient in the use of water, carbon dioxide, nutrients, and solar light (Kundiyana 1996, Serna-Saldívar 2010, Chuck-Hernandez et al. 2011). Woods (2000) considered sorghum as one of the most drought-resistant plants, making it suitable for semidesert regions spread throughout the globe, especially as the average world temperature is increasing due to the greenhouse effect and deforestation. The photosynthetic efficiency of sorghum is high due mainly to its C4 photosynthetic pathway that allows this plant to concentrate CO_2 and lower the photorespiration rate. Drought resistance is enhanced in sorghum because of the reduced loss of water across the stoma and the waste of carbon (Keeley and Rundel 2003). Furthermore, the leaves of sorghum are covered by a waxy coat that protects the plant from prolonged droughts. Other important advantage is the possibility of producing multiple crops per year, either from seed (replanting) or from ratoon (Saballos 2008, Turhollow et al. 2010).

The genus sorghum is incredibly diverse and has 25 recognized species that have been taxonomically divided into five sections: Eusorghum, Chaetosorghum, Heterosorghum, Parasorghum, and Stiposorghum (Garber 1950). These sections are distributed through Africa, Asia, and Australia with chromosome numbers of $2n = 10, 20, 30$, or 40 (Garber 1950, Lazarides et al. 1991).

The section Eusorghum includes the cultivated *S. bicolor* and several additional subspecies. Within this group, *S. bicolor* is grown primarily as an annual crop and most of the species are perennial. Within *S. bicolor*, the morphology is highly variable with plant height ranging from less than 1 to 6 m and the days to flowering ranging from 30 to 180 days. *S. bicolor* is further divided into five races: *guinea*, *caudatum*, *kafir*, *durra*, and *bicolor*. Each race has different varieties and cultivars generally planted to produce animal feed or for energy (Singh and Lohithaswa 2006). In addition to *S. bicolor*, another important species in this section is *Sorghum halepense* (Pers.), which is a perennial species with rhizomes that originated in south Europe and then spread to India and China and is now well known as a troublesome weed.

8.3 SORGHUM PRODUCTION AND SUPPLY LOGISTICS

Based on grain production, sorghum is the world's fifth largest grain cereal crop behind maize, wheat, rice (*Oryza sativa*), and barley (*Hordeum vulgare*). It is grown on 40.5 million hectares in practically all agricultural lands around the globe. The United States, India, Nigeria, China, Mexico,

Sudan, and Argentina are the leading producers (FAOSTAT 2012). While the reported statistics are for grain production, additional sorghum production area is devoted to forage and sweet sorghum. There is potential for additional growth. According to ICRISAT scientists, an estimated 50% of the grain sorghum area in Asia and sub-Saharan Africa could be sown with sweet sorghums that would produce both grain and sugars.

Sweet sorghums are defined by the presence of high sugar yields, which is the product of high sugar content and juicy stalks. True sweet sorghums have traditionally been grown primarily for syrup and chewing (Reddy and Sanjana 2003, Serna-Saldivar et al. 2012). This short-rotational crop is adapted to many regions of the world that are not conducive to other sugar-producing crops. Given that water availability has become a major constraint to agricultural production in recent years (Ryan and Spencer 2001), there is the opportunity for sweet sorghum to complement sugarcane production in many regions of the world and possibly substitute for sugarcane when water availability is limited. In addition to complementing sugarcane production, sweet sorghum can use the same industrial infrastructure for harvesting and processing as sugarcane. Moreover, sorghum can be harvested for high sugar yields when sugarcane feedstocks are scarce.

In terms of the economics of ethanol production from sweet sorghum, Vencor Green (2012) reported that the costs for the production of ethanol from the sweet sorghum were 20% lower than that of either sugarcane or maize. Energy balance analysis showed that sweet sorghum produced 8 units of energy for every unit of energy input in its cultivation and production, compared with 8.3 units for sugarcane and 1.8 for maize (vencorgreen.com). There is the potential to produce up to three crops per year for sweet sorghum, which is, however, contingent on available moisture, fertility, and day lengths as not all environments are conducive to production on a year-round basis. For example, in most temperate climates, sorghum is usually produced as a single crop on an annual basis.

Because sugar in stalks is not a shelf-stable commodity, continual harvest and processing of the crop is essential. This is possible for sugarcane, but the production logistics must be developed for a specific environment. It will be accomplished by planting sweet sorghum hybrids with different maturities at different dates to maximize yield and distribute maturity overtime. This will allow *just-in-time* harvesting and provide a constant supply of the feedstock for a biorefinery. Given that sweet sorghum and sugarcane are usually at peak sugar yields in different seasons, this allows the two crops to complement each other and extend to the harvest season beyond either of them alone.

8.4 CHEMICAL COMPOSITION AND ETHANOL PRODUCTION FROM SWEET SORGHUM JUICE

8.4.1 Juice from Sweet Sorghum

Mature stems of sweet sorghum contain about 73% moisture and the solids are divided in structural and nonstructural carbohydrates (Table 8.1). Approximately 13% of the solids are nonstructural carbohydrates composed of sucrose, glucose, and fructose, in variable amounts according to cultivar, harvesting season, and maturity stage, among other agronomic factors (Mamma et al. 1996, Phowchinda et al. 1997). Anglani (1998) classified sweet sorghums based on sugar fractions in the juice. The sugar types were high in sucrose, while syrup types had a larger monosaccharide (glucose and fructose) component.

The environment has variable effects on sugar composition and production. Smith et al. (1987) evaluated the juices extracted from the six sweet sorghum varieties at nine different locations for 4 years and did not find significant differences in sugar content or composition. The average sugar composition was approximately 70% sucrose and the rest glucose and fructose in equal parts. However, Davila Gomez et al. (2011) analyzed the sugar composition of sweet sorghum juices harvested at northeast Mexico and concluded that the Rio, M81E, and Della varieties contained only

TABLE 8.1
Chemical Composition of Sweet Sorghum Stalks

Chemical Component	% (As Is Basis)	% (Solid Basis)
Water	73	0
Sugars	13	48.1
Cellulose	5	18.5
Hemicellulose	4	14.8
Lignin	3	11.1
Others	2	7.4
Total	100	100

Source: Data from Prasad, S. et al., *Energy Fuels*, 21, 2415, 2007.
The sweet sorghum was harvested at optimum maturity (post-anthesis).

from 1.7% to 9.9% sucrose and 51% to 54% and 39% to 42% glucose and fructose, respectively. These syrup-type juices were converted into bioethanol with an efficiency of nearly 90%. On a stem dry basis, Woods (2000) reported fermentable sugar content between 41% and 44% in Keller and Wray varieties with 80% and 63% represented by sucrose and the rest by glucose and fructose. A fiber variety analyzed by the same author (H173) reached only 20% fermentable sugars based on the dry stem weight; sucrose, glucose, and fructose were found in equivalent amounts (around 7% for each sugar). Compared to sugarcane, the main difference is that the sucrose content in cane is significantly higher compared to glucose and fructose (90%, 4%, and 6%, respectively) and the total content sugar was 49% of the dry stem weight. In general terms, the composition of simple sugars in sweet sorghum juice is 53%–85%, 9%–33%, and 6%–21% for sucrose, glucose, and fructose, respectively (Mamma et al. 1996, Phowchinda et al. 1997, Gnansounou et al. 2005, Prasad et al. 2007, Serna-Saldivar et al. 2012).

Beyond the proportion of soluble sugars in sweet sorghum plants, the yield of total sugars per harvested area is a better guide in the analysis for fuel production. According to Woods (2000), Keller, Wray, and H173 cultivars yielded an average of 7, 10, and 4 tons of fermentable sugars ha^{-1}, significantly lower than 17 tons ha^{-1} for sugarcane reported by the same author. The sweet sorghum varieties studied by Davila-Gomez et al. (2011) planted in northeast Mexico yielded an average of 1.85–2.03 tons of sugar ha^{-1}. In an extensive study performed in multiple locations across the United States, Smith et al. (1987) reported sugar yields between 4.5 and 10.6 tons ha^{-1}. In China, the yields as high as 18 tons ha^{-1} were reported (Zhang et al. 2010).

Based on these sugar concentrations and yields, ethanol production is easily estimated. These highly fermentable sugars are capable of yielding about 46% ethanol after 48 h fermentation taking into consideration a theoretical yield of 51.1 g ethanol for 100 g glucose metabolized and a conversion efficiency of 90%. Therefore, the potential yield of anhydrous ethanol varies from 1000 to 8000 L. Likewise, Luhnow and Samor (2006) reported that sugar accumulated in sweet sorghum stalks had the potential to yield up to 8000 L ha^{-1} or about twice the ethanol yield potential of maize grain and 30% greater than the average Brazilian sugarcane productivity of 6000 L ha^{-1}. In order to achieve these yields with maize, it will be necessary to harvest 20 tons ha^{-1} and employ a more expensive process due to the necessity of using amylolytic enzymes indispensable to convert the starch into fermentable sugars.

The storage of either sweet sorghum biomass or the extracted juice would be very desirable as it minimizes mill shutdown time due to weather and extend to harvest and/or process schedule. However, sugars are very sensitive to microbial contamination especially after the plant material is either chopped or crushed. Davila-Gomez et al. (2011) reported the amounts of sugars, expressed as °Brix, dropped from 24% to 11% in 8 h from harvest and crushing to conversion. The microbial contamination and high environmental temperatures (32°C) were the most obvious explanations of these losses. Besides, the sucrose proportion in fermented juices was lower in relation to glucose and fructose (0%–10% of total), which is a strong indicator of degradation by invertase activity of contaminating wild yeasts that hydrolyzed the disaccharide into its monomers. Wu et al. (2010), working with cultivars with 16%–18% of fermentable sugars, concluded that as much as 20% of substrate can be lost after a 3-day storage at 25°C. This loss corresponds to approximately 375 L ethanol ha^{-1} when a yield of 50 tons of sorghum stems ha^{-1} with 15°Bx is considered. Daeschel et al. (1981) reported that juices can be preserved at 4°C for 14 days without detectable changes or deterioration. These authors also reported that the dominant spoilage microorganisms associated with sweet sorghum juice are *Leuconostoc mesenteroides* and *Lactobacillus plantarum* and recommends processing the juice within 5 h after extraction.

8.4.2 Sugar Extraction and Conversion to Ethanol from Sweet Sorghum Juice

After harvest, the stalks of sweet sorghum are processed using procedures optimized for sugarcane. Soluble sugars and significant proportions of the water are removed from the biomass by milling and washing or the diffusion processes. In milling, sugars and juice are extracted by mechanical pressing of the biomass with a series of roller mills composed of a set of cylinders, similar if not identical to sugarcane mills (Serna-Saldivar et al. 2012). At various stages, the biomass is washed to extract residual soluble sugars. According to the literature published by Wu et al. (2010), extractable sweet sorghum juice yields are approximately 50% juice in relation to the initial stem weight. However, these authors described an extraction process by pressing, which resulted in lower yields compared to roller mills. Furthermore, pressing is a batch process, which is difficult to optimize under industrial conditions. In the diffusion system, which is common in sugarcane processing systems, ground biomass is washed in a series of hot water treatments to remove solubles; after extensive washing, the biomass passes through a three-roller mill to remove residual water. Both of these methods are efficient for sugarcane and initial tests indicate that efficiency in sweet sorghum is similar and could be improved with testing to identify optimum conditions for sweet sorghum.

Gnansounou et al. (2005) and Almodares and Hadi (2009) reported that approximately 90% of fermentable sugars from sorghum stalks could be obtained after conventional roller milling, yielding an extraction ratio of 0.7 in relation to the initial plant weight. Gnansounou et al. (2005) reported extraction ratios ranging from 0.59 to 0.65 for the sweet sorghum cultivars Kelley, Wray, Rio, and Tianza. Kundiyana (1996) observed that extraction percentages varied between 47% and 58%. Similar juice extraction efficiencies were also observed by Davila-Gomez et al. (2011) in northeast Mexico.

After extraction, the sweet sorghum juice is fermented and distilled and the semipurified ethanol is dehydrated or subjected to azeotropic distillation (Figure 8.1). This is the simplest way to produce fuel ethanol because the grain and fiber processes require the prior hydrolysis of starch and fiber components into fermentable sugars. These steps are considered relatively expensive because they demand more processing time, energy, and reagents (i.e., enzymes, chemicals). Despite these benefits, some challenges must be solved in order to efficiently convert the sweet sorghum crop into fuel ethanol. The main setbacks are the relatively higher rate of sugar degradation at ambient temperature and the low nitrogen content and availability for yeast growth (Mei et al. 2009, Wu et al. 2010). Thus, the logistics of *just-in-time* harvesting and the storage of the feedstock in facilities that can retard the decomposition and degradation of fermentable carbohydrates should be considered and stressed. In relation to nitrogen availability, if there is a disadvantage, it can be overcome with the supplementation of urea, ammonia, or yeast extract in order to avoid sluggish fermentation.

Davila-Gomez et al. (2011) tested three cultivars of sweet (Rio, M81E, and Della) and two forage sorghums (Tato and Thor) in northeast Mexico with the goal to determine optimum harvesting

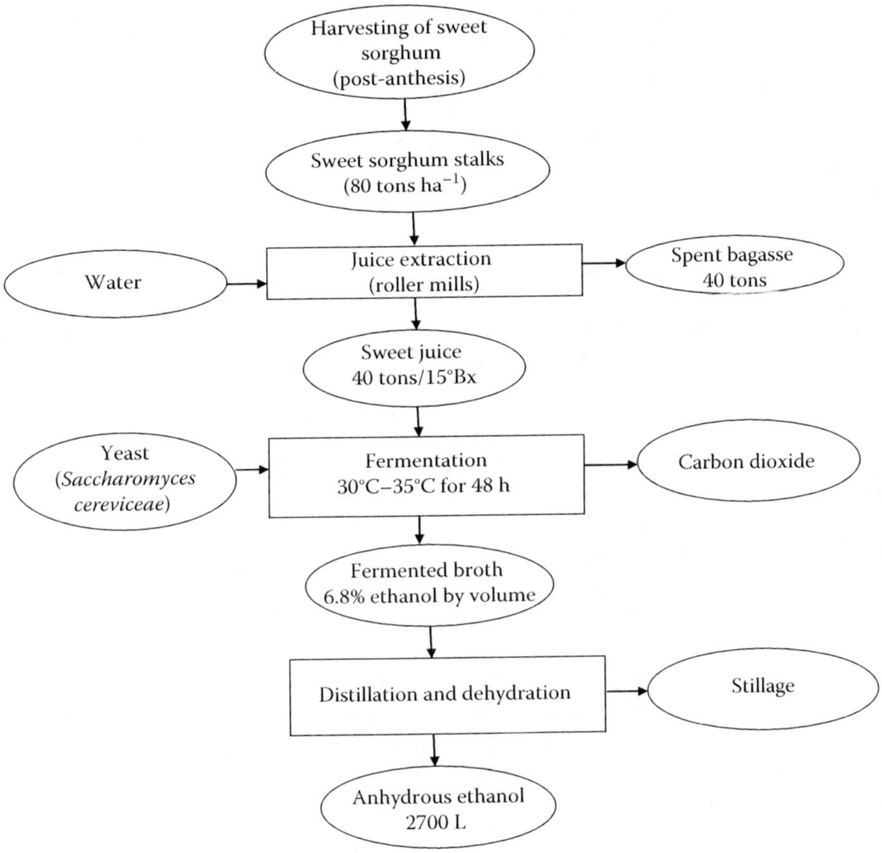

FIGURE 8.1 Flowchart and mass balance of first-generation ethanol production from sweet sorghum juice.

time for the first cut, sugar production, biomass composition, and ethanol yields. The juices were characterized in terms of sugar composition, free amino nitrogen (FAN), and phenolics and then fermented into ethanol using yeast. As expected, the cultivars yielded different volumes of juice and total sugars and had different optimum harvesting time. Glucose was the most abundant sugar in raw juices, followed by fructose and sucrose. FAN concentration ranged from 19 to 36 mg L^{-1}; therefore, nitrogen supplementation was required for adequate fermentation. After 18 h fermentation, there were no differences in efficiencies among cultivars, but the sweet sorghums yielded more ethanol per hectare compared to the two forage sorghums (approximately 1000 versus 770 L ha^{-1}). Della was the cultivar with the highest productivity that can produce 1051 L ha^{-1} ethanol for the first cut.

The sweet sorghum juice is not commonly used for crystallized sugar production because of the presence of significant amounts of inverted sugars (glucose and fructose) and antiquality compounds such as aconitic acid that makes crystallization difficult and expensive. However, sweet sorghum juice is easily bioconverted into ethanol with average fermentation efficiencies from 85% to 90% (Woods 2000, Prasad et al. 2007, Almodares and Hadi 2009, Wang et al. 2009, Wu et al. 2010, Chuck-Hernandez et al. 2011, Davila-Gomez et al. 2011, Serna-Saldivar et al. 2012). Besides sugar and nitrogen content, fermentation performance of sweet sorghum juice can also be affected with processing parameters and bioreactor configuration. In normal gravity fermentation, the initial sugar concentration of 15°Bx–20°Bx can achieve ethanol concentrations of approximately 7%–9% (v v^{-1}). The other alternative is to process the so-called very high gravity (VHG) substrates that contain initial sugar concentrations between 25°Bx and 28°Bx. The main constraint of this technology is that the high contents of saccharides in the fermentation medium cause an increase in the osmotic pressure, which has a detrimental effect on yeast cell growth (Bafrncova et al. 1999) and stress, resulting in sluggish fermentations. The advantage of adopting this technology is the higher rate of fermentation, ethanol concentration, and production output with the savings in energy and the optimization of existing plant facilities (Bvochora et al. 2000, Bayrock and Ingledew 2001, Bai et al. 2004, Laopaiboon et al. 2009). Nuanpeng et al. (2011) observed in a repeated-batch study that VHG fermentation was a good alternative to produce ethanol at high concentrations from sweet sorghum juice when adequate levels of yeast cell concentration, nitrogen, and agitation were used. Laopaiboon et al. (2007) reported that fed-batch fermentation achieved better results than batch configuration in terms of ethanol concentration and product yield but not in productivity (measured as grams of ethanol generated L^{-1} h^{-1}). These findings indicate that there is a need to optimize parameters of feeding and withdrawn rates in order to optimize yield. Thomas et al. (1996) concluded that under appropriate environmental and nutritional conditions, *Saccharomyces cerevisiae* can effectively produce and tolerate ethanol at high concentrations. Minteer (2006) reported that the yeast is capable of resisting extreme environmental stresses and osmolality (25%–30% solids w v^{-1}) and high ethanol concentrations (12%–18%, v v^{-1}). On the other hand, Laopaiboon et al. (2008) researched three high-ethanol-producing strains of *S. cerevisiae* (TISTR 5048, TISTR 5339, and NP 01) and concluded that NP 01 yielded the maximum ethanol concentration under batch fermentation in an ethanol production medium containing 280 g L^{-1} glucose. The same authors (Laopaiboon et al. 2009) studied ethanol yields from VHG sweet sorghum juice fermented with yeast NP01 and various carbon adjuncts and nitrogen sources. When sucrose was used as an adjunct, the sweet sorghum juice containing total sugar of 280 g L^{-1}, 3 g yeast extract L^{-1}, and 5 g peptone L^{-1} gave the maximum ethanol yield and efficiency. The ethanol concentration, productivity, and yield were 120.7 g L^{-1}, 2.0 g L^{-1} h^{-1}, and 51%, respectively. In addition, ammonium sulfate was not suitable for use as a nitrogen supplement since it reduced ethanol yields.

Liu et al. (2008) and Liu and Shen (2008) reported a novel fermentation technology by using immobilized yeast in a fluidized bed reactor that shortened process time and increased conversion efficiency. These results can be improved if the operating parameters such as temperature, agitation rate, particles stuffing rate, and pH are optimized.

8.4.3 Yeast Modification to Improve Ethanol Yields from Sweet Sorghum

Besides yeast, other microorganisms, such as bacteria, are especially designed for ethanol fermentation. *Escherichia coli* is a typical modified microorganism for ethanol production because it can metabolize a wide spectrum of carbohydrates, and it has a well-known genetic makeup and the easiness of manipulation. *Zymomonas mobilis*, a rod-shaped, gram-negative, nonspore-forming bacteria, is naturally ethanologenic, which has higher rates of glucose uptake than yeast. *Z. mobilis* also has a higher ethanol productivity and increased yield and tolerance, which makes it a good option for VHG fermentation. Kesava et al. (1995), working with *Z. mobilis*, reported 95% conversion rates after 35 h fermentation and ethanol yields of approximately 70 g L^{-1} when fermenting mashes containing 150 g glucose L^{-1}. The bacterium was able to ferment mashes containing 200 g glucose L^{-1} in a step-fed system. Perez-Carrillo et al. (2011) observed that *Z. mobilis* had lower nitrogen requirements than *S. cerevisiae* when fermenting sorghum grain mashes adjusted to 20° Plato. This bacterium has potential and advantages for commercial use in biorefineries.

8.5 SECOND-GENERATION ETHANOL FROM SORGHUM BAGASSE

8.5.1 Chemical Composition of Sweet Sorghum Bagasse

Sorghum bagasse is the fraction remaining after juice extraction. Bagasse is a fiber-rich material with some variation in composition depending on intrinsic and extrinsic factors such as cultivar

TABLE 8.2
Fiber and Structural Carbohydrates Composition Analyses of Sweet Sorghum Spent Bagasse

Component	Amount (%) ± SD[a]
Water extractives	19.1 ± 0.8
Ethanol extractives	4.3 ± 0.6
Glucans	42.4 ± 2.9
Xylans	12.4 ± 0.9
Arabinans	10.5 ± 0.1
Acetyl groups	0.05 ± 0.004
Acid insoluble lignin	13.5 ± 1.8
Acid soluble lignin	0.5 ± 0.2
Acid insoluble ash	2.4 ± 0.2

Source: Heredia-Olea, E. et al., *Bioresour. Technol.*, 119, 216, 2012.

[a] Composition is expressed on dry matter basis. The sweet sorghum bagasse contained approximately 54% water before drying.

type, maturity or lignification, and climatic conditions. On average, 15% of total plant dry weight is a fibrous component, which ranges from 12% to 17% (Woods 2000). Typically, the fresh sweet sorghum harvested at optimum time yields 50% juice and 50% bagasse after milling or crushing the stalks. This bagasse contains around 52% moisture, 5.4% residual sugar, 17% cellulose, 11.9% hemicellulose, and 8.5% lignin (Gnansounou et al. 2005). The main structural carbohydrates of sweet sorghum bagasse were glucans derived from cellulose and xylans and arabinans derived from hemicellulose as shown in Table 8.2 (Heredia-Olea et al. 2012). The amount of lignin expressed on dry matter basis was 13.5%, which was similar to the value previously determined by Gnansounou et al. (2005) and slightly higher than 11.1% reported by Prasad et al. (2007) (Table 8.1). This specific component is directly related to plant maturity and inversely related to the susceptibility of the fiber to hydrolysis and ethanol yield.

8.5.2 Second-Generation Ethanol Production from Sweet Sorghum Bagasse

By definition, the second-generation ethanol production uses biomass feedstocks that are not starch or simple sugars. Thus, the second-generation ethanol production is typically associated with the conversion of cellulose and hemicellulose. There is an abundance of structural carbohydrates, which are the building block of plant cell walls and the most abundant source of biomass on the planet (Hanchar et al. 2007, Viuchard et al. 2009). More importantly, the ethanol produced from lignocellulosic feedstocks is a viable option to decrease the perceived competition between production of foods and bioenergy (Chakracorty et al. 2009, Lin et al. 2011).

There are five key steps for the production of lignocellulosic ethanol: milling of the feedstock, chemical or physical pretreatment, enzymatic hydrolysis or saccharification, fermentation of hexose (C6) and pentose (C5) carbohydrates, and distillation–dehydration of the ethanol. While cellulose and hemicellulose can be deconstructed and fermented, it must be first extracted from lignin, which is not fermentable (Salvi et al. 2010). Hydrolysis of cellulose and hemicelluloses yield the C6 and C5 fermentable sugars (Martinez et al. 2001). The cell wall barrier that is constituted of hemicellulose and lignin and the crystalline structure of the cellulose

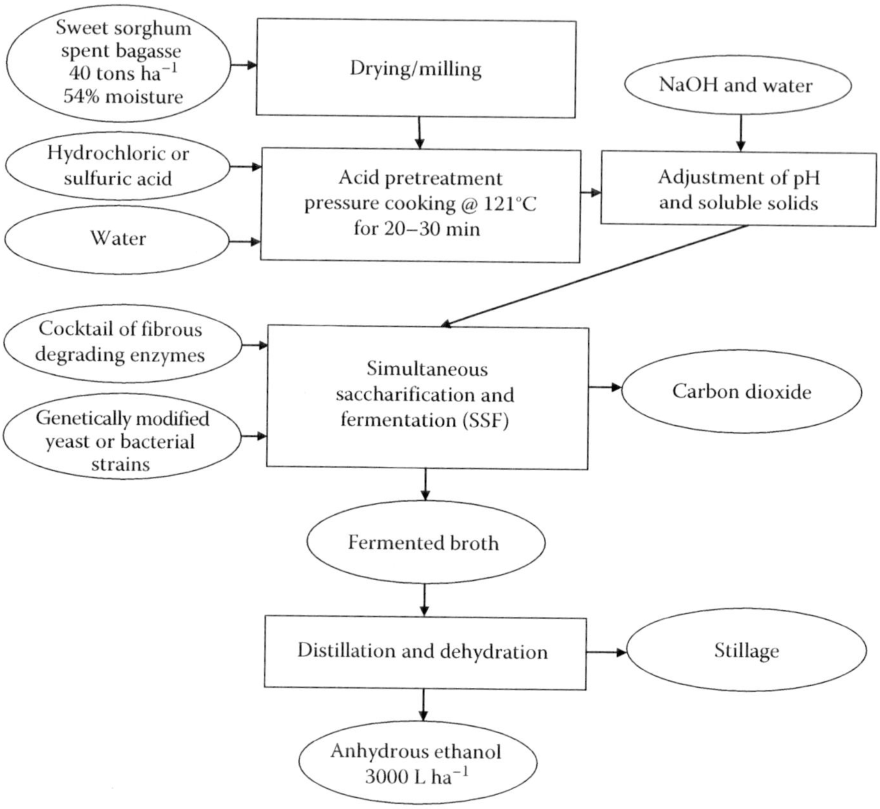

FIGURE 8.2 Flowchart and mass balance of ethanol production from sweet sorghum spent bagasse. Average composition of sweet sorghum bagasse: water 54%, simple fermentable sugars 5.4%, cellulose 17%, hemicellulose 12%, and lignin 11.7%.

must be disrupted (Sun 2010). The hydrolyses of hemicellulose and cellulose represent approximately 30% and 28% of the processing cost, respectively, as shown in Figure 8.2 (Jacobsen and Wyman, 2000).

8.5.2.1 Milling

After the extraction of juice, the lignocellulosic residue is chopped, milled, and dried at 50°C–60°C to reduce the moisture content to about 10%–15%. There are many options to reduce particle size. Among them, hammer or rotary milling is the most frequently used method. These inexpensive milling technologies can be used on both dry and wet feedstocks. The milling operation aims toward the reduction of the particle size of biomass to a fine powder so it is more susceptible to subsequent chemical, enzymatic, and fermentation treatments (Herrera et al. 2003, Mizuno et al. 2009, Sipos et al. 2009).

8.5.2.2 Physical and Chemical Pretreatments

One of the most significant problems in ethanol production from lignocellulose is production cost because the fiber conversion requires high-energy investments in order to obtain high concentration of fermentable sugars from the insoluble polymers (Mamma et al. 1996, Mizuno et al. 2009, Kurian et al. 2010). A prehydrolysis step is aimed toward the liberation of both the hemicellulose and cellulose associated to cell walls at a relatively low cost (Herrera et al. 2003). The main pretreatment methods for the production of ethanol from sorghum biomass are the chemical- (acid, alkaline, ammonia) and/or enzyme-catalyzed hydrolyses. Generally, the chemical or physical hydrolysis precedes the enzymatic in order to maximize production of C6 and C5 fermentable sugars (Mamma et al. 1996, Sipos et al. 2009).

8.5.2.2.1 Physical Pretreatments

The aim of physical pretreatments is to expose cellulose and hemicellulose from lignified cell walls for the subsequent step of enzymatic hydrolysis.

Extrusion cooking: The thermoplastic extrusion is the mechanical treatment that has been previously used to treat fiber-containing materials, especially free hemicellulose (Sun 2010) without the use of large amounts of water. Heredia-Olea (2011) investigated the use of a thermoplastic twin extruder to disrupt previously ground sweet sorghum bagasse. The dry feedstock was subjected to six different extrusion conditions varying in the screw rotational speed, torque, and temperature at the last section of the extruder barrel. The extrusion process further reduced the particle size of the ground material and exposed cellulose that was more prone to enzymatic hydrolysis as previously documented by Lamsal et al. (2010) and Shen and Liu (2009). The extruded feedstock treated with the enzymes had a conversion efficiency of 60% and 77.5% of the hemicellulose and cellulose, respectively. These values were very similar to the ones achieved by the combined acid and enzymatic hydrolyses. The main potential advantages of the extrusion pretreatment are that it is continuous and environment friendly and generates hydrolyzates practically free of yeast inhibitors such as acetic acid, furfural, and hydroxymethylfurfural (HMF) (Heredia-Olea 2011).

Steam explosion: In this mechanical process, raw or SO_2-impregnated bagasse is placed in a sealed reactor equipped with a blown down valve where steam is injected. The operation temperature is in the range of 170°C–210°C (Stenberg et al. 1998, Öhgren et al. 2005, Sipos et al. 2009). After a 2–10 min holding period, the valve suddenly opens and the resulting pressure change disrupts the fiber matrix of the feedstock, which is collected in a cyclone (Stenberg et al. 1998). Ballesteros et al. (2003) used steam explosion pretreatment and obtained around 50% of solids recovery and only 20% solubilization of the cellulose. On the other hand, the hemicellulose was extensively solubilized and more prone to this specific treatment. Sipos et al. (2009) achieved an extraction of 89%–92% of cellulose after steam explosion. The hydrolyzates contained 18 g glucose, 23 g xylose, and 5.5 g arabinose L^{-1}. They also used steam explosion to treat ground sorghum bagasse impregnated with up to 3% w w^{-1} SO_2 in plastic bags for 20–30 min (Stenberg et al. 1998, Öhgren et al. 2005, Sipos et al. 2009). The SO_2-impreganted bagasse prior to steam pretreatment improved the subsequent enzymatic saccharification of the feedstock (Stenberg et al. 1998).

8.5.2.2.2 Chemical Pretreatments

Dilute acid hydrolysis: Acid hydrolysis is the most common chemical pretreatment used to produce second-generation bioethanol (Ban et al. 2008) because it yields significant amounts of sugars, makes the fiber more susceptible to the subsequent enzymatic step, and is relatively inexpensive compared to the use of other chemicals (Gnansounou et al. 2005, Ban et al. 2008, Serna-Saldivar et al. 2012). The main advantage of diluted acid hydrolysis is that it enhances cell wall delignification, while its main disadvantage is to generate significant amounts of yeast inhibitors. Many acids, among which the sulfuric, hydrochloric, hydrofluoric, or acetic acids are the most prevalent, have been tested and used as catalysts (Herrera et al. 2003). The process involves adding diluted acid solution (0.1%–10% w v^{-1}) to the milled feedstock and the mix is pressure-cooked in an autoclave. During this pretreatment, the fiber is disrupted and the hemicellulose liberates into simpler sugars. The major controls of this process are the strength of the acid solution, the hydrolysis time, and the pressure or temperature (121°C).

Several researchers have investigated acid hydrolyses of sweet sorghum bagasse. Kurian et al. (2010) pressure-cooked sorghum bagasse for 30 min with sulfuric acid at a concentration of 5 g kg^{-1} and achieved an extract with 92 g L^{-1} of total sugars. Ban et al. (2008) treated the same feedstock for 80 min at a solid–liquid mass ratio of 10% with phosphoric acid at 80 g L^{-1} and 120°C. The yield

TABLE 8.3
Effect of Acid Pretreatment and Detoxification with NaOH on the Sugar and Inhibitor Composition of Hydrolyzed Sweet Sorghum Spent Bagasse[a,b]

	Treatment			Detoxified Treatment		
Compound, mg/g	HCl	H_2SO_4	HCl/H_2SO_4	HCl Detoxified	H_2SO_4 Detoxified	HCl/H_2SO_4 Detoxified
Sugars						
Cellobiose	35.6±0.6 a	37.4±0.2 a	36.5±0.9 a	33.6±5.5	35.6±0.2	35.2±0.5
Glucose	247.1±8.0 a	253.4±13.4 a	234.9±26.3 a	255.3±13.7	248.2±8.9	247.2±13.7
Xylose	75.8±1.5 a	75.9±1.5 a	72.7±3.8 a	75.3±1.9	72.8±1.2	73.8±2.3
Arabinose	50.8±0.7 a	50.3±0.3 a	48.2±1.2 b	48.9±0.7	47.7±0.3	47.6±1.2
Total sugars	409.2±10.8 a	415.5±15.1 a	390.8±31.7 a	413.6±16.6	398.7±10.2	404.4±12.2
Inhibitors						
Acetic acid	47.7±1.5 a,b	49.5±3.9 a	34.9±6.5 a	38.8±0.9	33.4±0.6	33.0±0.6
5-HMF	6.7±0.3 a	8.1±0.5 b	4.4±1.5 a,b	4.2±0.1	4.5±0.3	4.1±0.1
Furfural	7.1±0.2 a	5.2±0.2 b	4.7±0.2 c	6.5±0.2	4.8±0.1	4.4±0.2
Total inhibitors	61.5±1.6 a	52.6±7.7 b	44.2±8.1 b	49.7±1.0	42.8±0.5	41.7±0.6

[a] Means with different letter(s) within rows are significant different ($P \leq 0.05$).
[b] Results are expressed on dry matter basis.

of sugars was 302 g kg^{-1} bagasse. Recently, Heredia-Olea et al. (2012) used response surface methodologies to optimize the pretreatment of crushed and milled sweet sorghum bagasse using sulfuric and hydrochloric acids individually and their mixture. The hydrolyzates were detoxified with lime. The feedstock contained 41.2% and 24.5% cellulose and hemicellulose, respectively. The response variables were C5 and C6 sugars and inhibitors production (acetic acid, 5-HMF, and furfural). Results indicated that the three different pretreatments yielded similar amounts of total fermentable sugars. The proposed acid hydrolysis schemes liberated 56%–57% of total sugars available in the sweet sorghum bagasse (390–415 mg sugar g^{-1} bagasse) and 44–61 mg total inhibitors g^{-1} bagasse. The hydrochloric acid treatment was a good alternative due to its relatively lower hydrolysis time and adequate generation of C5 and C6 fermentable sugars (Table 8.3).

Alkali pretreatment: Unlike other pretreatments, the use of strong alkali can delignify biomass by disrupting the ester bonds of cross-linked lignin and xylans, resulting in a cellulose and hemicellulose enriched fraction. The main commercial alkalis used are sodium hydroxide, ammonia, and lime or calcium hydroxide. The alkaline pretreatments are generally performed at lower temperatures, pressures, and residence times than other technologies (McIntosh and Vancov 2010). Generally, two temperature conditions are used for hydrolysis: mild (60°C) or high (121°C).

8.5.2.2.3 Inhibitors Produced during Chemical Pretreatments

Chemical pretreatments may yield large amounts of potential microbial inhibitors. These inhibitors are categorized into three groups: organic acids (i.e., acetic, formic, and levulinic), furans (furfural and 5-HMF), and phenolics released from lignin (*p*-hydroxybenzoic acid) (Lopez et al. 2004, Ban et al. 2008). These water-soluble inhibitors at high concentrations can affect cell physiology and viability, the efficiency of ethanologenic microorganisms during the critical step of fermentation, and ethanol yields (Nilvebrant et al. 2001, Almeida et al. 2008). Amartey and Jeffries (1996) found that the removal of these compounds prior to fermentation can reduce fermentation cost by about 25%. The acetic acid is produced after hemicellulose hydrolysis

(Pamqvist et al. 1999), whereas furfural and HMF are yielded after dehydration of pentoses and hexoses, respectively (Gamez et al. 2006).

Heredia-Olea et al. (2012) reported on the amounts of acetic acid, furfural, and HMF generated after different acid hydrolysis schemes (Table 8.3). These inhibitors were generated in higher amounts in those hydrolyzates treated with higher acid concentrations and processing times. The main inhibitor obtained was acetic acid released from hemicellulose covalently linked by acetic groups of lignin (Rodríguez-Chong et al. 2004). The different kinds of acids hydrolyzed these linkages and generated acetic acid.

If inhibitors are produced at levels that affect fermentation, detoxification must be used. An effective detoxification process applied after acid hydrolyses is the supplementation with lime to increase the pH to 9–10. This alkali treatment precipitates inhibitors in the form of insoluble salts and improves yeast performance during fermentation (Kurian et al. 2010). Other suggested detoxification strategies are the use of activated carbon that traps lignin-derived phenolics (Gamez et al. 2006), ion-exchange resins, and degrading enzymes such as laccase (Carvalho et al. 2002, Lopez et al. 2004). Heredia-Olea et al. (2012) successfully detoxified different types of acid hydrolyzates with calcium hydroxide and achieved concentrations of acetic acid and HMF of 49.72 mg g^{-1} equivalent to 0.59 g L^{-1} containing 6% solids. This concentration is considered low to inhibit ethanologenic inoculum.

8.5.2.3 Enzyme Conversion

The enzymatic hydrolysis of fiber components is considered as one of the key steps in second-generation alcohol production. Generally, this biocatalysis process is done in chemically predigested or mechanically disrupted fiber components (Reddy and Yang 2005, Heredia-Olea et al. 2013) although some processes employ only this treatment especially when the feedstock is low in lignin. The advantage of the sole enzymatic conversion is that it is more environmental friendly and could be applied simultaneously with fermentation saving energy and processing time. According to Gnansounou et al. (2005), the enzyme degrading approach appears to be the best for continued improvements because it can significantly reduce production costs and energy. There are several enzymes generally used to convert cellulose and hemicellulose into soluble sugars. They are a mixture of cellulases, hemicellulases, pectinases, xylanases, and other known fiber-degrading enzymes (Reddy and Yang 2005, Lin et al. 2011). Cellulose can be effectively hydrolyzed by the synergistic action of endo-acting enzymes, known as endoglucanases, and exo-acting enzymes, known as exoglucanases (Lin et al. 2011). Today, it is common to employ enzyme complexes consisting of seven or more degrading enzymes that act synergistically. Recently, Heredia-Olea (2011) and Heredia-Olea et al. (2013) analyzed the effectiveness of two different doses of a fiber-degrading enzyme cocktail added directly to ground sorghum bagasse and concluded that the high dose generated approximately 20% more potentially fermentable carbohydrates. The main sugars generated during biocatalysis were glucose and xylose (Table 8.4). The same author suggested the treatment of extruded feedstock with the fiber-degrading enzyme complex in a simultaneous saccharification and fermentation (SSF) scheme. This combination yields hydrolyzates with high amounts of fermentable sugars and saves energy, processing time, and the use of chemicals required for pretreatment.

8.5.2.4 Fermentation

The hydrolyzate solutions obtained from the sorghum fiber by chemical pretreatment and/or enzymatic hydrolysis are rich in both C6 and C5 (Kurian et al. 2010). The production of ethanol from these hydrolyzate mashes is possible only with the use of osmotolerant and pentose-fermenting yeast or bacterial strains. There are many genetically engineered yeast and bacteria capable of fermenting worts rich in both C5 and C6 carbohydrates. The most relevant ones are different strains of *Saccharomyces cerevisiae*, *Kluyveromyces marxianus*, *Pichia stipitis*, *Klebsiella oxytoca*, *Klebsiella planticola*, *Candida shehatae*, *Flammulina velutipes*, and *Issatchenkia orientalis*.

TABLE 8.4
Effect of Fiber-Degrading Enzyme Dosage on the Direct Conversion of Sweet Sorghum Spent Bagasse into C5 and C6 Carbohydrates

	Low[a]	High[b]
Cellobiose	11.7±0.5 a	12.4±0.4 a
Glucose	243.2±7.1 a	227.7±14.5 a
Xylose	19.61±0.1 b	23.4±0.6 a
Arabinose	32.3±11.6 b	107.5±5.6 a
Total sugars	299.3±6.8 b	366.6±17.7 a

Source: Heredia-Olea, E., Evaluación del proceso de obtención de bioetanol a partir de bagazo de sorgo dulce pretratado mediante extrusión, auto hidrólisis o hidrólisis acido diluida, hidrolizado enzimáticamente fermentado con tres diferentes levaduras, Maestría en Ciencias en Biotecnología, Instituto Tecnológico y de Estudios Superiores de Monterrey, Monterrey, Mexico, 2011.

Data expressed on a dry matter basis. Means with different letters within row are significantly different ($P < 0.05$).

[a] Low enzyme dosage: 1% cellulase complex, 0.2% b-glucosidase, 0.03% enzyme complex, 0.05% xylanase, 0.4% hemicellulase, and 0.06% glucoamylase.

[b] High enzyme dosage: 5% cellulase complex, 0.6% b-glucosidase, 0.4% enzyme complex, 0.25% xylanase, 2% hemicellulase, and 0.1% glucoamylase.

These have been used alone or in combination at temperatures ranging from 38°C to 36°C and initial pH ranging from 5.2 to 6 (Lee 1997, Ballesteros et al. 2007, Balat et al. 2008, Mizuno et al. 2009, Kurian et al. 2010, Serna-Saldivar et al. 2012).

Heredia-Olea et al. (2013) fermented detoxified worts obtained after acid- and/or enzyme-treated sweet sorghum bagasse with a genetically engineered *S. cerevisiae* and *I. orientalis*, which are capable of metabolizing both C5 and C6 sugars. These two types of yeast generated 18.43 and 20.9 g ethanol 100 g^{-1} dry sweet sorghum bagasse, respectively. An estimation of the conversion efficiencies indicated that these yeasts were capable of fermenting 64.4% and 73.3% of the potentially fermentable sugars into ethanol. Interestingly, most of the ethanol was generated during the first 24 h of fermentation (Table 8.5). Ballesteros et al. (2003) obtained 16.2 g ethanol L^{-1} when hydrolyzates obtained from sweet sorghum bagasse were fermented with *K. marxianus*. Kurian et al. (2010) obtained 38.7 g ethanol L^{-1} with a theoretical conversion of 82.5% using *P. stipitis*.

Given these reported conversions, sweet sorghum bagasse can be important for second-generation biofuel production. Assuming 40 tons ha^{-1} of sweet sorghum bagasse, 158 L ethanol ton^{-1} biomass (wet basis) can be obtained after a sulfuric acid hydrolysis (Figure 8.2). The process also yielded 110 kg of lignin and other nonfermentable by-products. According to Almodares and Hadi (2009) and Gnansounou et al. (2005), the cellulase used in SSF can be added directly or from material previously deviated from the pretreatment and inoculated along with *Trichoderma reesei* or other fungi such as *Neurospora crassa* and *Fusarium oxysporum*. These microorganisms were capable of directly fermenting cellulose (Mamma et al. 1996). *F. oxysporum* was used in a SSF along with *S. cerevisiae*, yielding 5.2–8.4 g ethanol 100 g^{-1} of fresh sorghum. Nevertheless, the efficiency was calculated based on soluble sugars and not on total polysaccharides (Mamma et al. 1996).

TABLE 8.5
Fate of Fermentable Sugars and Inhibitors and Ethanol Yields during Fermentation of Sweet Sorghum Bagasse Previously Hydrolyzed with Acid and Fiber-Degrading Enzyme Treatments

Type of Yeast	*Saccharomyces cerevisiae*			*Issatchenkia orientalis*		
Fermentation, h	**0**	**24**	**48**	**0**	**24**	**48**
Sugars, mg/g						
Cellobiose	17.0	9.4	8.8	18.4	10.5	9.8
Glucose	411.2	20.8	17.9	447.3	21.1	13.1
Xylose	60.8	50.9	48.6	64.5	51.9	45.8
Arabinose	129.6	28.3	27.3	145.1	122.6	108.1
Inhibitors, mg/g						
Acetic acid	27.8	24.0	24.9	30.0	24.1	22.6
HMF	4.8	2.2	1.8	5.0	ND	ND
Furfural	4.1	ND	ND	3.2	ND	ND
Ethanol yield						
Ethanol, mg/g	ND	183.9	177.4	ND	209.2	188.2
Ethanol yield, %	ND	64.4	62.1	ND	73.3	65.9

Sources: Heredia-Olea, E., Evaluación del proceso de obtención de bioetanol a partir de bagazo de sorgo dulce pretratado mediante extrusión, auto hidrólisis o hidrólisis acido diluida, hidrolizado enzimáticamente fermentado con tres diferentes levaduras, Maestría en Ciencias en Biotecnología, Instituto Tecnológico y de Estudios Superiores de Monterrey, Monterrey, Mexico, 2011; Heredia-Olea, E. et al., *Bioresour. Technol.*, 134, 386, 2013.

8.6 ESTIMATED ETHANOL YIELDS FROM AN INTEGRATED PROCESS

Sweet sorghum stalk yields are reported between 40 and 130 tons ha^{-1} for the first cut. Sweet sorghum stalk contains 73% moisture, 13% soluble sugars, 5.3% cellulose, 3.7% hemicelluloses, and 2.7% lignin (Table 8.1) (Prasad et al. 2007, Almodares and Hadi 2009). The stalks yield up to 70% sweet juice and 15.33 tons ha^{-1} of spent bagasse. Water added during extraction is considered part of the sweet juice yield, which contained around 14% soluble sugars. The sweet sorghum juice can produce 3450 L ethanol ha^{-1} with a fermentation efficiency of 95%, similar to the result reported by Kim and Day (2010) (3296 L ha^{-1}). The small difference might be that Kim and Day (2010) did not consider the juice losses that negatively affect fermentation efficiencies. Almodares and Hadi (2009) reported a yield of 1400–4600 L ethanol ha^{-1} (average of approximately 3000 L ethanol ha^{-1}) when processing juice extracted from varieties that yielded from 39 to 128 tons stalks ha^{-1}. Experimental data obtained from sweet sorghums cultivated in Central Mexico indicated that it is possible to achieve a yield of 6.38 tons of sugar ha^{-1} cut^{-1}, which has the potential to produce 4132 L ethanol. Although Wu et al. (2010) did not report ethanol yields per hectare, the calculated yield of ethanol produced from the amount of total fermentable sugars extracted from a high-yielding M81E cultivar that was planted at two different locations at a 95% of fermentation efficiency was in the range of 4750–5220 L ha^{-1}. These potential ethanol yields are equivalent to the bioconversion of 12–13 tons of maize kernels.

After sugar extraction, the spent bagasse can potentially yield up to 2400 L ethanol considering a yield of 15.3 tons ha^{-1}, which contains 29% cellulose and hemicellulose and 5.4% of remaining unextracted soluble sugars (Figure 8.2). This yield represents almost half of the 4058 L ha^{-1} described by Kim and Day (2010) as theoretical ethanol.

In central Mexico, 42.5 tons of bagasse ha^{-1} with 50% fermentable sugars is commonly obtained. This biomass is capable of yielding 6375 L ethanol with high conversion efficiency. However, experimental data show that the efficiency was only 60% for the fermentation of acid-treated biomass with *I. orientalis* and the ethanol yield from the bagasse was then 3865 L ha^{-1}. These results show that there are still many areas for potential improvements especially when processing spent biomass.

8.7 YIELD ENHANCEMENT AND PRODUCTIVITY

Ultimately, the yield of biomass per area (Figure 8.1) is the most important factor that affects ethanol yields. Thus, further improvements rely on plant breeding for the development of enhanced cultivars and agronomic research to minimize fertility and water input costs. The new biomass cultivars should adapt to marginal lands in order to minimize their competition with cereal grain production. The potential to obtain ethanol yields of 6,630, 7,000, and 10,000 L ha^{-1} (with 95% of extraction and fermentation efficiency) can be achieved because yields of 50–120 tons of biomass ha^{-1} have been reported. Comparatively, Kim and Day (2010) indicated that the theoretical yield of maize kernels can be as high as 5100 L ha^{-1} and up to 8625 L ha^{-1} when the whole plant (grain + corn stover) is bioconverted into ethanol.

One of the most important factors to be addressed during yield calculation is the energy input required for ethanol production. Biomass and starch require more energy for hydrolysis compared to sweet sorghum juice. The technologies for starchy kernels and sweet juice have little room for improvement, whereas the conversion and estimation of energy balances when processing lignocellulosic material will be critically important for the evaluation of economic advisability.

8.8 FUTURE TRENDS

Sweet sorghum has the potential for significant improvement in all four phases of its utilization: genetics, production, processing, and conversion. As described herein, there have already been significant improvements in sweet sorghum varieties, and there will be a rapid transition to hybrid production that will further increase yields. The breeding program should be aimed toward the productivity and other relevant traits such as plant height, total soluble solids, juice production, and fiber yield and quality, especially the lignin/cellulose/hemicellulose ratio. As sweet sorghum production becomes more popular, agronomic production practices will be optimized to maximize the efficiency of nitrogen and water use as well as harvest schedules and sugar utility. It is also of top priority to develop machinery to harvest sweet sorghums because the use of existing sugarcane equipment reduces yields and efficiencies.

Further optimization of milling and extraction methodologies, specifically for sweet sorghum, should produce sugar extraction rates comparable to those currently available in sugarcane. Similar improvement should be expected for fermentation of sugars from sweet sorghum. For the conversion of bagasse, there are still significant hurdles in the logistics and economics of conversion, which however apply to all sources of lignocellulosic biomass. As these systems become more common and the logistics and economics are improved, sweet sorghum bagasse will be an important energy crop.

REFERENCES

Almeida, J.R., T. Modig, A. Röder, G. Lidén, and M.F. Gorwa-Grauslund. 2008. *Pichia stipitis* xylose reductase helps detoxifying lignocellulosic hydrolysate by reducing 5-hydroxymethyl-furfural (HMF). *Biotechnology for Biofuels* 1(12): 1–9.

Almodares, A. and M.R. Hadi. 2009. Production of bioethanol from sweet sorghum: A review. *African Journal of Agricultural Research* 5(9): 772–780.

Amartey, S. and T. Jeffries. 1996. An improvement in *Pichia stipitis* fermentation of acid-hydrolysis hemicellulose achieved by overliming (calcium hydroxide treatment) and strain adaptation. *World Journal of Microbiology and Biotechnology* 12: 281–283.

Anglani, C. 1998. Sorghum carbohydrates-A review. *Plant Foods for Human Nutrition* 52(1): 77–83.

Arbatti, S.V. 2001. Brief review of alcohol industry. *Bharatiya Sugar* March: 119–121.

Bafrncová, P., D. Šmogrovičová, I. Sláviková, J. Pátková, and Z. Dömény. 1999. Improvement of very high gravity ethanol fermentation by media supplementation using *Saccharomyces cerevisiae*. *Biotechnology Letters* 21: 337–341.

Bai, F.W., L.J. Chen, Z. Zhang, W.A. Anderson, and M. Moo-Young. 2004. Continuous ethanol production and evaluation of yeast cell lysis and viability loss under very high gravity medium conditions. *Journal of Biotechnology* 110: 287–293.

Balat, M., H. Balat, and C. Öz. 2008. Progress in bioethanol processing. *Progress in Energy and Combustion Science* 34: 551–573.

Ballesteros, M., J.M. Oliva, M.J. Negro, P. Manzanares, and I. Ballesteros. 2003. Ethanol from lignocellulosic materials by a simultaneous saccharification and fermentation process (SFS) with *Kluyveromyces marxianus* CECT 10875. *Process Biochemistry* 39(12): 1843–1848.

Ban, J., J. Yu, X. Zhang, and T. Tan. 2008. Ethanol production from sweet sorghum residual. *Frontiers of Chemical Engineering in China* 2(4): 452–455.

Bayrock, D.P. and W.M. Ingledew. 2001. Application of multistage continuous fermentation for production of fuel alcohol by very-high-gravity fermentation technology. *Journal of Industrial Microbiology and Biotechnology* 27: 87–93.

Berg, C. 2004. World fuel ethanol, analysis and outlook. Available from http://www.distill.com/World-Fuel-Ethanol-A&O-2004.html. Accessed on January 5, 2014.

Bvochora, J.M., J.S. Read, and R. Zvauya. 2000. Application of very high gravity technology to the cofermentation of sweet stem sorghum juice and sorghum grain. *Industrial Crops and Products* 11: 11–17.

Carvalho, W., S.S. Silva, A. Converti et al. 2002. Use of immobilized *Candida* yeast cells for xylitol production from sugarcane bagasse hydrolysate: Cell immobilization conditions. *Applied Biochemistry and Biotechnology* 0489(02): 98–100.

Chakravorty, U., M.H. Hubert, and L. Nostbakken. 2009. Fuel versus food. *Annual Review of Resource Economics* 1: 645–663.

Chuck-Hernández, C., E. Pérez-Carrillo, E. Heredia-Olea, and S.O. Serna-Saldívar. 2011. Sorgo como un cultivo multifacético para la producción de bioetanol en México; tecnologías, avances y áreas de oportunidad. *Revista Mexicana de Ingeniería Química* 10(3): 529–549.

Daeschel, M.A., J.O. Mundt, and I.E. McCarty. 1981. Microbial changes in sweet sorghum (*Sorghum bicolor*) juices. *Applied Environmental Microbiology* 42(2): 381–382.

Davila-Gomez, F.J., C. Chuck-Hernandez, E. Perez-Carrillo, W.L. Rooney, and S.O. Serna-Saldivar. 2011. Evaluation of bioethanol production from five different varieties of sweet and forage sorghums (*Sorghum bicolor* (L.) Moench.). *Industrial Crops and Products* 33(3): 611–616.

FAOSTAT. 2012. Cereal production. In FAOSTAT, Available from http://faostat.fao.org/. Accessed on December 12, 2013.

Gamez, S., J.J. González-Cabriales, J.A. Ramírez, G. Garrote, and M. Vázquez. 2006. Study of the hydrolysis of sugar cane bagasse using phosphoric acid. *Journal of Food Engineering* 74: 78–88.

Garber, E.D. 1950. Cytotaxonomic studies in the genus *Sorghum*. *University of California Publications in Botany* 23: 283–361.

Gnansounou, E., A. Dauriat, and C.E. Wyman. 2005. Refining sweet sorghum to ethanol and sugar: Economic trade-offs in the context of North China. *Bioresource Technology* 96(9): 985–1002.

Grassi, G. 2001. Sweet sorghum. Published by ETA Florence and WIP-Munich in the framework of Latin America Thematic Network on Bioenergy (LAMNET). Online: http://web2.etaflorence.it/cms/uploads/ETA/pubblicazioni/brochures_online/LAMNET_sweet_sorghum_01.pdf. Accessed on December 12, 2013.

Hanchar, R.J., F. Teymouri, C.D. Nielson, D. Mccalla, and M.D. Stowers. 2007. Separation of glucose and pentose sugars by selective enzyme hydrolysis of AFEX-treated corn fiber. *Applied Biochemistry and Biotechnology* 7: 136–140, 313–326.

Heredia-Olea, E. 2011. Evaluación del proceso de obtención de bioetanol a partir de bagazo de sorgo dulce pretratado mediante extrusión, auto hidrólisis o hidrólisis acido diluida, hidrolizado enzimáticamente fermentado con tres diferentes levaduras. Maestría en Ciencias en Biotecnología. Instituto Tecnológico y de Estudios Superiores de Monterrey, Monterrey, Mexico.

Heredia-Olea, E., E. Perez-Carrillo, and S.O. Serna-Saldivar. 2012. Effects of different acid hydrolyses on the conversion of sweet sorghum bagasse into C5 and C6 sugars and yeast inhibitors using response surface methodology. *Bioresource Technology* 119: 216–223.

Heredia-Olea, E., E. Perez Carrillo, and S.O. Serna Saldivar. 2013. Production of ethanol from sweet sorghum bagasse pretreated with different chemical and physical processes and saccharified with fiber degrading enzymes. *Bioresource Technology* 134: 386–390.

Herrera, A., S.J. Téllez-Luis, J.A. Ramírez, and M. Vázquez. 2003. Production of xylose from sorghum straw using hydrochloric acid. *Journal of Cereal Science* 37(3): 267–274.

Jacobsen, S.E. and C.E. Wyman. 2000. Cellulose and hemicellulose hydrolysis models for application to current and novel pretreatment processes. *Applied Biochemistry and Biotechnology* 84–86(81): 81–96.

Jessup, R. 2009. Development and status of dedicated energy crops in the United States. *In Vitro Cellular and Developmental Biology* 45(3): 282–290.

Keeley, J.E. and P.W. Rundel. 2003. Evolution of CAM and C4 carbon-concentrating mechanisms. *International Journal of Plant Science* 164(3): S55–S77.

Kesava, S.S., S.K. Rakshit, and T. Panda. 1995. Production of ethanol by *Zymomonas mobilis*: The effect of batch step-feeding of glucose and relevant growth factors. *Process Biochemistry* 30(1): 41–47.

Kim, M. and D. Day. 2010. Composition of sugar cane, energy cane, and sweet sorghum suitable for ethanol production at Louisiana sugar mills. *Journal of Industrial Microbiology and Biotechnology* : 1–5.

Kundiyana, D.K. 1996. Sorganol: In-field production of ethanol from sweet sorghum. Available from http://digital.library.okstate.edu/etd/umi-okstate-1974.pdf. Accessed on January 5, 2014.

Kurian, J.K., A.K. Minu, A. Banerji, and V.V.N. Kishore. 2010. Bioconversion of hemicellulose hydrolysate of sweet sorghum bagasse to ethanol by using *Pichia stipitis* NCIM 3497 and *Debaryomyces hansenii* sp. *Bioresources* 5(4): 2404–2416.

Lamsal, B., J. Yoo, K. Brijwani, and S. Alavi. 2010. Extrusion as a thermo-mechanical pre-treatment for lignocellulosic ethanol. *Biomass and Bioenergy* 34: 1703–1710.

Laopaiboon, L., S. Nuanpeng, P. Srinophakun, P. Klanrit, and P. Laopaiboon. 2008. Selection of *Saccharomyces cerevisiae* and investigation of its performance for very high gravity ethanol fermentation. *Biotechnology* 7: 493–498.

Laopaiboon, L., S. Nuanpang, P. Srinophakun, P. Klanrit, and P. Laopaiboon. 2009. Ethanol production from sweet sorghum juice using very high gravity technology: Effects of carbon and nitrogen supplementations. *Bioresource Technology* 100: 4176–4182.

Laopaiboon, L., P. Thanonkeo, P. Jaisil, and P. Laopaiboon. 2007. Ethanol production from sweet sorghum juice in batch and fed-batch fermentations by *Saccharomyces cerevisiae*. *World Journal of Microbiology and Biotechnology* 23(10): 1497–1501.

Lazarides, M., Hacker, J.B., and Andrew, M.H. 1991. Taxonomy, cytology and ecology of indigenous Australian sorghums (*Sorghum* Moench: Andropogoneae: Poaceae). *Australian Systematic Botany* 4(4): 591–635.

Lee, J. 1997. Biological conversion of lignocellulosic biomass to ethanol. *Journal of Biotechnology* 56(1): 1–24.

Lin, Z.X., H.M. Zhang, X.J. Ji, J.W. Chen, and H. Huang. 2011. Hydrolytic enzyme of cellulose for complex formulation applied research. *Applied Biochemical Biotechnology* 164(1): 23–33.

Liu, R., J. Li, and F. Shen. 2008. Refining bioethanol from stalk juice of sweet sorghum by immobilized yeast fermentation. *Renewable Energy* 33(5): 1130–1135.

Liu, R. and F. Shen. 2008. Impacts of main factors on bioethanol fermentation from stalk juice of sweet sorghum by immobilized *Saccharomyces cerevisiae* (CICC 1308). *Bioresource Technology* 99(4): 847–854.

López, M.J., N.N. Nichols, B.S. Dien, J. Moreno, and R.J. Bothast. 2004. Isolation of microorganisms for biological detoxification of lignocellulosic hydrolysates. *Applied Biochemistry and Biotechnology* 64: 125–131.

Luhnow, D. and G. Samor. 2006. As Brazil fills up on ethanol, it weans off energy imports. *The Wall Street Journal*, January 16, 2006, p. A1.

Mamma, D., D. Koullas, G. Fountoukidis et al. 1996. Bioethanol from sweet sorghum: Simultaneous saccharification and fermentation of carbohydrates by a mixed microbial culture. *Process Biochemistry* 31(4): 377–381.

Martinez, A., M.E. Rodriguez, M.L. Wells et al. 2001. Detoxification of dilute acid hydrolysates of lignocellulose with lime. *Biotechnology Process* 17: 287–293.

McCutchen, B.F., R.V. Avant, and D. Baltensperger. 2008. High-tonnage dedicated energy crops: The potential of sorghum and energy cane. *Proceedings of the Twentieth Annual Conference of the National Agricultural Biotechnology Council*, Columbus, OH, pp. 119–122.

McIntosh, S. and T. Vancov. 2010. Enhanced enzyme saccharification of *Sorghum bicolor* straw using dilute alkali pretreatment. *Bioresource Technology* 101(17): 6718–6727.

Mei, X., R. Liu, F. Shen, and H. Wu. 2009. Optimization of fermentation conditions for the production of ethanol from stalk juice of sweet sorghum by immobilized yeast using response surface methodology. *Energy and Fuels* 23(1): 487–491.

Minteer, S.D. 2006. *Alcoholic Fuels*. London, U.K.: CRC Press/Taylor & Francis Group.

Mizuno, R., H. Ichinose, M. Honda et al. 2009. Use of whole crop sorghums as a raw material in consolidated bioprocessing bioethanol production using *Flammulina velutipes*. *Bioscience, Biotechnology, and Biochemistry* 73(7): 1671–1673.

Nilvebrant, N.-O., A. Reimann, S. Larsson, and L.J. Jönsson. 2001. Detoxification of lignocellulose hydrolysates with ion-exchange resins. *Applied Biochemistry and Biotechnology* 91–93: 35–49.

Nuanpeng, S., L. Laopaiboon, P. Srinophakun et al. 2011. Ethanol production from sweet sorghum juice under very high gravity conditions: Batch, repeated-batch and scale up fermentation. *Electronic Journal of Biotechnology* 14(1). http://dx.doi.org/10.2225/vol14-issue1-fulltext-2.

Öhgren, K., M. Galbe, and G. Zacchi. 2005. Optimization of steam pretreatment of SO_2-impregnated corn stover for fuel ethanol production. *Applied Biochemistry and Biotechnology* 124: 1055–1068.

Palmqvist, E., H. Grage, N.Q. Meinander, and B. Hahn-Hägerdal. 1999. Main and interaction effects of acetic acid, furfural, and *p*-hydroxybenzoic acid on growth and ethanol productivity of yeasts. *Biotechnology and Bioengineering* 63(1): 46–55.

Perez-Carrillo, E., M.L. Cortés-Callejas, L.E. Sabillón-Galeas et al. 2011. Detrimental effect of increasing sugar concentrations on ethanol production from maize or decorticated sorghum mashes fermented with *Saccharomyces cerevisiae* or *Zymomonas mobilis*. *Biotechnology Letters* 33(2): 301–307.

Phowchinda, O., M.L. Delia-Dupuy, and P. Strehaiano. 1997. Alcoholic fermentation from sweet sorghum: Some operating problems. Available from http://www.aseanbiotechnology.info/Abstract/23005950.pdf. Accessed on January 5, 2014.

Prasad, S., A. Singh, N. Jain, and H.C. Joshi. 2007. Ethanol production from sweet sorghum syrup for utilization as automotive fuel in India. *Energy and Fuels* 21(4): 2415–2420.

Reddy, B.V.S., S. Ramesh, P.S. Reddy et al. 2005. Sweet sorghum—A potential alternate raw material for bio-ethanol and bio-energy. International Crops Research Institute for the Semi-Arid Tropics, Patancheru, Andhra Pradesh, India. Available from http://www.icrisat.org/Biopower/BVSReddySweetSorghumPotentialAlternative.pdf. Accessed March 28, 2011.

Reddy, B.V.S. and P. Sanjana. 2003. Sweet sorghum: Characteristics and potential. *International Sorghum and Millet Newsletter* SICNA and ICRISAT No. 44: 26–28.

Reddy, N. and Y. Yang. 2005. Biofibers from agricultural by products for industrial applications. *Trends in Biotechnology* 23(1): 22–27.

Renewable Fuels Association. 2012. The ethanol industry-statistics. Available from http://www.ethanolrfa.org. Accessed on January 5, 2014.

Rodriguez-Chong, A., J.A. Ramírez, G. Garrote, and M. Vázquez. 2004. Hydrolysis of sugar cane bagasse using nitric acid: A kinetic assessment. *Journal of Food Engineering* 61: 143–152.

Rooney, W., J. Blumenthal, B. Bean, and J.E. Mullet. 2007. Designing sorghum as a dedicated bioenergy feedstock. *Biofuels, Bioproducts and Biorefining* 1(2): 147–157.

Ryan, J.G. and D.C. Spencer. 2001. Future challenges and opportunities for agricultural R&D in the semi-arid tropics. International Crops Research Institute for the Semi-Arid Tropics, Patancheru, Andhra Pradesh, India, 83pp.

Saballos, A. 2008. Development and utilization of sorghum as a bioenergy crop. In *Genetic Improvement of Bioenergy Crops*, ed. W. Vermerris, pp. 211–248. New York: Springer.

Salvi, D.A., G.M. Aita, D. Robert, and V. Bazan. 2010. Ethanol production from sorghum by a dilute ammonia pretreatment. *Journal of Industrial Microbiology and Biotechnology* 37: 27–34.

Serna-Saldivar, S.O. 2010. *Cereal Grains: Properties, Processing and Nutritional Attributes*. Boca Raton, FL: CRC Press.

Serna-Saldivar, S.O., C. Chuck Hernandez, E. Perez Carrillo, and E. Heredia-Olea. 2012. Sorghum as a multifunctional crop for the production of fuel ethanol: Current status and future trends. In *Bioethanol*, eds. M.A. Pinheiro Lima and A.P. Policastro Natalence, pp. 55–74. London, U.K.: In Tech.

Shen, F. and R. Liu. 2009. Research on solid-state ethanol fermentation using dry sweet sorghum stalks particles with active dry yeast. *Energy and Fuels* 23: 519–525.

Singh, H.P. and H.C. Lohithaswa. 2006. Sorghum. In *Genome Mapping and Molecular Breeding in Plants*, Vol. 1, Cereals and Millets, ed. C. Kole, pp. 257–302. Berlin, Germany: Springer Verlag.

Sipos, B., J. Réczey, Z. Somorai et al. 2009. Sweet sorghum as feedstock for ethanol production: Enzymatic hydrolysis of steam-pretreated bagasse. *Applied Biochemistry and Biotechnology* 153: 151–162.

Smith, G.A., M.O. Bagby, R.T. Lewellan et al. 1987. Evaluation of sweet sorghum for fermentable sugar production potential. *Crop Science* 27(4): 788–793.

Stenberg, K., C. Tengborg, M. Galbe, and G. Zacchi. 1998. Optimisation of steam pretreatment of SO_2-impregnated mixed softwoods for ethanol production. *Journal of Chemical Technology and Biotechnology* 71(4): 299–308.

Sun, R.C. 2010. *Cereal Straw as a Resource for Sustainable Biomaterials and Biofuels: Chemistry, Extractives, Lignins, Hemicelluloses and Cellulose*, 1st edn. Amsterdam, the Netherlands: Elsevier.

Thomas, K.C., S.H. Hynes, and W.M. Ingledew. 1996. Practical and theoretical considerations in the production of high concentration of alcohol by fermentation. *Process Biochemistry* 31: 321–331.

Turhollow, A.F., E.G. Webb, and M.E. Downing. 2010. Review of sorghum production practices: Applications for bioenergy. Available from http://info.ornl.gov/sites/publications/files/Pub22854.pdf. Accessed on January 5, 2014.

Vencor Green LLC. 2012. Feedstocks sugarcane and sweet sorghum. Online: www.vencorgreen.com. Accessed on December 12, 2013.

Viuchard, N., P. Ciais, and A. Wolf. 2009. Soil carbon sequestration of biofuel production: New land use opportunities for mitigating climate over abandoned soviet farmlands. *Environmental Science and Technology* 43(22): 8678–8683.

Wang, M.L., C. Zhu, N.A. Barkley et al. 2009. Genetic diversity and population structure analysis of accessions in the US historic sweet sorghum collection. *Theoretical and Applied Genetics* 120(1): 13–23.

Woods, J. 2000. Integrating sweet sorghum and sugarcane for bioenergy: Modelling the potential for electricity and ethanol production in SE Zimbabwe. Doctor of Philosophy Thesis, King's College London, University of London, London, U.K.

Wu, X., S. Staggenborg, J.L. Propheter et al. 2010. Features of sweet sorghum juice and their performance in ethanol fermentation. *Industrial Crops and Products* 31(1): 164–170.

Zhang, C., G. Xie, S. Li, L. Ge, and T. He. 2010. The productive potentials of sweet sorghum ethanol in China. *Applied Energy* 87(7): 2360–2368.

9 Production and Supply Logistics of Sugarcane as an Energy Feedstock

Jose Goldemberg, Suani T. Coelho, Plinio M. Nastari, and Patricia Guardabassi

CONTENTS

9.1 INTRODUCTION

Brazil can be considered a special case in terms of the share of renewable in its energy matrix mainly due to the electric energy supply mostly based on hydro and the Brazilian initiatives related to biofuels, which have been intensively implemented since 1975. By that time, public policies designed to promote the large-scale production and consumption of biofuels were introduced in the country to reduce the external dependence on imported oil. Since then, several experiences transformed Brazil into a world leader in bioenergy mainly due to sugarcane ethanol production, distribution, and use in the transportation sector.

Brazil has a very sustainable energy matrix due to the use of hydraulic energy for power production, which represents more than 74% of the electricity supply, and to biofuels initiatives. Bioenergy has been part and parcel of the Brazilian energy mix and policy for a long time and certain policy choices. This is the reason why greenhouse gas (GHG) emissions from energy production in Brazil are relatively low, ranked in the 18th position among the world's largest country emissions, when excluded CO_2 emissions resulting from deforestation in the Amazon.

The share of biofuels in the total energy used in transportation is significant, with 17.9% from sugarcane ethanol (used as pure ethanol in flexible vehicles and blended with gasoline), against 25.3% for gasoline. The consumption of diesel oil, however, is still high, since diesel oil is still highly subsidized in the country. For this reason also, the diesel fuel is not allowed to be used in light vehicles, only in trucks, agricultural machines, and urban transportation. Aiming at reducing the fuel diesel consumption in the country, a program to incentivize the use of biodiesel was implemented in 2003, and, since 2010, all diesel oil in the country is blended with 5% biodiesel in volume.

Anhydrous ethanol, with minimum 99.50% total alcoholic content by volume at 15°C, is used as a blend component in gasoline at a ratio that, according to current legislation, can vary from 18% to 25% by volume. This fuel can be referred to as E18–E25.

Hydrous ethanol (E100), with average 95.54% alcoholic content by volume at 15°C, is used mainly by the flexible fuel fleet. The flexifleet was technically developed to use any proportion of gasoline and ethanol blends.

Ethanol has substituted 44.6% of the Otto cycle fuels, or referred to as gasoline in most countries in Brazil in 2010 (Nastari, 2012a). This was a remarkable achievement in terms of substitution. Since 2006, the United States has overcome Brazil as the world's largest ethanol producer. Most of the ethanol in the United States is produced from corn and ethanol substituted 9.5% of the gasoline used in the United States in 2010 (Nastari, 2012a). The target defined by the 2007 Energy Independence and Security Act (EISA) in the United States, commonly known as phase 2 of the Renewable Fuels Standard (RFS-2), is to achieve the substitution of 20% of all transportation fuels by 2022.

The Renewable Energy Directive (RED) issued by the European Commission established a target to substitute 10% of all transportation fuels with renewable fuels in Europe by 2020. In 2010, the actual proportion of bioethanol and biodiesel in fuel consumption in Europe was only 3.4% (Nastari, 2012a).

9.2 GENERAL OVERVIEW ON SUGARCANE AS AN ENERGY FEEDSTOCK

The use of bioenergy in the Brazilian energy matrix has been a reality for a long time. Ethanol production in Brazil was incentivized through a subsidized program in 1975. However, as the improvements in technology and economies of scale have reduced production costs, ethanol had already become economically competitive with gasoline without subsidies since 2004 (Goldemberg et al., 2008).

Nowadays, Brazil is the world's second largest producer of ethanol and the largest producer of sugarcane ethanol. Brazil produced 22.7 billion liters of ethanol from sugarcane while the United States produced 52.1 billion liters of ethanol in 2011. In the last harvesting season, there were 440 mills that produced ethanol and sugar in a planted area of 9.2 million hectares of sugarcane.

Brazil has a remarkable historical performance on agroindustrial yields, measured in terms of liters of hydrous ethanol per hectare. The average agroindustrial yield was 2024 L of hydrous ethanol equivalent per hectare in 1975 but had currently evolved to 6831 L of hydrous ethanol equivalent per hectare. This represents a compound average growth rate (CAGR) of 3.54% per year over the past 35 years (Nastari, 2012b). The average agricultural yield, measured in metric tons of cane per hectare in the 5-year period from 2005 to 2009 was 85.5 metric tons per hectare in the more productive central-south region. However, due to the severe droughts in 2010 and 2011 and the aging of cane fields caused by renewals at a lower rate than normal, the average agricultural yield has fallen to 67 tons/ha in 2011 (DATAGRO, 2012).

The main advantage of sugarcane ethanol is its positive net energy balance in comparison to ethanol from corn or other crops. This energy balance* is around 8.3 on average, with the best cases showing a balance of 10.2 (Macedo et al., 2008). This positive balance is due to the fact that the tropical grass crop such as sugarcane is highly efficient in converting solar energy into sugar-rich biomass and that the bagasse supplies all the energy needed (heat and electricity) for the industrial process of ethanol production.

Initially, ethanol was available for ethanol-dedicated engines (hydrous ethanol, with an average alcoholic content of 95.54% in volume, at 15°C) or as an octane enhancer (anhydrous ethanol, with a minimum 99.5% alcoholic content in volume, at 15°C), replacing lead and methyl tert-butyl ether (MTBE) that are commonly used as an octane enhancer. Nowadays, instead of ethanol-dedicated vehicles, hydrous ethanol is used in flex-fuel vehicles, which accounts for more than 90% of all new cars sold in Brazil. These flex-fuel vehicles can run on any blend of gasoline or ethanol, allowing drivers to make price-driven fuel choices (Anfavea, 2011). Projections are expected to increase

* Energy balance is defined here as the ratio of the energy content of 1 L of ethanol to the energy from fossil fuels used in the production of the ethanol (including transportation of sugarcane in diesel-run trucks, fertilizers, and pesticides).

ethanol production to almost 57 billion liters in Brazil over the next 10 years (Conab, 2011), due to the expected increase of 51.7% in light vehicle fuel consumption for transportation.

Besides ethanol, the sugarcane bagasse, which is the residue from sugarcane crushing, is used in combined heat and power generation systems (cogeneration) in the mills to supply the thermal and electromechanical energy within the mills and sell the surplus electricity to the commercial grid. In June 2012, there were 352 cogeneration systems in sugar and ethanol plants, registered by the National Electricity Regulatory Agency in Brazil, with a total installed capacity of 7588 MW.*

The use and production of ethanol in Brazil has been facing many challenges since the introduction of the National Alcohol Program in 1975. At the beginning of the program, the main barriers were the high production costs, low productivity, small production scale, the inexistence of dedicated engines, and low levels of consumer acceptance. All of these barriers have been overcome by the introduction of proper public policies and successful investments in research and development. The current challenges include the sustainable production of ethanol, the guarantee of decent labor conditions in the sugarcane industry and the demystification of the relationship between sugarcane ethanol, the price escalation of staple commodities and the deforestation in the Amazon forest.

The production of ethanol from sugarcane has shown good sustainability indicators, considering its local impacts and its larger, more general implications. The Environmental Protection Agency (EPA) in the United States has recognized Brazilian sugarcane ethanol as an *advanced biofuel* (EPA, 2010). In December of 2011, 118 Brazilian sugarcane mills were allowed to export sugarcane ethanol to the United States after they were registered with EPA by complying with the EPA's requirements. Furthermore, 14 Brazilian mills, which represent 1.4% of the world's sugarcane cropped area, have been certified by the Bonsucro Initiative.

Sugarcane can produce large amounts of liquid and electric energy per hectare compared with other food crops. Due to its high productivity, sugarcane has proven to be an effective source of biomass for energy production, without disrupting food production systems. The so-called food crisis, with high prices and relative scarcity, has multiple causes such as the rapid expansion of Asiatic demand, the rising oil prices, and climate change. The efficient production of bioenergy constitutes a second-order factor in this scenario. The deforestation in Amazon cannot be directly or indirectly associated with ethanol production. The methodologies for estimating the indirect land use change still lack clarity. However, well-conducted and detailed studies in Brazil indicate that the sugarcane crop has relatively little association with the deforestation of Amazon.

The jobs created in the sugarcane industry, according to several indicators, are better than those observed on average agricultural sectors and in most cultures with economic expressiveness in Brazil. In the case of cogeneration, the lack of a regulatory framework is partially solved, and the remaining gaps include the interconnection costs and the price offered to biomass energy in the public auctions.

9.3 POTENTIAL ENERGY PRODUCTION FROM SUGARCANE RESIDUES

The use of sugarcane residues as feedstock in steam cycles to supply the energy demand of sugarcane mills and provide surplus electricity to the grid has been in place in Brazil since the middle of 1990s.

In the last decade, the development of a regulatory framework that motivated the trade of electricity surplus induced the investments in more efficient cogeneration technology (higher-pressure boilers), hence increasing the amount of surplus electricity generated. Despite the significant increase in this surplus production, the introduction of mechanical harvesting and the consequent availability of cane trash, previously burnt in preparation of manual harvesting, can contribute to further increasing the surplus electricity available.

* ANEEL. Generation database. Available at http://www.aneel.gov.br/aplicacoes/capacidadebrasil/CombustivelPorClasse.asp?Classe=Biomassa. Accessed on June 15, 2012.

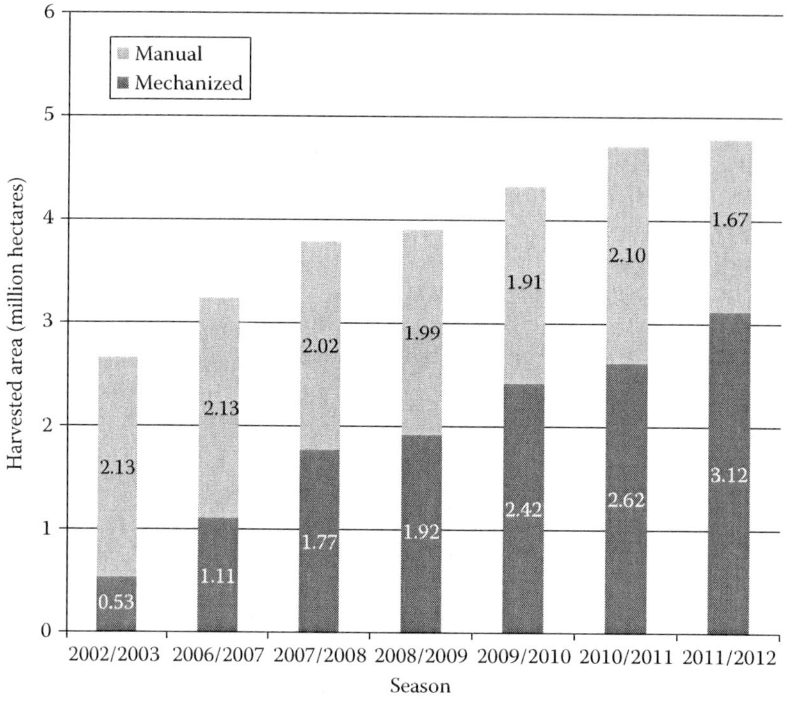

FIGURE 9.1 Evolution of mechanical harvesting in the state of São Paulo.

Preharvest burning of cane leaves, which results in intense local air pollution, is being phased out. The cane leaves, which account for an increase of 30% in biomass availability after collected, are used to produce more electricity (State Law 11,241/2002). This phaseout is controlled, and the remaining burning must be authorized by the Secretary for the Environment in the state of São Paulo, according to atmospheric conditions in each region of the state.

In 2007, the Secretariats of the Environment and of Agriculture and Food Supply in the state of São Paulo and UNICA (the Sugarcane Agroindustry Association) signed a voluntary environmental agreement, which aims at rewarding good practices in the sugarcane sector. As of June 2012, 173 out of the 192 ethanol plants operating in the state have adhered to this agreement.* One of the main guidelines of this agreement is to anticipate the timetable for phasing out of burning sugarcane. In the 2011/2012 harvesting season, 65.2% of the sugarcane was harvested in a green way in the State of São Paulo, and until 2014, the use of fire during sugarcane harvesting must be banished in all mechanizable areas. For other areas, the legal deadline of 2031 will be anticipated to 2017. Figure 9.1 shows the elimination of harvest burning in the state.

Despite the high investment costs for harvesting machinery as a harvesting machine costs approximately R$ 1 million (or US$ 600,000), operational costs of mechanical harvesting are lower than those of manual harvesting and the productivity of mechanical harvesting is also higher.

Mechanical harvesting will prevent the release of 3.9 thousand (metric) tons of particulates (~28% of emissions from diesel vehicles in the São Paulo Metropolitan Region—SPMR), 45.3 thousand tons of carbon monoxide (12% of diesel emissions in SPMR), and 6.5 thousand tons of hydrocarbons (11% of diesel in SPMR). There are other initiatives for the banishment of sugarcane burning in many states. In the central-west region, local environmental agencies require that all new sugar/alcohol mills to be licensed must have 100% green technology to harvest sugarcane. Only in

* São Paulo State Government. The Green Ethanol Program. Available at http://www.ambiente.sp.gov.br/etanolverde/resultadoSafras.php. Accessed on June 14, 2012.

TABLE 9.1
Potential for Electricity Surplus from Sugarcane Residues (Bagasse and Leaves)

Season	Sugarcane Production (Tons of Cane/Season)	Potential for 65 Bar Boilers (MW_{med}) 125 kWh/ton of cane	Potential for 92 Bar Boilers (MW_{med}) 145 kWh/ton of cane
2011/2012	558,350,000	8387	9729
2012/2013[a]	562,600,000	8449	9801

Notes: Theoretical potential using all sugarcane bagasse produced (30% of sugarcane crushed, 50% wet) and 40% of tops and leaves. Electricity production all year around.

[a] Estimates DATAGRO.

the northeastern region, due to declivity higher than 12%, the existing machines cannot be used, and manual harvesting of (burnt) cane will still be used.

As green or mechanical harvesting collects an additional amount of biomass, more electricity can be produced from the additional amount of biomass. Studies have recommended that 50% of cane leaves should be left in the fields to protect the soil, which depends on soil and climate conditions.

Currently, as electricity generation technology became more efficient by using high-pressure boilers (65–100 bar) and condensed extraction steam turbines and the amount of steam used to crush each ton of sugarcane was reduced from 500 to 340 kg of steam at 2.5 bar, the surplus electricity generated in a sugarcane mill increased to 120–145 kWh/ton cane each year. Table 9.1 shows the data for the 2011/2012 and 2012/2013 cropping seasons.

It is necessary to note that it is important to generate electricity year round beyond the harvesting season only. The power generated from cane complements hydro power generation due to the fact that the electricity generation from sugarcane bagasse happens mostly in the dry period when dam levels are lower. The generation of electricity all year around is a more interesting proposition as it corresponds to firm energy.

In the 2009/2010 harvesting season, the total electricity produced from sugarcane bagasse was 20,031 GWh. In this scenario, 28.2% of the mills sold their surplus electricity to the grid. Nonetheless, there was a potential to produce 39,950 GWh (Conab, 2011). Over the next 10 years, in the best scenario, considering higher-pressure boilers installed (99 bar) in all mills, the amount of electricity produced from the bagasse of 1.04 billion tons of sugarcane produced in Brazil each year is expected to increase up to 68,730 GWh (Conab, 2011).

9.4 SUGARCANE PRODUCTION: MAIN TECHNICAL ASPECTS

Sugarcane is a semiperennial crop with a typical production cycle in Brazil of 5–6 years. Due to the variation in climate and agronomical practice, the production cycle can be as high as 9–11 years for some cases. The length of the production cycle varies from one country to another. In India, which is the second largest sugarcane producer in the world after Brazil, the production cycle is only 2 years.

Sugarcane is usually planted at the end of the wet season and is harvested in the dry season. In Brazil, approximately 90% of all cane produced for industrial uses to produce sugar and ethanol without including these for forage and production of spirits is produced in the central-south region. The remaining is produced in states located in the north and northeastern regions of the country. In the central-south, sugarcane is planted in the months of February to April, and the first harvest happens 18 months later. After the first harvest (cane plant), sugarcane ratoons are harvested every year for another 4–5 years, therefore yielding a total of 5–6 cuts. Productivity, in tons of cane per hectare,

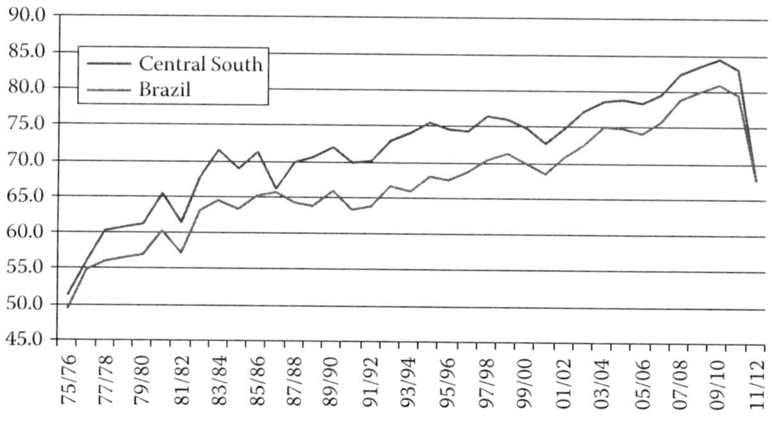

FIGURE 9.2 Average agricultural yield in the central-south region and in Brazil, from 1975/1976 to 2011/2012. (From Datagro, Informativo Datagro, São Paulo, Brazil, n. 02P, February 2012.)

is higher for the first cuts and decreases at every successive cut. Typical yields for the cane planted in the central-south region are 122, 105, 97, 87, 78, and 65 tons/ha for each subsequent cut. However, actual productivity is a function of many different variables including cane variety, climate, soil fertility, intensity of inputs application, weed control, pest and disease controls, age of cane fields, and agricultural practices, among other factors.

In 1975/1976, the average agricultural yield in the central-south region of Brazil was 51.4 tons of cane per hectare (DATAGRO, 2012). In 2010/2011, the average yield had grown to 83.1 tons/ha, mostly because of higher productive cane varieties that are adapted to the soil and climate conditions observed in each microregion and the development of agronomical technologies for weed and pest control and other agricultural management practices (DATAGRO, 2012). In 2011/2012, the average yield fell to 67.84 tons/ha due to the impact of drought, frosts, flowering, and aging of cane fields, all of which happened in 1 year (DATAGRO, 2012). The average agricultural yield for Brazil evolved from 49.5 to 79.6 tons/ha between 1975/1976 and 2010/2011 and reduced to 67.72 tons/ha in 2011/2012 (DATAGRO, 2012) (Figure 9.2).

Sugarcane varieties have been developed by independent, mostly privately funded research centers. Among them, the most noteworthy ones include Rede Interuniversitária para o Desenvolvimento do setor Sucroenergético (Ridesa), a group of universities whose research activities are funded by the private sector; Centro de Tecnologia Canavieira (CTC), controlled by a large group of mills in the central-south region of Brazil; CanaVialis, a company owned by Monsanto; and Instituto Agronomico de Campinas (IAC), linked to the Secretariat of Agriculture in the state of São Paulo. Typically, a modern sugar mill and its sugarcane suppliers grow between 15 and 20 different cane varieties simultaneously for the reasons of phytosanitary control and adapting different cane varieties to each soil/climate condition.

There are 440 mills in operation in Brazil: 153 produce ethanol only; 20 produce sugar only; and 267 produce both sugar and ethanol (Nastari, 2012a). The average scale of crushing in the industry is 1.3 million tons of cane per harvest. However, the scale considered to yield the least average cost is between 3.5 and 4 million tons of cane crushed per year. Larger mills, with projected scales of 6–8 million tons of cane-crushing capacity per year, during a crop season of 240 days, are being installed in pioneer regions in the states of Minas Gerais, Goias, Mato Grosso do Sul, and Mato Grosso.

Land rental costs are measured in a value pegged to the price of cane. Land rental agreements are normally made for a price that is negotiated in tons of cane per hectare, or per alqueire, equal to 2.42 ha. In the traditional sugarcane-producing regions, land rental costs have risen from the competition of sugar mills to the available areas owned by cane suppliers and, in 2011, are observed in the range of 30–35 tons of cane per hectare per year. In the pioneer and new frontier areas, land rental is much lower, ranging from 8 to 18 tons of cane per hectare. However, in these pioneer and frontier areas, logistics costs are higher, and there are not yet too many cane varieties already developed to capture the full production potential that exists in those areas. Contractual arrangements between cane suppliers and sugar mills are made for two or three cycles, of 5–6 years each, therefore lasting for 10–18 years, on average. These agreements establish the basis in which cane produced by cane suppliers will be delivered and paid for. Until 1999, the federal government determined the prices of cane that millers would pay to cane suppliers. Since then, prices have been liberalized and are negotiated freely between cane suppliers and mill owners. Based on the mutual interdependence of cane millers and suppliers and the fact that both of them should share the benefits and costs of market fluctuations, both sides negotiated a private agreement called CONSECANA. According to this agreement, which is not obligatory, the price of cane will be set in proportion to the price of the end products of sugar and ethanol that are produced and sold by the mill. The price of cane is determined by its value that corresponds to its share in the total costs of producing and processing cane for the end products such as sugar and ethanol. This modern and fair system for establishing the value of the feedstock of cane, which is delivered by cane suppliers to the mills, has brought a relative stability in the relationship between farmers and mills. Over time, there have been negotiations carried out by both sides to incorporate the impact of productivity gains and other by-product credits.

9.5 LOGISTICS: CANE TRANSPORT COSTS

A typical cane-crushing operation is actually a logistics operation by transporting a large volume of biomass to a central point of processing within a short radius of transport. The economical radius of transport for cane is in the range of 18–32 km distance to the mill. Transportation of cane from distances longer than 32 km is costly, which captures most of the value generated from its processing. Over time, cane transportation has evolved to the use of combined trucks with two or three load trucks (bitrem and treminhao) in order to optimize operations and reduce costs.

Because most of the agricultural operations, planting, cultivation, harvesting, and transport of cane are carried out using equipment and vehicles fueled by diesel, the substitution of diesel in cane production is a target that will be increasingly pursued by the industry in the coming years. The average consumption of diesel in the industry is 2.5 L of diesel per ton of cane crushed at the mill. For mills that use irrigation, the amount of diesel used rises to 4 L per ton of cane (DATAGRO, 2012). Considering that the amount of cane crushed for industrial uses was 558 million tons in 2011/2012, the total amount of diesel used was estimated to be 1.4 billion liters per year, which was 2.67% of all 52.26 billion liters of diesel consumed in Brazil in 2011 (DATAGRO, 2012).

Most of the sugar and ethanol produced is also transported by trucks using diesel, and only in recent years, the use of rail systems has been significantly increased in order to reduce costs. The most noteworthy initiatives have been made to transport sugar with railcars by Rumo Logística owned by Grupo Cosan, which is the largest independent cane processor in the world and crushed 52 million tons of cane in 2011, and Grupo São Martinho, which crushed 18 million tons of cane in 2011. For more efficient logistics to transporting ethanol, a company called Logum was formed as a joint venture of Copersucar, Cosan, Petrobras, ETH Bioenergia, and other independent producers to construct and operate a dedicated ethanol pipeline with a capacity of 21.8 billion liters per year when it becomes full operation. The first section of this pipeline, linking Ribeirao Preto to Paulinia and then to Ilha d'Água terminal in Rio de Janeiro, will be commissioned in 2013. The cost of pipeline transportation is estimated to be 1/8 of the cost of transporting ethanol by trucks, thus significantly reducing the logistics costs of bringing ethanol to the ports for cabotage operations in coastal Brazil and exports to other countries.

As previously mentioned, producers in the state of São Paulo, and the government of the state of São Paulo, have established a protocol in 2004 in which they agreed to the anticipated deadline of

2014 for banning fire as a preharvest activity during manual harvesting. The increased mechanization of harvesting will be able to cut cane in green and not to burn cane leaves during harvesting, which will also increase the availability of leaves left over in cane fields after the harvest. While the coverage of leaves is beneficial to the maintenance of soil humidity, it will also create a fire risk, which causes great damage to the lifespan of cane ratoons. For this agronomical reason, there is a great incentive to harvest part of the cane leaves to control the fire risk and use the bioenergy contained in the leaves.

A lot of studies that were conducted by CTC, Ridesa, and independently many sugar mills suggested the alternative methods to harvest cane leaves including bailing of cane leaves after harvest and harvesting and transporting of leaves along with canes to the mill and then separating at the mouth of the crushing tandems. The second alternative is the preferred one at this moment, and an increasing number of producers are bringing cane leaves to the mill to increase cogeneration. The transportation of cane leaves represents a new challenge to the industry. However, it also brings opportunities to create additional income from power and possibly second-generation ethanol produced from the leaves using the same cane production base.

Infrastructure at the ports has evolved more rapidly than inland infrastructure. In 1989, when the state monopoly on sugar exports was liberalized, Brazil was exporting 1.05 million tons of sugar, mostly from the northeast region and very little through the central-south ports. Brazil's share in world exports at that time was only 3.1%. In 2010, Brazil exported 28.0 million tons, corresponding to 51.0% of the world's sugar exports (Figure 9.3).

Obviously, a considerable effort was developed to create the necessary port infrastructure to allow such rapid and dramatic growth in sugar exports. Ethanol exports also grew but not in a consistent way, which have increased from 0.2 billion liters in 1989 to 5.1 billion liters in 2008 and then fell to 2.0 billion liters in 2011 (Figure 9.4).

Until recently, the use of biofuels was limited to local markets (in Brazil and in the United States) and played a marginal role in the global energy mix. However, today, biofuels have acquired a global dimension. In general, developing countries have a higher potential to produce biomass than industrialized countries due to more favorable climate conditions and lower labor costs. As a result, international trade in biofuels and/or feedstocks from developing to the developed

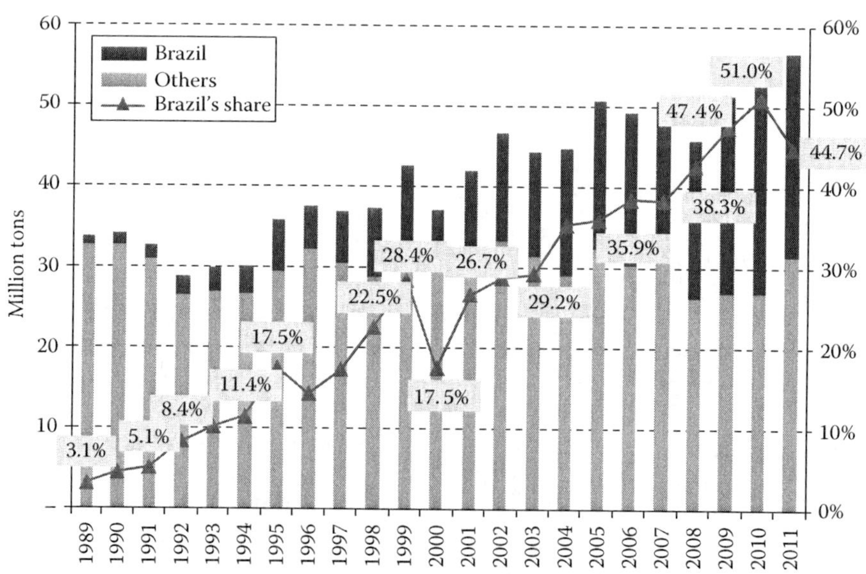

FIGURE 9.3 Brazilian exports of sugar. (From Datagro, Informativo Datagro, São Paulo, Brazil, n. 02P, February 2012.)

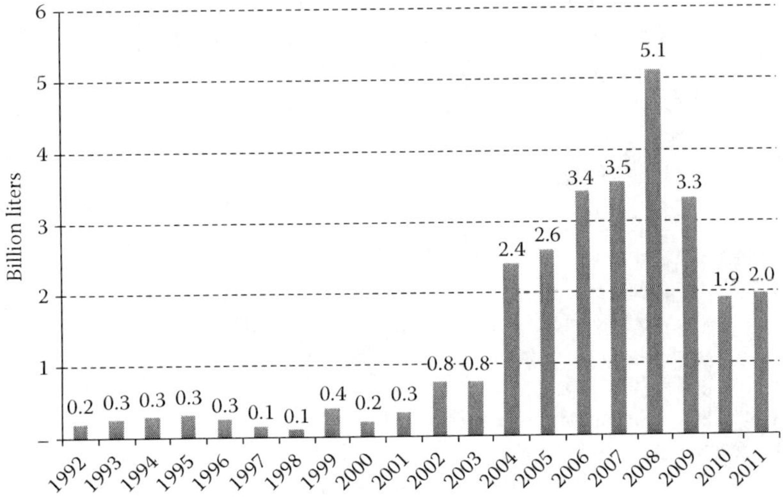

FIGURE 9.4 Brazilian ethanol exports time series. (From Datagro, Informativo Datagro, São Paulo, Brazil, n. 02P, February 2012.)

countries is expected to increase with significant positive implications for development (UNCTAD, 2009). According to Smeets et al. (2007), for the year 2050, sub-Saharan Africa and Latin America and the Caribbean regions have the greatest potential for the production of agricultural residues that could be used as bioenergy feedstock.

Furthermore, the *Global Agro-Ecological Zones* (GAEZ) system developed by FAO in conjunction with the International Institute for Applied Systems Analysis (IIASA) has identified and quantified areas with potential to produce raw materials for biofuels based on climate and soil conditions. In the case of rain-fed sugar cane, GAEZ identified 135 million suitable or very suitable hectares and 130 million moderately suitable hectares (of which only 22 million are currently planted with sugarcane). For unprotected areas, the overall potential is estimated at 87 million hectares that are suitable or very suitable for sugarcane production, of which 26 million hectares are located in Africa and 54 million hectares in South America. The regions with greater aptitude for the cultivation of rain-fed sugar cane are located in South and Central America, the Caribbean, Central Africa and some countries in West and South Africa, South India, and Southeast Asia (Fischer et al., 2008). However, these regions, which are suitable for the production of biofuels, currently produce small amounts of ethanol. Among the total 86.8 billion liters of ethanol produced around the world in 2010, the United States and Brazil were responsible for 76.2 billion. Despite their potential, the African continent produced only 0.16 billion liters of ethanol and Latin America and Caribbean produced 2.6 billion liters in 2010 (RFA, 2011).

The production and use of biofuels can improve energy security, power access, and rural development by accelerating both economic development and employment. It must also be noticed that, as mentioned from the Brazilian experience, electricity from residues presents a huge opportunity with enormous benefits for the increase of energy access in rural areas in developing countries. There are barriers that must be overcome to allow the development of a biofuels production sector. In order to surpass such barriers, it is necessary to settle a legal framework that ensures an appropriate institutional condition, which is economically attractive to investors, and to promote the ethanol production while minimizing environmental and social impacts.

The experience obtained by Brazil over the decades of ethanol production and use can assist these countries in developing sustainable biofuel programs. The most significant experience to get from the Brazilian lessons includes the adequate choice for biofuel crops, through the establishment of an agroenvironmental/agroeconomic zoning to define the best areas for food and fuel production, ensuring food security and contributing to the rural development (not only through the creation of jobs in rural areas but also through the increase of energy access from sugarcane bagasse, as already being done) (Goldemberg, 2009). The existing *Cogen for Africa Project*, funded by GEF-UNEP-AfDB,* is an extremely important experience that is being developed in sub-Saharan countries.

9.6 CONCLUSIONS

Many developing countries, mostly in Africa and Asia, have a small internal market for biofuels but have land and climate adequate for the production of biofuels. The production of biofuels to be exported to industrialized countries could stimulate rural development, generate jobs, increase energy access, and reduce poverty. Preconditions for the development of biofuels in developing countries include capacity building to master the technologies required both in the agricultural and industrial areas. Assistance from other developing countries, such as Brazil or India, which have important activities in sugarcane production (either for sugar or ethanol), could be very fruitful in this case, fostering South–South cooperation. The introduction of a bioethanol activity in developing countries should in all cases be preceded by a proper agronomic ecological zoning to identify producing areas, in order to respond to frequent criticism that biofuels production does not comply with appropriate certification criteria to local conditions. Such arguments can in reality be interpreted as nontariff protectionism barriers adopted by some countries in Europe and the United States, to protect noncompetitive agroindustrial activities in their countries. The experience of Brazil with sugarcane production and logistics, as discussed in this chapter, could be significant for other countries that are interested in developing local biofuel programs.

REFERENCES

Anfavea. 2011. *Anuário da Indústria Automobilística Brasileira*. Sao Paulo, Brazil: Anfavea, 158pp. Available from www.anfavea.com.br.

Conab. 2011. *A Geração Termoelétrica com a Queima do Bagaço de Cana-de-Açúcar no Brasil—Análise do Desempenho da Safra 2009–2010*. Brasília, Brazil: Conab.

DATAGRO. 2012. Informativo Datagro. São Paulo, Brazil, n. 02P, February.

EPA. 2010. 40 CFR Part 80. Regulation of Fuels and Fuel Additives: Changes to Renewable Fuel Standard Program; Final Rule, s.l.: s.n.

Fischer, G. et al. 2008. *Global Agro-Ecological Zones Assessment for Agriculture (GAEZ 2008)*. Laxenburg/Roma, Austria/Italy: IIASA/FAO.

Goldemberg, J. 2009. The Brazilian experience with biofuels. *Innovations Journal*, 4, 91–107.

Goldemberg, J., Coelho, S.T., and Guardabassi, P. 2008. The sustainability of ethanol production from sugarcane. *Energy Policy*, 36, 2086–2097.

Macedo, I.C, Seabra, J.E.A., and Silva, J.E.A.R. 2008. Green house gases emissions in the production and use of ethanol from sugarcane in Brazil: The 2005/2006 averages and a prediction for 2020. *Biomass and Bioenergy*, 32(7), 582–595.

Nastari, P.M. 2012a. When will Brazil be ready for another expansion spree? *Sixth ISO DATAGRO New York Sugar & Ethanol Conference*, The challenge of meeting global demand [for sugar & ethanol], The Starlight Roof of the Waldorf Astoria Hotel, New York, May 16, 2012, 53pp.

Nastari, P.M. 2012b. Sugar and ethanol production—Where Brazil Stands in 2020? *International Sugar Journal*, 2, 10. Online: https://www.agra-net.com/portal2/isj. Accessed on January 5, 2014.

RFA. 2011. World fuel ethanol production. Retrieved November 15, 2011, from Renewables Fuel Association: http://ethanolrfa.org/pages/World-Fuel-Ethanol-Production.

Smeets, E.M., Faaij, A.P., Lewandowski, I.M., and Turkenburg, W.C. 2007. A bottom-up assessment and review of global bio-energy potentials to 2050. *Progress in Energy and Combustion Science*, 33, 56–106.

United Nation Conference on Trade and Development (UNCTAD). 2008. *Making Certification Work for Sustainable Development: The Case of Biofuels*. Geneva, Switzerland: UNCTAD.

* Personal information S Coelho (suani@iee.usp.br). Mid Term Reviewer. http://cogen.unep.org/.

10 Crop Residues as an Energy Feedstock

Availability and Sustainability

Dong Jiang, Dafang Zhuang, and Yaohuan Huang

CONTENTS

10.1 INTRODUCTION

Biomass energy occupies a significant status in the world's energy consumption and in the fight against climate change. Sustainable bioenergy production can reduce energy poverty, contribute to rural development, and avoid the negative environmental impacts such as greenhouse gas emissions by fossil energy (Wicke et al. 2011). Although bioenergy accounts for only 10% of global energy consumption at present, the potential of bioenergy will be very great in the near future. A report from World Bioenergy Association (WBA) in 2010 stated that the reasonable and sustainable utilization of global biomass energy could meet global energy demand (World Bioenergy Association 2010). The US Biomass Roadmap by the US Department of Energy (US DOE) set forth a goal that, by the year 2030, biomass will supply energy approximately equivalent to 30% of the current petroleum consumption (Han et al. 2011). The European Commission (EC) has set mandatory targets for an overall share of 20% renewable energy and a 10% share of renewable energy in transport in the European Union (EU)'s consumption in 2020 (van Dam and Junginger 2011). For Romania, the target set by the Renewable Energy Directive on the promotion of the use of energy from renewable sources for 2020 is a 24.0% share of energy in gross final consumption (Scarlat et al. 2011). National estimates of the fraction of Australia's petrol usage that could be replaced by biofuels range from 10% to 140% (Herr and Dunlop 2011).

Rising energy price and emerging new technologies have renewed people's interest in converting crop residues to energy products (Scarlat et al. 2010; Valdez-Vazquez et al. 2010). Conversion of forest and agricultural residues to biomass feedstock for electricity generation and district heating is becoming a common form of bioenergy (van Dam and Junginger 2011). Crop residues such as the straws of cereal and corn, which can be converted into liquid or gaseous biofuels by thermochemical or biological techniques, are potential resources for the development of bioenergy, especially for rural areas in developing countries (Elmore et al. 2008). In addition to the primary agricultural crops residues such as straw or husks, there are several other types of residues, including organic material from excess production, animal manures, and slurries. However, these supplies are significantly less than the primary agricultural crops residue. In this chapter, we focus on the primary agricultural crops residues.

The increasing role of biomass in future energy supply requires the use of all available resources in a sustainable way, without causing directly or indirectly negative impacts (Scarlat et al. 2011). Some recent researches indicated that excessive straw reaping might have adverse impact on soil, environment, and crop yield. Thus, reasonable reaping (removal) ratio is needed for sustainability purpose. For example, Andrews et al. (Andrews 2006) discussed the effect of straw reaping on soil quality, pointing out that sustainable straw reaping rates vary depending on factors such as management, yield, and soil types. The US Conservation Technology Information Center (CTIC) provides the assessments of conservation tillage practices and pointed out that the nonconservation tillage (intensive/conventional and reduced tillage) leaves less than 30% crop residue cover (CRC), while conservation tillage leaves more than 30% CRC (Zheng et al. 2012). A study sponsored by the National Renewable Energy Laboratory (NREL) and the US DOE showed that the use of corn stover as a feedstock for ethanol production in Iowa, United States, could be done without adversely impacting soil health while providing significant energy and greenhouse gas benefits (Sheehan et al. 2003; Powers et al. 2011).

10.2 AVAILABILITY OF CROP RESIDUES

10.2.1 Estimation of the Total Amount of World Crop Residue Production

The accurate estimation of the amount of crop residues used for bioenergy purpose is very important for the sustainable supply of biomass. The official agricultural statistics data of yield and planting area at the county or province/state level for each type of crops are firstly needed, and the crop-to-residue ratios were mainly derived from this source (Statistics 2010). The regional straw production could be calculated in terms of the yield of each crop and the corresponding crop-to-residue ratio of each species (Lal 2005):

$$R_i = G_i * r_i \tag{10.1}$$

where

R_i is the residue production of the ith crop
G_i is grain production
r_i is the straw/residue ratio of the ith crop

The annual yields data of crops, including rice, wheat, maize, beans, tubers, oil-bearing crops (peanut, sesame), cotton, fiber crops, and sugar crops (sugarcane, sugar beet), can be derived from FAO statistical datasets (FAO 2012) (Figures 10.1 and 10.2, Table 10.1). Crop residues production for each country could be calculated with Equation 10.1. The results of the year 2010 were listed in Table 10.2.

As shown in Table 10.2, the total amount of crop residues of the world was estimated at 4.46 billion tons in 2010. The increasing rate of world crop residue production is approximately 18.8% in the recent 10 years, which is almost twice of the increasing rate from 1991 to 2001 (9.9%, presented by Lal 2005). The residues of cereals, legumes, oil crops, sugar crop, and tubers account for 72.6%, 8.8%, 5.2%, 11.0%, and 2.4%, respectively. Among all of these crop types, rice yielded the largest amount of residues (1.02 billion tons). Wheat and maize yielded about 0.98 and 0.84 billion tons of fibrous residues.

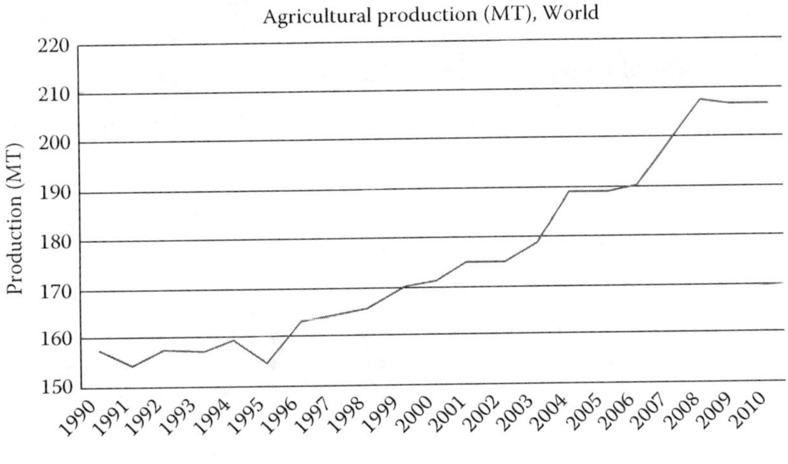

FIGURE 10.1 Temporal variations of world crop yield from 2000 to 2010.

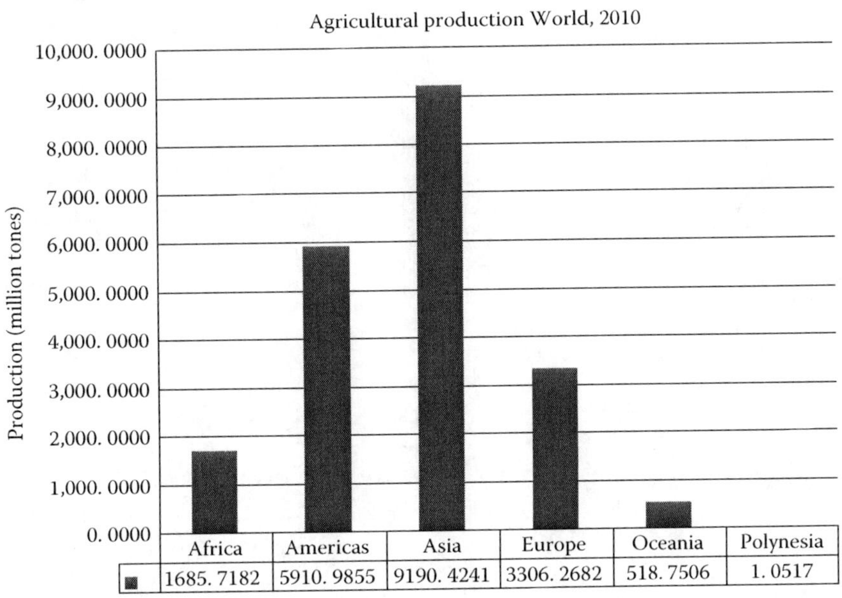

FIGURE 10.2 World crop yield of 2010.

Production of crop residues from individual cereals and certain groups of crops is given separately for each continent. Asia has the largest amount of crop residue products (2.14 billion tons). The main source of crop is rice, followed by wheat and maize. America is the leading region in the production of residues from maize (0.45 billion tons), soybeans, and sugarcane, and the total residue production (1.28 billion tons) ranked second in the world. Europe leads the residue fields from barley, rye, and sugar beet.

10.2.2 Estimates of Crop Residue Production in Selected Countries

Increasing attention has been paid to the quantification of bioenergy potentials of crop residues. Ruane et al. (2010) provided an overview of the current status of bioenergy development globally,

TABLE 10.1
Annual World Crop Yield Production in 2010 (10^4 Tons)

Crop	Africa	Americas	Asia	Europe	Oceania	World
			Grains			
Barley	668	1,603	1,975	7,349	760	12,354
Maize	6,358	44,534	24,575	8,510	53	84,031
Millet	1527	27	1,567	33	4	3,158
Oats	20	508	98	1,195	142	1,962
Rice	2,298	3,697	63,184	432	21	69,632
Rye	7	248	167	2,267	54	2,572
Sorghum	2,111	2,251	980	71	160	5,572
Wheat	2,210	11,273	29,252	20,371	2258	65,365
Total	15,198	64,141	121,798	40,229	3452	244,648
			Legumes			
Beans	473	775	2,892	128	8	4,276
Broad beans	127	17	151	90	20	406
Chick peas	49	36	936	8	60	1,089
Groundnuts	1,036	313	2,443	1	3	3,795
Lentils	16	235	188	7	14	461
Peas	101	439	1,497	524	40	2,601
Pulses	106	3	152	79	1	341
Soybeans	146	22,796	3,072	479	6	26,499
Total	2,054	24,614	11,330	1,317	153	39,468
			Oil crops			
Linseed	17	73	62	41	1	193
Rapeseed	11	1,320	2,044	2,308	218	5,901
Safflower	2	24	37	0	0	64
Seed cotton	362	1,387	4,910	78	94	6,830
Sesame	143	17	272	0	0	432
Sunflower	130	428	536	1,980	4	3,078
Total	663	3,249	7,861	4,408	318	16,498
			Sugar crop			
Sugar beet	1,028	3,103	3,663	15,051	0	22,845
Sugar cane	8,959	96,384	62,410	1	3355	171,109
Total	9,987	99,487	66,073	15,052	3355	193,954
			Tubers			
Potatoes	2,233	3,966	15,250	10,812	181	32,442
Sweet potatoes	1,522	0	94	5	74	10,764
Total	3,755	3,966	15,344	10,817	255	43,206
Grand total	31,658	195,457	222,406	71,823	7533	537,774

focusing on biomass energy and the potential contribution of agricultural biotechnologies in developing countries. Jingura and Matengaifa (2008) indicated that biomass provided 47% of the gross energy consumption in Zimbabwe. Using long-term crop yields data from 1970 to 1999, the total annual yield of crop residues was estimated at 7.8 Mt/year in Zimbabwe. Valdez-Vazquez et al. (2010) discussed the potentials of bioenergy resources used in agricultural activities in Mexico. It was estimated that 75.73 million tons of dry biomass was generated from 20 types of crops in Mexico each year. The spatial distribution of crop residues from different crops was also presented. Hiloidhari and Baruah (2011) presented the spatial assessment of crop residue and its power potential at village level in Sonitpur District of Assam, India. Sixteen types of

TABLE 10.2
Annual Estimate of World Crop Residue Production in 2010 (10^4 Tons)

Crops	Africa	Americas	Asia	Europe	Oceania	World
			Grains			
Barley	1,004	2,409	2,969	11,050	1143	18,576
Maize	6,358	44,534	24,575	8,510	53	84,031
Millet	2,316	41	2,378	50	6	4,792
Oats	20	508	98	1,195	142	1,962
Rice	3,449	5,549	94,830	649	31	104,507
Rye	10	377	254	3,450	81	3,914
Sorghum	3,166	3,376	1,469	106	240	8,358
Wheat	3,318	16,919	43,904	30,575	3389	98,104
Total	19,640	73,715	170,477	55,585	5086	324,244
			Legumes			
Beans	473	775	2,892	128	8	4,276
Broad beans	127	17	151	90	20	406
Chick peas	49	36	936	8	60	1,089
Groundnuts	1,036	313	2,443	1	3	3,795
Lentils	16	235	188	7	14	461
Peas	101	439	1,497	524	40	2,601
Pulses	106	3	152	79	1	341
Soybeans	146	22,796	3,072	479	6	26,499
Total	2,054	24,614	11,330	1,317	153	39,468
			Oil crops			
Linseed	17	73	0	41	1	193
Rapeseed	16	1,980	3,065	3,462	328	8,851
Safflower	3	41	62	0	1	106
Seed cotton	543	2,080	7,365	117	141	10,246
Sesame	286	33	544	0	0	863
Sunflower	130	428	536	1,980	4	3,078
Total	993	4,635	11,634	5,601	474	23,337
			Sugar crop			
Sugar beet	259	782	924	3,795	0	5,760
Sugar cane	2,242	24,115	15,615	0	839	42,811
Total	2,501	24,898	16,538	3,795	839	48,571
			Tubers			
Potatoes	558	991	3,812	2,703	45	8,111
Sweet potatoes	383	0	24	1	19	2,711
Total	942	991	3,836	2,704	64	10,821
Grand total	26,130	128,853	213,816	69,002	6617	446,443

crop residues were identified based on remote sensing data and geographic information system (GIS). The total amount of crop residue biomass and the power potential were estimated to be 0.17 million tons and 17 MW, respectively. Simon and Wiegmann (2009) modeled sustainable bioenergy potential from agriculture for Germany and Eastern European countries. Available land and residue potential were estimated up to 2030 for Germany, Poland, the Czech Republic, and Hungary. The results show that the sustainable energy potential from agricultural biomass is strongly restricted for Germany and the Czech Republic. But in Poland and Hungary, native biomass provides a much higher potential for energy supply, even if sustainability is comprehensively considered. Scarlat et al. (2011) assessed the availability of agricultural residues for bioenergy production in Romania. The amount of agricultural residues available for bioenergy in Romania was estimated at 154.1 PJ/a on average, of which 137.1 PJ/a was from annual crop residues and 17.3 PJ residues from permanent crops. Skoulou et al. (2011) presented an assessment of the sustainability of four energy crops and their residues in Greece, including rapeseed, soya, cotton, and sunflower. The study indicated that proper use and integrated management of the previously mentioned energy crops and their residues could result in increase of the domestic raw material production for the biodiesel industry and enhance the cohesion of agriculture, energy, and environmental policies in Greece.

Scarlat et al. (2010) assessed the utilization potentials of straws for bioenergy purpose in the EU. Sustainable removal rates were considered in order to protect soil fertility. The results show large spatiotemporal variations of available residues within EU27. The average amount of crop residues in EU27 was estimated at 1530 PJ/a, with a variation between 1090 and 1900 PJ/a. According to the 2011 US Billion-Ton Update report by the US DOE, there are about 111 million dry tons of primary crop residues in the United States under baseline assumption. More than 76% of this residue is the corn stover, and the remaining 24% is wheat and other grain residue. The total crop residue will reach 180 and 320 million dry tons in the United States by 2030, under low- and high-yield scenarios, respectively (Update 2011).

China is one of the largest agricultural countries in the world. Since the 1980s, with the increase of crop yield, the total yield of straw increased rapidly (Cao et al. 2008). In the past, the majority of these straws was used as fuels in rural areas or burned in the fields as the market for straws was limited. During the recent 10 years, for the requirements of air quality protection, straw burnings have sharply reduced, and the rational utilization of the straws has been required (Ren et al. 2008). A large number of studies have been conducted on assessment of the amount and distribution of crop residues for biomass in China (Liu and Shen 2007; Ming et al. 2008; Liwei et al. 2009; Xie et al. 2010). Previous efforts were mainly based on agricultural statistical data on the provincial level. The total amount of crop residues produced in China was estimated at 600–800 million tons/year. Dong et al. presented a new and detailed estimation of crop yields in China (Dong et al. 2012). It was estimated that the total yield of crop residues in China in 2009 was about 806.9 million tons, and the average yield was 716.0 million tons in the past 10 years. The three principal agricultural residues are straws of maize, wheat, and rice that accounted for 40.6%, 24.2%, and 15.7%, respectively (Table 10.3).

TABLE 10.3
Annual Estimate of Crop Residues in China in 2009

Crops	Yield (Million Tons)	Crop Residue (Million Tons)	%
Rice	195.1	195.1	24.2
Wheat	115.1	126.6	15.7
Maize	163.7	327.4	40.6
Beans	19.3	30.0	4.1
Tubers	30.0	30.0	3.7
Oil-bearing crops	31.5	63.1	7.8
Cotton	6.4	19.1	2.4
Fiber crops	0.4	0.6	0.1
Sugar crops	121.7	12.2	1.5
Total	530.8	806.9	/

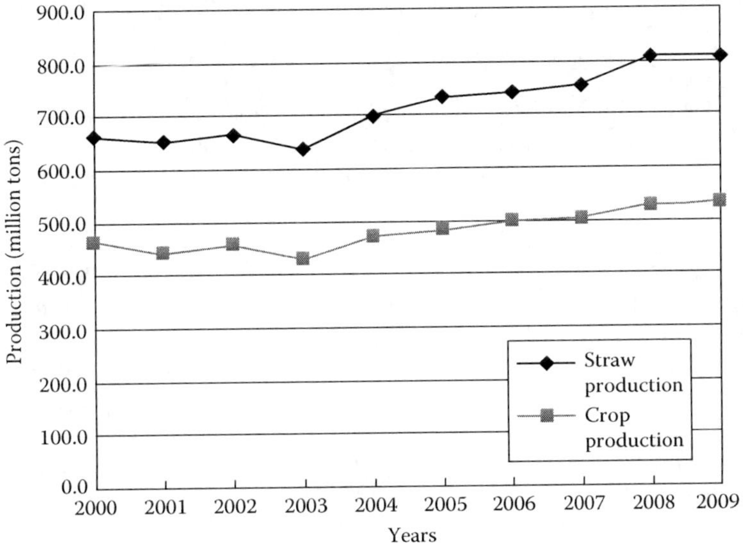

FIGURE 10.3 Temporal variations of crop yield and straws in China from 2000 to 2009.

Figure 10.3 shows the temporal variation of crop yield and straws in China from 2000 to 2009. The annual total yield of straws increased successively from about 660 million tons to more than 800 million tons in the past 10 years at an average rate of 2.1%/year.

The provincial statistical results indicate that the majority of straw productions are concentrated in Henan, Shandong, and Heilongjiang Province, with total straw yields of 8.61, 7.41, and 6.75 million tons in 2009, respectively (Figure 10.4).

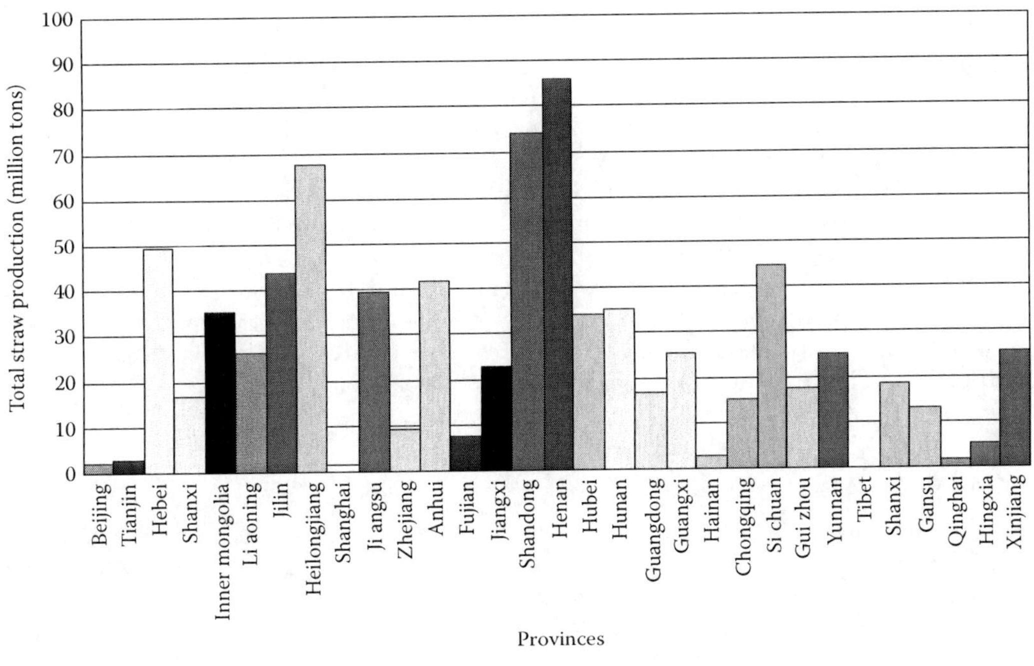

FIGURE 10.4 Total yields of straws in each province in China, 2009.

10.3 NET AVAILABILITY OF CROP RESIDUES FOR ENERGY PURPOSE

10.3.1 Net Availability of Crop Residues: A Sustainable Way

As mentioned earlier, not all straws are available for biomass use because some of them should be returned to the soil as residue to prevent erosion and enhance soil quality (Bi et al. 2008; Shi 2008). The previous methods remarkably overstated the crop residue yield since the part of cropland returning and reaping loss was not considered (Bi et al. 2008; Liu et al. 2008), and this part of straws should be left in the field to meet all the environmental and farming requirements (Andrews 2006; Blanco-Canqui and Lal 2009; Scarlat et al. 2010). Net available crop residue equals that of the total crop residue yield minus the amount of crop residues returning to the fields for conservation purpose.

The recommended amount of straws returning to fields depends on local climate conditions, soil corrodibility, nutrient demands, slopes of land, and governmental policies. A great deal of studies has presented that estimates on the sustainable removal rate of straws vary between 30% and 70% (Lemke et al. 2010; Statistics 2010; Zheng et al. 2010). Scarlat et al. (2010) used the sustainable reaping rate of 40% for wheat, rye, and barley and 50% for maize, rice, and sunflower, during the estimate of available straws in the EU. The National Crop Residue Management (CRM) Survey is the only survey in the United States to measure at the county level the type of tillage used by crop. The crops include corn, small grain (spring-seeded and fall-seeded), soybeans, and cotton. The ways of tillage were divided into two types: conservation tillage and other tillage. Conservation tillage is any cropland system that leaves at least 30% of the soil covered with crop residue after planting (CTIC 2010). Banowetz et al. (2008) assessed the availability of straw biomass energy in the Northwest Pacific in the United States, including Idaho, Oregon, and Washington. County-scale US Department of Agriculture (USDA) National Agricultural Statistical Service (NASS) data were used to quantify total grass seed and cereal straw production in each county. Over 15 million tons of straw is produced in those regions and approximately 5.73 million tons is available for bioenergy purpose after sufficient straw is returned to the soil for conservation.

In China, Wang et al. (2010) pointed out that 18.52% of straws were returned to the fields in China in 2005, and the recommended average sustainable reaping rate of straws was about 65%–71% in China. In many recent studies and projects, the sustainable straw reaping rate in China was determined according to requirements of National Fertile Soil Project launched in 2005 by China's Ministry of Agriculture (Li et al. 2006; Statistics 2010). A pixel-by-pixel analysis could be performed to obtain sustainable straw reaping rate, and net available straws are then calculated based on soil-type map at 1:100,000 scale, crop-zoning map, and information on crop types.

The cost of transportation, a major obstacle to the economic conversion of straws into energy, should also be considered in this study. Some low-density straws that are far away from road networks are considered unavailable for energy use and are thus dismissed in further statistics. With the GIS software, the aforementioned straw data layer (in raster format) is overlaid with the national road-network data layer (in vector format). Straws in pixels with a distance to main road greater than 15–20 km, according to the updated literatures (Bi et al. 2008; Dong et al. 2012), are excluded using the overlay and buffering functions of the GIS software.

10.3.2 Net Availability of Crop Residues and Its Spatial Distribution

A detailed assessment of spatial distribution of crop residues for different regions is needed for regional bioenergy planning (Gan and Yu 2008). Based on the GIS and remote sensing techniques, the spatial distribution of crop residues could be simulated. Besides information of the total amount of crop residues, several types of spatial datasets are needed: land-use data, road data, soil-type data, etc. Based on this map and the planting information, the yield of straws for each geographic unit (e.g., 100 m × 100 m pixel) was assigned according to the proportion of farmland area to each pixel.

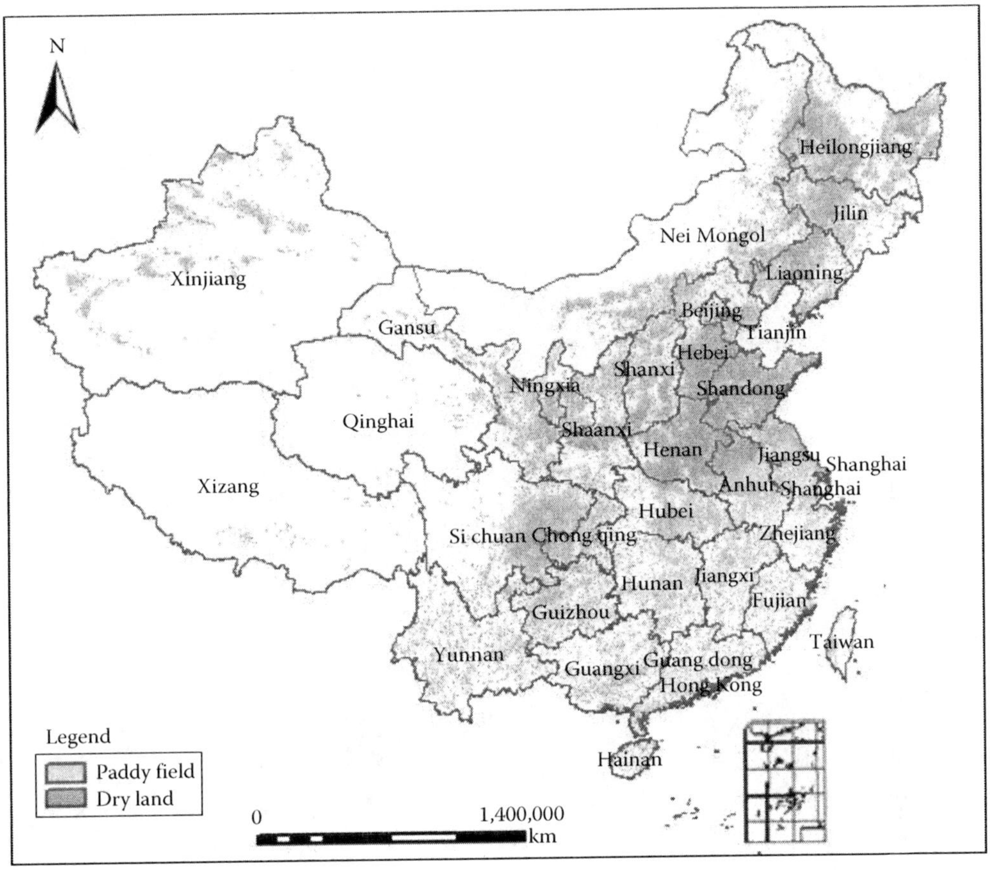

FIGURE 10.5 (Color figure at CRC website http://www.crcpress.com/product/isbn/9781466505520.) The map of farmland distribution in China, 2008.

Thus, the analysis on net availability could be performed by combining these data with other data such as road and soil-type data for further processing.

Figure 10.4 shows that the map of farmland distribution was derived from a national database of land use at a 1:100,000 scale, provided by the Resources and Environmental Sciences Data Center (RESDC) of the Chinese Academy of Sciences (Figure 10.5).

Based on the total yield of straws and the consideration of sustainability (sustainable reaping rate, cost analysis, etc.), net availability of straws for bioenergy purpose is then estimated. The estimates show that the amount of straws available for bioenergy production was about 505.5 million tons in China in 2009, with the average of 451.3 million tons/year in the past 10 years. For the convenience of further processing and analysis, the results were spread onto 100 m × 100 m pixels, making its spatial distribution much clearer than the map based on statistical data at the province or county level (Figure 10.6).

Figure 10.6 shows a significant regional unbalance of net availability of straws in China. The majority of crop residues are concentrated in the two principal grain-producing areas, which are the Huang-Huai-Hai region and Northeast Plain. Within the Huang-Huai-Hai region, crop residues are mainly distributed in the northern and the eastern part of Henan Province, western part of Shandong Province, and central belt region of Hebei Province. The Huang-Huai-Hai region is the principal grain-producing area of China, where wheat straws and maize straws are main productions. In the northern part of Anhui and Jiangsu Provinces, which are adjacent to the aforementioned areas, the yield of straws is also relatively higher. The main crop residues are straws of winter wheat, maize, and rice. In the Northeast Plain, the regions with higher density of crop residues per unit area are

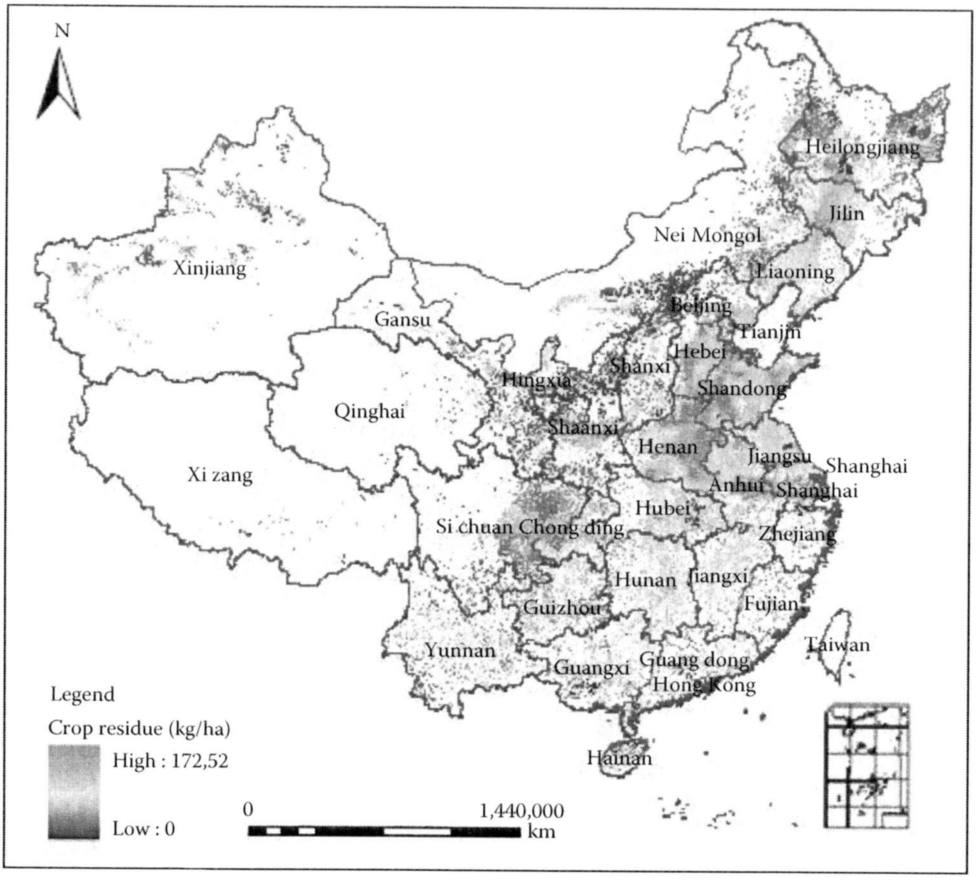

FIGURE 10.6 (Color figure at CRC website http://www.crcpress.com/product/isbn/9781466505520.) Spatial distributions of net crop residues in China, 2009 (100 m × 100 m).

located at the belt stretching from the north of Liaoning Province to the south of Heilongjiang Province, with main crop residues of straws of maize and rice. The crop residue resources are quite abundant in Sichuan Province and Hubei Province with a lower density of crop residues per unit area compared to the aforementioned regions.

10.3.3 Availability of Crop Residues for Bioenergy Purpose

Currently, crop residues can be converted to energy productions by using techniques such as combustion (direct and cofiring) and noncombustion methods (thermochemical and biochemical). The conversion efficiencies of different types of techniques vary and can be obtained from published literatures (Li 2006; Sha et al. 2010). Based on net biomass yield from straws and conversion efficiency (thermal values of each type of straws), the energy potential of crop residues, taking China as a case in point, could be estimated (Table 10.4).

The estimates indicate that the total bioenergy potential from straws in China reached 253.7 million tons standard coal (7.4 EJ) in 2009, which accounted for about 8.3% of total energy consumption of the country. According to the national mid- and long-term planning of renewable energy development, biomass energy is expected to account for about 4.0% of total energy consumption in China by 2020. In China, biomass energy resource mainly consists of crop residues, forest residues, grassland biomass, etc. Crop residues will play more and more important role in energy supply in the near future.

TABLE 10.4
Annual Estimates of Energy Potentials of Crop Residues in China in 2009

Crop	Crop Residue (Million Tons)	Conversion Ratio[a]	Equivalent Standard Coal (Million Tons)	Energy Potential (EJ)[b]
Rice	103.8	0.429	44.5	1.303
Wheat	75.6	0.500	37.8	1.106
Maize	223.0	0.529	117.9	3.452
Beans	20.5	0.543	11.1	0.326
Tubers	19.2	0.486	9.3	0.274
Oil-bearing crops	41.1	0.529	21.7	0.636
Cotton	13.3	0.543	7.2	0.211
Fiber crops	0.4	0.521	0.2	0.006
Sugar crops	8.7	0.441	3.8	0.112
Total	505.5		253.7	7.426

Notes: [a] Conversion ratio stands for the ratio from each type of crop residue to the equivalent standard coal (Andrews 2006).

[b] Lower heating value of standard coal is used for 29.27 MJ/kg (Blanco-Canqui and Lal 2009).

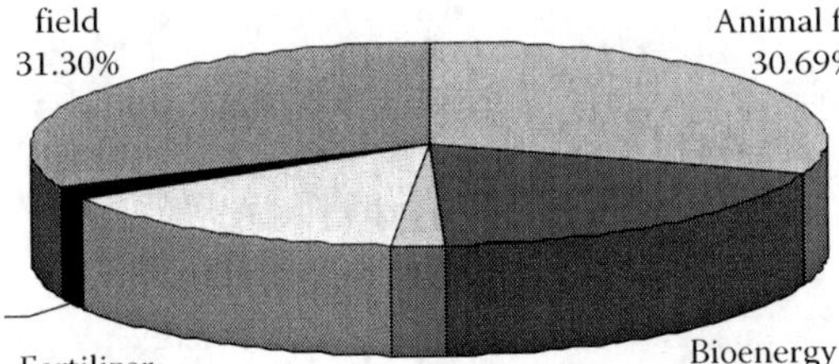

FIGURE 10.7 Utilization of crop residues in China (Data source: China's Ministry of Agriculture).

10.3.4 Competitive Use of Crop Residues

The net crop residues are available for bioenergy purpose. However, this type of resources is used for several alternative purposes currently, such as animal feed, industry materials, fertilizer, and mushroom production.

The utilization of crop residue in China is illustrated in Figure 10.7. It shows that 31.30% of crop residues were left in the field in China in 2009, 30.69% of crop residues were used as animal fodders, and only 18.72% of crop residues were used for bioenergy purpose. In recent years, the percentage of the straws burned directly in the fields has decreased dramatically in China. In general, crop residue resources have not been effectively exploited for the supply of bioenergy. Residue resources play an increasingly significant role in sustainable development of agriculture and economy in rural areas, and the comprehensive utilization of crop residue resources is still a challenging task.

10.4 CONCLUSIONS

Crop residues such as the straws of cereal and corn, which can be converted into liquid or gaseous biofuels by thermochemical or biological techniques, are potential resources for the development of bioenergy. With the development of agriculture, there are abundant crop residue resources in the world. The total amount of crop residue of the world was estimated at 4.46 billion tons in 2010. The increasing rate of world crop residue production was approximately 18.8% in the recent 10 years, which was almost twice of the increasing rate from 1991 to 2001. For sustainable development of bioenergy, the assessment of the current available crop residues should be conducted taking into account a number of issues: resources (total amount, spatial, and temporal distribution), economy influence (transportation costs), and environmental and technological conditions.

NOMENCLATURE

R_i: Residue production of the ith type of crop, kg
G_i: Grain production, kg
R_i: Straw/residue ratio of the ith type of crop

REFERENCES

Andrews, S. S. 2006. Crop residue removal for biomass energy production: Effects on soils and recommendations. White paper, USDA Natural Resources Conservation Service. Available from http://soils.usda.gov/sqi/management/files/AgForum_Residue_White_Paper.pdf.

Banowetz, G. M., A. Boateng, J. J. Steiner et al. 2008. Assessment of straw biomass feedstock resources in the Pacific Northwest. *Biomass & Bioenergy* 32: 629–634.

Bi, Y. Y., D. L. Wang, and C. Y. Gao. 2008. *Straw Resources Evaluation and Utilization in China*. Beijing, China: Agricultural Science and Technology Press.

Blanco-Canqui, H. and R. Lal. 2009. Crop residue removal impacts on soil productivity and environmental quality. *Critical Reviews in Plant Sciences* 28: 139–163.

Cao, G. L., X. Y. Zhang, Y. Q. Wang, and F. C. Zheng. 2008. Estimation of emissions from field burning of crop straw in China. *Chinese Science Bulletin* 53: 784–790.

CTIC. 2010. National survey of conservation tillage practices. West Lafayette, IN: Conservation Technology Information Center. Available from http://www.conservationinformation.org/pdf/TillageDefinitions.pdf. Retrieved on February 2010.

Dong, J., Z. Dafang, F. Jinying, and H.Yaohuan. 2012. Bioenergy potential from crop residues in China: Amount and availability. *Renewable & Sustainable Energy Reviews* 16:1377–1382.

Elmore, A. J., X. Shi, N. J. Gorence et al. 2008. Spatial distribution of agricultural residue from rice for potential biofuel production in China. *Biomass & Bioenergy* 32: 22–27.

FAO. 2012. FAOSTAT. Available from http://faostat.fao.org/site/339/default.aspx. Accessed on 1 June, 2012.

Gan, L. and J. Yu. 2008. Bioenergy transition in rural China: Policy options and co-benefits. *Energy Policy* 36: 531–540.

Han, F. X. X., R. L. King, J. S. Lindner et al. 2011. Nutrient fertilizer requirements for sustainable biomass supply to meet US bioenergy goal. *Biomass & Bioenergy* 35: 253–262.

Herr, A. and M. Dunlop. 2011. Bioenergy in Australia: An improved approach for estimating spatial availability of biomass resources in the agricultural production zones. *Biomass & Bioenergy* 35: 2298–2305.

Hiloidhari, M. and D. C. Baruah. 2011. Crop residue biomass for decentralized electrical power generation in rural areas (part 1): Investigation of spatial availability. *Renewable & Sustainable Energy Reviews* 15: 1885–1892.

Jingura, R. M. and R. Matengaifa. 2008. The potential for energy production from crop residues in Zimbabwe. *Biomass & Bioenergy* 32: 1287–1292.

Lal, R. 2005. World crop residues production and implications of its use as a biofuel. *Environment International* 31: 575–584.

Lemke, R. L., A. J. VandenBygaart, C. A. Campbell, G. P. Lafond, and B. Grant. 2010. Crop residue removal and fertilizer N: Effects on soil organic carbon in a long-term crop rotation experiment on a Udic Boroll. *Agriculture Ecosystems & Environment* 135: 42–51.

Li, W. G., Q. Li, and X. X. He. 2006. Progress in the study on returning crop stalks to the field. *Hunan Agricultural* 1: 46–48.

Li, Y. Y. 2006. *Feed Processing Techniques of Stalks of Agricultural Crops*. Beijing, China: Light Industry Press.

Liu, G. and L. Shen. 2007. Quantitative appraisal of biomass energy and its geographical distribution in China. *Journal of Natural Resources* 22: 9–19.

Liu, H., G. M. Jiang, H. Y. Zhuang, and K. J. Wang. 2008. Distribution, utilization structure and potential of biomass resources in rural China: With special references of crop residues. *Renewable & Sustainable Energy Reviews* 12: 1402–1418.

Liwei, G., M. Lin, Z. WeiFeng et al. 2009. Estimation of nutrient resource quantity of crop straw and its utilization situation in China. *Transactions of the Chinese Society of Agricultural Engineering* 25: 173–179.

Ming, C., Z. Lixin, T. Yishui et al. 2008. Analysis and evaluation on energy utilization of main crop straw resources in China. *Transactions of the Chinese Society of Agricultural Engineering* 24: 291–296.

Powers, S. E., J. C. Ascough, R. G. Nelson, and G. R. Larocque. 2011. Modeling water and soil quality environmental impacts associated with bioenergy crop production and biomass removal in the Midwest USA. *Ecological Modelling* 222: 2430–2447.

Ren, H., Z. A. Li, Q. F. Guo, and Q. Wang. 2008. Bioenergy: Future direction of China's energy and environment integrated strategy. *Ambio* 37: 136–138.

Ruane, J., A. Sonnino, and A. Agostini. 2010. Bioenergy and the potential contribution of agricultural biotechnologies in developing countries. *Biomass & Bioenergy* 34: 1427–1439.

Scarlat, N., V. Blujdea, and J. F. Dallemand. 2011. Assessment of the availability of agricultural and forest residues for bioenergy production in Romania. *Biomass & Bioenergy* 35: 1995–2005.

Scarlat, N., M. Martinov, and J. F. Dallemand. 2010. Assessment of the availability of agricultural crop residues in the European Union: Potential and limitations for bioenergy use. *Waste Management* 30: 1889–1897.

Sha, H., S. Tong, W. Zhang et al. 2010. Analysis on current situation of producing and comprehensive utilization of stalks of agricultural crops. *Journal of Jilin Agricultural Science* 4: 51–55.

Sheehan, J., A. Aden, K. Paustian et al. 2003. Energy and environmental aspects of using corn stover for fuel ethanol. *Journal of Industrial Ecology* 7: 117–146.

Shi, Y. C. 2008. *China Renewable Energy Development Strategic Research*. Beijing, China: Electric Power Press.

Simon, S. and K. Wiegmann. 2009. Modelling sustainable bioenergy potentials from agriculture for Germany and Eastern European countries. *Biomass & Bioenergy* 33: 603–609.

Skoulou, V., N. Mariolis, G. Zanakis, and A. Zabaniotou. 2011. Sustainable management of energy crops for integrated biofuels and green energy production in Greece. *Renewable & Sustainable Energy Reviews* 15: 1928–1936.

Statistics, D. o. I. a. T. 2010. *China Energy Statistics Yearbook 2009*. Beijing, China: China Statistics Press.

Update, U. S. B. T. 2011. *Biomass Supply for a Bioenergy and Bioproducts Industry*. Washington, DC: US Department of Energy.

Valdez-Vazquez, I., J. A. Acevedo-Benitez, and C. Hernandez-Santiago. 2010. Distribution and potential of bioenergy resources from agricultural activities in Mexico. *Renewable & Sustainable Energy Reviews* 14: 2147–2153.

van Dam, J. and M. Junginger. 2011. Striving to further harmonization of sustainability criteria for bioenergy in Europe: Recommendations from a stakeholder questionnaire. *Energy Policy* 39: 4051–4066.

Wang, Y. J., Y. Y. Bi, and C. Y. Gao. 2010. The assessment and utilization of straw resources in China. *Agricultural Sciences in China* 9: 1807–1815.

Wicke, B., E. Smeets, H. Watson, and A. Faaij. 2011. The current bioenergy production potential of semi-arid and arid regions in sub-Saharan Africa. *Biomass & Bioenergy* 35: 2773–2786.

World Bioenergy Association. 2010. Certification criteria for sustainable biomass for energy. Available from http://www.worldbioenergy.org/sites/default/files/Certification%20Criteria%20for%20Sustainable%20Biomass%20for%20Energy_Position%20Paper.pdf. Accessed on 1 September, 2012.

Xie, G., X. Wang, and L. Ren. 2010. China's crop residues resources evaluation. Sheng wu gong cheng xue bao. *Chinese Journal of Biotechnology* 26: 855.

Zheng, B., J. B. Campbell, and K. M. de Beurs. 2012. Remote sensing of crop residue cover using multi-temporal Landsat imagery. *Remote Sensing of Environment* 117: 177–183.

Zheng, Y. H., Z. F. Li, S. F. Feng et al. 2010. Biomass energy utilization in rural areas may contribute to alleviating energy crisis and global warming: A case study in a typical agro-village of Shandong, China. *Renewable & Sustainable Energy Reviews* 14: 3132–3139.

11 Woody Biomass as an Energy Feedstock

Availability and Sustainability

Birger Solberg

CONTENTS

11.1 INTRODUCTION

Wood is considered as humankind's very first source of energy. Today, it still is the most important single source of renewable energy providing over 9% of the total global primary energy supply. Wood energy is as important as all other renewable energy sources altogether (hydro, geothermal, wastes, biogas, solar, and liquid biofuels). More than two billion people depend on wood energy for cooking and/or heating, particularly in households in developing countries, where it often represents the only domestically available and affordable source of energy. Private households' cooking and heating with wood fuels represents one-third of the global renewable energy consumption, making wood the most decentralized energy in the world (FAO 2010c).

Wood fuels arise from multiple sources including forests, other wooded land and trees outside forests, coproducts from wood processing, postconsumer recovered wood, and processed wood-based fuels. Wood energy is also an important emergency backup fuel, and people at any socio-economic level will switch easily back to wood energy when encountering economic difficulties, natural disasters, conflict situations, or fossil energy supply shortages (FAO 2010c). Wood-based energy can be produced in many ways such as fuel wood, pellets, biodiesel, bioethanol, and electricity. Wood energy can also be used in combination with coal and gas in central heating and central heat and power plants.

The increased use of bioenergy from forests has implications for existing industries and land uses, and the question of what is sustainable bioenergy production has been an important topic both in the energy and global climate change arenas. During the last decades, we have witnessed a strong increase in the demand for wood-based bioenergy. Three factors have been the main driving forces here: first, to reduce the emission of greenhouse gases (GHG) to the atmosphere; second, to meet a steadily increasing energy demand; and, third, to provide income in rural areas. This has raised at least two main questions: first, how much wood resources are available in total and, second, what is the role of sustainability here?

This chapter addresses these questions by, first, giving a brief overview of the global forest resources. Then, the question of sustainable wood-based energy is discussed, and finally conclusions are drawn. The presentation takes a global and generic view. For more detailed analyses, information, and statistic analyses, the readers can refer to Chum et al. (2011), FAO (2001, 2007, 2010a–c), Dam et al. (2011), and FAOSTAT.

11.2 FOREST RESOURCES AND THEIR USE AT PRESENT: AN OVERVIEW

11.2.1 Forest Areas and Plantations

Table 11.1 gives an overview of the global forest resources. Forests cover just over 4.0 billion ha or 31% of the earth's land surface according to the FAO definition of forest lands, which have a tree crown cover of more than 10% of the land area, an area size of more than 0.5 ha, and trees with a minimum height of 5 m at maturity in situ (FAO 2010a). Europe has the largest forest land of 1005 million ha, followed by South America of 864 million ha, North and Central America of 705 million ha, Africa of 674 million ha, Asia of 593 million, and Oceania of 191 million ha. About 40% of the world's forests are in developed countries that have only 20% of the world's population, while 60% of the forests are in developing countries with 80% of the world population. The tropical forests of Latin America are the dominant contributor in developing countries, while Asia and Africa only account for 1.7 out of the total of about 4 billion ha (FAO 2010b).

The global forest area has decreased by 0.1% each year during the period 2005–2010. The forest areas in Europe and Asia increased by 0.1% and 0.3% each year, respectively. However, the forest areas decreased by 0.5% in Africa, 0.4% in South America, and 0.6% in Oceania each year during this period. The forest area per capita is 0.60 ha globally but varies from 0.15 ha in Asia to 5.48 ha in Oceania.

The global growing stock in these forest lands was 527 billion m^3 over bark in 2010, which included 177 billion m^3 in South America, 112 billion m^3 in Europe, 86 billion m^3 in North and Central America, 77 billion m^3 in Africa, 54 billion m^3 in Asia, and 21 billion m^3 in Oceania. Growing stock per capita was 131 m^3 on average globally, varying from 91 m^3 in Asia to 205 m^3 in South America.

TABLE 11.1
Estimated Area of Forest and Other Wooded Land and Volume of Growing Stock

	World	Europe	Asia	Africa	North and Central America	South America	Oceania
Forest area[a] (million ha)	4033	1005	593	674	705	864	191
Annual change in forest cover from 2005 to 2010 (1000 ha per year and % per year)	−5581 −0.1	770 0.1	1693 0.3	−3410 −0.5	19 0	−3581 −0.4	−1072 −0.6
Forest area per capita (ha)	0.60	1.37	0.15	0.68	1.32	2.25	5.48
Forest area in% of total land area	31	45	19	23	33	49	23
Growing stock (billion m^3 over bark)	527	112	53.7	77	86.4	177	20.9
Growing stock per ha (m^3 over bark)	131	111	91	114	123	205	109

Source: FAO, Global forest resources assessment 2010—Main report, FAO Forestry Paper No. 163, Rome, Italy, 2010b, www.fao.org/docrep/013/i1757e/i1757e00.htm, accessed on November 2012.

[a] The following definition of forest has been used: Land with tree crown cover of more than 10% and area of more than 0.5 ha. The tree should be able to reach a minimum height of 5 m at maturity in situ.

TABLE 11.2
Area of Planted Forests by Region in 1990 and 2010

	Area of Planted Forests (million ha)		Share of Total Forest Area (%)	
Region	**1990**	**2010**	**1990**	**2010**
World	178	264	4.3	6.5
Developed countries	90.9	117.7	5.1	6.5
United States	17.9	25.4	6	8.4
Canada	1.4	9	0.5	2.9
Europe	59	69.3	6	7
Japan	10.3	10.3	41.2	41.2
Australia and New Zealand	2.3	3.7	1.4	2.3
Developing countries	86.8	146.1	3.6	6.5
China	42	77.2	26.8	37.3
West–Central and East Asia (w/o China)	7.5	9.7	10.9	14.7
South and Southeast Asia	16.5	25.6	5.1	8.7
Africa	11.7	15.4	1.6	2.3
Latin America and Caribbean	9.1	18.2	0.9	1.9

Source: FAO, Global forest resources assessment 2010—Main report, FAO Forestry Paper No. 163, Rome, Italy, 2010b, www.fao.org/docrep/013/i1757e/i1757e00.htm, accessed on November 2012.

Planted forests become increasingly important to supply timber. Table 11.2 gives an overview of the development of planted forests and the share of total planted forest by region and country since 1990. The total area of planted forests around the world was 264 million ha in 2010, which was increased by 52% from 1990 to 2010. The developing countries, particularly China and Southeast Asia, have a large increase in the planted area, whereas the increase in developed countries is much less in terms of planted area but the percentage of increase is high. The areas of planted forests were increased by 22% in developed countries and 69% in developing countries. It is also seen that the share of planted forest area in Japan is as high as 41.2%, followed by China at 37.3% and West–Central Asia at 14.7%.

Although the planted forests are currently only about 6.5% of total global forest area, planted forests represent a major potential to increase industrial wood supply through the introduction of intensive management of high-yielding species and high genetic quality. Planted forests represent an important component of industrial timber production in the temperate zone and are a source of both industrial timber and nonindustrial products in the tropical zone. However, there are some challenges to estimate the total area of planted forests. There are various definitions of planted forests. Even though there is a unified definition available, it cannot be universally applicable. For temperate zone countries, the term planted forests is commonly used for forests that are established using artificial regeneration, in some cases using genetically improved plantation stock. The management of planted forests is generally more intensive than that of natural forests (e.g., shorter rotations, higher planting densities). In planted forests, the number of species is reduced, sometimes to a single one. A planted forest may be structurally and biologically less complex than natural forests in the same region.

Temperate zone planted forests are predominantly coniferous species, most of which are managed for industrial wood products as the consequence of reforestation rather than afforestation. Therefore, these plantations generally do not expand the existing forest area. Native species account for the majority of the planted forest area in most of the northern hemisphere temperate zone countries.

Growth rates of planted forests may be somewhat higher than those of natural forests as a result of the use of improved planting stock, or more intensive management practices. In the temperate zone of the southern hemisphere, most planted areas are fast-growing and exotic species with a shorter production cycle than native species. In the past decade, the production of wood from exotic softwood plantations in New Zealand, Chile, and Australia has increased significantly.

11.2.2 Wood Production and Consumption

Table 11.3 shows the harvest of round and fuel wood and the production of forest products in 1990, 2005, and 2010. The harvest of industrial wood and fuel wood was rather stable over time. The production of sawn wood was also rather stable in both developed and developing countries during this period. The total amount of wood harvested was 3.3 billion m^3 around the world in 1995, increasing to 3.6 billion m^3 in 2005 and then decreasing to 3.4 billion m^3 in 2010. The production of both panels and paper was increased globally and the developing countries contributed to most of these increases. The production of wood and the manufacture of wood products contributed about $400 billion to the global economy or around 2% of global GDP (FAO 2010b). Developed countries located in the temperate zone with a significant area of coniferous forests harvest primarily coniferous species (about 70% of total harvest) and use nearly all of the timber harvested for industrial purposes (about 85%). Developing countries harvest mainly nonconiferous species (about 80% of total harvest) and use this timber primarily for fuel (80%).

The volume of wood that is used as fuel is around 1.8 billion m^3 annually, more than one-half of the total harvested volume of timber. The wood contributes to 5% of the total world energy supply. Direct consumption of wood for energy in developed countries is 0.2 billion m^3. In developing countries, the fuel wood represents about 15% of their total energy use. However, the dependence on wood energy varies widely among regions and countries.

World production of industrial roundwood (1.5 billion m^3 per year in 2010) provides the raw material for the production of sawn wood, wood-based panels, and paper. Consumption has a similar distribution, and the level of per capita consumption reflects the level of economic well-being. Thus, the annual average consumption of sawn wood and panels was roughly about 0.3 m^3 per capita in developed countries and 0.03 m^3 in developing countries in 2000. The average consumption of paper was 150 kg per capita in developed countries and 12 kg per capita in developing countries, respectively.

TABLE 11.3
Harvest of Round Wood and Production of Forest Products in 1995, 2005, and 2010 (Billions m^3 Per Year)

	World			Developing Countries			Developed Countries		
	1995	2005	2010	1995	2005	2010	1995	2005	2010
Harvest									
Roundwood	3.3	3.6	3.4	2.1	2.2	2.2	1.2	1.4	1.2
Fuel wood	1.8	1.8	1.9	1.6	1.7	1.7	0.2	0.2	0.2
Industrial roundwood	1.5	1.7	1.5	0.5	0.5	0.5	1.1	1.2	1.0
Products									
Sawn wood	0.39	0.44	0.39	0.11	0.12	0.14	0.28	0.32	0.25
Panels	0.15	0.25	0.28	0.05	0.10	0.16	0.10	0.15	0.12
Paper (billion tons)	0.28	0.36	0.40	0.06	0.12	0.17	0.22	0.25	0.23

Source: FAOSTAT—FAO's online statistical database, Rome, Italy (available at http://faostat.fao.org).iv, accessed on November 2012.

In developed countries, about 60% of the 1.2 billion m^3 is used in manufacturing sawn wood and panels and 40% in paper manufacture. On average, 25%–30% of the wood used in manufacturing paper is supplied from wood residues from sawmills and panel mills. In addition, a very important source of fiber in the manufacture of paper is recovered paper, which currently contributes 35%–40% of fiber. The extension of the utilization of wood raw material by better recovery of wood from the forest, recovery of residues from wood manufacturing, and the recycling of used paper in paper manufacture have been important features in the evolution of the wood products sector over the past five decades.

11.3 CHALLENGE OF SUSTAINABLE WOOD-BASED ENERGY

There is some wood available that potentially could be used for bioenergy. But such utilization has to meet at least two challenges. The first challenge is that the use of wood for bioenergy will compete with other usages of the wood fiber—notably traditional fuel wood, sawn wood, and wood-based boards (particleboard, fiberboard, etc.)—used in the building and construction industry, and in particular the production of pulp and paper products. Secondly, it has to be sustainable. The importance of sustainability in biomass production and utilization has increased during the last years, which is discussed in Chapter 5. This chapter will focus on the special aspects that are inherent in relation to woody biomass for energy.

The first aspect to be emphasized is that sustainable wood-based bioenergy consumption and production are closely linked—sustainable bioenergy consumption requires sustainable bioenergy production. The second aspect is that wood-based bioenergy production has to be based on sustainable land use—that is, forest management should be sustainable. This leads directly to the question: what does the sustainability mean? The sustainability was introduced to the forestry more than 200 years ago. The sustainability in the forestry meant that the harvest today should not reduce future harvest possibilities—in fact, it meant that the future harvest possibilities should be the same as or higher than the present.

Centuries later, the Brundtland Report (WCED 1987) introduced the same concept of sustainability but now for sustainable development. Several definitions are presented, but the following is often referred to "Sustainable development is the development which covers the needs of the present generation without reducing the possibilities for the future generations to meet their needs." As such, sustainable development is often defined to have three pillars—ecological, economic, and social. In practice, to be operational, each of these pillars has to have specific criteria and indicators (C&I) for deciding the degree of sustainability or making it possible to determine whether one is moving toward the improvement or reduction of sustainability.

In forestry, such C&I have been discussed intensively, and it is a common understanding now that sustainable forestry has to not only secure volume sustainability but also consider

- Soil quality and nutrients as the resource base for woody biomass production
- Water as scarce resource under global change
- Biodiversity
- GHG impacts
- Recreation, wildlife, hunting
- Economic viability
- Risks (on all the aforementioned factors)

To estimate the impacts on these criteria, indicators have to be widely acceptable and measurable. This requires science-based silviculture, learning from recent research and practice, developing appropriate decision support systems, and monitoring. Several certification schemes have been developed, as reviewed by e.g. Dam et al. (2008).

Regarding wood-based bioenergy, there is a growing concern that in addition to secure a sustainable supply of wood fiber for bioenergy, it is equally important to secure a sustainable energy production from the wood fiber. This means, for example, to stimulate cascade use—that is, utilize the fiber optimally in linked production chains in order to keep the carbon as long as possible out of the atmosphere, provide high-energy efficiency technologies, avoid polluting gas emissions, and provide income. All of those imply that we should consider both forestry and fiber use together in various types of impact analyses such as life cycle analyses, carbon footprint analyses, and analyses of economic and social impacts.

It is also becoming increasingly understood that the choice of C&I for sustainability cannot be made without political considerations and political processes. This brings in an interesting issue about the role between politics and research. The main role of politics is to clarify the relative weights between the C&I for sustainability and provide adequate policy instruments.

The main roles of research here are at least two: (i) improve present bioenergy technologies and forest management systems and develop new technology options for both sustainable bioenergy production and consumption, and (ii) provide consistent appraisals of the impacts that each of these options have on the C&I for sustainable development and how various relative weights influence the results.

11.4 CONCLUSIONS

The bioenergy produced from forest resources is a very important forest product. Global production of fuel wood exceeds the production of industrial roundwood in terms of volume. Fuel wood and charcoal production are often the predominant uses of woody biomass in developing countries and economies in transition. Today, wood energy has entered into a new phase of high importance and visibility with climate change and energy security concerns.

Wood fuels have a large potential to increase, in particular through better utilization of planted forests and forest residues. However, to be classified as sustainable energy, both the production and consumption of the woody bioenergy have to fulfill politically decided C&I for the three main pillars of sustainable development: economic, ecological, and social. Such C&I have already been discussed and are available in many regions.

In many cases, it will be possible to find *win–win* solutions, that is, to gain on all these three pillars. But one should be aware of potential trade-offs creating conflicts, for example, trade-offs regarding social and economic indicators on one side and carbon sequestration, biodiversity, or recreational impacts on the other.

The use of woody biomass as energy feedstock is closely linked to the general utilization of forest resources, including the policy incentives implemented for securing sustainable utilization of the global forest resources.

REFERENCES

Chum, H., A. Faaij, J. Moreira, G. Berndes, P. Dhamija, H. Dong, B. Gabrielle et al. 2011. Bioenergy. In: *IPCC Special Report on Renewable Energy Sources and Climate Change Mitigation*, O. Edenhofer, R. Pichs-Madruga, Y. Sokona, K. Seyboth, P. Matschoss, S. Kadner, T. Zwickel et al. (eds.), Cambridge University Press, Cambridge, U.K./New York.

Dam, J., M. Junginger, and A.P.C. Faaij. 2011. From the global efforts on certification of bioenergy towards an integrated approach based on sustainable land use planning. *Renewable and Sustainable Energy Reviews* 14(9):2445–2472.

Dam, J. van, M. Junginger, A. Faaij, I. Jürgens, G. Beste, and U. Fritsche. August 2008. Overview of recent developments in sustainable biomass certification, *Biomass and Bioenergy* 32(8):749–780.

FAO. 2001. State of the world's forests 2001. Rome, Italy. ftp.fao.org/docrep/fao/003/y0900e/. Accessed on November 2012.

FAO. 2007. Forests and energy in OECD countries. Forest and energy working paper. 1ftp://ftp.fao.org/docrep/fao/010/k0966e/k0966e00.pdf. Accessed on November 2012.

FAO. 2010a. Criteria and indicators for sustainable woodfuels. FAO Forestry Paper No. 160. Rome, Italy. www.fao.org/docrep/012/i1673e/i1673e00.htm, Accessed on November 2012.

FAO. 2010b. Global forest resources assessment 2010—Main report. FAO Forestry Paper No. 163. Rome, Italy. www.fao.org/docrep/013/i1757e/i1757e00.htm. Accessed on November 2012.

FAO. 2010c. What woodfuels can do to mitigate climate change. Forestry Paper No. 162. Rome, Italy.

FAOSTAT—FAO's online statistical database. Rome, Italy (available at http://faostat.fao.org).iv. Accessed on November 2012.

WCED. 1987. *Our Common Future*. The World Commission on Environment and Development. Oxford University Press, Oxford, U.K.

12 Animal Wastes as an Energy Feedstock

Availability and Sustainability

Hui Wang and Lijun Wang

CONTENTS

12.1 INTRODUCTION

Confined animal-feeding operations (CAFOs) have been increased over the past decades. Without proper management, animal wastes can cause water and air pollution (Zhu, 2000; Krapac et al., 2002). Animal manure is one of the main sources of anthropogenic greenhouse gases (GHGs), mostly methane and nitrous oxide (Park et al., 2006). Animal manure also harbors pathogens that are harmful to livestock, wildlife, and people exposed to the manure (Vanotti et al., 2005).

Animal manure from small farms is usually spread onto cropland as a cheap and natural fertilizer (Choudhary et al., 1996). However, with more and more livestock being raised in industrial-style confinement facilities, the proper management of animal wastes has become an economic and environmental imperative. New technically and environmentally sound technologies are needed urgently, not only for the sustainable development of the livestock industry but also for the protection of our environment. With rising energy prices and concerns over petroleum supplies, there is an increasing interest in the conversion of animal wastes into energy products (He et al., 2000; Xiu et al., 2008). The conversion of animal waste into energy provides an opportunity for the agricultural sectors to reduce their reliance on imported fossil fuels while improving the soil, water, and air quality. This chapter is to review the availability and quality of animal manures, the technologies to convert animal manures into bioenergy products, and the economic and environmental impacts of the animal manure to energy technologies.

12.2 ANIMAL WASTES

12.2.1 Quantities and Characteristics of Livestock Wastes

Most dairy cows, chickens, pigs, and turkeys are now raised in CAFOs. A single operation can generate millions of gallons of liquid animal waste and/or large quantities of manure each year. Globally, CAFOs have increased more than fivefold in the last 50 years, and become the fastest growing industrialized model for animal production worldwide. CAFOs provide 74% of poultry, 50% of pork, and 43% of beef around the world. CAFOs are growing rapidly in many countries in Asia, Africa, and Latin America with relatively weak veterinary and public health infrastructure (Gilroyed et al., 2008).

A large quantity of animal wastes is produced as a result of the wide use of CAFOs. According to the US Department of Agriculture, confined food animals produced roughly 335 million tons of wastes (dry weight) in 2005, representing almost a third of the total municipal and industrial waste produced every year. In contrast, the amount of biosolids from residential and commercial sources, stormwater runoff, and residual compounds from wastewater treatment plants were estimated to be 7.6 million tons (dry wt.) in 2005 (USDA-ARS, 2006). The amount of manure produced by confined food animals each year is about 100 times more than that of human sewage sludge processed in US municipal wastewater plants. For instance, one dairy farm with 2,500 cows produces as much waste as a city with around 411,000 residents. The amounts of animal wastes produced in the United States are given in Table 12.1 (Carolina, 2012).

The physical and chemical properties of manure are important parameters in their conversion processes, especially the moisture content, heating value, fractions of fixed carbon and volatiles,

TABLE 12.1
Annual Energy Available in the United States from Manure, Sorted by Animal Category

Animal Type	Animal Units (Millions)	Biogas Energy per Animal Unit Day^{-1} (Thousand BTU)	Biogas Energy $Year^{-1}$ (Trillion BTU)
Fattened cattle	9.6	25.7	89.9
Milk cows	12.3	20.6	92.4
Other beef and dairy cattle	58.8	23.2	497
Swine	8.5	39.8	124
Poultry	6.1	56.0	125

TABLE 12.2
Chemical and Physicochemical Characterization of Poultry Manure

	Poultry Manure
Organic matter[a]	85.38
pH	8.8
Moisture[b]	48.69
Total nitrogen[a]	3.56
Inorganic nitrogen[a]	1.74
Ammonia nitrogen[a]	1.76
OCC/nitrogen ratio	10.89
TCC/nitrogen ratio	12.24
P_2O_5 (wt.%)	0.71
K_2O_5[a]	3.79
Paraffinic C[c]	12.4
C in OCH_3, sterols and amino acids[c]	6.7
C in carbohydrates and aliphatics with OH groups[c]	68.1
Aromatic and N-heterocyclic C[c]	3.2
Phenolic C[c]	3.1
C in CO_2H[c]	6.5
Total aliphatic C[c]	87.2
Total aromatic C[c]	6.3
Aromaticity[c]	6.7

Source: Font-Palma, C., *Energy Convers. Manage.*, 53, 92, 2012.
OCC: oxidizable carbon content, OCC=(TOMC − 15.356)/1.805.
TCC: total carbon content, TCC=(TOMC − 9.33)/1.745.
TOMC: total organic matter content, weight loss on ignition at 540°C for 16 h.
[a] In wt.% dry basis.
[b] In wt.% wet basis.
[c] Distribution of C as % of total C in poultry manure.

and ash content (Prapaspongsa et al., 2010). Animal manure is characterized by low fixed carbon and high moisture and volatile matter contents. The major organic components in animal manure are given in Table 12.2 (Font-Palma, 2012).

Nitrogen presents in animal wastes in several forms. It is continuously converted by microbial activity and changes in temperature, pH, moisture, and oxygen concentration. Poultry litter contains typically 60%–80% of nitrogen compounds in organic forms of urea, protein, and amino acids. A large percentage of organic nitrogen compounds (40%–90%) can be transformed to ammonia gas or ionized ammonia (NH_4^+) depending on environmental conditions. Ionized ammonia can then be converted to nitrate by microorganisms. Typical composition of poultry litter is shown in Table 12.3. As seen in Table 12.3, the composition of poultry litter varies significantly because of different origin and management practices of the farms. The poultry litter ash contains high concentrations of potassium (K), calcium (Ca), and phosphorous (P). The content of K in the ash is dependent on the type of bedding material used. For example, potassium content in the ash is very high, about 4%–6% if straw is used as the bed material, while wood shavings reduce the level of K to around 1.5% (Carolina, 2012).

TABLE 12.3
Composition of Poultry Litter from Different Farms

Components	Kirubakaran	Reardon	Antares Group	Whitely
HHV (MJ kg^{-1})	9.56	13.99 (dry)	14.79 (dry)	12.0
Proximate analysis (wt.%)				
Fixed carbon	10.2	14.0	9.8	8.11
Volatile matter	50.3	62.2	47.3	54.72
Moisture	8.2	25.5	27.4	10.59
Ash	28.8	23.9	15.7	26.58
Ultimate analysis (wt.%)				
C	24.84	35.6	27.22	29.09
H	1.9	4.6	3.72	5.11
O	33.76	29.8	23.1	
N	2.5	5.3	2.69	3.44
S	2.5	0.9	0.33	0.8
Cl	2.5	—	0.71	
Ash	28.8	23.9	15.7	26.58
Moisture	8.2	—	27.4	10.59
SiO_2	18.5	22.2	8.1	9.83
Al_2O_3	3.1	2.9	1.9	1.36
TiO_2	—	—	−0.2	0.06
Fe_2O_3	2.1	1.7	1.16	0.58
CaO	18.5	21.3	17.3	3.72
MgO	4.2	5.0	5.0	1.26
Na_2O	3.6	3.9	9.2	1.61
K_2O	—	—	16.3	3.34
P_2O_5	25.7	20.2	24.4	4.23
SO_3	5.2	7.0	6.7	1.91
BaO	—	—	—	—
SrO	—	—	—	—
CO_2/other	—	0.3	9.4	—
Ash fusion (°C)				
Initial deformation	1139	—	—	—
Softening temperature	1149	—	—	—
Hemispherical temperature	1161	—	—	—
Fusion temperature	1163	—	—	—

Source: Font-Palma, C., *Energy Convers. Manage.*, 53, 92, 2012.

12.2.2 Challenges in Livestock Wastes Management

12.2.2.1 Water Pollution

Because of the large amount of untreated animal waste that is disposed of on land, CAFOs have emerged as a major source of water pollution. The problem is exacerbated if operations are applied in particular locations and weather conditions, such as steep slopes, near waterways and natural drainage ways, snow-covered land, frozen land, saturated soil and/or poor soil porosity, and rainfall (Cox et al., 2005; Hodne, 2005). The US Environmental Protection Agency (EPA) (2003) has determined that 16% of more than 300,000 impaired river and stream miles in the United States are caused by animal-feeding operations. Forty percent of the US rivers and estuaries that fail to meet ambient water quality standards were caused by pathogens. In the Western United States, it is estimated that 80% of the impaired river and streams are associated with livestock.

12.2.2.2 Greenhouse Gas Emission

In the United States, livestock animals produce over one billion tons of manure annually. Manure emits two GHGs, methane and nitrous oxide. According to the standards of Intergovernmental Panel on Climate Change (IPCC), global warming potential of methane is 21 times of carbon dioxide and nitrous oxide is 310 times over a 100-year time span. Based on the data from EPA, the total GHG emission from US agricultural sector is about 536 million metric tons of carbon dioxide equivalent, among which at least 50.8 million metric tons of carbon dioxide emission equivalent comes from methane and nitrous oxide derived from livestock manure. Moreover, methane and nitrous oxide emissions from manure increased from 1990 to 2005. It is very important to find ways to decrease the emissions from manure in the agriculture production (Cuéllar and Webber, 2008).

12.2.2.3 Human Health Threat

Animal waste can emit harmful air pollutants and GHGs. Manure emits air pollutants, including ammonia, VOCs, hydrogen sulfide, and particulate matter. Many of them cause health problems in humans. In addition, some microbes and organic matters in the waste will threat human health (Cuéllar and Webber, 2008). The threats of animal manure to human health include microbial pathogens, antimicrobials, and hormones in the manure.

Pathogenic microorganisms in food-animal waste include a large number of species. *Salmonella* spp., *Campylobacter* spp., *Listeria monocytogenes*, *Cryptosporidium parvum*, *Giardia lamblia*, and *Escherichia coli* O157:H7 are the most common disease agents derived from animal-feeding operations (AFOs) (Smith and Perdek, 2004). Many of these organisms can survive from several days to several months in manure and wastewater and are often transported for long distances in the environment (Stehman et al., 1996; Hodne, 2005; Graham et al., 2009). Compared to bacteria and protozoa, there is relatively less knowledge about viruses in food-animal waste. According to some studies measuring the inactivation of animal viruses, it is known that viruses may survive a very long time (over 300 days in some cases) if no sufficient treatment is done (Pesaro et al., 1995). Animal wastes may also contain fungal organisms such as *Coccidioides immitis, Histoplasma capsulatum*, and various *Microsporum* and *Trichophyton* organisms (Krider, 1999). The use of antibiotics in food-animal production has led to the emergence of antibiotic resistance among pathogens and bacteria of humans (Graham et al., 2009). Antibiotic-resistant bacteria have been found in soil, surface and groundwater, air, and wild animal populations near AFOs (Stine et al., 2007).

Most antimicrobials are used for improving growth, feed efficiency, and disease prevention in the United States. It is estimated that approximately 25 million pounds of antimicrobials were used in US food-animal production annually (Mellon et al., 2001). About 25%–75% of antimicrobials in food animals is excreted to the waste and can be found in surface and groundwater located near swine and poultry operations (Kummerer, 2004). Roxarsone is an organoarsenical feed additive and used for improving the growth and feed efficiency and prevent coccidial infections (Han et al., 2003). It is excreted in the waste and degraded to arsenite and arsenate, most of which are water soluble (Garbarino et al., 2003; Jackson et al., 2003). Arsenate is associated with cancer and commonly found in drinking water (National Research Council, 2001). In addition, unmetabolized roxarsone has angiogenic potential in cultured human aortic and lung microvascular endothelial cells, increasing the risk of vascular disease from roxarsone exposure (Basu et al., 2008).

Hormone is widely used in food-animal operations in the United States to increase animal growth. It was estimated that 90% of beef cattle raised in animal-feeding operations receive hormone supplements to improve feed efficiency and animal growth (Raloff, 2002).The most commonly used hormones include androgens (testosterone and trenbolone acetate), estrogens, and progestogens (progesterone and melengestrol acetate) (Soto et al., 2004). Most of them are excreted to the waste and found in surface and groundwater near the feeding operation, lasting for a long time (Arai et al., 2003; Han et al., 2003; Lorenzen et al., 2004; Soto et al., 2004). Metabolites of trenbolone acetate have been found to survive in animal waste for more than 270 days (Schiffer et al., 2001). High levels of hormonally active chemicals were found in cattle wastewater. It has been estimated that estrogen excretions from cattle, pigs, and chickens in the United States are 45, 0.8, and 2.7 metric tons year^{-1}, respectively (Hanselman et al., 2003).

12.2.3 Policies to Improve Animal Waste Management

For small- (e.g., fewer than 300 animal units) and medium- (e.g., 300–999 animal units) sized CAFOs, livestock waste management depends on voluntary operation that is usually focused on nutrient management in the United States. The United States Department of Agriculture (USDA) has funded programs to assist small farmers in evaluating existing facilities and management systems and taking more voluntary actions. However, voluntary efforts to reduce negative environmental impacts associated with food-animal operation will be unsatisfactory because of lacking of incentives (Ribaudo et al., 2003; Hodne, 2005).

The rules for concentrated animal-feeding operations established by US EPA is the main US federal regulatory policy to manage the food-animal wastes. Approximately 15,500 out of the 238,000 AFOs in the United States are now considered as CAFOs. According to these rules, animal-feeding operations with more than 1000 animal units must apply for a water pollution permit and implement a nutrient management plan. The minimum number of animals for CAFO is approximately 1,000 heads of beef cattle, 700 dairy cattle, 2,500 hogs, or 125,000 broiler chickens (United States Environmental Protection Agency, 2003).

The National Pollutant Discharge Elimination System (NPDES) permit, issued by the EPA or federally approved states, is the key technology-based mechanism for regulating CAFOs. It requires technology-based restrictions on water pollution and includes (1) adequate storage of animal waste and wastewater, including operation and maintenance capability; (2) proper management of animal mortalities (disposal in waste is not allowed); (3) diversion of clean water from the operation; (4) prevention of contact between animals and US bodies of water; (5) proper chemical disposal (disposal in waste is not allowed); (6) identification of site-specific conservation practices to control runoff of pollutants into surface water; (7) annual nutrient analyses of waste, wastewater, and soil; (8) land application of waste in accordance with site-specific nutrient management plans; and (9) maintenance of records that will document implementation and management of the topics described earlier.

12.3 CONVERSION OF ANIMAL WASTES INTO ENERGY PRODUCTS

12.3.1 Recovery of Materials and Energy from Animal Wastes

There are a wide range of animal wastes that can be used as sources of bioenergy. The most common sources are animal and poultry manures. In the past, these wastes were recovered and sold as a fertilizer or simply spread onto agricultural land. However, waste management is now required because of tighter environmental controls on odor and water pollution, which provides a good opportunity for the implementation of waste-to-energy conversion technologies. Basically, there are three main ways to convert organic wastes to energy products including thermochemical, biochemical, and physicochemical conversion. Quantity and quality of animal waste are two main factors affecting the selection of a conversion method and the recovery efficiency of energy from the wastes.

Thermochemical conversion (TCC), including incineration, liquefaction, pyrolysis, and gasification, is operated at a high temperature and conversion rates. Animal wastes with low moisture content such as poultry litter are suitable feedstock for TCC. Incineration is controlled combustion of waste for recovering heat to produce steam and then power through steam turbines. Pyrolysis, liquefaction, and gasification are refined thermal conversion technologies and transform the waste into energy products such as syngas and bio-oil (Wang, 2008).

The biochemical conversion, such as anaerobic digestion (AD) and fermentation, is preferred for wastes with high contents of biodegradable matter and water. AD can be used to recover both nutrients and energy contained in organic wastes such as animal manure by converting the organic components into biogas with up to 70% of methane (Demirer and Chen, 2005). Fibers in dairy manure can be hydrolyzed into sugar and further fermented to alcohols by a series of biochemical reactions in microorganisms (Liao et al., 2006).

The physicochemical technology involves physical and chemical processes to improve properties of solid waste. Pelleting is a process to transform combustible fractions in the waste into high-energy fuel pellets that may be used in steam generation (Bernhart et al., 2010). Fuel pellets have several distinct advantages over coal and wood because of its lower ash and moisture contents, uniform size, cost-effectiveness, and ecofriendship.

12.3.2 Anaerobic Digestion of Animal Wastes into Biogas

AD breaks down complex organic fractions in animal wastes and produces biogas, mainly methane (CH_4) and carbon dioxide (CO_2), by a community of anaerobic microorganisms. The AD process has three main stages: hydrolysis, fermentation, and methanogenesis. Hydrolysis is the first stage to break down macromolecules into soluble components, which are easily used by fermentative bacteria (acidogenic and acetogenic) and converted into alcohols, acetic acid, other volatile fatty acids (VFAs), and off-gas containing H_2 and CO_2. These intermediate products are then converted into CH_4 (60%–70%), CO_2 (30%–40%), and other gases by methanogens. Efficiency of the methanogenesis process is sensitive to the composition of influent materials, pH, temperature, organic loading rate (OLR), and hydraulic retention time (HRT). As a result, the aforementioned parameters must be monitored and optimized in order to maximize efficiency (Cantrell et al., 2008).

Methanogens are sensitive to pH change but the VFAs generated during AD reduce the pH value of the slurry. Methanogenic population will be destroyed when the pH falls below 6.3 (Chen et al., 2002). Therefore, it is crucial to keep a balance among the acidogens, acetogens, and methanogens for an effective AD process to produce biogas. Temperature also affects the activity of the microorganisms that further affects the efficiency of digestion and methane production. Methane production occurs at temperatures from 4°C to 25°C and is dramatically improved with the increase of the temperature (Safley and Westerman, 1992; Umetsu et al., 2005). There are three common temperature ranges for AD: (1) a low temperature range (<20°C) is considered as psychrophilic, (2) a temperature range of 20°C–45°C is mesophilic, and (3) a temperature range of 45°C–60°C is thermophilic (Cantrell et al., 2008).

The OLR is the amount of volatile solids (VSs) or chemical oxygen demanding (COD) components fed per day per unit digester volume. The increase in OLR can reduce the size of the digesters and consequently save the capital cost. As microfloras require enough time to break down an organic material and convert it to biogas. The HRT is needed to be controlled. The HRT of a digester is the average time that the liquid is held in the digestion process and is calculated as the ratio of the digester volume to the volumetric flow rate of the effluent. If manure passes through a digester with too high flow rate, microorganisms do not have enough time to proliferate, which will lead to digestion failure.

AD can provide several benefits for the treatment of animal manure, which include (1) odor control, (2) reduction of gas emissions, (3) potential pathogen kill, (4) reduction of wastewater strength (oxygen demand), (5) conversion of organic nitrogen into plant available ammonia nitrogen, (6) preservation of plant nutrients (e.g., N, P, K), and (7) production of biofuel such as biogas (Kashyap et al., 2003; Wilkie et al., 2004; Beddoes et al., 2007). The liquid portion in the digestate can be separated and utilized as a fertilizer. Solid portion can be composted to stabilize them and produce more useful products. Part of the biogas can be used to supply heat for the AD via the combustion in a boiler or heater. Excess biogas can be used to generate electricity to be sold to local utility grids (Beddoes et al., 2007; Lazarus and Rudstrom, 2007).

12.3.3 Production of Hydrogen from Animal Wastes

Hydrogen is a promising alternative to fossil fuels because of the green nature of H_2 combustion, which only emits water during combustion. Hydrogen can be produced via dark fermentation, which utilizes principles similar to AD. However, methanogens, which would consume H_2 during the CH_4 generation, are inhibited (Kraemer and Bagley, 2007). Dark fermentation happens in a variety of approaches including (1) the selection of hydrogen-producing bacteria by proper heat treatment (Lay et al., 1999; Mu et al., 2007); (2) changes of the reactor pH to be below 6.3 or above 7.8 to deactivate the methanogenesis; (3) forced aeration to inhibit the methanogenesis (Ueno et al., 1995); and (4) addition of chemicals such as 2-bromoethanesulfonate (BES) (Wang et al., 2003), acetylene (Sparling et al., 1997), or chloroform (Liang et al., 2002). The digestion process can also be precisely controlled to produce both hydrogen and methane. This has been accomplished and demonstrated with two-stage reactors (Kyazze et al., 2007; Ueno et al., 2007; Zhu et al., 2007).

Factors affecting the production of hydrogen from animal wastes include the substrate concentration and characteristics, the type of inoculants used, pH, temperature, and the partial pressure of hydrogen in the fermentation vessel (Ginkel and Sung, 2001; Lin and Lay, 2004). Fermentation pH is extremely important because hydrogenase enzyme is pH sensitive and microbial metabolism tends to shift from acidogenesis to solventogenesis to produce alcohols instead of VFA at pH values below 4.5 (Dabrock et al., 1992). Besides, composition of microbial communities changes with pH and methanogenic activity is inhibited at $pH < 6.0$ (Zhang et al., 2005). Temperature affects not only the composition of the microbial community but also the rate of biochemical processes (Fang et al., 2002; Hernon et al., 2006; Levén et al., 2007). Although biodegradation is more active at thermophilic temperatures, the microbial community is more genetically diverse at mesophilic temperatures (Levén et al., 2007). While optimal conditions are different in published studies because of varied substrates and inoculums used, a pH value of 5.5 and temperatures from 35°C to 40°C are common (Mu et al., 2006; Zhang and Shen, 2006; Yokoyama et al., 2007).

Nonsterile substrate and readily available microbial inoculants are essential to make fermentative production of hydrogen at a commercial scale. Animal manures can be used in hydrogen production as both substrate (or cosubstrate) and inoculants. AD of animal waste can reduce environmental impact while simultaneously producing hydrogen energy. Stabilized biosolids can be produced through further subsequent AD or aerobic treatment (composting), which can be used as soil amendment for agricultural production.

12.3.4 Production of Methanol from Carbohydrate-Rich Animal Wastes

There is a great need for renewable methanol in the international market. 90% of the 29 million tons of current worldwide methanol demand is derived from methane in natural gas. A possible new way may be found for the biological production of methanol at ambient temperature and pressure. There are two pathways that are both based on the generation of either CO_2 or CH_4 for the biological production of methanol from animal waste (Cantrell et al., 2008).

12.3.4.1 Methane-Derived Methanol

The principle of the conversion from CH_4 to methanol is that CH_4 is oxidized by the enzyme methane monooxygenase (MMO) according to the following reaction:

$$CH_4 + O_2 + 2e^- + 2H^+ \rightarrow CH_3OH + H_2O \quad (12.1)$$

Methane MMO is found in two forms: a soluble MMO (sMMO) and a membrane-bound or particulate MMO (pMMO). With one or both enzymatic systems, methanotrophs can utilize CH_4 as their sole source for carbon and energy. However, methanol is so unstable as it is quickly oxidized by methanol dehydrogenase (MDH) to formate. Therefore, the presence of MDH inhibition is necessary. It was reported that *Methylosinus trichosporium* accumulated methanol at a maximum rate of 6 μ mol mg^{-1} h^{-1} at a cell density of 2 mg/ml using phosphate inhibition of MDH (Mehta et al., 1987). It was reported that 36.1 μmol methanol/ml was produced at an hourly stationary rate of 3.17 μmol mg^{-1} dry cell in a semicontinuous process operated over 6 h using cyclopropanol as an MDH inhibitor (Xin et al., 2004).

12.3.4.2 CO_2-Derived Methanol

Unlike one-step conversion from CH_4 to methanol, the conversion of CO_2 to methanol is a three-step reaction process as shown in Equation 12.2. Firstly, carbon dioxide is reduced to formate by formate dehydrogenase (F_{ate}DH). Then, formate is converted to formaldehyde by formaldehyde dehydrogenase (F_{ald}DH). Finally, formaldehyde is converted into methanol by alcohol dehydrogenase (ADH). For each reaction, reduced nicotinamide adenine dinucleotide (NADH) acts as a terminal electron donor for a ratio of NADH to CO_2 or methanol at 3:1:

$$CO_2 \xrightarrow[F_{ate}DH]{NADH} HCOOH \xrightarrow[F_{ald}DH]{NADH} HCHO \xrightarrow[ADH]{NADH} CH_3OH \qquad (12.2)$$

F_{ate}DH, F_{ald}DH, and ADH can be obtained from a variety of bacteria and yeast. Based on this principle, Obert and Dave (1999) realized the conversion of CO_2 to methanol in a silica sol–gel matrix. Methanol yield of 15.2 μmol was obtained for 50 μmol NADH used in the silica sol–gel matrix, which was over an 11-fold increase compared to the enzymes in solution (1.3 μmol methanol generated). Based on the work of Obert and Dave (1999), the enzymatic activities of sol–gel-mediated conversion of CO_2 to methanol has been successfully improved using polyethylene glycol (PEG) (Wu et al., 2004). These studies lay the groundwork for the biological production of methanol at ambient temperatures and pressures, which should eventually allow for more cost-effective production of methanol from CO_2.

12.3.5 Hydrothermal Liquefaction of Animal Manure into Bio-Oil

TCC is a high-temperature chemical process that breaks down the bonds of organic matter and converts these intermediates into char, syngas, and bio-oil. TCC of animal waste has a number of benefits and advantages: (1) conversion of organic portion to energy, (2) efficient nutrient recovery, (3) odor reduction, (4) short processing time on the order of minutes, (5) versatile application to a variety of feedstocks and blends, and (6) high-temperature elimination of pathogens and pharmaceutically active compounds (Cantrell et al., 2008).

Hydrothermal liquefaction (HTL), as an important method of TCC, is depolymerization and reforming reactions of organic matters occurring in a heated, pressurized, and oxygen-free or oxygen-deficient enclosure. HTL is similar to pyrolysis to produce oils but the heating values of bio-oil produced from HTL are usually higher than those of pyrolysis oils. Besides, high-pressure environment makes water in the reactor to remain in the liquid phase; predrying the livestock waste is not required for HTL. The carbonaceous materials are converted to liquefied products through a complex sequence of changes in physical structure and chemical bonds. Such liquefaction processes had been used to convert cellulosic wastes to bio-oil (He et al., 2000; Xiu et al., 2008).

Different liquid–solid separation (LSS) techniques were investigated for swine manure, such as stationary inclined screen, vibrating screen, rotary screen, belt press, centrifuge, and screw press (Zhang and Westerman, 1997; Chastain et al., 2001). These techniques were compared in terms of their efficiency at capturing the solids from the original manure slurry. The application of LSS improves the potential for HTL converting animal manure into oil (Minarick et al., 2011).

Xiu et al. (2009) investigated the principal structural and physicochemical changes of bio-oils produced from the liquefaction of swine manure with crude glycerol at 340°C and their key fraction, free fatty acids. The bio-oils were subjected to various physicochemical characterization methods. FTIR data indicated a reduction in aliphatic structures and an increase in more oxidized and, probably, more polycondensed aromatic components resulting from the addition of crude glycerol to swine manure. GC-MS data indicated that the addition of crude glycerol facilitated the esterification reaction in subcritical water to convert organic acids contained in bio-oil into various kinds of esters. The dynamic viscosity of bio-oil decreased dramatically by adding crude glycerol into the swine manure (Xiu et al., 2009).

12.3.6 Pyrolysis of Animal Manure into Bio-Oil

Pyrolysis converts the organic portion of animal waste into char and volatile gases containing both noncondensable vapors and condensable tars (oxygenated hydrocarbons), which form a combustible bio-oil (Bridgwater and Peacocke, 2000; Mohan et al., 2006). There are already lots of published findings for bio-oil from biomass (Bridgwater and Peacocke, 2000; Yaman, 2004; Mohan et al., 2006; Boateng et al., 2007); little research has been done to investigate bio-oil production from livestock waste. Slow pyrolysis converts animal wastes into char, which can be used as a green coal for existing coal combustion and gasification plants. Char can also be used as a soil amendment to improve fertility (Antal and Grønli, 2003). Char produced from animal waste can be upgraded to activated carbon (Bridgwater, 2003; Lima and Marshall, 2005; Koutcheiko et al., 2007; Sanchez et al., 2007). For instance, activated carbon can be obtained from poultry litter by pyrolyzing it into char at 700°C under nitrogen and then steam-activating the char (Lima and Marshall, 2005). Compared to commercial activated carbon produced from coal, biomass-based activated carbon has higher capacity of adsorbing copper ions, which shows promise in potential metal ion removal applications.

Shinogi and Kanri (2003) studied the pyrolysis of sugarcane bagasse and rice husks, cow biosolids, and treated municipal sludge at temperatures from 250°C to 800°C in closed containers. Physical properties (yield, surface area, density) and specific chemical properties (total carbon, total nitrogen, pH, fixed carbon, ash content, and volatility) were compared among the four waste products. The results indicated that surface area, total carbon, ash content, and pH increased as the carbonization temperature increased, while carbonization yield decreased with increasing temperature. Product density was not affected by temperature. Source materials had considerable influence on the physical and chemical properties of the carbonized products (Shinogi and Kanri, 2003).

Ro et al. (2010) used a commercial pilot-scale pyrolysis reactor to produce combustible gas and biochar at 620°C from chicken litter, swine solids, and a mixture of swine solids with ryegrass. The greatest higher heating value (HHV) of gas was obtained from swine solids followed by the mixture of swine solids with ryegrass and chicken litter. Gases with high sulfur content were produced. Biochar yield ranged from 43% to 49% based on dry weight with about 53% of carbon recovery. Manure biochars contained higher concentrations of P and K than the original manure feedstocks. Manure biochar could be used as a low-grade fertilizer to improve soil fertility and crop yields (Ro et al., 2010).

Broiler manure chars were characterized for their ability to adsorb selected metal ions (Lima et al., 2009). Char yields ranged between 29.6% and 33.5% when pelletized broiler manure was slowly pyrolyzed at 700°C and 800°C for 1 h. Surface areas of the chars from the broiler litter and cake decreased with pyrolysis temperature. Broiler manure chars absorbed more metal ions from solution (up to 0.91 mmol g^{-1} Cu^{2+} and 0.96 mmol g^{-1} Zn^{2+}) when compared to chars produced from coal, wood, or coconut shells (up to 0.0 mmol g^{-1} Cu^{2+} and 0.10 mmol g^{-1} Zn^{2+}). The noncondensable gas produced during pyrolysis of broiler manure included CO, CO_2, low, molecular, weight hydrocarbon gases, and H_2. Except CO, noncondensable gases increased with pyrolysis temperature (Lima et al., 2009).

12.3.7 Gasification of Animal Manure into Syngas

Gasification converts the organic portion of the feedstock into noncondensable gases of CO, CO_2, H_2 and low molecular weight hydrocarbon, and char using air, oxygen, or steam as a reaction medium at a high temperature (McKendry, 2002; Bridgwater, 2003). Primary procedures include drying, pyrolysis, reduction, and oxidation (McKendry, 2002; Bridgwater, 2003; Priyadarsan et al., 2004). In the drying stage, moisture is removed because of the heat generated by the later stages. At 250°C, the dried biomass was pyrolyzed to produce volatile compounds and char, which are then subjected to oxidative and reductive reactions to produce a mixture of gases including H_2, CO, CH_4, and CO_2.

Average HHVs of syngas produced from poultry litter (TS=92.5 wt.%) and feedlot manure (TS=92.4 wt.%) using air as an oxidizing agent were 4.5 MJ m^{-3} and 4.1 MJ m^{-3}, respectively. The product gases consist of approximately 5.8% H_2, 27.6% CO, and 1.0% CH_4. However, the product gases were severely diluted with nitrogen, thus decreasing the potential HHV by roughly 60% (Priyadarsan et al., 2004). Wu et al. (2012) studied the effects of key operating parameters on gasification. In this research, dairy manure from the University of Nebraska dairy farm was gasified in a fluidized-bed, laboratory-scale gasifier to generate syngas. Results showed that increase in the temperature favored the formation of H_2, CO, and CH_4, but the composition of each gas behaved differently according to the changing parameters. The heating value of the syngas varied from 2.0 to 4.7 MJ m^{-3}. Wu et al. (2012) studied an internally circulating fluidized-bed gasifier (ICFG) with bed material acting as the medium for char combustion and heat exchange by its internal circulation. The ICFG was used for the steam gasification of animal waste to produce fuel gas. Hydrogen-rich and low-tar product gas could be produced from the low-calorific feedstock after optimizing the process and using high-performance catalysts.

12.3.8 Combustion of Animal Wastes for Heat and Power

The cofiring of biomass and organic waste is a beneficial use of manure for power plants located near feedlots. Firstly, it utilizes biomass and/or waste locally generated as fuels. Secondly, co-firing in existing infrastructures avoids the technology transfer problem. The use of manure as a fuel source reduces not only CO_2 emissions from fossil fuel but also NO_x emissions from manure. Even though the effective N content of the coal–manure mixture is higher than that of coal, the NO_x emissions are lower because most of the N is in the form of NH_3 (Otero et al., 2011).

Cocombustion seems to be a promising way of producing renewable energy, because it makes use of existing technologies and infrastructure. The organic material can be fed into an existing plant together with the coal.

12.3.9 Growth of Energy Crops Using Animal Wastes

Biofertilizer is another important way to utilize the manure. The biofertilizer can be used to grow energy crops, such as algae, corn and soybean. Chen et al. (2012) found that nonfilamentous green algae, especially *Chlorella*, were able to tolerate high-nutrient loadings in a 5-month cultivation. A chemically pretreated AD effluent with 200 g m^{-3} of total nitrogen and 2.5 g m^{-3} of total dissolved phosphorus (TDP) provided an optimal nutrient concentration for the growth of selected algae. The selected algae fed with AD effluent in a pilot-scale semicontinuously fed raceway pond got a stable algal biomass productivity of 6.83 g m^{-2} day^{-1} (Chen et al., 2012).

Thelen et al. (2010) conducted experiments of corn–soybean rotation with whole-plant corn harvest over a 3-year period. Compost and manure supplied enough carbon to overcome the C debt associated with manure production, manure collection and storage, land application, and post-application field emissions. The net GWP in carbon dioxide equivalents for the manure-and compost-amended cropping systems was −934 and −784 g m^{-2} $year^{-1}$, respectively, compared to 52 g m^{-2} $year^{-1}$ for the non-manure-amended synthetic fertilizer (Thelen et al., 2010).

12.4 ENVIRONMENTAL AND ECONOMIC IMPACTS OF ANIMAL WASTES TO ENERGY TECHNOLOGIES

12.4.1 Environmental Impacts

The technologies of animal waste to energy are environmentally friendly because it does not increase GHG concentrations. Actually, climate neutrality cannot be derived from the emissions of biomass-fuelled installations, because it emits more CO_2 than fossil-fuelled installations in order to get the same power output. The view behind *climate neutrality* is that farm animals are grown on annual crops and grasses, which can fix the main GHG CO_2. CO_2 released by burning animal wastes can be counterbalanced by the photosynthesis of crops and grasses, reflecting climate neutrality (Reijinders and Huijbregts, 2005).

12.4.1.1 Positive Impact of an Effective Use of Manure for Energy Production

Animal manure has the potential for saving the fossil energy required in the industrial production of mineral fertilizers. This aspect has been neglected since fossil energy was widely used at low cost and damage of GHG emission was underestimated. Practical solutions and policy actions should be taken to reduce the dependence on industrial fertilizers by increasing the application of manure (Sangiorgi et al., 1996).

12.4.1.2 Negative Impact of Manure for Energy Production

The implicit condition of the previous statements about manure is that the energy necessary to deliver nutrients to the recycling crops should be as limited as possible. In fact, the energy required for transportation of manure, with a given content of N, can be compared with manufacturing an equivalent amount of N in the form of mineral fertilizer. It is clear that manure with a higher concentration of nutrients could be conveniently delivered over longer distances. Therefore, any solution that prevents the dilution of manure (e.g., by avoiding the addition of rainwater) can be regarded as economical. Currently, there is no economic advantage in paying for the transportation of manure instead of purchasing industrial fertilizers. For example, the typical cost of N fertilizer in Europe is 0.502 Euro kg^{-1} N. If pig slurry is used as a N fertilizer, the transportation cost of 44 tons of pig slurry by truck will be 99 Euro h^{-1}. As the pig slurry only contains 3 kg N ton^{-1}, 44 tons of the pig slurry contain 132 kg N. Therefore, the transportation cost of the pig slurry for 1 h distance to be used as an N fertilizer is $132/99 = 1.33$ Euro kg^{-1} N (Ceotto, 2005).

However, present economic conditions have been essentially driven by the market and policy measures like levies on the use of industrial fertilizers or direct subsidies for using manure in the place of industrial fertilizers. Technologies have been available to separate manure into clean water, NPK liquid concentrate, and dry matter. For instance, the Manura 2000® system separates pig slurry into clean water, dry organic fertilizer, liquid NPK fertilizer, and liquid N fertilizer. The procedure includes mechanical preseparation, concentration by evaporation, and purification of the distilled water. The power consumption of the system is approximately 50 kWh m^{-3} of slurry. The production of electricity has an emission factor of 0.834 kg CO_2 kWh. Therefore, the treatment of 1 ton involves an emission of $50 \times 0.834 = 41.7$ kg CO_2, which corresponds to 13.9 kg CO_2 kg^{-1} N for a diluted pig slurry with 3 kg N $tons^{-1}$, compared to 5.3 kg CO_2 kg^{-1} N mineral fertilizer (Smith et al., 2001). This result is based on the assumption that all electricity was produced by coal-fired power plants. In reality, only part of the electricity is produced by gas- or oil-fired power plants with lower CO_2 emissions. Furthermore, electricity generated by hydroelectric and nuclear power plants is free of CO_2 emissions (Ceotto, 2005).

12.4.2 Economic Impacts

There are a number of published studies that present a detailed economic analysis for the energy use of residues. Gomez et al. (2010) estimated the potential of electricity from combustion of animal waste. They found that the cost of the electricity generated depends greatly on the type of residue and the technology used for the transformation. Thus, the most economical option was found to be the incineration of municipal solid waste, with an entry cost of around 4 c€ kWh^{-1}. The generation entry costs from livestock manure and sewage sludge were higher than 8 c€ kWh^{-1} (Gomez et al., 2010).

Nolan et al. (2012) performed economic analysis on treatment options for pig manure in Ireland. A farm with 500 sows produces 10,500 m^3 of manure per year at 4.8% dry matter. The AD of pig manure and grass silage costs € 5.2 m^{-3} manure. Subsequent solid–liquid separation of the digestate would cost additional € 12.8 m^{-3} manure. Composting of the separated solid fraction would add € 2.8 to the total costs. The result of cost analysis showed that the technologies investigated are currently not cost effective in Ireland. Transport and spreading of raw manure, at € 4.9 m^{-3} manure (15 km maximum distance from farm), is the most cost-effective option (Nolan et al., 2012).

Minarick et al. (2011) investigated the impact of hydrothermal process on oil yields and fertilizer values. Maximum oil yield was obtained when using manure containing around 20% solids. Fertilizer was obtained as the by-product of the hydrothermal process, retaining most of its nitrogen with a reduced level of phosphorus. The economic analyses investigated feasibility for several different LSS methods based on the equipment and maintenance costs (Minarick et al., 2011). They found that conveyor-belt manure collection systems in conjunction with hydrothermal liquefaction are especially promising.

Lantz (2012) evaluated the economic feasibility of various technologies on different scales for the production of combined heat and power from manure-based biogas in Sweden. Overall, energy production from manure is not profitable under current conditions because of the gap between the calculated biogas production cost and the acceptable cost. In general, efficiency of large-scale plants is better than that of individual farm-scale ones. However, a centralized biogas plant is not always more cost efficient than a farm-scale plant because it needs to collect manure from lots of farms and increases the transportation cost. The results are useful for policy makers and companies involved in production and utilization of biogas.

12.5 CONCLUSIONS

Livestock waste-to-bioenergy technologies have been widely investigated. The biochemical process of AD is an established technology for the production of biogas from animal wastes. However, other biological processes for the production of biohydrogen and biomethanol from animal wastes are still in early research stages, but they have shown the promise to produce bioenergy from animal wastes in a sustainable way. Thermochemical technologies, such as gasification, pyrolysis, and liquefaction, have the potential to convert livestock wastes into combustible bio-oil or syngas. It will also be promising to integrate biological and thermal-based conversion technologies by recapturing the evolved CO_2 to promote algal growth and utilizing wet gasification as the algal energy recovery component.

REFERENCES

Antal, M.J. and Grønli, M. 2003. The art, science, and technology of charcoal production. *Industrial and Engineering Chemistry Research* 42: 1619–1640.

Arai, Y., Lanzirotti, A., Sutton, S., Davis, A., and Sparks, D.L. 2003. Arsenic speciation and reactivity in poultry litter. *Environmental Science and Technology* 37: 4083–4090.

Basu, P., Ghosh, R.N., Grove, L.E., Klei, L., and Barchowsky, A. 2008. Angiogenic potential of 3-nitro-4-hydroxy benzene arsenic acid (roxarsone). *Environmental Health Perspectives* 116: 520–523.

Beddoes, J.C., Bracmort, K.S., Burn, R.B., and Lazarus, W.F. 2007. An analysis of energy production costs from anaerobic digestion systems on US livestock production facilities. Technical Note No. 1. USDA, Natural Resources Conservation Service.

Bernhart, M., Fasina, O.O., Fulton, J., and Wood, C.W. 2010. Compaction of poultry litter. *Bioresource Technology* 101: 234–238.

Boateng, A.A., Banowetz, G.M., Steiner, J.J., Barton, T.F., Taylor, D.G., Hicks, K.B., El-Nashaar, H., and Sethi, V.K. 2007. Gasification of Kentucky bluegrass (*Poa pratensis* L.) straw in a farm-scale reactor. *Biomass and Bioenergy* 31: 153–161.

Bridgwater, A.V. 2003. Renewable fuels and chemicals by thermal processing of biomass. *Chemical Engineering Journal* 91: 87–102.

Bridgwater, A.V. and Peacocke, G.V.C. 2000. Fast pyrolysis processes for biomass. *Renewable and Sustainable Energy Reviews* 4: 1–73.

Cantrell, K.B., Ducey, T., Ro, K.S., and Hunt, P.G. 2008. Livestock waste-to-bioenergy generation opportunities. *Bioresource Technology* 99: 7941–7953.

Carolina, F.P. 2012. Characterization, kinetics and modeling of gasification of poultry manure and litter: An overview. *Energy Conversion and Management* 53: 92–98.

Ceotto, E. 2005. The issue of energy and carbon cycle: New perspectives for assessing the environmental impacts of animal waste utilization. *Bioresource Technology* 96: 191–196.

Chastain, J.P., Lucas, W.D., Albrecht, J.E., Pardue, J.C., Adams, I.J., and Moore, K.P. 2001. Removal of solids and major plant nutrients from swine manure using a screw press separator. *Applied Engineering in Agriculture* 17(3): 355–363.

Chen, C.C., Lin, C.Y., and Lin, M.C. 2002. Acid-base enrichment enhances anaerobic hydrogen production process. *Applied Microbiology and Biotechnology* 58(2): 224–228.

Chen, R., Li, R., Deitz, L., Liu, Y., Stevenson, J.R., and Liao, W. 2012. Freshwater algal cultivation with animal waste for nutrient removal and biomass production. *Biomass and Bioenergy* 39: 128–138.

Choudhary, M., Bailey, L.D., and Grant, C.A. 1996. Review of the use of swine manure in crop production: Effects on yield and composition and on soil and water quality. *Waste Management and Research* 14: 581–595.

Cox, P., Griffith, M., Angles, M., Deere, D., and Ferguson, C. 2005. Concentrations of pathogens and indicators in animal feces in the Sydney watershed. *Applied and Environmental Microbiology* 71(10): 5929–5934.

Cuéllar, D.A. and Webber, E.M. 2008. Cow power: The energy and emissions benefits of converting manure to biogas. *Environmental Research Letters* 3: 12–19.

Dabrock, B., Bahl, H., and Gottschalk, G. 1992. Parameters affecting solvent production by *Clostridium pasteurianum*. *Applied and Environmental Microbiology* 58: 1233–1239.

Demirer, G.N. and Chen, S. 2005. Two-phase anaerobic digestion of unscreened dairy manure. *Process Chemistry* 40: 3542–3549.

Fang, H.H.P., Zhang, T., and Liu, H. 2002. Microbial diversity of a mesophilic hydrogen-producing sludge. *Applied Microbiology and Biotechnology* 58: 112–118.

Font-Palma, C. 2012. Characterization, kinetics and modeling of gasification of poultry manure and litter: An overview. *Energy Conversion and Management* 53: 92–98.

Garbarino, I.R., Bednar, A., Rutherford, D.W., Beyer, R.S., and Wershaw, R.L. 2003. Environmental fate of roxarsone in poultry litter. *Environmental Science and Technology* 37(8): 1509–1514.

Gilroyed, B.H., Chang, C., Chu, A., and Hao, X.Y. 2008. Effect of temperature on anaerobic fermentative hydrogen gas production from feedlot cattle manure using mixed microflora. *International Journal of Hydrogen Energy* 33: 4301–4308.

Ginkel, S.V. and Sung, S. 2001. Biohydrogen production as a function of pH and substrate concentration. *Environmental Science and Technology* 35: 4726–4730.

Gomez, A., Zubizarreta, J., Rodrigues, M., Dopazo, C., and Fueyo, N. 2010. Potential and cost of electricity generation from human and animal waste in Spain. *Renewable Energy* 35: 498–505.

Graham, J.P., Price, L.B., Evans, S.L., Graczyk, T.K., and Silbergeld, E.K. 2009. Antibiotic resistant enterococci and staphylococci isolated from flies collected near confined poultry feeding operations. *Science of the Total Environment* 407(8): 2701–2710.

Han, F.X., Kingery, W.L., Selim, H.M., Gerard, P.D., Cox, M.S., and Oldham, J.L. 2003. Arsenic solubility and distribution in poultry waste and long-term amended soil. *Science of the Total Environment* 320: 51–61.

Hanselman, T.A., Graetz, D.A., and Wilkie, A.C. 2003. Manure-borne estrogens as potential environmental contaminants: A review. *Environmental Science and Technology* 37(24): 5471–5478.

He, B.J., Zhang, Y., Funk, T.L., Riskowski, G.L., and Yin, Y. 2000. Thermochemical conversion of swine manure: Alterative process for waste treatment and renewable energy production. *Transactions- American Society of Agricultural Engineers* 43: 1827–1833.

Hernon, F., Forbes, C., and Colleran, E. 2006. Identification of mesophilic and thermophilic fermentative species in anaerobic granular sludge. *Water Science Technology* 54: 19–24.

Hodne, C.J. 2005. *Concentrating on Clean Water: The Challenge of Concentrated Animal Feeding Operations*. Mount Vernon, IA, The Iowa Policy Project. Available from http://www.iowapolicyproject.org/2005docs/050406-cafo-fullx.pdf.

Jackson, B.P., Bertsch, P.M., Cabrera, M.L., Camberato, J., Seaman, C., and Wood, C.W. 2003. Trace element speciation in poultry litter. *Journal of Environmental Quality* 32(2): 535–540.

Kashyap, R.D., Dadhich, S.K., and Sharma, K.S. 2003. Biomethanation under psychrophilic conditions. *Bioresource Technology* 87: 147–153.

Koutcheiko, S., Monreal, C.M., Kodama, H., McCracken, T., and Kotlyar, L. 2007. Preparation and characterization of activated carbon derived from the thermo-chemical conversion of chicken manure. *Bioresource Technology* 98: 2459–2464.

Kraemer, J.T. and Bagley, D.M. 2007. Improving the yield from fermentative hydrogen production. *Biotechnology Letters* 29: 685–695.

Krapac, I.G., Dey, W.S., Roy, W.R., Smyth, C.A., Storment, E., Sargent, S.L., and Steele, J.D. 2002. Impacts of swine manure pits on groundwater quality. *Environmental Pollution* 120: 475–492.

Krider, J.N. 1999. *Agricultural Waste Management Field Handbook*. Washington, DC, National Resource Conservation Service.

Kummerer, K. 2004. Resistance in the environment. *Journal of Antimicrobial Chemotherapy* 54(2): 311–320.

Kyazze, G., Dinsdale, R., Guwy, A.J., Hawkes, F.R., Premier, G.C., and Hawkes, D.L. 2007. Performance characteristics of a two-stage dark fermentative system producing hydrogen and methane continuously. *Biotechnology and Bioengineering* 97: 759–770.

Lantz, M. 2012. The economic performance of combined heat and power from biogas produced from manure in Sweden-a comparison of different CHP technologies. *Applied Energy* 98: 502–511.

Lay, J.J., Lee, Y.J., and Noike, T. 1999. Feasibility of biological hydrogen production from organic fraction of municipal solid waste. *Water Research* 33: 2579–2586.

Lazarus, F.W. and Rudstrom, M. 2007. The economics of anaerobic digester operation on a Minnesota dairy farm. *Review of Agricultural Economics* 29: 349–364.

Levén, L., Eriksson, A.R.B., and Schnürer, A. 2007. Effect of process temperature on bacterial and archaeal communities in two methanogenic bioreactors treating organic household waste. *FEMS Microbiology Ecology* 59: 683–693.

Liang, T.M., Cheng, S.S., and Wu, K.L. 2002. Behavioral study on hydrogen fermentation reactor installed with silicone rubber membrane. *International Journal of Hydrogen Energy* 27: 1157–1165.

Liao, W., Liu, Y., Liu, C., Wen, Z., and Chen, S. 2006. Acid hydrolysis of fibers from dairy manure. *Bioresource Technology* 97: 1687–1695.

Lima, I. and Marshall, W.E. 2005. Utilization of turkey manure as granular activated carbon: Physical, chemical and adsorptive properties. *Waste Management* 25: 726–732.

Lima, I.M., Boateng, A.A., and Klasson, K.T. 2009. Pyrolysis of broiler manure: Char and product gas characterization. *Industrial and Engineering Chemistry Research* 48: 1292–1297.

Lin, C.Y. and Lay, C.H. 2004. Carbon/nitrogen effect on fermentative hydrogen production by mixed microflora. *International Journal of Hydrogen Energy* 29: 41–45.

Lorenzen, A., Hendel, J.G., Conn, K.L., Bittman, S., Kwabiah, A.B., Lazarovitz, G., Massé, D., McAllister, T.A., and Topp, E. 2004. Survey of hormone activities in municipal biosolids and animal manures. *Environmental Toxicology* 19(3): 216–225.

McKendry, P. 2002. Energy production from biomass (part 3): Gasification technologies. *Bioresource Technology* 83: 55–63.

Mehta, P.K., Mishra, S., and Ghose, T.K. 1987. Methanol accumulation by resting cells of Methylosinus trichosporium (I). *Journal of General and Applied Microbiology* 33: 221–229.

Mellon, M., Benbrook, C., and Benbrook, K.L. 2001. *Hogging It: Estimates of Antimicrobial Abuse in Livestock*. Cambridge, U.K., UCS Publications.

Minarick, M., Zhang, Y.H., Schideman, L., Wang, Z.C., Yu, G., Funk, T., and Barker, D. 2011. Product and economic analysis of direct liquefaction of swine manure. *Bioenergy Research* 4: 324–333.

Mohan, D., Pittman, C.U., and Steele, P.H. 2006. Pyrolysis of wood/biomass for bio-oil: A critical review. *Energy and Fuels* 20: 848–889.

Mu, Y., Yu, H.Q., and Wang, G. 2007. Evaluation of three methods for enriching H2-producing cultures from anaerobic sludge. *Enzyme and Microbial Technology* 40: 947–953.

Mu, Y., Zheng, X.J., Yu, H.Q., and Zhu, R.F. 2006. Biological hydrogen production by anaerobic sludge at various temperatures. *International Journal of Hydrogen Energy* 31: 780–785.

National Research Council. 2001. *Arsenic in Drinking Water: 2001 Update*. Washington, DC, National Academy Press.

Nolan, T., Troy, S.M., Gilkinson, S., Frost, P., Xie, S.H., Zhan, X.M., Harrington, C., Healy, M.G., and Lawlor, P.G. 2012. Economic analyses of pig manure treatment options in Ireland. *Bioresource Technology* 105: 15–23.

Obert, R. and Dave, C.B. 1999. Enzymatic conversion of carbon dioxide to methanol: Enhanced methanol production in silica sol-gel matrices. *Journal of American Chemical Society* 121(51): 12192–12193.

Otero, M., Sanchez, M.E., and Gomez, X. 2011. Co-firing of coal and manure biomass: A TG-MS approach. *Bioresource Technology* 102: 8304–8309.

Park, K.H., Thompson, G., Marinier, M., Clark, K., and Wagner-Riddle, G. 2006. Greenhouse gas emissions from stored liquid swine manure in a cold climate. *Atmospheric Environment* 40: 618–627.

Pesaro, F., Sorg, I., and Metzler, A. 1995. In situ inactivation of animal viruses and a coliphage in nonaerated liquid and semiliquid animal wastes. *Applied and Environmental Microbiology* 61(1): 92–97.

Prapaspongsa, T., Christensen, P., Schmidt, J.H., and Thrane, M. 2010. LCA of comprehensive pig manure management incorporating integrated technology systems. *Journal of Cleaner Production* 18: 1413–1422.

Priyadarsan, S., Annamalai, K., Sweeten, J.M., Mukhtar, S., and Holtzapple, M.T. 2004. Fixed-bed gasification of feedlot manure and poultry litter biomass. *Transaction of ASAE* 47: 1689–1696.

Raloff, J. 2002. Hormones: Here's the beef Environmental concerns reemerge over steroids given to livestock. *Science News* 161: 10–12.

Reijinders, L. and Huijbregts, M.A.J. 2005. Life cycle emissions of greenhouse gases associated with burning animal wastes in countries of the European Union. *Journal of Cleaner Production* 13: 51–56.

Ribaudo, M., Kaplan, J.D., Christensen, L.A., GoUehon, N., Johansson, R., Breneman, V.E., Aillery, M., Agapoff, J., and Peters, M. 2003. *Manure Management for Water Quality Costs to Animal Feeding Operations of Applying Manure Nutrients to Land*. Washington, DC, USDA Economic Research Service.

Ro, K.S., Cantrell, B.K., and Hunt, G.P. 2010. High-temperature pyrolysis of blended animal manures for producing renewable energy and value-added biochar. *Industrial and Engineering Chemistry Research* 49: 10125–10131.

Safley, M.L. and Westerman, W.P. 1992. Performance of a low temperature lagoon digester *Bioresource Technology* 41: 167–175.

Sanchez, M.E., Martinez, O., Gomez, X., and Moran, A. 2007. Pyrolysis of mixtures of sewage sludge and manure: A comparison of the results obtained in the laboratory (semi-pilot) and in a pilot plant. *Waste Management* 27: 1328–1334.

Sangiorgi, F., Provolo, G., Carton, O.T., Comby, T.R., Lorenz, F., Huijsmans, J.F.M., Eftichidis, G., McGechan, M.B., and Magette, W.L. 1996. Optimal use of animal slurries for input reduction and protection of the environment in sustainable agricultural systems (SWAMP). *Proceedings of the International Scientific Conference Faculty of Agronomy and Agricultural Development of Slovakia*, Nitra, Slovakia, pp. 101–105.

Schiffer, B., Daxenberger, A., Meyer, K., and Meyer, H.H.D. 2001. The fate of trenbolone acetate and melengestrol acetate after application as growth promoters in cattle: Environmental studies. *Environmental Health Perspectives* 109: 1145–1511.

Shinogi, Y. and Kanri, Y. 2003. Pyrolysis of plant, animal and human waste: Physical and chemical characterization of the pyrolytic products. *Bioresource Technology* 90: 241–247.

Smith, A., Brown, K., Ogilvie, S., Rushton, K., and Bates, J. 2001. Waste management options and climate change—report to the European Commission. Luxembourg: Office for Official Publications of the European Communities. Online: http://ec.europa.eu/environment/waste/studies/pdf/climate_change.pdf. Accessed on January 4, 2014.

Smith, J.E. and Perdek, J.M. 2004. Assessment and management of watershed microbial contaminants. *Critical Reviews in Environmental Science and Technology* 34: 109–139.

Soto, A.M., Calabro, J.M., Prechtl, N.V., Yau, A.Y., Orlando, E.F., Daxenberger, A., Kolok, A.S., Guillette, L.J., Bizec, B.L., Lange, I.G., and Sonnenschein, C. 2004. Androgenic and estrogenic activity in water bodies receiving cattle feedlot effluent in Eastern Nebraska, USA. *Environmental Health Perspectives* 112(3): 346–352.

Sparling, R., Risbey, D., and Poggi-Varaldo, H.M. 1997. Hydrogen production from inhibited anaerobic composters. *International Journal of Hydrogen Energy* 22: 563–566.

Stehman, S.M., Rossiter, C.R., McDonough, P., and Wade, S. 1996. Potential pathogens in manure. In: J.S. Popow (ed.) *Animal Agriculture and the Environment: Nutrients, Pathogens, and Community Relations*. Ithaca, NY, Northeast Regional Agricultural Engineering Service, pp. 47–55.

Stine, O.C, Johnson, J.A., Keefer-Norris, A., Perry, K.L., Tigno, J., Qaiyumi, S., Stine, M.S., and Morris, J.J.G. 2007. Widespread distribution of tetracycline resistance genes in a confined animal feeding facility. *International Journal of Antimicrobial Agents* 29(3): 348–352.

Thelen, K.D., Fronning, E.B., Kravchenko, A., Min, H.D., and Robertson, P.G. 2010. Integrating livestock manure with a corn-soybean bioenergy cropping system improves short-term carbon sequestration rates and net global warming potential. *Biomass and Bioenergy* 34: 960–966.

Ueno, Y., Kawai, T., Sato, S., Otsuka, S., and Morimoto, M. 1995. Biological production of hydrogen from cellulose by natural anaerobic microflora. *Journal of Fermentation and Bioengineering* 79: 395–397.

Ueno, Y., Tatara, M., Fukui, H., Makiuchi, T., Goto, M., and Sode, K. 2007. Production of hydrogen and methane from organic solid wastes by phase-separation of anaerobic process. *Bioresource Technology* 98: 1861–1865.

Umetsu, K., Kimura, Y., Takahashi, J., Kishimoto, T., Kojima, T., and Young, B. 2005. Methane emission from stored dairy manure slurry and slurry after digestion by methane digester. *Journal of Animal Science* 76: 73–79.

United States Environmental Protection Agency. 2003. National pollutant discharge elimination system permit regulation and effluent limitation guidelines and standards for concentrated animal feeding operations (CAFOS). In *Code of Federal Regulations*: *Title 40 Protection of Environment*. pp. 7176–7274. Online: http://www.ecfr.gov. Accessed on January 4, 2014.

USDA-ARS. 2006. 2005 annual report: Manure and byproduct utilization. Online: http://www.ars.usda.gov. Accessed on January 4, 2014.

Vanotti, M.B., Millner, P.D., Hunt, P.G., and Ellison, A.Q. 2005. Removal of pathogen and indicator microorganisms from liquid swine manure in multi-step biological and chemical treatment. *Bioresource Technology* 96: 209–214.

Wang, C.C., Chang, C.W., Chu, C.P., Lee, D.J., Chang, B.V., Liao, C.S, and Tay, J.H. 2003. Using filtrate of waste biosolids to effectively produce bio-hydrogen by anaerobic fermentation. *Water Research* 37: 2789–2793.

Wang, L.J. 2008. *Energy Efficiency and Management in Food Processing Facilities*. Boca Raton, FL: CRC Press.

Wilkie, C.A., Castro, F.H., Cubinski, R.K., Owens, M.J., and Yan, C.S. 2004. Fixed-film anaerobic digestion of flushed dairy manure after primary treatment: Wastewater production and characterization. *Biosystems Engineering* 89: 457–471.

Wu, H., Huang, S., and Jiang, Z. 2004. Effects of modification of silica gel and ADH on enzyme activity for enzymatic conversion of CO_2 to methanol. *Catalysis Today* 98: 545–552.

Wu, H.J., Hanna, A.M., and Jones, D.D. 2012. Fluidized-bed gasification of dairy manure by Box-Behnken design. *Waste Management and Research* 30: 506–511.

Xin, J.Y., Cui, J.R., Niu, J.Z., Hua, S.F., Xia, C.G., Li, S.B., and Zhu, L.M. 2004. Production of methanol from methane by methanotrophic bacteria. *Biocatalysis and Biotransformation* 22: 225–229.

Xiu, S.N., Zhang, Y., and Shahbazi, A. 2008. Swine manure solids separation and thermochemical conversion to heavy oil. *Bioresources Journal* 4: 458–470.

Xiu, S.N., Zhang, Y.H., and Shahbazi, A. 2009. Swine manure solids separation and thermochemical conversion to heavy oil. *Bioresources Technology* 4: 458–470.

Yaman, S. 2004. Pyrolysis of biomass to produce fuels and chemical feedstocks. *Energy Conversion and Management* 45: 651–671.

Yokoyama, H., Waki, M., Moriya, N., Yasuda, T., Tanaka, Y., and Haga, K. 2007. Effect of fermentation temperature on hydrogen production from cow waste slurry by using anaerobic microflora within the slurry. *Applied Microbiology and Biotechnology* 74: 474–483.

Zhang, R.H. and Westerman, W.P. 1997. Solid-liquid separation of animal manure for odor control and nutrient management. *Applied Engineering in Agriculture* 13(3): 385–393.

Zhang, T., Liu, Y., and Fang, H.H.P. 2005. Effect of pH change on the performance and microbial community of enhanced biological phosphate removal process. *Biotechnology and Bioengineering* 92: 173–182.

Zhang, Y. and Shen, J. 2006. Effect of temperature and iron concentration on the growth and hydrogen production of mixed bacteria. *International Journal of Hydrogen Energy* 31: 441–446.

Zhu, H., Stadnyk, A., Beland, M., and Seto, P. 2007. Co-production of hydrogen and methane from potato waste using a two-stage anaerobic digestion process. *Bioresource Technology* 99: 5078–5084.

Zhu, J. 2000. A review of microbiology in swine manure odor control. *Agriculture, Ecosystems and Environment* 78: 93–106.

13 Municipal Solid-Waste Management Strategies for Renewable Energy Options

Richa Kothari, Virendra Kumar, N.L. Panwar, and V.V. Tyagi

CONTENTS

13.1 INTRODUCTION

Waste is defined as any material that is not useful and does not represent any economic value to its owner, and the owner being the waste generator should be responsible for its management. Depending on the physical state of waste, wastes are categorized into solid, liquid, and gaseous. Solid wastes are categorized into municipal wastes, hazardous wastes, medical wastes, and radioactive wastes (Table 13.1). Managing solid waste generally involves planning, financing, construction, and operation of facilities for the collection, transportation, and management strategies for final settlement of the waste. Municipal solid-waste (MSW) generation is an issue of worldwide concern. It is defined as any waste generated by household, commercial, and/or institutional activities and is not hazardous. Depending upon the source, MSW is categorized into three types: residential or household waste, which arises from domestic areas from individual houses; commercial wastes and/or institutional wastes, which arise from individually larger sources of MSW like hotels, office buildings, and schools; and municipal services wastes, which arise from area sources like streets and parks. MSW usually contains food wastes, paper, cardboard, plastics, textiles, glass, metals, wood, street sweepings, landscape and tree trimmings, and general wastes from parks, beaches, and other recreational areas. Sometimes, other household wastes like batteries and consumer electronics also get mixed up with MSW due to the lack of separation practices at source. Hence, management of MSW is a complex task, which depends as much upon organization and cooperation between households, communities, private enterprises, and municipal authorities as it depends upon the

TABLE 13.1
Types of Solid Waste

Residential	Single and multifamily dwelling	Food waste, paper, cardboard, plastics, textile, leather, yard waste, wood, glass, metals, ashes, special waste
Industrial	Light and heavy manufacturing, fabrication, construction sites, power and chemical plants	Housekeeping waste, packing and food waste construction and demolition materials, hazardous waste, ashes, special waste
Commercial	Restaurants, markets, office buildings, etc.	Stones, bottles, papers, cardboard, plastics, wood, food wastes, glass, metals, special waste
Institutional	School, hospitals, prisons, governmental institutes	Same as commercial
Construction and demolition	New construction sites, road repair, renovation site, demolition of buildings	Wood, steel, concrete, dirt, etc.
Municipal sites	Street cleaning, land scoping parts, beaches, other recreational areas, water and waste water treatment plants	Street sweeping; landscape and tree trimming; general waste from parks, beaches, and other recreational areas; sewage sludge
Process	Heavy and light manufacturing, refineries, chemical plants, power plants, mineral extraction and processing	Industrial process waste, scrap materials, off-specification products, slag tailing
Agricultural	Crops, farms, dairies	Spoiled food waste, agriculture residues, pesticides

Source: World Bank, What a waste: Solid waste management in Asia Urban Development Sector Unit East Asia and Pacific Region, May 1999, http://www.worldbank.org/html/fpd/urban/publicat/whatawaste.pdf (accessed September, 2003).

selection and application of appropriate technical solutions for waste collection, transfer, recycling, and disposal. Furthermore, MSW management is an essential task also, which has important consequences for public health and well-being, the quality and sustainability of the urban environment, and the efficiency and productivity of the urban economy. In developed countries, the production of MSW per capita is higher and management of MSW is efficient than developing countries.

Tremendous increase in the amount of MSW has been reported in the cities of developing countries due to an improved lifestyle and social status (Sharholy et al. 2007). In most cities of developing countries, waste management is inadequate: a significant portion of the population does not have access to a waste-collection service and only a fraction of the generated waste is actually collected. MSW management is a major responsibility of local governments, and it requires an additional financial allocation to treat the waste typically consuming between 20% and 50% of municipal budgets in developing countries (Peter et al. 1996). Systems for transfer, recycling, and/or disposal of solid waste are unsatisfactory from the environmental, economic, and financial points of view. The urban areas of Asia spent about US$25 billion on solid-waste management per year and this figure will increase to at least US$50 billion in 2025. The daily waste generation rate is about 760,000 tons. By 2025, this rate will be increased to about 1.8 million tons per day (World Bank 1999).

In recent years, MSW management has attracted increasing attention from bilateral and multilateral development agencies, due to the mounting urgency of urban environmental problems identified. MSW management systems should therefore include special measures for preventing hazardous materials from entering the waste stream. Improper management of waste leads to several problems for the environment as well as for the living beings. These negative environmental impacts are only a result of improper solid-waste disposal. They do not include the substantial environmental degradation resulting from the extraction and processing of materials at the beginning of the product life cycle. Liquid waste could leach out and contaminate soil and groundwater. Uncontrolled burning of wastes and improper incineration contribute significantly to urban air pollution. Greenhouse gases

(GHGs) are generated from the decomposition of organic wastes in landfills. Therefore, it is required to dispose of the waste in such a way without posing any environmental losses and creating health hazards. The generation of energy from the MSW is a new approach of solid-waste management. It provides dual benefits through the treatment of waste and generation of energy. There are various technologies available to generate energy from MSW, but the use of an appropriate one is dependent on characteristics of waste. The implementation of MSW management practices benefits both public health and environmental quality directly and substantially.

This study focuses only on the sources that generate MSW and management strategies to reduce the amount of MSW on environment and energy crisis.

13.2 CLASSIFICATION, COMPOSITION, AND CHARACTERISTICS OF MSW

MSW, also called urban solid waste, is a type of waste that includes predominantly household waste (domestic waste) with sometimes the addition of commercial wastes, construction and demolition debris, sanitation residue, and waste from streets collected by a municipality within a given area. They are in either solid or semisolid form and generally exclude industrial hazardous wastes. Semisolid wastes such as sludge and night soil are considered to be liquid wastes. Figure 13.1 shows the classification of MSW.

The composition of MSW varies widely from place to place as shown in Table 13.2 (World Bank 1999). The most striking difference that can be seen is the difference in inorganic content of MSW in low-, medium-, and high-income countries.

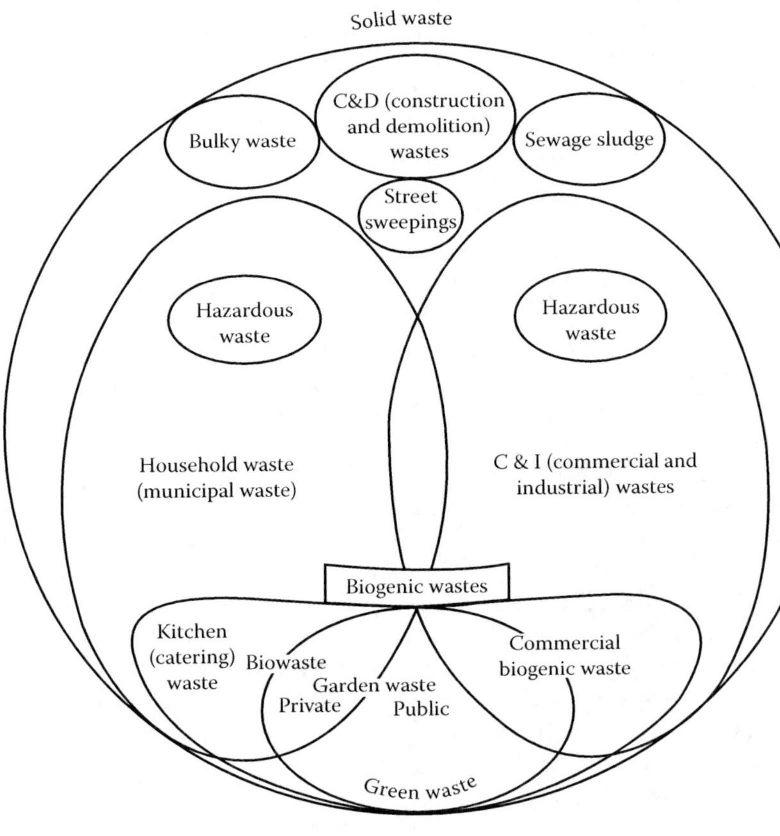

FIGURE 13.1 Typical classification of MSW.

TABLE 13.2
Relative Composition of Household Waste in Low-, Medium-, and High-Income Countries

Parameters	Low-Income Countries	Medium-Income Countries	High-Income Countries
Organic (%)	40–85	20–65	20–30
Paper (%)	1–10	15–30	15–40
Plastics (%)	1–5	2–6	2–10
Metals (%)	1–5	1–5	3–13
Glass (%)	1–10	1–10	4–10
Rubber, leather, etc. (%)	1–5	1–5	2–10
Others (%)	15–60	15–60	2–20
Moisture content (%)	40–80	40–60	5–20
Specific weight (kg/m^3)	250–500	170–330	100–170
Calorific value (kcal/kg)	800–1100	1000–1300	1500–2700

Source: Hassan, M.N., Solid waste management—What's the Malaysian position? In *Urban Sustainability in the Context of Global Change*, ed. R.B. Singh, Plymouth, MN, Science, 2004.

The total amount of MSW generated in the United States in 2010 was 250 million tons (US EPA 2011). Organic materials are the largest component of MSW. For the MSW generated in 2010, paper and paperboard accounted for 29% and yard trimmings and food scraps accounted for another 27%. Plastics were 12%, and rubber, leather, and textiles accounted for 8%. Wood was around 6%. The noncombustibles of metals and glass in MSW were 9% and 5%, respectively. Other miscellaneous wastes made up approximately 3% of the MSW. The pie chart depicts the percentages by which different materials contribute to the MSW stream (Figure 13 2). The compositions of MSW in selected developed and developing countries are given in Table 13.3 (Zhang et al. 2010). The composition of MSW determines the technology needed for waste processing prior to disposal.

13.3 GENERATION AND MANAGEMENT OF MSW WORLDWIDE

The amount and composition of solid waste generated in any country depend on the economic development, density of population, size of the urban habitation, and consumption rate of commercial goods. The increase of waste generation rates due to population growth, change of people's lifestyles, and development and consumption of products with materials that are less biodegradable has led to the various challenges for MSW management in cities around the world. The waste generation rate per capita, that is, the amount of waste generated by one person in 1 day in a country or region is strongly correlated to the gross domestic product (GDP) of the country or region. The waste generation rate generally increases with the increase in GDP. At global level, the world can be divided using two methods. First, the classification of developed and developing countries is based on economic growth, technology development, and infrastructure in the last 50 years. Second, the classification of low-, medium-, and high-income countries is based on the income per capita. In this chapter, we consider both categories according to data available to support the data analyzed.

High-income countries generate more waste per person compared to low-income countries due to the reasons discussed in following sections. As shown in Table 13.4, the average amount of waste generated per capita in India is 370 g/day as compared to 2200 g/day in Denmark, 2000 g/day in the United States, and 700 g/day in China (US EPA 2009; Liu and Wu 2011).

According to UNEP (2005), the rate of waste generation generally increases in direct proportion to that of a nation's advance in development. The failure to provide a proper MSW management system could result in greater environmental degradation and increase health risk to the urban

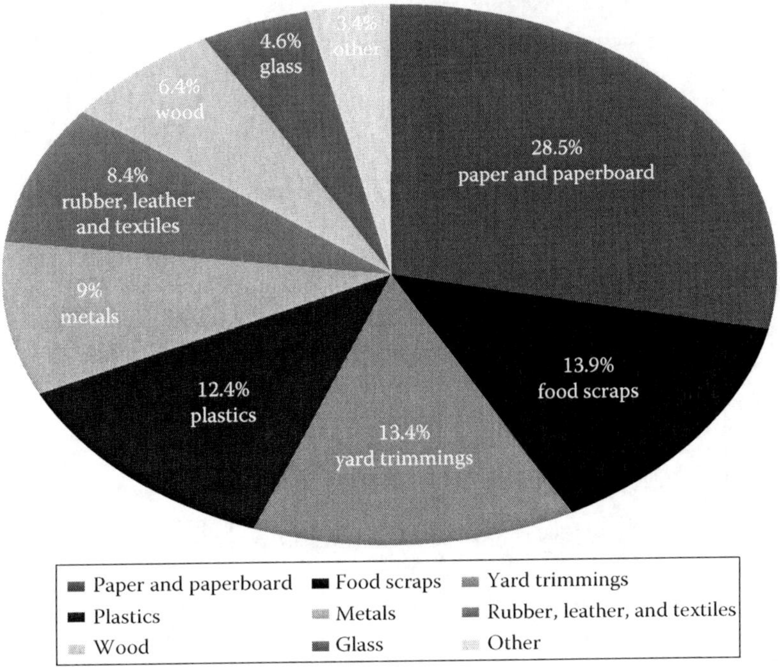

FIGURE 13.2 Compositional variation in MSW in the United States before recycling. (From US EPA, United States Environmental Protection Agency Solid Waste and Emergency Response, Washington, DC, U.S. Environmental Protection Agency Office of Resource Conservation and Recovery, December 2011.)

population. In most developed countries, focus is on optimization of waste management practices with a broader goal of resource conservation (McDougall et al. 2001; Wilson 2007). Table 13.5 highlights the amounts of MSW generated in developed and developing countries.

13.3.1 MSW Generation and Management in a Developed Country (United States)

The amount of MSW generated in the United States was 250 million tons in 2010, which was approximately tripled from 88 million tons in 1960. This increase was caused by the economic development and population growth. The amount of MSW generated by each person was 4.43 pounds per day in the United States in 2010, which was also doubled in the last 50 years from 1960 to 2010. Residential waste has been estimated to be 55%–65% of the total MSW generated in the United States as shown in Figure 13.3 according to the 2011 US EPA report. Waste from commercial and institutional locations, such as businesses, schools, and hospitals, amounted to the remaining 35%–45% (US EPA 2011).

The most part of the MSW is recovered by various methods, but a large part of it is disposed of without any further treatment (Table 13.6). The organic part of MSW is disposed of in landfill, which is a common practice in developed nations, but it requires land. The liquid part of MSW or the leachate in landfill percolates inside the soil, which contaminates the groundwater and soil. The recovery of energy from organic part of MSW is a new and promising approach. Thermal conversion or biochemical conversion can be used to recover energy from organic wastes.

13.3.2 MSW Generation and Management in Developing Countries

Rapid economic growth by the industrialization in the developing countries has created serious problems of waste disposal due to uncontrolled and unmonitored urbanization (ISWA and UNEP 2002).

TABLE 13.3
Comparison of Typical Distribution of MSW in Different Countries

Country	Year	Organic Garbage	Papers/ Cardboard	Plastics	Glass	Metals	Textiles	Wood	Ash	Construction Debris	Horticulture Waste	Others
China	2003	52.6	6.9	7.3	1.6	0.5	4.7	6.9	19.2	—	—	—
Singapore	2008	9.5	21.2	11.5	1.0	14.6	1.6	4.5	9.5	15.4	3.8	7.5
United States	2005	25.0	34.0	12.0	5.0	8.0	—	—	—	—	—	16.0
Japan	2002	34.0	33.0	13.0	5.0	3.0	—	—	—	—	—	12.0
Korea	2005	28.0	24.0	8.0	5.0	7.0	—	—	—	—	—	28.0
France	2005	32.0	20.0	9.0	6.0	13.0	—	—	—	—	—	8.0
Netherlands	2005	35.0	26.0	19.0	10.0	3.0	—	—	—	—	—	26.0
Germany	2005	14.0	34.0	22.0	4.0	4.0	—	—	—	—	—	12.0
Switzerland	2005	29.0	20.0	19.0	12.0	5.0	—	—	—	—	—	12.0
Australia	2005	47.0	23.0	15.0	4.0	3.0	—	—	—	—	—	29.0
Mexico	2005	51.0	15.0	4.0	7.0	5.0	—	—	—	—	—	13.0
Slavic	2005	38.0	13.0	6.0	6.0	3.0	—	—	—	—	—	18.0
Portugal	2005	34.0	21.0	7.0	8.0	3.0	—	—	—	—	—	31.0
Hungary	2005	29.0	15.0	11.0	7.0	4.0	—	—	—	—	—	32.0

Source: Zhang, D.Q. et al., *J. Environ. Manage.*, 91, 1623, 2010.

TABLE 13.4
Comparison between the Per Capita MSW Generation Rates in Low-, Medium-, and High-Income Countries

Country	Per Capita Urban MSW Generation (kg/Day)	
	1999	2025
Low-income countries	0.45–0.9	0.6–1.0
Medium-income countries	0.52–1.1	0.8–1.5
High-income countries	1.1–5.07	1.1–4.5

Source: Ranjith, K.A., Sustainable solid waste management in India by submitted in partial fulfillment of the requirements for the degree of Master of Science in earth resources engineering, Department of Earth and Environmental Engineering, Fu Foundation, School of Engineering and Applied Science, Columbia University in the City of New York, 2012.

TABLE 13.5
Waste Generations in Different Developed Countries

Countries	Year	Total Amount of MSW Generation (1000 Tons)	MSW Generation Rate (kg/Capita/Day)
United States	2005	222,863	2.05
France	2005	33,963	1.48
Germany	2005	49,563	1.64
Denmark	2005	3,900	2.03
Switzerland	2005	4,855	1.78
Australia	2005	—	—
Poland	2005	9,354	0.68
Portugal	2005	5,009	1.29
Hungary	2005	4,632	1.26
Mexico	2005	36,088	0.93
Japan	2005	51,607	1.10
Korea	2005	18,252	1.04
China	2006	212,100	0.98

Source: Zhang, D.Q. et al., *J. Environ. Manage.*, 91, 1623, 2010.

The amount of MSW generated per capita in the cities of developing countries was in the range of 0.2–1.7 kg/day (Visvanathan and Trankler 2003). MSW generation also depends on the urbanization and development of the country. For example, higher amounts of waste generation in the cities of Sri Lanka are caused by the increased consumption patterns as well as the movement of the people from the rural areas to urban centers. Similarly, in Thailand, over 23% of the population is urban, and its high rate of economic growth caused the higher amount of MSW generated per capita per day (Visvanathan and Trankler 2003). The problem is further aggravated by the lack of financial as well as human resources trained in waste management practices including collection,

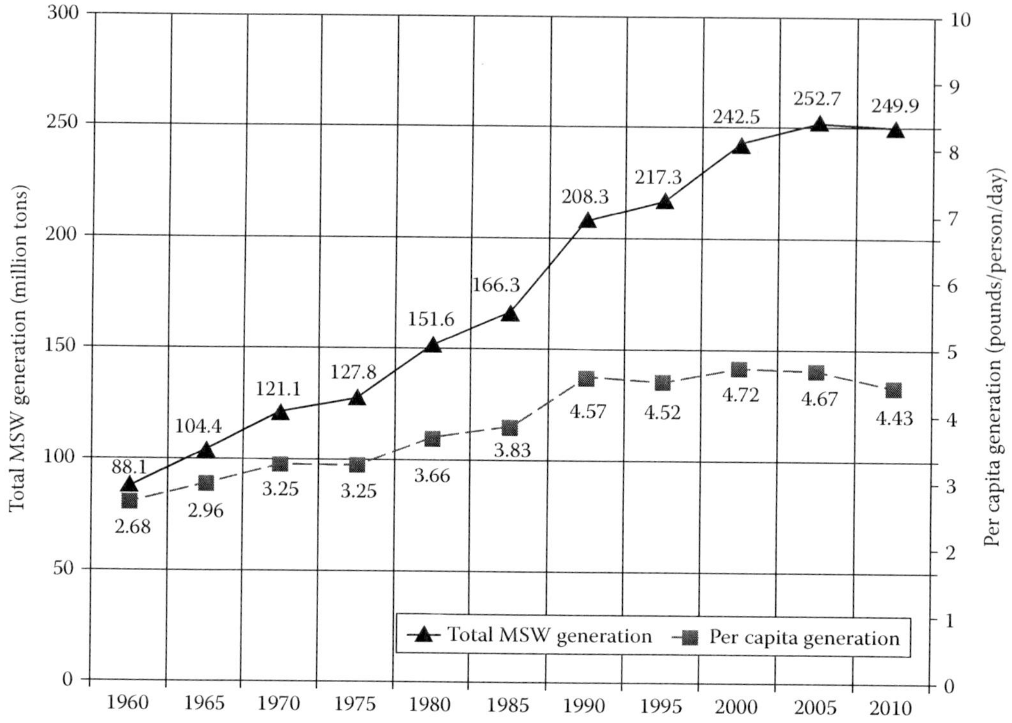

FIGURE 13.3 Total waste generation in the United States on per capita basis. (From US EPA, United States Environmental Protection Agency Solid Waste and Emergency Response, Washington, DC, U.S. Environmental Protection Agency Office of Resource Conservation and Recovery, December 2011.)

TABLE 13.6
Generation and Recovery of Materials in MSW in the United States, 2010 (in Millions of Tons and Percent of Generation of Each Material)

Material	Waste Generated	Waste Recovered	Remained Disposable	Remained Disposable in %
Paper and paperboard	71.31	44.57	26.47	37.12
Glass	11.53	3.13	8.4	72.85
Total metals	22.41	7.87	14.54	64.88
Plastics	31.04	2.55	28.49	91.78
Rubber and leather	7.78	1.17	6.61	84.96
Textiles	13.12	1.97	11.15	84.98
Wood	15.88	2.30	13.58	85.51
Other materials	4.79	1.41	3.38	70.56
Other waste				
Food, others	34.76	0.97	33.79	97.20
Yard trimming	33.40	19.20	14.20	42.51
Miscellaneous organic waste	72.00	20.17	51.83	71.98
Total MSW	249.86	85.14	164.72	65.92

Source: US EPA, United States Environmental Protection Agency Solid Waste and Emergency Response, Washington, DC, U.S. Environmental Protection Agency Office of Resource Conservation and Recovery, December 2011.

transportation, processing, and final disposal. A typical waste management system in a low- or medium-income country includes the following elements:

- Waste generation and storage
- Segregation, reuse, and recycling at the household level
- Primary waste collection and transport to a transfer station or community bin
- Street sweeping and cleansing of public places
- Management of the transfer station or community bin
- Secondary collection and transport to the waste disposal site
- Waste disposal in landfills
- Collection, transport, and treatment of recyclables at all points on the solid-waste pathway (collection, storage, transport, and disposal)

The management practices such as recycling, reuse, and recovery of the MSW are grossly driven by demand and supply or disorganized in most cases. The composition of the waste generated in the developing countries is similar, and the variation in the MSW composition among different regions is caused by the climatic, cultural, industrial, infrastructural, and legal factors. In considering the major technical operations involved in MSW management, the guidelines based on the methods used in the developed nations and their regulations may not be amenable to the conditions in the developing countries (Zurbrugg 1999). The associated environmental and public health risks play a key role in designing the various operations based on the technical, climatic, and socioeconomic setup in the developing countries. According to the composition of the waste generated in these developing countries, the biodegradable portion is mainly food and yard waste. Within the developing countries, the MSW in the developed regions has a higher paper and cardboard content (Zhang et al. 2010).

The waste composition from India shows a comparable lower food content but higher inorganic (ash and dust) content. The lower amounts of paper, glass, and plastic are due to the intensive collection and scavenging by informal waste collectors (Ministry of Finance, Government of India 2009). China has also higher ash content, which is due to the geographical location in the temperate latitudes and common use of raw coal. Thailand, on the other hand, has an increasing trend in plastic and paper—an impact of the progressing industrialization and urbanization with a growing GDP (National Research Institute [NRI] 2003). As indicated, the waste generated is primarily biodegradable that is usually dumped on barren land or nonengineered landfills.

13.4 MSW IN INDIA: CASE STUDY

Covering an area of 3,287,263 km^2, India is the seventh largest country in the world. Lying entirely in the northern hemisphere, the Indian mainland extends between latitudes 8°4′ and 37°6′ north and longitudes 68°7′ and 97°25′ east and measures about 3214 km from north to south between the extreme latitudes and about 2933 km from east to west between the extreme longitudes. It is the second most populated country in the world. The total Indian population according to the census of 2011 was 1.21 billion on the map of the world, which was 17.7% of the total world's population. It is as much as the combined population of the United States, Indonesia, Brazil, Pakistan, Bangladesh, and Japan. Based on estimates, more than 55 million tons of MSW is generated in India per year, and the yearly increase in MSW generation is estimated to be about 5%. It is estimated that the amounts of solid waste generated in small, medium, and large cities and towns in India are about 0.1, 0.3–0.4, and 0.5 kg per capita per day, respectively. Waste generation rate in Indian cities ranges between 200 and 870 g/day/person, depending upon the region's lifestyle and the size of the city. The estimated annual increase in per capita waste generation is about 1.3% per year (Ministry of Finance, Government of India 2009). The increase in the population and rapid income growth in India have changed the lifestyle of urban residents, thus changing the composition of the garbage generated. The presence of paper, plastic, and metal is on the rise, resulting in more disposal difficulty. The municipalities have not been able to collect and dispose of the enormous quantity of waste being generated.

13.4.1 Composition of MSW Generated in Indian Cities

With rising urbanization and change in lifestyle and food habits, the amount of MSW has been increasing rapidly and its composition has been changing. MSW contains recyclables (paper, plastic, glass, metals, etc.), toxic substances (paints, pesticides, used batteries, medicines), compostable organic matter (fruit and vegetable peels, food waste), soiled waste (blood-stained cotton, sanitary napkins, disposable syringes), and inerts (sand, pebbles, and gravels) (Khan 1994; Reddy and Galab 1998; Jha et al. 2003) (Figure 13.4a and b). In Indian cities, the total amount of solid waste generated annually has been increased from 6 to 48 million tons from year 1947 to 1997 with an annual growth rate of 4.25% and is anticipated to increase and reach up to 300 million tons by 2047 (CPCB 2004; Gupta and Garg 2009).

The quantity of MSW generated (CPCB 2000a) and the generation rate of MSW per capita (CPCB 2004) are shown in Table 13.7 and Figure 13.5, respectively. It can be seen from Table 13.7 and Figure 13.5 that the per capita generation rate is high in some states (Gujarat, Delhi, and Tamil Nadu) and cities (Madras, Kanpur, Lucknow, and Ahmedabad). This may be due to the high living standards, the rapid economic growth, and the high level of urbanization in these states and cities.

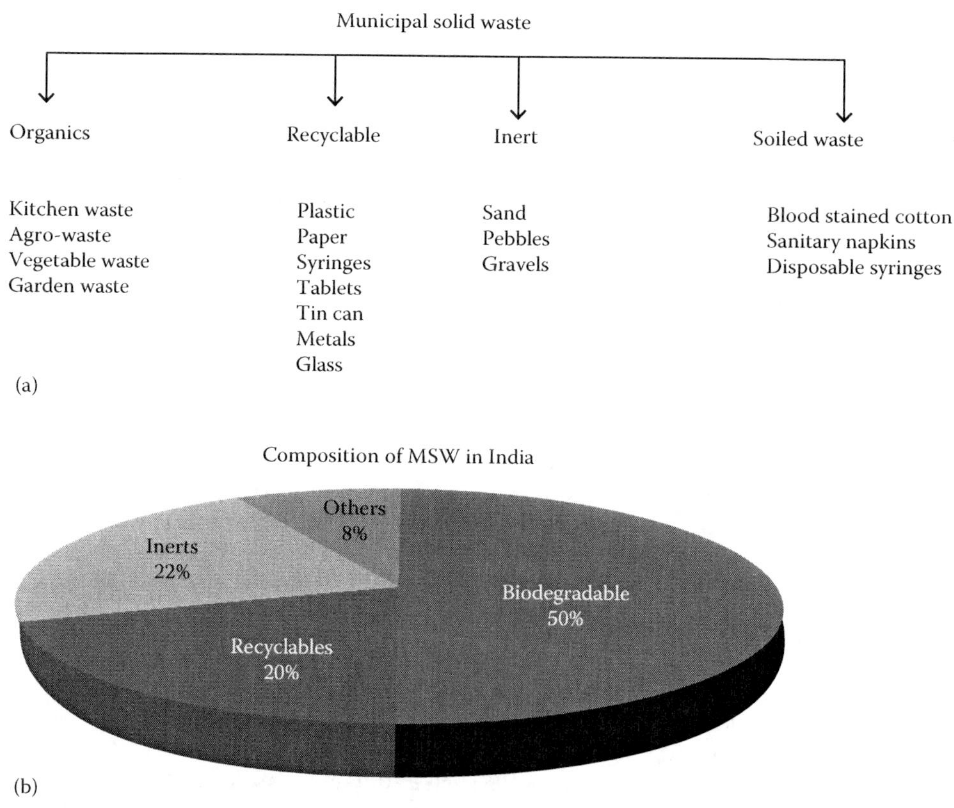

FIGURE 13.4 Contents (a) and percentage composition (b) of MSW in India. (From http://www.eai.in/ref/ae/wte/typ/clas/msw.htm46.)

TABLE 13.7
MSW Generation in Different States in India

S. No.	Name of State	No. of Cities	Municipal Population	MSW (Tons/Day)	Per Capita Generated (kg/Day)
1	Andhra Pradesh	32	10,845,907	3,943	0.364
2	Assam	4	878,310	196	0.223
3	Bihar	17	5,278,361	1,479	0.280
4	Gujarat	21	8,443,962	3,805	0.451
5	Haryana	12	2,254,353	623	0.276
6	Himachal Pradesh	1	82,054	35	0.427
7	Karnataka	21	8,283,498	3,118	0.376
8	Kerala	146	3,107,358	1,220	0.396
9	Madhya Pradesh	23	7,225,833	2,286	0.316
10	Maharashtra	27	22,727,186	8,589	0.378
11	Manipur	1	198,535	40	0.201
12	Meghalaya	1	223, 366	35	0.157
13	Mizoram	1	155,240	46	0.296
14	Orissa	7	1,766,021	646	0.366
15	Punjab	10	3,209,903	1,001	0.312
16	Rajasthan	14	4,979,301	1,768	0.355
17	Tamil Nadu	25	10,745,773	5,021	0.467
18	Tripura	1	157,358	33	0.210
19	Uttar Pradesh	41	14,480,479	5,515	0.381
20	West Bengal	23	13,943,445	4,475	0.321
21	Chandigarh	1	504,094	200	0.397
22	Delhi	1	8,419,084	4,000	0.297
23	Pondicherry	1	203,065	60	0.295
	Total	299	128,113,865	48,134	0.376

Sources: CPCB (Central Pollution Control Board), Management of municipal solid waste Delhi, New Delhi, India, Central Pollution Control Board, 2000a; CPCB, Status of municipal solid waste generation, collection, treatment and disposal in class I cities, Series: ADSORBS/31/1999–2000, 2000b; CPCB, Status of solid waste generation, collection, treatment and disposal in metrocities, Series: CUPS/46/1999–2000, 2000c.

It has been noticed that the physical and chemical characteristics of MSW also varies with population density in Indian cities, as shown in Tables 13.8 and 13.9 (Bhide and Shekdar 1998; CPCB 2000; Garg and Prasad 2003; Mufeed et al. 2008). It has been clear that the composition of MSW at generation sources and collection points consists mainly of a large organic fraction (40%–60%), ash and fine earth (30%–40%), paper (3%–6%), and plastic, glass, and metals (each less than 1%). The C/N ratio ranges between 20 and 30, and the lower calorific value ranges between 3.35 and 4.18 MJ/kg.

The composition dictates the technology needed for waste processing prior to disposal. Generally, MSW is disposed of in low-lying areas without taking any precautions or operational controls. Therefore, the management of MSW is going through a critical phase, due to the unavailability of suitable facilities to treat and dispose of the larger amount of MSW generated daily in metropolitan cities. Improper disposal causes an adverse impact on all components of the environment and human health (Kansal et al. 1998; Singh and Singh 1998; Kansal 2002; Ray et al. 2005; Sharholy et al. 2005; Rathi 2006). In the next section, an attempt has been made to provide a comprehensive analysis of MSW management practices for Indian cities to evaluate the current status.

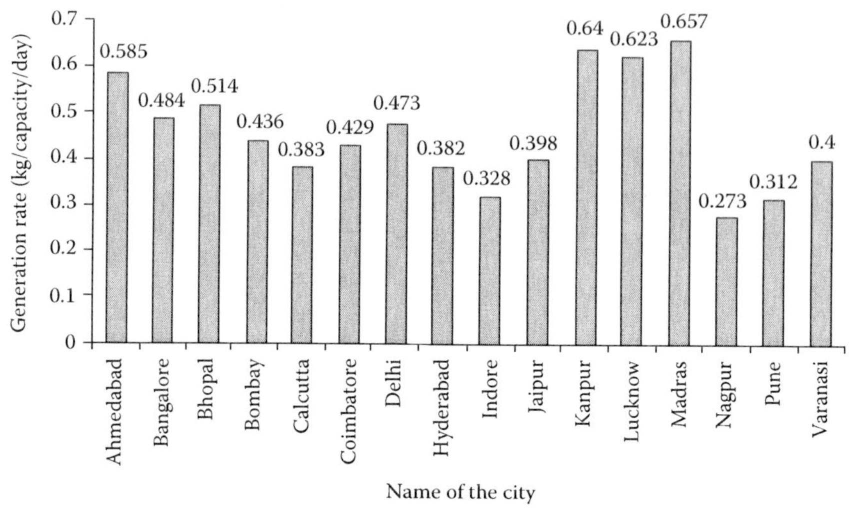

FIGURE 13.5 Per capita generation rate of MSW for Indian cities. (From Central Pollution Control Board (CPCB), Management of municipal solid waste, New Delhi, India, Ministry of Environment and Forests, 2004.)

TABLE 13.8
Physical Characteristics of MSW in Indian Cities Population Wise

Population Range (in Millions)	No. of Cities Surveyed	Paper	Rubber, Leather, and Synthetics	Glass	Metals	Compostable Matter	Inert Material
0.1–0.5	12	2.91	0.78	0.56	0.33	44.57	43.59
0.5–1.0	15	2.95	0.73	0.56	0.32	40.04	48.38
1.0–2.0	9	4.71	0.71	0.46	0.49	38.95	44.73
2.0–5.0	3	3.18	0.48	0.48	0.59	56.57	49.07
5.0 and above	4	6.43	0.28	0.94	0.8	30.84	53.9

Source: NEERI report strategy paper on SWM in India, August 1995.

TABLE 13.9
Chemical Charecteristics of MSW in Indian Cities Population Wise

Population Range (in Millions)	Nitrogen as Total Nitrogen	Phosphorous as P_2O_5	Potassium as K_2O	C/N Ratio	Calorific Value
0.1–0.5	0.71	0.63	0.83	30.94	1009.89
0.5–1.0	0.66	0.56	0.69	21.13	900.61
1.0–2.0	0.64	0.82	0.72	23.68	980.05
2.0–5.0	0.56	0.69	0.78	22.45	907.18
5.0 and above	0.56	0.52	0.52	30.11	800.70

Source: Sharholy, M. et al., *Waste Manage.*, 27(4), 490, 2007.

13.4.2 MSW Management Practices in India

MSW management is one of the major environmental problems of Indian megacities. It involves activities associated with generation, storage, collection, transfer and transport, processing, and disposal of solid wastes. But in most cities, the MSW management system comprises only four activities, that is, waste generation, collection, transportation, and disposal. The existing practices in solid-waste management can be classified at three levels, depending upon the quantity of solid waste and the physical area covered, that is, rural level, town level, and big city/metropolitan city level. Among these, a relatively well-defined system of collection, transportation, and disposal only exists at metropolitan city level, whereas at the rural and state level, no specific mode of collection, transportation, and disposal exists. At these levels, waste is generally dumped in low-lying areas and burned openly. Rag pickers can also be seen at waste-collection and disposal points. Normally, collection of MSW is the responsibility of corporations/municipalities. The predominant system of collection in most of the cities is through communal bins placed at various points along the roads, and sometimes, this leads to the creation of unauthorized open collection points. Efforts to organize house-to-house collection are just starting in many metro cities such as Delhi, Mumbai, Bangalore, Madras, and Hyderabad with the help of NGOs. It has been observed that many municipalities have employed private contractors for secondary transportation from the communal bins or collection points to the disposal sites. This type of public–private partnerships (PPPs) may be a good option for the management of MSW in the future. Figure 13.6 shows the flow for the case of MSW management in Mumbai (Rathi 2006).

After this, a fraction of MSW generated remains uncollected on streets due to nonavailability of waste-collection services at overcrowded low-income settlement areas. The reason is that these settlements are often illegal and the inhabitants are unwilling or unable to pay for the services. They throw away the waste near or around their houses at different times, which makes the collection and transportation of waste very difficult in these areas. The collection efficiency is the quantity of MSW collected and transported from streets to disposal sites divided by the total quantity of MSW generated during the same period. Many studies on urban environment have revealed that MSW collection efficiency is a function of two major factors: manpower availability and transport capacity (Nema 2004; Siddiqui et al. 2006). Hence, management of MSW requires proper infrastructure, maintenance, and upgrade for all activities. This becomes increasingly expensive and complex due to the continuous and unplanned growth of urban centers. The difficulties in providing the desired level of public service in the urban centers are often attributed to the poor financial status of the managing municipal corporations (MoEF 2000; Raje et al. 2001; Mor et al. 2006).

The major stakeholders in the management of MSW from the government sector include (a) the Ministry of Environment and Forests (MoEF), (b) Ministry of Urban Development (MoUD), (c) Central and State Pollution Control Boards, (d) Department of Urban Development, (e) state-level nodal agency, (f) urban local bodies, and (g) private formal and informal sectors. Many acts and regulations relating to protecting the environment have come up from time to time from major stakeholder's side. The rule pertaining to MSW management in India is the MSW (Management and Handling) Rules, 2000. These rules deal with the scientific management of MSW by ensuring proper collection, segregation, storage, transportation, processing, and disposal of MSW. These rules clearly indicated the following measures for improving solid-waste management practices: prohibit street littering, organize a waste-collection system, conduct awareness programs, provide adequate community storage facilities and color-coded bins, promote segregation at the source, cover transport vehicles, and process waste through appropriate technologies including the composting, recycling, and recovery of materials (APO 2007). So, proper disposal, reuse, and recovery from MSW are a necessity to minimize environmental health impacts and degradation of land resources.

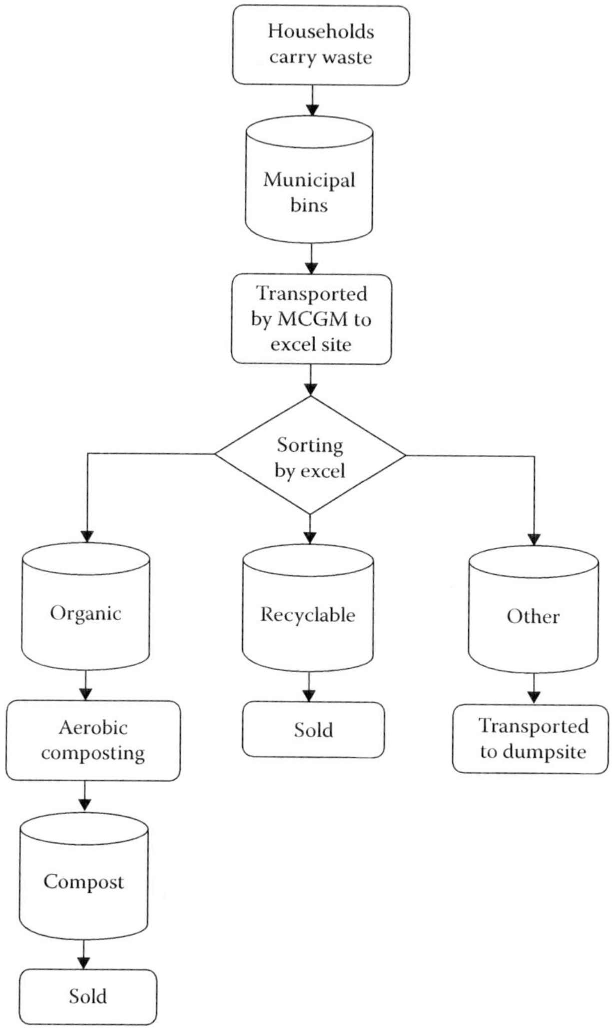

FIGURE 13.6 Flow sheet diagram used in PPP for Mumbai. (From Rathi, S., *J. Waste Manage.*, 26(10), 1192, 2006.)

13.4.3 Energy Recovery from MSW as a Renewable Option

There is an urgent need to work toward a sustainable solid-waste management system, which is environmentally, economically, and socially sustainable. An MSW-to-renewable-energy recovery option can be an alternative for sustainable management of this waste and will be helpful in tackling this huge quantity of waste of India (Figure 13.7). A variety of processes exist for MSW-to-energy conversion. The most used methods are thermal conversions (incineration, fast and slow pyrolysis, gasification, production of refuse-derived fuel [RDF]), biochemical (composting, vermicomposting, anaerobic digestion/biomethanation), and chemical conversions (transesterification and other processes to convert plant and vegetable oils to biodiesel). The choice of a conversion process depends on the type, property, and quantity of waste feedstock, the desired form of the energy, end user requirements, environmental standards, economic conditions, and project-specific factors (Table 13.10).

Thermochemical conversion entails thermal decomposition of organic matter to produce either heat energy or fuel oil or gas. The thermochemical conversion processes are useful for wastes

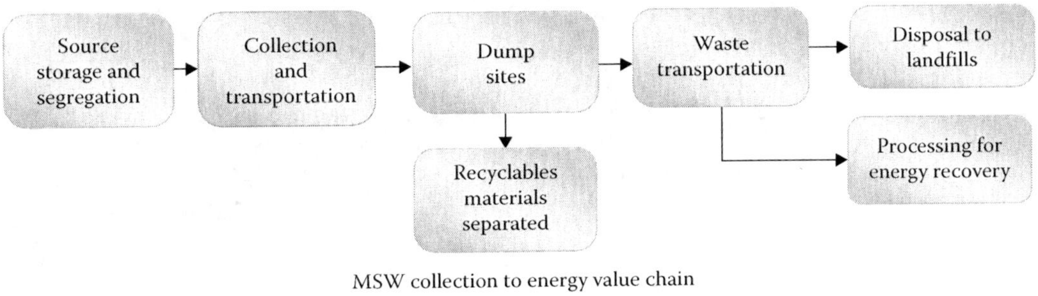

FIGURE 13.7 Schematic flow chart for renewable energy recovery from MSW. (From Santham, N., Sewage, http://www.eai.in/ref/ae/wte/typ/clas/msw.html.)

TABLE 13.10
Status of MSW in Developing Countries

	MSW Generation per Capita (kg/Day)		
Country	**2007**	**2010**	**2030 * (Expected)**
India	0.75	0.79	0.97
Nepal	0.32	0.35	0.44
Pakistan	0.39	0.44	0.59
Bangladesh	0.67	0.73	0.98
Sri Lanka	0.56	0.59	0.73

Sources: World fact sheet, http://www.worldfactsandfigures.com/gdp_country_desc.php, 2001; World Bank 2009.

Note: The data is significant and calculated on the basis of rate of generation of solid waste for developing countries. Expected data clearly indicate increase in solid waste generation per capita. "*" indicates expected outcome calculated on the basis of previous year data.

containing a high percentage of nonbiodegradable matter and low moisture content. The main technological options under this category include incineration, pyrolysis/gasification, and RDF.

Biochemical conversion is based on enzymatic decomposition of organic matter by microbial action to produce methane gas or alcohol. The biochemical conversion processes, on the other hand, are preferred for wastes having a high percentage of organic biodegradable (putrescible) matter and high level of moisture/water content, which aids microbial activity. The main technological options under this category are anaerobic digestion/biomethanation.

Due to the difference in resources (materials/energy) that can be recovered from MSW, only three important and economically viable processes, which are anaerobic digestion, RDF, and waste-to-energy combustion (WTEC), are discussed and reviewed in the Indian scenario in this chapter, which will give an idea for investors about the market potential, the maturity of the practicing technology, and the environmental aspects with renewable options.

13.4.3.1 Anaerobic Digestion

Anaerobic digestion is helpful in lessening the amount of organic solid waste and recovering energy. Basically, the organic fraction of MSW is a potential feedstock for anaerobic digestion (Braber 1995; Iglesias et al. 2000; Ambulkar and Shekdar 2004; Elango et al. 2007; Igoni et al. 2008). Typically, the organic fraction may be digested and the biogas may be utilized either for combined heat and power (CHP) or as a transport fuel. However, nonrecyclable nonorganic fraction from MSW may be either incinerated or gasified. During the decomposition process of the organic fraction of MSW, the temperature rises and may reach as high as 65°C but starts to fall after 1–2 months. However, the process of fermentation goes on for a long time and a number of gases are produced, including small amounts of CO and H_2S. Under anaerobic conditions, methane is produced. Methane (high-heat gas) can be efficiently converted into methanol. Figure 13.5 shows the steps involved in a typical anaerobic process (Metcalf Eddy 1996). It has been estimated that in controlled anaerobic digestion, 1 ton of MSW produces the amount of methane in 3 weeks as much as two to four times that 1 ton of waste in landfill will produce in 6–7 years (Ahsan 1999; Saxena et al. 2009).

The Indian government has enhanced the outlay for energy recovery from urban waste in the fiscal year 1999–2000 budget. This has given a boost to projects through the national program on energy recovery from urban (municipal solid) wastes and projects on the development of high-rate biomethanation processes as a means of reducing GHG emissions. Financial incentives to state nodal agencies, municipal corporations, financial institutions, project developers, and technology providers for setting up MSW-to-energy recovery projects are being provided by the Ministry of Nonconventional Energy Sources (MNES) under these programs (CPCB 2000; REN21 2009). In Uttar Pradesh State, the Nonconventional Energy Development Agency (NEDA), Department of Additional Sources of Energy, government of Uttar Pradesh, and the Maharashtra Energy Development Agency (MEDA) invited potential promoters to set up power projects based on MSW (Ramachandra 2006). Chennai City, the Tamil Nadu state capital, generates approximately 2500 tons of MSW per day. Two international companies, CGEA (Singapore) and Albucheri (Kuwait), are in a race along with two Indian companies, namely, Swacherita Corporation (Bangalore) and the state-owned Tamil Nadu Agro Engineering Federation, to apply the government approval for the collection and disposal of MSW. A 15 MW waste-to-energy (WTE) project has been established by Energy Development Ltd. (Australia) (Ramachandra and Varghese 2003). The State Bank of India and Canara Bank will finance the project, including a waste treatment plant. The chemical characteristics of MSW in India are given in Table 13.8. Table 13.9 shows zone-wise solid-waste generation in India (Ramachandra and Bachamanda 2007). Studies had been carried out on anaerobic digestion of MSW (Structured Urban and Industrial Database 2007), and it was found that biogas could be generated at a rate of 95 m^3/t of solid waste. The biogas generation potential for the MSW is estimated at 9.23 $m^3/t/day$ (Ravindranath et al. 2005).

13.4.3.2 Refuse-Derived Fuel

RDF refers to the segregated high calorific fraction of processed MSW. It can be defined as the final product from waste materials that have been processed mainly to achieve a high calorific value to be useful as secondary/substitute fuels in the solid fuel industry (Gendebien et al. 2003). It is a good alternative for coal in high-energy industrial processes like power production, cement kilns, and steel manufacturing, where RDF's use can be optimized to enhance economic performance (Gendebien et al. 2003). This process can be used for both organic and inorganic fractions of MSW. With organic fraction, the net GHG emissions are zero. The overall GHG emissions from RDF are, however, not zero. This is due to carbon emissions from burning the plastics fractions presented in RDF. The amount of GHG emissions from RDF depends upon the composition of organics and plastics in the MSW stream that the RDF is produced from. The use of RDF prevents GHG emissions from landfills, displaces fossil fuels, and reduces the volume of waste that needs to be landfilled, thus increasing their operating life. It's a good and sustainable option for energy recovery from MSW (Table 13.11).

In developing countries like India, as MSW has a low calorific value (7.3 MJ/kg compared to values greater than 10 MJ/kg in Europe, Japan, and the United States) (European Environment Agency 2011) and high percentage of inerts, it is necessary to process the MSW to make it suitable as a fuel. One of the less expensive and well-established technologies to produce RDF from MSW is mechanical biological treatment (MBT). An MBT plant separates out metals and inert materials,

TABLE 13.11
Important MSW Parameters for Technical Viability of Energy Recovery

Treatment Method	Basic Principal	Important Parameters
Thermochemical conversion	Decomposition of organic matter by action of heat	Organic content
Incineration		Volatile matter
Pyrolysis		Fixed carbon
Gasification		Calorific value
RDF		
Biochemical conversion	Decomposition of organic matter by microbial action	Moisture content
Anaerobic		Organic content
Digestion/biomethanation		Volatile matter
		C/N ratio

Source: CPHEEO, *Manual on Municipal Solid Waste Management*, New Delhi, India, Central Public Health and Environmental Engineering Organisation, Ministry of Urban Development, 2000.

screens out organic fractions (for stabilization using composting processes), and separates out high calorific fractions for RDF. RDF can also result from a *dry stabilization process* in which residual waste (after separating out metals and inert materials) is dried through a composting process, leaving the residual mass with a higher calorific value (Gendebien et al. 2003). The RDF thus produced is either used directly as floc/fluff or is compressed to make pellets. RDF fluff (as it is called in India) can be directly combusted, whereas making RDF pellets increases the marketability of the product as they can be used through cocombustion in various solid fuel industries like cement kilns and coal-fired power plants. There are six RDF plants in India, near Hyderabad, Vijayawada, Jaipur, Chandigarh, Mumbai, and Rajkot. The plant in Vijayawada used to serve the city of Guntur too. The Hyderabad and Vijayawada plants handled 700 and 500 tons/day of MSW to generate 6 MW of electricity, respectively. RDF produced in these plants was combusted in specifically designed boilers. The RDF plants near Jaipur and Chandigarh combust the RDF produced in cement kilns to replace fossil fuels. They handle 500 tons/day of MSW each.

13.4.3.3 Waste-to-Energy Combustion

WTEC is defined as a process of controlled combustion, using an enclosed device to thermally break down combustible solid waste to an ash residue that contains little or no combustible material and meanwhile produce electricity, steam, or other energy products (Bureau of Land and Waste Management 1999). The objective of WTEC process is to treat MSW to reduce its volume. Generating energy and electricity also adds value to this process. WTE is recognized as a renewable energy technology by the Government of India (GOI). Australia, Denmark, Japan, Netherlands, and the United States also recognize WTE as a renewable energy technology (Kothari et al. 2010).

Thermal WTE technologies are the only solutions to handling mixed wastes. WTE is a point source pollution control technology using thermal process, where the impurities in the input mixed waste are captured using extensive pollution control technologies and can be handled separately. The bottom ash from WTEC contains nothing but inert inorganic materials and minerals that could be used to make bricks and other construction material. The fly ash from WTE contains pollutants from the input stream and needs to be disposed of in sanitary landfills. By controlling the types of materials fed in to the boiler, European and Japanese WTE plants are known to have achieved nearly zero emissions in the fly ash too. WTEC decreases the volume of wastes by up to 90%. Such reduction in volume would prolong the life of a 20-year landfill to 200 years. However, MSW should be combusted after all possible recycling and composting have been done. Therefore, WTEC is placed after recycling and anaerobic digestion for sustainable waste management.

There are only two WTE mass burn combustion plants currently in operation in India. The latest one among them has finished the construction on the Okhla landfill site, New Delhi, and is about to start operations. An earlier WTEC plant, which was built in Timarpur, New Delhi, has the option of WTE for thermal treatment of 300 tons/day of waste with 3.75 MW of envisaged power generation (Nguyen et al., 2007). The plant was on trial run and was operated for a few months only and subsequently closed down in 1990 due to a mismatch of the quality of incoming refuse with the plant design. The plant was based on the technology developed on turnkey basis by M/s Volund of Denmark. A well-developed and well-controlled waste management system is considered a prerequisite for a waste incineration project to be successful and effective. WTE recovery projects are more complex than conventional fossil fuel–based plants and require skilled personnel.

The two more WTE plants in Hyderabad and Vijayawada are also recognized for mass burn combustion. But they combust RDF produced after considerable processing of MSW and the addition of secondary biomass fuels like rice husk. Therefore, they are known as RDF-WTE plants.

13.5 CONCLUSION

This chapter examined and discussed the quantity and quality of MSW and its management in developed and developing countries. There are many shortcomings in the existing practices for the management of MSW. These pertain mainly to inadequate manpower, financial resources, and implements/machinery required to effectively carry out various activities of MSW management. To overcome the deficiencies in the existing MSW management systems, three alternative approaches were considered in this chapter for sustainable management of MSW and energy recovery from MSW. All these alternatives are part of integrated waste management system with reduction in waste load and energy potential for both clean environment and energy crisis.

REFERENCES

Ahsan, N. 1999. Solid waste management plan for Indian megacities. *Indian Journal of Environmental Protection* 19(2):90–95.

Ambulkar, A.R. and A.V. Shekdar. 2004 Prospects of biomethanation technology in the Indian context: A pragmatic approach. *Resources Conservation Recycling* 40:111–128.

APO. 2007. Solid waste management: Issues and challenges in Asia. *Report of the APO Survey on Solid-Waste Management 2004–05*, ed. The Environmental Management Centre, Mumbai, India: The Asian Productivity Organization, Hirakawacho, Chiyoda-ku, Tokyo, Japan, p. 69.

Bhide, A.D. and A.V. Shekdar. 1998. Solid waste management in Indian urban centers. *International Solid Waste Association Times (ISWA)* 1:26–28.

Braber, K. 1995. Anaerobic digestion of municipal solid waste: A modern waste disposal option on the verge of breakthrough. *Biomass and Bioenergy* 9:365–376.

Bureau of Land and Waste Management, Division of Mining and Solid Waste Management. May 28, 1999. Regulation 67-107.12, SWM: Solid Waste Incineration and solid waste pyrolysis facilities. Columbia, SC: South Carolina Department of Health and Environmental Control. [Online].

Central Pollution Control Board (CPCB). 2004. Management of municipal solid waste. New Delhi, India: Ministry of Environment and Forests.

CPCB. 2000a. Management of municipal solid waste Delhi. New Delhi, India: Central Pollution Control Board.

CPCB. 2000b. Status of municipal solid waste generation, collection, treatment and disposal in class I cities, Series: ADSORBS/31/1999–2000.

CPCB. 2000c. Status of solid waste generation, collection, treatment and disposal in metrocities, Series: CUPS/46/1999–2000.

CPHEEO. 2000. Manual on Municipal Solid Waste Management. New Delhi, India: Central Public Health and Environmental Engineering Organisation, Ministry of Urban Development.

Department of Economic Affairs, Ministry of Finance, Government of India. November 2009. Position Paper on the Solid Waste Management Sector in India. Public Private Partnerships in India. [Online].

Elango, D., M. Pulikesi, P. Baskaralingam, V. Ramamurthi, and S. Sivanesan. 2007. Production of biogas from municipal solid waste with domestic sewage. *Journal of Hazardous Material* 141:301–304.

European Environment Agency. 2011. Municipal Waste Generation Per Capita in Western Europe (EU-15), New Member States (EU-12), EU Countries (EU-27) and total in Europe (EU-27 + Turkey, Croatia, Norway, Iceland, Switzerland). European Environment Agency. [Online].

Garg, S. and B. Prasad. 2003. Plastic waste generation and recycling in Chandigarh. *Indian Journal of Environmental Protection* 23(2):121–125.

Gendebien, A., K. Leavens, A. Godley, K. Lewin, K.J. Whiting, R. Davis, J. Giegrich, H. et al. July 2003. *Refuse Derived Fuel, Current Practice and Perspectives*. Environment, European Commission, Wiltshire, U.K. [Online].

Gupta, R. and V.K. Garg. 2009. Vermiremediation and nutrient recovery of non-recyclable paper waste employing Eisenia fetida. *Journal of Hazardous Materials* 162(1):430–439.

Hassan, M.N. 2004. Solid waste management—What's the Malaysian position? in *Urban Sustainability in the Context of Global Change*, Ed. R.B. Singh. Plymouth, MN: Science.

Iglesias, J.R., L. Castrilloa, N. Pelaez, E. Marana, O.N. Maison, and H.S. Andres. 2000. Biomethanization of municipal solid waste in a pilot plant. *Water Research* 34:447–454.

Igoni, A.H., M.J. Ayotamuno, C.L. Eze, S.O.T. Ogaji, and S.D. Probert. 2008. Designs of anaerobic digesters for producing biogas from municipal solid-waste. *Applied Energy* 85:430–438.

International Solid Waste Association and United Nations Environment Programme (ISWA and UNEP). 2002. Waste management, industry as a partner for sustainable development.

Jha, M.K., O.A.K. Sondhi, and M. Pansare. 2003. Solid waste management—A case study. *Indian Journal of Environmental Protection* 23(10):1153–1160.

Kansal, A. 2002. Solid waste management strategies for India. *Indian Journal of Environmental Protection* 22(4):444–448.

Kansal, A., R.K. Prasad, and S. Gupta. 1998. Delhi municipal solid waste and environment—An appraisal. *Indian Journal of Environmental Protection* 18(2):123–128.

Khan, R.R. 1994. Environmental management of municipal solid wastes. *Indian Journal of Environmental Protection* 14(1):26–30.

Kothari, R., V.V. Tyagi, and A. Pathak. 2010 Waste to energy: A way from renewable energy sources to sustainable development. *Renewable and Sustainable Energy Review* 14:3164–3170.

Liu, C. and X.W. Wu. 2011. Factors influencing municipal solid waste generation in China: A multiple statistical analysis study. *Waste Management and Research, International Solid Wastes and Public Cleansing Association* 29(4):371–378.

McDougall, F., P. White, M. Franke, and P. Hindle. 2001. *Integrated Waste Management: A Life Cycle Inventory*, 2nd edn. Oxford, U.K.: Blackwell Science.

Metcalf, E. and Eddy, I N C. 1996. *Waste Water Engineering Treatment Disposal Reuse*. New Delhi, India: Tata McGraw-Hill.

Ministry of Environment and Forests (MoEF). 2000. The Gazette of India. New Delhi, India: Municipal Solid Waste (Management and Handling) Rules.

Mor, S., K. Ravindra, A.D. Visscher, R.P. Dahiya, and A. Chandra. 2006. Municipal solid waste characterization and its assessment for potential methane generation: A case study. *Journal of Science of the Total Environment* 371(1):1–10.

Mufeed, S., K. Ahmad, G. Mahmood, and R.C. Trivedi. 2008. Municipal solid waste management in Indian cities—A review. *Waste Management* 28(2):459–467.

National Research Institute (NRI) 2003. China. Municipal solid waste management in China. Country Report.

NEERI report strategy paper on SWM in India, August 1995.

Nema, A.K., 2004. Collection and transport of municipal solid waste. In: *Training Program on Solid Waste Management*. Delhi, India: Springer.

Nguyen, P.H.L., P. Kuruparan, and C. Visvanathan. 2007. Anaerobic digestion of municipal solid waste as a treatment prior to landfill. *Bioresource Technology* 98:380–387.

Peter, S., K. Wehrle, and J. Christen. August 1996. *Conceptual Framework for Municipal Solid Waste Management in Low Income Countries*, 1st edn. SKAT (Swiss Centre for Development Cooperation in Technology and Management), St. Gallen, Switzerland.

Raje, D.V., P.D. Wakhare, A.W. Despande, and A.D. Bhide. 2001. An approach to assess level of satisfaction of the residents in relation to SWM system. *Journal of Waste Management and Research* 19:12–19.

Ramachandra, T.V. 2006. Management of municipal solid waste. New Delhi, India: Capital Publishing Company.

Ramachandra, T.V. and S. Bachamanda. 2007. Environmental audit of municipal solid waste management. *International Journal of Environmental Technology and Management* 7(3/4):369–391.

Ramachandra, T.V. and S.K. Varghese. 2003. Exploring possibilities of achieving sustainability in solid waste management. *Indian Journal of Environmental Health* 45(4):255–264.

Ranjith, K.A. 2012. Sustainable solid waste management in India by submitted in partial fulfillment of the requirements for the degree of Master of Science in earth resources engineering, Department of Earth and Environmental Engineering, Fu Foundation, School of Engineering and Applied Science, Columbia University in the City of New York.

Rathi, S. 2006. Alternative approaches for better municipal solid waste management in Mumbai, India. *Journal of Waste Management* 26(10):1192–1200.

Ravindranath, N.H., H.I. Somashekar, M.S. Nagaraja, P. Sudha, G. Sangeetha, and S.C. Bhattacharya. 2005. Assessment of sustainable non-plantation biomass resources potential for energy in India. *Biomass and Bioenergy* 29:178–190.

Ray, M.R., S. Roychoudhury, G. Mukherjee, S. Roy, and T. Lahiri. 2005. Respiratory and general health impairments of workers employed in a municipal solid waste disposal at open landfill site in Delhi. *International Journal of Hygiene and Environmental Health* 108(4):255–262.

Reddy, S. and S. Galab. 1998. An Integrated Economic and Environmental Assessment of Solid Waste Management in India—The Case of Hyderabad, India.

REN21, Renewables 2009 global status report. http://www.ren21.net/Portals/0/documents/activities/gsr/RE_GSR_2009_Update.pdf.

Santham, N. Sewage. http://www.eai.in/ref/ae/wte/typ/clas/msw.html.

Saxena, R.C., D.K. Adhikari, and H.B. Goyal. 2009. Biomass-based energy fuel through biochemical routes: A review. *Renewable and Sustainable Energy Reviews* 13:167–178.

Sharholy, M., K. Ahmad, G. Mahmood, and R.C. Trivedi. 2005. Analysis of municipal solid waste management systems in Delhi—A review. In: *Book of Proceedings for the second International Congress of Chemistry and Environment*, Indore, India, pp. 773–777.

Sharholy, M., K. Ahmad, R. Vaishya, and R. Gupta. 2007. Municipal solid waste characteristics and management in Allahabad, India. *Waste Management* 27(4):490–496.

Siddiqui, T.Z., F.Z. Siddiqui, and E. Khan. 2006. Sustainable development through integrated municipal solid waste management (MSWM) approach—A case study of Aligarh district. In: *Proceedings of National Conference of Advanced in Mechanical Engineering (AIME-2006)*, Jamia Millia Islamia, New Delhi, India, pp. 1168–1175.

Singh, S.K. and R.S. Singh. 1998. A study on municipal solid waste and its management practices in Dhanbad–Jharia coalifield. *Indian Journal of Environmental Protection* 18(11):850–852.

Structured Urban and Industrial Database. 2007. National Master Plan for Development of Waste-to-Energy in India,.

UNEP. 2005. http://www.cep.unep.org/publications-and-resources/marine-and-coastal-issues-links/solid-waste-and-marine-litter.

United States Environmental Protection Agency. 2009. Municipal solid waste generation, recycling, and disposal in the United States: Facts and Figures for 2009. Washington, DC: United States Environmental Protection Agency. [Online].

US EPA. December 2011. United States Environmental Protection Agency Solid Waste and Emergency Response. Washington, DC: U.S. Environmental Protection Agency Office of Resource Conservation and Recovery.

Visvanathan, C. and J. Trankler. 2003. Municipal Solid Waste Management in Asia: A comparative analysis workshop on sustainable landfill management. Chennai, India, pp. 3–15.

Wilson, D.C. 2007. Development drivers for waste management. *Waste Management and Research* 25:198–207.

World Bank. May 1999. What a waste: Solid waste management in Asia Urban Development Sector Unit East Asia and Pacific Region. http://www.worldbank.org/html/fpd/urban/publicat/whatawaste.pdf (accessed September, 2003).

World fact sheet. http://www.worldfactsandfigures.com/gdp_country_desc.php. (2001).

Zhang, D.Q., S.K. Tan, and R.M. Gersberg. 2010. Municipal solid waste management in China: Status, problems and challenges. *Journal of Environmental Management* 91:1623–1633.

Zurbrugg, C. 1999. The challenge of solid waste disposal in developing countries. *SANDEC News*, EAWAG, No. 4.

14 Food Processing Wastes as an Energy Feedstock
Availability and Sustainability

Lijun Wang

CONTENTS

14.1 INTRODUCTION

Large quantities of liquid and solid wastes are produced by the food processing industry. The total amount of food wastes around the world is 1.3 billion tons each year. In the developed nations, 40% of the food waste is generated at the retail and consumer levels, while 40% of the food waste is generated at the production and processing stages in developing nations (Curry and Pillay, 2012). The total amount of food wastes and by-products produced in the European Union has been estimated at approximately 222 million tons/year (Mahro and Timm, 2007). The US EPA estimated that 230 million metric tons of municipal solid waste (MSW) were generated in 2007, of which 77.1 million metric tons were recovered via recycling or composting and 28.8 million metric tons were food wastes. Based on the US EPA figures, only 3% of food waste was decomposed and the remaining 97% of food waste was buried in landfills in 2007 (US EPA, 2008; Levis et al., 2010). Food processing wastes contain principally biodegradable organic matter, and disposal of them may create serious environmental problems such as contamination of water and generation of odor (Hang, 2004). Various methods including landfill, incineration, and complete decomposition have been used for the treatment of food wastes at a high cost (Park et al., 2002). Life cycle cost analysis showed that the total costs of collection, transportation, and treatment for landfilling, incineration, and composting of food wastes in South Korea in 2008 were $77.24/ton, $119.84/ton, and $133.63/ton, respectively. The total revenues of by-products and environmental benefit for the landfilling, incineration, and composting methods were $2.94/ton, $31.06/ton, and $29.88/ton, respectively. Therefore, the benefit–cost ratio was only 0.04 for landfilling, 0.26 for incineration, and 0.22 for composting (Kim et al., 2011a).

Most of the wastes generated in food processing facilities can become a resource for producing bioenergy and biochemicals (Matteson and Jenkins, 2007). Conversion of food processing wastes into energy products such as bioethanol, biodiesel, bio-oil, biogas, hydrogen, syngas, steam, and electricity in a food processing facility can reduce the amount of energy purchased and waste disposal costs in the food manufacturing industry (Wang, 2008). It has been estimated that an anaerobic digestion unit at a processing capacity of 165 tons of food wastes per year could save a total of $30,168 annually including the savings of $21,615 in transportation and landfill costs and revenue of $8,553 from biogas for an anaerobic digestion unit (Curry and Pillay, 2012).

The types of raw food materials and the operations of a food processing facility greatly influence the kinds and amounts of wastes produced. Since the marginal profit in the food industry is usually small, the costs versus return on investment may not be favorable for some alternative waste treatment technologies (Kim et al., 2011b). However, the selection of a food waste treatment technology is affected not only by the net return on the investment but also by the emphasis of sustainable development and governmental environmental regulations. In this chapter, wastes from different food processing sectors are reviewed in terms of their availability, quality, and current utilizations. Technologies for value-added processing of different food wastes into bioenergy and biochemical products are then reviewed.

14.2 PROCESSING WASTES IN DIFFERENT FOOD PROCESSING SECTORS

14.2.1 Fruit and Vegetable Processing Wastes

The fruit and vegetable processing facilities generate large amounts of liquid and solid wastes. For comparative purposes, the International Finance Corporation (IFC, 2007) has established industrial benchmark values given in Table 14.1 for the amount of solid wastes produced per ton of fruit and vegetable products that are generally considered to be achievable in new processing facilities by existing technologies. As shown in Table 14.1, the amount of wastes can be as high as 20% of the original mass for processing fruits and vegetables.

Solid processing wastes from fruits and vegetables have high moisture and carbohydrate contents. The moisture content of fresh fruits and vegetables is as high as 95% on a wet basis. Pomace or press cakes from juice production may have moisture content below 50% depending on the method and efficiency of the press. Most of the carbohydrates in fruit and vegetable wastes are composed of soluble sugars and other easily hydrolysable polysaccharides. A major portion of the carbohydrates is dissolved or suspended in the processing wastewater. Appropriate disposal of the wastewater is a costly burden to the fruit and vegetable processing industry because wastewater must be properly treated to the degree necessary to comply with the regulatory standards for discharge into surface water or reuse for agricultural application (El-Kamah et al., 2010).

TABLE 14.1
Amounts of Wastes Generated to Process Selected Fruits and Vegetables

Fruits and Vegetables	Solid Waste Produced per Ton of Product (kg)
Maize	40
Peas	40
Potatoes	40
Broccoli	200
Carrots	200
Strawberries	60
Apples	90
Peaches	180

Source: International Finance Corporation (IFC), Environmental, health, and safety guidelines-food and beverage processing, Online: www.ifc.org/ifcext/enviro.nsf/Content/EnvironmentalGuidelines, 2007.

14.2.2 Oilseed Processing Wastes

Approximately 125 million tons of oils and fats were produced worldwide in 2003 (Hill, 2006). Soybean, palm, rapeseed, and sunflower are main oil sources as shown in Table 14.2. In the United States, the average production capacity of vegetable oils was 10.7 million tons/year between 1995 and 2000. Soybean was the dominant oil source in the United States, which was

TABLE 14.2
World Production of Oils and Fats in 2003

Oil and Fat	Amount (Million Tons)
Soybean	31.3
Palm oil	27.8
Rapeseed oil	12.5
Sunflower oil	8.9
Palm kernel oil/coconut oil	6.5
Other plant oils	14.9
Tallow	8.0
Butter	6.3
Other animal fats	8.2
Total	125

Source: Hill, K., Industrial development and application of biobased oleochemicals, in *Biorefineries-Industrial Processes and Products*, Vol. 2, pp. 291–314, B. Kamm, P.R. Gruber and M. Kamm eds., Weinheim, Germany, Wiley-VCH Verlag GmbH & Co, 2006.

TABLE 14.3
Composition of Oil Cakes

Oil Cake	Dry Matter (%)	Crude Protein (%)	Crude Fiber (%)	Ash (%)	Calcium (%)	Phosphorous (%)
Canola	90	33.9	9.7	6.2	0.79	1.06
Coconut	88.8	25.2	10.8	6.0	0.08	0.67
Cottonseed	94.3	40.3	15.7	6.8	0.31	0.11
Groundnut	95.6	49.5	5.3	4.5	0.11	0.74
Mustard	89.8	38.5	3.5	9.9	0.05	1.11
Olive	85.2	6.3	40.0	4.2	—	—
Palm kernel	90.8	18.6	37	4.5	0.31	0.85
Sesame	83.2	35.6	7.6	11.8	2.45	1.11
Soybean	84.8	47.5	5.1	6.4	0.13	0.69
Sunflower	91	34.1	13.2	6.6	0.30	1.30

Source: Ramachandran, S. et al., *Bioresour. Technol.*, 98, 2000, 2007.

8.3 million tons/year or about 78% of the total oil production in the United States and 27% of the total soybean production in the world (Canakci, 2007).

The oil contents of oilseeds are related to their varieties, genotype, and environmental influences. Sunflower seed typically has 40%–50% of oil by weight (Gercel, 2002). Rapeseed has 40% of oil (Onay et al., 2001). Soybean has only 15%–25% of oil, but it is a rich source of protein for foods and feeds. In oilseed milling facilities, oil cakes are a solid residues generated after oil extraction. There are two types of oil cakes: edible and nonedible (Ramachandran et al., 2007). Oil cakes are rich in protein and fiber. The composition of oil cakes depends on their variety, growing condition, and extraction methods. The chemical compositions of selected oil cakes are given in Table 14.3.

14.2.3 Grain Processing Wastes

Grains are an abundant renewable resource of starch for the production of foods, feeds, and fuels. Corn, wheat, and rice account for 85% of all grain production worldwide (Food and Agriculture Organization, 2011). Grain processing facilities extract carbohydrates, proteins, fats, and other materials from grains and convert them into multiple products including foods, fuels, and high-value chemicals and materials. Grains such as corn and wheat are processed through either a wet milling or dry milling process (Johnson, 2006). The wet milling has been used for many years in the starch industry. It has also been adapted and modified for ethanol production. In the corn wet milling process, corn is first soaked in water containing sulfur dioxide for 24–48 h. It is then ground, and the components are physically separated using a series of centrifuges, screens, and washes. Some of the separated fractions then undergo additional refining steps. The corn wet milling process converts corn into starch, oil, gluten meal, gluten feed, and germ meal. The need for freshwater in a corn wet milling plant is as high as 1.5 m^3/ton of corn (Kollacks and Rekers, 1988). Commercial manufacture of wheat starch and gluten involves physical separation of starch granules and gluten particles formed in a neutral aqueous system. The wet separation of gluten proteins and starch from wheat flour is based on their water insolubility, density, and particle size (Sayaslan, 2004). The dry milling process has been widely used in the ethanol industry. The dry milling process does not fractionate the different components of the grains. A dry milling process involves grinding of the grain, followed by addition of water and treatment with heat. Enzymes are added to the slurry, and the sugar that results from starch hydrolysis is fermented to ethanol by the addition of yeast. Fuel ethanol is recovered by distillation and evaporation. Distillers' grains and solubles, which are the processing wastes after recovering the ethanol, are traditionally used as animal feeds (Wang, 2008).

About 40.6% of the corn produced (or 5021.21 million bushels) was used to produce fuel ethanol in the United States in 2011 (USDA, 2012). Wheat is considered the primary raw material for ethanol production in Europe and Australia (Du et al., 2009). The type of coproducts that are produced in grain processing industry depends on a number of factors, including conversion technology, feedstock, and milling processing (Turhollow and Heady, 1986). Grain processing facilities may generate four main coproducts besides starch: (1) gluten with a high protein content, (2) bran with a high fiber content, (3) germ with high protein and lipid contents, and (4) distillers' grains and solubles in ethanol production.

Corn at a moisture content of 16% (wet basis) has about 62% starch, 8% protein, 4% fat, and 2% fiber by weight (Grull et al., 2006). Corn gluten feed (CGF) and corn gluten meal (CGM) are two major coproducts from the wet milling of corn (Rausch et al., 2003). The protein content in CGF and CGM varies from 22% to 26% (Rausch et al., 2003) and 60% to 70% (Di Gioia et al., 1999), respectively. Corn germ is another attractive coproduct because it contains high concentrations of protein and minerals and a number of high-value lipid compounds. A full-fat corn germ meal has 48.5% oil, 12.6% protein, 11.4% fiber, and 22.3% starch (Ronyai et al., 1998). Cereal bran has high fiber content. Fiber is composed mainly of nonstarch polysaccharides and lignin. The insoluble dietary fiber includes cellulose, lignin, and hemicelluloses, while the soluble fiber consists of pectins, beta-glucans, gums, and mucilages. In a corn dry mill, one-third of the original corn mass becomes corn distillers' dried grains and solubles (DDGS). The concentrations of dietary fiber, protein, and fats are three times of their concentrations in corn (Rasco et al., 1987; Wang et al., 2005).

Wheat at a moisture content of 13% has 60% starch, 13% protein, 3% oil, and 1.3% fiber by weight (Grull et al., 2006). About 80%–90% of the protein is made up of gluten (Weegels et al., 1992). Typically, wheat milling results in 78%–80% flour, 19% bran, and 1% wheat germ. The flour is washed with water to produce starch, and the starch processing residue is a commercially accepted feedstock for ethanol production (Nguyen, 2003). The wastewater from a wheat starch process was found to have 1.73% total solids (Verma et al., 2007). The distillery effluent from a starch-based ethanol plant was found to have 3.3% total solids (Ngyyen, 2003). These solids in the starch processing wastewater can be recovered as a potential renewable source (Nguyen, 2003; Verma et al., 2007).

14.2.4 Meat Processing Wastes

The consumption of pork, beef, and poultry in developed countries has been on the increase (Salminen and Rintala, 2002). In the meat industry, slaughterhouse is producing increasing amounts of organic solid by-products and wastes. The organic solid wastes generated in a poultry slaughterhouse include blood, feather, offal, feet, head, trimmings, and bones. A broiler is about 1.8–1.9 kg before it is slaughtered. During slaughtering and processing, about 40% of the broiler mass enters into the waste stream. Table 14.4 gives the quantities and characteristics of organic solid wastes produced in poultry slaughterhouses (Salminen and Rintala, 2002).

It can be seen from Table 14.4 that 39% of the feet and heart offals are solids and 95% of the solids are volatiles. Protein is the dominant solid matter in feather, which is 91% of the total solid matters. Fat is the dominant solid matter in offal, which is 54% of the total solid matters.

14.3 WASTE AUDITING AND MONITORING IN FOOD PROCESSING FACILITIES

14.3.1 Significance of Waste Auditing and Monitoring

Food processing wastes are commodities like other raw materials that should be measured regularly in facility operation. The audit of food processing waste not only quantifies the loss of food

TABLE 14.4
Quantities and Characteristics of Organic Solid Wastes Produced in Poultry Slaughterhouses

	Percentage of the Live Weight (%)	Total Solid (%)	Volatile Solid (%)	Protein (% of Total Solid)	Fat (% of Total Solid)
Blood	2	22	91	48	2
Feather	10	24.3	96.7	91	1–10
Offal, feet and heart	21	39	95	32	54
Trimmings and bone	7	22.4	68	51	22

Source: Salminen, E. and Rintala, J., *Bioresour. Technol.* 83, 13, 2002.

materials but also serves as a measure to determine the process-control effectiveness. Monitoring waste continuously and keeping records of the waste flows are also important to determine whether the effluent from a facility meets local, regional, and national pollution abatement regulations (Zall, 2004).

Monitoring waste is a management tool. Waste monitoring can be done to answer the following questions: (1) What is the waste source? (2) How much waste is produced? (3) What is its composition or concentration? and (4) where does it come from? The amount and quality of the waste are the important information to identify opportunities for recovering the unused resources into value-added products. The conversion of raw materials into finished food products involves a series of unit operations. Material loss can occur at each unit operation. Some food processing facilities first collect wastes from individual unit operations to a sump within the facility and then transport to a central waste treatment area. To quantify loss at only the central discharge of a facility neglects the ability to determine the loss from a specific unit operation. Therefore, it is necessary to establish waste sampling stations to gather and test representative samples of waste materials sent to the central discharge in a food processing facility (Zall, 2004).

14.3.2 Auditing the Amount of Food Processing Wastes

To manage food processing waste, the first step is to find out how much waste the facility produces. Most food processing facilities fail to measure their wastes as a separate identifiable by-product (Zall, 2004). For liquid food processing wastes, flow measurement is used to determine the quantity of the wastes. A simple technique for the flow measurement is to record the time that a pump of known capacity is in operation to deliver the wastes. A number of waste pumps in a large facility can be timed and their total output capacity is then calculated. The quantity of the liquid wastes is the flow rate of pump(s) times the operating time. A weir, which has a dam-like structure, is a common system to measure flow (Zall, 2004). During measurement, a flow is restricted through a weir and the height of flow above a discharge outlet is a measure of volume passing over or through the weir. Solid wastes in a food processing facility may be combined. The solid waste may contain food fragments, damaged packaged goods, and a mixture of raw materials, paper, plastics, and others. The mass of solid wastes can be weighted directly.

14.3.3 Auditing the Quality of Food Processing Wastes

For liquid food processing wastes, the common quality parameters are biological oxygen demand (BOD), chemical oxygen demand (COD), pH value, and temperature. The BOD to COD ratio determines the biodegradability of wastes. Other common analyses of liquid wastes include total solids, total suspended solids, volatile suspended solids, fixed suspended solids or ash, and dissolved solids.

Temperature of wastes should be measured because heat may be recovered from high-temperature liquid wastes. Hot water can be generated from high-temperature wastes. The hot water can be used as the feed water of a boiler and plant wash water. The record of the temperature of an effluent from a facility can also help to isolate the causes of fuel waste in the facility if an unusual temperature rise occurs (Zall, 2004).

Waste concentration or composition can be determined through sampling. For a small amount of liquid wastes, the best sampling method would be to contain waste in a single vessel, agitate the mass, and then collect a well-mixed representative portion of the waste. For a large amount of liquid waste, periodic sampling may be used. Periodic sampling is time and flow dependent. Proper sampling equipment is a unit where both sampling time and sample size can be varied (Zall, 2004). The composition of solid waste can be determined by grinding the waste and sampling homogeneous material. The food processing wastes contain carbohydrate, fat, proteins, and minerals besides waste. The different components have variable values depending on the ingredient sources (Wang, 2008).

14.4 CONVERSION OF FOOD PROCESSING WASTES INTO ENERGY PRODUCTS

14.4.1 Recovery of Materials and Energy from Food Wastes

Table 14.5 summarizes different technologies that have been investigated to recover materials and energy from various food wastes. Selection of an appropriate process is dependent upon (1) the characteristics of wastes such as moisture content and heat value, (2) the quantity of wastes available, (3) the desired form of energy products, (4) the efficiency of the energy conversion process, (5) energy demands for the processing facilities and in the reachable market, and (6) economical feasibility (Wang, 2008). Food processing wastes have traditionally been used as animal feeds (Lahmar et al., 1994; Ramachandran et al., 2007; Canakci, 2007) and organic fertilizers (Ritter and Chinside, 1995; Mohaibes and Heinonen-Tanski, 2004; Ramachandran et al., 2007). In recent years, there was an increasing interest in converting various food processing wastes into bioenergy and bioproducts. High-moisture-content food wastes (e.g., >50%) are usually better suited to biological processes such as anaerobic digestion, while low-moisture-content wastes are better for thermochemical conversion processes such as combustion, gasification, and pyrolysis. In many cases, it is the form of marketable bioenergy products that determines the process selection. Fermentation, transesterification, pyrolysis, and liquefaction produce liquid fuels including ethanol, biodiesel, and bio-oil suitable for the use as transportation fuels. Combustion, gasification, and anaerobic digestion produce gaseous energy products such as hot gas for steam generation, syngas, biogas, and hydrogen, which are suitable to be used at the production location. It is worth noting that high-value phytochemicals can be identified and extracted from the waste residuals in the food processing facilities before they are used for energy production for the increase of economic profitability (Wang et al. 2005; Wang and Weller, 2006; Wang, 2013).

14.4.2 Production of Ethanol from Carbohydrate-Rich Processing Wastes

Residues from fruit, vegetable, and grain processing sectors and food wastes from cafeterias are rich in carbohydrates, which can be used as a source for the production of fermentable sugars and bioethanol. The total sugar content of the juice extracted from pineapple was as high as 82.52 g/L (Nigam, 1999). Ethanol can be produced directly from the sugar juice of fruit pomace (Hang et al., 1986; Ngadi and Correia, 1992; Nigam, 2000). Starchy and cellulosic waste materials from fruit and vegetable processes, however, must be hydrolyzed first to fermentable sugars before yeast and bacteria such as *Zymomonas mobilis* and *Clostridium thermocellum* can ferment them to ethanol. Campo et al. (2006) and Fischer and Bipp (2005) used acidic hydrolysis of monosaccharides from several food processing residues including tomato residue, red pepper residue, sugar beet molasses,

TABLE 14.5
Value-Added Processing Food Wastes into Bioenergy Products

Value-Added Products	Selected Food Wastes	Major Technology	Challenges	Selected Reference
Bioethanol	(1) Fruit and vegetable processing residues (2) Grain processing residues (3) Cafeteria wastes	(1) Hydrolysis (2) Fermentation	(1) Increase process efficiency (2) Reduce production cost	Nigam (2000); Campo et al. (2006) and Fischer and Bipp (2005); Hang (2004); Choteborska et al. (2004)
Biogas	(1) Fish processing waste (2) Slaughterhouse wastes (3) Fruit and vegetable residues (4) Food processing wastewater	Anaerobic digestion	Reduce production cost	Gebauer (2004); Salminen and Rintala (2002); Kalia et al. (2000); Berardino et al. (2000)
Hydrogen	(1) Fish processing waste (2) Slaughterhouse wastes (3) Fruit and vegetable residues (4) Food processing wastewater	(1) Anaerobic digestion at controlled pH (e.g., 5.5) and temperature (e.g., 50°C) (2) Gasification	(1) Increase process efficiency (2) Reduce production cost	Pan et al. (2008); Redondas et al. (2012); Song et al. (2010) Muangrat et al. (2010a,b)
Heat and power	Dry food waste such as rice husk, bran, and bagasse	Incineration/combustion	(1) Require dry waste (2) Control pollutant emissions	Armesto et al. (2003); Digman and Kim (2008)
Syngas	Dry food waste such as rice husk, bran, and bagasse	Gasification	(1) Require dry waste (2) Increase the operational reliability (3) Reduce production cost	Mansaray et al. (1999); Gabra et al. (2001); Wang et al. (2008); Digman and Kim, (2008)
Bio-oil	Dry food waste such as oil cakes	Pyrolysis Liquefaction	(1) Upgrade the quality of bio-oil (2) Reduce production cost	Gercel (2002); Digman and Kim (2008); Bridgwater (1999)
Biodiesel	(1) Used cooking oil (2) Grease (3) Animal fats	Transesterification	Technical challenge to use feedstock with high FFA and moisture contents	Canakci (2007)

whey powder, wine yeast, potato peel sludge, spent hops, malt dust, and apple marc. The maximum yields of sugars including glucose and fructose in the liquid fraction from dilute sulfuric acid hydrolysis were 40.29% and 50.20% (w/w) of the total amounts of sugars that could be hydrolyzed from tomato and red pepper residues, respectively (Campo et al., 2006). However, to increase the overall sugar yield of the pretreatment process, the solid fraction from the dilute acid hydrolysis should be further hydrolyzed by enzymes.

Grain residuals are rich in cellulose and hemicellulose that can be a renewable source for enzymatic production of soluble sugars as feedstocks for ethanol fermentation (Hang, 2004). Wheat bran consists of three main components of residual starch, cellulose, and hemicellulose. Choteborska et al. (2004) produced a sugar solution from wheat bran for ethanol fermentation using a two-step hydrolytic process. Starch with the wheat bran was first hydrolyzed into glucose by enzymes including Termamyl 120 L (endoamylase) and AMG 300 L (amyloglucosidase) obtained from Novo Nordisk in Denmark. The starch-free wheat bran that mainly consists of cellulose and hemicelluloses was further hydrolyzed into sugars using dilute sulfuric acid. The hydrolysis of starch-free wheat bran with 1% of sulfuric acid at 130°C for 40 min could achieve a maximum yield of sugars at 52.1 g/100 g of the bran. Distillers' grains are rich in residual starch and fibers. Tucker et al. (2004) found that 77% of available carbohydrates in the distillers' grains could be hydrolyzed into sugars using 3.27% H_2SO_4 at 140°C for 20 min. The sugars were further fermented to ethanol using the yeast of *Saccharomyces cerevisiae* D_5A. The ethanol yield was 73% of the theoretical value from available glucans in the distillers' grains.

The carbohydrate in the food waste collected from the cafeterias in Institute of Plasma Physics, Chinese Academy of Sciences, and at the Seoul National University of Science and Technology was as high as 65% of the total solids on a dry basis (Yan et al., 2011; Kim et al., 2011b). It was found that the ethanol yield was 0.43 and 0.31 g ethanol/g total solid in the food waste collected from the cafeteria for the separate enzymatic hydrolysis and ethanol fermentation with *S. cerevisiae* and simultaneous saccharification and ethanol fermentation, respectively (Kim et al., 2011b).

Wastewater from fruit, vegetable, and starch processing is another potential feedstock for alcohol fermentation. The solids in the wastewater can economically be recovered by ultrafiltration (Nguyen, 2003). Moreover, the recovery of sugars from food processing wastewater can remove most of the BOD-contributing solids in the wastewater, which can reduce wastewater disposal costs and reuse the wastewater.

14.4.3 Anaerobic Digestion of Food Processing Wastes into Biogas

Anaerobic digestion is a commercially proven technology to degrade organic materials into biogas usually containing 50%–60% methane in the absence of oxygen and is widely used for treating organic wastes with a high moisture content (i.e., >80%–90% moisture) (McKendry, 2002). For the treatment of food processing wastes, anaerobic digestion can not only produce methane for energy but also destroy pathogenic bacteria present in the wastes and reduce pollutant emission from the wastes. Recent advances in anaerobic digestion technologies have made them competitive for the treatment of various food processing wastes. Anaerobic digestion has been used to treat solid fish wastes (Gebauer, 2004), slaughterhouse wastes (Salminen and Rintala, 2002), fruit and vegetable processing wastes (Kalia et al., 2000), and food processing wastewater (Berardino et al., 2000). Browne and Murphy (2013) reported that the biomethane potential of food wastes was found to be between 467 and 529 L CH_4/kg volatile solids added. As the heating value of methane is around 36 MJ/m^3 at 1 atm and 0°C, which is equivalent to 0.76 kg standard coal, 1 kg volatile solids digested can generate the amount of methane, which is equivalent to 0.35–0.4 kg standard coal. The solid concentration has a significant effect on the anaerobic digestion of food wastes. Among the total solid concentration at 20%, 25%, and 30%, the highest methane yield was found to be 0.49 L CH_4/g VS, which was achieved at the solid concentration of 20% (Forster-Carneiro et al., 2008).

Codigestion of wastes with different characteristics is an approach to dilute toxicants and supply required nutrients (Mata-Alvarez et al., 2000; Serrano et al., 2013). Fruit and vegetable wastes and animal manure are a promising combination to be used as codigestion feedstock. Callaghan et al. (2002) investigated the codigestion of fruit and vegetable wastes and chicken manure in a completely stirred tank reactor at 35°C, hydraulic retention time (HRT) of 21 days, and loading rate in the range of 3.19–5.01 kg volatile solid (VS)/m^3 day. They found increasing the fruit and vegetable wastes from 20% to 50%; the methane yield increased from 0.23 to 0.45 m^3/kg VS added. The increase of fruit and vegetable wastes in the feedstock can decrease the ammonia inhibition from chicken manure. Fruit and vegetable wastes and fish waste are also used in codigestion (Serrano et al., 2013). Strawberry waste can increase the organic matter concentration in the digestion mixture and dilute the inhibitory compounds such as chloride, nitrogen, and phosphorus contained in the fish waste. The mean methane yield was 120 L/kg total volatile solids for an organic loading rate from 22.8 to 50.6 kg waste mixture/(m^3 day) during codigestion of fish and strawberry wastes (Serrano et al., 2013). Nonedible oil cakes have been used to adjust the carbon-to-nitrogen ratio of feedstocks to anaerobic digestion processes for maximum microbiological activity and fuel output (Lingaiah and Rajasekaran, 1986).

Anaerobic digestion effluents are generally not suitable to directly be disposed of on the land because they are too wet, contain some phytotoxic volatile fatty acids, and are not hygienized if the digestion does not occur at a thermophilic temperature (Mata-Alvarez et al., 2000). Digestate from food waste derived from a mixed waste source (including food waste and other wastes) may not be applied to agricultural land according to Department of Agriculture Regulations in EU countries (Browne and Murphy, 2013). Therefore, the segregation of municipal waste must be fully implemented to make full use of food waste.

14.4.4 Production of Hydrogen from Food Processing Wastes

Both biological and thermochemical methods have been used to convert food wastes into H_2. Controlling the pH value and temperature at proper levels is a feasible approach to enhance H_2 production during anaerobic digestion of food wastes (Pan et al., 2008; Kim et al., 2011c,d; Yasin et al., 2011). Solid adsorbents such as activated carbon can be used to remove CO_2 in the H_2-rich gas stream generated from food wastes to produce highly concentrated H_2 (Redondas et al., 2012). Food wastes may lack some proteinaceous nutrients that are essential for H_2 production. Codigestion of food wastes and sewage sludges can be complementary in the concentrations of carbohydrates, total nitrogen, phosphate, and some metals for biological hydrogen production (Zhu et al., 2008). *S. cerevisiae* (yeast) was found to be a good cooperator with hydrogen-producing bacteria in promoting hydrogen production from protein-rich food wastes such as defatted milk powder because it could not only provide a better anaerobic condition for H_2-producing bacteria but also produce some proteinase that facilitates the degradation of lactoprotein in the defatted milk powder for further H_2 production (Song et al., 2010).

The pH value has a significant effect on the biological production of H_2 from food wastes (Kim et al., 2011c; Yasin et al., 2011). If the pH value is too low (e.g., pH=5), alcohols and organic acids may be dominant products during fermentation of food wastes (Kim et al., 2011c; Yasin et al., 2011). The pH value between 5.5 and 6.0 had lower lactic acid production than the pH of 5.0, which could result in effective H_2 production during anaerobic fermentation of food wastes at a thermophilic temperature of 55°C (Yasin et al., 2011). However, during fermentation of food wastes, the pH value continues to decrease due to the generation of organic acids (Kim et al., 2011c). Temperature is another factor that affects the anaerobic digestion of food wastes for producing H_2. At a high temperature, the activity of indigenous lactic acid bacteria is suppressed, while H_2-producing bacteria, such as *Clostridium sp.*, *Acetanaerobacterium elongatum*, and *Caloramater indicus*, are predominantly cultivated. It was found that without any inoculum addition or pretreatment, food wastes could be naturally decomposed and converted to H_2 when cultivated at 50°C–60°C in an anaerobic state (Kim et al., 2011d). The highest H_2 yield was 57 mL H_2/g VS during anaerobic digestion of kitchen wastes at a temperature of 50°C and pH of 5.9–6.6 (Pan et al., 2008). For a continuous process, the H_2 yields and specific production rate from the fermentation of food wastes were found to decrease with the decrease in the HRT due to the incomplete hydrolysis of the food wastes into soluble substrates for H_2-producing microflora at a low HRT (Redondas et al., 2012).

H_2 production from food waste is usually accompanied by the production of organic acids that are suitable substrates for methane production. Two-stage anaerobic digestion including acidogenic and methanogenic stages with properly controlling pH value, temperature, and HRT can effectively separate H_2-producing bacteria from methanogenic archaea for simultaneously generating both H_2 and methane from food wastes (Wang and Zhao, 2009; Zhu et al., 2009; Chu et al., 2010; Lee et al., 2010; Song et al., 2010; Zhu et al., 2011). The first stage can achieve hydrolysis, acidification, and H_2 gas production. The second stage can further produce methane. The hydrolysis and acidification in the first stage can also reduce the residence time requirements in the second stage. Furthermore, hydrogenesis under thermophilic condition at the first stage can replace the pretreatment of feedstock to kill pathogens involved in waste materials and prevent the indigenous microorganisms to prevail (Chu et al., 2010). Research found that the cogeneration of H_2 and methane from defatted milk powers could significantly increase the energy conversion efficiency. The energy conversion efficiency of the cogeneration process was 55.58%–61.96% compared to 10.85%–11.75% for the H_2 production only (Song et al., 2010). The production of H_2 could be significantly improved by adding a phosphate buffer solution such as 1 M K_2HPO_4 and 1 M KH_2PO_4 in the first stage (Song et al., 2010). The maximum hydrogen production was found to be 120 mL/g VS added when the initial pH was 6.5 and buffer addition was in the range of 15–20 mmol/g VS (Zhu et al., 2009).

Thermochemical methods can also be used to convert food wastes into H_2. During thermochemical conversion, food wastes are first gasified into syngas mainly consisting of H_2, CO, CO_2, CH_4, and other short-chain hydrocarbons. The CO and CH_4 can be further converted to H_2 via water–gas shift reaction ($H_2O + CO \leftrightarrow CO_2 + H_2$) and water-reforming reaction ($2H_2O + CH_4 \leftrightarrow CO_2 + 4H_2$). Catalysts are usually used to enhance the thermochemical production of H_2. Among the three catalysts, NaOH, Ni/Al_2O_3, and Ni/SiO_2, the catalyst NaOH was found to be the most effective because the NaOH can react with CO_2 to promote the water–gas shift and water-reforming reactions in the forward direction, which results in the increase of H_2 and decrease of CO, CO_2, and hydrocarbons such as CH_4 (Muangrat et al., 2010a). The H_2O_2 can be used to partially oxidize the food wastes to enhance the gasification of the wastes. The addition of both NaOH and H_2O_2 was found to give higher H_2 product than that of NaOH or H_2O_2 alone (Muangrat et al., 2010b).

14.4.5 Thermochemical Conversion of Food Processing Wastes into Biofuels

Thermochemical conversion provides a competitive way to produce chemical and energy products from low-value and highly distributed biomass resources with large variations in properties (Wang, 2008). Combustion, pyrolysis, gasification, and thermochemical liquefaction are four main thermochemical conversion methods (Wang, 2008). Combustion is to convert the chemical energy stored in an organic matter into heat, generating carbon dioxide and water as final products. Combustion usually produces hot gas at temperatures around 800°C–1000°C. Pyrolysis is the conversion of biomass to liquid, solid, and gaseous fractions by heating the biomass in the absence of air or oxygen to around 500°C. Gasification process falls between complete combustion and pyrolysis. Gasification is the partial oxidation of organic matter at a high temperature to convert the organic matter into a combustible gas mixture called syngas, which mainly consists of CO, H_2, CH_4, and CO_2. Gasifiers are operated at approximately 800°C–900°C although noncatalytic entrained flow gasifies could be operated at a temperature as high as 1300°C. Solvents such as water and alcohols at an elevated temperature and pressure can convert solid biomass into liquid fuels via a thermochemical liquefaction process. Thermochemical conversion technologies have been used for the reduction of environmental impact of food processing wastes and the recovery of energy from the wastes

(Shinogi and Kanri, 2003). It is possible to convert any type of organic food wastes into liquid and gaseous fuels, heat, and power in a thermochemical process. However, it could be practical and economical to thermochemically convert a feedstock with a low moisture content (e.g., <50%–60%) if the biomass is not dried before the thermochemical conversion except hydrothermal liquefaction. It is possible to dry the wet food wastes using the waste heat from the thermochemical conversion (Wang et al., 2009).

Incineration or combustion has been widely used to generate energy from waste materials and reduce the volume of waste substantially (e.g., 80%) (Digman and Kim, 2008; Khoo et al., 2010). The heat generated by burning the food wastes can be used to operate steam turbines for the generation of electricity or as process heat used in food processing facilities. Rice husk was used as a fuel in a 30 kWh bubbling fluidized bed combustor (Armesto et al., 2003). Food wastes usually have lower oxygen content, higher nitrogen and ash content, and higher energy content than wood (Caton et al., 2010). Caton et al. (2010) measured and compared the temperatures and emissions from combustion of wood pellets, dried pelletized food waste, and dried nonpelletized food waste using a modified residential pellet stove. They found that the temperatures from the combustion of food wastes were higher due to the higher energy content. The emissions of NO, HC, and soot were slightly higher for food waste. However, most food processing wastes are not proper for burning because of the high moisture content. An increasing concern of adverse environmental impact from emissions such as dioxins is another reason that restricts the use of incineration in the food industry (Digman and Kim, 2008).

Food wastes with low moisture content are also suitable for gasification. Gasification can have better control of pollutant emissions and higher energy efficiency than incineration. The syngas can be used to generate steam and electricity and synthesize biochemicals and biofuels (Wang et al., 2008). Mansaray et al. (1999) used a fluidized bed gasifier for air gasification of rice husk at a fluidization velocity from 0.22 to 0.33 m/s, air equivalence ratio (the ratio of O_2 required for gasification to O_2 required for full combustion of a given amount of biomass) from 0.25 to 0.35, and temperature from 665°C to 830°C. The gas yield and carbon conversion were in the range of 1.30–1.98 Nm^2/kg and 55%–81%, respectively. The higher heating value of the syngas was 3.09–5.03 MJ/Nm^3. A sugar factory produces nearly 30% of bagasse out of its total crushing. Many research efforts have been attempted to use the bagasse as a renewable feedstock for power generation and the production of bio-based materials. A cyclone gasifier was used to gasify bagasse powder at 39–53 kg/h and 820°C–850°C. The heating values of the syngas at oxygen equivalence ratio from 0.18 to 0.25 were in the range of 3.5–4.5 MJ/Nm^3 dry gas (Gabra et al. 2001). The consumption of thermal energy and electricity is significant in the grain and oilseed milling facilities. In the United States, the grain and oilseed milling sector is the largest energy consumer, which consumed 18.8% of the total energy input and generated 13.3% of the total shipment value in the whole food industry in 2011. The total costs for purchasing fuels and electricity in this sector were $2.08 billion in 2011 (U.S. Census Bureau, 2011). It is a promising way to supply heat and power from the grain processing residues. However, gasification of food processing wastes still has difficulty in satisfying the energy profitability and operational reliability (Digman and Kim, 2008).

Pyrolysis can convert dry food wastes into bio-oil. As bio-oil can be easily stored and transported compared to bulky biomass and gaseous fuels, pyrolysis offers significant logistical and economic advantages over other thermal conversion processes such as combustion and gasification (Bridgwater, 1999). Pyrolysis has been used to convert oilseed cakes into a liquid fuel of bio-oil (Ozbay et al., 2001; Yorgun et al., 2001; Gercel, 2002; Ozcimen and Karaosmanoglu 2004). Gercel (2002) used a fixed-bed tubular reactor to pyrolyze sunflower oil cake. The maximum bio-oil yield was 48.89% by weight obtained at a pyrolysis temperature of 550°C and a heating rate of 5°C/s. The calorific value of bio-oil produced from sunflower cake was measured at 32.15 MJ/kg, which is very close to those of petroleum fractions. Direct solvent liquefaction at high pressure and moderate temperature provides another method for the production of liquid fuels from solid biomass (Wang, 2008). Bio-oils are complex mixtures of volatile hydrocarbons, alcohols, organic acids, aldehydes, ketones, ethers, furans, phenols, and other nonvolatile compounds. Bio-oil also contains a substantial amount of water. The unstable fragments in bio-oil can rearrange through condensation, cyclization, and polymerization to form new compounds such as aromatics. It is necessary to refine and upgrade bio-oils to transportation fuels by reducing the viscosity, oxygen content, and acidity, removing water, and increasing the heating value and stability (Zhang et al., 2007).

14.4.6 Conversion of Used Cooking Oil and Animal Fats into Biodiesel

Biodiesel is currently produced from food-grade vegetable oils such as soybean oil in the United States and rapeseed in Europe. Since food-grade vegetable oils are expensive, biodiesel produced from food-grade vegetable oil is not economically feasible. Animal fats, waste cooking oils, and restaurant grease are potential feedstocks for biodiesel production.

The food processing industry generates a large amount of waste cooking oils and animal fats. The production of inexpensive nonedible feedstocks including grease and animal fats as shown in Table 14.6 represents one-third of the US total oil and fat production (Canakci, 2007). In France, fatty residues from both plants and animals represent an overall production of 0.55 million tons/year. The food industry generates 29% of the total fatty residues. The catering industry, wastewater treatment plants, and autonomous sanitation have shares of about 32%, 23%, and 16%, respectively (Mouneimne et al., 2003).

Grease is produced from vegetable oils or animal fats that are heated and used for cooking food products. Yellow grease is required to have a free fatty acid (FFA) level of less than 15%. Brown grease or trap grease has an FFA level higher than 15%. The price of yellow grease varied widely from \$0.09 to \$0.20/lb in 2000, compared to \$0.35/lb for soybean oil. Brown grease is usually discounted \$0.01–\$0.03/lb below the price of yellow grease. Brown grease is often considered as a potential feedstock for biodiesel production because of its low price (Canakci, 2007). Animal fats are mainly from meat processing facilities. Another source of animal fats is the collection and processing of animal mortalities by rendering companies.

One pound of most fats and oils can be converted to a pound of biodiesel. If all of the 5.284 million tons/year of greases and animal fats in the United States were converted to biodiesel, it would replace about 1.5 billion gallons of diesel fuel (Canakci, 2007). Research showed that the content of FFAs in nonedible oils and fats varied from 0.7% to 41.8% and moisture content from 0.01% to 55.38% (Canakci, 2007). The FFAs and water contents have significant effects on the transesterification of nonedible oils and fats (glycerides) with alcohols using alkaline in a traditional biodiesel production process. They also interfere with the separation of fatty acid esters and glycerol.

TABLE 14.6
Average Production Capacity of Nonedible Oils and Fats in the United States between 1995 and 2000

Animal Fats and Grease	Amount (Million Tons)
Tallow	2.489
Yellow grease	1.195
Lard and grease	0.593
Poultry fat	1.006
Total	5.284

Source: Canakci, M., *Bioresour. Technol.*, 98, 183, 2007.

FFAs and moisture reduce the efficiency of transesterification reaction to convert a feedstock into biodiesel using traditional alkaline catalysts. Therefore, an efficient process for converting waste grease and animal fats must tolerate a wide range of feedstock properties (Ma and Hanna, 1999; Wang, 2008).

14.5 ENVIRONMENTAL, ECONOMICAL, AND SUSTAINABLE IMPACTS OF FOOD WASTES TO ENERGY TECHNOLOGIES

14.5.1 Environmental Impacts

There are numerous environmental factors to be considered in evaluating alternatives for organic waste management. Without proper treatment, food wastes create odor and hygiene concerns and cause negative environmental impacts. Large amounts of solid food processing wastes are currently disposed of as wastes in landfills, on the land, or burned on site. Solid organic wastes from slaughterhouses may contain several species of microorganisms including potential pathogens (Salminen and Rintala, 2002). In addition, animals may accumulate various metals, drugs, and other chemicals added in their feed for nutritional and pharmaceutical purposes (Haapapuro et al., 1997). The disposal and processing of animal wastes should destroy potential pathogens present in the wastes for preventing animal and public health. Fatty wastes are a major source of pollution and landfilling of fatty wastes is not acceptable in some developed countries. In most developing countries, a large proportion of the waste oils and fats are disposed of in landfills or discharged to public sewers inappropriately (Al-Widyan and Al-Shyoukh, 2002). The requirement for the treatment of the wastewater from food processing facilities is determined by the specified discharge limitations defined in the municipal sewer use ordinance. If the liquid wastes contain a high amount of suspended solids, some treatments are required before being released to the environment. If the liquid wastes contain toxic constituents such as salt brines and lye, they must be detoxified prior to disposal (Sargent and Steff, 1986). The wastewater from the grain processing facilities is usually disposed of by spray irrigation. The problems caused by the spray irrigation include strong odor, insect invasion, increase of soil acidity, salt leaching, buildup of sulfates, and putrescity (Nguyen, 2003).

A food treatment process can both result in emissions with negative environmental impacts and affect the potentials for nutrient and energy recovery. Organic food wastes produce methane when they decompose anaerobically in a landfill. As methane has a global warming potential (GWP) 25 times that of CO_2, landfills become the major contributor to GWP from the waste management sector (Themelis, 2003). N_2O is another gas that causes global warming. The GWP of N_2O is 310 times that of CO_2. Research shows that every ton of MSWs processed can avoid 1 ton of CO_2 equivalent (Themelis, 2003). High transport and landfill costs and strict governmental regulations have led to alternative utilization of the food processing wastes (Laufenberg et al., 2003).

Life cycle assessment (LCA) is an emerging tool to measure and compare the environmental impacts of human activities. The selection of impact indicators and the analysis of inventory data for emissions are two important steps for LCA. The impact indicators used in the LCA may include GWP, acidification, eutrophication, photochemical oxidation, and energy use (Lundie and Peters, 2005). The global warming score of a system can be determined by

$$GW = \sum_{s} GWP_s \times m_s \tag{14.1}$$

where

GW is the global warming score

s is the substances (the different greenhouse gases such as CO_2, CH_4)

GWP_s is the GDP of the sth substance

m_s is the emitted amount of substance s in kg

Kim and Kim (2010) and Khoo et al. (2010) evaluated different food waste treatment technologies from the perspective of global warming and resource recovery using LCA. Kim and Kim's research showed that 200 kg of CO_2 equivalent could be produced from dry animal feeding process, 61 kg of CO_2 equivalent from wet animal feeding process, 123 kg of CO_2 equivalent from composting process, and 1010 kg of CO_2 equivalent from landfilling for processing 1 ton of food wastes. Khoo et al. (2010) used LCA to identify and quantify the potential environmental impacts of different food waste management methods including incineration, anaerobic digestion, and composting. The LCA shows that anaerobic digestion of food wastes could significantly reduce the global warming impacts of food wastes, compared to incineration and composting. However, it should be noted that the global warming scores for different treatment alternatives vary not only by the actual difference in environmental impacts from different treatment alternatives but also largely by the difference in system boundary settings, process parameters, and input data such as the characteristics of treated food wastes, losses of carbon and other compounds during treatment, and potential energy recovery (Bernstad and Jansen, 2012). The following text is the case study adopted from the literature published by Khoo et al. (2010).

Case study 1: Life cycle analysis of combined anaerobic digestion and composting of food processing wastes

Anaerobic digestion combined with composting has been widely used to treat food wastes. In the combined anaerobic digestion and composting process used in Singapore, food wastes are pretreated and then fed into an anaerobic digester. The biogas from the anaerobic digester is transferred into gas engines to generate energy. The solid residues from the anaerobic digester are converted into biocompost (Khoo et al. 2010). The whole system produces two products: electrical energy and biocompost. The electricity generated will be used to replace fossil-based electricity on the existing grid. The biocompost material can be used as a replacement of mineral fertilizers. The nutrient contents of the biocompost are 0.0076 kg N/kg biocompost and 0.0011 kg P/kg biocompost. Greenhouse gas savings in terms of CO_2 equivalence are 5.3 kg/kg N mineral fertilizer and 0.52 kg/kg P mineral fertilizer (Khoo et al., 2010). The process information is as follows (Khoo et al. 2010):

- Treatment capacity: 109,500 tons/year
- Amount of compost produced: 66,000 tons/year
- Electrical energy consumption for pretreatment: 25 kWh/ton
- Electrical energy consumption for anaerobic digestion and composting: 32 kWh/ton
- Electrical energy output: 260.82 kWh/ton

LCA is used to investigate the environmental performance of this combined anaerobic digestion and compositing process for the treatment of food processing wastes. The LCA considers the positive environmental impacts of this process including (1) useful energy of electricity from the process and (2) carbon dioxide mitigation from the biocompost and negative environmental impacts including (1) air emissions of the process and (2) energy consumption by the process. The electricity grid emissions and the emissions to air for combined anaerobic digestion and composting are given in Table 14.7. Main assumptions for the LCA include the following:

- Transportation emissions are not included.
- Construction materials for the waste treatment plant and chemical used are not included.
- Both CO_2 and CH_4 contribute to global warming, and the GWP of CH_4 is 25 times that of CO_2.

The LCA has two steps. The first step is to select the impact indicators. In this case study, GWP is selected as an impact indicator. The second step is to analyze the inventory data. The LCA of the inventory data for the GWP is given as follows:

TABLE 14.7
Emissions to Air by an Electricity Grid and the Combined Anaerobic Digestion and Composing Process for the Treatment of Food Wastes

Pollutant	CO	CO_2	CH_4	N_2O	NO_x	SO_x	PM	NMVOC
Electricity grid emission (kg/kWh)	1.78×10^{-4}	5.00×10^{-1}	8.15×10^{-6}	3.03×10^{-6}	1.19×10^{-4}	7.34×10^{-4}	1.16×10^{-5}	2.17×10^{-5}
Waste conversion (kg/kg)	1.13×10^{-6}	6.00×10^{-3}	1.64×10^{-5}	8.15×10^{-5}	4.50×10^{-5}	3.19×10^{-5}	1.65×10^{-7}	2.70×10^{-5}

Source: Khoo, H.H. et al., *Sci. Total Environ.*, 408, 1367, 2010.

1. *Annual saving of CO_2 equivalence by electricity generated*

$$GW_1 = 109,500 \text{ tons/year} \times 260.82 \text{ kWh / ton} \times 5.00\times10^{-1} \text{ kg}\,CO_2/\text{kWh}$$

$$= 1.43\times10^7 \text{ kg}\,CO_2/\text{year} \left(CO_2 \text{ emission}\right)$$

$$GW_2 = 109,500 \text{ tons/year} \times 260.82 \text{ kWh/ton} \times 8.15\times10^{-6} \text{ kg}\,CH_4/\text{kWh} \times 25 \text{ kg}\,CO_2/CH_4$$

$$= 5.82\times10^3 \text{ kg}\,CO_2/\text{year} \left(CO_2 \text{ equivalence from } CH_4 \text{ emission}\right)$$

2. *Annual saving of CO_2 equivalence by bio-composting generated*

$$N: GW_3 = 66,000 \text{ tons/year} \times 1000 \text{ kg/ton} \times 0.0076 \text{ kg N/kg} \times 5.3 \text{ kg}\,CO_2/\text{kg N}$$

$$= 2.66\times10^6 \text{ kg}\,CO_2/\text{year}$$

$$P: GW_4 = 66,000 \text{ tons/year} \times 1000 \text{ kg/ton} \times 0.0011 \text{ kg P/kg} \times 0.52 \text{ kg}\,CO_2/\text{kg P}$$

$$= 3.78\times10^4 \text{ kg}\,CO_2/\text{year}$$

3. *Annual cost of CO_2 equivalence for the electricity consumption*

$$GW_5 = 109,500 \text{ tons/year} \times (25+32) \text{ kWh/ton} \times 5.00\times10^{-1} \text{ kg}\,CO_2/\text{kWh}$$

$$= 3.12\times10^6 \text{ kg}\,CO_2/\text{year} \quad \left(CO_2 \text{ emission}\right)$$

$$GW_6 = 109,500 \text{ tons/year} \times (25+32) \text{ kWh/ton} \times 8.15\times10^{-6} \text{ kg}\,CH_4/\text{kWh} \times 25 \text{ kg}\,CO_2/CH_4$$

$$= 1.27\times10^3 \text{ kg}\,CO_2/\text{year} \left(CO_2 \text{ equivalence from } CH_4 \text{ emission}\right)$$

4. *Annual cost of CO_2 equivalence for the combined AD and composting of food wastes*

$$GW_7 = 109,500 \text{ tons/year} \times 1000 \text{ kg/ton} \times 6.00\times10^{-3} \text{ kg}\,CO_2/\text{kg waste}$$

$$= 6.57\times10^5 \text{ kg}\,CO_2/\text{year} \left(CO_2 \text{ emission}\right)$$

$$\begin{aligned} GW_8 &= 109{,}500\ \text{tons/year} \times 1000\,\text{kg/ton} \times 1.64 \times 10^{-5}\ \text{kg CH}_4/\text{kg waste} \times 25\,\text{kg CO}_2/\text{CH}_4 \\ &= 4.49 \times 10^4\ \text{kg CO}_2/\text{year} \quad (\text{CO}_2 \text{ equivalence from CH}_4 \text{ emission}) \end{aligned}$$

5. *Net saving of global warming due to the combined AD and composting conversion*

$$\begin{aligned} GW &= (GW_1 + GW_2 + GW_3 + GW_4) - (GW_5 + GW_6 + GW_7 + GW_8) \\ &= (1.43 \times 10^7 + 5.82 \times 10^3 + 2.66 \times 10^6 + 3.78 \times 10^4) \\ &\quad - (3.12 \times 10^6 + 1.27 \times 10^3 + 6.57 \times 10^5 + 4.49 \times 10^4)\ \text{kg CO}_2/\text{year} \\ &= 1.32 \times 10^7\ \text{kg CO}_2/\text{year} \end{aligned}$$

Therefore, the combined anaerobic digestion and composting process with a loading capacity to treat 109,500 tons of food wastes per year will reduce the CO_2 emission by 1.32×10^7 kg CO_2 each year, or it can reduce 120 kg CO_2 for each ton of food wastes treated.

14.5.2 Economical Impacts

The tipping fees at the composting facilities in the United States are ranged from about $20 to $50/ton of food wastes (Levis et al., 2010). The food wastes to energy technologies may reduce the costs of treating food wastes. The investments can be broadly categorized into two classifications of capital costs and operating costs. Capital investments for land, building, and equipment are generally more strategic and have long-term effects. The capital investment is usually a function of the capacity or the scale of the project. Operating costs include labor, power, repairs, taxes, and other expenses. The costs versus return on investment may not be favorable for some alternative waste treatment technologies. However, the selection of a food waste treatment technology is affected not only by the net return on the investment but also by the emphasis of sustainable development and governmental regulations. As the costs of landfill, which is widely used to treat solid food wastes, are rising higher, alternative methods such as pyrolysis and gasification to treat food processing wastes become more favorable. Many governments have focused on diverting waste away from landfill through regulation, taxation, and public awareness (Mena et al., 2011). The economics of the treatment technologies for recovering energy from food processing wastes also depends on energy costs.

There are a number of methods for evaluating economic performance. These methods include the simple payback period, life cycle cost method, net benefit or net savings method, benefit/cost ratio method, internal rate of return method, overall rate of return method, and payback method. Usually, several methods are used to provide a better understanding of an investment's worth. The simple payback period is commonly used by businesses. However, the primary criterion mandated for assessing the effectiveness of waste management technologies may be the minimization of life cycle costs (Wang, 2008). A life cycle cost analysis is to quantify costs over the entire life cycle of the project investment. The life cycle cost method sums the costs of acquisition, maintenance, repair, replacement, energy, and other costs such as salvage value that are affected by the investment decision. All amounts are usually measured either in a present value or annual value. The time value of money must be taken into account for all amounts over the relevant period. Two important parameters for a life cycle cost analysis are the lifetimes of equipment and the interest rate. The life cycle cost method is particularly useful for decisions that are made primarily on the basis of cost-effectiveness to determine whether an investment will lower total cost. However, it cannot be used to find the best investment in general. Numerous alternatives may be compared.

The alternative with the lowest life cycle cost that meets the investor's objectives and constraints is the preferred investment (Wang, 2008).

A formula for calculating the life cycle costs at the present value is given by (Wang, 2008)

$$\mathrm{LCC}_p = I_p + \mathrm{CF}_p \tag{14.2}$$

where

I_p is the present value of investment costs
CF_p is the present value of expected total cash flow

The expected annual cash flow is given by (Gebrezgabher et al., 2010)

$$\mathrm{CF}_a = \sum_{i=1}^{n} p_i C_i + \mathrm{FC} - \sum_{j=1}^{m} p_j R_j \tag{14.3}$$

where

p_i is the unit cost of ith item
C_i is the total input of ith item
p_j is the unit price of jth item
R_j is the total output of jth item
i is the index of input items such as feedstock, labor, and maintenance
j is the index of output items such as electricity and heat from a CHP unit
FC is the fixed cost such as labor cost and maintenance cost

The present value of expected total cash flow is calculated by (Wang, 2008)

$$\mathrm{CF}_p = \mathrm{CF}_a \frac{(1+r)^t - 1}{r(1+r)^t} \tag{14.4}$$

where

r is the internal rate of return
t is the life of the investment

The capital investment cost consists of equipment cost and installation cost. The equipment cost is a function of its size, which is calculated by (Wang, 2008)

$$\frac{\mathrm{Cost}}{\mathrm{Cost}_{\mathrm{ref}}} = \left(\frac{\mathrm{Size}}{\mathrm{Size}_{\mathrm{ref}}}\right)^n \tag{14.5}$$

where

$\mathrm{Cost}_{\mathrm{ref}}$ is the reference cost at a reference unit size of $\mathrm{Size}_{\mathrm{ref}}$
n is the cost scaling factor

The cost scaling factor is usually between 0.6 and 0.8 (Wang, 2008).

Case study 2: Economic analysis of combined anaerobic digestion and composting of food processing wastes

A life cycle cost method is used to analyze the economics of the combined anaerobic digestion and composting process described in case study 1.

1. *Total capital investment*
The capital investment for an anaerobic digestion–based power plant at a loading capacity of 70,000 tons/year is about US$ 9 million given in literature (Gebrezgabher et al., 2010). Using a cost scaling factor of 0.6, the capital investment for a similar unit at a loading capacity of 109,500 tons/year is

$$I_p = 9 \times \left(\frac{109{,}500}{70{,}000} \right)^{0.6} = \$\ 11.8 \text{ million}$$

2. *Expected annual cash flow*
The total revenue of this investment consists of the revenues of generated electricity, compost fertilizer, and the saving of disposal costs for the food wastes.
Electricity: The unit price of electricity is assumed to be \$0.1/kWh. The annual revenue of the electricity generated is 109,500 tons/year × 260.82 kWh/ton × \$0.1/kWh = \$2.86 million/year.
Compost: The price of compost as fertilizer is assumed to be \$10/ton excluding transportation costs. The annual revenue of the compost is 66,000 tons/year × \$10/ton = \$0.66 million/year.
Saving of landfill cost: The cost for the disposal of food processing wastes is assumed to be \$25/ton. The total revenue for the saving of disposal costs is 109,500 tons/year × \$25/ton = \$2.74 million/year.
The operating cost includes the costs for feedstock and energy, personnel, and facility maintenance. There is no cost for the feedstock of food processing wastes because it requires paying \$25/ton to dispose of the wastes in landfills. The saving of landfill cost using the anaerobic digestion method is considered as revenue.
Cost for electricity consumption: The annual cost of the electricity consumed is 109,500 tons/year × (25 + 32) kWh/ton × \$0.1/kWh = \$0.62 million/year.
Labor cost: It is assumed that six people are hired to work for three shifts and the average cost for salary and benefits is \$100,000/person/year. The total annual labor cost is \$0.6 million/year.
Maintenance cost: The annual maintenance cost is calculated at 1.5% of the total capital investment, which is \$11.8 million × 1.5%/year = \$0.18 million/year.
The expected net annual cash flow is given by

$$\mathrm{CF}_a = \sum_{i=1}^{n} p_i C_i + \mathrm{FC} - \sum_{j=1}^{m} p_j R_j = (0.62 + 0.60 + 0.18)$$

$$-(2.86 + 0.66 + 2.74) = \$ - 4.86 \text{ million/year}$$

3. *Present value of expected total cash flow*
It is assumed that the life of the plant, *t*, is 20 years and the internal rate of return is 10%. The present value of expected total cash flow is calculated by

$$\mathrm{CF}_p = \mathrm{CF}_a \frac{(1+r)^t - 1}{r(1+r)^t} = -4.86 \times \frac{(1+0.1)^{20} - 1}{0.1(1+0.1)^{20}} = \$ - 41.38 \text{ million}$$

Therefore, the present value of the life cycle cost of the project is

$$LCC_p = I_p + CF_p = 11.8 - 41.38 = \$ - 29.58 \text{ million}$$

That is, the net present value of the project is $29.58 million over 20 years.

14.5.3 Sustainable Development in Food Processing Facilities

The food processing industry depends heavily on raw materials and energy. Most of the energy sources used in food processing facilities have a limited availability. The food industry is under increasing pressure from governments and environmental groups to improve the sustainability of its process and create sustainable business practices. Gerbens-Leenes et al. (2003) used three indicators that address the global environmental sustainability of a food production system: (1) the total energy from both fossil and renewable sources, (2) the water, and (3) land requirements per kilogram of available foods. This method can be used to compare the trends of production sustainability over time.

Energy resources, energy consumption, and the utilization of food processing wastes are intimately related to the sustainable development of the food industry. One of the major factors undermining the sustainability of a food production process is the depletion of resources it uses. The difference between renewable and nonrenewable resources is that renewable resources are created at least as fast as they are consumed while nonrenewable resources are consumed faster than they are created (de Swaan Arons et al., 2004). The food processing facilities driven by nonrenewable resources should be transformed into one based on renewable resources. The food industry can adapt the zero-waste concept in its processing facilities. The main idea of the zero-waste concept is that the waste generated in one industry could become a raw material for other industries (Pauli, 1998). Each food processing facility produces a large mass of waste streams, which can be reused or recycled (Wang, 2008).

14.6 CONCLUSIONS

The food processing facilities generate large amounts of organic wastes in a solid or liquid form. Large amounts of solid food processing wastes are buried in landfills at a cost, while liquid food processing wastes are released untreated into rivers, lakes, and oceans and disposed in public sewer systems. It has become important to develop alternative methods for treating food processing wastes due to the greenhouse gas emission and leachate into groundwater during traditional disposal of food wastes, the availability of landfill sites, and high landfill disposal cost. Conversion of food processing wastes into bioenergy and bioproducts can reduce the disposal costs of the wastes and generate revenues from the products in the food processing facility. Food processing waste can be converted to biofuels, heat, and power using a biological, thermochemical, or chemical conversion process. Wet food processing waste is usually better suited for biological processes such as anaerobic digestion, while dry waste is better for thermochemical conversion processes such as combustion, gasification, and pyrolysis. Thermochemical liquefaction can also be used to convert wet food processing wastes at high pressure and moderate temperature into a bio-oil. Fermentation, transesterification, pyrolysis, and liquefaction produce liquid fuels for use as transportation fuels. Combustion, gasification, and anaerobic digestion produce gaseous energy products, which are suitable to be used at the production location. High-value phytochemicals can be identified and extracted from the waste residuals in the food processing facilities before they are used for energy production for the increase of economic profitability.

REFERENCES

Al-Widyan, M.I. and A.O. Al-Shyoukh. 2002. Experimental evaluation of the transesterification of waste palm oil into biodiesel. *Bioresour. Technol.* 85: 253–256.

Armesto, L., A. Bahillo, K. Veijonen, A. Cabanillas, and J. Otero. 2003. Combustion behavior of rice husk in a bubbling fluidized bed. *Biomass Bioenerg.* 23: 171–179.

Berardino, S.D., S. Costa, and A. Converti. 2000. Semi-continuous anaerobic digestion of a food industry wastewater in an anaerobic filter. *Bioresour. Technol.* 71: 261–266.

Bernstad, A. and J.C. Jansen. 2012. Review of comparative LCAs of food waste management systems-current status and potential improvements. *Waste Manag.* 32: 2439–2455.

Bridgwater, A.W. 1999. Principles and practice of biomass fast pyrolysis processes for liquids. *J. Anal. Appl. Pyrol.* 51: 3–22.

Browne, J.D. and J.D. Murphy. 2013. Assessment of the resource associated with biomethane from food waste. *Appl. Energ.* 104: 170–177.

Callaghan, F.J., D.A.J. Wase, K. Thayanithy, and C.F. Forster. 2002. Continuous co-digestion of cattle slurry with fruit and vegetable wastes and chicken manure. *Biomass Bioenerg.* 27: 71–77.

Campo, I., I. Alegria, M. Zazpe, M. Echeverria, and I. Echeverria. 2006. Diluted acid hydrolysis pretreatment of agri-food wastes for bioethanol production. *Ind. Crop. Prod.* 24: 214–221.

Canakci, M. 2007. The potential of restaurant waste lipids as biodiesel feedstocks. *Bioresour. Technol.* 98: 183–190.

Caton, P.A., M.A. Carr, S.S. Kim, and M.J. Beautyman. 2010. Energy recovery from waste food by combustion or gasification with the potential for regenerative dehydration: A case study. *Energ. Convers. Manag.* 51:1157–1169.

Choteborska, P., B. Palmarola-Adrados, M. Galbe, G. Zacchi, K. Melzoch, and M. Rychtera. 2004. Processing of wheat bran to sugar solution. *J. Food Eng.* 61: 561–565.

Chu, C.F., Y. Ebie, K.Q. Xu, Y.Y. Li, and Y. Inamori. 2010. Characterization of microbial community in the two-stage process for hydrogen and methane production from food waste. *Int. J. Hydrogen Energ.* 35: 8253–8261.

Curry, N. and P. Pillay. 2012. Biogas prediction and design of a food waste to energy system for the urban environment. *Renew. Energ.* 41: 200–209.

Di Gioia, L., B. Cuq, and S. Guilbert. 1999. Thermal properties of corn gluten meal and its protein components. *Int. J. Biol. Macromol.* 24: 341–350.

de Swaan Arons, J., H. van der Kooi, and K. Sankaranarayanan. 2004. *Efficiency and Sustainability in the Energy and Chemical Industries.* New York: Marcel Dekker, Inc.

Digman, B. and D.S. Kim. 2008. Review: Alternative energy from food processing wastes. *Environ. Prog.* 27: 524–537.

Du, C., G.M. Campbell, N. Misailidis, F. Mateos-Salvador, J. Sadhukhan, M. Mustafa, and R.M. Weightman. 2009. Evaluating the feasibility of commercial arabinoxylan production in the context of a wheat biorefinery principally producing ethanol. Part 1. Experimental studies of arabinoxylan extraction from wheat bran. *Chem. Eng. Res. Des.* 87: 1232–1238.

El-Kamah, H., A. Tawfik, M. Mahmoud, and H. Abdel-Halim. 2010. Treatment of high strength wastewater from fruit juice industry using integrated anaerobic/aerobic system. *Desalination.* 253: 158–163.

Fischer, K. and H.P. Bipp. 2005. Generation of organic acids and monosaccharides by hydrolytic and oxidative transformation of food processing residues. *Bioresour. Technol.* 96: 831–842.

Food and Agriculture Organization (FAO). 2011. FAOSTAT. Online: http://faostat3.fao.org. Accessed on January 3, 2014.

Forster-Carneiro, T., M. Perez, and L.I. Romero. 2008. Influence of total solid and inoculum contents on performance of anaerobic reactors treating food waste. *Bioresour. Technol.* 99: 6994–7002.

Gabra, M., E. Pettersson, R. Backman, and B. Kjellstrom. 2001. Evaluation of cyclone gasifier performance for gasification of sugar cane residue-Part 1: Gasification of bagasse. *Biomass Bioenerg.* 21: 351–369.

Gebauer, R. 2004. Mesophilic anaerobic treatment of sludge from saline fish farm effluents with biogas production. *Bioresour. Technol.* 93: 155–167.

Gebrezgabher, S.A., M.P.M. Meuwissen, B.A.M. Prins, and G.J.M.O. Lansink. 2010. Economic analysis of anaerobic digestion—A case of green power biogas plant in the Netherlands. *NJAS-Wagen. J. Life Sci.* 57: 109–115.

Gercel, H.F. 2002. The production and evaluation of bio-oils from the pyrolysis of sunflower-oil cake. *Biomass Bioenerg.* 23: 307–314.

Gerbens-Leenes, P.W., H.C. Moll, and A.J.M. Schoot Uiterkamp. 2003. Design and development of a measuring method for environmental sustainability in food production systems. *Ecol. Econ.* 46: 231–248.

Grull, D.R., F. Jetzinger, M. Kozich, and M.M. Wastyn. 2006. Industrial starch platform-status quo of production, modification and application. In *Biorefineries-Industrial Processes and Products*, pp. 61–95. B. Kamm, P.R. Gruber and M. Kamm, eds. Weinheim, Germany: Wiley-VCH Verlag GmbH & Co.

Haapapuro, E.R., N.D. Barnard, and M. Simon. 1997. Review-animal waste used as livestock feed: Danger to human health. *Prev. Med.* 26: 599–602.

Hang, Y.D. 2004. Management and utilization of food processing wastes. *J. Food Sci.* 69: 104–107.

Hang, Y.D., C.Y. Lee, and E.E. Woodams. 1986. Solid state fermentation of grape pomace for ethanol production. *Biotechnol. Lett.* 8: 53–56.

Hill, K. 2006. Industrial development and application of biobased oleochemicals. In *Biorefineries-Industrial Processes and Products*. Vol. 2, pp. 291–314. B. Kamm, P.R. Gruber, and M. Kamm eds. Weinheim, Germany: Wiley-VCH Verlag GmbH & Co.

International Finance Corporation (IFC). 2007. Environmental, health, and safety guidelines-food and beverage processing. Online: www.ifc.org/ifcext/enviro.nsf/Content/EnvironmentalGuidelines. Accessed on January 3, 2014.

Johnson, D.L. 2006. The corn wet milling and corn dry milling industry-a base for biorefinery technology developments. In *Biorefineries-Industrial Processes and Products*, pp. 344–352. B. Kamm, P.R. Gruber, and M. Kamm eds. Weinheim, Germany: Wiley-VCH Verlag GmbH & Co.

Kalia, V.C., V. Sonakya, and N. Raizada. 2000. Anaerobic digestion of banana stem waste. *Bioresour. Technol.* 73: 191–193.

Khoo, H.H., T.Z. Lim, and R.B.H. Tan. 2010. Food waste conversion options in Singapore: Environmental impacts based on an LCA perspective. *Sci. Total Environ.* 408: 1367–1373.

Kim, M.H. and J.W. Kim. 2010. Comparison through a LCA evaluation analysis of food waste disposal options from the perspective of global warming and resource recovery. *Sci. Total Environ.* 408: 3998–4006.

Kim, D.H., S.H. Kim, K.W. Jung, M.S. Kim, and H.S. Shin. 2011c. Effect of initial pH independent of operational pH on hydrogen fermentation of food waste. *Bioresour. Technol.* 102: 8646–8652.

Kim, D.H., J. Wu, K.W. Jeong, M.S. Kim, and H.S. Shin. 2011d. Natural inducement of hydrogen from food waste by temperature control. *Int. J. Hydrogen Energ.* 36: 10666–10673.

Kim, J.H., J.C. Lee, and D. Pak. 2011b. Feasibility of producing ethanol from food waste. *Waste Manag.* 31: 2121–2125.

Kim, M.H., Y.E. Song, H.B. Song, J.W. Kim, and S.J. Hwang. 2011a. Evaluation of food waste disposal options by LCC analysis from the perspective of global warming: Jungnang case, South Korea. *Waste Manag.* 31: 2112–2120.

Kollacks, W.A. and C.J.N. Rekers. 1988. Five years of experience with the application of reverse osmosis on light middlings in a corn wet milling plant. *Starch.* 40: 88–94.

Lahmar, M., V. Fellner, R.L. Belyea, and J.E. Williams. 1994. Increasing the solubility and degradability of food processing biosolids. *Bioresour. Technol.* 50: 221–226.

Laufenberg, G., B. Kunz, and M. Nystroem. 2003. Transformation of vegetable waste into value added products: (A) the upgrading concept; (B) practical implementations. *Bioresour. Technol.* 87: 167–198.

Lee, D.Y., Y. Ebie, K.Q. Xu, Y.Y. Li, and Y. Inamori. 2010. Continuous H_2 and CH_4 production from high-solid food waste in the two-stage thermophilic fermentation process with the recirculation of digester sludge. *Bioresour. Technol.* 101: S42–S47.

Levis, J.W., M.A. Barlaz, N.J. Themelis, and P. Ulloa. 2010. Assessment of the state of food waste treatment in the United States and Canada. *Waste Manag.* 30: 1486–1494.

Lingaiah, V. and P. Rajasekaran. 1986. Biodigestion of cow dung and organic wastes mixed with oil cake in relation to energy. *Agr. Wastes.* 17: 161–173.

Lundie, S. and G.M. Peters. 2005. Life cycle assessment of food waste management options. *J. Clean. Prod.* 13: 275–286.

Ma, F. and M.A. Hanna. 1999. Biodiesel production: A review. *Bioresour. Technol.* 70: 1–15.

Mahro, B. and M. Timm. 2007. Potential of biowaste from the food industry as a biomass resource. *Eng. Life Sci.* 7: 457–468.

Mansaray, K.G., A.E. Ghaly, A.M. Al-Taweel, F. Hamdullahpur, and V.I. Ugursal. 1999. Air gasification of rice husk in a dual distributor type fluidized bed gasifier. *Biomass Bioenerg.* 17: 315–332.

Mata-Alvarez, J., S. Mace, and O. Llabres. 2000. Anaerobic digestion of organic solid wastes. An overview of research achievements and perspectives. *Bioresour. Technol.* 74: 3–16.

Matteson, G.C. and B.M. Jenkins. 2007. Food and processing residues in California: Resource assessment and potential for power generation. *Bioresour. Technol.* 98: 3098–3105.

McKendry, P. 2002. Energy production from biomass (part 2): Conversion technologies. *Bioresour. Technol.* 83: 47–54.

Mena, C., B. Adenso-Diaz, and O. Yurt. 2010. The causes of food waste in the supplier–retailer interface: Evidences from the UK and Spain. *Resour. Conserv. Recy.* 55(6): 648–658.

Mohaibes, M. and H. Heinonen-Tanski. 2004. Aerobic thermophilic treatment of farm slurry and food wastes. *Bioresour. Technol.* 95: 245–254.

Mouneimne, A.H., H. Carrere, N. Bernet, and J.P. Delgenes. 2003. Effect of saponification on the anaerobic digestion of solid fatty residues. *Bioresour. Technol.* 90: 89–94.

Muangrat, R., J.A. Onwudili, and P.T. Williams. 2010a. Influence of NaOH, Ni/Al2O3 and Ni/SiO_2 catalysts on hydrogen production from the subcritical water gasification of model food waste compounds. *Appl. Catal. B: Environ.* 100: 143–156.

Muangrat, R., J.A. Onwudili, and P.T. Williams. 2010b. Reaction products from the subcritical water gasification of food wastes and glucose with NaOH and H_2O_2. *Bioresour. Technol.* 101: 6812–6821.

Ngadi, M.O. and L.R. Correia. 1992. Kinetics of solid-state ethanol fermentation from apple pomace. *J. Food Eng.* 17: 97–116.

Nguyen, M.H. 2003. Alternatives to spray irrigation of starch waste based distillery effluent. *J. Food Eng.* 60: 367–374.

Nigam, J.N. 1999. Continuous ethanol production from pineapple cannery waste. *J. Biotechnol.* 72: 197–202.

Nigam, J.N. 2000. Continuous ethanol production from pineapple cannery waste using immobilized yeast cells. *J. Biotechnol.* 80: 189–193.

Onay, O., S.H. Beis, and O.M. Kochar. 2001. Fast pyrolysis of rape seed in a well-swept fixed-bed reactor. *J. Anal. Appl. Pyrol.* 58–59: 995–1007.

Ozbay, N., A.E. Putun, and E. Putun. 2001. Biocrude from biomass: Pyrolysis and steam pyrolysis of cottonseed cake. *J. Anal. Appl. Pyrol.* 60: 89–101.

Ozcimen, D. and F. Karaosmanoglu. 2004. Production and characterization of bio-oil and biochar from rapeseed cake. *Renew. Energ.* 29: 779–787.

Pan, J., R. Zhang, H.M. El-Mashad, H. Sun, and Y. Ying. 2008. Effect of food to microorganism ratio on biohydrogen production from food waste via anaerobic fermentation. *Int. J. Hydrogen Energ.* 33: 6968–6975.

Park, J.I., Y.S. Yun, and J.M. Park. 2002. Long-term operation of slurry bioreactor for decomposition of food wastes. *Bioresour. Technol.* 84: 101–104.

Pauli, G. 1998. *Upsizing: The Road to Zero Emissions—More Jobs, More Income and No Pollution*. London, U.K.: GreenLeaf Publishing.

Ramachandran, S., S.K. Singh, C. Larroche, C.R. Soccol, and A. Pandey. 2007. Oil cakes and their biotechnological applications-a review. *Bioresour. Technol.* 98: 2000–2009.

Rasco, B.A., F.M. Dong, A.E. Hashisaka, S.S. Gazzaz, S.E. Downey, and M.L. San Buenaventura. 1987. Chemical composition of distillers' dried grains with solubles (DDGS) from soft white wheat, hard red wheat and corn. *J. Food Sci.* 52: 235–237.

Rausch, K.D., C.I. Thompson, R,L. Belyea, and M.E. Tumbleson. 2003. Characterization of light gluten and light steep water from a corn wet milling plant. *Bioresour. Technol.* 90: 49–54.

Redondas, V., X. Gómez, S. García, C. Pevida, F. Rubiera, A. Morán, and J.J. Pis. 2012. Hydrogen production from food wastes and gas post-treatment by CO2 adsorption. *Waste Manag.* 32: 60–66.

Ritter, W.F. and A.E.M. Chinside. 1995. Impact of dead bird disposal pits on groundwater quality on the Delmarva Peninsula. *Bioresour. Technol.* 53: 105–111.

Ronyai, E., B. Simandi, S. Tomoskozi, A. Deak, L. Vigh, and Z. Weinbrenner. 1998. Supercritical fluid extraction of corn germ with carbon dioxide-ethyl alcohol mixture. *J. Supercrit. Fluid.* 14: 75–81.

Salminen, E. and J. Rintala. 2002. Anaerobic digestion of organic solid poultry slaughterhouse waste—A review. *Bioresour. Technol.* 83: 13–26.

Sargent, S.A. and J.E. Steffe. 1986. Energy generation from direct combustion of solid food processing wastes. In *Energy in Food Processing*, pp. 247–266. R.P. Singh, ed. New York: Elsevier Science Publishing Company Inc.

Sayaslan, A. 2004. Wet-milling of wheat flour: Industrial processes and small-scale test methods. *Lebensm.-Wiss. U.-Technol.* 37: 499–515.

Serrano, A., J.A. Siles, A.F. Chica, and M.A. Martin. 2013. Agri-food waste valorization through anaerobic co-digestion: Fish and strawberry residues. *J. Clean. Prod.* 54: 125–132.

Shinogi, Y. and Y. Kanri. 2003. Pyrolysis of plant, animal and human waste: Physical and chemical characterization of the pyrolytic products. *Bioresour. Technol.* 90: 241–247.

Song, W., J. Cheng, J. Zhou, B. Xie, H. Su, and K. Cen. 2010. Cogeneration of hydrogen and methane from protein-mixed food waste by two-phase anaerobic process. *Int. J. Hydrogen Energ*. 35: 3141–3146.

Themelis, N.J. 2003. An overview of the global waste-to-energy industry. *Waste Manag. World*. 7–8: 40–47.

Tucker, M.P., N.J. Nagle, E.W. Jennings, K.N. Ibsen, A. Aden, Q.A. Nguyen, K.H. Kim, and S.L. Noll. 2004. Conversion of distiller's grain into fuel alcohol and a higher-value animal feed by dilute-acid pretreatment. *Appl. Biochem. Biotechnol*. 113–116: 1139–1159.

Turhollow, A.F. and E.O. Heady. 1986. Large-scale ethanol production from corn and grain sorghum improving conversion technology. *Energ. Agr*. 5: 309–316.

US Census Bureau. 2011. 2011 Annual survey of manufactures. Online: http://factfinder.census.gov. Accessed on January 3, 2014.

USDA. 2012. The USDA economic research service feed grain database. Online: http://www.ers.usda.gov/data/feedgrains/FeedYearbook.aspx. Accessed on January 3, 2014.

US EPA. 2008. Municipal solid waste in the United States: 2007 facts and figures, EPA 530-R-08-010. Office of Solid Waste and Emergency Response, Washington, DC. Online: http://www.epa.gov/osw/nonhaz/municipal/pubs/msw07-rpt.pdf.

Verma, M., S.K. Brar, R.D. Tyagi, R.Y. Surampalli, and J.R. Valero. 2007. Starch industry wastewater as a substrate for antagonist Trichoderma viride production. *Bioresour. Technol*. 98: 2154–2162.

Wang, L.J. 2008. *Energy Efficiency and Management in Food Processing Facilities*. Boca Raton, FL: CRC Press.

Wang, L.J. 2013. Production of bioenergy and biochemicals from food processing wastes: A review. *Trans. ASABE*. 56(1): 217–230.

Wang, L.J., M.A. Hanna, C.L. Weller, and D.D. Jones. 2009. Technical and economical analyses of combined heat and power generation from distillers grains and corn stover in ethanol plants. *Energ. Convers. Manag*. 50: 1704–1713.

Wang, L.J. and C.L. Weller. 2006. Recent advances in extraction of natural products from plants. *Trends Food Sci. Technol*. 17: 300–312.

Wang, L.J., C.L. Weller, M.A. Hanna, and D.D. Jones. 2008b. Contemporary issues in thermal gasification of biomass and application to electricity and fuel production. *Biomass Bioenerg*. 32: 573–581.

Wang, L.J., C.L. Weller, and K.T. Hwang. 2005. Extraction of lipids from grain sorghum DDG. *Trans. ASABE*. 48: 1883–1888.

Wang, X. and Y.C. Zhao. 2009. A bench scale study of fermentative hydrogen and methane production from food waste in integrated two-stage process. *Int. J. Hydrogen Energ*. 34(1): 245–254.

Weegels, P.L., J.P. Marseille, and R.J. Hamer. 1992. Enzymes as a processing aid in the separation of wheat flour into starch and gluten. *Starch*. 44: 44–48.

Yan, S., J. Li, X. Chen, J. Wu, P. Wang, J. Ye, and J. Yao. 2011. Enzymatical hydrolysis of food waste and ethanol production from the hydrolysate. *Renew. Energ*. 36: 1259–1265.

Yasin, N.H.M., N.A.A. Rahman, H.C. Man, M.Z.M. Yusoff, and M.A. Hassan. 2011. Microbial characterization of hydrogen-producing bacteria in fermented food waste at different pH values. *Int. J. Hydrogen Energ*. 36: 9571–9580.

Yorgun, S., S. Sensoz, and O.M. Kockar. 2001. Flash pyrolysis of sunflower oil cakes for production of liquid fuels. *J. Anal. Appl. Pyrol*. 60: 1–12.

Zall, R.R. 2004. *Managing Food Industry Waste*. Ames, IW: Blackwell Publishing.

Zhang, Q., J. Chang, T. Wang, and Y. Xu. 2007. Review of biomass pyrolysis oil properties and upgrading research. *Energ. Convers. Manag*. 48: 87–92.

Zhu H., W. Parker, R. Basnar, A. Proracki, P. Falletta, M. Béland, and P. Seto. 2008. Biohydrogen production by anaerobic co-digestion of municipal food waste and sewage sludges. *Int. J. Hydrogen Energ*. 33: 3651–3659.

Zhu, H., W. Parker, R. Basnar, A. Proracki, P. Falletta, M. Béland, and P. Seto. 2009. Buffer requirements for enhanced hydrogen production in acidogenic digestion of food wastes. *Bioresour. Technol*. 100: 5097–5102.

Zhu, H., W. Parker, D. Conidi, R. Basnar, and P. Seto. 2011. Eliminating methanogenic activity in hydrogen reactor to improve biogas production in a two-stage anaerobic digestion process co-digesting municipal food waste and sewage sludge. *Bioresour. Technol*. 102: 7086–7092.

Section III

Sustainable Bioenergy Processing

15 Fermentation in Transportation Alcohols

Jonathan Moncada, Luis A. Toro, and Carlos A. Cardona

CONTENTS

15.1 INTRODUCTION

Nowadays, the world is facing the progressive depletion of its energy resources mainly based on nonrenewable fuels. At the same time, energy consumption grows at rising rates. In addition, the intensive utilization of fossil fuels has led to the increase in the generation of polluting gases released into the atmosphere, which have caused changes in the global climate. The solution depends on how the development and implementation of technologies based on alternative sources of energy will be undertaken. Through the use of renewable energetic resources, humankind can find part of the solution to their energy requirements in an environmentally and friendly way. Global potential of bioenergy is represented in energy crops and lignocellulosic residues. Conversion of these feedstocks into biofuels is an important choice for the exploitation of alternative energy sources and reduction of polluting gases. In addition, the utilization of biofuels has important economic and social effects.

Since the late 1970s and early 1980s, the international community, led by several governments, showed a great interest in fuel alcohols from renewable resources (Qureshi and Blaschek 2000).

Currently, ethanol is one of the most important renewable resources that contribute significantly to the reduction of negative environmental impact generated by fossil fuels (Cardona and Sánchez 2007). In 2001, fuel ethanol production was 31 billion liters, with a growth up to 39 billion liters in 2006, and is expected to reach 100 billion liters in 2015 (Sarkar et al. 2011). The United States (the first ethanol producer in the world) contributed with 18 billion liters from corn while Brazil produced 17 billion liters from sugarcane (Siqueira et al. 2008). Anhydrous ethanol can be blended directly with gasoline in several countries, with the aim of increasing fuel's oxygenation. Moreover, many vehicles known as flexible ones can use gasoline–alcohol blends or 100% alcohols. It is estimated that Brazil has 3 million of this kind of cars (Siqueira et al. 2008).

Most of the fuel ethanol produced in the world is currently obtained from starch or sucrose (molasses or cane juice), but the technology for ethanol production from nonfood plant sources is being developed rapidly so that large-scale production will be a reality in the coming years (Lin and Tanaka 2006). The production of ethanol from lignocellulosic biomass involves the different steps of pretreatment, hydrolysis (saccharification), fermentation, and ethanol recovery (van Zessen et al. 2003).

Raw materials containing sugars, or materials that can be transformed into sugars, can be used as substrates in fermentation. The fermentable raw materials can be grouped as directly fermentable sugary materials; starchy, lignocellulosic materials; and urban/industrial wastes. Direct fermentation of sugarcane, sugar beet, and sweet sorghum to produce ethanol has also been reported (Bryan 1990; Ganesh et al. 1995). Sugar-containing materials require the least costly pretreatment, while starchy, lignocellulosics, and urban/industrial wastes need costly pretreatment to be converted into fermentable substrates (Sun and Cheng 2002). Sugar-containing materials, which can be transformed into glucose, can be used directly as fermentation substrates under anaerobic condition. Under this condition, glucose is converted to ethanol and carbon dioxide by glycolysis. The phosphorylation of carbohydrates is carried out through the metabolic pathway and the end products are two moles of ethanol and carbon dioxide (Ingram et al. 1998).

On the other hand, butanol is emerging as an alternative to traditional fuels (Rakopoulos et al. 2010). Similar to ethanol fermentation, the study of viable substrates for butanol production has been focused on renewable agricultural crops, mainly corn. Other commercial substrates that can be used for this bioconversion include molasses, wood waste, and several agricultural and food industry wastes (Mariano et al. 2009; Foteinis et al. 2011). Recent studies have been focused on wheat fiber (Qureshi et al. 2008). Another abundant agricultural waste is the corn fibers, which is produced by corn processing industry. Butanol production in the world is estimated at 0.98 million of m^3, maintaining a ratio of 3:6:1 acetone–butanol–ethanol (Ezeji et al. 2007).

The objective of this chapter is to discuss the feedstocks, microorganisms, and operating conditions of fermentation processes for the production of transportation alcohols, with special focus on fuel ethanol. One study case for fuel ethanol production is then discussed. A techno-economical analysis is presented to compare several scenarios with different microorganisms and operating conditions used in Colombia.

15.2 FEEDSTOCKS

One of the most important aspects to be considered in fermentation to produce transportation alcohols is feedstock. Due to the nature of raw materials, it may require different steps such as hydrolysis, neutralization, and detoxification before fermentation. For the production of fuel ethanol, the selection of feedstock directly affects the design of cost-effective processes in order to meet product specifications (Cardona and Sánchez 2007).

Alcohols can be produced from raw materials containing fermentable sugars such as sugarcane and beet that are rich in sucrose. In addition, alcohols may also be produced from some polysaccharides that can be hydrolyzed to obtain fermentable sugars. Starch in grains is the major polysaccharide that is currently used for ethanol production (Cardona and Sánchez 2007; Quintero et al. 2008; Cardona et al. 2010). Lignocellulosic biomass is the most promising feedstock due to its great availability and low cost, but the large-scale commercial production of fuel ethanol from lignocellulosic materials has still not been implemented. In this way, feedstocks can be classified into three main groups: sucrose-rich, starchy, and lignocellulosic biomass. A brief description of these groups is shown as follows.

15.2.1 Sucrose-Rich Materials

Sugarcane is one of the most important crops in tropical countries such as Brazil and India. It is used either in the form of cane juice or cane molasses to produce fuel ethanol. In Colombia, sugarcane is mainly produced in zones close to the Cauca Valley river. In 2010, the planted area in the Cauca Valley region (about the) reached nearly 218,000 ha and produced about 20,272,500 tons of sugarcane, which was about the 99% of total production in Colombia (Asocaña 2011). In European countries, beet molasses are the most used sucrose-rich raw materials for ethanol fermentation. Also, sweet sorghum has become an interesting raw material with high content of sucrose in its juices, starch from grains, and lignocellulosic biomass present in bagasse (Gibbons et al. 1986; Quintero et al. 2008).

The potential of fodder beets juice for fermentation to ethanol was studied by Kosaric et al. (Kosaric et al. 1983) with two strains of yeast of *Saccharomyces cerevisiae* and *S. diastaticus* and *Kluyveromyces marxianus*, and they found that each ton of fodder beet could produce 119.77 L absolute alcohol. Smith and Buxton (Smith and Buxton 1993) reported that the average ethanol yields obtained from a 2-year study were above 3100 L/ha and the maximum yield was up to 5235 L/ha from different sweet sorghum varieties. On the other hand, Reddy and Reddy (2003) reported that the ethanol yield was 6106 L/ha for sweet sorghum while only 4680 L/ha for sugarcane. This crop is ideally suited for semiarid agroclimatic regions and it gives reasonably good yield with minimum requirement of irrigation and fertilizer (Prasad et al. 2007). Lengthy growing period and high water requirement are the disadvantage in sugarcane (*Saccharum officinarum*) and sugar beet (*Beta vulgaris*), in those areas where irrigation facilities are not available. Due to these factors along with the comparative disadvantage of molasses (higher price and water and air pollution), it is expected to increase the interest in sweet sorghum (Reddy and Reddy 2003). Sorghum stalks are ideal for ethanol production, as the ethanol from the sorghum is significantly cleaner than that from sugarcane (Prasad et al. 2007).

One of the most important advantages of sucrose-rich feedstocks is its easy conversion to fuel ethanol because disaccharide can be broken by some microorganism, avoiding the requirement of hydrolysis. Nevertheless, the conditioning of sugar-rich juice or molasses (hydrolysis) led to higher fermentation yields, therefore more productivity in terms of entire technological scheme (Quintero et al. 2008).

15.2.2 Starchy Materials

Starch is very important and abundant natural solid substrate. Many microorganisms are capable to hydrolyze starch. However, generally efficient hydrolysis requires gelatinization. Some recent work investigated the hydrolysis of the raw (crude or native) starch as it occurs naturally. Starch is composed of two related polymers in different proportions according to its source: amylose (16%–30%) and amylopectin (65%–85%). Amylose is a polymer of glucose linked by α-1,4 bonds, mainly in linear chains. Amylopectin is a large highly branched polymer of glucose including α-1,6 bonds at the branch points. Within the plant, cell starch is stored in the form of granules located in amyloplasts, which are intracellular organelles surrounded by a lipoprotein membrane. Starch granules are highly variable in size and shape depending on plant materials. Granules contain both amorphous and crystalline internal regions in proportions of about 30/70. During gelatinization, starch granules swell when heated in the presence of water, which involves the breaking of hydrogen bonds, especially in the crystalline regions (Raimbault 1998). Many microorganisms can hydrolyze starch, especially fungi that are then suitable for simultaneous saccharification and fermentation (SSF) of starchy substrates. Glucoamylase, α-amylase, β-amylase, pullulanase, and isoamylase are involved in the starch degradation. Mainly α-amylase and glucoamylase are of importance for SSF. α-Amylase is an endo-amylase attacking α-1,4 bonds in a random fashion, which rapidly reduces the molecular size of starch and consequently its viscosity of liquefied starch. Glucoamylase occurs almost exclusively in fungi including *Aspergillus* and *Rhizopus* groups. This exo-amylase produces glucose units from amylose and amylopectin chains. Microorganisms generally prefer gelatinized starch. But large quantity of energy is required for the gelatinization so it would be attractive to use organisms growing well on raw (ungelatinized) starch. Different works have been dedicated to isolate fungi producing enzymes that can degrade raw starch (Bergmann et al. 1988; Soccol et al. 1994).

Starchy materials require a hydrolysis (or saccharification) step with water to break down the starch into fermentable sugars. Typically, the starch is mixed with water to form a slurry. The slurry is then stirred and heated to rupture the cell walls. Specific enzymes that will break the chemical bonds are added at various times during heating (Prasad et al. 2007).

Ethanol production from grains involves milling of grains, hydrolysis of starch into fermentable sugar, followed by inoculation with yeast. Yeast cannot use starch directly to produce ethanol. Therefore, grain starch has to be completely broken down to glucose by using two enzymes, amylase and amyloglucosidase, before it is fermented by yeast to produce ethanol. The remaining stillage in the fermented broth during alcohol production is processed to dried distillers grain with soluble (DDGS), which is an excellent ingredient for animal feed (Prasad et al. 2007). Starchy grains such as corn and wheat are the most utilized for the production of fuel ethanol, particularly in the United States and part of Europe (Dien et al. 2002; Quintero et al. 2008). In some countries, other crops with high content of starch, such as tubers, have been used to produce fuel ethanol. Currently in Colombia, there is an ethanol distillery that uses cassava as feedstock with important contribution to the gasoline oxygenation program (FedeBiocombustibles 2012).

15.2.3 Lignocellulosic Materials

The importance of a particular type of biomass depends on the chemical and physical properties of the large molecules that the biomass consists of. Biomass contains varying amounts of cellulose, hemicellulose, lignin, and a small amount of extractive (Bridgewater 1999). The presence of lignin in lignocellulosic biomass limits the usage of cellulose and hemicellulose. Cellulose is a polymer of six-carbon sugar of glucose. Hemicelluloses are complex polysaccharides that exist in association with cellulose in the cell wall. It is a mixture of polysaccharides, composed almost entirely of sugars such as glucose, mannose, xylose and arabinose, and methlyglucoronic and galacturonic acids, with an average molecular weight of 30,000. Lignins are highly branched, substituted, mononuclear aromatic polymers in the cell walls of certain biomass, especially woody species, and are often adjacent to cellulose fibers to form a lignocellulosic complex (Ana-Rita and Ian 1996). Cellulose is generally the largest fraction, about 40%–50% by weight, and hemicellulose about 20%–40% (McKendry 2002; Saxena et al. 2009). For example, the sugarcane bagasse contains 40%–50% cellulose, 20%–30% hemicellulose, 20%–25% lignin, and 1.5%–3.0% ash.

To convert cellulose and hemicellulose molecules into sugar, it is necessary to remove the lignin from lignocellulosic materials. A number of pretreatment methods such as concentrated acid hydrolysis (Liao et al. 2006), dilute-acid hydrolysis (Cara et al. 2008), alkali treatment (Carrillo et al. 2005), sodium sulfite treatment (Kuhad et al. 1999; Kapoor et al. 2008), sodium chlorite treatment (Sun et al. 2004), steam explosion (Ohgren et al. 2005), ammonia fiber explosion (Teymouri et al. 2005), lime treatment (Kim and Holtzapple 2005), and organic solvent treatment (Xu et al. 2006) have frequently been used to remove lignin and improve the saccharification of the cell wall carbohydrates.

The production of ethanol from lignocellulosic material consists of mainly five different steps, namely, pretreatment, (enzymatic) hydrolysis, fermentation, product separation, and posttreatment of the liquid fraction. The pretreatment is necessary to improve the rate of production and the total yield of monomeric sugars at the hydrolysis step. The conversion of hemicellulose and cellulose to monomeric sugars can be done chemically by acids or enzymatically by addition of cellulases that are the enzymes responsible for the hydrolysis of cellulose. The produced monomeric hexoses (six-carbon sugars) can be fermented to ethanol quite easily, while the fermentation of pentoses (five-carbon sugars) is only done by few strains. The product ethanol during fermentation is an inhibitor for the yeasts/bacteria that perform the fermentation, which puts a limit on the concentration of fermentable sugars during fermentation (Hendriks and Zeeman 2009). Furfural and other inhibitors such as soluble lignin compounds are also a problem for the fermentation step, because such compounds can inhibit or even stop the fermentation (Laser et al. 2002). After the fermentation, the ethanol must be recovered from the fermentation broth by distillation.

15.3 MICROORGANISMS

The microorganisms of fungi, bacteria, and yeast can be used for fermentation. Specific yeast (*S. cerevisiae* also known as bakers' yeast) is often used to ferment glucose to ethanol. Theoretically, 100 g of glucose will produce 51.4 g of ethanol and 48.8 g of carbon dioxide. However, in practice, the microorganisms use some of the glucose for their growth and the actual yield is less than 100%. A brief description of yeast, bacteria, and fungi is shown as follows.

15.3.1 Yeasts

Metabolic pathway engineering is constrained by the thermodynamic and stoichiometric feasibility of enzymatic activities of introduced genes. Engineering of xylose metabolism in *S. cerevisiae* has focused on the introduction of genes for the initial xylose assimilation from *Pichia stipitis*, a xylose-fermenting yeast, into *S. cerevisiae*, a yeast traditionally used in the production of ethanol from hexose. However, recombinant *S. cerevisiae* strains created in several laboratories have used xylose oxidatively rather than in a fermentative manner as the yeast metabolizes glucose (Jin and Jeffries 2004). D-xylose is a major hydrolysate from the hemicellulose of biomass. Therefore, ethanol production from xylose is essential for efficient utilization of lignocellulose (Jeffries 1985). Many bacteria, yeast, and fungi assimilate xylose, but only a few metabolize it to ethanol (Skoog and Hahn-Hagerdahl 1988).

Xylose-fermenting yeasts, such as *P. stipitis, Pachysolen tannophilus*, and *Candida shehatae*, require precisely to regulate the oxygenation for the maximum ethanol production (Ligthelm et al. 1988; Skoog et al. 1992) and detoxify the hydrolysate, because they tolerate the inhibitory environment of lignocellulose hydrolysates poorly (van Zyl et al. 1991; Hahn-Hagerdal et al. 1994). These factors increase the cost of xylose fermentation. *S. cerevisiae* has an efficient anaerobic sugar metabolism, tolerates inhibitory industrial substrates better than other microorganisms (Olsson, Linden et al. 1992; Olsson and Hahn-Hagerdal 1993), and ferments hexoses abundantly present in lignocellulosic hydrolysates, such as glucose, mannose, and galactose with high yield and productivity. Table 15.1 shows some yeast that is capable of producing ethanol.

15.3.2 Bacteria

Ethanol-producing bacteria have attracted much attention in recent years because their growth rate is substantially higher than that of the *Saccharomyces* currently used for industrial production of fuel alcohol. With the recent advances in biotechnology, they have the potential to play a key role to make the ethanol production more economic (Dien et al. 2003). Among such ethanol-producing bacteria, *Zymomonas mobilis* is a well-known organism used historically in tropical areas to make

TABLE 15.1
Description of Some Yeast for Ethanol Production

Yeast	Remarks	References
S. cerevisiae	Limited to glucose, fructose, and sucrose as substrates. Tolerates inhibitory industrial substrates better than other microorganisms.	Olsson et al. (1992); Olsson and Hahn-Hagerdal (1993); Dien et al. (2003)
Recombinant *S. cerevisiae*	Strain with introduced genes for the initial xylose assimilation steps from *P. stipitis*. Can use xylose oxidatively rather than in the fermentative manner.	Jin and Jeffries (2004)
P. stipitis	Xylose-fermenting yeast. Require well-regulated oxygenation for maximal ethanol production. Require detoxification of the hydrolysate. Ethanol production rate from glucose is at least five times less than that observed for *S. cerevisiae*.	Ligthelm et al. (1988); Skoog et al. (1992)
P. tannophilus	Xylose-fermenting yeast. Require well-regulated oxygenation for maximal ethanol production. Require detoxification of the hydrolysate. Ethanol production rate from glucose is at least five times less than that observed for *S. cerevisiae*.	Ligthelm et al. (1988); Skoog et al. (1992)
C. shehatae	Xylose-fermenting yeast. Require well-regulated oxygenation for maximal ethanol production. Require detoxification of the hydrolysate. Ethanol production rate from glucose is at least five times less than that observed for *S. cerevisiae*.	Ligthelm et al. (1988); Skoog et al. (1992)
K. marxianus	Uses directly cellulose for ethanol production. Has high ethanol productivity from cellulose (0.5 g of ethanol/g cellulose in 78 h).	Ballesteros et al. (1991)
K. fragilis	Uses directly cellulose for ethanol production. Has high ethanol productivity from cellulose.	Ballesteros et al. (1991)
C. acidothermophilum	Thermotolerant yeast. Produces 80% of the theoretical ethanol yield.	Kadam and Schmidt (1997)

alcoholic beverages from plant sap (Skotnicki et al. 1982). The advantages of *Z. mobilis* are its high growth rate and specific ethanol production. Unfortunately, its fermentable carbohydrates are limited to glucose, fructose, and sucrose. On the other hand, the Gram-negative strain *Zymobacter palmae*, which was isolated by Okamoto et al. (Okamoto et al. 1993), can use a broad range of carbohydrate substrates. *Z. palmae* is a facultative anaerobe that ferments hexoses, α-linked di- and trisaccharides, and sugar alcohols (fructose, galactose, glucose, mannose, maltose, melibiose, sucrose, raffinose, mannitol, and sorbitol). This strain produces approximately 2 mol of ethanol per mole of glucose without the accumulation of by-products and shows similar productivity to that of *Z. mobilis* (Okamoto et al. 1994). Table 15.2 shows some bacteria with its characteristics to produce ethanol.

15.3.3 Fungi

The filamentous fungus *Fusarium oxysporum* is known for its ability to produce ethanol by SSF of cellulose. However, the conversion rate is low and significant amounts of acetic acid are produced as a by-product (Panagiotou et al. 2005). A few microbial species such as *Neurospora*, *Monilia*,

TABLE 15.2
Description of Some Bacteria for Ethanol Production

Bacteria	Remarks	References
Z. mobilis	Has high growth rate and specific ethanol production. Its fermentable carbohydrates are limited to glucose, fructose, and sucrose. Has a high ethanol tolerance.	Skotnicki et al. (1982); Okamoto et al. (1993)
Z. palmae	Facultative anaerobe bacterium. Uses a broad range of carbohydrate substrates (hexoses, α-linked di- and trisaccharides, and sugar alcohol). Shows productivity similar to that of *Z. mobilis*.	Okamoto et al. (1993)
Recombinant *E. coli*	Strain capable of assimilating both hexoses and pentoses.	Park et al. (2001)
Clostridium thermohydrosulfuricum	These bacteria may transform pentoses and amino acids into ethanol.	McMillan (1997)
Thermoanaerobacterium thermosaccharolyticum	Can synthesize up to 2 mol EtOH/mol hexose. Main drawback lies in their very low ethanol tolerance.	
Clostridium thermocellum	These bacteria may transform pentoses and amino acids into ethanol. Can synthesize up to 2 mol EtOH/mol hexose. Can even directly convert lignocellulosic materials into ethanol. Main drawback lies in their very low ethanol tolerance.	McMillan (1997)

Paecilomyces, and *Fusarium* have been reported to hold the ability to ferment cellulose directly to ethanol (Singh et al. 1992). *F. oxysporum* produces a broad range of cellulases and xylanases, which has been characterized (Christakopoulos et al. 1996). Acetic acid was the major fermentation product of *Neocallimastix* sp., another ethanol-producing fungus (Dijkerman et al. 1997). Table 15.3 shows description for some fungi strains that are capable to produce alcohol.

TABLE 15.3
Description of Some Fungi for Ethanol Production

Fungi	Remarks	References
F. oxysporum	Produce ethanol from cellulose but significant amounts of acetic acid are produced too	Panagiotou et al. (2005)
Neurospora, *Monilia*, *Paecilomyces*, and *Fusarium*	Has the ability to ferment cellulose directly to ethanol	Singh et al. (1992)
Neocallimastix sp.	Ethanol-producing fungus but produce acetic acid as the major fermentation product	Dijkerman et al. (1997)
T. reesei	Use cellulose directly and is implemented in the SSF process usually	Sun and Cheng (2002)
Mucor indicus	Capable of fermenting a great variety of sugars including hexoses and pentoses Has an ethanol yield of 0.46 g/g glucose when cultivated under anaerobic conditions Assimilates the inhibitors present in dilute-acid hydrolysates	Sues et al. (2005)
Chalara parvispora	Capable of fermenting pentose-containing materials	Holmgren and Sellstedt (2006)

15.4 PROCESS OPERATING MODES

Ethanol can be produced by four main types of industrial operations: batch, continuous, fed-batch, and semicontinuous. In batch fermentation, substrate and yeast culture are charged into the bioreactor together with nutrients. Most of the ethanol produced today is done by the batch operation since the investment costs are low, it does not require much control, and it can be accomplished with unskilled labor (Caylak and VardarSukan 1998). Complete sterilization and management of feedstocks in a batch process are easier than in the other processes. The other advantage of batch operation is the greater flexibility that can be achieved by using a bioreactor for various product specifications. In the continuous process, feed, which contains substrate, culture medium, and other required nutrients, is pumped continuously into an agitated vessel where the microorganisms are active. The product, which is taken from the top of the bioreactor, contains ethanol, cells, and residual sugar (Maiorella et al. 1981).

The fed-batch operation, which may be regarded as a combination of the batch and continuous operations, becomes very popular in the ethanol industry. In this operation, the feed solution, which contains substrate, yeast culture, and the required minerals and vitamins, is fed at constant intervals while effluent is removed discontinuously. The main advantage of the fed-batch system is that intermittent feeding of the substrate prevents inhibition and catabolite repression. If the substrate has an inhibitory effect, intermittent addition improves the productivity of the fermentation by maintaining a low substrate concentration. In semicontinuous processes, a portion of the culture is withdrawn at intervals and a fresh medium is added to the system. This method has some of the advantages of the continuous and batch operations. There is no need for a separate inoculums vessel, except at the initial start-up. Time is also not wasted in nonproductive idle time for cleaning and resterilization. Another advantage of this operation is that not much control is required. However, there is a high risk of contamination and mutation due to long cultivation periods and periodic handling. Furthermore, since larger reactor volumes are needed, slightly higher investment costs are required (Caylak and VardarSukan 1998). It is essential to keep the culture volume constant in continuous operation, whereas there is volume variation in the fed-batch and semicontinuous processes.

15.5 PROCESS CONFIGURATIONS

15.5.1 Separate Hydrolysis and Fermentation (SHF)

The classic configuration employed for fermenting biomass hydrolysates involves a sequential process where the hydrolysis of cellulose and the fermentation are carried out in different units. This configuration is known as separate hydrolysis and fermentation (SHF). When sequential process is utilized, solid fraction of pretreated lignocellulosic material undergoes hydrolysis (saccharification). This fraction contains the cellulose in an accessible form to acids or enzymes. Once the hydrolysis is completed, the resulting cellulose hydrolysate is fermented into ethanol. One of the main features of the SHF process is that each step can be performed at its optimal operating conditions. The most important factors to be taken into account for the saccharification step are reaction time, temperature, pH, enzyme dosage, and substrate load (Cardona and Sánchez 2007).

15.5.2 Simultaneous Saccharification and Fermentation (SSF)

During SSF, the hydrolysis and fermentation are performed in a single unit. The SSF process has been extensively studied to reduce the inhibition of end products of hydrolysis (Saxena et al. 1992; Zheng et al. 1998). In the process, reducing sugars produced in cellulose hydrolysis or saccharification is simultaneously fermented to ethanol, which greatly reduces the product inhibition to the hydrolysis. However, the need of employing more dilute media to maintain suitable rheological properties makes the final product concentration to be low. In addition, this process operates at nonoptimal conditions for the hydrolysis and thus requires higher enzyme dosage, which increases the substrate conversion but also increases the process costs. Considering that enzymes account for an important part of production costs, it is necessary to find methods that can reduce the cellulases doses (Cardona and Sánchez 2007).

The microorganisms used in the SSF are usually the fungus *Trichoderma reesei* and *S. cerevisiae* (Sun and Cheng 2002). Hydrolysis is usually the rate-limiting process in SSF (Philippidis and Smith 1995). Thermotolerant yeasts and bacteria have been used in the SSF to raise the temperature close to the optimal hydrolysis temperature. Ballesteros et al. (Ballesteros et al. 1991) have identified *K. marxianus* and *Kluyveromyces fragilis* that have the highest ethanol productivity at 42°C from a number of yeast strains. *K. marxianus* has an ethanol yield of 0.5 g/g cellulose using Solka-Floc 200 as substrate at 42°C. Kadam and Schmidt (Kadam and Schmidt 1997) found that thermotolerant yeast *Candida acidothermophilum* produced 80% of the theoretical ethanol yield at 40°C using dilute-acid pretreated poplar as the substrate. Compared to the two stage of hydrolysis–fermentation process, SSF has some advantages such as the increase of hydrolysis rate by the conversion of sugars that inhibit the cellulase activity, lower enzyme requirement, higher product yield, lower requirements for sterile conditions since glucose is removed immediately and ethanol is produced, shorter process time, and less reactor volume because a single reactor is used. However, ethanol may also exhibit inhibition to the cellulase activity in the SSF process (Sun and Cheng 2002). According to Sun and Cheng (Sun and Cheng 2002), this configuration has some disadvantages, which include incompatible temperature of hydrolysis and fermentation, ethanol tolerance of microbes, and inhibition of enzymes by ethanol. These disadvantages need to be considered for the improvement of SSF.

15.5.3 Pentose Fermentation

One of the main problems in bioethanol production from lignocellulosics is that *S. cerevisiae* can only ferment certain mono- and disaccharides like glucose, fructose, maltose, and sucrose. This microorganism is not able to assimilate cellulose and hemicellulose directly. In addition, pentoses, mainly xylose, obtained during hemicellulose hydrolysis cannot be assimilated by this yeast. A way to overcome this obstacle is the use of recombinant DNA technology (genetic engineering). Another approach is the use of pentose-fermenting microorganisms like some species of yeasts and bacteria. In this case, configurations involving the separate fermentation of pentoses and hexoses have been proposed. Yeasts such as *P. stipitis, C. shehatae*, and *P. tannophilus* can assimilate pentoses but their ethanol production rate from glucose is at least five times less than that observed for *S. cerevisiae*. Moreover, their culture requires oxygen and their ethanol tolerance is 2–4 times lower (Claassen et al. 1999). Pentose-fermenting yeasts require a careful control to maintain low oxygen levels in the culture medium needed for their oxidative metabolism. Additionally, these yeasts successfully ferment pure xylose but not the aqueous hemicellulose streams generated during the biomass pretreatment, probably due to the presence of different inhibitors (Chandrakant and Bisaria 1998).

15.5.4 Simultaneous Saccharification and Cofermentation

Another promising process configuration is the inclusion of the pentose fermentation in the SSF, which is called as simultaneous saccharification and cofermentation (SSCF) (Cardona and Sánchez 2007). In an initial stage, the cofermentation of mixed cultures was studied. For example, the co-culture of *P. stipitis* and *Brettanomyces claussenii* has been utilized for the SSCF of aspen at 38°C and pH of 4.8, which yielded 369 L ethanol per ton of aspen for a 48 h batch process, as reported by Olsson and Hahn-Hägerdal (Olsson and Hahn-Hägerdal 1996). In this configuration, both fermenting microorganisms should be compatible in terms of operating pH and temperature. Chandrakant and Bisaria (Chandrakant and Bisaria 1998) suggested that a combination of *C. shehatae* and *S. cerevisiae* is suitable for this kind of process. Similarly, a system including the isomerization of xylose and the fermentation with *S. cerevisiae* in a simultaneous way can be utilized. This system has been proven in non-pretreated spent sulfite liquor and in pretreated acid-hydrolyzed wheat straw (Lindén and Hahn-Hägerdal 1989). Among the drawbacks of this configuration are the high by-product formation in the form of CO_2 and xylitol, poor enzyme stability, incompatible pH and optimal temperature (pH of 7.0 and 70°C for the isomerization process), and the reversibility of the enzymatic transformation (Chandrakant and Bisaria 1998).

15.6 STUDY CASE: FUEL ETHANOL PRODUCTION FROM SUGARCANE MOLASSES IN COLOMBIAN CONTEXT

Colombian sugarcane is the main feedstock especially for three products in food and energy sectors: sugar as food and food additive, fuel ethanol as indirect energy for the Colombian gasoline oxygenation program, and recently electricity as direct energy from cogeneration projects of sugarcane bagasse (Mañosca-Ruiz 2003. Cardona and Sanchez 2007; Quintero et al. 2008; Cardona et al. 2010; Asocaña 2011).

Nowadays, sugar is the main product in sugarcane factories in Colombia, producing raw and refined sugar for national supply and exportation. Between 2001 and 2010, Colombia produced an annual average of 2.4 million tons of sugar per year, among which 1.5 million tons stayed in the national market and an average of 0.9 tons was exported (Asocaña 2011). Because Cauca Valley area is one of the three regions in the world where sugarcane can be cropped and harvested in the whole year, production is constant, which ensures the stability of sugar supply in the national market (Arbeláez et al. 2010; Asocaña 2011).

Fuel ethanol is another important product obtained from sugarcane with special care due to the national ethanol program for gasoline oxygenation. In April of 2010, with the inclusion of the Atlantic coast, the ethanol program covered all national territory with an E8 mixture (8% of ethanol in gasoline). However, five ethanol distilleries are currently working in Cauca Valley River region with an ethanol production capacity of 1,250,000 L/day (FedeBiocombustibles 2012). It is predicted that the annual ethanol production capacity in 2012 is 412 millions of liters, which can be enough to oxygenate 9.2% gasoline of the country (Asocaña 2011; FedeBiocombustibles 2012). The aim of this study case is to show the comparison on fuel ethanol production by two strains used and three process configurations based on fermentation stage. This procedure is shown for ethanol production in the Colombian context and projections on future scenarios.

15.6.1 General Process Descriptions

In the case of ethanol production from sugarcane molasses, the first process scheme includes the selection of widely used conventional technologies for each one of the processing steps, as the Colombian case (Quintero et al. 2008). Also, pretreatment step is included with acid hydrolysis where the sucrose is hydrolyzed mainly into glucose that is more assimilable by the microorganism (Quintero et al. 2008). The scheme includes a conventional batch fermentation using the yeast *S. cerevisiae*. The separation of the cell biomass from the culture broth is carried out by a centrifugation technology. In order to reduce the large organic charge that the broth may have due to cell growth and reduce environmental impact of stillage, additional separation is done to consider the possibility of recycling cells. No treatment of the cells was included. The broth has an ethanol concentration of around 10% by mass. The broth is sent to a distillation column to obtain a concentrated stream of ethanol at 40%–50%. Then, this stream is directed to a rectification column where the ethanol content is elevated to 90%–95% under an atmospheric pressure. To utilize the ethanol as a fuel, the water should be almost totally removed. Thus, it is necessary to dehydrate the ethanol up to 99.7% purity in molecular sieves. The main liquid effluent of the process, the stillage, is evaporated up to a 30% concentration of solids. This procedure is done because there may be many problems in soil content when all stillage is directly spread in fields due to the supersaturation of phosphorus soil content as reported previously for Colombia (Quintero et al. 2008).

The second process configuration is different from the first one in the fermentation step. In this case, the operation of a cascade of three fermentors in series, using conventional yeasts (*S. cerevisiae*) in a continuous process for the cultivation of the microorganisms, is used, a more advanced technology with a higher productivity. The continuous fermentation consists of a process for cell cultivation in a bioreactor where fresh culture medium is continuously added and an outlet stream (effluent) is continuously removed. The microorganisms grow and proliferate inside the bioreactor at such a rate that it offsets the removal rate of the cells. As a result, the bioreactor reaches an equilibrium or steady state where the concentrations of cells, substrate, and target product in the culture broth (and in the effluent as well) do not change over time. To ensure the system homogeneity and reduce the concentration gradients in the broth, continuous stirred tank reactors (CSTR) are used in the continuous operation. In this way, a permanent production of fermented broth can be achieved without the need to stop the bioreactor operation for its load and discharge. This leads to a significant increase on the volumetric productivity compared to the batch (discontinuous) processes. High productivity of ethanol implies best economic indexes as the raw material consumption increases and the concentration of the broth including distillation steps will use less energy as discussed by Cardona and Sánchez (2006). The second process scheme also contemplates the operation of centrifugation for the separation of the yeast cells from culture broth. The purification includes an adsorption process with molecular sieves for ethanol dehydration. The liquid effluents from the process are treated by combination of evaporation of the stillage up to a 30% solids concentration.

In the third process configuration, the ethanol-producing bacteria *Z. mobilis* is chosen. The use of this bacterium is more advantageous than the conventional yeast *S. cerevisiae*. For instance, it has a higher conversion yield to ethanol and consumes lower amounts of nutrients to produce cell biomass (Wooley and Putsche 1996; Cazetta et al. 2007; Rogers et al. 2007). High-productivity bacterial ethanol-producing strains can be selected without the need to use genetically modified microorganisms (developed by the means of recombinant DNA technology or genetic engineering). Currently, these are used in pilot-scale plants where the production of ethanol by continuous fermentation using these bacteria has been demonstrated. One of the current limiting factors for the utilization of this microorganism at commercial level is the adaptation of bacterial strains to industrial conditions. It should be noted that the industry has not inclined to substitute the use of conventional yeasts with bacterium strains.

15.6.2 Simulation Procedure

Each of the proposed technological schemes corresponding to each feedstock and each level of technological development was simulated. The objective of this procedure was to generate the mass and energy balances from which the requirements for raw materials, consumables, service fluids, and energy needs are calculated. Prior to the simulations, extensive information was reviewed and analyzed to define the operating parameters for each one of the processing steps.

The simulation and modeling were performed using different commercial packages as well as specialized software. The simulation for each technological flow sheets included all the processing steps for the conversion of feedstocks into bioethanol. For this, the main simulation tool used was the package Aspen Plus Version v7.2 (Aspen Technology, Inc., United States). Additional data on the physical properties of components required for the simulation were obtained from the work of Wooley and Putsche (Wooley and Putsche 1996). One of the most important issues to be considered during the simulation is the appropriate selection of the thermodynamic models that describe the liquid and vapor phases. The nonrandom two-liquid (NRTL) thermodynamic model was applied to calculate the activity coefficients of the liquid phase, and the Hayden–O'Connell equation of state was used for the description of the vapor phase.

The estimation of the energy consumption was performed based on the results of the mass and energy balances generated by the simulation. For this, the thermal energy required in the heat exchangers and reboilers was taken into account, as well as the electric energy needs of the pumps, compressors, and other equipments. The capital and operating costs were calculated using the software Aspen Icarus Process Evaluator (Aspen Technologies, Inc., United States). However, specific parameters regarding some Colombian conditions such as the costs of the raw materials, income tax, and labor salaries, among others, were incorporated in order to calculate the production costs of one liter of fuel for the Colombian case.

15.6.3 Simulation Results

Simulations of the different technological schemes were used to generate their respective mass and energy balance sheets, which are the basic inputs for the techno-economic analysis. The simulations were validated by comparing the composition of by-products generated in each of the proposed technological schemes with reported values for actual commercial processes.

One of the best alternatives to reduce production cost is to reduce the energy consumption of the production process by implementing more energy-efficient and better-performing technologies. The energy consumption results from the simulations were used to assess the implication of technology choices on energy consumption and the effect of energy savings on production cost. This in turn has a positive impact on internal energy consumption and lowers production costs. Table 15.4 shows the energy requirements for each process scheme. As seen in Table 15.4, technological conversion pathway (technology configuration) shows different productivities, as well as the strains themselves. For example, for process schemes 1 and 2, the strain is the same (*S. cerevisiae*), but global yields for fuel ethanol production are different, due to configuration. In the same way, global yield of ethanol is higher when *Z. mobilis* is used because of the use of different technical configuration and strain.

The cost of producing fuel-grade ethanol from sugarcane molasses is reported in Table 15.5, which clearly shows the importance of raw material cost. The raw material cost (including transport)

TABLE 15.4
Yield and Energy Consumption of Ethanol Production Based on Sugarcane Molasses

Technology	Yield (L/ton)[a]	Utility Requirements (MJ/L)[b]	Electricity (kWh/ton)[c]
Level 1	258.23	26.14	39.09
Level 2	260.43	26.38	3.46
Level 3	287.77	24.16	3.52

[a] Liters of ethanol per ton of molasses.
[b] The utility requirements make reference to the required cooling and heating utilities, expressed in MJ, to produce 1 L of ethanol. Most of the time, these utilities are low- and mid-pressure steam and cooling water. This does not consider any level of energy integration. If any level is considered, the utility requirements will be lower.
[c] Electricity requirement expressed in kWh per ton of molasses.

TABLE 15.5
Estimated Production Costs for Fuel-Grade Alcohol Obtained from Sugarcane Molasses under Colombian Conditions

Category	Level 1		Level 2		Level 3	
	US$/L	Share of Total (%)	US$/L	Share of Total (%)	US$/L	Share of Total (%)
1. Raw materials[a]	0.4435	65.90	0.4350	66.33	0.4022	67.28
2. Services	0.0633	9.41	0.0816	12.44	0.0660	11.04
3. Labor	0.0007	0.10	0.0006	0.09	0.0007	0.12
4. Maintenance	0.0225	3.34	0.0142	2.17	0.0128	2.14
5. Operating charges	0.0002	0.03	0.0002	0.03	0.0002	0.03
6. Indirect plant expenses	0.0137	2.04	0.0074	1.13	0.0068	1.14
7. General and administrative costs	0.0495	7.36	0.0495	7.55	0.0450	7.53
8. Capital depreciation[b]	0.0796	11.83	0.0673	10.26	0.0641	10.72
Total	0.6730	100	0.6558	100	0.5978	100

[a] Price of molasses: US$ 0.10/kg (based on market price as provided by Colombia Sugarcane Board).
[b] Calculated using the straight-line method.

comprises 65.90% of the total production cost for the process scheme 1 and 66.33% and 67.28% for process schemes 2 and 3, respectively.

Another important expenditure is the services and capital cost. Overall, the lowest-cost technology per liter of ethanol produced is the process scheme 3. The cost of per liter ethanol produced from molasses in Colombia is very high, due to the high cost of molasses that was estimated to be around US$ 0.1/kg, considering no integration with sugar refineries. As molasses are a by-product of the sugar industry, it is expected that the value of this raw material should make it possible to produce ethanol at a similar cost to that obtained from the base case. Cost estimates can be compared with actual values of sugarcane distilleries, as the base case for Colombia (using batch fermentation and *S. cerevisiae* as strain). Nowadays, Colombian distilleries are integrated with sugar refineries and cogeneration units from cane bagasse, using cane molasses to produce fuel-grade ethanol and contribute to the Colombian Oxygenation Program. Production cost for fuel-grade ethanol in current distilleries at Colombian conditions is reported in Table 15.6. It is important to consider that 26.2 L of ethanol per ton of cane is obtained in the presented Colombian base case. After the distilleries are integrated with sugar refineries and cogeneration units, and raw material cost represents about 68% of total production cost when no credit for by-/coproducts such as sugar and electricity is considered, the production cost of ethanol is 2.79 US$/L, which is very high compared with the sale price of fuel ethanol in Colombia (1.21–1.31 US$/L). Nevertheless, when it is considered as the credit of coproduct sales (i.e., sugar and electricity), ethanol's production cost is reduced to US$ 0.61/L, which is significantly lower than the cost shown for scenarios using juice or molasses as raw material. This is relevant since it is demonstrated that production cost can be decreased when fuel ethanol is integrated within sugar refineries (see Table 15.4). In the Colombian case, different products share the costs of the entire plant, therefore costs can be easily allocated to show individual production costs, therefore integrated processes might have a higher potential than stand-alone processes. For the Colombian case, the savings in production cost are covered by coproducts sales even when sugarcane price is higher in the Colombian context, and process schemes do not include coproducts sales or coproducts do not represent important saving in total production cost.

TABLE 15.6
Estimation of Production Costs for Fuel-Grade Alcohol Obtained from Sugarcane Molasses Based on Distilleries Integrated with Sugar Refineries and Cogeneration Unit in Colombia (US$/L) (Base Case)

Category	Base Case
1. Raw materials[a]	1.9162
2. Services	0.2042
3. Labor	0.0037
4. Maintenance	0.0308
5. Operating charges	0.0009
6. General plant costs	0.0172
7. General and administrative costs	0.5139
8. Capital depreciation[b]	0.1080
9. Coproducts sale[c]	2.2981
Total without coproduct credit	2.7949
Total with coproduct credit	0.4968

[a] Price of sugarcane: US$ 0.035/kg at Colombian conditions.
[b] Calculated using the straight-line method.
[c] Refined sugar and electricity considered as coproducts. Refined sugar yield: 0.11 ton of sugar per ton of cane. Electricity generated: 151.74 kWh/ton of cane. Electricity for sale: 101.13 kWh/ton of cane.

15.7 CONCLUSIONS

Fermentation of transportation alcohols is an important subject in current and future biorefineries. As mentioned in the chapter, there are several parameters that can affect directly the performance of a fermentation process. These parameters include raw materials, microorganism, process modes, and configurations. The proper selection of each parameter can lead to a processing system with higher performance and better economics. Combined use of process engineering and metabolic engineering tools can increase the productivity of ethanol during fermentation for better exploitation of resources and minimizing wastes.

ACKNOWLEDGMENTS

The authors express their acknowledgments to the National University of Colombia at Manizales and Orinoquia for financial support.

REFERENCES

Ana-Rita, F. D. and W. D. Ian (1996). Pyrolysis of sugar cane bagasse in a wire mesh reactor. *Industrial and Engineering Chemistry Research* 35: 1263–1268.

Arbeláez, M. A., A. Estacio, and M. Olivera (2010). Impacto socioeconómico del sector azucarero colombiano en la economía nacional y regional. *Cuadernos de Fedesarrollo* 31: 3–10.

Asocaña (2011). Informe Anual 2010–2011. Sector Azucarero Colombiano. *Asociación de cultivadores de caña de azúcar de Colombia* 1: 1–102.

Ballesteros, I., M. Ballesteros, A. Cabanas, J. Carrasco, C. Martin, and M. J. Negro (1991). Selection of thermotolerant yeasts for simultaneous saccharification and fermentation (SSF) of cellulose to ethanol. *Applied Biochemistry and Biotechnology* 28–29: 307–315.

Bergmann, F. W., J. I. Abe, and S. Hizukuri (1988). Selection of microorganisms which produce raw-starch degrading enzymes. *Applied Microbiology and Biotechnology* 27: 443–446.

Bridgewater, A. V. (1999). Principles and practice of biomass fast pyrolysis processes for liquids. *Journal of Analytical and Applied Pyrolysis* 51: 3–22.

Bryan, W. L. (1990). Solid-state fermentation of sugars in sweet sorghum. *Enzyme and Microbial Technology* 12: 437–442.

Cara, C., E. Ruiz, J. M. Oliva, F. Sáez, and E. Castro (2008). Conversion of olive tree biomass into fermentable sugars by dilute acid pretreatment and enzymatic saccharification. *Bioresource Technology* 99: 1869–1876.

Cardona, C. A., J. A. Quintero, and I. C. Paz (2010). Production of bioethanol from sugarcane bagasse: Status and perspectives. *Bioresource Technology* 101(13): 4754–4766.

Cardona, C. A. and Ó. J. Sánchez (2006). Energy consumption analysis of integrated flowsheets for production of fuel ethanol from lignocellulosic biomass. *Energy* 31(13): 2447–2459.

Cardona, C. A. and Ó. J. Sanchez (2007). Fuel ethanol production: Process design trends and integration opportunities. *Bioresource Technology* 98(12): 2415–2457.

Carrillo, F., M. J. Lis, X. Colom, M. Valldeperas, and J. Valldeperas (2005). Effect of alkali pretreatment on cellulase hydrolysis of wheat straw: Kinetic study. *Process Biochemistry* 40: 3360–3364.

Caylak, B. and F. Vardar Sukan (1998). Comparison of different production processes for bioethanol. *Turkish Journal of Chemistry* 22: 351–359.

Cazetta, M., M. Celligoi, J. Buzato, and I. Scarmino (2007). Fermentation of molasses by *Zymomonas mobilis*: Effects of temperature and sugar concentration on ethanol production. *Bioresource Technology* 98(15): 2824–2828.

Chandrakant, P. and V. S. Bisaria (1998). Simultaneous bioconversion of cellulose and hemicellulose to ethanol. *Critical Reviews in Biotechnology* 18(4): 295–331.

Christakopoulos, P., D. Kekos, B. J. Macris, M. Claeyssens, and M. K. Bhat (1996). Purification and characterization of a major xylanase with cellulase and transferase activities from *Fusarium oxysporum*. *Carbohydrate Research* 289: 91–104.

Claassen, P. A. M., J. B. van Lier, A. M. López Contreras, E. W. J. van Niel, L. Sijtsma, A. J. M. Stams, S. S. de Vries, and R. A. Weusthuis (1999). Utilisation of biomass for the supply of energy carriers. *Applied Microbiology and Biotechnology* 52: 741–755.

Dien, B. S., R. J. Bothast, N. N. Nichols, and M. A. Cotta (2002). The US corn ethanol industry: An overview of current technology and future prospects. *International Sugar Journal* 104(1241): 204–208.

Dien, B. S., M. A. Cotta, and T. W. Jeffries (2003). Bacteria engineered for fuel ethanol production: Current status. *Applied Microbiology and Biotechnology* 63: 258–266.

Dijkerman, R., J. Ledeboer, H. J. M. Opden Camp, R. A. Prins, and C. van der Drift (1997). The anaerobic fungus *Neocallimastix sp.* strain L2: Growth and production of (hemi)cellulotic enzymes on a range of carbohydrate substrates. *Current Microbiology* 34: 91–96.

Ezeji, T. C., N. Qureshi, and H. P. Blaschek (2007). Bioproduction of butanol from biomass: From genes to bioreactors. *Current Opinion in Biotechnology* 18(3): 220–227.

FedeBiocombustibles (2012). Ethanol anhidro de caña. Federación Nacional de Biocombustibles de Colombia. *Cifras Informativas del Sector Biocombustible* 2012: 1–6.

Foteinis, S., V. Kouloumpis, and T. Tsoutsos (2011). Life cycle analysis for bioethanol production from sugar beet crops in Greece. *Energy Policy* 39: 4834–4841.

Ganesh, S., A. K. Fazlullah-Khan, M. Suresh, and N. Senthil (1995). Character associated for ethanol yield in sweet sorghum. *Madras Agricultural Journal* 82(5): 361–363.

Gibbons, W. R., C. A. Westby, and T. L. Dobbs (1986). Intermediate scale, semicontinuous solid phase fermentation process for production of fuel ethanol from sweet sorghum. *Applied and Environmental Microbiology* 51(1): 115–122.

Hahn-Hagerdal, B., H. Jeppsson, L. Olsson, and A. Mohagegi (1994). An interlaboratory comparison of the performance of ethanol producing microorganisms in a xylose-rich acid hydrolysate. *Applied Microbiology and Biotechnology* 41: 62–72.

Hendriks, A. T. W. M. and G. Zeeman (2009). Pretreatments to enhance the digestibility of lignocellulosic biomass. *Bioresource Technology* 100: 10–18.

Holmgren, M. and A. Sellstedt (2006). Producing ethanol through fermentation of organic starting materials, involves using fungus e.g., *Chalara parvispora*, capable of metabolizing pentose compounds. Patent#: SE527184. Sweden. Available online: http://europepmc.org/abstract/PAT/SE527184. Accessed on January 5, 2014.

Ingram, L. O., P. F. Gomez, X. Lai, M. Moniruzzaman, and B. E. Wood (1998). Metabolic engineering of bacteria for ethanol production. *Biotechnology and Bioengineering* 58: 204–214.

Jeffries, T. W. (1985). Emerging technology for fermenting d-xylose. *Trends in Biotechnology* 3: 208–212.

Jin, Y. S. and T. W. Jeffries (2004). Stoichiometric network constraints on xylose metabolism by recombinant *Saccharomyces cerevisiae. Metabolic Engineering* 6: 229–238.

Kadam, K. L. and S. L. Schmidt (1997). Evaluation of *Candida acidothermophilum* in ethanol production from lignocellulosic biomass. *Applied Microbiology and Biotechnology* 48: 709–713.

Kapoor, M., L. M. Nair, and R. C. Kuhad (2008). Cost-effective xylanase production from free and immobilized Bacillus pumilus strain MK001 and its application in saccharification of *Prosopis juliflora. Biochemical Engineering Journal* 38: 88–97.

Kim, S. and M. T. Holtzapple (2005). Lime pretreatment and enzymatic hydrolysis of corn stover. *Bioresource Technology* 96: 1994–2006.

Kosaric, N., A. Wieczorek, and S. Kliza (1983). Ethanol production from fodder beets. *Biotechnology and Bioengineering* 25: 1737–1745.

Kuhad, R. C., M. Manchanda, and A. Singh (1999). Hydrolytic potential of extracellular enzymes from a mutant strain of *Fusarium oxysporum. Bioprocess Engineering* 20: 47–125.

Laser, M., D. Schulman, S. G. Allen, J. Lichwa, M. J. Antal Jr, and L. R. Lynd. (2002). A comparison of liquid hot water and steam pretreatments of sugar cane bagasse for bioconversion to ethanol. *Bioresource Technology* 81: 33–44.

Liao, W., Y. Liu, C. Liu, Z. Wen, and S. Chen (2006). Acid hydrolysis of fibers from dairy manure. *Bioresource Technology* 97: 1687–1695.

Ligthelm, M. E., B. A. Prior, and J. C. du Preez (1988). The oxygen requirements of yeasts for the fermentation of d-xylose and d-glucose to ethanol. *Applied Microbiology and Biotechnology* 28: 63–68.

Lin, Y. and S. Tanaka (2006). Ethanol fermentation from biomass resources: Current state and prospects. *Applied Microbiology and Biotechnology* 69: 627–642.

Lindén, T. and B. Hahn-Hägerdal (1989). Fermentation of lignocellulose hydrolysates with yeasts and xylose isomerase. *Enzyme and Microbial Technology* 11: 583–589.

Maiorella, B., C. R. Wilke, and H. W. Blanch (1981). Alcohol production and recovery. *Advances in Biochemical Engineering/Biotechnology* 20: 9–31.

Mañosca-Ruiz, M. H. (2003). Study of Cogeneration as a Clean Development Mechanism Option in Colombia, MSc thesis, University of Calgary, Faculty of Graduate Studies.

Mariano, A. P., R. C. Tomasella, C. Di Martino, M. H. R. Seleghim, J. Contiero, and D. F. de Angelis (2009). Aerobic biodegradation of butanol and gasoline blends. *Biomass and Bioenergy* 33(9): 1175–1181.

McKendry, P. (2002). Energy production from biomass (part 1): Overview of biomass. *Bioresource Technology* 83: 37–46.

McMillan, J. D. (1997). Bioethanol production: Status and prospects. *Renewable Energy* 10: 295–302.

Ohgren, K., M. Galbe, and G. Zacchi (2005). Optimization of steam pretreatment of SO2-impregnated corn stover for fuel ethanol production. *Applied Biochemistry and Biotechnology* 124: 1055–1067.

Okamoto, H., H. Taguchi, K. Nakamura, and H. Ikenaga (1994). Production of ethanol from maltose by *Zymobacter palmae* fermentation. *Agricultural and Biological Chemistry* 58: 1328–1329.

Okamoto, T., H. Taguchi, K. Nakamura, H. Ikenaga, H. Kuraishi, and K. Yamasato (1993). *Zymobacter palmae gen. nov., sp. nov.*, a new ethanol-fermenting Peritrichous bacterium isolated from palm sap. *Archives of Microbiology* 21: 333–337.

Olsson, L. and B. Hahn-Hagerdal (1993). Fermentative performance of bacteria and yeasts in lignocellulose hydrolysates. *Process Biochemistry* 28: 249–257.

Olsson, L. and B. Hahn-Hägerdal (1996). Fermentation of lignocellulosic hydrolysates for ethanol production. *Enzyme and Microbial Technology* 18: 312–331.

Olsson, L., T. Linden, and B. Hahn-Hagerdal (1992). Performance of microorganisms in spent sulfite liquor and enzymatic hydrolysate of steam-pretreated Salix. *Applied Biochemistry and Biotechnology* 34–35: 359–367.

Panagiotou, G., S. G. Villas-Boas, P. Christakopoulos, J. Nielsen, and L. Olsson (2005). Intracellular metabolite profiling of *Fusarium oxysporum* converting glucose to ethanol. *Journal of Biotechnology* 115(4): 425–434.

Park, E. Y., A. Michinaka, and N. Okuda (2001). Enzymatic hydrolysis of waste office paper using viscosity as operating parameter. *Biotechnology Progress* 17: 379–382.

Philippidis, G. P. and T. K. Smith (1995). Limiting factors in the simultaneous saccharification and fermentation process for conversion of cellulosic biomass to fuel ethanol. *Applied Biochemistry and Biotechnology* 51–52: 117–124.

Prasad, A., S. Singh, and H. C. Joshi (2007). Ethanol as an alternative fuel from agricultural, industrial and urban residues. *Resources, Conservation and Recycling* 50: 1–39.

Quintero, J. A., M. I. Montoya, Ó. J. Sánchez, O. H. Giraldo, and C. A. Cardona (2008). Fuel ethanol production from sugarcane and corn: Comparative analysis for a Colombian case. *Energy* 33(3): 385–399.

Qureshi, N. and H. Blaschek (2000). Economics of butanol fermentation using hyper-butanol Producing *Clostridium beijerinckii* BA101. *Food and Bioproducts Processing* 78(3): 139–144.

Qureshi, N., B. C. Saha, R. E., Hector, S. R. Hughes, and M. A. Cotta (2008). Butanol production from wheat straw by simultaneous saccharification and fermentation using *Clostridium beijerinckii*: Part I—Batch fermentation. *Biomass and Bioenergy* 32(2): 168–175.

Raimbault, M. (1998). General and microbiological aspects of solid substrate fermentation. *Electronic Journal of Biotechnology* 1(3): 174–188.

Rakopoulos, D., C. Rakopoulos, E. Giakoumis, A. Dimaratos, and D. Kyritsis (2010). Effects of butanol–diesel fuel blends on the performance and emissions of a high-speed DI diesel engine. *Energy Conversion and Management* 51(10): 1989–1997.

Reddy, B. V. S. and P. S. Reddy (2003). Sweet sorghum: Characteristics and potential. *International Sweet Sorghum and Millets Newsletter* 44: 26–28.

Rogers, P., Y. Jeon, K. Lee, and H. Lawford (2007). *Zymomonas mobilis* for fuel ethanol and higher value products. *Biofuels* 108: 263–288.

Sarkar, N., S. K. Ghosh, S. Bannerjee, and K. Aikat (2011). Bioethanol production from agricultural wastes: An overview. *Renewable Energy* 37(1): 19–27.

Saxena, A., S. K. Garg, and J. Verma (1992). Simultaneous saccharification and fermentation of waste newspaper to ethanol. *Bioresource Technology* 39: 13–15.

Saxena, R. C., D. K. Adhikari, and H. B. Goya (2009). Biomass-based energy fuel through biochemical routes: A review. *Renewable and Sustainable Energy Reviews* 13: 167–178.

Singh, A., P. K. R. Kumar, and K. Schugerl (1992). Bioconversion of cellulosic materials to ethanol by filamentous fungi. *Advances in Biochemical Engineering/Biotechnology* 45: 29–55.

Siqueira, P. F., S. G. Karp, J. C. Carvalho, W. Sturm, J. A. Rodríguez-León, J. L. Tholozan, R. R. Singhania, A. Pandey, and C. R. Soccol (2008). Production of bio-ethanol from soybean molasses by *Saccharomyces cerevisiae* at laboratory, pilot and industrial scales. *Bioresource Technology* 99(17): 8156–8163.

Skoog, K. and B. Hahn-Hagerdahl (1988). Xylose fermentation. *Enzyme and Microbial Technology* 10: 66–90.

Skoog, K., H. Jeppsson, and B. Hahn-Hagerdal (1992). The effect of oxygenation on glucose fermentation with *Pichia stipitis. Applied Microbiology and Biotechnology* 34–35: 369–375.

Skotnicki, M. L., R. G. Warr, A. E. Goodman, K. J. Lee, and P. L. Rogers (1982). High-productivity alcohol fermentations using *Zymomonas mobilis. Biochemical Society Symposia* 19: 53–86.

Smith, G. A. and D. R. Buxton (1993). Temperate zone sweet sorghum ethanol production potential. *Bioresource Technology* 43: 71–75.

Soccol, C., B. Marin, M. Raimbault, and J. M. Lebeault (1994). Breeding and growth of Rhizopus in raw cassava by solid state fermentation. *Applied Microbiology and Biotechnology* 41: 330–336.

Sues, A., R. Millati, L. Edebo, and M. J. Taherzadeh (2005). Ethanol production from hexoses, pentoses, and dilute-acid hydrolyzate by *Mucor indicus. FEMS Yeast Research* 5: 669–676.

Sun, J. X., X. F. Sun, H. Zhao, and R. C. Sun (2004). Isolation and characterization of cellulose from sugarcane bagasse. *Polymer Degradation and Stability* 84: 331–339.

Sun, Y. and J. Cheng (2002). Hydrolysis of lignocellulosic materials for ethanol production: A review. *Bioresource Technology* 83(1): 1–11.

Teymouri, F., L. Laureano-Perez, H. Alizadeh, and B. E. Dale (2005). Optimization of the ammonia fiber explosion (AFEX) treatment parameters for enzymatic hydrolysis of corn stover. *Bioresource Technology* 96: 2014–2018.

van Zessen, E., M. Weismann, R. R. C. Bakker, H. W. Elbersen, J. H. Reith, and H. den Uil (2003). *Lignocellulosic Ethanol, A Second Opinion*. Netherlands Agency for Energy & Environment. Available online: http://gave.novem.nl. Accessed on January 5, 2014.

van Zyl, C., B. A. Prior, and J. C. Du Preez (1991). Acetic acid inhibition of d-xylose fermentation by *Pichia stipitis. Enzyme and Microbial Technology* 13: 82–86.

Wooley, R. and V. Putsche (1996). *Development of an ASPEN PLUS Physical Property Database for Biofuels Components*. Golden, CO, National Renewable Energy Laboratory, p. 38.

Xu, F., J. X. Sun, C. F. Liu, and R. C. Sun (2006). Comparative study of alkali- and acidic organic solvent-soluble hemicellulosic polysaccharides from sugarcane bagasse. *Carbohydrate Research* 341: 253–261.

Zheng, Y., H. M. Lin, and G. T. Tsao (1998). Pretreatment for cellulose hydrolysis by carbon dioxide explosion. *Biotechnology Progress* 14: 890–896.

16 Biodiesel Production from Waste Feedstock

Dennis Y.C. Leung and Xuan Wu

CONTENTS

16.1 INTRODUCTION

Biodiesel, a monoalkyl ester of long-chain fatty acid, is a clean diesel substitute that can be used in most diesel engines without the need of engine modification. The combustion of biodiesel in engines emits much less air pollutants and greenhouse gases than petrodiesel (Leung, 2001). Most of the feedstocks for biodiesel production comes from vegetable oils, both edible and nonedible, because they are renewable in nature and can be cultivated on a large scale (Patil and Deng, 2009). At present, more than 95% of biodiesel production feedstock comes from edible oils (Gui et al., 2008). However, the use of virgin edible oil has caused some problems in our society, such as competition with the edible oil market, hence raises the market price of both edible oils and biodiesel (Kansedo et al., 2009). Thus, researchers have growing interest in using nonedible oils that are not suitable for human consumption, such as *Jatropha curcas* seeds, okra seeds, and algae oils (Berchmans and Hirata, 2008; Sahoo and Das, 2009; Anwar et al., 2010; Gao et al., 2010). Nonedible oil crops can be grown in wastelands that are not suitable for food crops, and the cost of cultivation is much lower than normal food crops because these crops can still sustain reasonably high yield even without

TABLE 16.1
Amount of Waste Cooking Oil Produced in Major Countries

Country	Quantity (million tons/year)
United States	10.0
China	4.5
Europe	0.7–1.0
Japan	0.45–0.57
Malaysia	0.5
Canada	0.12
Taiwan	0.07

Source: Gui, M.M. et al., *Energy*, 33, 1646, 2008.

intensive care (Tiwari et al., 2007; Gui et al., 2008). However, most nonedible oils (such as jatropha and rubber seed oil) contain high free fatty acids (FFAs) that may require multiple chemical steps or alternate approaches to produce biodiesel. This inevitably increases the production cost, and the ester yield of biodiesel may be lower than the standard specifications for biodiesel (Haas, 2005; Patil and Deng, 2009; Sahoo and Das, 2009). Low-cost animal fats (with high saturated fatty acids) and used cooking oil (UCO) or used frying oil (UFO) can be feasible feedstocks of biodiesel production. The consumption of oils and fats in the world is enormous, estimated to be about 154 million tons/year (MPOC, 2007). A great amount of it will become UCO/UFO after using, which is not suitable for human consumption. Table 16.1 shows the quantity of waste cooking oil produced in major countries (Gui et al., 2008). The use of these waste oils can significantly reduce the cost of biodiesel production due to its low price in acquiring.

As mentioned earlier, biodiesel can be produced from various feedstocks such as raw vegetable oils, UFOs, animal fats, and even yellow grease from grease trap (Guo and Leung, 2003; Chakrabarti et al., 2008). These oils contain triglycerides that can be converted to methyl ester through the process of transesterification. Table 16.2 shows the physicochemical properties of biodiesel produced from virgin vegetable oils, animal fats, and waste oils (Leung et al., 2010). As seen in Table 16.2, while there are some variations in their chemical compositions, the physicochemical properties of the biodiesel produced from different feedstocks are more or less similar and can be used as a diesel substitute.

16.2 ADVANTAGES OF USING WASTE OILS FOR BIODIESEL PRODUCTION

Biodiesel has been developed commercially for more than three decades, but up to now, it is still not widely used in the world and most of its usage is blended with petrodiesel at 2%–5% (B2–B5) only. One of the main obstacles to biodiesel commercialization is its high cost. In almost all countries that use virgin oil–based biodiesel, its cost is much higher than petrodiesel unless subsidized by local government. Various ways have been adopted to reduce the cost of and add value to biodiesel production.

16.2.1 Reduction of Biodiesel Production Cost

According to literature, the cost of raw materials accounts about 60%–80% of the total cost of biodiesel production (Gui et al., 2008; Singh and Singh, 2009). As can be seen from Table 16.3, the

TABLE 16.2
Feedstocks for Biodiesel Production and Their Physiochemical Properties

Type of Oil		Species	Density (g/cm³)	Flash Point (°C)	Kinematic Viscosity (cst, at 40°C)	Acid Value (mg KOH/g)	Heating Value (MJ/kg)
		Soybean	0.91	254	32.9	0.2	39.6
		Rapeseed	0.91	246	35.1	2.92	39.7
		Sunflower	0.92	274	32.6	—	39.6
		Palm	0.92	267	39.6[a]	0.1	—
	Edible oil	Peanut	0.90	271	22.72	3	39.8
		Corn	0.91	277	34.9[a]	—	39.5
		Camelina	0.91	—	—	0.76	42.2
Vegetable oil		Canola			38.2	0.4	
		Cotton	0.91	234	18.2		39.5
		Pumpkin	0.92	>230	35.6	0.55	39
		J. curcas	0.92	225	29.4	28	38.5
		Pongamia pinnata	0.91	205	27.8	5.06	34
	Nonedible oil	Sea mango	0.92	—	29.6	0.24	40.86
		Palanga	0.90	221	72.0	44	39.25
Animal fats		Tallow	0.92	—	—	—	40.05
		Nile tilapia	0.91	—	32.1[b]	2.81	—
		Poultry	0.90	—	—	—	39.4
Others		UCO	0.90	—	44.7	2.5	—

[a] Kinematic viscosity at 38°C, mm2/s.
[b] Kinematic viscosity at 37°C, mm2/s.

TABLE 16.3
Average International Prices from Common Vegetable Oil and Waste Oil Feedstocks for Biodiesel Production in 2007 (US$/ton)

Feedstock	Cost (US$/Ton)
Crude palm oil	543
Rapeseed oil	824
Soybean oil	771
Crude corn oil	802
Crude peanut oil	891
Waste cooking oil	224
Yellow grease	412
Poultry fat	256

Source: Demirbas, A., *Appl. Energy*, 86, S108, 2009a.

cost of some waste oil feedstock (i.e., UCO, yellow grease, poultry fat) is much lower than that of crude vegetable oils. Escobar et al. (2009) indicated that the overall production costs of biodiesel can be reduced by more than half compared to virgin vegetable oil if UFO is used as feedstock, while Demirbas (2009a) estimated that the cost of biodiesel produced from waste grease is about 35% less than that of virgin vegetable oil. Therefore, choosing suitable feedstocks can reduce the cost of biodiesel production that can ultimately reduce its market price and make it more competitive and economically viable.

16.2.2 Solving Problems of Waste Oil Disposal and Treatment

One obvious potential environmental benefit of using the waste oils as a feedstock for biodiesel production is the reduction in the amount of waste oils to be treated. Every day, a large amount of these waste edible oils is generated from restaurants, fast-food shops, food processing industries, and abattoirs. Most of these oils have poor quality for reconsumption and many countries, including the European Union, have banned their use as animal feed. Therefore, these waste oils need to be disposed of in landfills or treated in wastewater treatment plants. This induces tremendous pressure to the waste treatment facilities of a modernized city. So the reuse of waste oils to produce biodiesel has the benefits of both waste recycling and reducing of biodiesel production cost, which can normally receive support from local government. It can also help to reduce some of the waste oils from entering the human food chain in some countries such as China.

16.3 WASTE FEEDSTOCK FOR BIODIESEL PRODUCTION

As mentioned earlier, many plants have considered or already used waste oils for producing biodiesel, mainly due to cost consideration. Table 16.4 shows a list and characteristics of waste oil feedstock that can be used for producing biodiesel.

TABLE 16.4
Different Waste Feedstocks for Biodiesel Production

Waste Feedstocks	Characteristics	Reference for Biodiesel Production
UFO based on vegetable oils	Most commonly used waste feedstock for biodiesel production; composition varies with types of base oil used; contain impurities due to the interaction with other food during cooking	Zhang et al. (2003), Kulkarni and Dalai (2006), Canakci (2007), Lam et al. (2010), and Patil et al. (2010)
Animal fats (lard, poultry fat, tallow, ghee)	Contained high saturated FFA; concern in usage under cold weather; normally exist in solid form at room temperature; higher CV and cetane number but easier oxidation	Ma et al. (1998), Tashtoush et al. (2004), Canakci (2007), Bhatti et al. (2008), Dias et al. (2009), Guru et al. (2010), and Ito et al. (2012)
Yellow grease	UCO originated from vegetable oils or animal fats; FFA <15%; main composition is triglycerides; yellow color; liquid or solid form at room temperature	Canakci and Van Gerpen (2001), Cao et al. (2008), and Diaz-Felix et al. (2009)
Brown grease	Trap grease with FFA >15%; possesses higher proportion of mono- and diglycerides than triglycerides; brown color and solid form at room temperature; strong and offensive odor smell	Canakci and Van Gerpen (2001), Lotero et al. (2005), Kim et al. (2011), and Yan et al. (2012)

16.3.1 Used Frying Oil or Used Cooking Oil

UFO or UCO is the main source of waste oil for producing biodiesel. A lot of studies have been conducted in this area over the past two decades. The earliest study can be traced back to 1983 by Nye et al. (1983) who converted UFO to biodiesel and tested its properties. A number of studies were subsequently conducted using waste vegetable oils for cooking (Zhang et al., 2003; Leung and Guo, 2006; Phan and Phan, 2008; Meng et al., 2008; Demirbas, 2009a,b; Hingu et al., 2010; Patil et al., 2010; Wan Omar and Amin, 2011; Chen et al., 2012; Noshadi et al., 2012). Apart from traditional waste vegetable oils based on rapeseed, corn, or peanut oils, other types of UFO have also been studied. Al-Widyan and Al-Shyoukh (2002) studied the conversion characteristics of palm oil–based UFO using acid catalysts, while Lertsathapornsuk et al. (2008) successfully converted the same type of waste oil to biodiesel with a yield of 97% using alkaline catalyst and microwave power.

16.3.2 Animal Fats

Animal fats include lard, poultry fat (i.e., duck fat and chicken fat), tallow, and ghee. Animal fats contain high saturated FFA that can be converted to biodiesel. Most animal fats are wastes from animal meat-processing facilities and rendering companies for processing animal mortalities. The properties of animal fats were found to be suitable for biodiesel production, while the inherent high FFA levels can be reduced by using an acid catalyst process (see Section 16.4.3) (Canakci 2007). Ma et al. (1998) studied the transesterification of beef tallow to biodiesel under different conditions. The properties of the biodiesel produced from tallow were assessed to determine its suitability in engine operations (Alcantara et al., 2000; Moraes et al., 2008). Since the animal fats have relatively high levels of saturated fats, the cold flow properties of biodiesel produced from them may cause concern in usage under cold weather. Compared with biodiesel produced from vegetable oils, animal fats–based biodiesel has higher calorific value and cetane number (Lebedevas et al., 2006) but is easier to be oxidized because of the absence of natural antioxidants (Sendzikiene et al., 2005).

16.3.3 Yellow Grease

Yellow grease refers to those oils generated from UCO and other fats and oils with FFAs less than 15%. It is generally collected from commercial or industrial cooking operations. Typically, the main composition of yellow grease is triglycerides and it has a yellow color and exists in either liquid or solid form at room temperature. According to the report of the National Renderers Association, about 7.2 billion pounds of nonedible tallow and grease are annually produced in commercial US restaurants (Diaz-Felix et al., 2009). With the help of a well-developed collection system, yellow grease has the potential to be an effective feedstock for biodiesel production because of lower cost (half price of soybean oil) (Groschen, 2002; Lai et al., 2005). However, since yellow grease from different collection sources varies widely in the levels of FFA, the production process of biodiesel from this waste feedstock becomes more complicated. The technical challenge is to find an efficient method to reduce the FFA level of yellow grease. Cao et al. (2008) successfully produced biodiesel from yellow grease in a continuous membrane reactor under base-catalyzed conditions. Diaz-Felix et al. (2009) investigated the relationship between the composition of yellow grease and process variables of biodiesel production. They found that the FFA level of yellow grease could be successfully reduced by using an acid-catalyzed esterification process, and the amount of methanol used plays a significant role in this process.

16.3.4 Brown Grease (Trap Grease)

Brown grease refers to those waste grease materials with FFA over 15%. It is usually collected from grease traps, which are installed in commercial, industrial, or municipal sewage facilities to separate grease and oil from wastewater (Groschen, 2002). Generally, brown grease has a higher proportion of monoglycerides and diglycerides than triglycerides and is typically brown in color and solid at room temperature. Moreover, it may have a very strong and offensive odor because of bacterial-produced compounds, septage introduced from collection or processing, and volatile short-chain FFAs produced from biodegradation (Grueb and Gatfield, 1989). Since the commercial viability of brown grease might be limited unless its inherent odor could be mitigated, a headspace solid-phase microextraction (HS-SPME) analysis method has been developed by Thompson et al. (2012) to provide a rapid and quantitative identification method of odor compounds in brown grease and its biodiesel product. The significant technical difficulty of converting brown grease to biodiesel is the reduction of the high FFA level. Recently, Yan et al. (2012) successfully developed a novel and efficient method for the production of biodiesel from brown grease, which is using tandem lipases to catalyze the conversion of grease to biodiesel via esterification and transesterification in one pot with methanol in a solvent-free system. Furthermore, since high concentrations of sulfur and metal compounds are also found in brown grease, most catalytic approaches for biodiesel production from brown grease are easily poisoned. To solve this problem, a new class of zirconia-supported metal oxide catalysts has been developed to produce biodiesel with high FAME yields through esterification of FFA while simultaneously performing desulfurization and demetallization functions (Kim et al., 2011).

16.4 PRODUCTION PROCESSES

Many biodiesel studies have been conducted with a focus on the optimization of the production processes using different waste oil feedstocks. The main problem generally associated with using waste oil feedstock is the presence of a high level of FFA and water. The alkali catalyst used in the alkali transesterification will react with the FFA in the oils to form soap (saponification reaction). Equation 16.1 shows an example of the saponification reaction in which the alkaline catalyst (e.g., sodium hydroxide) reacts with the FFAs to form soap and water:

$$\underset{(\text{FFA})}{R_1-\text{COOH}} \quad + \quad \underset{(\text{Sodium hydroxide})}{\text{NaOH}} \quad \longrightarrow \quad \underset{(\text{Soap})}{R_1-\text{COONa}} \quad + \quad \underset{(\text{Water})}{H_2O} \tag{16.1}$$

The earlier saponification reaction is undesirable because the soap would lower the biodiesel yield and inhibit the separation of the esters from the glycerol after the transesterification reaction. Furthermore, it would bind with the alkaline catalyst, thus increasing its catalyst consumption (Van Gerpen et al., 2004). On the other hand, water, which either originated from the original oils or formed during the saponification reaction, would retard the transesterification reaction through the hydrolysis reaction. Typical hydrolysis reaction is shown in Equation 16.2 in which water hydrolyzes the triglycerides to diglycerides and forms FFAs:

$$\begin{array}{ccccccc} CH_2-O-CO-R_1 & & & & CH_2-OH & & \\ | & & & & | & & \\ CH-O-CO-R_2 & + & H_2O & \longrightarrow & CH-O-CO-R_2 & + & R_1-COOH \\ | & & & & | & & \\ CH_2-O-CO-R_3 & & & & CH_2-O-CO-R_3 & & \\ (\text{Triglycerides}) & & (\text{Water}) & & (\text{Diglyceride}) & & (\text{FFA}) \end{array} \tag{16.2}$$

This hydrolysis reaction is also undesirable since it would reduce the biodiesel yield and cause the saponification reaction. Therefore, the feedstock for producing biodiesel with high FFA and water needs to be pretreated before the transesterification process to avoid its decrement in biodiesel yield. Several processes have been developed to cater for these special characteristics of waste oils, such as alkaline transesterification, acid-catalyzed transesterification, acid- and alkali-catalyzed two-stage transesterification, enzyme-catalyzed transesterification, and supercritical transesterification (see Table 16.5). These different processes have been reviewed by Zhang et al. (2003), Kulkarni and Dalai (2006), and Banerjee and Chakraborty (2009) and are briefly discussed in the following sections.

16.4.1 Alkaline-Catalyzed Transesterification

In this method, an alkaline is first dissolved in a methanol solution to form methoxide that is then added to the oil to carry out the transesterification reaction. This method is well developed and commonly used with virgin vegetable oil feedstock that can also be applied to those waste oils with low FFA content (<5%) (Leung et al., 2001). Studies in this area mainly were focused on the way to improve the process efficiency and reaction conditions. Guo (2005) investigated alkaline-catalyzed transesterification kinetics of UFO and the study revealed that the overall reaction followed a pseudo second-order kinetic model. Lertsathapornsuk et al. (2008) and Chen et al. (2012) adopted a microwave heating system to improve the conversion efficiency and reduce the time of reaction. Hingu et al. (2010) found a significant improvement in the reaction rate using a sonochemical reactor with the help of ultrasound. Omar and Amin (2011) used a zirconia catalyst (a heterogeneous catalyst) to enhance the conversion performance using waste cooking oil with high FFA content.

Although the technology of alkaline transesterification is well developed, the product yield of this method is sensitive to both water and FFA content in the raw material due to the formation of soap. As most waste cooking oils contain water and have high FFA content, the yield of production will be affected under this process. Under this situation, other methods of production as discussed in the following would be more effective.

16.4.2 Acid-Catalyzed Transesterification

Certain types of waste oil feedstock contain water and high levels of FFA that will react with the alkaline catalyst to form soap during the alkaline-catalyzed transesterification process. This reaction is undesirable because it will lower the yield of the biodiesel and inhibit the separation of esters from the glycerol formed. With the use of an acidic catalyst (such as concentrated sulfuric acid and hydrochloric acid), the oil can react with alcohol to form ester (biodiesel) directly. This process is insensitive to any FFA in the oil and therefore suitable even when there is a high level of FFA (Freedman et al., 1984). The major disadvantage of this process is the long processing time and the potential corrosion problem due to the use of homogeneous acid catalysts. Lam et al. (2010), after reviewing different catalysts for the transesterification of waste cooking oil of high FFA content, concluded that a heterogeneous acid catalyst is one of the best options as compared to the current commercial homogeneous alkali-catalyzed and acid-catalyzed processes. However, a suitable heterogeneous acid catalyst needs to be developed with special attention on catalyst deactivation and regeneration.

16.4.3 Acid- and Alkali-Catalyzed Two-Stage Transesterification

Since both acid and alkali catalysts have their own merits and demerits, acid- and alkali-catalyzed two-stage transesterification converts the two processes into a two-stage process. An acidic catalyst is used initially to convert FFA to esters and decrease the FFA level to <1%. In the second stage, the transesterification of the remaining oil is performed using an alkaline catalyst (Kulkarni and Dalai, 2006). Jain et al. (2011) studied the kinetic of the two-stage transesterification and obtained a maximum yield of 21.5% and 90.6%, respectively, from the esterification and transesterification

TABLE 16.5
Characteristics of Biodiesel Produced from Various Processes

Process Type	Catalyst Type	Catalyst Used (wt.%)	Reaction Time (h)	Reaction Temperature (°C)	Alcohol to Oil Molar Ratio	FAME Yield (%)	Comment	References
One step	Alkaline	NaOH, 1.1	0.33	60	Methanol, 7:1	88.8	Produce high FAME yield in shorter reaction time; essential for low FFA content;	Leung and Guo (2006) and Patil et al. (2010)
	Acid	H_2SO_4, 4	20	95	Methanol, 20:1	≥90	not sensitive to FFA; long reaction time; high alcohol used	Wang et al. (2006) and Lam et al. (2010)
Two step	Acid followed by alkaline	Ferric sulfate, 2	Acid, 2	Acid, 100	Acid, 9:1	96	Not sensitive to FFA; low product yield; low recovery of catalyst	Patil et al. (2010) and Lam et al. (2010)
		KOH, 0.5	Alkaline, 1	Alkaline, 100	Alkaline, 7.5:1			
Supercritical methanol	No catalyst	—	0.5	560	Methanol, 41:1	99.6	No catalyst used; short reaction time; high FAME yield; high temperature and pressure requirement; high energy consumption	Patil et al. (2010)
Ultrasonic-assisted	Alkaline	KOH, 1	0.67	45	Methanol, 6:1	93.5	Short reaction time; high conversion rate	Hingu et al. (2010)
Microwave heating	Alkaline	CH_3ONa, 0.75	0.05	—	Methanol, 6:1	97.9	Short reaction time; high energy efficiency	Chen et al. (2012)

of pretreated waste cooking oil with an FFA content of 21.8%. Dorado et al. (2002) studied the conversion of waste cooking oils from different sources with different FFA contents that vary from 1.3% to 5.1%, using a two-stage transesterification process. They found that this process is a good method for biodiesel production from waste cooking oils of different origins with an FFA <=3%, and a conversion efficiency of 88%–95% can be achieved.

16.4.4 Enzyme-Catalyzed Transesterification

This process is similar to alkali- or acid-catalyzed transesterification except that a special enzyme is used to replace the acid or alkali catalyst. The enzyme is a type of biocatalyst but its synthesis is rather costly. Nevertheless, the biodiesel production process using an enzyme catalyst has the merits of no by-products, easy product recovery, and catalyst recycling, and the reaction is insensitive to the FFA and water content in the waste oils (Hsu et al., 2004; Kulkarni and Dalai, 2006). Lam et al. (2010) found that the enzyme-catalyzed transesterification is one of the best options to produce biodiesel from waste cooking oil with a high FFA level, which, however, suffers from serious heat and mass transfer limitation. More advanced research works on process scale up and enzyme development are required.

16.4.5 Supercritical Methanol Method

When a fluid or gas is subjected to temperatures and pressures in excess of its critical point, there are a number of unusual properties exhibited. Under such conditions, distinct liquid and vapor phases no longer exist and a single fluid phase is formed (Van Gerpen et al., 2004). In this method, the temperature and pressure inside the reactor are raised to above 350°C and 450 bar, respectively, such that the thermophysical properties of methanol (solvent) are changed, including the reduction in viscosity and polarity (Kulkarni and Dalai, 2006). This enhances the solubility of oil in methanol and hence the rate of methyl ester (biodiesel) formation. This method is a noncatalytic method that eliminates the needs of catalyst recovery.

The transesterification of waste cooking oil using supercritical methanol was studied by Warabi et al. (2004) and Kusdiana and Saka (2004) with different water and FFA levels. Patil et al. (2010) converted waste cooking oil to biodiesel through supercritical methanol method and found that the yield increased to 50%–65% in the first 15 min and reached 80% after 20 min. Also, high temperature has a favorable effect on the biodiesel yield. Biodiesel produced by this method can be scaled up, resulting in high purity of methyl ester (99.6%) and pure glycerol by-product (96.5%) (Demirbas, 2009b). The major advantages of this process include the fast reaction rate, high yield, catalyst-free operation, and the ability handle to waste oils with high FFA content. On the other hand, the main disadvantages of this method include the high energy requirement (high temperature and pressure) for the reactor and high molar ratio of oil to alcohol (>1:40), all of which will increase the cost significantly.

16.4.6 Pretreatment

Pretreatment operation is essential for biodiesel production using waste oils, particularly those oils with high FFA and moisture content. The operation can be a physical pretreatment (such as filtration for removing suspended solids and water washing for removing water-soluble impurities as found in most UFO and trap grease) or chemical pretreatment (such as acidic esterification in the two-stage transesterification and distillation) (Banerjee and Chakraborty, 2009). In addition, deodorization is another important step in pretreatment, particularly for UCO or grease. During this step, low-pressure steam is injected into the oils at a temperature of about 500 K in order to eliminate FFA, aldehydes, unsaturated hydrocarbons, and ketones, all of which cause undesirable odors and flavors in the waste oils (Demirbaşa and Karaa, 2006).

16.4.7 Analytical Methods

The quality of the biodiesel produced governed its usability particularly in the internal combustion engine application. Knothe (2006) has reviewed various analytical methods to evaluate biodiesel quality to ensure meeting international standards (i.e., ASTM, EN14214), which was updated by Monteiro et al. (2008).

16.5 PROSPECT OF BIODIESEL PRODUCED FROM WASTE FEEDSTOCKS

Due to the needs to improve the air quality and combat global climate change, there is a worldwide growing trend in the use of biofuels, including biodiesel. The needs to reduce the cost of biodiesel production and the pressure in waste oil disposal make the use of waste oil feedstock to produce biodiesel an attractive option for biodiesel producers. As shown in Table 16.1, countries like the United States and China produced an enormous amount of waste cooking oil every year (Gui et al., 2008), which are 10 and 4.5 million tons/year (excluding other types of waste oils such as animal fats and tallow), respectively. Technologies are also developed that can economically convert trap grease to biodiesel (Chakrabarti et al., 2008; Wang et al., 2008), which greatly increase the amount of waste oil feedstock for biodiesel production. A commercial biodiesel plant with a capacity of 100,000 tons, which can produce biodiesel using the oil and fat recovered from trap grease waste, is under construction and will be commissioned in Hong Kong in 2013. This project can act as a demonstration plant for the enormous market in mainland China. It can be envisaged that more and more biodiesel will be produced from waste oil feedstock in the near future.

16.6 CONCLUSIONS

Due to air pollution and global climate change issues, it is clear that the demand of biofuels would increase significantly in the coming decades. Biodiesel is a type of biofuel and a diesel alternative that has grown steadily over the past two decades. A major factor that affects the widespread development of biodiesel is the high production cost due to the high cost of raw oil feedstock. On the other hand, with the growth in world economy and population, more and more oily wastes are produced in modern cities that provide a golden opportunity for economic biodiesel production. Various types of oil and grease waste have been found to be a suitable alternative feedstock for biodiesel production. Technologies have been developed that can produce biodiesel economically using UFO/UCO, animal fats, tallow, oil, and grease recovered from grease trap. Various processes have been developed for the conversion of different types of waste oils to biodiesel. The selection of a suitable process will depend on several factors, but the most important ones are the water and FFA contents in the oil samples. It is anticipated that more and more plants will be commissioned that can produce more competitive biodiesel fuel and reduce the pressure of treating and disposing the waste oils in our society.

ACKNOWLEDGMENT

This study is supported by the ICEE and CRCG grant of the University of Hong Kong.

REFERENCES

Alcantara, R., Amores, J., Canoira, L., Fidalgo, E., Franco, M.J., and Navarro, A. 2000. Catalytic production of biodiesel from soy-bean oil, used frying oil and tallow. *Biomass and Bioenergy* 18: 515–527.

Al-Widyan, M.I. and Al-Shyoukh, A.O. 2002. Experimental evaluation of the transesterification of waste palm oil into biodiesel. *Bioresource Technology* 85: 253–256.

Anwar, F., Rashid, U., Ashraf, M., and Nadeem, M. 2010. Okra (Hibiscus esculentus) seed oil for biodiesel production. *Applied Energy* 87: 779–785.

Banerjee, A. and Chakraborty, R. 2009. Parametric sensitivity in transesterification of waste cooking oil for biodiesel production-A review. *Resources, Conservation and Recycling* 53: 490–497.

Berchmans, H.J. and Hirata, S. 2008. Biodiesel production from crude *Jatropha curcas* L. seed oil with a high content of free fatty acids. *Bioresource Technology* 99: 1716–1721.

Bhattia, H.M., Hanifa, M.A., Qasima, M., Rehmanb, A. 2008. Biodiesel production from waste tallow. *Fuel* 87 (13–14): 2961–2966.

Canakci, M. 2007. The potential of restaurant waste lipids as biodiesel feedstocks. *Bioresource Technology* 98: 183–190.

Canakci, M. and Van Gerpen, J. 2001. Biodiesel production from oils and fats with high free fatty acids. *Transactions ASAE* 44: 1429–1436.

Cao, P., Dubé, M.A., Tremblay, A.Y. 2008. High-purity fatty acid methyl ester production from canola, soybean, palm, and yellow grease lipids by means of a membrane reactor. *Biomass and Bioenergy* 32(11): 1028–1036.

Chakrabarti, A.R., Hake, J.M., Zarchi, I., and Gray, D.M.D. 2008. Waste grease biodiesel production at a wastewater treatment plant. *Proceedings of WEFTEC.08, 81st Annual Technical Exhibition and Conference*, October 18–22, 2008, Chicago, IL. pp. 2770–2789.

Chen, K.S., Lin, Y.C., Hsu, K.H., and Wang, H.K. 2012. Improving biodiesel yields from waste cooking oil by using sodium methoxide and a microwave heating system. *Energy* 38: 151–156.

Demirbas, A. 2009a. Political, economic and environmental impacts of biofuels: A review. *Applied Energy* 86: S108–S117.

Demirbas, A. 2009b. Biodiesel from waste cooking oil via base-catalytic and supercritical methanol transesterification. *Energy Conversion and Management* 50: 923–927.

Demirbaşa, A., Karaa, H. 2006. New options for conversion of vegetable oils to alternative fuels. *Energy Sources, Part A: Recovery, Utilization, and Environmental Effects* 28(7): 619–626.

Dias, J.M., Alvim-Ferraz, M.C.M., and Almeida, M.F. 2009. Production of biodiesel from acid waste lard. *Bioresource Technology* 100: 6355–6361.

Diaz-Felix, W., Riley, M.R., Zimmt, W., and Kazz, M. 2009. Pretreatment of yellow grease for efficient production of fatty acid methyl esters. *Biomass and Bioenergy* 33: 558–563.

Dorado, M.P., Ballesteros, E., de Almeida, J.A., Schellert, C., Lohrelein, H.P., and Krause, R. 2002. An alkali-catalyzed transesterification process for high free fatty acid waste oils. *Transactions ASAE* 45: 525–529.

Escobar, J.C., Lora, E.S., Venturini, O.J., Yáñez, E.E., Castillo, E.F., and Almazan, O. 2009. Biofuels: Environment, technology and food security. *Renewable Sustainable Energy Reviews* 13: 1275–1287.

Freedman, B., Pryde, E.H., and Mounts, T.L. 1984. Variables affecting the yields of fatty esters from transesterified vegetable oils. *Journal of the American Oil Chemical Society* 61: 1638–1643.

Gao, C., Zhai, Y., Ding, Y., and Wu, Q. 2010. Application of sweet sorghum for biodiesel production by heterotrophic microalga *Chlorella protothecoides. Applied Energy* 87: 756–761.

Groschen, R. 2002. Overview of: The feasibility of biodiesel from waste/recycled greases and animal fats. Report prepared for the Legislative Commission on Minnesota Resource. Available online: http://www.mda.state.mn.us. Accessed on January 5, 2014.

Grueb, H. and Gatfield, J.L. 1989. Generation of flavor components by microbial fermentation and enzyme technology, In: *Flavor Chemistry of Lipid Foods*. Min, D.B. and Smouse, T.H., eds., Champaign, IL.: American Oil Chemists Society, pp. 367–375.

Gui, M.M., Lee, K.T., and Bhatia, S. 2008. Feasibility of edible oil vs. non-edible oil vs. waste edible oil as biodiesel feedstock. *Energy* 33: 1646–1653.

Guo, Y. 2005. Alkaline-catalyzed production of biodiesel fuel from virgin canola oil and recycled waste oils. PhD thesis, Department of Mechanical Engineering, The University of Hong Kong, Hong Kong.

Guo, Y. and Leung, D.Y.C. 2003. Analysis on the biodiesel production using grease trap oils from restaurants. *Macro Review* 16: 421–426.

Guru, M., Koca, A., Can, O., Cinar, C., and Sahin, F. 2010. Biodiesel production from waste chicken fat based sources and evaluation with Mg based additive in a diesel engine. *Renewable Energy* 35: 637–643.

Haas, M.J. 2005. Improving the economics of biodiesel production through the use of low value lipids as feedstocks: Vegetable oil soapstock. *Fuel Processing Technology* 86: 1087–1096.

Hingu, S.M., Gogate, P.R., and Rathod, V.K. 2010. Synthesis of biodiesel from waste cooking oil using sonochemical reactors. *Ultrasonics Sonochemistry* 17: 827–832.

Hsu, A., Jones, K.C., Foglia, T.A., and Marmer, W.N. 2004. Continuous production of ethyl esters of grease using an immobilized lipase. *Journal of the American Oil Chemical Society* 81: 749–752.

Ito, T., Sakurai, Y., Kakuta, Y., Sugano, M., and Hirano, K. 2012. Biodiesel production from waste animal fats using pyrolysis method. *Fuel Processing Technology* 94: 47–52.

Jain, S., Sharma, M.P., and Rajvanshi, S. 2011. Acid base catalyzed transesterification kinetics of waste cooking oil. *Fuel Processing Technology* 92: 32–38.

Kansedo, J., Lee, K.T., and Bhatia, S. 2009. Cerbera odollam (sea mango) oil as a promising non-edible feedstock for biodiesel production. *Fuel* 88(6): 1148–1150.

Kim, M., DiMaggio, C., Yan, S., Wang, H., Salley, S.O., and Simon Ng, K.Y. 2011. Performance of heterogeneous ZrO_2 supported metaloxide catalysts for brown grease esterification and sulfur removal. *Bioresource Technology* 102: 2380–2386.

Knothe, G. 2006. Analyzing biodiesel: Standards and other methods. *Journal of the American Oil Chemical Society* 83: 873–877.

Kulkarni, M.G. and Dalai, A.K. 2006. Waste cooking oil an economical source for biodiesel: A review. *Industrial & Engineering Chemistry Research* 45(9): 2901–2913.

Kusdianna, D. and Saka, S.2004. Effects of water on biodiesel fuel production by supercritical methanol treatment. *Bioresource Technology* 91: 289–295.

Lai, C.C., Zullaikah, S., Vali, S.R., and Ju, Y.H. 2005. Lipase-catalyzed production of biodiesel from rice bran oil. *Journal of Chemical Technology and Biotechnology* 80: 331–337.

Lam, M.K., Lee, K.T., and Mohamed, A.R. 2010. Homogeneous, heterogeneous and enzymatic catalysis for transesterification of high free fatty acid oil (waste cooking oil) to biodiesel: A review. *Biotechnology Advances* 28: 500–518.

Lebedevas, S., Vaicekauskas, A., Lebedeva, G., Makareviciene, V., Janulis, P., and Kazancev, K. 2006. Use of waste fats of animal and vegetable origin for the production of biodiesel fuel: Quality, motor properties, and emissions of harmful components. *Energy and Fuels* 20:2274–2280.

Lertsathapornsuk, V., Pairintra, R., Aryusuk, K., and Krisnangkura, K. 2008. Microwave assisted in continuous biodiesel production from waste frying palm oil and its performance in a 100 kW diesel generator. *Fuel Processing Technology* 89: 1330–1336.

Leung, D.Y.C. 2001. Development of a clean biodiesel fuel in Hong Kong using recycled oil. *Journal Water, Air & Soil Pollution* 130: 277–282.

Leung, D.Y.C. and Guo, Y. 2006. Transesterification of neat and used frying oil: Optimization for biodiesel production. *Fuel Processing Technology* 87: 883–890.

Leung, D.Y.C., Wu, Xuan, and Leung, M.K.H. 2010. A review on biodiesel production using catalyzed transesterification. *Applied Energy* 87: 1083–1095.

Lotero, E., Liu, Y., Lopez, D.E., Suwannakarn, K., Bruce, D.A., and Goodwin, Jr J.G. 2005. Synthesis of biodiesel via acid catalysis. *Industrial and Engineering Chemistry Research* 44: 5353–5363.

Ma, F., Clements, L.D., and Hanna, M.A. 1998. Biodiesel fuel from animal fat. Ancillary studies on transesterification of beef tallow. *Industrial and Engineering Chemistry Research* 37: 3768–3771.

Malaysia Palm Oil Council (MPOC). 2007. *The Palm Oil*. Selangor, Malaysia: MPOC. Available from http://www.mpoc.org.my.

Meng, X., Chen, G., Wang, Y. 2008. Biodiesel production from waste cooking oil via alkali catalyst and its engine test. *Fuel Processing Technology* 89(9): 851–857.

Monteiro, M.R., PepeAmbrozin, A.R., Lião, L.M., and Ferreira, A.G. 2008. Critical review on analytical methods for biodiesel characterization. *Talanta* 77: 593–605.

Moraes, M.S.A., Krause, L.C., da Cunha, M.E., Faccini, C.S., de Menezes, E.W., Veses, R.C., Rodrigues, M.R.A., and Caramão, E.B. 2008. Tallow biodiesel: Properties evaluation and consumption tests in a diesel engine. *Energy Fuels* 22: 1949–1954.

Noshadi, I., Amin, N.A.S., and Parnas, R.S. 2012. Continuous production of biodiesel from waste cooking oil in a reactive distillation column catalyzed by solid heteropolyacid: Optimization using response surface methodology (RSM). *Fuel* 94: 156–164.

Nye, M.J., Williamson, T.W., Deshpande, S., Schrader, J.H., Snively, W.H., Yurkerwich, T.P., and French, C.L. 1983. Conversion of used frying oil to diesel fuel by transesterification: Preliminary tests. *Journal of the American Oil Chemistry Society* 60: 1598–1601.

Omar, W.N.N.W. and Amin, N.A.S. 2011. Biodiesel production from waste cooking oil over alkaline modified zirconia catalyst. *Fuel Processing Technology* 92: 2397–2405.

Patil, P. and Deng, S. 2009. Optimization of biodiesel production from edible and non-edible vegetable oils. *Fuel* 88: 1302–1306.

Patil, P., Deng, S., Rhodes, J., and Lammers, P.J. 2010. Conversion of waste cooking oil to biodiesel using ferric sulfate and supercritical methanol processes. *Fuel* 89: 360–364.

Phan, A.N. and Phan, T.M. 2008. Biodiesel production from waste cooking oils. *Fuel* 87: 3490–3496.

Sahoo, P.K. and Das, L.M. 2009. Process optimization for biodiesel production from Jatropha, Karanja and Polanga oils. *Fuel* 88: 1588–1594.

Sendzikiene, E., Makareviciene, V., and Janulis, P. 2005. Oxidation stability of biodiesel fuel produced from fatty wastes. *Polish Journal of Environmental Studies* 14: 335–339.

Singh, S.P. and Singh, D. 2009. Biodiesel production through the use of different sources and characterization of oils and their esters as the substitute of diesel: A review. *Renewable and Sustainable Energy Reviews* 14: 200–216.

Tashtoush, G.M., Al-Widyan, M.I., and Al-Jarrah, M.M. 2004. Experimental study on evaluation and optimization of conversion of waste animal fat into biodiesel. *Energy Conversion and Management* 45: 2697–2711.

Thompson, J.G., Bertman, S., Hill, S., Kushner, L., and Miller, J.B. 2012. Using headspace solid phase microextraction to evaluate the odor compounds in trap grease feedstock for biodiesel. *Biomass and Bioenergy* 43: 36–41.

Tiwari, A.K., Kumar, A., and Raheman, H. 2007. Biodiesel production from jatropha oil (*Jatropha curcas*) with high free fatty acids: An optimized process. *Biomass and Bioenergy* 31: 569–575.

Van Gerpen, J., Shanks, B., Pruszko, R., Clement, D., and Knothe, G. 2004. Biodiesel production technology: August 2002–January 2004. National Renewable Energy Laboratory subcontractor report NREL SR-510-36244. Available from http://www.nrel.gov/docs/fy04osti/36244.pdf. Accessed on January 5, 2014.

Wan Omar, W.N.N. and Amin, N.A.S. 2011. Biodiesel production from waste cooking oil over alkaline modified zirconia catalyst. *Fuel Processing Technology* 92: 2397–2405.

Wang, Y., Ou, S., Liu, P., Xue, F., and Tang, S. 2006. Comparison of two different processes to synthesize biodiesel by waste cooking oil. *Journal of Molecular Catalysis A: Chemical* 252(1–2): 107–112.

Wang, Z.M., Lee, J.S., Park, J.Y., Wu, C.Z., and Yuan, Z.H. 2008. Optimization of biodiesel production from trap grease via acid catalysis. *Korean Journal of Chemical Engineering* 25: 670–674.

Warabi, Y., Kusdianna, D., and Saka, S. 2004. Reactivity of triglycerides and fatty acids of rapeseed oil in supercritical alcohols. *Bioresource Technology* 91: 283–287.

Yan, J., Li, A., Xu, Y., Ngo, T.P.N., Phua, S., and Li, Z. 2012. Efficient production of biodiesel from waste grease: One-pot esterification and transesterification with tandem lipases. *Bioresource Technology* 123: 332–337.

Zhang, Y., Dube, M.A., McLean, D.D., and Kates, M. 2003. Biodiesel production from waste cooking oil: 1. Process design and technological assessment. *Bioresource Technology* 89: 1–16.

17 Biomass Gasification and Syngas Utilization

Ajay Kumar, Ashokkumar Sharma, and Pushpak Bhandari

CONTENTS

17.1 INTRODUCTION

Gasification is a robust technology, which converts diverse biomass feedstocks into syngas, a consistent product stream. Gasification is a well-demonstrated technology to convert low-grade energy of solid biomass into high-value gaseous product, called producer gas or syngas. Gasification process occurs under high temperature (600°C–1000°C) in the presence of a limited oxidizing agent.

TABLE 17.1
Major Reactions Involved in a Gasification Process

Biomass decomposition		
$CH_xO_yN_zS_s \rightarrow a\,CO + b\,CO_2 + c\,CH_4 + d\,H_2 + e\,NH_3 + f\,H_2S + g\,H_2O + h$ tar		R(1)
Char combustion		
Partial combustion of char	$C + \frac{1}{2}O_2 \rightarrow CO$	R(2)
Complete combustion of char	$C + O_2 \rightarrow CO_2$	R(3)
Char gasification		
Boudouard reaction	$C + CO_2 \leftrightarrow 2CO$	R(4)
Water–gas reaction	$C + H_2O \leftrightarrow CO + H_2$	R(5)
Methanation reaction	$C + 2H_2 \leftrightarrow CH_4$	R(6)
Shift reaction		
Water gas shift reaction	$CO + H_2O \rightarrow CO_2 + H_2$	R(7)
Reforming reaction		
Methane reforming	$CH_4 + H_2O \leftrightarrow CO_2 + 3H_2$	R(8)
Volatile oxidation		
CO oxidation	$CO + \frac{1}{2}O_2 \rightarrow CO_2$	R(9)
H_2 oxidation	$H_2 + \frac{1}{2}O_2 \rightarrow H_2O$	R(10)
CH_4 oxidation	$CH_4 + 2O_2 \rightarrow CO_2 + 2H_2O$	R(11)

High temperature and oxidizing agent break down high molecular weight compounds of biomass, such as cellulose, hemicellulose, and lignin, into lighter and stable gaseous molecules, such as CO, H_2, CO_2, and CH_4. The process involves four major types of reactions: drying, devolatilization (pyrolysis), oxidation, and reduction. Table 17.1 lists major reactions occurring during biomass gasification (Kumar et al. 2009a). Biomass properties, gasifier type, and operating variables affect the extent of these reactions and hence determine yield and composition of syngas and bioproducts obtained (Figure 17.1). The following sections describe how these factors influence the product yields and qualities.

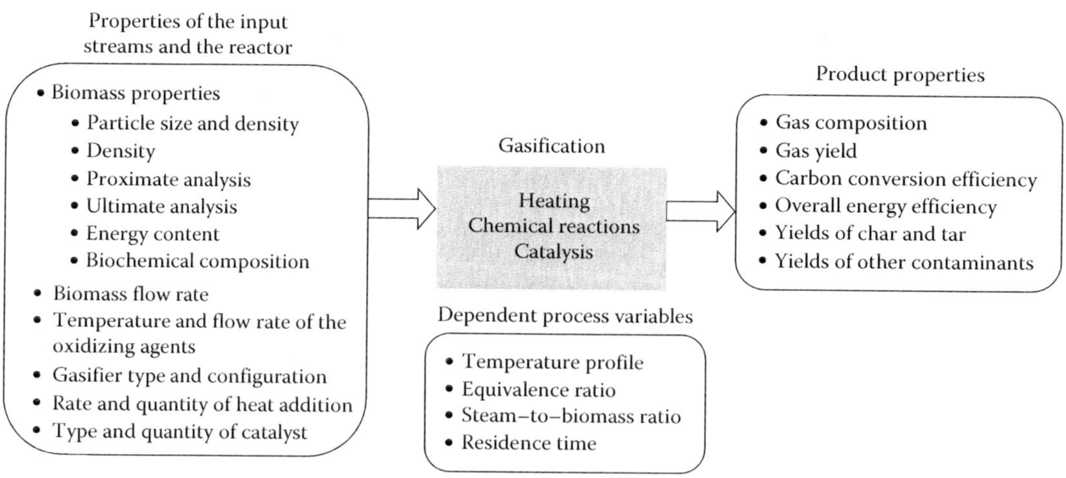

FIGURE 17.1 Gasification process variables.

17.2 GASIFIER TYPES

Several gasifier types, such as fixed bed, fluidized bed, entrained bed, and cyclone, have been successfully demonstrated and commercialized (Bridgwater 2006). Figure 17.2 shows the main gasifier types classified based on the bed configuration and flow directions of biomass and gasification agents.

17.2.1 Fixed-Bed Gasifier

Fixed-bed gasifier consists of a fixed bed of feed material (biomass) at the reactor bottom through which the gasification agent flows and thus reacts with biomass particles. Depending on the flow direction of gasifying agents, the fixed-bed gasifiers are subclassified into three: countercurrent (updraft), concurrent (downdraft), and crosscurrent (cross draft) gasifiers.

In an updraft gasifier, biomass and gasification agent flow in a countercurrent direction wherein the biomass is fed from top and flows from top to bottom of the reactor and the gasifying agent flows from bottom to top of the reactor. The reactions occurring in the gasifier from top to bottom can be subdivided into four major zones: drying, devolatilization, reduction, and combustion. In the drying zone, the drying of biomass takes place, removing moisture from biomass in the form of steam. In the devolatilization zone, the volatile matter of the dry biomass releases. The devolatilization process occurs at moderate temperature between 300°C and 500°C and produces char, gases, and tar. The devolatilization products further move down to the reduction and combustion zones of the gasifier. In the combustion zone, the gasifying agent, such as air, entering at the bottom of the reactor reacts with the char and produces hot combustion gases at temperature around 1000°C. In the reduction zone, the hot combustion gases from the combustion zone are reduced to a product gas called producer gas or syngas. In the updraft gasifier, the upward flow of product gas carries considerable amount of tar from the devolatilization zone and exits the reactor from the low-temperature zone at the top of the gasifier. As a result, the tar content of the producer gas from an updraft gasifier is higher than that from other types of gasifiers (Reed 1981; Klass 1998).

In a downdraft gasifier, both biomass and gasifying agent flow in the downward direction and undergo similar sequence of reactions: drying, devolatilization, combustion, and reduction.

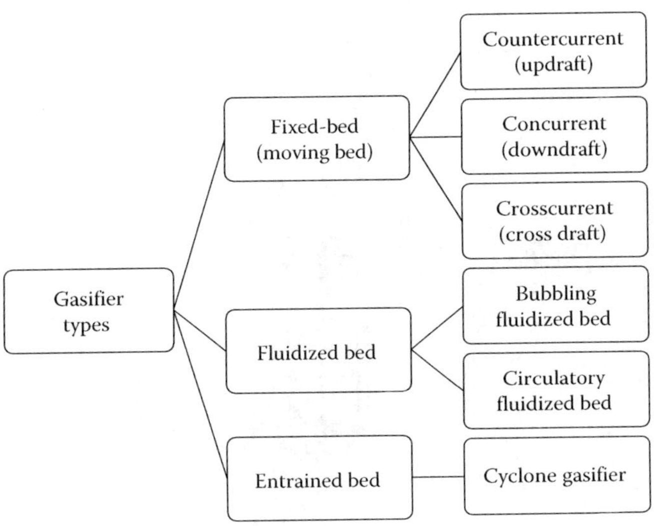

FIGURE 17.2 Types of gasifier.

The combustion zone is above the reduction zone in this gasifier. Due to downward flow of the product gas, most of the tars crack into lighter permanent gases at high temperature (about 1000°C) in the combustion zone resulting in low tar content in the producer gas. In a cross draft gasifier, the biomass is fed from the top of the reactor, while the gasifying agent is introduced from one sidewall of the reactor, and the product gas exits from the opposite sidewall of the reactor. The cross flow arrangement provides low residence time for tar cracking in the combustion zone, which results in a high tar content in the producer gas.

17.2.2 Fluidized-Bed Gasifier

Fluidized bed is one of the most commonly used gasifiers because of its high throughput and efficient heat and mass transfer rates. The fluidizing medium includes inert material such as silica sand. The biomass is injected into the fluidized-bed reactor at the bottom. The gasifying agent is supplied below the bed of the fluidizing material. The same gas acts as both fluidizing and gasifying agent. The velocity of the gasifying agent must be above the minimum fluidization velocity at which all solid particles are in a suspended position. Under such suspended condition, the fluidized bed of biomass behaves like a fluid with much higher particle movement for the enhancement in heat and mass transfers. Initially, the inert material bed is preheated to temperatures above 400°C, and then the biomass particles are injected to initiate the gasification process. The resulted product exits from the top of the gasifier. The fluidized-bed gasifier is subcategorized into bubbling fluidized-bed and circulatory fluidized-bed gasifiers. Compared to other gasifier types, the fluidized-bed gasifier has several benefits such as flexibility in accepting biomass with variable properties such as particle size, efficient heat and mass transfer, efficient gas–solid-phase reactions, and high conversion efficiency. However, the product gas from a fluidized-bed gasifier contains high level of particulates (Kumar et al. 2009a).

In the bubbling fluidized-bed gasifier, the gas superficial velocity, which is the ratio of gas flow rate to the cross-sectional area of the reactor, is maintained between minimum fluidization velocity and terminal velocity of the inert material particles. Figure 17.3 shows the lab-scale bubbling fluidized-bed gasifier and its components (Sharma et al. 2011).

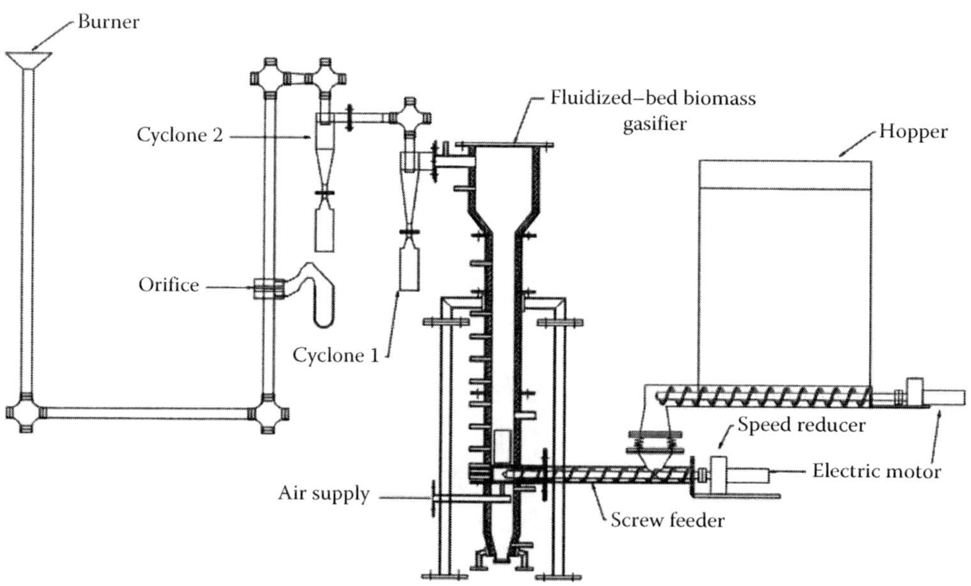

FIGURE 17.3 Schematic of a bubbling fluidized-bed gasifier at Oklahoma State University.

17.2.3 Entrained-Bed Gasifier

The entrained-bed, also called entrained-flow, gasifier has been originally developed to use coal with very fine particle size of 80–100 μm (Basu 2006) and later demonstrated for biomass materials. In the entrained-bed gasifier, both the material and gasification agent (usually oxygen) flow together in a concurrent direction. The gasification process occurs inside the dense-cloud phase of the feed particles. The entrained-bed gasifier operates at a high temperature and pressure and has the highest throughput capacity. The entrained-bed gasifier requires a very fine particle size, which limits the use of gasifier for fibrous biomass. A cyclone gasifier is an example of an entrained-flow gasifier, which simultaneously gasifies the biomass and removes particulates from the product gas.

17.3 BIOMASS PROPERTIES

The reactivity of biomass materials to participate in gasification reactions depends heavily on their physical and chemical properties (Basu 2006). Primary biomass properties affecting gasification include biomass type, particle size and shape, density, angle of repose and flowability, proximate and ultimate analyses, and energy content (Mckendry 2002). The effects of these biomass properties are discussed in the following.

17.3.1 Type, Particle Size, and Shape

Physical and chemical properties of biomass vary with the biomass type such as wood, crop residues, and grasses. Literature shows gasification studies on the effects of several biomass types, such as wood, groundnut shell, wheat straw, rice hulls, sawdust, almond shells, and switchgrass, on gasifier performance (Boateng et al. 1992; Ergudenler and Ghaly 1993; Rapagnà and Latif 1997; Natarajan et al. 1998; Bhoi et al. 2006; Sharma et al. 2011).

The particle size of biomass feedstocks plays a crucial role in the gasification process. A smaller particle provides a higher surface area for heat and mass transfers, leading to a quicker release of volatiles from biomass. Particle shape is described using sphericity, which is the ratio of the surface area of a spherical particle to the surface area of a given particle with the same volume. A biomass particle with low sphericity has a high surface-to-volume ratio that offers better heat and mass transfer than the particle with high sphericity (de Souza-Santos 2004).

17.3.2 Density

Biomass feedstock is characterized using two densities: particle density and bulk density. Particle density is further classified as apparent density and real density. Particle real density is the ratio of the particle mass and particle volume without considering the inner porous volume within the particle. Apparent particle density is the ratio of the particle mass and particle volume including the inner porous volume within the particle but excluding the void volume between the particles. High apparent density offers better biomass flowability to facilitate biomass flow through hopper, feeder, and gasification reactor. On the contrary, high apparent density shows low inner porosity, limiting the gas–solid interactions during gasification. Bulk density is the ratio of the mass of biomass to the total volume of biomass including the void spaces between and within the biomass particles. A high bulk density biomass results in a high energy density per cubic meter volume and improved flowability. Bulk density of biomass is influenced by the size, shape, and type of the biomass.

17.3.3 Flowability

Flowability, the ability of the biomass material to flow, depends on the size, shape, density, and angle of repose of the biomass. The angle of repose is the maximum angle of an inclined face of a stable biomass heap with the horizontal plane. The angle of repose also depends upon the size, shape, and density of the biomass particle and strongly influences the biomass flow in the gasifier hopper. Compared to granular biomass, loose and fibrous biomass materials are difficult to feed, resulting in a lower gasification performance. This is because loose and fibrous particles have the tendency to get interlocked with each other and bridge inside the gasifier hopper that interrupts material flow into the gasifier.

17.3.4 Proximate and Ultimate Analyses and Energy Content

Moisture, volatile, fixed carbon, and ash contents constitute the biomass proximate analysis, which influence gas composition, product yields, and impurities. Generally, a biomass of up to 15% moisture (on a wet basis) is suitable for an efficient gasification process (Reed 1981; Klass 1998). High moisture content results in H_2-rich syngas (Narvaez et al. 1996); however, as it draws more heat energy toward sensible heating and moisture evaporation, the reactor temperature (in autothermal gasification) decreases, which affects the gasifier performance and syngas quality.

Biomass organic matter that volatilizes releases at temperature between 125°C and 500°C is known as the volatile content of biomass. Volatile matter occupies a significant portion, generally about 70%–85% by mass, of biomass. Increase in volatile level improves biomass reactivity and results in high conversion efficiency. On the contrary, biomass with low volatile content exhibits low reactivity and poses difficulties during gasification, resulting in an inefficient conversion and high char production (Basu 2006).

Ash, inorganic compounds of biomass, is composed of the oxides of several minerals such as Fe, Si, K, Na, Mg, and Zn. Ash may have a catalytic role during gasification; however, the available literature contains only little information on the catalytic effects of ash. Usually, at high temperature, ash particles melt, fuse together (agglomerate), and form a clinker, a stonelike solid porous structure, in the gasifier bed and halt the gasification process (Ergudenler and Ghaly 1993; Fryda et al. 2008; Sharma et al. 2011). Less ash content is preferred in biomass feedstocks since ash does not contain any energy. In a fluidized-bed gasification, high ash content, such as in cattle manure and poultry litter (shown in Table 17.2), causes additional operational challenges due to agglomeration leading to decrease in fluidization.

Fixed carbon is the solid residue excluding ash that remains after biomass devolatilization in an inert atmosphere. Biomass with high fixed carbon content has high energy density resulting in high energy throughput of the gasifier. The gasification of biomass with higher fixed carbon content also yields syngas with less tar but more char. Proximate analyses of several biomass feedstocks are shown in Table 17.2.

TABLE 17.2
Proximate Analysis of Biomass Feedstocks

Biomass	Moisture (% w.b.)	Ash (% d.b.)	Volatile (% d.b.)	Fixed Carbon (% d.b.)	References
Switchgrass	8.0	4.36	79.21	16.43	Lemus et al. (2002), Mani et al. (2006)
Corn stover	10.6	3.7	78.7	17.6	Demirbaş (1997)
Wheat straw	4.1	6.3	76.4	17.3	Bridgeman et al. (2008)
Woody residue/ wood bark	45–50	1.6	75–80	20–25	Baker (1982)
Cattle manure	15.3–36.7	23.5–29.2	70.27	13.86	Halligan et al. (1975)
Chicken litter	43.01	16.42	38.91	1.66	Henihan et al. (2003)

TABLE 17.3
Ultimate Analysis of Biomass Feedstocks

Biomass	C (% d.b.)	H (% d.b.)	O (% d.b.)	N (% d.b.)	S (% d.b.)	References
Switchgrass	480	54	414	4.1	1.6	Lemus et al. (2002)
Corn stover	43.7	5.56	43.3	0.61	0.01	Kumar et al. (2008)
Wheat straw	40.78	5.84	52.92	0.18	0.28	Mani et al. (2010)
Woody residue/ wood bark	53.1	6.1	40.6	0.2	1.6	Baker (1982)
Cattle manure	35.1–39.6	5.3–5.9	30.98	2.5–3.1	0.4–0.6	Halligan et al. (1975)
Chicken litter	39.57	5.11	48.27	5.31	0.77	Henihan et al. (2003)

Carbon, hydrogen, oxygen, sulfur, and nitrogen contents constitute the ultimate analysis of biomass. Biomass with high carbon and hydrogen is desirable for energy applications because most of the biomass energy is derived from the chemical bonds of C and H. As shown in Table 17.3, biomass contains low sulfur and nitrogen that form H_2S and NH_3, respectively, as impurities in the syngas.

The energy content of biomass, expressed as higher heating value (HHV), depends on its contents of carbon, hydrogen, and oxygen. The HHV of biomass feedstocks usually falls between 14.6 and 23.3 MJ/kg (dry basis) (Ebeling and Jenkins 1985). Biomass with a high energy content results in an increased gasification performance such as the heating value of the resulted syngas.

17.4 GASIFICATION OPERATING VARIABLES

Operating variables such as temperature, pressure, gasifying agent, and flow rates of biomass and gasification agents (equivalence ratio [ER] and steam-to-biomass ratio [SBR]) play key roles in the process and affect composition, heating value and yield of syngas, and conversion efficiency. These operating variables, described in the following sections must be optimized to achieve high gasification performance (Han and Kim 2008).

17.4.1 Temperature

The gasifier performance is heavily influenced by the gasification temperature. High temperature leads to more gas yield, less tar yield, and high carbon conversion (Kumar et al. 2009b). High temperature also favors endothermic reactions R(4) and R(5) and increases H_2 and CO concentrations. Literature shows that the increase in temperature from 700°C to 900°C increased H_2 and CO concentrations while it reduced CH_4 and CO_2 concentrations (González et al. 2008). The operating temperature range of fluidized-bed gasifiers is from 700°C to 900°C (Han and Kim 2008), while that of fixed-bed gasifiers is 900°C–1000°C.

17.4.2 Pressure

A gasifier operated at an elevated pressure directly affects partial pressures of participating species in the reactor and thus influences gas composition. Low-pressure gasification results in high CO and H_2 and low CO_2 and CH_4 concentrations (Basu 2006). The operating pressure is even more influential in a fluidized-bed gasification because an appropriate pressure of the fluidizing medium is needed to maintain fluidization of the bed material (Sharma et al. 2013) and create a pressure gradient across the fluidized-bed reactor. However, a high-pressure operation may also lead backflow of gas from the fluidized-bed reactor into the biomass feeding unit.

17.4.3 Gasification Agent (Air, Steam, and Oxygen)

Gas composition and heating value, gas tar content, and gasifier conversion efficiency vary with the change in gasifying agent. A gasifying agent helps to break down heavier biomass molecules into lighter molecules and provides necessary heat (through exothermic reactions) for endothermic reactions. Air, steam, and oxygen are various gasifying agents that are conventionally used for biomass gasification.

The use of air as a gasification agent has been very common because air does not need extensive preprocessing; it is available free of cost, and oxidation reactions with air provide heat for other endothermic reactions. However, the gasification with air results in syngas that is highly diluted with nitrogen (up to 65%) and with low caloric value (Delgado et al. 1997; Seo et al. 2010). The nitrogen dilution increases the reactor sizes of the downstream processes by more than twofold for producing the same amount of power, fuels, and chemicals. Air as an oxidizing medium results in less tar compared to O_2–steam mixture and pure steam as oxidizing agents (Devi et al. 2003).

The use of steam, as a gasifying agent, results in syngas with high concentrations of H_2 and H_2/CO ratio because of shift R(5) and reforming R(8) reactions with H_2O that adds to H_2 production (Bridgwater 2006; Weerachanchai et al. 2009). This makes syngas more suitable for the conversions into fuels and chemicals. However, the gasification with only steam as a gasifying agent requires much higher amount of steam than its stoichiometric requirement to maintain gas flow and bed material fluidization (in fluidized-bed gasifiers). As high as 90% of the total steam supplied into the gasifier remains unreacted, resulting in substantial energy loss (Corella et al. 2007). By reducing or avoiding the excess steam, net energy efficiency can be significantly improved in steam gasification.

The use of O_2, as a gasifying agent, results in syngas with high CO and H_2 concentrations (Zhou et al. 2009). However, it requires oxygen generation facilities, which can make the operation uneconomical at large scale and add energy penalty (Maniatis 2001; Campoy et al. 2009).

17.4.4 Biomass Feed Rate

Optimum biomass feed rate is influenced by scale and design of the gasifier and feeding units, and biomass properties. High biomass feed rate results in plugging and reduction in conversion efficiency and may halt the gasification process (Kumar et al. 2009a), whereas low feed rate leads to a decrease in the gasification temperature and gas yield. Several feeding types, such as continuous feeding or batch feeding, also affect the gasifier performance. Continuous feeding provides consistent energy throughput to the reactor resulting in a stable and smooth operation of the gasifier, while batch-type feeding leads to a variation in energy throughput to the reactor and may cause unstable operation and poor performance of the gasifier.

17.4.5 Gasification Agent Flow Rate (Equivalence Ratio and Steam-to-Biomass Ratio)

The flow rate of the gasifying agent injected into the gasifier is one of the best available controls to tune gas composition. For example, once a gasifier is in operation, the gasifier design cannot be changed easily to influence the gas composition. However, a gasifying agent can be changed even during the operation, which has a significant effect on gas quality. The flow rate of air (or oxygen) and steam as gasification agents is expressed using parameters: ER and SBR, respectively. ER is a ratio of the quantity of actual air supplied per unit mass of biomass to the stoichiometric quantity of air required per unit mass of biomass for a complete combustion. The ER for gasification typically falls between 0.2 and 0.4 to achieve the maximum energy efficiency (Reed 1981). ER directly influences the gasification temperature, residence time, gas yield and composition, and carbon conversion and energy efficiencies. As ER increases, an additional oxygen supply into the gasifier increases the degree of biomass oxidation and causes rise in the gasification temperature leading to a higher biomass conversion. Gasifier operation, such as ER, is optimized to achieve maximum energy efficiency. However, gasifier operation above the optimum ER results in oxidation of generated syngas and decreases the amount of energy available in syngas. Biomass gasification at an optimum ER of 0.32 resulted in the highest syngas heating value and energy efficiency (Sharma et al. 2011). SBR is the ratio of mass of steam supplied to the mass of biomass supplied. Steam as a gasifying agent favors H_2-forming reactions R(5), R(7), and R(8) due to the increase in the partial pressure of H_2O within the gasifier reactor and results in an H_2-rich gas (Kumar et al. 2009a). However, H_2-forming reactions are highly endothermic that require gasification temperature above 750°C to proceed in a forward direction to yield H_2-rich syngas. Steam also acts as a tar reforming agent and decreases tar level in the syngas.

Catalysts, such as char, dolomite, olivine, limestone, calcined dolomite, $CaCO_3$, Ni/Al catalysts, metal oxides, magnesites, zeolite, $Rh/CeO_2/SiO_2$ catalyst, and ϒ-alumina catalyst, have been used in the gasifier bed to enhance gasification reactions. The properties of the catalysts such as attrition resistance, availability, and temperature stability are considered for the selection of appropriate catalysts (Devi et al. 2003).

17.5 PRODUCT PROPERTIES

The products of biomass gasification include syngas (the gaseous product also called producer gas), tar (liquid product accompanied with the syngas), and char and ash (the solid products). The properties of these products highly depend on the biomass and gasification operating variables. Typically, syngas is the main product of biomass containing the highest amount of energy. Tar is considered a contaminant and must be removed from the syngas. Tar and impurity removal are discussed in the syngas conditioning section. Char and ash are considered by-products, which contain minerals that can be utilized as a soil amendment. New high-value applications of char, such as for catalyst synthesis, are also being discovered. The properties of the main product, syngas, are described in the following.

Syngas is mainly composed of CO, H_2, CH_4, CO_2, and N_2 (if air is used as a gasifying agent). Hydrocarbons such as C_2H_2, C_2H_4, and C_2H_6 are also found in the syngas at a low concentration (below 5% v/v). Gasification studies mostly focus on increasing CO and H_2 concentrations of the syngas because CO and H_2 hold approximately 75%–95% energy of the gas and are the building blocks for the production of many liquid fuels and chemicals. In addition, several processes for fuels and chemical production, such as a Fischer–Tropsch (FT) process, require a specific H_2/CO ratio for an effective conversion. The heating value of the syngas is determined using the gas composition. The heating value of biomass-derived syngas using steam and oxygen as gasifying agents varies between 12 and 18 MJ/N m^3, while using air as a gasifying agent varies between 4 and 6 MJ/N m^3 (McKendry 2002).

Syngas also contains unavoidable impurities, such as tar, particulate, alkali compounds, NH_3, and H_2S, which require an extensive gas cleaning step. Tars, the most prevalent gas impurity, are organic compounds that are heavier than benzene, condense at room temperature, and cause pipe blockage and equipment fouling (Klass 1998). Particulate, a mixture of char and ash, mixes with the tar and clogs pipelines, which intensifies maintenance problems. Alkali compounds, such as CaO, K_2O, P_2O_5, and SiO_2, present in the syngas condense at a temperature below 650°C and cause surface corrosion in downstream process units (Kumar et al. 2009a).

17.6 GASIFICATION PERFORMANCE

Gasifier performance can be expressed using various efficiencies such as carbon conversion efficiency (CCE), cold gas efficiency (CGH), and hot gas efficiency (HGE). All these efficiencies are determined based on the product yields and properties. CCE is the amount of biomass carbon that is converted into useful gaseous products and is calculated using Equation 17.1 (Sharma et al. 2011):

$$\mathrm{CCE}(\%) = 12 \times \mathrm{Y}_{\text{dry gas}} \times \frac{\left[(\mathrm{CO}+\mathrm{CO_2}+\mathrm{CH_4})+2\times(\mathrm{C_2H_2}+\mathrm{C_2H_4}+\mathrm{C_2H_6})\right]}{(22.4\times\mathrm{C})} \tag{17.1}$$

where
- CO, CO_2, CH_4, C_2H_2, C_2H_4, and C_2H_6 are the components of product syngas in volume percentage (%)
- C is the percentage of carbon in the dry biomass
- $\mathrm{Y}_{\text{dry gas}}$ is the dry gas yield per kg of dry biomass (N m^3/kg)

CGE and HGE are the cold and hot gas efficiencies, respectively, which are determined using Equations 17.2 and 17.3 (Sharma et al. 2011):

$$\mathrm{CGE}(\%) = \left(\frac{\mathrm{CE}}{\mathrm{DBE}+\mathrm{SE}}\right) \times \mathrm{Y}_{\text{dry gas}} \times 100\% \tag{17.2}$$

$$\mathrm{HGE}(\%) = \left(\frac{\mathrm{CE}+\mathrm{PE}}{\mathrm{DBE}+\mathrm{SE}}\right) \times \mathrm{Y}_{\text{dry gas}} \times 100\% \tag{17.3}$$

where
- CE is the chemical energy content of dry syngas per unit volume (MJ/Nm^3)
- DBE is the energy content of dry biomass (MJ/kg)
- SE is the sensible energy of the gasification agent (if preheated) in MJ/kg
- PE is the sensible energy of hot product gas (MJ/m^3)

The effects of the major gasification variables on gasifier performances are shown in Table 17.4.

17.7 SYNGAS CLEANING AND CONDITIONING

Various impurities in producer gas such as tars, ammonia (NH_3), hydrogen sulfide (H_2S), nitrogen oxides (NOx), sulfur oxides (SOx), alkalis, and particulates cause problems in the downstream applications of the gas. These contaminants must be removed from the producer gas before the gas can be used in its downstream applications (see Figure 17.4). This section provides a brief overview of the four major contaminants present in producer gas, that is, tar, ammonia, hydrogen sulfide, and CO_2. Table 17.5 shows various problems caused due to producer gas contaminants and the popular techniques for removing these contaminants. Reviews on various physical and chemical techniques for upgrading producer gas and quality are available in literature (Milne et al. 1998; Devi et al. 2003). However, information on carbon-based catalysts/sorbents, a novel catalytic removal approach, is very limited. The following section provides a brief overview of some of the advancements in the use of carbon-based catalysts for the removal of contaminants generated during biomass gasification.

Tar can be defined as all syngas organic contaminants having a molecular weight higher than benzene (Neeft et al. 1999). Tar causes problems such as clogging of fuel process lines, deactivation of catalysts, and disruption of engine and turbine functioning (Milne et al. 1998; Devi et al. 2003; Li and Suzuki 2009). Various physical and chemical removal techniques, such as scrubbers and electrostatic precipitators (ESPs), have been used to reduce the levels of tar from producer

TABLE 17.4
Effects on Gasifier Performance

Biomass	Gasifier Type	Gasification Agent(s)	ER	SBR	Temperature, °C	CCE,%	CGE,%	HGE,%	Reference
Switchgrass	Bubbling fluidized bed	Air	0.20	—	801	51	31	33	Sharma et al. (2011)
			0.32	—	825	89	71	75	
			0.45	—	907	96	47	56	
Pine sawdust	Bubbling fluidized bed	Air and steam	0.19	1.56	800	76	—	—	Lv et al. (2004)
			0.23	1.56	800	91	—	—	
			0.27	1.56	800	70	—	—	
Hemlock	Circulatory fluidized bed	Air	0.34	—	718	92	71	—	Li et al. (2004)
Sugarcane bagasse	Cyclone gasifier	Air and steam	0.18	0.34	793	84	—	—	Gabra et al. (2001)
			0.21	0.43	809	83	—	—	
			0.25	0.42	827	88	—	—	
Wood chips	Downdraft	Air	0.38	—	1000[a]	—	80	—	Li et al. (2004)
Rice hull	Bubbling fluidized bed	Steam	—	18.5	700	31	37	—	Boateng et al. (1992)
				to	750	43	51		
				13.3	800	46	57		
Distiller grains	Bubbling fluidized bed	Air and steam	0.29	7.3	650	62	70	—	Kumar et al. (2009b)
			0.29	7.3	750	71	82	—	
			0.29	7.3	850	82	96	—	

[a] Combustion zone temperature.

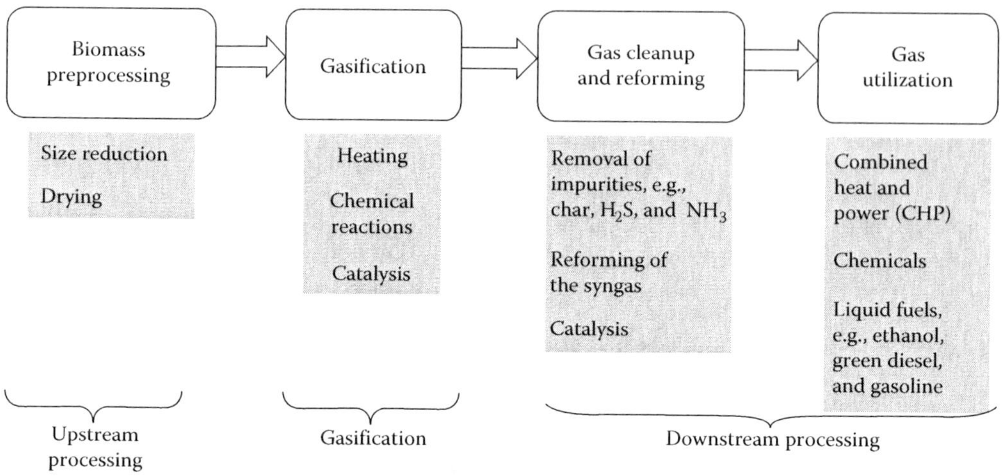

FIGURE 17.4 Operations in biomass conversion through gasification.

TABLE 17.5
Overview of Problems Caused due to Producer Gas Contaminants and Popular Cleanup Techniques

NH_3	H_2S	Particulates	Alkalis	Tars
Problems Caused				
• Poison catalysts • Precursors for NOx and photochemical smog • Promote corrosion	• Poison catalysts	• Valve functioning • Clog fuel lines • Foul equipment surfaces	• Deposition of vapors • Fouling, slagging, and corrosivity • Deactivate catalysts	• Handling and disposal problems • Plugging and fouling • Dehydration to form char and coke
Cleanup Techniques				
• Dolomite- or Ni/Fe-based catalyst • Mixed metal oxide catalysts	• Wet scrubbers • Adsorb on metal oxides	• Cyclonic or barrier filters • ESPs	• Product condensation • ESPs	• Wet ESPs • Ni/dolomite/biochar type catalysts • Thermal treatment

gas (Jothimuragesan et al. 1996; Zhang et al. 2004). Ammonia (NH_3) is formed primarily from the nitrogen-containing compounds present in the biomass feedstocks. This fuel-bound nitrogen is released at high temperatures inside the gasifier. Ammonia is a precursor for NOx emissions in downstream applications of the syngas such as turbines, gas engines, and burners. NH_3 and other nitrogen-containing compounds are also undesirable for catalytic upgrading and conversions as they poison catalysts. The syngas ammonia content has been found to vary between a few hundred ppm to 3% by volume (Zhou et al. 2000). Sulfur in biomass feedstock usually leads to the formation of hydrogen sulfide (H_2S), carbonyl sulfide (COS), and other gaseous sulfur compounds. These sulfur compounds are toxic and result in various problems such as the deactivation of catalysts, acid rain precursors, and pollution. These sulfur contaminants must be removed before the producer gas can be used further (Torres et al. 2007).

Various physical and chemical removal techniques such as scrubbers and ESPs have been utilized for reducing the levels of contaminants from the syngas. However, these techniques also end up reducing the gas energy content since these are cold treatment techniques. Also, the disposal of the solvents used as well as the high equipment and operation costs makes these processes uneconomical and inefficient. On the contrary, catalysts offer a more cost-efficient and effective approach to the removal of these contaminants (Jothimuragesan et al. 1996).

17.7.1 Carbon-Based Catalysts for Syngas Upgrading

Carbon-based catalysts include, but are not limited to, materials such as char, activated carbon, carbon fibers, and carbon molecular sieves. Carbon-based catalysts or sorbents have many applications such as energy storage, energy conversion, and sensors. Carbon catalysts can be synthesized from gas-phase, solid-phase, or liquid-phase reactions from carbon-containing precursors. Hydrocarbons are the most widely used precursors for the synthesis of carbon-based catalysts/sorbents. The porosities of carbon-based catalysts are crucial in determining their effectiveness for removing contaminants. The general characteristics of the various types of porosities in carbon adsorbents are shown in Table 17.6. Microporous and mesoporous carbons are generally suitable for the removal of ammonia (molecule radius = 2.6 Å approx.), hydrogen sulfide (molecule radius = 4 A° approx.), and CO_2 (molecule radius = 3.4 Å approx.). Tars are higher in molecular weight as well as have a larger molecule size (5–6 Å), and hence, mesoporous and macroporous carbons may be more effective for the removal of tar. Table 17.7 shows the effectiveness of carbon-based catalysts for contaminant removal from syngas.

Char, a by-product of biomass gasification, has been found to be promising for synthesizing carbon-based catalysts especially for syngas cleaning. Char is a nongraphitizable material and is defined as a product of a carbonization process of carbonaceous precursors such as coal and

TABLE 17.6
General Characteristics of the Types of Porosities in Carbon Adsorbents

Type of Porosity	Pore Radii (nm)	Pore Volume (cm^3/g)	Function in Adsorption
Micropore	<2	0.15–0.70	Volume filling
Mesopore	2–50	0.2–0.65	Filled through capillary condensation
Macropore	500–2000	0.2–0.4	Channel for adsorbate to reach inside particle

TABLE 17.7
Effectiveness of Carbon-Based Catalysts for Contaminant Removal

Contaminant	Approximate Molecule Radius (°A)	Carbon-Based Catalysts Effective for Removal	Remarks
Sulfur	4	Activated carbon, activated carbon cloth, coke, activated carbon fibers	The number of acidic groups, surface pH, and amount of surface oxygen groups on catalyst are critical for sulfur removal. The optimum temperature is 600°C–800°C.
Ammonia	2.6	Activated carbon, activated carbon fibers	Ammonia adsorption improves with strong acidic groups and oxide groups on sorbent surface.
Tar	5–6	Char, activated carbon	Carbon catalysts generally act as sorbents below reaction temperature of 600°C. Higher temperatures (>700°C) improve the removal of tar.

biomass (Bottani and Tascón 2008). The incomplete combustion or thermochemical treatment of a carbonaceous precursor gives rise to a black solid material rich in carbon content known as char. Char derived from biomass is popularly known as biochar; if derived from wood, it is known as wood char, whereas coal-derived char is known as charcoal or coal char. Various thermochemical processes such as gasification, pyrolysis, and combustion produce char. It is mostly considered as a by-product or waste material. However, being rich in carbon, char is being used as a source of fuel in some instances. Char has also shown promise as an organic fertilizer to increase the overall plant growth (Chan et al. 2009). Char can be synthesized from a variety of feedstocks. Biochar can be synthesized from perennial grasses (switchgrass) (Boateng 2007), *Miscanthus* (Melligan et al. 2010), forestry waste, crop residues, animal manure, sewage, and waste, among others. Wood char can be obtained from pinewood (Janse et al. 1998), maple (DiBlasi et al. 1999), birch (Chen et al. 1997), poplar (Scott and Piskorz 1982), and others, whereas coal char can be derived from either peat (Arpiainen and Lappi 1989), lignite (Suuberg et al. 1978), bituminous (Wu et al. 2000), or anthracite (Desypris et al. 1982) types of coal. The reactivity of char can be increased with high heating rates, small biomass particle sizes, and short residence times (Zanzi et al. 1996).

17.7.2 Sulfur Compounds

Activated carbon is a promising catalyst for sulfur adsorption, and its low cost and easy synthesis offer an added advantage to it. It is suggested that the choice of a sulfur sorbent should be made based on the acidity of the sorbent and associated parameters such as the number of acidic groups, surface pH, and the amount of surface oxygen groups (Adib et al. 2000). The reaction mechanism of H_2S adsorption on the carbon surfaces can be basically summarized into three types of reactions:

$$C + H_2S \rightarrow C - S + H_2 \tag{17.4}$$

$$C - O + H_2S \rightarrow C - S + H_2O \tag{17.5}$$

$$C - M + H_2S \rightarrow C - M - S + H_2 \tag{17.6}$$

Generally, activated carbon can be synthesized by either chemical activation, physical activation, or a combination of chemical and physical activation of a carbon precursor (Lee et al. 2006). This activation process produces a highly porous carbon material showing high adsorptive properties. The carbon precursor can be any material with a high carbon content. Various agricultural waste precursors like wheat (Schroder et al. 2007), rice husks (Ahmedna et al. 2000), sugarcane bagasse (Ahmedna et al. 2000), corn stover (Fan et al. 2004), hazelnut (Sayan 2006), and biochar (Azargohar and Dalai 2006) have been used to synthesize this activated carbon. It can be prepared in either granular (Toles et al. 1997), powdered (Bansal 2005), fibrous (Suzuki 1994), or cloth (Huidobro and Pastor 2001) form. Large internal surface area and small pores are the characteristics of a granular activated carbon, whereas powdered forms have large pore diameters and small internal surface area. Large surface area and wide pores are the characteristics of activated carbon cloth and fibers (Bansal 2005).

17.7.3 Ammonia and HCN

Ammonia removal using catalysts/sorbents is thought to proceed through one of the following two reaction mechanisms:

$$2NH_3 \rightarrow N_2 + 3H_2 \tag{17.7}$$

$$4NH_3 + 3O_2 \rightarrow 2N_2 + 6H_2O \tag{17.8}$$

Reaction 17.7 is also called as catalytic decomposition reaction, whereas Reaction 17.8 is known as selective catalytic decomposition and requires oxygen for the reaction to proceed. Chemical decomposition of ammonia is not feasible since syngas has a high concentration of H_2. Research is currently underway on using high surface area acidic sorbents for capturing (as an alternative to decomposing) ammonia (Jothimuragesan et al. 1996; Devi et al. 2003). Activated carbon sorbents (Radovic and Rodriguez-Reinoso 2004) and activated carbon fibers (Mangun et al. 1999) seem to be promising catalysts/sorbents for the removal of ammonia from producer gas.

17.7.4 Tar

Char, obtained through biomass gasification, has been found to be promising for tar removal. In a downdraft gasifier, the product gas is passed through a hot bed of char to decrease syngas tar content (Patil et al. 2011). Dogru et al. (2002) highlighted the activity of char for lowering tar contents in a downdraft gasifier system. Zanzi et al. (1996) found a high tar removal efficiency when the gasifier product gas was passed through the partial oxidation zone followed by a bed of char. Brandt et al. (2000) carried out a study on a 100 kWth gasifier where the gases were contacted with a charcoal bed. A high tar reduction of the aromatic hydrocarbons including PAH was obtained using the charcoal bed down to 15 mg/N m^3 tar content. Donnot et al. (1991) found that the residence time for an efficient tar removal is inversely proportional to the char surface area. They also investigated the carbon deposition phenomena on the char catalyst and found that at temperatures below 750°C, a decrease in activity occurs after a short time due to coking. Chembukulam et al. (1981) reported a complete decomposition of tar and pyroligneous liquor into gas with low calorific value over semi-coke and charcoal made from teakwood sawdust at 950°C. Seshadri and Shamsi (1998) reported tar cracking over a char–dolomite mixture. At a high pressure (17 atm), the conversion of tar increased due to secondary reactions of the tar in the pores of the sorbent material. An increase in temperature from 535°C to 560°C also resulted in an increase of tar conversion rates by 15%–25%. Abu (2008) compared biomass-derived char to other catalysts such as dolomite, olivine, nickel, FCC catalysts, and ash using model tar compounds of phenol and naphthalene in a fixed-bed reactor. Biomass char performed satisfactorily for naphthalene conversion among the low-cost catalysts. In a review of tar removal catalysts carried out by the same group of authors, it was concluded that char has a high potential of being a low-cost and efficient tar removal catalyst (El-Rub et al. 2004). Griffiths and Mainhood (1967) evaluated activated carbon for tar removal.

17.8 SYNGAS UTILIZATION

Syngas is a very well-known building block to produce many fuels and chemicals such as methanol, ethanol, ammonia, and hydrocarbons. Currently, syngas with 6000 PJ (equivalent to the energy of 1100 million barrels of petroleum oil) is generated primarily by steam reforming of natural gas (Drift et al. 2004). Biomass-generated syngas has the potential to supplement the need of syngas. The following section described the major applications of the syngas.

17.8.1 Syngas Conversion into Hydrocarbons

The most common routes available for the conversion of syngas into hydrocarbon fuels are FT process and syngas to methanol to gasoline (MTG). FT conversion of syngas into hydrocarbons is one of the most recognized technologies with the first plant operated in Germany in 1938. Currently, three plants use the FT process to produce gasoline, diesel, and chemicals from coal and natural gas. FT process uses Fe- or Co-based catalysts for the conversion. The resulting product is a wide range of primarily linear hydrocarbons from C_1 compounds to high molecular mass waxes, which need reprocessing to obtain hydrocarbons in the range of diesel or gasoline. Diesel is the most appropriate FT fuel because the FT product contains mostly linear hydrocarbon that results in diesel with high cetane number (Dry 2004). However, due to the need of an additional processing of long-chain hydrocarbons (waxes), the capital cost is high (Spath and Dayton 2003). An FT process requires an optimum syngas H_2/CO of 2, which is difficult to achieve in biomass-derived syngas without steam reforming. The conversion of syngas to methanol is also a well-known process. However, since methanol cannot be used directly because of its toxicity, methanol is converted into gasoline through an MTG process developed by Mobil Oil Corporation in the 1970s. The process uses zeolite-based catalysts (ZSM-5) resulting in higher than 85% selectivity to gasoline-range hydrocarbons (Spath and Dayton 2003).

17.8.2 Syngas Conversion to Alcohols and Other Fuels and Chemicals

Syngas can be converted into alcohols using microbial or chemical catalysts. Syngas fermentation research using microorganisms such as strains of *Clostridium ljungdahli*, *Clostridium autoethanogenum*, *Clostridium carboxidivorans*, *Clostridium ragsdalei*, and *Alkalibaculum bacchi* yielded ethanol, butanolisopropanol, and acetic acid. Recent advances in the syngas fermentation include developing new strains of microorganisms, improved reactor design, and optimized conditions such as temperature, pH, buffer presence, and media to increase yield and reduce cost for the production of alcohols (Kundiyana et al. 2011a,b, 2010; Maddipati et al. 2011; Liu et al. 2012).

Chemical catalysts have also been used to convert syngas into mixed alcohols. The process takes place at high pressure and low temperature in the presence of catalysts with the function of hydrogenation, C–O bond breaking, and CO insertion. Catalysts based on both noble and non-noble metals have been used for the synthesis of mixed alcohols. Noble metal-based catalysts containing Rh, Ru, or Re and supported on oxides such as SiO_2 and Al_2O_3 have high alcohol selectivity but are not economical for commercial applications. Major non-noble metal-based catalysts for mixed-alcohol synthesis contain MoS_2, Cu–Co, Cu–Zn–Al, and Zn–Cr–K (Fang et al. 2009). Recent advances in mixed-alcohol production using chemical catalysts include synthesis and development of new catalysts and optimization of reaction conditions to increase yield and reduce cost of alcohol production. However, the alcohol synthesis process still suffers from the low yield and poor selectivity of the desired alcohol product (Subramani and Gangwal 2008). Syngas is also considered as the building block for the production of many chemicals, such as aldehydes and acetic acid, produced through catalytic and microbial conversions. Hydrogen can also be separated from syngas for producing ammonia or refining hydrocarbon fuels.

17.8.3 Biopower

The use of syngas for electrical power and heat production is more adaptable and direct as compared to its use for fuel and chemical productions. Several demonstrations of heat and power generation using syngas produced from biomass and wastes are available in literature (Young and Pian 2003; Bengtsson 2011; Son et al. 2011). The technologies used for heat and power productions include external combustion engines such as steam engine, internal combustion engines such as reciprocating gas engines and gas turbines, and fuel cells. However, the challenges remain in using existing equipment because of low volumetric energy content of biomass-generated syngas (4–15 MJ/m^3) as compared to that of fossil fuels such as natural gas (38 MJ/m^3) (Wang et al. 2008), for which the engines are designed for. The use of gas turbines was demonstrated in a weeklong test for power production from wood chip-generated syngas in Sweden at the scale of 18 MW (thermal) (Bengtsson 2011). In developing countries, the gasification of locally available biomass and wastes has the potential to provide electricity (Abe et al. 2007). However, the challenges in the commercialization of biomass gasification for power generation are several: robust syngas cleaning technologies are needed, power is a lower-value product than liquid fuels, and the infrastructure and equipment to use biomass is not well established unlike fossil-based feedstocks such as coal and natural gas.

17.9 CONCLUSIONS

Biomass gasification and syngas conditioning and utilization are reviewed. The use of biomass for the production of fuels, chemicals, and power through gasification and gas utilization technologies is very promising because of the well-known technologies and products. However, the commercialization of these technologies remains challenging because of the low energy and bulk densities of biomass, complex and uneconomical syngas cleaning technologies, and availability of cheaper, robust, and infrastructure-ready petroleum fuels. Carbon-based materials, especially activated carbon catalysts and sorbents, might be effective for the removal of tars, ammonia, and H_2S. New research and demonstration aimed to make the gasification, gas conditioning and gas utilization technologies robust and economical are needed.

REFERENCES

Abe, H., A. Katayama, B.P. Sah et al. 2007. Potential for rural electrification based on biomass gasification in Cambodia. *Biomass and Bioenergy* 31(9):656–664.

Abu, E.R. 2008. Experimental comparison of biomass chars with other catalysts for tar reduction. *Fuel* 87(10–11):2243–2252.

Adib, F., A. Bagreev, and T.J. Bandosz. 2000. Analysis of the relationship between H2S removal capacity and surface properties of unimpregnated activated carbons. *Environmental Science & Technology* 34(4):686–692.

Ahmedna, M., W.E. Marshall, and R.M. Rao. 2000. Production of granular activated carbons from select agricultural by-products and evaluation of their physical, chemical and adsorption properties1. *Bioresource Technology* 71(2):113–123.

Arpiainen, V. and M. Lappi. 1989. Products from the flash pyrolysis of peat and pine bark. *Journal of Analytical and Applied Pyrolysis* 16(4):355–376.

Azargohar, R. and A.K. Dalai. 2006. Biochar as a precursor of activated carbon. *Applied Biochemistry and Biotechnology* 131(1–3):762–773.

Baker, A.J. 1982. Wood fuel properties and fuel products from woods. In *Proceedings of the Fuel Wood Management and Utilization Seminar* (East Lansing, MI, November 9–11, 1982) pp. 14–25.

Bansal, R.C. 2005. *Activated Carbon Adsorption*. Boca Raton, FL: CRC.

Basu, P. 2006. *Combustion and Gasification in Fluidized Beds*. Boca Raton, FL: CRC.

Bengtsson, S. 2011. VVBGC demonstration plant activities at Varnamo. *Biomass and Bioenergy* 35:S16–S20.

Bhoi, P.R., R.N. Singh, A.M. Sharma, and S.R. Patel. 2006. Performance evaluation of open core gasifier on multi-fuels. *Biomass and Bioenergy* 30(6):575–579.

Boateng, A.A. 2007. Characterization and thermal conversion of charcoal derived from fluidized-bed fast pyrolysis oil production of switchgrass. *Industrial & Engineering Chemistry Research* 46(26):8857–8862.

Boateng, A.A., W.P. Walawender, L.T. Fan, and C.S. Chee. 1992. Fluidized-bed steam gasification of rice hull. *Bioresource Technology* 40(3):235–239.

Bottani, E.J. and J.M.D. Tascón. 2008. *Adsorption by Carbons*. Amsterdam, the Netherlands: Elsevier.

Brandt, P., E. Larsen, and U. Henriksen. 2000. High tar reduction in a two-stage gasifier. *Energy & Fuels* 14(4):816–819.

Bridgeman, T.G., J.M. Jones, I. Shield, and P.T. Williams. 2008. Torrefaction of reed canary grass, wheat straw and willow to enhance solid fuel qualities and combustion properties. *Fuel* 87(6):844–856.

Bridgwater, T. 2006. Biomass for energy. *Journal of the Science of Food and Agriculture* 86(12):1755–1768.

Campoy, M., A. Gomez-Barea, F.B. Vidal, and P. Ollero. 2009. Air-steam gasification of biomass in a fluidised bed: Process optimisation by enriched air. *Fuel Processing Technology* 90(5):677–685.

Chan, K.Y., Z.H. Xu, J. Lehmann, and S. Joseph. 2009. Biochar–nutrient properties and their enhancement. *Biochar for Environmental Management: Science and Technology*. London, U.K.: Earthscan, pp. 67–84.

Chembukulam, S.K., A.S. Dandge, N.L.K. Rao, K. Seshagiri, and R. Vaidyeswaran. 1981. Smokeless fuel from carbonized sawdust. *Industrial & Engineering Chemistry Product Research and Development* 20(4):714–719.

Chen, G., Q. Yu, and K. Sjostrom. 1997. Reactivity of char from pyrolysis of birch wood. *Journal of Analytical and Applied Pyrolysis* 40:491–499.

Corella, J., J.M. Toledo, and G. Molina. 2007. A review on dual fluidized-bed biomass gasifiers. *Industrial & Engineering Chemistry Research* 46(21):6831–6839.

Delgado, J., M.P. Aznar, and J. Corella. 1997. Biomass gasification with steam in fluidized bed: Effectiveness of CaO, MgO, and CaO-MgO for hot raw gas cleaning. *Industrial & Engineering Chemistry Research* 36(5):1535–1543.

Demirbaş, A. 1997. Calculation of higher heating values of biomass fuels. *Fuel* 76(5):431–434.

de Souza-Santos, M.L. 2004. *Solid Fuels Combustion and Gasification: Modeling, Simulation, and Equipment Operations*. Vol. 174, Boca Raton, FL: CRC Press (Taylor and Francis Group).

Desypris, J., P. Murdoch, and A. Williams. 1982. Investigation of the flash pyrolysis of some coals. *Fuel* 61(9):807–816.

Devi, L., K.J. Ptasinski, and F.J.J.G. Janssen. 2003. A review of the primary measures for tar elimination in biomass gasification processes. *Biomass and Bioenergy* 24(2):125–140.

Di Blasi, C., G. Signorelli, C. Di Russo, and G. Rea. 1999. Product distribution from pyrolysis of wood and agricultural residues. *Industrial & Engineering Chemistry Research* 38(6):2216–2224.

Dogru, M., C.R. Howarth, G. Akay, B. Keskinler, and A.A. Malik. 2002. Gasification of hazelnut shells in a downdraft gasifier. *Energy* 27(5):415–427.

Donnot, A., P. Magne, and X. Deglise. 1991. Kinetic parameters of the cracking reaction of tar from wood pyrolysis; comparison of dolomite with industrial catalysts. *Journal of Analytical and Applied Pyrolysis* 22(1–2):47–59.

Drift, A. van der, R. van Ree, H. Boerrigter, and K. Hemmes. 2004. Bio-syngas: Key intermediates for large scale production of green fuels and chemicals. In *Second World Conference and Technology Exhibition on Biomass for Energy, Industry and Climate Protection*. Rome, Italy.

Dry, M.E. 2004. Present and future applications of the Fischer–Tropsch process. *Applied Catalysis A: General* 276(1–2):1–3.

Ebeling, J.M. and B.M. Jenkins. 1985. Physical and chemical properties of biomass fuels. *Transactions of the ASAE (American Society of Agricultural Engineers)* 28(3):898–902.

El-Rub, Z.A., E.A. Bramer, and G. Brem. 2004. Review of catalysts for tar elimination in biomass gasification processes. *Industrial & Engineering Chemistry Research* 43(22):6911–6919.

Ergudenler, A. and A.E. Ghaly. 1993. Agglomeration of silica sand in a fluidized bed gasifier operating on wheat straw. *Biomass and Bioenergy* 4(2):135–147.

Fan, M., W. Marshall, D. Daugaard, and R.C. Brown. 2004. Steam activation of chars produced from oat hulls and corn stover. *Bioresource Technology* 93(1):103–107.

Fang, K.G., D.B. Li, M.G. Lin, M.L. Xiang, W. Wei, and Y.H. Sun. 2009. A short review of heterogeneous catalytic process for mixed alcohols synthesis via syngas. *Catalysis Today* 147(2):133–138.

Fryda, L.E., K.D. Panopoulos, and E. Kakaras. 2008. Agglomeration in fluidised bed gasification of biomass. *Powder Technology* 181(3):307–320.

Gabra, M., E. Pettersson, R. Backman, and B. Kjellström. 2001. Evaluation of cyclone gasifier performance for gasification of sugar cane residue—Part 1: Gasification of bagasse. *Biomass and Bioenergy* 21(5):351–369.

González, J.F., S. Román, D. Bragado, and M. Calderón. 2008. Investigation on the reactions influencing biomass air and air/steam gasification for hydrogen production. *Fuel Processing Technology* 89(8):764–772.

Griffiths, D.M.L. and J.S.R. Mainhood. 1967. The cracking of tar vapour and aromatic compounds on activated carbon. *Fuel* 46(3):167–176.

Halligan, J.E., K.L. Herzog, and H.W. Parker. 1975. Synthesis gas from bovine wastes. *Industrial & Engineering Chemistry Process Design and Development* 14(1):64–69.

Han, J. and H. Kim. 2008. The reduction and control technology of tar during biomass gasification/pyrolysis: An overview. *Renewable & Sustainable Energy Reviews* 12(2):397–416.

Henihan, A.M., M.J. Leahy, J.J. Leahy, E. Cummins, and B.P. Kelleher. 2003. Emissions modeling of fluidised bed co-combustion of poultry litter and peat. *Bioresource Technology* 87(3):289–294.

Huidobro, A. and A.C. Pastor. 2001. Preparation of activated carbon cloth from viscous rayon: Part IV. Chemical activation. *Carbon* 39(3):389–398.

Janse, A.M.C., H.G. de Jonge, W. Prins, and W.P.M. van Swaaij. 1998. Combustion kinetics of char obtained by flash pyrolysis of pine wood. *Industrial & Engineering Chemistry Research* 37(10):3909–3918.

Jothimuragesan, K., A.A. Adeyiga, and S.K. Gangwal. 1996. Simultaneous removal of H_2S and NH_3 in coal gasification processes. Final report, DE-FG-22-93MT93005.

Klass, D.L. 1998. *Biomass for Renewable Energy, Fuels, and Chemicals*. San Diego, CA: Academic Press.

Kumar, A., K. Eskridge, D.D. Jones, and M.A. Hanna. 2009b. Steam-air fluidized bed gasification of distillers grains: Effects of steam to biomass ratio, equivalence ratio and gasification temperature. *Bioresource Technology* 100(6):2062–2068.

Kumar, A., D.D. Jones, and M.A. Hanna. 2009a. Thermochemical biomass gasification: A review of the current status of the technology. *Energies* 2(3):556–581.

Kumar, A., L. Wang, Y.A. Dzenis, D.D. Jones, and M.A. Hanna. 2008. Thermogravimetric characterization of corn stover as gasification and pyrolysis feedstock. *Biomass and Bioenergy* 32(5):460–467.

Kundiyana, D.K., R.L. Huhnke, and M.R. Wilkins. 2010. Syngas fermentation in a 100-L pilot scale fermentor: Design and process considerations. *Journal of Bioscience and Bioengineering* 109(5):492–498.

Kundiyana, D.K., R.L. Huhnke, and M.R. Wilkins. 2011b. Effect of nutrient limitation and two-stage continuous fermentor design on productivities during "*Clostridium ragsdalei*" syngas fermentation. *Bioresource Technology* 102(10):6058–6064.

Kundiyana, D.K., M.R. Wilkins, P. Maddipati, and R.L. Huhnke. 2011a. Effect of temperature, pH and buffer presence on ethanol production from synthesis gas by "*Clostridium ragsdalei*". *Bioresource Technology* 102(10):5794–5799.

Lee, J., J. Kim, and T. Hyeon. 2006. Recent progress in the synthesis of porous carbon materials. *Advanced Materials* 18(16):2073–2094.

Lemus, R., E.C. Brummer, K.J. Moore, N.E. Molstad, C.L. Burras, and M.F. Barker. 2002. Biomass yield and quality of 20 switchgrass populations in southern Iowa, USA. *Biomass and Bioenergy* 23(6):433–442.

Li, C. and K. Suzuki. 2009. Tar property, analysis, reforming mechanism and model for biomass gasification—An overview. *Renewable and Sustainable Energy Reviews* 13(3):594–604.

Li, X.T., J.R. Grace, C.J. Lim, A.P. Watkinson, H.P. Chen, and J.R. Kim. 2004. Biomass gasification in a circulating fluidized bed. *Biomass and Bioenergy* 26(2):171–193.

Liu, K., H.K. Atiyeh, R.S. Tanner, M.R. Wilkins, and R.L. Huhnke. 2012. Fermentative production of ethanol from syngas using novel moderately alkaliphilic strains of *Alkalibaculum bacchi*. *Bioresource Technology* 104:336–341.

Lv, P.M., Z.H. Xiong, J. Chang, C.Z. Wu, Y. Chen, and J.X. Zhu. 2004. An experimental study on biomass air-steam gasification in a fluidized bed. *Bioresource Technology* 95(1):95–101.

Maddipati, P., H.K. Atiyeh, D.D. Bellmer, and R.L. Huhnke. 2011. Ethanol production from syngas by *Clostridium* strain P11 using corn steep liquor as a nutrient replacement to yeast extract. *Bioresource Technology* 102(11):6494–6501.

Mangun, C.L., R.D. Braatz, J. Economy, and A.J. Hall. 1999. Fixed bed adsorption of acetone and ammonia onto oxidized activated carbon fibers. *Industrial & Engineering Chemistry Research* 38(9):3499–3504.

Mani, S., L.G. Tabil, and S. Sokhansanj. 2006. Effects of compressive force, particle size and moisture content on mechanical properties of biomass pellets from grasses. *Biomass and Bioenergy* 30(7):648–654.

Mani, T., P. Murugan, J. Abedi, and N. Mahinpey. 2010. Pyrolysis of wheat straw in a thermogravimetric analyzer: Effect of particle size and heating rate on devolatilization and estimation of global kinetics. *Chemical Engineering Research and Design* 88(8):952–958.

Maniatis, K. 2001. Progress in biomass gasification: An overview, In *Progress in Thermochemical Biomass Conversion*, Vol. 1, A.V. Bridgwater (Ed.) Oxford, U.K.: Blackwell Science Ltd.

McKendry, P. 2002. Energy production from biomass (Part 3): Gasification technologies. *Bioresource Technology* 83(1):55–63.

Melligan, F., R. Auccaise, E.H. Novotny, J.J. Leahy, M.H.B. Hayes, and W. Kwapinski. 2010. Pressurised pyrolysis of *Miscanthus* using a fixed bed reactor. *Bioresource Technology* 102(3):3466–3470.

Milne, T.A., N. Abatzoglou, R.J. Evans, Inc Kemestrie, and National Renewable Energy Laboratory. 1998. *Biomass Gasifier"Tars": Their Nature, Formation, and Conversion*, National Renewable Energy Laboratory, Golden, CO.

Narvaez, I., A. Orio, M.P. Aznar, and J. Corella. 1996. Biomass gasification with air in an atmospheric bubbling fluidized bed. Effect of six operational variables on the quality of the produced raw gas. *Industrial & Engineering Chemistry Research* 35(7):2110–2120.

Natarajan, E., A. Nordin, and A.N. Rao. 1998. Overview of combustion and gasification of rice husk in fluidized bed reactors. *Biomass and Bioenergy* 14(5–6):533–546.

Neeft, J.P.A., H.A.M. Knoef, S.E.C. Nederland, and BTG Biomass Technology Group. 1999. *Behaviour of Tar in Biomass Gasification Systems: Tar Related Problems and Their Solutions*, Utrecht, the Netherlands.

Patil, K., P. Bhoi, R. Huhnke, and D. Bellmer. 2011. Biomass downdraft gasifier with internal cyclonic combustion chamber: Design, construction, and experimental results. *Bioresource Technology* 102(10):6286–6290.

Radovic, L.R. and F. Rodriguez-Reinoso. 2004. Carbon materials in catalysis. In *Chemistry and Physics of Carbon*, Vol. 29, L.R. Radovic (Ed.), pp. 243–358. Boca Raton, FL: CRC Press.

Rapagnà, S. and A. Latif. 1997. Steam gasification of almond shells in a fluidised bed reactor: The influence of temperature and particle size on product yield and distribution. *Biomass and Bioenergy* 12(4):281–288.

Reed, T.E. 1981. *Biomass Gasification: Principles and Technology*. Park Ridge, NJ: Noyes Data Corporation.

Sayan, E. 2006. Ultrasound-assisted preparation of activated carbon from alkaline impregnated hazelnut shell: An optimization study on removal of Cu2+ from aqueous solution. *Chemical Engineering Journal* 115(3):213–218.

Schroder, E., K. Thomauske, C. Weber, A. Hornung, and V. Tumiatti. 2007. Experiments on the generation of activated carbon from biomass. *Journal of Analytical and Applied Pyrolysis* 79(1–2):106–111.

Scott, D.S. and J. Piskorz. 1982. The flash pyrolysis of aspen-poplar wood. *The Canadian Journal of Chemical Engineering* 60(5):666–674.

Seo, M.W., J.H. Goo, S.D. Kim, S.H. Lee, and Y.C. Choi. 2010. Gasification characteristics of coal/biomass blend in a dual circulating fluidized bed reactor. *Energy & Fuels* 24:3108–3118.

Seshadri, K.S. and A. Shamsi. 1998. Effects of temperature, pressure, and carrier gas on the cracking of coal tar over a char-dolomite mixture and calcined dolomite in a fixed-bed reactor. *Industrial & Engineering Chemistry Research* 37(10):3830–3837.

Sharma, A.M., A. Kumar, K.N. Patil, and R.L. Huhnke. 2011. Performance evaluation of a lab-scale fluidized bed gasifier using switchgrass as feedstock. *Transactions of the ASABE* 56(6):2259–2266.

Sharma, A.M., A. Kumar, K.N. Patil, and R.L. Huhnke. 2013. Fluidization characteristics of a mixture of gasifier solid residues, switchgrass and inert material. *Powder Technology* 235(0):661–668.

Son, Y.I., S.J. Yoon, Y.K. Kim, and J.G. Lee. 2011. Gasification and power generation characteristics of woody biomass utilizing a downdraft gasifier. *Biomass and Bioenergy* 35(10):4215–4220.

Spath, P.L. and D.C. Dayton. 2003. Preliminary screening—technical and economic assessment of synthesis gas to fuels and chemicals with emphasis on the potential for biomass-derived syngas. NREL/TP-510-34929. Golden, CO: National Renewable Energy Laboratory.

Subramani, V. and S.K. Gangwal. 2008. A review of recent literature to search for an efficient catalytic process for the conversion of syngas to ethanol. *Energy & Fuels* 22(2):814–839.

Suuberg, E.M., W.A. Peters, and J.B. Howard. 1978. Product composition and kinetics of lignite pyrolysis. *Industrial & Engineering Chemistry Process Design and Development* 17(1):37–46.

Suzuki, M. 1994. Activated carbon fiber: Fundamentals and applications. *Carbon* 32(4):577–586.

Toles, C.A., W.E. Marshall, and M.M. Johns. 1997. Granular activated carbons from nutshells for the uptake of metals and organic compounds. *Carbon* 35(9):1407–1414.

Torres, W., S. Pansare, and J. Goodwin. 2007. Hot gas removal of tars, ammonia, and hydrogen sulfide from biomass gasification gas. *Catalysis Reviews* 49(4):407–456.

Wang, L., C.L. Weller, D.D. Jones, and M.A. Hanna. 2008. Contemporary issues in thermal gasification of biomass and its application to electricity and fuel production. *Biomass and Bioenergy* 32(7):573–581.

Weerachanchai, P., M. Horio, and C. Tangsathitkulchai. 2009. Effects of gasifying conditions and bed materials on fluidized bed steam gasification of wood biomass. *Bioresource Technology* 100(3):1419–1427.

Wu, H., G. Bryant, K. Benfell, and T. Wall. 2000. An experimental study on the effect of system pressure on char structure of an Australian bituminous coal. *Energy & Fuels* 14(2):282–290.

Young, L. and C.C.P. Pian. 2003. High-temperature, air-blown gasification of dairy-farm wastes for energy production. *Energy* 28(7):655–672.

Zanzi, R., K. Sjostrom, and E. Bjornbom. 1996. Rapid high-temperature pyrolysis of biomass in a free-fall reactor. *Fuel* 75(5):545–550.

Zhang, R., R.C. Brown, A. Suby, and K. Cummer. 2004. Catalytic destruction of tar in biomass derived producer gas. *Energy Conversion and Management* 45(7–8):995–1014.

Zhou, J., Q. Chen, H. Zhao et al. 2009. Biomass-oxygen gasification in a high-temperature entrained-flow gasifier. *Biotechnology Advances* 27(5):606–611.

Zhou, J., S.M. Masutani, D.M. Ishimura, S.Q. Turn, and C.M. Kinoshita. 2000. Release of fuel-bound nitrogen during biomass gasification. *Industrial & Engineering Chemistry Research* 39(3):626–634.

18 Biomass Pyrolysis and Bio-Oil Utilization

Lijun Wang and Dan Cheng

CONTENTS

18.1 INTRODUCTION

Transportation fuel is the second largest energy consumption in the United States. Each year over 200 billion gallons of gasoline and diesel are consumed for light-duty and heavy-duty vehicles. Currently the dominating alternative biofuels in the market are ethanol for gasoline and biodiesel for diesel, respectively. Ethanol is currently produced from corn starch in the United States. The growth of corn needs tractors, fertilizers and pesticides that are heavily dependent on petroleum. Therefore, corn ethanol does not greatly reduce the net greenhouse gas (GHG) emission from the environmental perspective (Kim and Dale, 2005). Biodiesel market in the United States is much smaller than ethanol market. Currently biodiesel is mostly produced from soybean oil. The difficulty for the market to grow up is the fluctuation of the upstream material price, as up to 70% or more of the biodiesel production cost comes from soybean oil price (Haas et al., 2006). Furthermore, both corn and soybean are food sources. The production of ethanol from corn and biodiesel from soybean competes with food supply.

Lignocellulosic biomass such as agricultural and forest residues is an abundant renewable energy source. Lignocellulosic biomass can be pyrolyzed into bio-oil in the absence of oxygen (Bridgwater, 2012). One of the main advantages for the conversion of biomass into bio-oil is that it is an effective method for the densification of voluminous biomass into bio-oil locally that can be easily transported and refined at a centralized refinery facility. The density of bio-oil is typically 1200 kg/m^3, which is almost 10 times of the bulk density of a typical biomass material such as crop residues at 150 kg/m^3 (Bridgwater, 2012).

Crude bio-oil is a dense and dark liquid, which is highly considered as a substitute of petroleum oil. However, as crude bio-oil has poor physical and chemical qualities, research has been conducted on upgrading crude bio-oil for the applications in engines and boilers. This chapter is focused on the review of biomass pyrolysis, bio-oil utilization, bio-oil upgrading, and the economical and environmental impacts of biomass pyrolysis.

18.2 BIOMASS PYROLYSIS

18.2.1 Principles and Processes of Biomass Pyrolysis

Biomass consists of cellulose, hemicellulose, lignin, and other organic materials. Biomass pyrolysis is the thermochemical decomposition of biomass at an elevated temperature in the absence of air (or oxygen). Biomass pyrolysis can generate liquid bio-oil, syngas, and solid biochar. At a high temperature, both bio-oil and syngas are in a gaseous form. A dark and dense bio-oil, which has almost half of the heating value of crude petroleum oil, is formed after the hot vapor is rapidly condensed. The syngas that mainly consists of CO, H_2, CO_2, and CH_4 is noncondensable. The proportion and quality of these three products from biomass pyrolysis heavily depend on the characteristics of biomass and the pyrolytic conditions such as the heating rate, temperature, pressure, and residence time. For a fast pyrolytic process, the yield of bio-oil can be as high as 75% of the mass of feedstock. The biochar is typically about 15% of the original mass of the biomass but about 25% of the energy of the biomass feedstock. The remaining 10% is the syngas, which only contains about 5% of the energy in the feedstock (Bridgwater, 2012).

During slow pyrolysis, biomass is heated up at a low rate over a long period, leading to the maximum yield of solid product of biochar. In contrast, a fast pyrolysis process, which requires a much larger heating rate (e.g., 10^3°C/s–10^5°C/s) and short vapor residence time (e.g., <2 s), produces more liquid product. Biomass particle size and moisture content have significant effect on its heating rate. Although biomass particle size is expected to be quite small (e.g., less than 3 mm) to achieve a high heating rate, the reduction of biomass particle size will increase the energy consumption and capital costs. Due to low thermal conductivity of biomass, biomass with a large particle size decreases the heating rate and ends up forming more char. Secondary reactions of chemicals in the bio-oil within large biomass particles become more obvious. An ablative process is another way to achieve fast heat transfer between biomass particles and heat source. Ablative heat transfer between biomass and hot solids peels the char off the particle that continues to facilitate fresh biomass to react (Lede, 2003). This reduces the restriction of particle size but increases the difficulty of removing micropored carbons from vapor as the ablation generates a lot of fine char particles. The vaporization of water in biomass during pyrolysis decreases the heating rate. Biomass should be dried to reduce the moisture content below 10% prior to pyrolysis. Furthermore, water from biomass or generated by reactions during pyrolysis may stay in bio-oil. The water in bio-oil cannot be easily

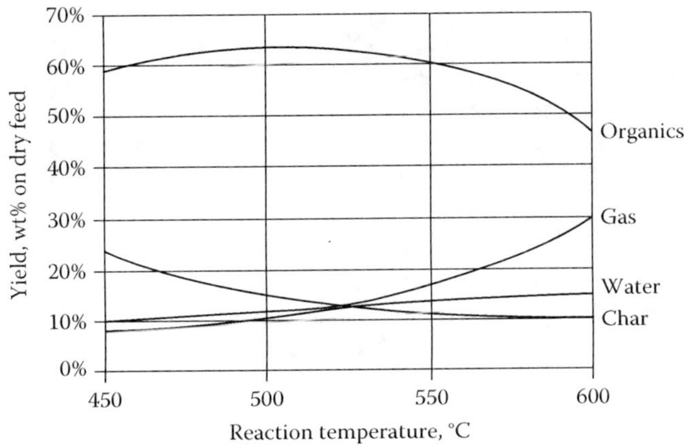

FIGURE 18.1 Product compositions of wood fast pyrolysis at different temperatures. (Adapted from Toft, A.J., A comparison of integrated biomass to electricity systems. PhD thesis, Aston University, Birmingham, U.K., 1996.)

removed by conventional distillation and affects the bio-oil quality such as pH, viscosity, and thermochemical stability. Fast pyrolysis usually cannot reach the thermodynamic equilibrium because the bio-oil is produced at a short reaction time and rapidly cooled from the pyrolysis temperature. The chemical composition of the bio-oil from fast pyrolysis thus changes toward thermodynamic equilibrium during storage (Qian et al., 2007).

Temperature affects the bio-oil yield. The hot vapor generated in biomass pyrolysis can be further decomposed at a temperature higher than 500°C to generate more syngas, while the hot vapor can experience the condensation reaction at a temperature lower than 400°C to generate more biochar. Therefore, the pyrolysis temperature should be carefully controlled in order to achieve a high bio-oil yield (Bridgwater, 1999). The effect of temperature on the yields of bio-oil, syngas, and biochar during the fast pyrolysis of wood is shown in Figure 18.1 (Toft, 1996). As shown in Figure 18.1, the highest bio-oil yield was around 65%, which was achieved at approximately 500°C. Butler et al. (2011) also observed that the bio-oil yield can reach up to 80% on a dry weight basis from wood pyrolyzed around 500°C for less than 1 s. However, it is still not clear how temperature affects the quality of bio-oil.

The residence time of the vapor generated from biomass during pyrolysis should be as short as possible to avoid secondary reactions such as cracking reaction, which converts the bio-oil to noncondensable gas with smaller molecular weight and thus reduces the bio-oil yield (Zheng et al., 2008). Typically, vapor should be quenched within 5 s if temperature is below 500°C (Bridgwater, 1999). Very short residence time of biomass would cause insufficient lignin depolymerization (Mckinley, 1989).

The yield of bio-oil depends on biomass type, temperature, hot vapor residence time, char separation, and biomass ash content. Bridgwater et al. (1999) and Bridgwater (2012) summarized that a successful fast pyrolysis process must have the following:

- High heating rate and heat transfer rate (e.g., 10^{3}°C/s–10^{5}°C/s)
- Strictly controlled temperature around 500°C
- Short hot vapor residence time of typically less than 2 s to minimize secondary reactions
- Fast cooling of the pyrolysis vapor to produce bio-oil (e.g., <5 s)
- Rapid removal of product char to minimize catalytically cracking of bio-oil vapor

18.2.2 Biomass Pyrolysis Mechanisms and Kinetics

Cellulose pyrolysis follows two pathways: gradual degradation, decomposition, and char formation at a temperature as low as 325 K, and fast volatilization as well as levoglucosan generation at a higher temperature (Shafizadeh, 1982). The bio-oil that is produced from cellulose at an isothermal temperature between 623 and 643 K contains 62%–73% aqueous distillate, mainly consisting of water, large proportion of levoglucosan, and small proportions of acetaldehyde, propionaldehyde, butyraldehyde, acrolein, crotonaldehyde, furan, acetone, butanedione, and methanol (Wegener and Fengel, 1978). Hemicellulose is more easily to be decomposed than cellulose. Research found that the yield of bio-oil from the pyrolysis of cellulose at 580°C was 72%, compared to 45% for the pyrolysis of hemicellulose (Shen et al., 2010).

Lignin is a complex phenolic polymer that is left as residues after a fermentation or pulping process. Usually lignin is discarded as waste or burned directly in boilers as a low-grade fuel. Lignin that is a very recalcitrant matrix has a stronger thermal stability than cellulose and hemicelluloses during pyrolysis. Lignin decomposes within a wider temperature range than cellulose and hemicelluloses (Bridgwater, 1999). Lignin does not decompose completely until a temperature reaches up to 973 K, compared to a small temperature interval between 573 and 673 K for the pyrolysis of cellulose and hemicelluloses. The thermogravimetric curves show that it requires higher temperature to decompose woody biomass with a higher amount and different types of lignin than straws (Burhenne et al., 2013). Herbaceous biomass such as wheat straw with higher cellulose and hemicelluloses contents decomposes faster and produces a larger fraction of gaseous products than woody biomass with higher lignin content. Woody biomass produces larger fraction of solid products than herbaceous biomass during pyrolysis (Burhenne et al., 2013). Bio-oil from lignin has lower oxygen content and therefore a higher energy density than conventional bio-oil (Nowakowski et al., 2010). It was also found that pyrolysis of woody biomass with high lignin content is an overall endothermic reaction, while the pyrolysis of herbaceous biomass with lower lignin content is an exothermic reaction (Yang et al., 2006).

The ash in biomass and biochar produced during pyrolysis can catalyze the cracking reactions to convert bio-oil to syngas (Butler et al., 2011). Agricultural residues and grasses generally have higher ash contents than woody biomass. The application of high levels of nitrogen to grow biomass crop could increase the ash content of the crop (Hodgson et al., 2010). The biomass with high ash contents is not desirable as the feedstock in pyrolysis. The ash in biomass feedstocks can be washed out using water or an acid solution prior to pyrolysis. The washing pretreatment of empty fruit bunches increased the yield of bio-oil from 52% to 72% (Abdullah and Gerhauser, 2008). Char removal is another key factor in a reactor environment to maximize the bio-oil yield (Butler et al., 2011).

The thermal degradation behavior of biomass during pyrolysis primarily depends on the chemical composition and structure of the biomass. Cellulose, hemicelluloses, and lignin are three dominant components in lignocellulosic biomass. Thermogravimetric analysis is widely used to characterize the decomposition reactions during the pyrolysis of biomass and determine kinetic parameters (Fisher et al., 2002). First-order kinetic equation was used to determine the pyrolysis rates of cellulose, hemicelluloses, and lignin, which was given by (Burhenne et al., 2013)

$$\frac{dX_i}{dt} = -k_i(T)X_i = -k_i(T)\frac{m_i - m_{ash}}{m_{0,i} - m_{ash}} \tag{18.1}$$

where the temperature-dependent rate coefficient for each component is determined by Arrhenius equation

$$k_i(T) = k_{0,i}\exp\left[\frac{-E_{A,i}}{(RT)}\right] \tag{18.2}$$

where

$E_{A,i}$ is the activation energy

$k_{0,i}$ is the pre-exponential factor, which is given in Table 18.1

TABLE 18.1
Parameters for the Devolatilization Kinetics of Cellulose, Hemicelluloses, and Lignin from Spruce

	$E_{A,i}$ (kJ/mol)	$k_{0,i}$ (1/s)	References
Cellulose	236	$10^{17.43}$	Gronli et al. (2002)
Hemicellulose	100	$10^{6.4}$	Gronli et al. (2002)
Lignin	46	$10^{0.58}$	Gronli et al. (2002)
Xylan, the first step	168	$10^{15.4}$	Fisher et al. (2002)
Xylan, the second step	104	$10^{7.3}$	Fisher et al. (2002)

It was found that fast pyrolysis (25–70 K/s) of straws at temperature between 400 and 648 K followed two stages. The first step was 10 times faster than the second one, releasing most of the volatiles and forming solid intermediates. In the second step, solid intermediates were further degraded into char and volatiles (Demirbas, 2000). A two-step consecutive first-order reaction model given in Equation 18.1 can be used to fit the experimental data better (Fisher et al., 2002; Kumar et al., 2008; Wang et al., 2009a, 2011). The activation energies and pre-exponential factors in Equation 18.2 for the two steps of first-order reaction kinetics during pyrolysis of xylan are given in Table 18.1 (Fisher et al., 2002).

18.2.3 Pyrolysis Reactors

A successful fast pyrolysis reactor should be able to create high heating rate, moderate temperature, and short quenching time, while certain environmental issues, energy, and capital costs should be minimized. There are various designs of pyrolytic reactors including fixed bed, entrained bed, fluidized beds including circulating and bubbling beds, rotating cone, ablative reactor, screw and augur kilns, and microwave pyrolysis. Different types of pyrolysis reactors are summarized in Table 18.2 (Bridgwater, 1999, 2012). Fluid beds and auger pyrolysis reactors are attractive for fast pyrolysis. High-density heat carriers such as sand are usually used to supply heat to biomass (Butler et al., 2011).

The most technical challenge in the design of a fast pyrolytic reactor is heat transfer in the reactor. Pyrolysis is an endothermic process, which requires a significant amount of heat input. There are two ways to transfer heat in pyrolysis reactors: gas–solid heat transfer in which heat is transferred from gas to biomass by convection and solid–solid heat transfer by conduction. For good mixing of particles in a reactor, 90% heat can be transferred by conduction through solid–solid interaction, while the remaining 10% heat is transferred by convection through gas–solid interaction (Bridgwater, 2012). Radiation also somewhat contributes to the heat transfer in reactors. Combustion of product char and syngas is two common ways to supply heat for a pyrolysis process. Char should be removed from the bio-oil vapor rapidly because the char acts as a cracking catalyst to convert bio-oil to syngas. Char is usually separated from the bio-oil vapor and syngas by cyclones. Because the char and pyrolytic lignin product easily form a gel-like phase that can rapidly block the filter, pressure filtration is not a proper way to remove particles from liquid bio-oil. Hot gas filtration has not yet been demonstrated over a long-term operation. Rapid cooling of bio-oil vapor is essential to minimize secondary reactions in the bio-oil. As lignin-derived components in the bio-oil can cause deposition on the heat exchanger surface, quenching the bio-oil vapor in liquid bio-oil or an immiscible hydrocarbon solvent is widely used (Bridgwater, 2012).

TABLE 18.2
Reactor Types for Fast Pyrolysis

Reactor Type	Heating Method	Heating Mode	Advantages/Disadvantages/Features
Ablative	Wall heating	95% conduction 4% convection 1% radiation	Large-size feedstock acceptation Compact design Heat supply problematical High mechanical char abrasion from biomass Heat transfer gas not required Particulate transport gas not always required
Circulating fluidized bed	In-bed gasification of char to heat sand	95% conduction 4% convection 1% radiation	High heat transfer rates High char abrasion from biomass and char erosion Leading to high char product Char heat carrier separation required Solid recycle required Maximum particle size up to 6 mm Possible catalytic activity from hot char Possible liquids cracking by hot solids Greater reactor wear possible
Fluidized bed	Heated recycled gas	95% conduction 4% convection 1% radiation	High heat transfer rates High supply to fluidizing gas or to bed directly Limited char abrasion Very good solid mixing Particle size limit <2 mm in smallest dimension Simple reactor configuration
Entrained flow	Char combustion products	95% conduction 4% convection 1% radiation	Low heat transfer rates Particle size limit <2 mm Limited gas/solid mixing
Screw auger	Hot sand	N/A	For feedstocks that are difficult to handle and heterogeneous

18.2.4 Fast Hydropyrolysis and Catalytic Pyrolysis

The hydropyrolysis was developed by GTI, which includes three stages: (1) biomass hydropyrolysis under a medium hydrogen pressure in the presence of a novel glass ceramic catalyst, (2) hydrodeoxygenation of hydropyrolysis vapor with a conventional sulfide HDS catalyst under mild conditions, and (3) reforming of C_1–C_3 hydrocarbon gases into hydrogen for the supply of hydrogen. The yield of hydrocarbon liquid that is highly deoxygenated and contains no polynuclear aromatics and olefins is 24%–28% (Butler et al., 2011).

Catalytic pyrolysis is investigated for the production of bio-oil with high amounts of aromatics. Pine wood that was catalytically pyrolyzed using the ZSM-5 catalyst at 600°C produced 14% aromatics (Carlson et al., 2009). Zeolite-based catalysts are widely used for the catalytic pyrolysis of biomass. Among 40 zeolite catalysts, the ZSM-5 catalyst was found to perform better than other zeolite catalysts. The maximum hydrocarbon yield was 16%, which was achieved with a nickel-, cobalt-, iron-, and gallium-substituted ZSM-5 catalyst (French and Czernik, 2010). Torri et al. (2010) found CuO-mixed metal oxide catalysts and ZnO could reduce the content of heavy fraction in the bio-oil while slightly reducing the bio-oil yield.

18.2.5 Case Studies of Biomass Pyrolysis

Cassava is an annual crop extensively planted in tropical and subtropical areas. The tuberous root is edible and considered as the third largest food carbohydrates source. About 234 million tons of

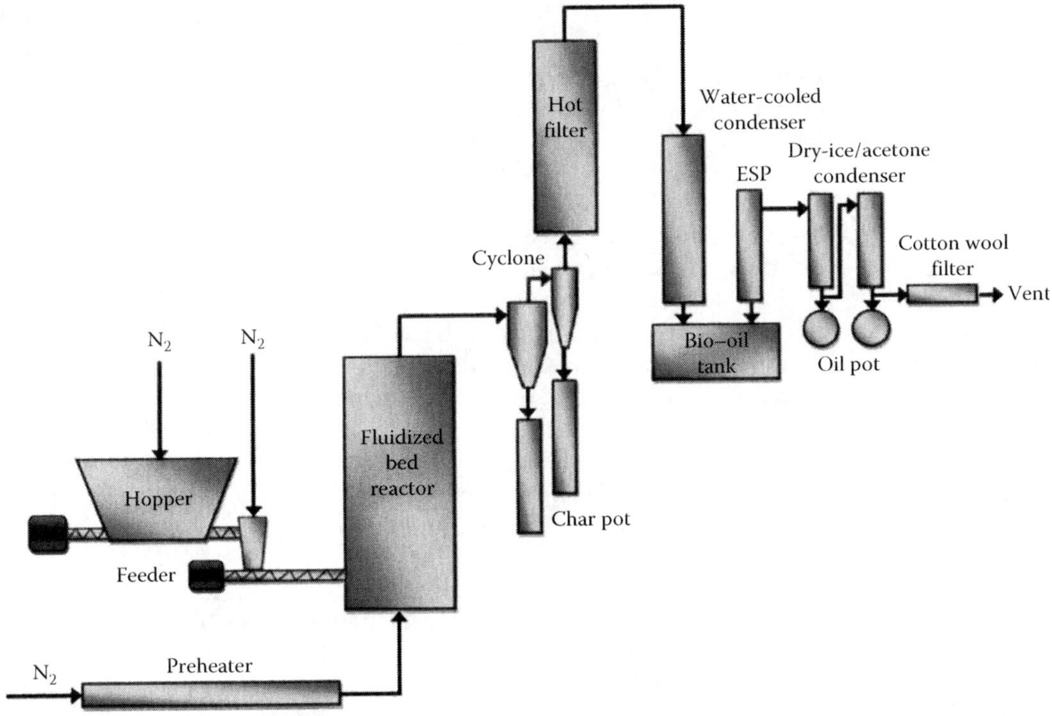

FIGURE 18.2 Schematic diagram of the fast pyrolysis unit. (Adapted from Pattiya, A. and Suttibak, S., *J. Anal. Appl. Pyrolysis*, 95, 227, 2012.)

cassava is produced worldwide each year, while approximately 115 million tons of cassava residues are generated, which is equivalent to 17,000 million US dollars petroleum oil at the petroleum price of $81/barrel. Pattiya and Suttibak (2012) conducted experiments on fast pyrolysis of cassava residues in a fluidized reactor with a hot vapor filter. The schematic diagram is shown in Figure 18.2. The cassava residues were transported from the hopper to the reactor by a twin-screw feeder. The temperature was monitored and controlled by K-type thermocouples and PID temperature controllers, respectively. Two cyclone separators and a hot filter unit were used to clean the vapor after which the pyrolysis vapor was collected by a series of condensers and heat exchangers.

The highest bio-oil yields for the fast pyrolysis of rhizome and stalk were 63.23% and 61.39%, respectively, which were obtained at 475°C and 469°C, respectively. As temperature increased, thermal cracking of bio-oil vapors and thermal decomposition of char resulted in the reduction of bio-oil and char yields and increase of syngas yield. It was also observed that the installation of a hot filter decreased the bio-oil yield by 6%–7% because of secondary vapor cracking reaction through the filter. However, a hot vapor filter was necessary because the filter could reduce the alkali content in bio-oil to an acceptable level. It was also observed that cassava with a particle size of 337 mm generated the highest oil yield as particles with smaller size were blown away before the completion of pyrolysis, while larger particles would restrict the heat transfer into the particles and lead to the formation of more char.

18.3 BIO-OIL CHARACTERIZATION

18.3.1 Bio-Oil Chemical Compositions

Crude bio-oil that is produced from depolymerization and fragmentation of cellulose, hemicelluloses, and lignin in biomass is a complex mixture composed of acids, alcohols, aldehydes, esters, ethers, ketones, sugars, phenols, guaiacols, syringols, furans, and extractible terpene with multifunctional compounds (Wang et al., 2004). There are more than 300 organic compounds that have been identified in bio-oil. The aldehydes and phenols in bio-oil cause an objectionable smell, while they are not really harmful to human beings. The molecular weights of those compounds vary from 18 for water to 5000 or higher for pyrolytic lignins (Lu et al., 2009). The chemical composition of bio-oil can be analyzed by a GC–MS. Identification of chromatographic peaks can be done using the National Institute of Standards and Tests (NIST) Mass Spectrometry (MS) library (Wang et al., 2004; Peng et al., 2009; Lu et al., 2010). The analysis of the chemical composition showed that crude bio-oil has abundant amounts of aldehydes and ketones, which make the bio-oil extremely hydrophilic (Wang et al., 2004).

The organic components of crude bio-oil produced by the fast pyrolysis of rice husk are given in Table 18.3 (Peng et al., 2009). Acetic acid, 1,6-Anhydro-beta-D-glycopyranose, and 2,3,4,6-tetrakis-O-gluconic acid are the three major compounds in the bio-oil that are 17.64%, 15.38%, and 11.69% of the total mass, respectively. The yield and concentration of major chemicals in bio-oil are affected by the type of the biomass feedstock, the type of pyrolytic reactors, and pyrolysis conditions. For the same type of biomass feedstock and pyrolysis system, different pyrolysis conditions may cause large variation of concentrations of the major chemicals in the crude bio-oil (He et al., 2009).

Table 18.4 shows the comparison of selected chemical and physical properties between wood-derived bio-oil and heavy fuel oil (Oasmaa and Czernik, 1999). However, it should be noted that the physical and chemical properties of bio-oil vary widely, which depend on the pyrolysis methods, variety of biomass feedstocks, type of reactors, and catalysts used in the pyrolysis. The oxygen content of bio-oil is usually 45%–50% by mass, which is distributed in more than 300 compounds. Proximate analysis of bio-oil shows that the chemical formula of bio-oil can be expressed as $CH_{1.9}O_{0.7}$, which represents about 46% of oxygen in the bio-oil (Oasmaa and Czernik, 1999). Bio-oil contains various oxygenated hydrocarbons. The oxygenized compounds are immiscible with hydrocarbon fuels (Zhang et al., 2007).

18.3.2 Bio-Oil Physical Properties

The bio-oil has 15%–30% of water by mass as shown in Table 18.4, which comes from the biomass itself that may contain a lot of water and is formed during the pyrolytic reactions (Zhang et al., 2007; Lu et al., 2009). The water content of bio-oil can be measured by a Karl Fischer titration method, which uses a Hydranal solvent to dissolve the sample and then Hydranal composite 5 to titrate the water in the bio-oil (Mahfud et al., 2007). Water in bio-oil can reduce the viscosity of bio-oil and thus increase its fluidity that facilitates the atomization in engines (Bridgwater, 1999).

Ash content of bio-oil is higher than 0.1% by mass, which causes poisoning of catalysts that are used for upgrading bio-oil, erosion and corrosion of vessels, and slag formation and damage to turbines. Sodium and potassium in bio-oil cause erosion at a high temperature. A vapor filter can be installed to reduce the ash content of bio-oil to an acceptable level (Pattiya and Suttibak, 2012). As bio-oil comprises substantial amounts of carboxylic acids such as acetic and formic acids, its pH value is as low as 2–3 (Wang et al., 2009b; Majhi et al., 2012). The acids in bio-oil make them very corrosive. The organic acids in bio-oil can react with alcohols to form esters through the esterification reaction (Xu et al., 2008).

Due to the high oxygen and water content, the heating value of crude bio-oil is 16–19 MJ/kg, which is less than half of those of conventional hydrocarbon fuels at 40 MJ/kg. The density of crude bio-oil is about 1200 kg/m^3, compared to 800–1000 kg/m^3 for petroleum fuels. The volumetric energy density of bio-oil is thus only 50%–60% that of petroleum-based fuels (Lu et al., 2009). The viscosity of bio-oil varies in a large range from 25 cP up to 1000 cP, depending on the water content and the variety of biomass feedstock. Viscosity of bio-oil is usually high, which causes high pumping cost and poor atomization (Zheng and Kong, 2010).

TABLE 18.3
Organic Components of Crude Bio-Oil Produced by the Fast Pyrolysis of Rice Husk

Components in Crude Bio-Oil	Content (%)
Acids	
Acetic acid	17.64
2-Propenoic acid, 2-methyl, ethyl	1.09
Benzeneaceticacid,4-hydroxy-3-methyl-	1.31
Benzoic acid, 4-hydroxy-3-methoxy-	1.44
Gluconic acid, 2,3,4,6-tetrakis-*O*-	11.69
Phenols	
Phenol	0.99
Phenol, 4-methyl-	0.67
Phenol, 2-methoxy-	4.58
Phenol, 4-ethyl-	1.12
Phenol, 2-methoxy-4-methyl-	2.05
1,2-Benzenediol	2.40
Phenol,4-ethyl-2-methoxy-	0.80
2-Methoxy-4-vinylphenol	1.47
3-Allyl-6-methoxyphenol	0.96
Phenol,2,6-dimethoxy-	1.02
Esters	
Acetic acid, methyl ester	5.93
Butanedioic acid, dimethyl ester	0.92
Ketones	
2-Propanone, 1-hydroxy-	9.87
2-Cyclopenten-1-one	0.35
2-Propanone, 1-(acetyloxy)-	0.19
2-Cyclopenten-1-one, 3-methyl-	0.76
2(5H)-Furanone	1.31
2-Cyclopenten-1-one,2-hydroxy-3-methyl-	3.10
2(3H)-Furanone, dihydro-5-propyl-	1.44
Ethanone, 1-(4-hydroxy-3-methoxyph)-	0.80
Aldehydes	
Furfural	2.08
Vanillin	2.37
Others	
Methane, diethoxy-	2.14
1,6-Anhydro-beta-D-glycopyranose	15.38

Source: Adapted from Peng, J. et al., *Bioresour. Technol.*, 100, 3415, 2009.

18.3.3 Bio-Oil Distillation Curves

Bio-oil cannot be completely vaporized once it has been condensed from the vapor phase. If the liquid bio-oil is heated to 100°C or higher, the bio-oil rapidly reacts and eventually produces a solid char (Bridgwater, 2012). The yield of distillate obtained in vacuum distillation will be higher than

TABLE 18.4
Comparison of Selected Properties of Bio-Oil Produced from Wood and Heavy Fuel Oil

Property	Bio-Oil	Heavy Fuel Oil
Moisture content (wt%)	15–30	0.1
pH	2.5	—
Specific gravity	1.2	0.94
Elemental composition (wt%)		
C	54–58	85
H	5.5–7.0	11
O	35–40	1.0
N	0–0.2	0.3
Ash	0–0.2	0.1
HHV (MJ/kg)	16–19	40
Viscosity at 50°C (cP)	40–100	180
Solids (wt%)	0.2–1	1
Distillation residue (wt%)	Up to 50	1

Source: Adapted from Zhang, Q. et al., *Energy Convers. Manage.*, 48, 87, 2007.

that obtained in atmospheric distillation from the same bio-oil because heating bio-oil at higher temperature during atmospheric distillation causes significant polymerization reactions (Lu et al., 2009). The residues left during bio-oil distillation are the char that includes the char presented in the crude bio-oil itself and the char formed during the distillation process.

Ott et al. (2008) developed an advanced distillation apparatus to measure the distillation curves of bio-oil produced by hydrothermal liquefaction of swine manure. The distillation curve is a graphical depiction of the boiling temperature of a fluid mixture against the volumetric fraction distilled. It was found that the yield of the distillate obtained at 500°C was 65% of the crude bio-oil by volume. However, the further increase of the distillation temperature beyond 500°C would not increase the volume of the distillate. The remaining bio-oil was converted to char (Ott et al., 2008).

18.3.4 Bio-Oil Stability

Bio-oil is not thermodynamically stable. Many compounds in bio-oil may take part in different reactions such as polymerization and esterification during storage. Esterification converts highly polar organic acids and alcohol molecules to esters at a low polarity. Polymerization generates large molecules that have poor solubility in the bio-oil. Aldehydes and compounds with unsaturated carbon bonds in crude bio-oil are very active for polymerization and condensation reactions. The properties of bio-oil such as water content and viscosity are changed as a result of these reactions (Diebold, 1999). It was observed that the viscosity of bio-oil during storage slightly increased (Oasmaa and Kuoppala, 2003). The aggregation of pyrolytic lignin, which was eventually separated as the heavy lignin-rich fraction and formed a viscous sludge, was found to play a key role to cause the aging of bio-oil (Fratini et al., 2006). At elevated temperatures, the aging reactions will be accelerated. Therefore, the increase of storage temperature will accelerate the increase of the viscosity of bio-oil during storage (Boucher et al., 2000; Chaala et al., 2004). It was reported that the properties of bio-oil treated at 80°C were significantly altered, while no significant variations were observed for the bio-oil at 50°C for a week (Boucher et al., 2000). Some inhibitors such as hydroquinone can be used to suppress the polymerization reaction of bio-oil during storage. Research from the National Renewable Energy Laboratory (NREL) in the United States showed the viscosity of bio-oil increased from 20 to 22 cp during 4-month storage at 20°C with the addition of 10% methanol (Zheng et al., 2008).

18.4 UPGRADING BIO-OIL INTO TRANSPORTATION FUELS

Although bio-oil produced either from hydrothermal liquefaction or from fast pyrolysis has been considered as an alternative liquid fuel, the low heating value due to high moisture and oxygen contents, high corrosive potential due to substantial amounts of carboxylic acids, high viscosity and poor mobility due to the presence of large molecules in the bio-oil, and poor stability have restricted the use of crude bio-oil as a transportation fuel. It is essential to upgrade the bio-oil for the use of the bio-oil as a transportation fuel. Current upgrading techniques include esterification, hydrogenation, catalytic cracking, emulsification, and reactive distillation. Catalysts may be added to a pyrolytic reactor to simultaneously produce and upgrade bio-oil (Williams and Nugranad, 2000) or in a separate upgrading reactor (Zhang et al., 2006). Integration of catalysis and pyrolysis requires operation at a single temperature and sufficiently robust catalyst (Bridgwater, 2012). Different catalysts used in crude bio-oil upgrading are summarized in Table 18.5. Coking of the catalysts used in catalytic biomass pyrolysis and bio-oil upgrading remains a technical issue. Fractionation of bio-oil prior to catalytic upgrading is considered as an approach to address this technical issue (Butler et al., 2011).

18.4.1 ESTERIFICATION

The crude bio-oil is very corrosive due to large amounts of organic acids in the bio-oil. The organic acids can react with alcohols to produce less corrosive esters through esterification. Zhang et al. (2006) upgraded crude bio-oil by the esterification over solid acid such as 40 $SiO_2/TiO_2–SO_4^{2-}$ and solid base 30 K_2CO_3/Al_2O_3–NaOH catalysts. They found that the density of the upgraded bio-oil

TABLE 18.5
Catalysts for Bio-Oil Upgrading

Reactions	Catalysts	Operating Condition	Main Results	References
Esterification	Solid acids (e.g., $40SiO_2/TiO_2–SO_4^{2-}$ or Amberlyst 15); solid base (e.g., $30K_2CO_3/Al_2O_3$–NaOH)	Temperature usually lower than alcohol boiling point (e.g., around 80°C for ethanol); supercritical conditions make reaction faster.	Heating value increased up to 50%; esters greatly increased, but acid-catalyzed bio-oil needs to be washed.	Zhang et al. (2006), Calvar et al. (2006) and Peng et al. (2009)
Hydrogenation	Transitional metal catalyst (e.g., Co–Mo–P/Al_2O_3)	Relative high H_2 pressure and temperature (e.g., 360°C, 2 MPa).	Oxygen content reduced from 41.8% to 3%.	Zhang et al. (2005)
Catalytic cracking	Zeolite (e.g., HZSM-5) or ZnO	Starts cracking around 300°C; 400°C gives the best performance.	Water soluble sugars decomposed; viscosity increased much more slowly than oil treated without catalyst.	Nokkosmaki et al. (2000) and Adjaye and Bakhshi (1995a,b)
Steam reforming	Transitional metal catalyst (e.g., Ni- or Pt-based catalysts)	High temperature up to 1000°C.	Yield up to 80% of bio-oil.	Czernik et al. (2004) and Seyedeyn-Azad et al. (2011)

was reduced from 1.24 to 0.96 kg/m^3 and the heating value of the upgraded bio-oil was doubled. However, they also found that the acidity of the upgraded bio-oil was alleviated by the solid base catalyst but intensified by the solid acid catalyst due to its strong acidification.

Under supercritical condition, the polarity and acidity of methanol will be increased. Therefore, supercritical methanol (T_c=239°C and P_c=8.09 MPa) can serve as both an acidic catalyst and a reactant for the esterification reaction. Li et al. (2011) used supercritical methanol at a temperature of 239°C and pressure of 8.09 MPa to esterify the light fraction of bio-oil obtained by the distillation of crude bio-oil at a reduced pressure. They found that all the acids in the bio-oil were converted into esters after 6 h esterification without a catalyst. HZSM-5 is usually used to catalyze the thermal cracking reaction for upgrading bio-oil. Acidic HZSM-5 (low Si/Al ratio) can also facilitate esterification in supercritical ethanol (T_c=243.1°C and P_c=6.37 MPa) to convert acids in crude bio-oil into esters and meanwhile crack heavy components of crude bio-oil (Peng et al., 2009).

18.4.2 Hydrogenation

Crude bio-oil has a high oxygen content that results in a low heating value. Hydrogenation can deoxygenize the bio-oil by catalytic reaction with hydrogen to significantly reduce the oxygen content and increase the heating value of the bio-oil. During hydrogenation, the oxygen in the bio-oil compounds is removed as H_2O, CO_2, and CO. The hydrogenation process typically occurs at a high pressure up to 20 MPa and moderate temperature up to 400°C. Hydrogen can be directly used in a process of hydropyrolysis to improve the quality of bio-oil (Butler et al., 2011). Rocha et al. (1999) pyrolyzed cellulose, sugarcane bagasse, and eucalyptus wood in a fixed-bed reactor at heating rates from 5°C/min to 300°C/min under hydrogen pressured up to 10 MPa. They found that the oxygen content of bio-oil was reduced from 40% to below 20% when hydrogen pressure was increased from an atmospheric pressure up to 10 MPa without any catalysts. If the FeS catalyst at a loading amount of 5% (w/w) was used, the oxygen content of the bio-oil was as low as 10%. The NMR analysis showed that the pyrolysis oil with reduced oxygen content was more aromatic. However, increased hydrogen pressure actually decreased the aromatization because of the low char yield under a high pressure. It was found that the hydrodeoxygenation of bio-oil derived from catalytic pyrolysis of biomass was easier than bio-oil from fast pyrolysis (Wang et al., 2012).

The common catalysts used for hydrogenation include Co–Mo, Ni–Mo, and their oxides loaded on Al_2O_3 (Zhang et al., 2007). Using the catalyst of Co–Mo–Pt/Al_2O_3 in a hydrogenation process at 360°C and 2 MPa, Zhang et al. (2005) found that the oxygen content of bio-oil was reduced from 41.8% to 3%. Hydrogen can be directly generated from bio-oil over platinum catalysts such as Pt supported on Al_2O_3. However, coke was formed by thermal catalytic reactions and needed to be removed to avoid catalyst clogging (Fisk et al., 2009). Another technical challenge is that the catalyst support of alumina or aluminosilicates is unstable in the bio-oil with high water content (Bridgwater, 2012). Precious metal catalysts on less susceptible supports such as ruthenium on carbon (Wildschut et al., 2009) and palladium on zirconium (Tang et al., 2009) have received increasing attention recently.

18.4.3 Catalytic Cracking

As crude bio-oil has a very high viscosity due to the presence of large molecules, cracking reaction has been used to convert heavy components in the crude bio-oil into light distillate. For example, rice husks were pyrolyzed in a fixed-bed reactor at 400°C, 450°C, 500°C, 550°C, or 600°C with a ZSM-5 catalyst, followed by molecular weight distribution tests of the oils determined by size exclusion chromatography (Williams and Nugranad, 2000). It was found that the number average (MWn) and weight average (MWw) were significantly reduced as recorded by the RI and UV detectors. A catalytic cracking process at 400°C could convert as high as 91.67% of the crude bio-oil into gasoline type of fuel within 15 min. The remaining bio-oil had a potential to be upgraded to other types of fuels such as kerosene and diesel (Hew et al., 2010).

Traditional zeolites such as HZSM-5 and HY are widely used to catalyze the cracking reactions to convert highly oxygenated compounds to hydrocarbons (Vitolo et al., 1999; Jackson et al., 2009; Peng et al., 2009). Peng et al. (2009) upgraded rice husk-derived bio-oil in sub- and supercritical ethanol over a HZSM-5 catalyst. GC–MS analysis showed that the acid content of upgraded bio-oil dropped to 3.78%, compared to the acid content of 17.64% for crude bio-oil. It was also found that HZSM-5 has catalytic effect on thermal cracking and catalysts with stronger acidity (lower Si/Al ratio) would facilitate thermal cracking more effectively in this process. Although catalytic cracking can convert oxygenated bio-oil to lighter fractions, high yield of coke generated during catalytic cracking, which is usually 8%–25% of the bio-oil, restricts the wide use of the catalytic cracking method (Zhang et al., 2007). The presence of oligomers that are derived mainly from lignin is partly responsible for the catalyst deactivation and low organic yield. At the catalytic cracking temperature, the oligomers are rapidly polymerized to form coke on catalyst surface and subsequently cause the deactivation of the catalyst (Lu et al., 2010). The catalysts should have high conversion efficiency and little coking tendency. TiO_2 and ZrO_2/TiO_2 were also used to catalyze the cracking of bio-oil. Mild cracking over base catalysts such as ZnO is an alternative to zeolite-based cracking (Nokkosmaki et al., 2000).

18.4.4 Emulsification

Unlike other crude bio-oil upgrading methods, emulsification of crude bio-oil with other biofuels or fossil fuels is a physical process that can enhance both physical and chemical properties of the pyrolytic fuel.

Crude bio-oil can be used as a transportation fuel in diesel engine when blended with diesel or biodiesel (Garcia-Perez et al., 2010; Alcala and Bridgwater, 2013). Garcia-Perez et al. (2010) found that the carboxylic acids, monophenols, furans, and lignin-derived oligomers were the most soluble compounds in biodiesel, while sugars had poor solubility. Therefore, biodiesel may be used to extract some bio-oil compounds such as phenols as fuel additives (Garcia-Perez et al., 2010). Alcala and Bridgwater (2013) used an alcohol such as ethanol, 1-butanol, and 2-propanol as a cosolvent for the formation of the bio-oil and biodiesel blend. They found that the type and quantity of the alcohol were critical for the blend formation and stability. Among ethanol, 1-butanol, and 2-propanol, 1-butanol blend could have the largest proportion of bio-oil. 1-Butanol-based blends could have a maximum of 60% bio-oil by mass without phase separation.

Biomass-derived pyrolytic oil is not miscible with hydrocarbons. However, with the aid of surfactants, the bio-oil can be emulsified with diesel fuel (Ikura et al., 2003). Ikura et al. (2003) emulsified pyrolytic oil with No. 2 diesel and found that bio-oil concentration, surfactant concentration, and power input per unit volume of the mixture had critical effect on stability of the mixture. By emulsification, cetane number of crude bio-oil was found to be 5.6 by linear interpolation of emulsified fuel, which was a lot easier to be tested. The viscosity and corrosivity of emulsion fuel were found to be substantially lower than those of crude bio-oil itself. However, developing an economically feasible surfactant is very crucial in bringing down the price of the emulsified fuel. It was also found that an addition of 5% methanol into bio-oil can decrease the viscosity of the bio-oil by 35% (Sipilae et al., 1998).

18.4.5 Distillation

Distillation is another way to upgrade bio-oil. However, crude bio-oil especially produced from hydrothermal liquefaction of biomass is a very condensed material that is hardly fluidable and distillable. On the other hand, distillation temperature should not be too high to avoid char formation and condensation among reactive components. Zheng and Wei (2011) studied reduced pressure distillation of crude bio-oil at 80°C with a residual pressure of 2 kPa. The yield of distillate was up to 61% and the higher heating value of the distillate was 34.2 MJ/kg. The distillate had low acidity and viscosity and increased thermal stability.

Reactive distillation is a more complicated process that integrates both upgrading chemical reactions and separation. A reactive distillation process is to combine reaction and separation in a single unit to reduce both operating and capital costs (Xu et al., 2008; Avami et al., 2012). Organic acids and aldehydes can react with alcohols to esters and acetals, respectively. Xu et al. (2008) upgraded bio-oil in a reactive rectification process using ethanol and solid acidic catalysts. The pH values of light and heavy fractions were 7.06 and 5.35, respectively, compared to 2.82 for the crude bio-oil. The heat values of the light and heavy fractions were 21.5 and 24.5 MJ/kg, respectively, compared to 14.3 MJ/kg for the crude bio-oil. The dynamic viscosity was reduced from 10.5 mm^2/s for the crude bio-oil to 0.46 and 3.65 mm^2/s for the light and heavy fractions, respectively. Mahfud et al. (2007) proposed a reactive distillation process to upgrade bio-oil using a high-boiling alcohol such as n-butanol (boiling point, 118°C), ethylene glycol (boiling point, 197.3°C), and 2-ethylhexyl alcohol (boiling point, 186°C) in the presence of a solid acid catalyst, Nafion® SAC-13 from Aldrich.

18.5 OTHER BIO-OIL UTILIZATIONS

Bio-oil can be directly used in boilers and furnaces for electricity and heat generation, reformed to produce hydrogen and used to extract high-value chemicals. Czernik and Bridgwater (2004) did a review on applications of biomass fast pyrolysis oil.

18.5.1 Combustion for Heat and Power

Bio-oil, which has a lower heating value compared to petro-diesel, can be used in boilers, diesel engines, and gas turbines for heat, power, or combined heat and power (Czernik and Bridgwater, 2004; Zheng and Kong, 2010). Furnaces and boilers that are common devices used for heat and power generation are usually less efficient than engines and turbines. As furnaces and boilers can operate with various fuels, bio-oil has been used in furnaces and boilers for heat and power generation. The bio-oil can be delivered to the combustion chamber by a nozzle and atomized with air. Emissions from burning bio-oil were found to be lower than from burning heavy fuel oil except for particulates (Czernik and Bridgwater, 2004). Pyrolysis bio-oil has also been used in diesel engines. Without blending with other fuel additives, the use of pyrolysis bio-oil should be limited to low-speed diesel engines with relatively high compression ratios. The blend of pyrolysis oil and methanol can be used in high-speed diesel engines (Czernik and Bridgwater, 2004).

18.5.2 Chemicals Extracted from Bio-Oil

Phenols in bio-oil can be extracted to produce resins. Volatile organic acids can be used to make deicers, levoglucosan, hydroxyacetaldehyde, and some additives used in pharmaceutical, biber, and fertilizer industries (Bridgwater, 1999; Bridgwater and Peacocke, 2000). Chemicals that have been produced from bio-oil include polyphenols for resins, fertilizer, and emission control agents, magnesium acetate for biodegradable chemicals, a range of flavors for the food industry, and asphalt for transportation and civil industry. The challenges for the extraction of chemicals from bio-oil include the lack of low cost and efficient separation and refining techniques, and well-established markets for the less commonly used chemicals from bio-oil. Currently the only mature market for the chemicals produced from bio-oil is food flavors.

18.5.3 Steam Reforming of Bio-Oil into Hydrogen

Steam reforming reaction can be used to produce hydrogen from bio-oil (Wu et al., 2008). Steam reforming reactions usually take place at a temperature of 600°C–800°C with high space velocities and a Ni catalyst. Medrano et al. (2011) used magnesium and calcium to promote the Ni–Al catalyst for steam reforming of bio-oil in a fluidized bed reactor. They found that the magnesium-modified catalyst had the best performance for the highest carbon conversion to gas and hydrogen yield. The calcium-modified catalyst shows a poorer performance with lower H_2 yield and higher CO yield. It was reported that about 6 kg of hydrogen was produced from 100 kg of wood via the pyrolysis–steam reforming process (Czernik et al., 2002). Research showed that biomass-derived oxygenated hydrocarbons could be selectively converted into hydrogen fuel and methane by supported metal catalysts that are made from group VIII metals such as Pt and Ni (Huber and Dumesic, 2006). The catalysts supported on activated carbon, TiO_2, and Al_2O_3 were found to be very effective in cleaving C–C bonds as well as C–H and O–H bonds, resulting in production of H_2 and CO_2.

18.6 LIFE CYCLE ASSESSMENT AND ECONOMIC ANALYSIS

18.6.1 Energy Efficiency

The heat required to raise the temperature of biomass to 500°C and decompose it during fast pyrolysis was measured at average value of 1.5 MJ/kg of dry biomass. Additional 0.375 and 0.6 MJ/kg feed of heat was required to remove moisture at every 10% of total mass and increase the temperature of the purging gas (Daugaard and Brown, 2003). Therefore, without heat loss, it requires about a total amount of heat at 2 MJ/kg of dry biomass to pyrolyze the biomass with 10% moisture. As the heating value of dry biomass is usually around 15–20 MJ/kg, it thus requires about 10%–15% of the energy stored in the biomass for the pryolysis process.

The wet bio-oil yields are in the range of 60%–79% of the dry biomass, while the dry bio-oil yields are in the range of 41%–65% because the bio-oil usually contains 17%–35% water. If the heating values of the dry bio-oil and biomass are 23 and 19 MJ/kg, respectively, the energy yield of bio-oil is 51%–77% of the energy stored in the biomass (Rogers and Brammer, 2012). The by-products of char and syngas typically contain about 25% and 5% of the energy in the biomass feedstocks, respectively (Rogers and Brammer, 2012). The gases and char generated in biomass pyrolysis can be used to supply the heat. The electricity is used to supply purging gas, mill biomass, and convey biomass. The total electrical consumption for bubbling fluidized bed pyrolysis was estimated at 240 kWh/dry ton of biomass (Rogers and Brammer, 2012). It was estimated that about 18% of the bio-oil was required to be used in a diesel engine at an efficiency of 40% to supply the amount of electricity (Rogers and Brammer, 2012). The energy and carbon conversion efficiencies of fast pyrolysis coupled with upgrading technologies were as low as 30%–40% (Butler et al., 2011).

18.6.2 Cost of Biomass Pyrolysis

Economic evaluation of bio-oil production is very critical, because its production cost decides if bio-oil price is competitive and easy to popularize in the biofuel market. Bio-oil production cost typically includes raw biomass cost, capital costs of a biomass preprocessing and fast pyrolysis plant, labor cost, and cost of electricity utilization. The estimated capital cost is a function of plant size. Bridgwater (2012) developed the following two equations to determine the capital and productions cost for a fast pyrolysis plant. The capital cost was estimated by the power equation, which was given by

$$C_i = 6.98 \times F^{0.67} \tag{18.3}$$

where

C_i is the capital cost in million euros in 2011

F is the biomass feed rate in dry t/h

The production cost was estimated by

$$C_p = 1.1 \times \left[C_b + \left(H \times 16{,}935 \times F^{-0.33} \right) \right] Y^{-1} \tag{18.4}$$

where

C_p is the production cost of bio-oil in euros/t in 2011
C_b is the biomass cost, euro dry/t
H is the capital and capital-related charges and the default value is 0.18
F is the biomass feed rate in dry t/year
Y is the fractional bio-oil yield and the default value for wood and grasses are 0.75 and 0.6, respectively

Equation 18.4 was used to estimate the production costs of bio-oil in a fast pyrolysis process as a function of production scale and biomass supply costs, which were given in Figure 18.3 (Bridgwater, 2012).

For the biofuels produced through fast pyrolysis followed by hydrotreating and hydrocracking, a different study showed that the minimum selling prices are \$0.51/L–\$0.53/L of fuel in 2007 dollar value (Jones et al., 2009; Anex et al., 2010; Wright et al., 2010).

Roger et al. (2012) conducted the economic analysis of fast pyrolysis bio-oil by taking account of the sales of biochar and electricity consumption of the plant. They found that char sales could reduce the production cost of bio-oil by up to 18%. They concluded that it should be able to economically produce bio-oil at relatively large plants if the bio-oil could be sold at a price between distillate fuels and residual fuel oil. However, the breakeven price cannot be simply considered to be equal to the real market price because potential operational risks and certain amount of profit need to be calculated and covered. Those risks may involve the increases of electricity cost, labor cost, raw biomass cost, and interest rate and reduction of bio-oil yield and capacity factor. Gebreslassie et al. (2013) optimized the design and operation of an integrated process including fast pyrolysis, hydrotreating, and hydrocracking to refine biomass to hydrocarbon fuels under both economic and environmental criteria. They found that the fuel production cost was \$0.61/L

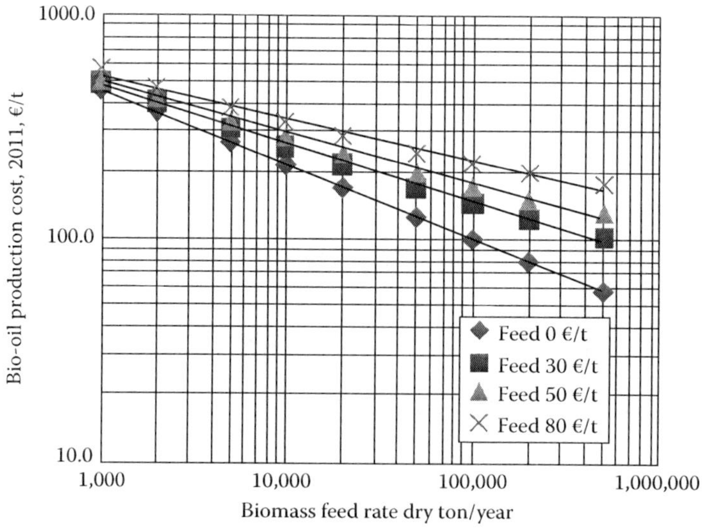

FIGURE 18.3 Bio-oil production costs affected by production scale and feedstock price. (Adapted from Bridgwater, A.V., *Biomass Bioenergy*, 38, 68, 2012.)

of gasoline equivalent for the maximum net present value and \$0.97/L for the minimum global warming potential design.

18.6.3 Life Cycle Assessment

Life cycle assessment (LCA) of a product usually includes the analyses of the energy and materials flow and environmental impact of a product from being produced from raw material, through transportation, handling, to being disposed. LCA becomes very important in the current and future economy because the limited natural resources are being depleted, environment is facing a huge challenge from manufacturing of human products, and both renewable and environment-friendly products are desirable and competitive in the market.

Cumulative energy demand, global warming, ozone layer depletion, photochemical oxidant formation, land competition, acidification, and eutrophication were evaluated for the production and upgrading of bio-oil from poplar in a circulating fluidized bed reactor (Iribarren et al., 2012). It was found that poplar pretreatment, pyrolysis process, and hydrogen production for bio-oil upgrading had high negative environmental impact.

An LCA for the production of gasoline and diesel via biomass pyrolysis at a load of 2000 dry ton/day and upgrading assumes that grid electricity is used and supplemental natural gas is supplied to the hydrogen plant (Hsu, 2012). Gasoline and diesel produced from pyrolysis are estimated to have GHG emissions of CO_2 equivalent of 117 and 98 g/km, respectively. The GHG emissions of CO_2 equivalent for the gasoline produced from pyrolysis are 65% lower than that of conventional gasoline at 300 g/km in 2005. The net energy values (NEVs) of gasoline and diesel produced from pyrolysis are 1.09 and 0.92 MJ/km, respectively, compared to conventional of −1.2 MJ/km in 2005. However, gasoline produced from pyrolysis does not perform as well as ethanol produced by the same feedstock from gasification. Grid electricity and natural gas used account for 37% and 51% of the net GHG emissions in the base case. Iribarren et al. (2012) and Zhong et al. (2010) also found that hydrogen and electricity input have significant impacts on GHG emission for the biofuels produced from biomass pyrolysis. The power and hydrogen should be better generated from biomass resources rather than fossil fuels.

18.7 CONCLUSIONS

Raw biomass is not appropriate to be used as a fuel directly in existing equipment since it has high moisture content and low bulk density. Biomass can be converted into bio-oil in the absence of air. Fast pyrolysis at a temperature of 450°C–550°C, high heating rate of 10^3°C/s–10^4°C/s, and short gas residence time can achieve a high bio-oil yield. Bio-oil produced from thermochemical conversion of biomass can be used as an alternative fuel in engines, boilers, and electricity generation. However, crude bio-oil is not appropriate to be used as a direct fuel because of its unfavorable physical and chemical properties such as high viscosity, thermal instability, and corrosiveness. Different upgrading techniques including esterification, catalytic cracking, hydrogenation, emulsification, and reactive distillation can be used to improve both chemical and physical properties of bio-oil. Crude bio-oil can be further catalytically upgraded and distilled into gasoline and diesel similar products. Crude bio-oil can also be catalytically reformed into hydrogen. The challenging issues related to catalytic upgrading include economic catalyst regeneration, prevention of catalyst clog, and development of continuous system.

The use of hydrogen and electricity for upgrading crude bio-oil has a high negative environmental impact. To minimize the environmental impact, hydrogen that is used during bio-oil upgrading can be produced from the fast pyrolysis process and electricity can be generated from biomass resources. A large portion of the bio-oil production costs come from the purchase of raw biomass and electricity. For a relatively large plant size, the prices of bio-oil-based fuels might be as competitive as that of diesel fuel or gasoline. However, increasing processing capabilities may also impose challenge for the supply of raw biomass.

ACKNOWLEDGMENTS

This chapter was partially supported by two funds provided by the U.S. Department of Agriculture-National Institute of Food and Agriculture (grant number: USDA NIFA 2010-38821-21512 and NC.X-269-5-12-130-1). Its contents are solely the responsibility of the authors and do not necessarily represent the official views of the funding agency. Mention of a trade name, a company name, or proprietary products is for presentation clarity and does not imply endorsement by the authors or the university.

REFERENCES

Abdullah N. and Gerhauser H. 2008. Bio-oil derived from empty fruit bunches. *Fuel* 87: 2606–2613.

Adjaye J.D. and Bakhshi N.N. 1995a. Production of hydrocarbons by catalytic upgrading of a fast pyrolysis bio-oil. Part II: Comparative catalyst performance and reaction pathways. *Fuel Processing Technology* 45(3): 185–202.

Adjaye J.D. and Bakhshi N.N. 1995b. Production of hydrocarbons by catalytic upgrading of a fast pyrolysis bio-oil. Part I: Conversion over various catalysts. *Fuel Processing Technology* 45(3): 161–183.

Alcala A. and Bridgwater A.V. 2013. Upgrading fast pyrolysis liquids: Blends of biodiesel and pyrolysis oil. *Fuel* 109: 417–426.

Anex R.P., Aden A., Kazi F.K. et al. 2010.Techno-economic comparison of biomass-to-transportation fuels via pyrolysis, gasification, and biochemical pathways. *Fuel* 89: S29–S35.

Avami A., Marquardt W., Saboohi Y., and Kraemer K. 2012. Shortcut design of reactive distillation columns. *Chemical Engineering Science* 71: 166–177.

Boucher M.E., Chaala A., Pakdel H., and Roy C. 2000. Bio-oils obtained by vacuum pyrolysis of softwood bark as a liquid fuel for gas turbines. Part II: Stability and ageing of bio-oil and its blends with methanol and a pyrolytic aqueous phase. *Biomass and Bioenergy* 19: 351–361.

Bridgwater A.V. 1999. Principles and practices of biomass fast pyrolysis processes for liquids. *Journal of Analytical and Applied Pyrolysis* 51: 3–22.

Bridgwater A.V. 2012. Review of fast pyrolysis of biomass and product upgrading. *Biomass and Bioenergy* 38: 68–94.

Bridgwater A.V., Meier D., and Radlein D. 1999. An overview of fast pyrolysis of biomass. *Organic Geochemistry* 30: 1479–1493.

Bridgwater A.V. and Peacocke G.V.C. 2000. Fast pyrolysis processes for biomass. *Sustain Renew Energy Reviews* 4: 1–73.

Burhenne L., Messmer J., Aicher T., and Laborie M.P. 2013. The effect of the biomass components lignin, cellulose and hemicelluloses on TGA and fixed bed pyrolysis. *Journal of Analytical and Applied Pyrolysis* 101: 177–184.

Butler E., Devlin G., Meier D., and McDonnell K. 2011. A review of recent laboratory research and commercial developments in fast pyrolysis and upgrading. *Renewable and Sustainable Energy Reviews* 15: 4171–4181.

Calvar N., Gonzalez B., and Dominguez A. 2006. Esterification of acetic acid with ethanol: Reaction kinetics and operation in a packed bed reactive distillation column. *Chemical Engineering and Processing* 46: 1317–1323.

Carlson T., Tompsett G., Sauvanaud L., and O'Connor P. 2009. Aromatic production from catalytic fast pyrolysis of biomass-derived feedstocks. *Topics in Catalysis* 52: 241–252.

Chaala A., Ba Y., Perez M.G., and Roy C. 2004. Colloidal properties of bio-oils obtained by vacuum pyrolysis of softwood bark: Ageing and thermal stability. *Energy Fuel* 18: 1135–1142.

Czernik S. and Bridgwater A.V. 2004. Overview of application of biomass fast pyrolysis oil. *Energy Fuels* 18: 590–598.

Czernik S., French R., Feik C. et al. 2002. Hydrogen by catalytic steam reforming of liquid byproducts from biomass thermoconversion process. *Industrial & Engineering Chemistry Research* 41: 4209–4215.

Daugaard D.E. and Brown R.C. 2003. Enthalpy for pyrolysis for several types of biomass. *Energy and Fuels* 17: 934–939.

Demirbas A. 2000. Mechanisms of liquefaction and pyrolysis reactions of biomass. *Energy Conversion & Management* 41: 633–646.

Diebold J.P. 1999. A review of the chemical and physical mechanisms of the storage stability of fast pyrolysis bio-oils. NREL/SR-570-27613. Subcontractor report.

Fisher T., Hajaligol M., Waymack B., and Kellogg D. 2002. Pyrolysis behavior and kinetics of biomass derived materials. *Journal of Analytical and Applied Pyrolysis* 62: 331–349.

Fisk C.A., Morgan T., Ji Y. et al. 2009. Bio-oil upgrading over platinum catalysts using in situ generated hydrogen. *Applied Catalysis A: General* 358: 150–156.

Fratini E., Bonini M., Oasmaa A. et al. 2006. SANA analysis of the microstructural evolution during the ageing of pyrolysis oils from biomass. *Langmuir* 22: 306–312.

French R. and Czernik S. 2010. Catalytic pyrolysis of biomass for biofuels production. *Fuel Processing Technology* 91: 25–32.

Garcia-Perez M., Shen J., Wang X.S., and Li C.Z. 2010. Production and fuel properties of fast pyrolysis oil/bio-diesel blends. *Fuel Processing Technology* 91: 296–305.

Gebreslassie B.H., Slivinsky M., Wang B., and You F. 2013. Life cycle optimization for sustainable design and operations of hydrocarbon biorefinery via fast pyrolysis, hydrotreating and hydrocracking. *Computers and Chemical Engineering* 50: 71–91.

Gronli M.G., Varhegyi G., and Di Blasi C. 2002. Thermogravimetric analysis and devolatilization kinetics of wood. *Industrial and Engineering Chemistry Research* 41: 4201–4208.

Haas M.J., McAloon A.J., Yee W.C., and Foglia T.A. 2006. A process model to estimate biodiesel production costs. *Bioresource Technology* 97: 671–678.

He R., Ye X.P., English B.C., and Satrio J.A. 2009. Influence of pyrolysis condition on switchgrass bio-oil yield and physicochemical properties. *Bioresource Technology* 100: 5305–5311.

Hew K.L., Tamidi A.M., Yusup S., Lee K.T., and Ahmad M.M. 2010. Catalytic cracking of bio-oil to organic liquid product (OLP). *Bioresource Technology* 101: 8855–8858.

Hodgson E.M., Fahmi R., Yates N., Barraclough T. et al. 2010. Miscanthus as a feedstock for fast-pyrolysis: Does agronomic treatment affect quality? *Bioresource Technology* 101: 6185–6191.

Hsu D.D. 2012. Life cycle assessment of gasoline and diesel produced via fast pyrolysis and hydroprocessing. *Biomass and Bioenergy* 45: 41–47.

Huber G.W. and Dumesic J.A. 2006. An overview of aqueous-phase catalytic processes for production of hydrogen and alkanes in a biorefinery. *Catalysis Today* 111: 119–132.

Ikura M., Stanciulescu M., and Hogan E. 2003. Emulsification of pyrolysis of pyrolysis derived bio-oil in diesel fuel. *Biomass and Bioenergy* 24: 221–232.

Iribarren D., Peters J.F., and Dufour J. 2012. Life cycle assessment of transportation fuels from biomass pyrolysis. *Fuel* 97: 812–821.

Jackson M.A., Compton D.L., and Boateng A.A. 2009. Screening heterogeneous catalysts for the pyrolysis of lignin. *Journal of Analytical and Applied Pyrolysis* 85: 226–230.

Jones S.B., Valkenburg C., Walton C. et al. 2009. Production of gasoline and diesel from biomass via fast pyrolysis, hydrotreating and hydrocracking: A design case. Richland, WA: Pacific Northwest National Laboratory, 76pp. Report No.: PNNL-18284 Rev. 1.

Kim S. and Dale B. 2005. Environmental aspects of ethanol derived from no-tilled corn grain: Nonrenewable energy consumption and greenhouse gas emissions. *Biomass and Bioenergy* 28: 475–489.

Kumar A., Wang L.J., Dzenis Y.A., Jones D.D., and Hanna M.A. 2008. Thermogravimetric characterization of corn stover as gasification and pyrolysis feedstock. *Biomass and Bioenergy* 32: 460–467.

Lede J. 2003. Comparison of contact and radiant ablative pyrolysis of biomass. *Journal of Analytical and Applied Pyrolysis* 70: 601–618.

Li W., Pan C., Zhang Q. et al. 2011. Upgrading of low-boiling fraction of bio-oil in supercritical methanol and reaction network. *Bioresource Technology* 102: 4884–4889.

Lu Q., Li W.Z., and Zhu X.F. 2009. Overview of fuel properties of biomass fast pyrolysis oils. *Energy Conversion and Management* 50: 1376–1383.

Lu Q., Zhang Y., Tang Z., Li W., and Zhu X. 2010. Catalytic upgrading of biomass fast pyrolysis vapors with titania and zirconia/titania based catalysts. *Fuel* 89: 2096–2103.

Mahfud F.H., Melian-Cabrera I., Manurung R., and Heeres H.J. 2007. Biomass to fuels: Upgrading of flash pyrolysis oil by reactive distillation using a high boiling alcohol and acid catalysts. *Process Safety and Environmental Protection* 85: 466–472.

Majhi A., Sharma Y.K., and Naik D.V. 2012. Blending optimization of Hempel distilled bio-oil with commercial diesel. *Fuel* 96: 264–269.

McKinley J. 1989. Biomass liquefaction: Centralised analysis, Final Report., DSS File No. 23216-4-6192, Energy, Mines and Resources Ministry: Ottawa, Ontario, Canada.

Medrano J.A., Oliva M., Ruiz J., Garcia L., and Arauzo J. 2011. Hydrogen from aqueous fraction of biomass pyrolysis liquids by catalytic steam reforming in fluidized bed. *Energy* 36: 2215–2224.

Nokkosmaki M.I., Kuoppala E.T., Leppamaki E.A., and Krause A.O.I. 2000. Catalytic conversion of biomass pyrolysis vapors with zinc oxide. *Journal of Analytical and Applied Pyrolysis* 55: 119–131.

Nowakowski D.J., Bridgwater A.V., Elliott D.C., Meier D., and de Wild P. 2010. Lignin fast pyrolysis: Results from an international collaboration. *Journal of Analytical and Applied Pyrolysis* 88: 53–72.

Oasmaa A. and Czernik S. 1999. Fuel oil quality of biomass pyrolysis oils-state of the art for the end-users. *Energy and Fuels* 13: 914–921.

Oasmaa A. and Kuoppala E. 2003. Fast pyrolysis of forestry residue. 3. Storage stability of liquid fuels. *Energy Fuel* 17: 1075–1084.

Ott L.S., Smith B.L., and Bruno T.J. 2008. Advanced distillation curve measurement: Application to a bio-derived crude oil prepared from swine manure. *Fuel* 87: 3379–3387.

Pattiya A. and Suttibak S. 2012. Production of bio-oil via fast pyrolysis of agricultural residues from cassava plantations in a fluidised-bed reactor with a hot vapour filtration unit. *Journal of Analytical and Applied Pyrolysis* 95: 227–235.

Peng J., Chen P., Lou H., and Zheng X.M. 2009. Catalytic upgrading of bio-oil by HZSM-5 in sub- and super-critical ethanol. *Bioresource Technology* 100: 3415–3418.

Qian Y.J., Zuo C.J., Tan J., and He J. 2007. Structural analysis of bio-oils from sub-and supercritical water liquefaction of woody biomass. *Energy* 32(2007): 196–202.

Rocha J.D., Luengo C.A., and Snape C.E. 1999. The scope for generating bio-oils with relatively low oxygen contents via hydropyrolysis. *Organic Geochemistry* 30: 1527–1534.

Rogers J.G. and Brammer J.G. 2012. Estimation of the production cost of fast pyrolysis bio-oil. *Biomass and Bioenergy* 36: 208–217.

Seyedeyn-Azad F., Salehi E., Abedi J., and Harding T. 2011. Biomass to hydrogen via catalytic steam reforming of bio-oil over Ni-supported alumina catalysts. *Fuel Processing Technology* 92(3): 563–569.

Shafizadeh F. 1982. Introduction to pyrolysis of biomass. *Journal of Analytical and Applied Pyrolysis* 3: 283–305.

Shen D.K., Gu S., and Bridgwater A.V. 2010. The thermal performance of the polysaccharides extracted from hardwood: Cellulose and hemicellulose. *Carbohydrate Polymers* 82: 39–45.

Sipilae K., Kuoppala E., and Fagernaes L. 1998. Characterization of biomass-based flash pyrolysis oils. *Biomass and Bioenergy* 14: 103–113.

Tang Z., Lu Q., Zhang Y., Zhu X., and Guo Q. 2009. One step bio-oil upgrading through hydrotreatment, esterification, and cracking. *Industrial and Engineering Chemistry Research* 48: 6923–6929.

Toft A.J. 1996. A comparison of integrated biomass to electricity systems. PhD thesis, Aston University, Birmingham, U.K.

Torri C., Reinikainen M., Lindfors C., Fabbri O., Oasmaa A., and Kuoppala E. 2010. Investigation on catalytic pyrolysis of pine sawdust: Catalyst screening by Py-GC-MIP-AED. *Journal of Analytical and Applied Pyrolysis* 88: 7–13.

Vitolo S., Seggiani M., Frediani P., Ambrosini G., and Politi L. 1999. Catalytic upgrading of pyrolytic oils to fuel over different zeolites. *Fuel* 78: 1147–1159.

Wang L.J., Kumar A., Weller C.L., Hanna M.A., and Jones D.D. 2009a. Thermal degradation kinetics of distillers grains in nitrogen and air. *Energy Sources, Part A: Recovery, Utilization, and Environmental Effects* 31: 797–806.

Wang L.J., Shahbazi A., and Hanna M.A. 2011. Characterization of corn stover, distiller grains and cattle manure for thermochemical conversion. *Biomass Bioenergy* 35: 171–178.

Wang S., Gu Y.L., Liu Q. et al. 2009b. Separation of bio-oil by molecular distillation. *Fuel Processing Technology* 90: 738–745.

Wang S.R., Luo Z.Y., Tan H. et al. 2004. The analyses of characteristics of bio-oil produced from biomass flash pyrolysis. *Journal of Engineering Thermophysics* 25: 1049–1052.

Wang Y., He T., Liu K., Wu J., and Fang Y. 2012. From biomass to advanced bio-fuel by catalytic pyrolysis/hydro-processing: Hydrodeoxygenation of bio-oil derived from biomass catalytic pyrolysis. *Bioresource Technology* 108: 280–284.

Wegener G. and Fengel D. 1978. Rapid extraction of lignin from ball-milled spruce wood by the use of ultrasonics. *Ultrasonics* 16: 186.

Wildschut J., Arentz J., Rasrendra C.B., Venderbosch R.H., and Heeres H.J. 2009. Catalytic hydrotreatment of fast pyrolysis oil: Model studies on reaction pathways for the carbohydrate fraction. *Environmental Progress and Sustainable Energy* 28: 450–460.

Williams P.T. and Nugranad N. 2000. Comparison of products from the pyrolysis and catalytic pyrolysis of rice husks. *Energy* 25: 493–513.

Wright M.M., Daugaard D.E., Satrio J.A., and Brown R.C. 2010. Technoeconomic analysis of biomass fast pyrolysis to transportation fuels. *Fuel* 89(Suppl 1): S2–S10.

Wu C., Huang Q., Sui M., Yan Y., and Wang F. 2008. Hydrogen production via catalytic steam reforming of fast pyrolysis bio-oil in a two-stage fixed bed reactor system. *Fuel Processing Technology* 89: 1306–1316.

Xu J.M., Jiang J.C., Sun Y.J., and Lu. Y.J. 2008. Bio-oil upgrading by means of ethyl ester production in reactive distillation to remove water and to improve storage and fuel characteristics. *Biomass and Bioenergy* 32: 1056–1061.

Yang H., Yan R., Chen H., Zheng C., Lee D.H., and Liang D.T. 2006. In-depth investigation of biomass pyrolysis based on three major components: Hemicelluloses, cellulose and lignin. *Energy & Fuels* 20: 388–393.

Zhang Q., Chang J., Wang T., and Xu Y. 2006. Upgrading bio-oil over different solid catalysts. *Energy & Fuels* 20: 2717–2720.

Zhang Q., Chang J., Wang T., and Xu Y. 2007. Review of biomass pyrolysis oil properties and upgrading research. *Energy Conversion and Management* 48: 87–92.

Zhang S.P., Yang, Y., Li, T. et al. 2005. Upgrading of liquid fuel from the pyrolysis of biomass. *Bioresource Technology* 96: 545–550.

Zheng J.L. and Kong Y.P. 2010. Spray combustion properties of fast pyrolysis bio-oil produced from rice husk. *Energy Conversion and Management* 51: 182–188.

Zheng J.L. and Wei Q. 2011. Improving the quality of fast pyrolysis bio-oil by reduced pressure distillation. *Biomass and Bioenergy* 35: 1804–1810.

Zheng J.L., Yi W.M., and Wang N.N. 2008. Bio-oil production from cotton stalk. *Energy Conversion and Management* 49: 1724–1730.

Zhong Z.W., Song B., and Zaki M.B.M. 2010. Life-cycle assessment of flash pyrolysis of wood waste. *Journal of Cleaner Production* 18: 1177–1183.

19 Hydrothermal Liquefaction of Biomass

Nor Aishah Saidina Amin, Javaid Akhtar, and Selhan Karagoz

CONTENTS

19.1 INTRODUCTION

Biomass is commonly referred to any plant-derived organic substance. Wood, energy crops, agricultural crops, and aquatic flora can be considered as an important resource of biomass feedstocks. Wood wastes in the form of sawdust, agricultural residue, pulp process wastes, and other wastes from wood processing industries also produce significant amount of waste biomasses. Cellulose, hemicelluloses, and lignin are the major components of biomass. Both cellulose and lignin are the most abundant renewable carbon resource on earth. Major buildup of cellulose is glucose units, which has a degree of polymerization of ~500–100,000. Lignin is a complex copolymer composed of phenyl propane units containing mainly guaiacyl propane and hydroxyphenylpropane monomers. Hemicellulose comprises of glucose, mannose, and xylose monomers and has a lower polymerization than cellulose (Kelly-Yong et al. 2007; Goyal et al. 2008). Plant biomass is a chemical energy buildup produced by the conversion of water and CO_2 into carbohydrates using sunlight during a photosynthetic process.

Petroleum and its allied petrochemicals have served the world as the cheapest source of energy over the years, and this has encouraged the development of relevant industries for millions of petroleum-based products. However, the long-term supply of fossil fuel has been questioned due to ever-increasing demand all over the globe. The increasing global demand for oil and exhaustion of available crude supply has increased the prices of petroleum in recent years. The uncertainties in continuous supply of energy resources have opened doors for alternative renewable energy. Soaring fuel demand and depletion of fossil reserves have demanded the substitution of energy resources supply with greener and renewable sources.

Alternative energy sources such as biomass, nuclear, and solar have attracted considerable attention from researchers around the world. The degree of success has been on the course for harnessing these renewable energy sources from nature. The role of biomass-based renewable energy is crucial for the supply of energy to transportation sector and for power generation. Synthesis of biomass-based chemical products emphasizes the broad application of biomass other than conventional combustion process and highlights it as a replacement for depleting fossil fuel. The utilization of biomass for various applications has the potential to enhance social welfare of developing regions in the world. Biomass is also regarded as a greenhouse gas reducer from the environmental aspect. The alternative to existing fossil fuels with lower greenhouse gases emission, renewable source, social welfare, and NO_2 and CO_2 reduction is a major advantage for utilizing biomass as a fuel source. Plants help in the reduction of greenhouse gas during photosynthesis. The concept of zero emissions is plausible by maintaining a balance between production and utilization of plant biomass. Greenhouse gas emission is the lowest for biomass at 17–27 g/kWh when compared with coal at 995 g/kWh, natural gas at 446 g/kWh, and furnace oil at 818 g/kWh. Better technological development may result in greenhouse gas reduction to about 15 g/kWh from biomass (Bauen 2004), which highlights its green nature. The conversion of biomass to fuel oil and other related chemicals may reduce the oil import bill and uphold the economic growth in areas that constitute abundant supply of biomass such as Indonesia, Malaysia, and Thailand. The conversion of such biomass resources into chemicals and/or transportation fuels can distribute the loads of these products from conventional sources to achieve cheap transportation fuels, electricity, and other biochemicals such as bioplastics, cosmetics, and biopolyesters. Relevant technological development is also necessary to propel the research and social activities for biomass applications and reduce the reliance on conventional energy.

Pyrolysis is the major technique for the production of bio-oil from biomass. However, the presence of water in biomass has a negative impact on overall pyrolysis yield, since vaporization of water requires excessive heat supply. This disturbs the overall economy of the operation and restricts the applications of biomass for bio-oil production. Although the biomass with moisture content up to 40 wt% is acceptable as a feedstock for pyrolysis, the moisture content of biomass below 10 wt% is highly recommended. However, the water content of plant wood is normally higher than 50 wt% and even up to 80 wt% for tropical grasses and 90 wt% for aquatic plants. The preprocessing of wet biomass is necessary for pyrolysis applications. The drying of wet biomass can be achieved by mechanical, solar, or atmospheric drying (Heinz et al. 2001). The drying of biomass using solar radiation is cheap but at the expense of longer time duration. On the other hand, the mechanical drying of biomass is costly and requires a separate preprocessing unit in pyrolysis industry.

Hydrothermal liquefaction provides an alternative method for producing bio-oil from biomass with high moisture content. The usual operating conditions for hydrothermal liquefaction involve temperature between 280°C and 370°C and pressure within 10–25 MPa (Behrendt et al. 2008). Water exits in a liquid state under hydrothermal operating conditions. The reactions among biomass and solvent components are complicated and depend upon many factors including biomass type. The major product of hydrothermal liquefaction is a biocrude that contains significant portion of water-soluble and high-calorific-value components. The addition of alkalis suppresses the formation of char residue and thus enhances the oil yield (Toor et al. 2011). There are a number of advantages to liquefy biomass under hydrothermal conditions, which include high energy and separation efficiencies, the ability to use mixed feedstock, and the suitability to utilize biomasses with high moisture content (Peterson et al. 2008). Due to high reactivity of a hydrothermal solvent, the fast decomposition of biomass occurs without the need of a drying step. In addition, hydrothermal liquefaction of biomass requires relatively lower temperature than that of the pyrolysis process. Any kind of biomass including grasses, aquatic or tropical plants, and wood crops are liquefiable efficiently via hydrothermal processing as the high solvation power of a hydrothermal fluid can extract the bio-oil or chemicals from biomass. The fluid under high temperature and pressure achieves high solvent density, efficient mass transfer capabilities, and faster extraction rate during a hydrothermal process. However, corrosion may cause major hindrance during catalytic sub-/supercritical fluid operations. The high pressure and temperature coupled with corrosion problems require expensive alloys for the fabrication of reactors and process lines (Toor et al. 2011), which consequently increases the investment for the operation. Major research conducted so far is on the laboratory or bench scale.

This chapter examines the mechanism of biomass degradation. The decomposition reactions and resulting fragmentations from woody biomass are then discussed. The effects of major operating conditions including temperature, particle size, biomass feedstock, biomass heating rates, pressure, residence time, reducing agents, and biomass to liquid ratio on optimum bio-oil yield are evaluated. Industrial applications, efficiency, and economics of the hydrothermal process are also discussed in this chapter.

19.2 DECOMPOSITION MECHANISM

19.2.1 Degradation Mechanism in Hydrothermal Liquefaction

The woody biomass is comprised of 40%–50% of oxygen content on ash-free basis. The basic objective of liquefaction is to reduce the oxygen content and increase carbon and hydrogen contents of biomass fragments. Two major reactions, that is, decarboxylation and dehydration, reduce oxygen contents of biomass by converting oxygen heteroatom into CO_2 or H_2O, respectively. Dehydration occurs when hydrogen at high temperature attacks oxygen at nucleophilic sites and converts oxygen into H_2O. Decarboxylation can reduce the oxygen content in biomass by forming CO_2 through thermal cracking of carboxylic acid groups in biomass. As the oxygen in biomass is fully oxidized into H_2O and CO_2, dehydration and decarboxylation reactions are the effective methods to decrease oxygen in bio-oil. The biomass is converted into pure carbon-like charcoal by dehydration reaction, while decarboxylation reaction increases H/C ratio of bio-oil. Two reactions of dehydration and decarboxylation are given in Figure 19.1a and b.

The hydrothermal decomposition of biomass relies mainly on depolymerization and solvolysis for the production of bio-oil (Behrendt et al. 2008). During solvolysis, biomass components are decomposed via nucleophilic substitution or the fragmented species are stabilized. The stabilization of fragmented components helped to reduce the amount of char or coke (Jakab et al. 1997).

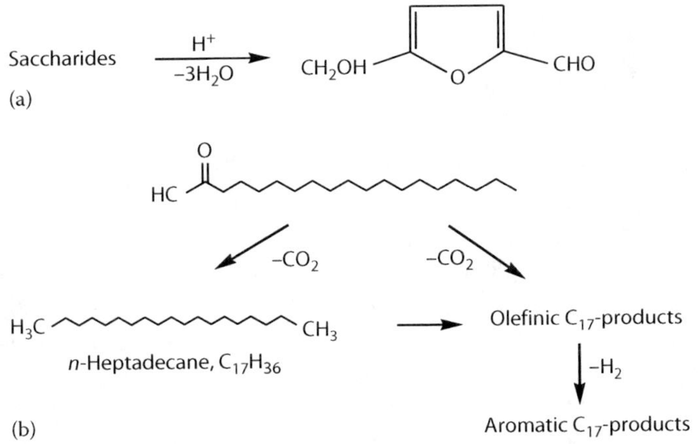

FIGURE 19.1 (a) Dehydration of saccharides to 5-hydroxymethylfurfural (From Asghari, F.S. and Yoshida, H., *Carbohydr. Res.*, 341, 2379, 2006) and (b) simplified decarboxylation network of stearic acid to diesel products. (From Snåre, M. et al., *Chem. Eng. J.*, 134, 29, 2007.)

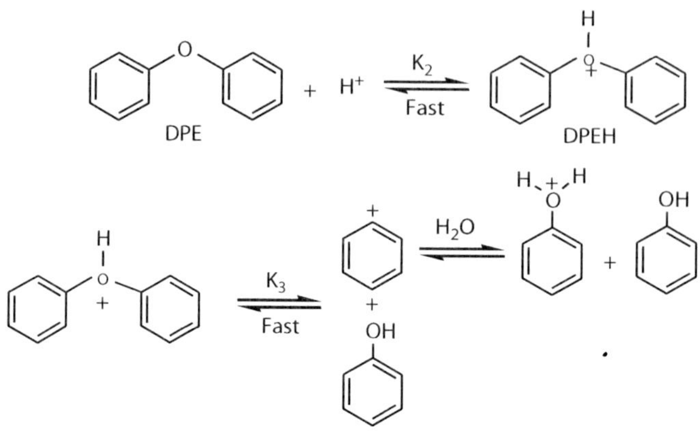

FIGURE 19.2 Mechanism for supercritical water hydrolysis of DPE. (From Penninger, J.M.L. et al., *J. Supercrit. Fluids*, 16, 119, 1999.)

Hydrolysis is the solvolysis when water is used as the solvent during hydrothermal liquefaction. The thermal cracking of biomass is plausible due to high temperature in hydrolysis. Hot water breaks the biomass components at heteroatoms sites, and the fragments are hydrolyzed subsequently. Acid and thermal hydrolyses are different from each other.

The focus here is to study the mechanism for thermal hydrolysis. Different researchers have studied the reaction pathways for biomass liquefaction. Sasaki et al. (2000) hydrolyzed the cellulose biomass at 320°C, 350°C, and 400°C under a pressure of 25 MPa. At near-supercritical conditions, cellulose was hydrolyzed into cellotriose-, cellohexaose-, cellobiose-, cellopentaose-, fructose-, and glucose-type components. However, long residence times were found to cause the decrease in the overall yield of hydrolyzed products and variation in composition of liquid components. This reflected the influence of both temperature and time on overall yield of the process. Figure 19.2 discusses the SN_1 and SN_2 mechanism for hydrolysis of diphenylether (DPE) in supercritical conditions. It is also true that hydrolysis and thermal decomposition occur side by side in hydrothermal processes (Minowa et al. 1998). Initially, the biomass is hydrolyzed and later followed by thermal decomposition of biomass components. The products from the decomposition are thus stabilized by hydrogen donor species like tetralin and other oxygenates compounds. Tetralin-type donors enhance the yield of bio-oil. The primary products are produced as a result of hydrolysis and further decomposed to secondary products via thermal decomposition. Condensation, isomerization, and dehydration are major secondary reactions. Further decomposition of secondary products produces components such as aldehydes, acids, and phenolic compounds that are collectively known as biocrude.

19.2.2 Degradation Mechanism for Cellulose in Supercritical Water

Cellulose is studied as a model biomass compound to explore about the liquefaction yield and product composition in hydrothermal processes (Schwald and Bobleter 1989; Jakab et al. 1997; Minowa and Fang 1998; Sasaki et al. 1998). Linear chains of glucose monomers form the building blocks of cellulose structure. The monomers in cellulose structure are connected by means of β-(1–4) D-glucose units. The crystallinity in the structure of cellulose is influenced by β-(1–4) D-glucose linkages that are resistant to enzymatic or acid hydrolysis (Puri 1984). The liquefaction in supercritical water has the ability to break the crystalline nature of cellulose by decomposing the hydrogen bonding of intramolecules. The mechanism for the decomposition of cellobiose, a disaccharide bonded by β-(1–4), in supercritical water is investigated by Kabyemela et al. (1998). They found that the hydrolysis and thermal decomposition of cellobiose occurred simultaneously. Two major intermediates, glucose and fructose, were further depolymerized to glucosyl-glycolaldehyde and glucosyl-erythrose intermediates. These intermediates were then decomposed to erythrose, pyruvaldehyde, 1, 6-anhydroglucose, and acids.

Sınağ et al. (2009) observed the decomposition mechanism for supercritical degradation of cellulose and the effect of temperatures (325°C, 300°C, 375°C) and catalysts (K_2CO_3, HZSM-5, Ni/SiO_2) on the product's composition. Methyl furfural, 5-hydroxymethylfurfural, glucose, fructose, formic acid, acetic acid, formaldehyde, acetaldehyde, and phenols were observed in the product. Carbohydrate intermediates such as fructose and glucose were detectable at a low operating temperature, while no such intermediates were found at higher temperatures. Such observation confirmed the effective decomposition of cellulose biomass at high temperatures. The formation of low molecular weight products at high temperatures was similar to the previous study (Kabyemela et al. 1998). The cellulose decomposition under supercritical conditions is common in many studies (Jakab et al. 1997).

Sakaki et al. (1996) proposed that the liquefaction of cellulose preceded saccharification and carbonization. Water-soluble sugars were major intermediates in saccharification, and these sugars were further decomposed in carbonization via decarboxylation, dehydration, and aromatization mechanisms. Ehara and Saka (2002) used flow-type reactors for the decomposition of cellulose in supercritical water. The operating conditions for both reactors were 40 MPa, 380°C, and 25 min. The reaction pathway was the hydrolysis of the cellulose to polysaccharides with a polymerization degree of 13–100. Further saccharification produced glucose and fructose as oligosaccharides and monosaccharide, respectively. Decomposition was found to be active only after the saccharification step. Glyceraldehydes and methylglyoxal were the main intermediates for the initial 10 min. Final products from the decomposition of intermediates were erythrose, levoglucosan, and 5-hydroxymethyl furfural.

19.3 EFFECT OF PROCESS CONDITIONS ON YIELD OF BIO-OIL

19.3.1 Temperature

The liquefaction temperature has a significant effect on bio-oil yields from biomass. It is due to the fact that the temperature induces fragmentation in biomass into smaller fractions. Temperature helps in the depolymerization of biomass by providing necessary activation energy for the breakdown of carbon bonds. The excessive decomposition of biomass into fragments occurs at a very high temperature, resulting in the generation of free radicals and large fragments. Therefore, the probability of recombination of fragmented biomass components is possible at high concentrations. The role of temperature is evident during the hydrolysis of fragmented species and their possible recombination. There is a slight range of suitable temperature that can give sufficient fragmentation at minimum loss of bio-oil fragments to gaseous products. In general, the decomposition is a dominant process at low temperatures at 250°C–300°C. As the temperature increases, the recombination of species becomes quite significant and the amount of char increases. The temperature in range of 250°C–330°C is suitable for high yield of bio-oil. In fact, it is an experimental task to select a suitable range of temperature for a given biomass type. The properties of hydrothermal fluids change very quickly near the supercritical temperature. The operating temperature higher than the supercritical point leads to the formation of gas products (Zhong and Wei 2004).

According to the research reported by Karagoz and coworkers (2006), the yield of liquefied sawdust increased from 3.7% to 8.5% as a function of temperature 180°C–280°C, that is, an indication of synergic effect. Sakaki et al. (1996) also found the maximum concentration of water-soluble compounds of 45% at operating temperature of 250°C. In another study by Xu and Etcheverry (2008), the oil yield was increased by approximately 25% when the temperature increased from 220°C to 350°C. However, they found that maximum rate of bio-oil conversion was in range of 250°C–300°C. The rate of bio-oil production decreased when the temperature was further increased from 300°C to 350°C. It is also possible that small fragmented species are easily converted to gaseous products at high temperatures. Therefore, the decrease in the decomposition rate is plausible at high temperatures.

The sequential effect of temperature on bio-oil yield is highly likely to occur. Initially, the increase in temperature enhances the bio-oil production. As maximum bio-oil yield is reached, further increase in temperature retards the overall liquefaction. The higher temperature increases gasification or charification activity, depending on the dominance of biomass composition. For example, if volatile components of the biomass still exist in the biomass, high temperature leads to gasification of volatile matter. Similarly, if excessive fragmented components are available for longer time period during the liquefaction, the probability of charification becomes higher (Abu El-Rub et al. 2004). These two mechanisms are very much dominant at high temperatures and reduce the overall bio-oil yield. On the contrary, hydrothermal process carried out at a temperature <280°C led to incomplete decomposition of biomass, hence low conversion yield of gas and liquid products. It is also found that the fragmentation of cellulose or lignin is excessive as temperature increased from 250°C during hydrothermal liquefaction (Yu et al. 2007). Therefore, a temperature that can give sufficient fragmentation of biomass and at the same time avoids any gasification or charification is a suitable one for given biomass. The effect of temperature on liquid oil yield is described in Figure 19.3. Low temperature was insignificant to induce any fragmentation in biomass, that is, low oil yield. It was found in many studies that the temperature from 300°C to 315°C was very much effective to achieve maximum oil yield (Zhang et al. 2009; Yin et al. 2010).

Several biomass types such as cattle manure, grassland perennials, and *Enteromorpha prolifera* (Sugano et al. 2008; Zhang et al. 2009; Yin et al. 2010) are considered as waste biomass. Zhang et al. (2009) observed 76% of bio-oil yield from softwood species over 300°C–330°C temperature. The extended oil yield was due to the presence of acetone as the extraction fluid that increased the amount of acetone-soluble components effectively. The operating temperature of 330°C was considered to be suitable for obtaining a high oil yield. The bio-oil yield increased as a function of temperature from 260°C to 340°C using manure as biomass feed (Xiu et al. 2010). The hydrothermal processing temperature of 360°C was found sufficient to obtain maximum oil yield from microalgae feedstock (Zhang et al. 2011). It was determined from those previous reports that the temperatures between 300°C and 350°C can be considered as an appropriate temperature for the

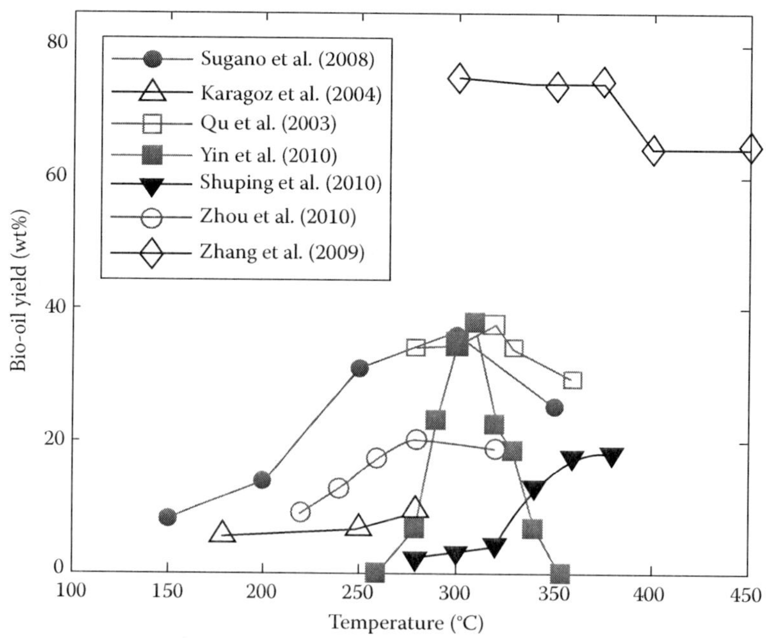

FIGURE 19.3 Influence of temperature on percent yield of bio-oil.

hydrothermal liquefaction of biomass. Sugano et al. (2008) liquefied eucalyptus biomass in hydrothermal media and found that the temperature exerted considerable influence on bio-oil yields. The amount of biochar was reduced sharply from 70 wt% at 150°C to 10 wt% at 300°C. The amount of bio-oil yield increased from 30 to 68 wt% with the increase of temperature from 150°C to 300°C. It is an indication about the effect of temperature on product distributions. The formation of gaseous products at temperatures higher than 350°C was prominent. Therefore, the optimal temperature for the liquefaction of biomass was reported at 300°C by Sugano et al. (2008).

19.3.2 Particle Size

The particle size determines the exposed surface area for heat and mass transfer during the hydrothermal liquefaction. The effect of temperature is pronounced for fine particle size. However, it is very costly to reduce particle sizes. Therefore, it is necessary to optimize the particle size of feedstock to minimize size reduction cost. Once a reasonable degree of particle size is achieved, further reduction in the particle size may not be able to significantly enhance the liquefaction. In a hydrothermal medium, the resistance of heat transfer at the particle–medium interface might not be significant because the particle is surrounded by a fluid medium.

The influence of particle sizes of perennial grass feedstock at 25.4, 2.0, and 0.5 mm on the performance of liquefaction at 350°C was investigated, and no significant effect of this parameter was observed on bio-oil yield (Zhang et al. 2009). The effect of particle size is not discussed in most literatures as it is less important than other variables. Overall, particle size that gives minimum amount of operating cost is sufficient for most practical purposes. The cost of energy consumption is low in case of coarse grinding, while fine grinding leads to excessive operational cost. The size reduction cost for fine grinding of switchgrasses, barley straw, and wheat straw was almost doubled when reducing the size from 3.2 to 0.8 mm (Mani et al. 2004). It is normally suggested that the feedstock size of approximately 4–10 mm is suitable for any practical hydrothermal liquefaction to offset any heat or mass transfer resistance.

19.3.3 Biomass Feedstock

Biomass exhibits compositional variability that leads to changes in the composition and yield of bio-oil. The yield of bio-oil is usually high when the major portion of biomass consists of cellulose or hemicellulose. Lignin leads to the formation of char components at low operating temperature. The hydrothermal liquefaction of cherrywood produced more bio-oil compared to cypress mainly due to more cellulose in the cherrywood (Bhaskar et al. 2008). In another study by Zhong and Wei (2004), *Populus tomentosa carr*, *Fraxinus mandschurica*, *Cunninghamia lanceolata*, and *Pinus massoniana lamb* were liquefied in a hydrothermal process. The yield of oil was the highest for the hydrothermal liquefaction of *F. mandschurica* mainly because it contains less amount of lignin. The yields of oil and char can be estimated from the content of lignin in biomass using Equations 19.1 and 19.2, respectively (Zhong and Wei 2004):

$$\text{Maximum heavy oil yield} = 40.525 - 0.583 \times (\text{lignin percentage}) \tag{19.1}$$

$$\text{Maximum amount of char residue} = \text{lignin percentage} - 5.477 \tag{19.2}$$

In Equation 19.1, it is clear that the amount of oil will decrease with the increase in amount of lignin in biomass feedstock. Similarly, the amount of char residue increases with the increase of lignin. The investigations have found that final temperature of hydrothermal medium also increases at high lignin composition. According to Zhong and Wei (2004), the final temperature for *C. lanceolata* with lignin content of 32.4 wt% was 340°C compared to 300°C in case of *F. mandschurica* with lignin content of 21 wt%. The bio-oil yield can be estimated from the cellulose or hemicelluloses content.

Thermal decomposition behaviors of hemicellulose, cellulose, and lignin depend upon the structure. Amorphous hemicelluloses decompose rapidly compared to the other two components. Cellulose contains β-(1–4) glycosidic linkages and its polymerization degree varies from ~500 to 10,000, making it a little harder to decompose. Molecular weight of lignin is higher than 10,000. Thus, lignin is the most difficult component in biomass to be decomposed during liquefaction and mainly appears as solid residue (Peterson et al. 2008).

The composition of bio-oil depends on the types of biomass. The portion of oxygenates is higher if major portion of biomass is hemicelluloses or cellulose. The presence of oxygenates decreases the quality and thermal stability of bio-oil but lowers its viscosity (Grange et al. 1996). Compact feedstocks like coal produce oil with better quality. The results of a few studies are presented here to highlight the importance of feedstock on the composition of bio-oil. It was reported that the bio-oil obtained from the hydrothermal liquefaction of monosaccharide (i.e., galactose, mannose, and glucose) at 340°C and 27.5 MPa constituted mostly of acetone, glycolic acid, acetaldehyde, acetic acid, acrylic acid, formic acid, lactic acid, glyceraldehydes, glycolaldehyde, and hydroxymethylfurfural (Srokol et al. 2004). In another study, the bio-oil produced from cellulose feedstock at 350°C contained more of cellobiose, erythrose, anhydroglucose, and hydroxymethylfurfural (Sasaki et al. 1998). These two studies show that the composition of bio-oil did not vary mainly due to the similar feedstock types. However, Wahyudiono et al. (2008) found that the hydrothermal decomposition of lignin at 350°C produced a variety of products in bio-oil that included phenoxymethane, phenol, cyclopentanone, *o*-cresol methyl ether, octamethyltetrasiloxane, 2-methylindane, cyclopentasiloxane, *p*-ethylguaiacol, and 1, 3-bis (trimethylsiloxy) benzene. This discussion highlights the importance of biomass feedstock on the yield and composition of bio-oil.

19.3.4 Biomass Heating Rate

The heating rates influence the decomposition pattern of biomass. Extensive fragmentation of biomass is possible at high heating rates. Rapid heating followed by quick cooling gives more quantity of bio-oil. However, the effect of heat rates is less significant on the composition of bio-oil. This is mainly due to the high stabilization and solvation property of solvent medium during hydrothermal liquefaction. The solvent quickly stabilizes the decomposition by-products, and therefore, changes in the composition are unlikely to happen by variations in heating rates. This is true in case of pyrolysis, where biomass with high moisture content produces little amount of char (Demirbas 2004).

The previous study has found positive influence of heating rate on bio-oil yield for grassland perennial feedstock (Zhang et al. 2009). They reported that the bio-oil yield increased from 63% to 73% as a function of heating rates from 5°C/min to 140°C/min. Equation 19.3 represents the influence of heating rates on the bio-oil yield in case of grassland perennials.

$$\text{Liquid yield } (\%) = [0.0042 \times \ln(\text{heating rate}) + 0.5514] \times 100 \tag{19.3}$$

The behavior of heating rates on char residue in hydrothermal medium is similar to that in pyrolysis. Char is the major component at low heating rates and the process is called carbonization. As heating rates increase, the amount of char decreases. The amount of gaseous by-product may increase at extremely high heating rates.

19.3.5 Pressure

The major role of pressure is to maintain the single-phase medium during liquefaction; otherwise the phase change would require large energy inputs. Pressure can control biomass decomposition and rate of the hydrolysis during liquefaction to produce high yield of liquid oil or gas through thermodynamically favorable reactions. Solvent density of the liquefaction process increases with pressure. The denser solvents dissolve and extract the decomposed components of biomass and stabilize the smaller fractions. However, pressure exerts little influence on the production of liquid oils once the supercritical condition is achieved (Kersten et al. 2006; Sangon et al. 2006). This is mainly due to the properties of solvent medium or water that vary slightly by the pressure variations at supercritical conditions of water.

Similar results were observed by Sangon et al. (2006), where the liquid oil yield was increased slightly under supercritical conditions by changing the pressure from 7 to 12 MPa. They also found that the yield of oil actually decreased by increasing pressure for catalytic liquefaction because the active sites of catalysts might be blocked by a denser medium. Similarly, Kabyemela et al. (1997) observed the decrease in decomposition rate at high pressure. The change of pressure from 30 to 40 MPa at 400°C did not improve the hydrolysis rate for the liquefaction of cellobiose. In fact, the overall rate for the decomposition of cellobiose was decreased. The reduction in overall hydrothermal treatment of biomass at high pressure may be attributed to the cage effect. At high pressure, the local solvent density increased, which might induce the cage effect on C–C linkages. The cage effect reduces the decomposition of C–C linkages and thus the fragmentation of biomass. It might be followed from the discussion that the pressure may exert insignificant or negative influence on overall bio-oil yield once a supercritical condition is achieved.

19.3.6 Residence Times

Reaction time is an important factor in hydrothermal liquefaction. The overall biomass decomposition and the composition of bio-oil and other products depend upon residence time (Qu et al. 2003). There has been plenty of research dedicated to determine the effect of residence times on the liquefaction of biomass in hydrothermal processes. It was found in the literature (Sasaki et al. 2003) that short residence times are suitable for supercritical hydrothermal processes due to the fast decomposition and hydrolysis of biomass components. The literature has reported the importance of short residence times for the process (Boocock and Sherman 1985). It was observed that the yield of bio-oil was suppressed at longer residence times except for higher ratios of biomass to water. Similarly, Yan et al. (1999) observed insignificant change in oil yield by varying residence times. However, longer residence times were suitable during hydrothermal liquefaction of sawdust at low temperatures (~150°C) (Karagöz et al. 2004). At high temperatures (250°C–280°C), the effect of residence times was slightly diminished for the conversion and bio-oil yield. Qu et al. (2003) ended up with similar conclusion that long residence times decreased the overall yield of heavy oils, while short residence times were desirable. The residence times influence the primary, secondary, and tertiary reactions. In hydrothermal decomposition, the fragmentation of heavy intermediates via the secondary and tertiary reactions led to the formation of smaller fractions like liquids and gases. As the process proceeds, the fragmentation reaches saturation and changed to either gaseous or solids via repolymerization.

The composition of bio-oils can change depending on the residence times. The decomposition products were different for short and long residence times for both high temperature (250°C) and low temperature (180°C) (Karagöz et al. 2004). They found that the 60 min hydrothermal liquefaction produced 4-hydroxy- 3-methoxybenzoic acid, bis-(2-ethylyhexyl) phthalate, 4-hydroxy-3-methoxybenzeneacetic acid, and 5-(hydroxymethyl)-2-furan carboxaldehyde that were not observed in the 15 min process. The decomposition of asphaltene and pre-asphaltene into light fractions is plausible at longer time period that may increase the liquid oil yield (Boocock and Sherman 1985). At the same time, it is also necessary to inhibit the decomposition of smaller products or repolymerization because these types of reactions decrease the overall oil yield. Generally, the amount of oil reaches a maximum and then decreases at longer residence times.

19.3.7 Reducing Agents

The reducing agents supply necessary hydrogen for the reduction of fragmented species during liquefaction. Reducing agents can be hydrogen donors or other reducing gases with basic purpose

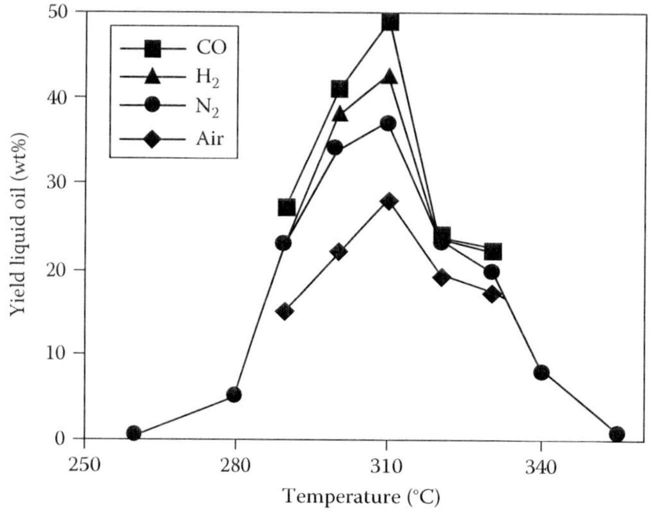

FIGURE 19.4 Effect of reducing gases on liquid oil yield as a function of liquefaction temperature. (From Yin, S. et al., *Bioresour. Technol.*, 101, 3657, 2010.)

to furnish the hydrogen supplies. Reducing gases limit the repolymerization and condensation of fragmented radicals, that is, inhibiting the charring of such fragments (Xu and Etcheverry 2008). Fragmented aromatic species ($Ar^{\bullet}$) can be stabilized to useful products with the help of hydrogen supply as given in Equations 19.4 and 19.5 (Shuping et al. 2010). The other options for reducing gases include steam, argon gas, synthesis gas (CO/H_2), and N_2. Although N_2 gas did not react, it was used for sweeping fragments from the reaction chamber:

$$Ar^{\bullet} + H_2 \rightarrow ArH + H^{\bullet} \tag{19.4}$$

$$Ar^{\bullet} + H^{\bullet} \rightarrow ArH \tag{19.5}$$

Figure 19.4 represents the influence of reducing environment on bio-oil yield during hydrothermal liquefaction of cattle manure. The reactive gases (CO and H_2) gave better yield of bio-oil compared to less reactive air or N_2 gases. Moreover, air gave the minimum amount of oil mainly due to its ability to combust the mixture. Both CO and H_2 gases are reactive and possess better affinity for fragmented components and are able to give a high amount of oil. Low quantity of bio-oil was reported in presence of N_2 gas due to its inert nature. The discussion evidently describes that reactive gases are much better option as reducing gases compared to inert ones. H_2 and mixture of H_2/CO are recommendable in many cases. The sole usage of CO can be avoided as reducing gas due to its noxious nature.

However, reducing gases have the tendency to bubble through reactors that led to its misdistribution within reactor and gas channeling. Hydrogen donors are the alternative option that can hydrogenate the fragmented species. Hydrogen donors transport the hydrogen by exchanging H^+ ions with fragmented species (Wang et al. 2007). The strength of a hydrogen donor is defined by its ability to donate hydrogen. Some of the popular hydrogen donor solvents are tetralin, octahydrophenanthrene, tetrahydroacenaphthene, tetrahydrophenanthrene, and hexahydrofluorene. An example of donating hydrogen by a hydroaromatic compound is given in the following equation (Neavel et al. 1981):

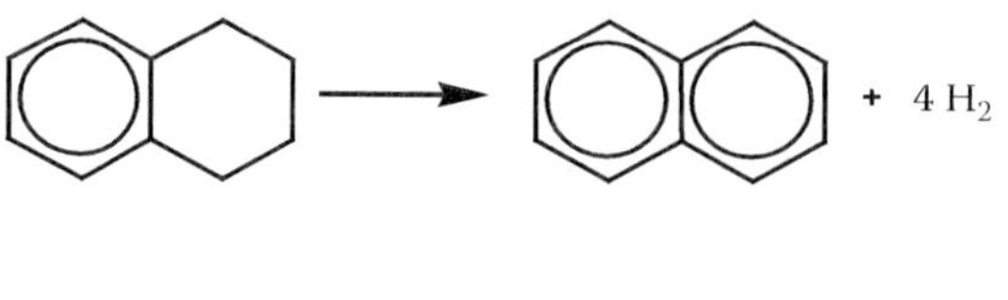

(19.6)

TABLE 19.1
Effect of Hydrogen Donor Solvents and Reducing Gas on the Yield of Liquefaction Products

Feed	Hydrogen Donor Solvent	Reducing Gas	Bio-Oil Yield (%) With Reducing Gas or Donor Solvents	Without Reducing Gas or Donor Solvents	References
Popular	Tetralin/ Tetrahydronaphthalene	Air/inert	56.6	23.3	Phillips and Granatstein (1981)
Cellulose	—	H_2/CO	38.6	5	Shinya et al. (1985)
Municipal	Tetralin	H_2	61	—	Benn and McAuliffe (1986)
Wood	Ethyl acetate	Ar	68.4	33	Silvestri and Naro (1975)
Sawdust	Tetralin	—	40	—	Amen-Chen et al. (2001)
Bituminous	Coal tar	H_2	59	48	Silvestri and Naro (1975)

Literatures have elucidated the role of hydrogen donors on the composition of bio-oil. Wang et al. (2007) investigated the role of reducing gases and donor solvents on the product composition during the hydrothermal liquefaction of saw dust. Tetralin was found to suppress the amount of asphaltenes and gas, thereby increasing oil yield. The bio-oil yield was 58.9% in presence of tetralin and 42% for a nonhydrogen donor toluene solvent. Tetralin enhanced the production of oil by suppressing the amount of asphaltenes and gases due to their hydrogen-shuttling ability. The shuttling ability of tetralin overcame the mass transfer limitations and hydrogen distribution unlike reducing gases (H_2 or syngas). Tetralin furnishes the quick supply of hydrogen to enhance the hydrogenation of fragmented asphaltenes fractions and restrict their polymerization. Hydrogen donors also supplied necessary hydrogen to separate oxygenates from fragmented species and converted it to water, thus increasing the quality of bio-oil by increasing the heating value of bio-oil (Li et al. 2009). Table 19.1 shows the effect of hydrogen donor solvents and reducing gas on the yield of liquefaction products.

19.3.8 Biomass to Water Ratio

The optimization of biomass to liquid ratio is necessary in hydrothermal treatment processes. Optimum ratio enables hydrothermal process to obtain higher yields of liquids at low energy inputs. It is possible that the variation in the biomass to water ratio can change (1) overall liquid yield, (2) gas formation, (3) bio-oil composition, and (4) char residues. In general, at high biomass to water ratios, the amount of gaseous products and char residues are likely to increase. The addition of water in biomass reduces the gaseous compounds and increases water solubility in the oil (Karagöz et al. 2004). It is assumable that the increase in water to biomass ratio leads to more liquid products. However, the excess of water in biomass may increase the energy inputs, and the process may not be economically viable.

Akhtar et al. (2010) reported the effect of biomass to water ratio during hydrothermal liquefaction of biomass. They found that high biomass to water ratio lowered liquid oil yield and decreased the amount of lignin degradation. The maximum oil yield was 68%, which was obtained at a biomass to water ratio of 2:10, 270°C, and 20 bar and with 1.0 M K_2CO_3. The biomass to water ratio of 8:10 produced about 50% of oil and EPFB conversion (Figure 19.5). The decrease in bio-oil yield at high biomass to water ratios may be attributed to the decrease in dissolution power of a hydrothermal solvent.

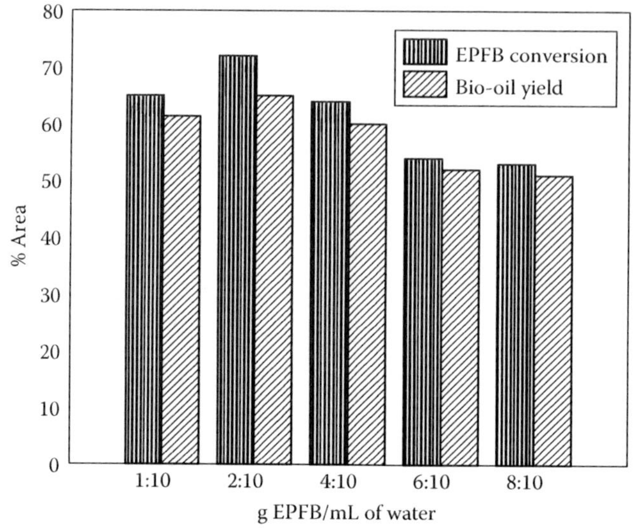

FIGURE 19.5 Effect of biomass to water ratios on liquid oil yield in hydrothermal liquefaction of empty palm fruit bunch (EPFB). (Redrawn after Akhtar, J. et al., *Renew. Energy*, 35, 1220, 2010. With permission.)

The effect of biomass to water ratio during the liquefaction of *C. lanceolata* in water was studied (Qu et al. 2003). They reported that the maximum yield of heavy oil of 24 wt% was obtained at 8 g of biomass to 100 mL of water and 320°C. The amount of heavy oil continuously decreased to 17 wt% for 13 g of biomass to 100 mL of water. The decrease in heavy oil may not be attributed to total bio-oil yield.

The amounts of water-soluble components increased significantly at high water to biomass ratios, which may be the reason for low heavy oil contents (Karagöz et al. 2006). The study reported the positive effect of the presence of water in biomass at high biomass to liquid ratios. They found that the amount of total oil increased from 6.4 to 8.6 wt% by increasing wood/water ratios from 5:30 (g/mL) to 10:30 (g/mL). They also observed significantly high gas yield in product composition without using any water that may suggest that a hydrothermal process favors the formation of liquid products. Hydrothermal process produced approximately 40 wt% of water-soluble compounds, compared to 7 wt% for pyrolysis. It is due to the ability of hydrothermal solvents to extract components from decomposing biomass fragments. The presence of hydrothermal solvents may be helpful to obtain the bio-oil fractions of desired components.

The power of solvent to dissolve the biomass fragments is linked with the amount of water molecules available per unit mass of biomass. Larger amount of water is expected to dissolve more biomass fragment. Water molecules also inhibit the recombination of fragmented biomass components. The amount of both char and gaseous product will be low in hydrothermal liquefaction. Therefore, high volumes of solvent per unit biomass probably give more water solubility and liquid oil yield. But it is not often helpful to use high volume of solvent due to the high energy inputs required for heating the solvents to achieve sub- or supercritical conditions (Qu et al. 2003). The optimum amount of water contents required can be found by performing few experiments. Figure 19.5 indicates that 2:10 ratio as the optimum point for obtaining higher bio-oil yield.

19.4 INDUSTRIAL APPLICATIONS OF HYDROTHERMAL LIQUEFACTION PROCESSES

One of the most common industrial applications of biofuels is biodiesel, which is obtained by chemically reacting fatty acids and alcohol, commonly in the presence of a basic catalyst. The production of biodiesel from vegetable or waste cooking oil is simple. Thus, there are many small and large producers all over the world. Biodiesel is currently used as an alternative to fossil fuels. It is used either directly or blended with diesel fuel at different concentrations from 2 up to 100% by weight. There are two main drawbacks to use biodiesel. One of them is the challenge in its transportation and storage. Another one is the use of biodiesel that could cause problems in cold climates due to its poor cold flow properties.

A process named HTU (hydrothermal upgrading) was originally developed by Shell company in the 1980s (Peterson et al. 2008). Experimental conditions for the HTU process were carried out at temperatures between 330°C and 350°C with holding times 5–20 min and under pressures of 10–18 MPa. The heating value of biocrude obtained from HTU process was approximately 35 MJ/kg. Hydrothermal treatment of crude algae oils was performed at the temperatures of 350°C, 400°C, and 450°C for a residence time of 60 min (Roussis et al. 2012). It was reported that the removal of oxygen and metals from crude algae oils was achieved without using any catalysts and hydrogen. It was also concluded that an increase in the temperature led to a decrease in boiling point of hydrocarbons in bio-oils from hydrothermal treatment. The hydrothermal process works very well not only in removing oxygen from bio-oils but also in removing nitrogen (Cranford et al. 2012). Biomass such as algae contains significant amounts of nitrogen. Thus, this process is very suitable for algal biomass in terms of removing nitrogen from bio-oils from hydrothermal treatment.

When oils containing fatty acids (i.e., crude algae oils, raw vegetable oils, waste cooking oils) were subjected to hydrothermal processing, it is possible to increase the yield of bio-oil and improve the quality by the removal of oxygenates. On the other hand, it is difficult to achieve yields for the hydrothermal liquefaction of biomass such as wood, agricultural residue, and fruit processing wastes without using any catalysts. For this reason, researchers are making great efforts to use catalysts and/or additives in the hot-compressed water liquefaction of biomass. The uses of alkaline solutions as catalysts in the hydrothermal liquefaction of biomass increased the yields of oils (Karagöz et al. 2006). However, the use of base or basic salts on the hydrothermal liquefaction of biomass might cause the separation problem (Karagöz et al. 2005). At an industrial scale, separation and corrosion are big concerns when a base or basic catalyst is used in the process.

In contrast to the HTU process, fast pyrolysis of biomass to produce biocrude is more developed in industries (Bridgewater 2004). The production of biocrude from biomass via fast pyrolysis is carried out at either a pilot or demonstration scale. Some of examples for these industrial applications are Red Arrow Products Co. (a nominal capacity of 50 t/day; Inc. in Wisconsin), DynaMotive Co. (a capacity of 10 t/day in Vancouver, Canada), BTG (a nominal capacity of 51 t/day in the Netherlands), and Fortum (a capacity of 12 t/day in Finland) (Bridgwater 2004). The heating values and the oxygen content of bio-oils from the pyrolysis of wood biomass are usually 18 MJ/kg and 45%–50% by mass, respectively. Although the processing costs are high and the products seem not cost competitive in the market, industries are making a great effort to produce bio-oils and/or bio-based chemicals at low cost.

19.5 ENERGY EFFICIENCY AND ECONOMICS OF A HYDROTHERMAL LIQUEFACTION PROCESS

Bioethanol is the strongest candidate among other bio-oils in terms of energy efficiency. Bioethanol production from corn is energy efficient (Shapouri et al. 2002). However, the production of biofuels from biomass using either pyrolysis or hydrothermal processes is currently not energy efficient. It is necessary to conduct a full economic evaluation of biofuels based on capital and operating costs. It is very difficult to estimate the economics of biofuel production as there are a number of factors such as availability of raw materials, transportation, and storage that affect the economics of biofuels. Hydrothermal liquefaction of biomass is unfavorable from the economics point of view at the current prices of fossil fuels. The major hurdle for the commercialization of hydrothermal liquefaction is (1) high cost of transportation of biomass from field to plant site, (2) low conversion efficiency of biomass feedstock into desirable bio-oil product, and (3) lack of research on the understanding of complex reactions during hydrothermal liquefaction that are retardant to the overall liquid yield (Midgett et al. 2012). The quality of bio-oil that contains significant amounts of oxygen is low, and it cannot be used as a transportation fuel without upgrading (Amen-Chen et al. 2001).

The economic issues mainly arise from the production of biomass and its allied expenditure until it converts into bio-oil. The production of bio-oil from oilseeds or other available oils might have some economic compatibility with fossil fuels. However, this chapter is focused on the economics of biomass feedstock that can be converted into liquid fuels directly using a hydrothermal process. The overall costs for the hydrothermal liquefaction include the cost of biomass feedstock and the cost of liquefaction. Biomass feedstock is the major component of any bio-oil production using any of the available conversion methods. Forest, agricultural, and logging residues, urban wood wastes, short-rotation woody corps, and sugarcane wastes may be the sources of biomass used in liquefaction. All of the crops or agricultural activities for the production of biomass require farmland. This may trigger the competition for the available land between food supplies and biofuel production. The prices of land are likely to increase in such case. Biomass obtained in the form of crops or forest residues is a scattered source of energy. The collection of such scattered biomass and management for an organized supply chain from feedstock in field to a biomass conversion facility will require significant amounts of capital investment, especially for transportation. Nevertheless, all of the residues, in most of the cases, are being used for other purposes. For example, forest residues are a source of burning fire in less developed areas. The agricultural residues are being used as cattle fodder or crushed to be buried in the agricultural lands to increase the fertility. Therefore, it is expected that with the increase in demand of such biomass, the cost of biomass feedstock will increase. The growth of biomass either in the form of agricultural residues or as energy crops requires farmland, energy inputs in agricultural machinery, fertilizers, and water supply. The cost of feedstock is expected to increase with its demand for bio-oil production.

The hydrothermal liquefaction is an energy-intensive process. The capital investments and the operational cost of producing bio-oil are also considerable. It is difficult to find a generalized model to determine energy inputs for the production of bio-oil from different biomass materials due to the variability in feedstock types that require different amounts of energy inputs. Few studies have discussed in detail the energy analysis of selected biomass types used in hydrothermal liquefaction (Minarick et al. 2011; Miao et al. 2012; Midgett et al. 2012). Miao et al. (2012) provided the energy analysis during hydrothermal liquefaction of algae, while Minarick et al. (2011) conducted the energy analysis for the hydrothermal liquefaction of swine manure.

Miao et al. (2012) used direct hydrothermal liquefaction (DHTL) and sequential hydrothermal liquefaction (SEHTL) to compare their energy efficiency for the production of bio-oil from algae. DHTL process was a semibatch steel autoclave (Parr reactor). SEHTL method was the repetition of DHTL method with an extra step. At the end of DHTL, the solid residue left after filtration of reaction mixture was fed back to the reactor for the second step. The second step was performed under similar conditions as the first step. Net energy balance (ΔH_c), energy recovery rate (%), and thermal efficiency (η) were the three major approaches to analyze the efficiency of both processes. They observed that the net energy balance in many cases was on the negative side, which showed that the required input energy was more than that obtained in heating value of bio-oil. From their study, the SEHTL method was comparatively a better technique than the DHTL method. However, these HTL processes were found to be less desirable for the commercialization due to their low efficiency.

Minarick et al. (2011) estimated the economics of liquefaction of swine manure using a batch reactor in a nitrogen environment. The energy requirement to increase the temperature of 1 kg of water from 22°C to 303°C was 1258 kJ. Therefore, careful measurement was required to properly estimate the water for liquefaction. The manure with a high concentration of solids produced 43% of liquid oil. The amount of bio-oil decreased with the addition of water. The energy output to input ratio increased to 4.43 in case of fresh manure. The fertilizer value of manure was not disturbed by the hydrothermal process, and they recommended a farm-based hydrothermal process for large production facilities.

19.6 SUMMARY

The bio-oil from hydrothermal liquefaction depends very much on input conditions and biomass types. The fragmentation reactions and type of recombination also depend upon the operating parameters. The major parameters that govern the liquefaction process include temperature, solvent density, biomass type, heating rates, the size of biomass particles, biomass to water ratio, and pressure. Among these, temperature is the most dominant parameter that has significant influence on fragmentation. The normal operation temperature of 280°C–374°C is recommended for a hydrothermal process that depends upon biomass type and the desirable composition of bio-oil. The amount of gas composition is higher for temperature higher than 350°C. Similarly, the temperatures lower than 280°C lead to smaller fragmentations, thus reducing the amount of bio-oil production. For the biomass types such as softwood, algae, grasses, cellulose, and hemicelluloses, the operating temperature range of 300°C–330°C is recommendable. The biomass types such as lignocelluloses and hardwood may require higher operating temperature.

The biomass to liquid ratio and solvent density are another two influencing parameters. The solvents help decomposition, dehydration, and solvolysis reactions. Suitable solvents support biomass fragmentation and stabilization of free radicals. Water is among the abundantly available solvents along with methanol, acetone, benzene, tetralin, and others. Biomass to liquid ratio is the measure of availability of solvent molecules per unit mass available. High solvent to biomass ratios are not recommended due to the increase in energy inputs for heating solvent and downstream treatment problems. The biomass types cause significant influence on attaining hydrothermal conditions and final composition of bio-oil. It is due to the different degradation behaviors of cellulose, lignin, and hemicelluloses.

The heating rates and the size of biomass particles are two insignificant operating variables. The dominant properties of hydrothermal solvents for reaction and heat transfer nullify the influence of particle size or heating rates. However, large particle size still hinders the heat transfer.

Residence times and reducing agents exert moderate to low influence on bio-oil yield. Short residence times are practically useful for a hydrothermal process in near-critical conditions. Longer residence times might be helpful at low-temperature conditions. Among reducing agents, hydrogen donor solvents are recommended due to their efficient reducing properties. Donor solvent in liquid form can stabilize free radicals easily due to effective mass transfer. Donor solvents work on hydrogen-shuttling principle that sustains the uniform distribution of hydrogen. Reducing gases such as CO and H_2 are helpful to enhance the oil yield. The role of pressure in hydrothermal process is to maintain a single phase during liquefaction. The pressure thereafter does not have any recognized influence on liquid yield.

Many efforts have been made to utilize biomass via thermochemical conversion due to the growing energy demand and environmental considerations. Most of the industrial applications of thermochemical conversion of biomass are concerned with the upgrading of vegetable and/or algal oils for possible production of biocrude, which can be done either by classical biodiesel production via transesterification or hydrothermal processing of vegetable oils. The upgrading of vegetable and/or algal oils via hydrothermal processing is possible without using any catalyst, which provides economically competitive production of biocrude. However, the use of the other biomass types such as wood, agricultural residues, and pulp processing wastes for the hydrothermal processing needs to be developed at a full scale in terms of industrial applications and economic aspects. Although the energy efficiency of hydrothermal processing of biomass needs to improve, the process is an important step toward providing a system for producing sustainable bioenergy.

ACKNOWLEDGMENT

The authors would like to express their sincere gratitude to the Ministry of Higher Education (MOHE), Malaysia, and Universiti Teknologi Malaysia for providing financial support.

REFERENCES

Abu El-Rub, Z., Bramer, E. A., Brem, G. Review of catalysts for tar elimination in biomass gasification processes. *Industrial & Engineering Chemistry Research* **43**, 6911–6919 (2004).

Akhtar, J., Kuang, S. K., Amin, N. S. Liquefaction of empty palm fruit bunch (EPFB) in alkaline hot compressed water. *Renewable Energy* **35**, 1220–1227 (2010).

Amen-Chen, C., Pakdel, H., Roy, C. Production of monomeric phenols by thermochemical conversion of biomass: A review. *Bioresource Technology* **79**, 277–299 (2001).

Asghari, F. S., Yoshida, H. Dehydration of fructose to 5-hydroxymethylfurfural in sub-critical water over heterogeneous zirconium phosphate catalysts. *Carbohydrate Research* **341**, 2379–2387 (2006).

Bauen, A., Woods, J., and Hailes, R. A biomass blueprint to meet 15% of OECD electricity demand by 2020. (2004). Available online: http://assets.panda.org/downloads/biomassreportfinal.pdf. Accessed on January 5, 2014.

Behrendt, F., Neubauer, Y., Oevermann, M., Wilmes, B., Zobel, N. Direct liquefaction of biomass. *Chemical Engineering & Technology* **31**, 667–677 (2008).

Benn, F. R., McAuliffe, C. A. Conversion of a municipal waste to fuel. US 46187361 (1986).

Bhaskar, T., Sera, A., Muto, A., Sakata, Y. Hydrothermal upgrading of wood biomass: Influence of the addition of K_2CO_3 and cellulose/lignin ratio. *Fuel* **87**, 2236–2242 (2008).

Boocock, D. G. B., Sherman, K. M. Further aspects of powdered poplar wood liquefaction by aqueous pyrolysis. *The Canadian Journal of Chemical Engineering* **63**, 627–633 (1985).

Bridgewater, A. V. Biomass fast pyrolysis. *Thermal Science* **8**, 21–49 (2004).

Cranford, R. J., Aravanis, A.M., and Roussis, S.G. Process for the recovery of oleaginous compounds from biomass. USA patent application US 8,192,628 B2 (2012).

Demirbas, A. Effect of initial moisture content on the yields of oily products from pyrolysis of biomass. *Journal of Analytical and Applied Pyrolysis* **71**, 803–815 (2004).

Ehara, K., Saka, S. A comparative study on chemical conversion of cellulose between the batch-type and flow-type systems in supercritical water. *Cellulose* **9**, 301–311 (2002).

Goyal, H. B., Seal, D., Saxena, R. C. Bio-fuels from thermochemical conversion of renewable resources: A review. *Renewable and Sustainable Energy Reviews* **12**, 504–517 (2008).

Grange, P., Laurent, E., Maggi, R., Centeno, A., Delmon, B. Hydrotreatment of pyrolysis oils from biomass: Reactivity of the various categories of oxygenated compounds and preliminary techno-economical study. *Catalysis Today* **29**, 297–301 (1996).

Heinz, A., Kaltschmitt, M., Stülpnagel, R., Scheffer, K. Comparison of moist vs. Air-dry biomass provision chains for energy generation from annual crops. *Biomass and Bioenergy* **20**, 197–215 (2001).

Jakab, E., Liu, K., Meuzelaar, H. L. C. Thermal decomposition of wood and cellulose in the presence of solvent vapors. *Industrial & Engineering Chemistry Research* **36**, 2087–2095 (1997).

Kabyemela, B. M., Adschiri, T., Malaluan, R. M., Arai, K. Kinetics of glucose epimerization and decomposition in subcritical and supercritical water. *Industrial & Engineering Chemistry Research* **36**, 1552–1558 (1997).

Kabyemela, B. M., Takigawa, M., Adschiri, T., Malaluan, R. M., Arai, K. Mechanism and kinetics of cellobiose decomposition in sub- and supercritical water. *Industrial & Engineering Chemistry Research* **37**, 357–361 (1998).

Karagöz, S., Bhaskar, T., Muto, A., Sakata, Y. Hydrothermal upgrading of biomass: Effect of K_2CO_3 concentration and biomass/water ratio on products distribution. *Bioresource Technology* **97**, 90–98 (2006).

Karagöz, S., Bhaskar, T., Muto, A., Sakata, Y., Oshiki, T., Kishimoto, T. Low-temperature catalytic hydrothermal treatment of wood biomass: Analysis of liquid products. *Chemical Engineering Journal* **108**, 127–137 (2005).

Karagöz, S., Bhaskar, T., Muto, A., Sakata, Y., Uddin, M. A. Low-temperature hydrothermal treatment of biomass: Effect of reaction parameters on products and boiling point distributions. *Energy & Fuels* **18**, 234–241 (2004).

Kelly-Yong, T. L., Lee, K. T., Mohamed, A. R., Bhatia, S. Potential of hydrogen from oil palm biomass as a source of renewable energy worldwide. *Energy Policy* **35**, 5692–5701 (2007).

Kersten, S. R. A., Potic, B., Prins, W., Van Swaaij, W. P. M. Gasification of model compounds and wood in hot compressed water. *Industrial & Engineering Chemistry Research* **45**, 4169–4177 (2006).

Li, H., Yuan, X., Zeng, G., Tong, J., Yan, Y., Cao, H., Wang, L., Cheng, M., Zhang, J., Yang, D. Liquefaction of rice straw in sub- and supercritical 1,4-dioxane–water mixture. *Fuel Processing Technology* **90**, 657–663 (2009).

Mani, S., Tabil, L. G., Sokhansanj, S. Grinding performance and physical properties of wheat and barley straws, corn stover and switchgrass. *Biomass and Bioenergy* **27**, 339–352 (2004).

Miao, C., Chakraborty, M., Chen, S. Impact of reaction conditions on the simultaneous production of polysaccharides and bio-oil from heterotrophically grown *Chlorella sorokiniana* by a unique sequential hydrothermal liquefaction process. *Bioresource Technology* **110**, 617–627 (2012).

Midgett, J. S., Stevens, B., Dassey, A., Spivey, J., Theegala, C. S. Assessing feedstocks and catalysts for production of bio-oils from hydrothermal liquefaction. *Waste and Biomass Valorization* **3**, 1–10 (2012).

Minarick, M., Zhang, Y., Schideman, L., Wang, Z., Yu, G., Funk, T., Barker, D. Product and economic analysis of direct liquefaction of swine manure. *BioEnergy Research* **4**, 324–333 (2011).

Minowa, T., Fang, Z. Hydrogen production from cellulose in hot compressed water using reduced nickel catalyst: Product distribution at different reaction temperatures. *Journal of Chemical Engineering Japan* **31**, 488–491 (1998).

Minowa, T., Fang, Z., Ogi, T., Varhegyi, G. Decomposition of cellulose and glucose in hot-compressed water under catalyst-free conditions. *Journal of Chemical Engineering of Japan* **31**, 131–134 (1998).

Neavel, R. C., Knights, C. F., Schulz, H. Exxon donor solvent liquefaction process. *Mathematical Physical Engineering Science* **300**, 141–156 (1981).

Penninger, J. M. L., Kersten, R. J. A., Baur, H. C. L. Reactions of diphenylether in supercritical water—Mechanism and kinetics. *The Journal of Supercritical Fluids* **16**, 119–132 (1999).

Peterson, A. A., Vogel, F., Lachance, R. P., Froling, M., Antal, J. M. J., Tester, J. W. Thermochemical biofuel production in hydrothermal media: A review of sub- and supercritical water technologies. *Energy & Environmental Science* **1**, 32–65 (2008).

Phillips, C. R., Granatstein, D. L. Production of hydrogen liquids from solid carbonaceous materials. CA Patent number 1100532 (1981).

Puri, V. P. Effect of crystallinity and degree of polymerization of cellulose on enzymatic saccharification. *Biotechnology and Bioengineering* **26**, 1219–1222 (1984).

Qu, Y., Wei, X., Zhong, C. Experimental study on the direct liquefaction of *Cunninghamia lanceolata* in water. *Energy* **28**, 597–606 (2003).

Roussis, S. G., Cranford, R., Sytkovetskiy, N. Thermal treatment of crude algae oils prepared under hydrothermal extraction conditions. *Energy & Fuels* **26**, 5294–5299 (2012).

Sakaki, T., Shibata, M., Miki, T., Hirosue, H., Hayashi, N. Reaction model of cellulose decomposition in near-critical water and fermentation of products. *Bioresource Technology* **58**, 197–202 (1996).

Sangon, S., Ratanavaraha, S., Ngamprasertsith, S., Prasassarakich, P. Coal liquefaction using supercritical toluene–tetralin mixture in a semi-continuous reactor. *Fuel Processing Technology* **87**, 201–207 (2006).

Sasaki, M., Adschiri, T., Arai, K. Production of cellulose ii from native cellulose by near- and supercritical water solubilization. *Journal of Agricultural and Food Chemistry* **51**, 5376–5381 (2003).

Sasaki, M., Fang, Z., Fukushima, Y., Adschiri, T., Arai, K. Dissolution and hydrolysis of cellulose in subcritical and supercritical water. *Industrial Engineering Chemistry Research* **39**, 2883–2890 (2000).

Sasaki, M., Kabyemela, B., Malaluan, R., Hirose, S., Takeda, N., Adschiri, T., Arai, K. Cellulose hydrolysis in subcritical and supercritical water. *The Journal of Supercritical Fluids* **13**, 261–268 (1998).

Schwald, W., Bobleter, O. Hydrothermolysis of cellulose under static and dynamic conditions at high temperatures. *Journal of Carbohydrate Chemistry* **8**, 565–578 (1989).

Shapouri, H., Duffield, J. A., and Wang, M. The energy balance of corn ethanol: An update. U.S. Department of Agriculture, Agricultural Economics Report 813 (2002). Available online: http://www.ers.usda.gov/topics/crops/corn/readings.aspx. Accessed on January 5, 2014.

Shinya, Y., Tomoko, O., Katsuya, O. Liquefaction of wooden biomass. JP Patent number 60262888 (1985).

Shuping, Z., Yulong, W., Mingde, Y., Kaleem, I., Chun, L., Tong, J. Production and characterization of bio-oil from hydrothermal liquefaction of microalgae *Dunaliella tertiolecta* cake. *Energy* **35**, 5406–5411 (2010).

Silvestri, A. J., Naro, P. A. Liquefaction of coal. US Patent number 3923634 (1975).

Sınağ, A., Gülbay, S., Uskan, B., Güllü, M. Comparative studies of intermediates produced from hydrothermal treatments of sawdust and cellulose. *The Journal of Supercritical Fluids* **50**, 121–127 (2009).

Snåre, M., Kubičková, I., Mäki-Arvela, P., Eränen, K., Wärnå, J., Murzin, D. Y. Production of diesel fuel from renewable feeds: Kinetics of ethyl stearate decarboxylation. *Chemical Engineering Journal* **134**, 29–34 (2007).

Srokol, Z., Bouche, A.-G., van Estrik, A., Strik, R. C. J., Maschmeyer, T., Peters, J. A. Hydrothermal upgrading of biomass to biofuel; studies on some monosaccharide model compounds. *Carbohydrate Research* **339**, 1717–1726 (2004).

Sugano, M., Takagi, H., Hirano, K., Mashimo, K. Hydrothermal liquefaction of plantation biomass with two kinds of wastewater from paper industry. *Journal of Materials Science* **43**, 2476–2486 (2008).

Toor, S. S., Rosendahl, L., Rudolf, A. Hydrothermal liquefaction of biomass: A review of subcritical water technologies. *Energy* **36**, 2328–2342 (2011).

Wahyudiono, V., Sasaki, M., Goto, M. Recovery of phenolic compounds through the decomposition of lignin in near and supercritical water. *Chemical Engineering and Processing: Process Intensification* **47**, 1609–1619 (2008).

Wang, G., Li, W., Li, B., Chen, H. Direct liquefaction of sawdust under syngas. *Fuel* **86**, 1587–1593 (2007).

Xiu, S., Shahbazi, A., Shirley, V., Cheng, D. Hydrothermal pyrolysis of swine manure to bio-oil: Effects of operating parameters on products yield and characterization of bio-oil. *Journal of Analytical and Applied Pyrolysis* **88**, 73–79 (2010).

Xu, C., Etcheverry, T. Hydro-liquefaction of woody biomass in sub- and super-critical ethanol with iron-based catalysts. *Fuel* **87**, 335–345 (2008).

Yan, Y., Xu, J., Li, T., Ren, Z. Liquefaction of sawdust for liquid fuel. *Fuel Processing Technology* **60**, 135–143 (1999).

Yin, S., Dolan, R., Harris, M., Tan, Z. Subcritical hydrothermal liquefaction of cattle manure to bio-oil: Effects of conversion parameters on bio-oil yield and characterization of bio-oil. *Bioresource Technology* **101**, 3657–3664 (2010).

Yu, Y., Lou, X., Wu, H. Some recent advances in hydrolysis of biomass in hot-compressed water and its comparisons with other hydrolysis methods. *Energy & Fuels* **22**, 46–60 (2007).

Zhang, B., von Keitz, M., Valentas, K. Thermochemical liquefaction of high-diversity grassland perennials. *Journal of Analytical and Applied Pyrolysis* **84**, 18–24 (2009).

Zhang, L., Champagne, P., Xu, C. Bio-crude production from secondary pulp/paper-mill sludge and waste newspaper via co-liquefaction in hot-compressed water. *Energy* **36**, 2142–2150 (2011).

Zhong, C., Wei, X. A comparative experimental study on the liquefaction of wood. *Energy* **29**, 1731–1741 (2004).

Zhou, D., Zhang, L., Zhang, S., Fu, H., and Chen, J. Hydrothermal liquefaction of macroalgae *Enteromorpha prolifera* to bio-oil, *Energy & Fuels* **24** (7), 4054–4061 (2010).

20 Anaerobic Digestion of Organic Wastes

Bo Zhang and Lijun Wang

CONTENTS

20.1 INTRODUCTION

Anaerobic digestion (AD) is a versatile technology platform that has experienced several waves of technological developments. It was originally developed as a wastewater treatment system in the 1970s. AD showed a great promise as an alternative energy source particularly from animal wastes in the 1980s, and soon it became a standard procedure for treating organic-matter-rich industrial wastewater. With the growing concern around limited availability and increasing demand for fossil fuels, the need for national energy independence and safety, and the need for reduction in greenhouse gas (GHG) emissions, recently, AD returned to our sight for its energy recovery potential, making use of a variety of organic wastes.

AD is a biological process in which organic matters are degraded to a gaseous mixture of biogas in the absence of oxygen. Biogas, which mainly consists of methane and carbon dioxide, can be used as an energy source to replace fossil natural gas. The methane in the biogas could be as high as 50%–60% by volume (Gebauer, 2004). When biogas produced by AD of biomass is used for electricity generation, the overall conversion efficiency from biomass to electricity is about 10%–16% (McKendry, 2002).

AD technology has previously been reviewed by Nallathambi Gunaseelan (1997), Mata-Alvarez et al. (2000), Rajeshwari et al. (2000), and Li et al. (2011). This chapter reviews the AD of organic wastes. First, basic principles, operating parameters, and monitoring technologies are introduced. Then the applications of AD for the treatment of organic wastes are reviewed. Finally, life cycle assessment (LCA) applications for the AD treatment of organic wastes are discussed.

20.2 PRINCIPLES OF ANAEROBIC DIGESTION

20.2.1 Basic Principles

The AD of organic material basically follows the four stages: hydrolysis, acidogenesis, acetogenesis, and methanogenesis, as shown in Figure 20.1. Although each of these stages could be a rate-limiting step, hydrolysis is generally considered as rate limiting. AD is a complex process, which requires strict anaerobic conditions (i.e., oxidation-reduction potential [ORP]<-200 mV) to proceed, and depends on the coordinated activity of a complex microbial association to transform organic material into mostly methane (CH_4) and CO_2 (Appels et al., 2008).

The hydrolysis step degrades both insoluble organic material and high molecular weight compounds such as lipids, polysaccharides, proteins, and nucleic acids into soluble organic substances (e.g., sugars, amino acids, and fatty acids). The compounds formed during hydrolysis are further digested during the second step, that is, acidogenesis. Volatile fatty acids (VFAs) are produced by acidogenic (or fermentative) bacteria along with ammonia (NH_3), CO_2, H_2S, and other by-products. The third step in the AD process is acetogenesis, in which the higher organic acids and alcohols produced by acidogenesis are further split by acetogens to produce mainly acetic acid as well as CO_2 and H_2. This conversion is controlled to a large extent by the partial pressure of H_2 in the mixture. The last step of AD, methanogenesis, produces methane by two groups of methanogenic microorganisms: the first group splits acetate into methane and carbon dioxide, and the second group uses hydrogen as electron donor and carbon dioxide as acceptor to produce methane.

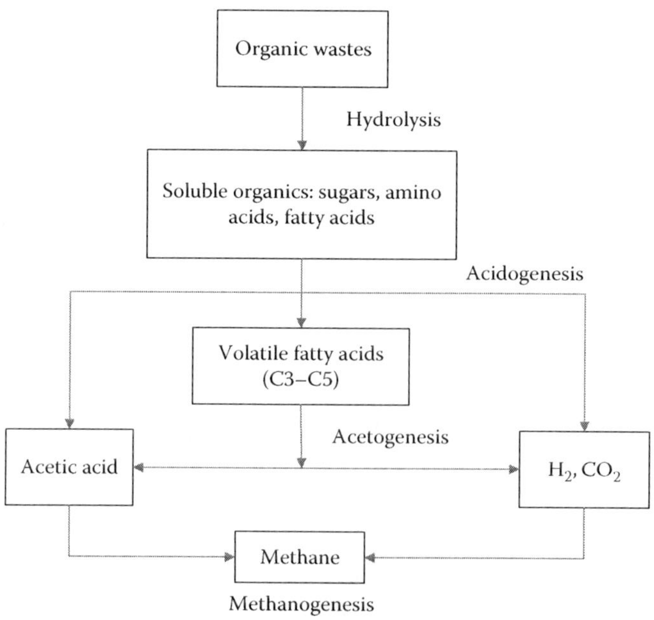

FIGURE 20.1 Successive steps in the AD process.

20.2.2 Operating Parameters

20.2.2.1 Volatile Fatty Acids

Major intermediates in the AD reactions are VFAs. Many authors have suggested VFAs as control parameters as these acids are indicative of the activity of the methanogenic consortia (Ahring et al., 1995; Pind et al., 2003). In case of VFA accumulation, this can be interpreted as either organic overloading or the inhibition of methanogenic microbial communities due to the influence of other factors. Once VFA accumulation happens, pH value is reduced, and action should be taken to avoid reactor failure. This pH reduction could normally be countered by the activity of the methanogenic bacteria, which also produce alkalinity in the form of carbon dioxide, ammonia, and bicarbonate. Relationship between specific VFAs is still unknown. Though it is unclear which VFAs should be included in the control strategies, simple short-chained VFAs have been the favorites of many researchers (Nielsen et al., 2007).

20.2.2.2 pH Value

The value of pH is an indicator of the amount of VFAs generated during AD. Methanogenic bacteria are extremely sensitive to pH. If a high concentration of VFAs is formed, pH will be decreasing to levels that the methanogenic bacteria are severely inhibited. A digester usually runs at the pH of 6.5–7.2, which is optimum for methanogenic bacteria. Therefore, it is important to have a buffering capacity in the system, and a buffering capacity of 70 mEq $CaCO_3$/L or a molar ratio of at least 1.4:1 of bicarbonate/VFA should be maintained for a stable and well-buffered digestion process (Hobson et al., 1981).

20.2.2.3 Solid Content

Solid concentration in a feedstock will significantly affect the digester design and operation cost. In a continuously stirred tank digester (CSTD), an influent substrate concentration of 3%–8% total solids (TSs) is added daily and an equal amount of effluent is withdrawn. Solid-state anaerobic digestion (SS-AD) generally can use a TS concentration of higher than 15% (Li et al., 2011). Animal manure, sewage sludge, and food waste are usually treated by regular liquid AD, while organic fractions of municipal solid waste (OFMSW) and lignocellulosic biomass such as crop residues and energy crops can be processed through SS-AD.

20.2.2.4 Temperature

AD of organic wastes is usually operated at mesophilic and thermophilic temperatures (Boubaker and Cheikh Ridha, 2007). Different bacterial populations are established within these two temperature ranges, which are from about 20°C to 50°C and from about 55°C to 70°C (Hobson et al., 1981). Relatively few studies have been conducted in the digestion of organic wastes at psychrophilic temperatures (i.e., below 20°C) (Lo and Liao, 1986). Psychrophilic digestion requires less energy input for heating influents and maintaining the temperature of digesters. The main disadvantage of psychrophilic digestion is the low biogas production rate. The rate of thermophilic digestion at a temperature of 50°C–55°C is normally faster than mesophilic digestion at 37°C–40°C (Kalia et al., 2000), and thermophilic digestion was usually more effective than mesophilic digestion for destroying pathogens (Shih, 1993). However, the application of high thermophilic temperatures increases the fraction of free ammonia and pKa of the VFAs, which play inhibiting roles to the microorganisms. Thus, control is a more sensitive issue for thermophilic than mesophilic.

AD process failure can occur at temperature changes in excess of 1°C/day; and changes in temperature of more than 0.6°C/day should be avoided (Turovskiy and Mathai, 2006). The AD of organic wastes at a high temperature may be economically sustainable if inexpensive external heat is available to warm the waste feedstock to the processing temperature. Digesters should be designed to insure more uniform digester temperatures. Obtaining this uniform temperature profile will require a better understanding of the modes of heat transfer that occur within and around the digester and the fluid dynamics in the digester.

20.2.2.5 Solids and Hydraulic Retention Time

The solids retention time (SRT) is the average time that the solid organic wastes are in the system, whereas the hydraulic retention time (HRT) is the average time the sludge liquid is held in the digester. The subsequent steps of the digestion process are directly related to the SRT. The SRT is an important digester design and operation factor for the AD process and is usually expressed in days. Each time solids are removed, a fraction of the bacterial population is also lost, thus suggesting that the cell growth must at least balance the cell removal to ensure a steady AD process.

20.2.3 Inhibition Factors on Anaerobic Digestion

The chemical composition of the organic wastes and the reaction intermediates are important for an AD process. The AD of organic wastes may be inhibited by presenting compounds including long-chain fatty acids, ammonia, and salts. The AD process may also be restricted by the ratio of carbon and nitrogen.

The growth of bacteria for the degradation of long-chain fatty acids is slow and the degradation of long-chain fatty acids requires low partial pressure of hydrogen. The accumulating long-chain fatty acids are toxic to anaerobic microorganisms of acetogens and methanogens (Angelidaki and Ahring, 1995). The inhibition of AD by long-chain fatty acids depends on the type of microorganism, the specific surface area of sludge, the carbon chain length, and the saturation of the long-chain fatty acids (Salminen and Rintala, 2002). Gram-positive microorganisms and methanogens are more vulnerable to fatty acids than gram-negative microorganisms (Roy et al., 1985). Saturated long-chain fatty acids with 12–14 carbons and unsaturated long-chain fatty acids with 18 carbons are more inhibitory (Komatsu et al., 1991). Some substances including albumin, starch, bile acids, cholesterol, bentonite, and calcium have the ability to reduce the toxicity and prevent the inhibition of long-chain fatty acids to AD.

Ammonia primarily produced from the protein degradation may inhibit methanogens, because unionized ammonia can diffuse across the cell membrane to directly inhibit the activity of cytosolic enzymes or affect intracellular pH and other cations (Kadam and Boone, 1996). Small amount of sulfide at 23 mg/L may increase ammonia inhibition, while activated carbon (2.5% by weight) or $FeCl_2$ (4.4 mM) could relieve this inhibition by reduction of sulfide concentration (Hansen et al., 1999). Several approaches including air stripping, dilution of digester content with water, adjustment of feedstock C/N ratio, adding magnesium and orthophosphate, and addition of phosphorite ore have been attempted and are able to successfully prevent ammonia inhibition (Kayhanian, 1999).

Salts such as sodium present in the feedstock may also inhibit an AD. It was found that the AD process was strongly inhibited by sludge from saline fish farm effluents, which contains sodium of 10–10.5 g/L. When diluting this sludge 1:1 with tap water to reduce the sodium concentration to 5.3 g/L, the inhibition effect could be relieved (Gebauer, 2004).

20.2.4 Anaerobic Digester Design

There are five typical types of anaerobic digesters, which are completely mixed digesters (i.e., continuously stirred tank digesters, CSTD), fixed-film digesters, plug-flow digester, anaerobic filter, and fluidized bed digester (Nishio and Nakashimada, 2007). CSTD is the most popular design. In the CSTD, the sludge is continuously fed for maximum efficiency. In a plug-flow digester, a volume of the medium with a suitable inoculum enters at one end of the tube and the digestion is completed when the medium reaches the other end. Unlike completely mixed digesters where the substrate concentration and temperature distribution are forced to uniform values with mechanical mixing, the performance of the plug-flow digester is strongly dependent on how well the fluid dynamics and fermentation kinetics are integrated. The anaerobic filter is primarily used for the digestion of wastewaters produced in large quantities. Microbes are attached to a solid support such as stones packed inside a tank. Wastewater flows upward through the tank and the retention time is only a few hours.

Fluidized bed digester is a modified form of anaerobic filter. In the fluidized bed digester, the bacteria are attached to small glass spheres that are freely suspended in the upflowing feed (Nallathambi Gunaseelan, 1997).

20.2.5 Monitoring Technologies

The development of a monitoring unit for an AD process is similar to that of the established fermentation industry sector. Most monitoring techniques focus on quantifying the key metabolites and parameters in the AD process: VFAs, bicarbonate, phosphate, nitrate, nitrite, ammonium, and TKN, pH, COD, TOC, total and partial alkalinity, redox potential (NADH, NADPH), coenzyme F_{420} from methanogenic bacteria, and biogas flow rate. Recent advancements in AD process analytical technologies (PATs) including fluorescence spectroscopy, infrared spectroscopy, near-infrared spectroscopy, ultraviolet and visual spectroscopy, electronic tongues and noses, microwaves, gas chromatography, titration, and acoustic chemometrics have been reviewed by Madsen et al. (2011).

The VFAs, as the relevant indicator of an AD process, have been widely studied. Among quantification methods developed, titration and chromatographic methods are viewed as the efficient and cheap AD process monitoring technologies. While the titration method cannot provide the relative abundance of individual VFAs, chromatographic methods are commonly used in industry and research laboratories (Brondz, 2002).

20.3 ANAEROBIC DIGESTION OF ORGANIC WASTES

AD is a versatile technology to treat a variety of organic wastes and produce fuel. Essentially all organic materials can be digested, except for stable woody materials since the anaerobic microorganisms are unable to degrade lignin. Existing excessive amount of lignin in the AD reactor may cause the process failure. The following section will discuss the applications of AD for the treatment of organic wastes: renewable feedstock (algae, aquatic plants, agricultural residues, and woody biomass), industrial wastewater, food processing wastes, and municipal solid wastes (MSWs).

20.3.1 Algae

Algae are prokaryotic or eukaryotic photosynthetic microorganisms, which have unicellular or simple multicellular structure, and perform oxygenic photosynthesis like higher plants. Examples of prokaryotic microorganisms are Cyanobacteria (*Cyanophyceae*) and eukaryotic microalgae are green algae (*Chlorophyta*) and diatoms (*Bacillariophyta*) (Li et al., 2008). Algae are essential to global carbon, nitrogen, and sulfur cycling. Approximately 45% of photosynthetic carbon is assimilated by algae. Algae are present in all existing habitat where light is available. Algae have close associations with many other organisms (e.g., lichens, coral, numerous protozoans). It is estimated that more than 50,000 species exist, but only a limited number of around 30,000 have been studied and analyzed (Richmond, 2004). The first use of microalgae by humans was dated back 2000 years to the Chinese, who used *Nostoc* to survive during famine. However, algal biotechnology only really began to develop in the middle of the last century. Nowadays, there are numerous commercial applications of microalgae. For example, (1) microalgae can be used to enhance the nutritional value of food and animal feed owing to their chemical composition; (2) they play a crucial role in aquaculture; (3) they can be incorporated into cosmetics; (4) they can be used in wastewater treatment; and (5) they can be used in biofuel production.

Algae, especially waste-grown algae, are viewed as next-generation biofuel feedstocks because of their superior photosynthetic efficiencies and higher carbon capturing capabilities compared to terrestrial plants. Algae are often used in the secondary wastewater treatment. Without proper treatment, excess nitrogen and phosphorus discharged in wastewater can lead to ecosystem damage (Correll, 1998). The negative effects of such nutrient overloading of receiver systems include low dissolved oxygen concentrations and fish kills, undesirable pH shifts, and cyanotoxin production. Chemical- and physical-based technologies are on hand to remove these nutrients, but they consume significant amounts of energy and chemicals, making them costly processes. Chemical treatment often leads to secondary contamination of the sludge by-products as well, creating additional troubles of safe disposal. The energy and cost required for treatment of wastewater remain a problem for industries and municipalities. Compared to physical and chemical treatment processes, algae-based treatment can potentially achieve nutrient deduction in a less expensive and ecologically safer way with the added benefits of resource recovery and recycling (Christenson and Sims, 2011).

However, according to the current practice for wastewater treatment ponds systems in the United States, algae were not even harvested. Algae biomass was returned to the ponds, in which it decomposes on the pond floor, releasing methane to the air and decreasing water quality. Instead, the algae biomass could be processed for lipid extraction to be used in transportation fuel, or it can be anaerobically digested to make biogas. Algae are usually composed of lipids, proteins, nucleic acids, and noncellulosic carbohydrates. Typical water content of plants is between 80% and 90%. Due to the growth environment of algae, the water content of algae is even higher, which is ~90%–95%. Conventional *algae to biodiesel* technology requires drying of the algal biomass followed by solvent and/or mechanical extraction. The energy consumed in drying process is more than 75% of the total energy consumption. Using AD technology, the wet algae can be directly employed to reduce the need for energy-intensive drying operations, which is almost inevitable in other biorefinery technologies.

However, algae typically yield less methane than wastewater sludge (~0.3 vs. 0.4 L CH_4/g volatile solids (VSs) introduced). Ammonia toxicity and recalcitrant cell walls are the common causes of the low biogas yield. Ammonia toxicity might be counteracted by codigesting algae with high-carbon organic wastes. Carbon-rich organic wastes that are available near major wastewater pond systems include primary and secondary municipal sludge, sorted municipal organic solid waste, waste fats–oils–greases (FOGs), food industry waste, waste paper, and various agricultural residues. Acclimation of the digester microbial community to algae digestion may also improve the yield (Salerno et al., 2009).

Besides wastewater treatment, there are hundreds of algae fuel producers throughout the world. In the perspective of setting up massive cultures, the management of large quantities of residual biomass and the high amounts of fertilizers must be considered. AD is a promising process that can solve this waste issue as well as the economical and energetic balance of such a promising technology. There are three main issues on directly digesting algae: (1) the biodegradability of algae can be low depending on both the biochemical composition and the nature of the cell wall; (2) the high protein content ranging from 30%–75% of cell dry weight results in ammonia release, which can lead to potential toxicity; and (3) the presence of sodium in marine species can also affect the digester performance.

In one study, green microalgae *Chlorella vulgaris* production was coupled with an AD unit. Digestion of *C. vulgaris* was studied under 16 and 28 days HRT, with a corresponding organic loading rate (OLR) of 1 g COD/L. However, 50% of the biomass did not undergo AD, even under long retention times. This points out the need for pretreating and intensifying microalgae biomass digestion (Ras et al., 2011). Physicochemical pretreatment, codigestion, and control of gross composition are comment strategies that can significantly and efficiently enhance the biomass conversion (Zhang and Shahbazi, 2011). In a recent study, a mixture of cellulose and green microalgal biomass of *Dunaliella tertiolecta* (marine species) or *C. vulgaris* was treated using a microbial consortium of TC60, which is known to degrade cellulose and other plant polymers. After five enrichment steps at 60°C, hydrogen yields increased at least 10% under all conditions. AD of *D. tertiolecta* and cellulose by TC60 produced 7.7 mmol H_2/g VSs, which were higher than the results (2.9–4.2 mmol/g VS) obtained with cellulose and *C. vulgaris*. The biomass-fed AD resulted in large shifts in short-chain fatty acid concentrations and increased ammonium levels. To date, this is the first study to examine H_2 production by AD of microalgal biomass. The results indicate that H_2 production is feasible but higher yields could be achieved by the optimization of the bioprocess conditions including biomass pretreatment (Carver et al., 2011).

The unbalanced nutrients of algal sludge (low C/N ratio) were also considered as a significant limitation factor to an AD process. In order to have a balanced C/N ratio in algal sludge, high carbon content of waste paper was added. The addition of 50% (VS based) of waste paper in algal sludge increased the methane production rate to 1170 mL/L/day, as compared to 573 mL/L day of algal sludge digestion alone, both operated at 4 g VS/L/day, 35°C, and 10 days HRT. The maximum methane production rate of 1607 mL/L/day was achieved at a 5 g VS/L day loading rate with 60% (VS based) of paper. Results suggested that an optimum C/N ratio for codigestion of algal sludge and waste paper was in the range of 20–25 to 1 (Yen and Brune, 2007).

Another study showed that AD of algal residues after lipid extraction is a favorable process that can recover more energy than the energy from the cell lipids. The lipid content of algae has a wide range of 10%–85% of cell dry weight, and most algae do not contain over 25% of lipid. When the cell lipid content does not exceed 40%, AD of the whole biomass appears to be the optimal strategy on an energy balance basis, for the energetic recovery of cell biomass (Sialve et al., 2009). When evaluating the potential of algae as feedstock for methane production from a process technical and economic point of view, production of mixed algae culture in raceway ponds on nonagricultural sites, such as landfills, was identified as a preferred approach (Zamalloa et al., 2011).

20.3.2 Aquatic Plants

Similar to algae, aquatic plants such as water hyacinth, duckweed, and cattails can be good candidates as feedstocks for AD due to the high yields of biomass per hectare and year, low fertilization and pesticide requirements, broad adaptability, and greater ability to sequester carbon in the oil than most other plants. Aquatic plants have been used for phytoremediation in constructed wetlands (Suda et al., 2009). Aquatic plants use some pollutants as nutrients. Aquatic plant farms near sewage treatment plants could clean troublesome nitrogen and phosphorus from effluents. Unlike nuclear power and fossil fuels, aquatic plants do not add heat and carbon dioxide to the earth but recycle them. The plants use the sun's energy and the atmosphere's carbon dioxide to produce starches and sugars through photosynthesis. Thus, using aquatic plants as a feedstock for biogas production will add value to the land with little or positive environmental impact.

Like algae, water contents in aquatic species can be as high as ~90%. When applying the AD technology to aquatic plants, the need for energy-intensive drying operations can be avoidable. In addition to the bioenergy benefits provided by substitution of biogas for fossil fuels, such as natural gas, a major advantage of biogas production is that residual material from anaerobic digesters can be readily removed and used as an organic fertilizer.

The basic operations for biogas production from aquatic plants include processing the wet biomass into a slurry, followed by loading the biomass slurry into the anaerobic digester (Malik, 2007). However, using the aquatic plants in single-stage slurry reactors can often cause major problems with clogging. In order to avoid such clogging issues, two-phase and three-phase reactors, which provide greater flexibility for managing spent biomass, have been developed as an alternative (Chanakya et al., 1992). Some research suggests that digestion efficiency and gas production can be greatly improved in multistage reactors when aquatic plant biomass is codigested with manure (Kivaisi and Mtila, 1997). Factors that generally influence biogas yields from aquatic plant biomass include particle size, VS content, plant composition (nitrogen, carbon, and P), trace nutrients, and inoculum volume (Moorhead and Nordstedt, 1993). Targeted sampling to measure these variables can be used to make accurate estimates of expected biogas yield from local aquatic plant populations.

There is an extensive literature describing the use of ADs to produce energy from aquatic plants. While many research has been directed toward water hyacinth (Moorhead and Nordstedt, 1993; Singhal and Rai, 2003), a variety of other aquatic plants have also received some attention for their biogas production potential (Abbasi et al., 1990, 1991; Verma et al., 2007; Alvarez and Lidén, 2008). Yields from different aquatic plant species, and even the same aquatic plant species harvested from different conditions, can be quite inconsistent. One comparative study showed that water hyacinth produced greater long-term biogas yields than both *Azolla* spp. and *Hydrilla* spp. (Taheruzzaman and Kushari, 1989). In an early study of eight common aquatic weeds (*Salvinia molesta, Hydrilla verticillata, Nymphaea stellata, Azolla pinnata, Ceratopteris* sp., *Scirpus* sp., *Cyperus* sp., and *Utricularia reticulata*), an almost tenfold difference in biogas yields was found between *S. molesta* (the highest) and *Cyperus* spp. (the lowest) (Abbasi et al., 1990). When comparing biogas production potentials of water hyacinth (*Eichhornia crassipes*) and water chestnut (*Trapa bispinosa*), the biogas production rate was quicker (maximum from 8 to 12 days) in water hyacinth than in water chestnut (maximum from 12 to 16 days). The qualitative and quantitative variations in biogas production were correlated with COD, C, N, C/N ratio, and toxic metal contents of the slurry used (Verma et al., 2007).

In a series of present studies, potentials of lignocellulosic aquatic plants, cattail (*Typha* spp.) and *Canna indica L.* (canna), were examined in terms of biogas generation by using rumen microorganisms in batch cultures. The maximum VS conversion and the total VFA yield were found to be a function of the substrate concentration and pH. The total VFA yields were about 0.56 and 0.362 g/g VS digested, respectively (Hu and Yu, 2006). The by-products including acetate, propionate butyrate, I-butyrate, and valerate were detected. And the VS conversion was significantly affected by pH value (Yue et al., 2007). The researchers developed a mathematical model to describe the anaerobic conversion of lignocellulose-rich cattails by rumen microbes. The hydrolysis of insoluble components was the rate-limiting step in the AD of cattail (Zhao et al., 2009).

20.3.3 Agricultural Residues

Agricultural residues such as corn stover, wheat straw, and rice straw are cellulosic materials, which are the structural portion of plants, including complex sugars that cannot directly be used for fermentation substrates. At the present time, rising oil prices and uncertainty over the security of existing fossil fuel reserves, combined with concerns over global climate change, have created the great interests for new cellulosic biofuels and bioproducts to substitute for fossil carbon-based materials.

A study conducted in early 1980 has showed that the addition of corn stover to swine manure could enhance biogas production in both mesophilic and thermophilic ADs. Adding corn stover had little effect on solids conversion efficiency, but only 50% of the carbon in corn stover was converted to biogas (Fujita et al., 1980). Most recent AD studies using cellulosic materials suggest or focus on pretreatment. Pretreatment is a key step for efficient utilization of biomass for biogas or biofuel production. The goal of the pretreatment is to increase the surface accessibility of carbohydrate polymers, preextract sugars, disrupt the lignin seal, and liberate the cellulose from the plant cell wall matrix. Based on the application and type of pretreatment catalysts, pretreatment techniques have generally been divided into three categories: physical, chemical, and biological pretreatment. Each pretreatment method has its advantages and disadvantages, and none of the pretreatment approaches is suitable for all biomass species. Pretreatment has been considered as the most expensive processing step in cellulosic ethanol processes, representing about 18% of the total cost (Lynd et al., 1996; Mosier et al., 2005; Yang and Wyman, 2008).

Many studies have been done to investigate the efficient pretreatment process for AD of agricultural residues. Among chemical pretreatment reagents, sodium hydroxide (NaOH) was often used because its alkalinity is much higher than others (Pavlostathis and Gossett, 1985). When NaOH loadings of 1%, 2.5%, 5.0%, and 7.5% (w/w) were applied to corn stover in an SS-AD, the highest biogas yield of 372.4 L/kg VS was obtained with 5% NaOH-pretreated corn stover, which was 37.0% higher than that of the untreated corn stover. However, a higher NaOH loading of 7.5% caused faster production of VFAs during the hydrolysis and acidogenesis stages, which inhibited the methanogenesis (Zhu et al., 2010). Zhang and Zhang (1999) combined physical grinding, heating, and ammonia treatment methods to treat rice straw. Pretreated rice straw resulted in a high biogas yield of 0.47 L/g/VS fed, which is 17.5% higher than the biogas yield of untreated whole straw. Pretreatment temperature was found to have a significant effect on the digestibility of straw (Zhang and Zhang, 1999).

Biological pretreatment by adding enzymes to cellulosic wheat grass (*Jose tall*) showed positive effects on the solubilization of wheat grass (Romano et al., 2009). But the cost of cellulase, hemicellulase, and β-glucosidase has limited the feasibility of applying biological pretreatment in the AD. And this pretreatment did not show significant improvements in biogas and methane yields and VS reduction.

AD was also proposed as an effective way to treat the whole corn stillage from the dry grinding corn-based ethanol plant. At whole corn stillage concentrations of 6,348–50,786 mg total chemical oxygen demand (TCOD)/L, at standard temperature (0°C) and pressure (1 atm), the biogas yields were 88 ± 8 L (49 ± 5 L CH_4) and 96 ± 19 L (65 ± 14 L CH_4) biogas per L stillage from mesophilic and thermophilic digesters, respectively. Continuous-flow studies for the full-strength stillage (TCOD = 254 g/L) at organic loadings of 4.25, 6.30, and 9.05 g TCOD/L/day indicated unstable performance for the thermophilic digester. And the sludge retention time of 60 days does not represent a practical operation time for a large-scale bioethanol plant (Eskicioglu et al., 2011).

20.3.4 Industrial Wastewater

AD is viewed as the most suitable option for the treatment of high-strength organic effluents (Speece, 1983). The presence of biodegradable components in the effluents coupled with the advantages of anaerobic process over other treatment methods makes AD an attractive option.

For this purpose, a number of anaerobic reactors for the digestion of organic effluents from sugar and distillery, pulp and paper, slaughterhouse, dairy units, etc., were well developed (Lettinga, 1995). Rajeshwari et al. (2000) provided a comprehensive review of anaerobic reactors for industrial wastewater treatment (Rajeshwari et al., 2000). In summary, although most of the high-rate reactors have proved their applicability for different high-strength wastewaters over a range of OLRs, there exist certain differences in the preference of a particular type of digester over others in terms of various factors such as requirement of pretreatment, dilution, and control of operating conditions. There are no governing factors that dictate the suitability of any particular reactor design for a specific effluent. By suitable modifications in the reactor designs and by altering the effluent characteristics, the existing high-rate digesters can be accommodated for the treatment of organic effluents.

20.3.5 Food Processing Wastes

Food wastes such as fruit and vegetable processing wastes (FVWs), meat processing wastes, and slaughterhouse wastes can be treated either aerobically or anaerobically to reduce pollutant and pathogen risk and recover materials from the wastes. Food processing wastes have high moisture content, which makes their conversion difficult in a thermochemical conversion process without predrying. AD is a commercially demonstrated technology and is widely used for treating organic wastes with a high moisture content (i.e., >80%–90% moisture) (McKendry, 2002). For the treatment of food processing wastes, AD can not only produce methane for energy but also destroy pathogenic bacteria presenting in the wastes and reduce pollutant emission from the wastes. AD has been used to treat solid fish wastes (Gebauer, 2004), slaughterhouse wastes (Salminen and Rintala, 2002), and food processing wastewater (Di Berardino et al., 2000).

Gebauer (2004) used a CSTD for the mesophilic anaerobic treatment of sludge from saline fish farm effluents with TSs of 8.2%–10.2% by weight, chemical oxygen demand (COD) of 60–74 g/L, and sodium of 10–10.5 g/L. After AD treatment, the COD content of the sludge was decreased from 60% to 36% and methane yields were between 0.114 and 0.184 L/g COD loaded. However, the process was strongly inhibited, presumably by sodium, and unstable, with propionic acid being the main compound of the VFAs. When diluting the sludge 1:1 with tap water (Na content: 5.3 g/L), the inhibition could be overcome and a stable process with low VFA concentrations was achieved (Gebauer, 2004). In another study, codigestion of fish wastes and sisal pulp has been shown to improve the digestibility of both materials and biogas yields (Mshandete et al., 2004).

AD has been used to treat solid slaughterhouse wastes such as the paunch wastes, blood, and settlement tank solids of poultry, pig, and cattle (Banks, 1994; Salminen and Rintala, 2002; Hejnfelt and Angelidaki, 2009). Hejnfelt and Angelidaki (2009) investigated the methane potential from pig meat and bone flour, fat, blood, hair, meat, ribs, and raw waste in batch and semicontinuously fed reactor experiments. The methane potential measured by batch assays is between 225 and 619 dm^3/kg, corresponding to 50%–100% of the calculated theoretical methane potential. Dilution of the by-products had a positive effect on the specific methane yield, and high concentrations of long-chain fatty acids and ammonia in the by-products inhibit the biogas process at concentrations higher than 5 g lipids/dm^3 and 7 g N/dm^3, respectively. Codigestion of 5% pork by-products mixed with pig manure at 37°C showed 40% higher methane production compared to digestion of manure alone (Hejnfelt and Angelidaki, 2009). Banks (1994) treated cattle and lamb paunch contents, blood, and process wastewaters with an organic loading of 0.36 kg COD/m^3/day in a 105 m^3 CSTD. A biogas yield of 0.18 m^3/kg COD was observed at a HRT of 43 days (Banks, 1994). Bank and Wang (1999) further investigated the treatment of cattle blood and rumen paunch contents in a two-phase process with uncoupled solids and liquid retention times. The two-phase process allowed a higher overall loading at 3.6 kg TSs/m^3/day. In the first phase of the process, hydrolysis was apparently enhanced owing to the hydraulic flush, which prevented the accumulation of intermediates, the potential inhibitors of hydrolysis (Banks and Wang, 1999). Salminen and Rintala (2002) treated solid poultry slaughterhouse waste at a loading rate of 0.8 kg (VS) VS/m^3/day and HRT of 50 days in a 2 L CSTD. A methane yield of 0.55 m^3/kg VS was observed (Salminen and Rintala, 2002).

FVWs contain 8%–18% TSs, with a total VS content of 86%–92%. The organic fraction includes about 75% easy biodegradable matter (sugars and hemicellulose), 9% cellulose, and 5% lignin. AD of FVW has been studied under different operating conditions using different types of bioreactors (Bouallagui et al., 2005). Overall, AD permits the conversion of 70%–95% of organic matter to methane, with a volumetric OLR of 1–6.8 g versatile solids/L/day. The major limitation of this process is the rapid acidification, resulting in a low pH value in the reactor and a higher VFA production. VFAs further stress and inhibit the activity of methanogenic bacteria. Continuous two-phase systems appear as more highly efficient technologies for AD of FVW. The greatest advantage lies in the buffering of the OLR taking place in the first stage, allowing a more constant feeding rate of the methanogenic in the second stage. Using a two-stage system consisting of a thermophilic liquefaction reactor and a mesophilic anaerobic filter, over 95% VSs were converted to methane at a volumetric loading rate of 5.65 g VS/L/day. The average methane production yield was about 420 L/kg added VS.

20.3.6 Municipal Solid Wastes

AD of MSW is considered as the modern waste disposal option (Braber, 1995). It has captured a significant share of the European market for the biological treatment of the OFMSW. Almost 4 million ton/year in digestion capacity has been installed through the construction of more than 120 full-scale plants. Though not all plants have been fully successful, due to poor planning, design, or bad operation, AD will most likely play an even more important role in the future (De aere, 2006).

The OFMSW could be anaerobically digested to biogas in the digestion treatment system (Hartmann and Ahring, 2005). Initializing MSW AD normally requires the inoculation with sewage sludge or the codigestion with manure, following a successively higher concentration of MSW (Sosnowski et al., 2003; Macias-Corral et al., 2008). The process variables of the AD of the OFMSW include solid content (high-solids vs. low-solids), processing temperature (mesophilic vs. thermophilic), and the reactor configuration (single-stage vs. multistage processes). The different feedstock characteristics such as composition of biodegradable fractions, C/N ratio, and particle size also have the influence on the biogas production. Generally, source-sorted OFMSW containing a high content of food waste leads to higher biogas yields than the use of mechanically sorted OFMSW. Thermophilic processes are more efficient than mesophilic processes in terms of higher biogas yields at different OLRs. Highest biogas yields are achieved by means of wet thermophilic processes at OLRs lower than 6 kg VS/m^3/day. High-solids processes appear to be relatively more efficient when OLRs higher than 6 kg VS/m^3/day are applied. Multistage systems show in some investigations a higher reduction of recalcitrant organic matter than single-stage systems, but they are seldom applied in full scale. One study indicated that continuously mixing was not necessary for good performance of the MSW AD and was even inhibitory at higher loading rates. In addition, reduction of mixing levels may be used as an operational tool to stabilize unstable digesters (Stroot et al., 2001). An extended cost-benefit calculation shows that the highest overall benefit of the process is achieved at an OLR that is lower and a HRT that is longer than those values of OLR and HRT, at which the highest biogas production is achieved (Hartmann and Ahring, 2006).

The OFMSW is also digestible by hydrogen-producing bacteria enriched from soybean meal silo. A simple model developed from the Gompertz equation was suitable for estimating the hydrogen production potential and rate (Lay et al., 1999). Through response surface methodology, empirical equations for specific hydrogen production potential and rate were fitted and plotted as contour diagrams in order to facilitate examination of experimental results. The contour plots showed the highest hydrogen production potentials of 180 mL H_2/g TVS.

20.4 LIFE CYCLE ASSESSMENT OF ANAEROBIC DIGESTION

LCA is an emerging tool to measure and compare the environmental impacts of human activities. LCA consists of two procedures: the selection of impact indicators and the analysis of inventory data for emissions. The impact indicators used in the LCA may include global warming potential, acidification, eutrophication, photochemical oxidation, and energy use. The energetic balance and the potential environmental impacts of AD are often selected as the indicators in the comparative LCA case studies.

Most of LCA studies focus on the waste treatment in different countries or areas. Finnveden et al. (2005) evaluated different strategies for treating solid waste in Sweden in a life cycle perspective. The whole process includes landfilling, incineration, recycling, AD, and composting. The waste hierarchy suggesting the environmental preference of recycling over incineration over landfilling is often put forward and used in waste policy making. Using LCA, their results indicate that the waste hierarchy is valid as a rule of thumb. The results as well suggest that a policy promoting recycling of paper and plastic materials, preferably combined with policies promoting the use of plastics replacing plastics made from virgin materials, leads to decreased use of total energy and emissions of gases contributing to global warming (Finnveden et al., 2005).

Chaya and Gheewalab (2007) performed LCA to evaluate environmental impacts of two MSW-to-energy schemes currently practiced in Thailand: incineration and AD. AD scheme was found to have more advantages: (1) Potential impacts such as global warming, acidification, stratospheric ozone depletion, and photooxidant formation can be avoided due to net electricity production and fertilizer production and (2) the AD resulted in the higher net energy output compared to the incineration scheme. But the incineration had less potential impact for nutrient enrichment. LCA analysis provides invaluable information to policy makers and suggests an integrated method of management (Chaya and Gheewala, 2007).

Cherubini et al. (2009) applied LCA to assess four waste management strategies in Rome (Italy): landfill without biogas utilization, landfill with biogas combustion to generate electricity, sorting plant that splits the inorganic waste fraction, and direct incineration of waste. The indicators consist of global and local emissions, total material demands, total energy requirements, and ecological footprints. The results show that the traditional landfill systems are the worst waste management option and significant environmental savings at global scale are achieved from undertaking energy recycling (Cherubini et al., 2009).

Tilche and Galatola (2008) looked into the potential of AD in terms of reduction of GHG emissions in Europe. The LCA was performed for the 27 European countries on the basis of their 2005 Kyoto declarations. They considered two different possible biogas applications: electricity production from manure waste and upgraded methane production for light goods vehicles. Their work shows that biogas may noticeably contribute to GHG reductions, and its use as a biofuel may allow for true negative GHG emissions, showing a net advantage with respect to other biofuels (Tilche and Galatola, 2008).

Though the LCA case analysis was performed individually in different countries and cities, the conclusions are reliable for most of the big cities. AD as a waste treatment technology has minor or negative impacts to human life, reduces GHG emissions, and generates high net energy output.

20.5 CONCLUSIONS

AD is a versatile technology that has the potential to recover energy from a variety of organic wastes. Almost all organic wastes, such as renewable feedstock (algae, aquatic plants, agricultural residues, and woody biomass), industrial wastewater, MSWs, and food processing wastes, can be digested using AD process. The chemical composition of the organic wastes, the reaction intermediates, and the operating parameters are important for the performance of an AD process. AD PATs including fluorescence spectroscopy, infrared spectroscopy, near-infrared spectroscopy, ultraviolet and visual spectroscopy, electronic tongues and noses, microwaves, gas chromatography, titration, and acoustic chemometrics have been well developed. Meanwhile, the LCA shows that AD as a waste treatment technology has minor or negative impacts to human life, reduces GHG emissions, and generates high net energy output.

ACKNOWLEDGMENTS

This publication was made possible by Grant Number NC.X-269-5-12-130-1 from the National Institute of Food and Agriculture, the United States Department of Agriculture. Its contents are solely the responsibility of the authors and do not necessarily represent the official views of the National Institute of Food and Agriculture.

REFERENCES

Abbasi, S. A., P. C. Nipaney, and M. B. Panholzer, 1991. Biogas production from the aquatic weed Pistia (*Pistia stratiotes*), *Bioresource Technology* 37, 211–214.

Abbasi, S. A., P. C. Nipaney, and G. D. Schaumberg, 1990. Bioenergy potential of eight common aquatic weeds, *Biological Wastes* 34, 359–366.

Ahring, B. K., M. Sandberg, and I. Angelidaki, 1995. Volatile fatty acids as indicators of process imbalance in anaerobic digestors, *Applied Microbiology and Biotechnology* 43, 559–565.

Alvarez, R. and G. Lidén, 2008. Anaerobic co-digestion of aquatic flora and quinoa with manures from Bolivian Altiplano, *Waste Management* 28, 1933–1940.

Angelidaki, I. and B. K. Ahring, 1995. Establishment and characterization of an anaerobic thermophilic (55(deg) C) enrichment culture degrading long-chain fatty acids, *Applied and Environmental Microbiology* 61, 2442–2445.

Appels, L., J. Baeyens, J. Degrève, and R. Dewil, 2008. Principles and potential of the anaerobic digestion of waste-activated sludge, *Progress in Energy and Combustion Science* 34, 755–781.

Banks, C. J., 1994. Anaerobic digestion of solid and high nitrogen content fractions of slaughterhouse wastes, environmentally responsible food processing, *AIChE Symposium Series* 90, 48–55.

Banks, C. J. and Z. Wang, 1999. Development of a two phase anaerobic digester for the treatment of mixed abattoir wastes, *Water Science and Technology* 40, 67–76.

Bouallagui, H., Y. Touhami, R. Ben Cheikh, and M. Hamdi, 2005. Bioreactor performance in anaerobic digestion of fruit and vegetable wastes, *Process Biochemistry* 40, 989–995.

Boubaker, F. and B. Cheikh Ridha, 2007. Anaerobic co-digestion of olive mill wastewater with olive mill solid waste in a tubular digester at mesophilic temperature, *Bioresource Technology* 98, 769–774.

Braber, K., 1995. Anaerobic digestion of municipal solid waste: A modern waste disposal option on the verge of breakthrough, *Biomass and Bioenergy* 9, 365–376.

Brondz, I., 2002. Development of fatty acid analysis by high-performance liquid chromatography, gas chromatography, and related techniques, *Analytica Chimica Acta* 465, 1–37.

Carver, S., C. Hulatt, D. Thomas, and O. Tuovinen, 2011. Thermophilic, anaerobic co-digestion of microalgal biomass and cellulose for H2 production, *Biodegradation* 22, 805–814.

Chanakya, H. N., S. Borgaonkar, M. G. C. Rajan, and M. Wahi, 1992. Two-phase anaerobic digestion of water hyacinth or urban garbage, *Bioresource Technology* 42, 123–131.

Chaya, W. and S. H. Gheewala, 2007. Life cycle assessment of MSW-to-energy schemes in Thailand, *Journal of Cleaner Production* 15, 1463–1468.

Cherubini, F., S. Bargigli, and S. Ulgiati, 2009. Life cycle assessment (LCA) of waste management strategies: Landfilling, sorting plant and incineration, *Energy* 34, 2116–2123.

Christenson, L. and R. Sims, 2011. Production and harvesting of microalgae for wastewater treatment, biofuels, and bioproducts, *Biotechnology Advances* 29, 686–702.

Correll, D. L., 1998. Role of phosphorus in the eutrophication of receiving waters: A review, *Environmental Quality* 27, 261–266.

De Baere, L., 2006. Will anaerobic digestion of solid waste survive in the future? *Water Science and Technology* 53(8), 187–194.

Di Berardino, S., S. Costa, and A. Converti, 2000. Semi-continuous anaerobic digestion of a food industry wastewater in an anaerobic filter, *Bioresource Technology* 71, 261–266.

Eskicioglu, C., K. J. Kennedy, J. Marin, and B. Strehler, 2011. Anaerobic digestion of whole stillage from dry-grind corn ethanol plant under mesophilic and thermophilic conditions, *Bioresource Technology* 102, 1079–1086.

Finnveden, G., J. Johansson, P. Lind, and Å. Moberg, 2005. Life cycle assessment of energy from solid waste—Part 1: General methodology and results, *Journal of Cleaner Production* 13, 213–229.

Fujita, M., J. M. Scharer, and M. Moo-Young, 1980. Effect of corn stover addition on the anaerobic digestion of swine manure, *Agricultural Wastes* 2, 177–184.

Gebauer, R., 2004. Mesophilic anaerobic treatment of sludge from saline fish farm effluents with biogas production, *Bioresource Technology* 93, 155–167.

Hansen, K. H., I. Angelidaki, and B. K. Ahring, 1999. Improving thermophilic anaerobic digestion of swine manure, *Water Research* 33, 1805–1810.

Hartmann, H. and B. K. Ahring, 2005. Anaerobic digestion of the organic fraction of municipal solid waste: Influence of co-digestion with manure, *Water Research* 39, 1543–1552.

Hartmann, H. and B. K. Ahring, 2006. Strategies for the anaerobic digestion of the organic fraction of municipal solid waste: An overview, *Water Science and Technology* 53, 7–22.

Hejnfelt, A. and I. Angelidaki, 2009. Anaerobic digestion of slaughterhouse by-products, *Biomass and Bioenergy* 33, 1046–1054.

Hobson, P. N., S. Bousfield, and R. Summers, 1981. *Methane Production from Agricultural and Domestic Wastes*. New York: John Wiley & Sons, Inc.

Hu, Z.-H. and H.-Q. Yu, 2006. Anaerobic digestion of cattail by rumen cultures, *Waste Management* 26, 1222–1228.

Kadam, P. C. and D. R. Boone, 1996. Influence of pH on ammonia accumulation and toxicity in halophilic, methylotrophic methanogens, *Applied Environmental Microbiology* 62, 4486–4492.

Kalia, V. C., V. Sonakya, and N. Raizada, 2000. Anaerobic digestion of banana stem waste, *Bioresource Technology* 73, 191–193.

Kayhanian, M., 1999. Ammonia inhibition in high-solids biogasification-an overview and practical solution, *Environmental Technology* 20, 355–365.

Kivaisi, A. K. and M. Mtila, 1997. Production of biogas from water hyacinth (Eichhornia crassipes) (Mart) (Solms) in a two-stage bioreactor, *World Journal of Microbiology and Biotechnology* 14, 125–131.

Komatsu, T., K. Hanaki, and T. Matsuo, 1991. Prevention of lipid inhibition in anaerobic processes by introducing a two-phase system, *Water Science and Technology* 23, 1189–1200.

Lay, J.-J., Y.-J. Lee, and T. Noike, 1999. Feasibility of biological hydrogen production from organic fraction of municipal solid waste, *Water Research* 33, 2579–2586.

Lettinga, G., 1995. Anaerobic digestion and wastewater treatment systems, *Antonie van Leeuwenhoek* 67, 3–28.

Li, Y., M. Horsman, N. Wu, C. Q. Lan, and N. Dubois-Calero, 2008. Biofuels from microalgae, *Biotechnology Progress* 24(4), 815–820.

Li, Y., S. Y. Park, and J. Zhu, 2011. Solid-state anaerobic digestion for methane production from organic waste, *Renewable and Sustainable Energy Reviews* 15, 821–826.

Lo, K. V. and P. H. Liao, 1986. Psychrophilic anaerobic digestion of screened dairy manure, *Energy in Agriculture* 5, 339–345.

Lynd, L., R. Elamder, and C. Wyman, 1996. Likely features and costs of mature biomass ethanol technology, *Applied Biochemistry and Biotechnology* 57–58, 741–761.

Macias-Corral, M., Z. Samani, A. Hanson, G. Smith, P. Funk, H. Yu, and J. Longworth, 2008. Anaerobic digestion of municipal solid waste and agricultural waste and the effect of co-digestion with dairy cow manure, *Bioresource Technology* 99, 8288–8293.

Madsen, M., J. B. Holm-Nielsen, and K. H. Esbensen, 2011. Monitoring of anaerobic digestion processes: A review perspective, *Renewable and Sustainable Energy Reviews* 15, 3141–3155.

Malik, A., 2007. Environmental challenge vis a vis opportunity: The case of water hyacinth, *Environment International* 33, 122–138.

Mata-Alvarez, J., S. Macé, and P. Llabrés, 2000. Anaerobic digestion of organic solid wastes. An overview of research achievements and perspectives, *Bioresource Technology* 74, 3–16.

McKendry, P., 2002. Energy production from biomass (part 2): Conversion technologies, *Bioresource Technology* 83, 47–54.

Moorhead, K. K. and R. A. Nordstedt, 1993. Batch anaerobic digestion of water hyacinth: Effects of particle size, plant nitrogen content, and inoculum volume, *Bioresource Technology* 44, 71–76.

Mosier, N., C. Wyman, B. Dale, R. Elander, Y. Y. Lee, M. Holtzapple, and M. Ladisch, 2005. Features of promising technologies for pretreatment of lignocellulosic biomass, *Bioresource Technology* 96, 673–686.

Mshandete, A., A. Kivaisi, M. Rubindamayugi, and B. Mattiasson, 2004. Anaerobic batch co-digestion of sisal pulp and fish wastes, *Bioresource Technology* 95, 19–24.

Nallathambi Gunaseelan, V., 1997. Anaerobic digestion of biomass for methane production: A review, *Biomass and Bioenergy* 13, 83–114.

Nielsen, H. B., H. Uellendahl, and B. K. Ahring, 2007. Regulation and optimization of the biogas process: Propionate as a key parameter, *Biomass and Bioenergy* 31, 820–830.

Nishio, N. and Y. Nakashimada, 2007. Recent development of anaerobic digestion processes for energy recovery from wastes, *Journal of Bioscience and Bioengineering* 103, 105–112.

Pavlostathis, S. G. and J. M. Gossett, 1985. Alkaline treatment of wheat straw for increasing anaerobic biodegradability, *Biotechnology and Bioengineering* 27, 334–344.

Pind, P. F., I. Angelidaki, and B. K. Ahring, 2003. Dynamics of the anaerobic process: Effects of volatile fatty acids, *Biotechnology and Bioengineering* 82, 791–801.

Rajeshwari, K. V., M. Balakrishnan, A. Kansal, K. Lata, and V. V. N. Kishore, 2000. State-of-the-art of anaerobic digestion technology for industrial wastewater treatment, *Renewable and Sustainable Energy Reviews* 4, 135–156.

Ras, M., L. Lardon, S. Bruno, N. Bernet, and J.-P. Steyer, 2011. Experimental study on a coupled process of production and anaerobic digestion of *Chlorella vulgaris*, *Bioresource Technology* 102, 200–206.

Richmond, A., 2004. *Handbook of Microalgal Culture: Biotechnology and Applied Phycology*. Oxford, U.K.: Blackwell Science Ltd.

Romano, R. T., R. Zhang, S. Teter, and J. A. McGarvey, 2009. The effect of enzyme addition on anaerobic digestion of Jose Tall Wheat Grass, *Bioresource Technology* 100, 4564–4571.

Roy, F., G. Albagnac, and E. Samain, 1985. Influence of calcium addition on growth of highly purified syntrophic cultures degrading long-chain fatty acids, *Applied and Environmental Microbiology* 49, 702–705.

Salerno, M., Y. Nurdogan, and T. J. Lundquist, 2009. Biogas production from algae biomass harvested at wastewater treatment ponds, ASABE Paper Number: Bio098023. Available at http://digitalcommons.calpoly.edu/cgi/viewcontent.cgi?article=1181&context=cenv_fac (accessed 30 December 2013).

Salminen, E. and J. Rintala, 2002. Anaerobic digestion of organic solid poultry slaughterhouse waste—A review, *Bioresource Technology* 83, 13–26.

Shih, J. C. H., 1993. Recent development in poultry waste digestion and feather utilization—A review, *Poultry Science* 72, 1617–1620.

Sialve, B., N. Bernet, and O. Bernard, 2009. Anaerobic digestion of microalgae as a necessary step to make microalgal biodiesel sustainable, *Biotechnology Advances* 27, 409–416.

Singhal, V. and J. P. N. Rai, 2003. Biogas production from water hyacinth and channel grass used for phytoremediation of industrial effluents, *Bioresource Technology* 86, 221–225.

Sosnowski, P., A. Wieczorek, and S. Ledakowicz, 2003. Anaerobic co-digestion of sewage sludge and organic fraction of municipal solid wastes, *Advances in Environmental Research* 7, 609–616.

Speece, R. E., 1983. Anaerobic biotechnology for industrial wastewater treatment, *Environmental Science & Technology* 17, 416A–427A.

Stroot, P. G., K. D. McMahon, R. I. Mackie, and L. Raskin, 2001. Anaerobic codigestion of municipal solid waste and biosolids under various mixing conditions—I. Digester performance, *Water Research* 35, 1804–1816.

Suda, K., A. Shahbazi, and Y. Li, 2009. *The Feasibility of Using Cattails from Constructed Wetlands to Produce Bioethanol*. New York: Springer.

Taheruzzaman, Q. and D. P. Kushari, 1989. Evaluation of some common aquatic macrophytes cultivated in enriched water as possible source of protein and biogas, *Hydrobiological Bulletin* 23, 207–212.

Tilche, A. and M. Galatola, 2008. The potential of bio-methane as bio-fuel/bio-energy for reducing greenhouse gas emissions: A qualitative assessment for Europe in a life cycle perspective, *Water Science Technology* 57, 1683–1692.

Turovskiy, I. S. and P. K. Mathai, 2006. *Wastewater Sludge Processing*. New York: Wiley.

Verma, V. K., Y. P. Singh, and J. P. N. Rai, 2007. Biogas production from plant biomass used for phytoremediation of industrial wastes, *Bioresource Technology* 98, 1664–1669.

Yang, B. and C. E. Wyman, 2008. Pretreatment: The key to unlocking low-cost cellulosic ethanol, *Biofuels, Bioproducts and Biorefining* 2, 26–40.

Yen, H.-W. and D. E. Brune, 2007. Anaerobic co-digestion of algal sludge and waste paper to produce methane, *Bioresource Technology* 98, 130–134.

Yue, Z.-B., H.-Q. Yu, H. Harada, and Y.-Y. Li, 2007. Optimization of anaerobic acidogenesis of an aquatic plant, *Canna indica L.*, by rumen cultures, *Water Research* 41, 2361–2370.

Zamalloa, C., E. Vulsteke, J. Albrecht, and W. Verstraete, 2011. The techno-economic potential of renewable energy through the anaerobic digestion of microalgae, *Bioresource Technology* 102, 1149–1158.

Zhang, B. and A. Shahbazi, 2011. Recent developments in pretreatment technologies for production of lignocellulosic biofuels, *Journal of Petroleum Environmental Biotechnology* 2, 108.

Zhang, R. and Z. Zhang, 1999. Biogasification of rice straw with an anaerobic-phased solids digester system, *Bioresource Technology* 68, 235–245.

Zhao, B.-H., Z.-B. Yue, B.-J. Ni, Y. Mu, H.-Q. Yu, and H. Harada, 2009. Modeling anaerobic digestion of aquatic plants by rumen cultures: Cattail as an example, *Water Research* 43, 2047–2055.

Zhu, J., C. Wan, and Y. Li, 2010. Enhanced solid-state anaerobic digestion of corn stover by alkaline pretreatment, *Bioresource Technology* 101, 7523–7528.

21 Combined Heat and Power Generation from Biomass

Hao Liu and Rabah Boukhanouf

CONTENTS

21.1 INTRODUCTION

The process of fuel energy conversion to mechanical energy is governed by well-understood laws of thermodynamics. These impose severe limitations on the thermal performance of heat-to-mechanical conversion devices, or heat engines, which depends greatly on the available heat source and the heat sink temperatures. Often, this means that huge quantities of low-temperature heat energy are rejected from industrial-scale power plants such as coal-fired steam power plants. Even in modern combined cycle power plants where the heat source temperature can be as high as 1300°C, the maximum electrical efficiencies are only about 50%. Hence, the impossibility of converting all fuel energy into useful work has led many energy consumers both large and small scale to adopt energy-efficient techniques such as combined heat and power (CHP) to reduce energy consumption and alleviate the environmental impact of burning fossil fuels.

CHP, also known as cogeneration, is the simultaneous generation of power and heat from a single fuel source with a view to using both products. CHP has long been accepted as an energy-efficient technology in many manufacturing and service industries where energy bills represent a significant share of the overall operating cost of a site. CHP in service industry market is mainly driven by small-scale packaged plants to supply low-temperature hot water and power in buildings such as leisure centers, hotels, and hospitals (Boukhanouf 2011). On the other hand, in energy-intensive industries with material transformation processes such as steel, petrochemical, paper, and pulp, large-scale CHP systems are exclusively adopted. These types of industries present particularly high confidence in investing in CHP because of consistent demand for electrical power and heat from continuous processes with a defined load profile throughout the year.

CHP is not a new technology. Power plants supplying heat to a pressurized hot water network system, which is then distributed to public and domestic buildings for space heating, also known as district heating, have been commonplace in many cities of Northern and Eastern Europe from as early as the turn of the last century. Despite this early success, CHP industry encountered widespread resistance from monolithic state-owned electricity companies in many countries. It is the energy crisis of the 1970s that had shifted opinions around the world in favor of CHP. In the United States, for example, the Congress passed in 1979 the Public Utility Regulatory Policies Act (PURPA) that guarantees CHP operators' electricity market prices for electricity fed into the grid (Doty and Turner 2009, Simons 2009). This had initially benefitted industries such as paper mills and petrochemical plants with large steam consumption. In other developed countries, many examples of small-scale CHP schemes supplying a specific site such as a hospital or a leisure center were becoming popular. The full advantage of CHP was not realized, however, until after the deregulation of electricity industry in major developed countries in the early 1990s. This had introduced a marked rethinking and organization of the power utilities by separating power generation, transmission, and distribution, creating a dynamic and competitive energy market that allowed new independent power generators to enter the different segments of the new energy market. This also encouraged new and established power generators to build customized CHP units in the proximity of customers' manufacturing plants to export heat (usually as steam) and feed the power to the grid network or sell it directly to the customer through an agreed contract. However, in the United Kingdom, the opening up of the electricity market and the subsequent introduction of the governing competition regulations in 2001, the New Electricity Trading Agreement (NETA), had led to a huge depression in electricity prices. This made the spark spread (the difference between fuel prices and electricity prices) unattractive for potential investors in CHP schemes, leading to a temporary decline in installed CHP capacity (Kelly and Pollitt 2010). The electricity prices have since recovered and the many economic incentives available to CHP operators have also encouraged the CHP market to expand again through new entrants.

The driving energy source for CHP plants can come from a variety of primary energy sources including traditional fossil fuels (coal, natural gas, and oil) or renewable energy sources such as solar energy, geothermal energy, and biomass. When a chemical fuel (e.g., coal, gas, oil, and biomass) is used as the primary energy source of a CHP plant, its energy is usually released through combustion, that is, by burning the fuel with air to produce high-temperature combustion products. While natural gas has been widely used in CHP systems since the 1990s, virtually all other types of fuels can be used to power CHP including coal, landfill gas, solid waste, waste gases from landfill and refineries, fuel and gas oils, biomass, and biogas.

21.2 CHP TECHNOLOGIES

A CHP plant consists essentially of a heat engine, also referred to as prime mover, and associated heat recovery and control equipment. The prime mover converts the energy released from the primary energy source (e.g., via burning the primary fuel) into mechanical energy in the form of a

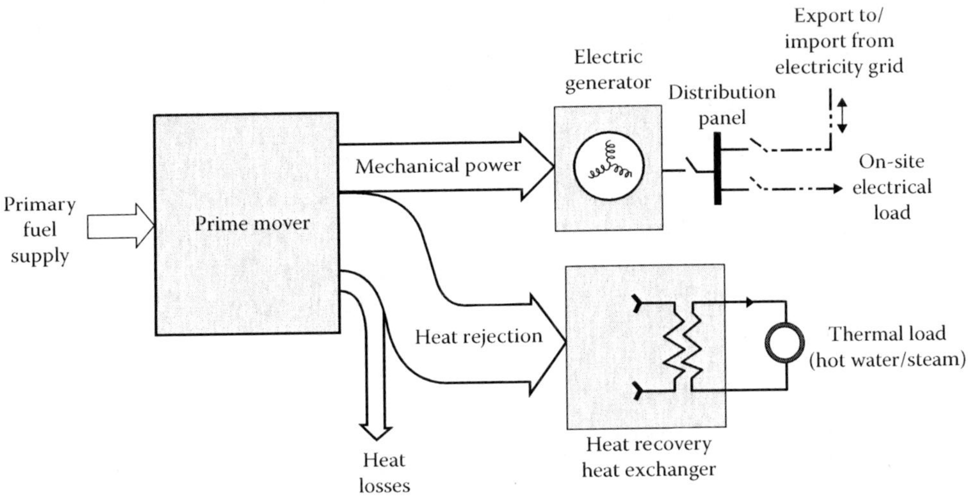

FIGURE 21.1 General representation of main components of a CHP system.

rotating or reciprocating shaft that is usually coupled to an electric generator for electrical energy generation or used to drive directly an air compressor or a hydraulic pump. The superior energy performance of a CHP is realized from recovering some of the thermal energy liberated in the prime mover, otherwise wasted in power-only generating plants, for useful heating and cooling applications. Figure 21.1 shows a schematic of a CHP plant.

21.2.1 CHP Performance Characterization

Recovery of a portion of the heat rejected in a power generating plant to operate on-site heat processes or a district heating system can increase overall fuel utilization efficiency of the plant to above 80%. From the energy conservation principle, the total energy conversion efficiency of a CHP system could be expressed as follows:

$$\eta_{CHP} = \frac{P + Q_{th}}{Q_{in}} \tag{21.1}$$

where
 P is the electrical power output
 Q_{th} is the thermal output of the CHP plant

The total thermal efficiency η_{CHP} gives a quantitative measurement of the CHP performance in terms of energy utilization. However, it does not distinguish between the quality of power and heat output. Therefore, electrical energy conversion efficiency, η_e, is often specified as a quality parameter of a CHP prime mover that needs maximizing as electricity is more valuable than heat with the price of electricity up to 10 times that of heat at peak load demand (ETSU 1999). The electrical energy conversion efficiency, η_e, is defined as

$$\eta_e = \frac{P}{Q_{in}} \tag{21.2}$$

Most CHP designers aim to maximize η_e. In practice, most CHP plants are employed primarily to satisfy on-site heat demand and generate power as a by-product. It becomes important when selecting the size of a prime mover that would match the heat and power demand of a site. Hence, there is need for a further design parameter that is specific to a prime mover technology, the heat-to-power ratio, expressed as

$$\kappa = \frac{Q_{th}}{P} \tag{21.3}$$

A well-sized CHP system would consume less primary energy for similar heat and power outputs compared to supplying heat and power from separate sources, resulting in savings in energy cost and carbon and pollutants emissions. The level of savings though depends both on technical parameters of the prime mover and prices of fuel and grid power. Assuming that all heat recovered from a CHP system is used with no provision for damping, the savings in cost that could be realized per kWh of electricity generated can be estimated from the following equation (CIBSE 1999):

$$\Delta c = C_E + C_m + \left(\frac{\kappa}{\eta_b} - \frac{1}{\eta_e}\right) C_F \tag{21.4}$$

where

C_E is the unit price of imported electricity from grid
C_m is the maintenance cost per unit of power generated
C_F is the unit of fuel cost
η_b is the efficiency of a boiler displaced by the CHP

21.2.2 Prime Mover Selection

In practice, the selection of a CHP prime mover is exclusively site dependent and the economic viability of a CHP scheme can only be realized if the full energy requirement of the site is well defined. This mainly includes detailed electrical and thermal energy load profiles, a breakdown of the grades of the heat energy requirement, fuel flexibility, and availability. Under optimized conditions, the prime mover will operate at nominal power capacity for maximum number of running hours and all recoverable heat is fully used. This also means that the prime mover's heat-to-power ratio is appropriately selected to match that of the site's energy demand.

Most existing CHP schemes are dominated by the traditional prime movers technologies such as steam turbines, gas turbines, and reciprocating internal combustion (IC) engines. Other devices such as organic Rankine cycle (ORC), Stirling engines, fuel cells, and thermoelectric generators that have found applications in CHP are either still in development or confined to serving niche markets.

21.2.2.1 Steam Turbines

Steam turbines are one of the oldest heat engine technologies that are used to convert thermal energy from high-pressure steam into mechanical power. Most of centralized electricity generation plants in the world operate on the steam cycle principle known as the Rankine cycle. In a Rankine cycle, high-pressure steam is raised in a steam boiler or heat recovery steam generator (HRSG) using energy from fossil fuels, fission reactions of radioactive materials, or concentrated solar energy. Depending on application, steam turbines can be designed to operate over a wide range of steam pressures. Modern high-efficiency steam turbines have been designed to operate at supercritical vapor pressure and condenser pressure near absolute vacuum conditions (Bugge et al. 2006). A simple schematic of the components of a steam cycle is shown in Figure 21.2a.

In the steam boiler, the energy released from the combustion of the fuel in the boiler combustion furnace is transferred to the working fluid (water) through the boiler metal tubes. Water circulating through the boiler tubes is heated up to generate saturated steam, which in turn is heated further in steam superheaters to become superheated steam, which is represented by point 3 on a temperature—entropy diagram of an ideal Rankine cycle of Figure 21.2b. The high-temperature

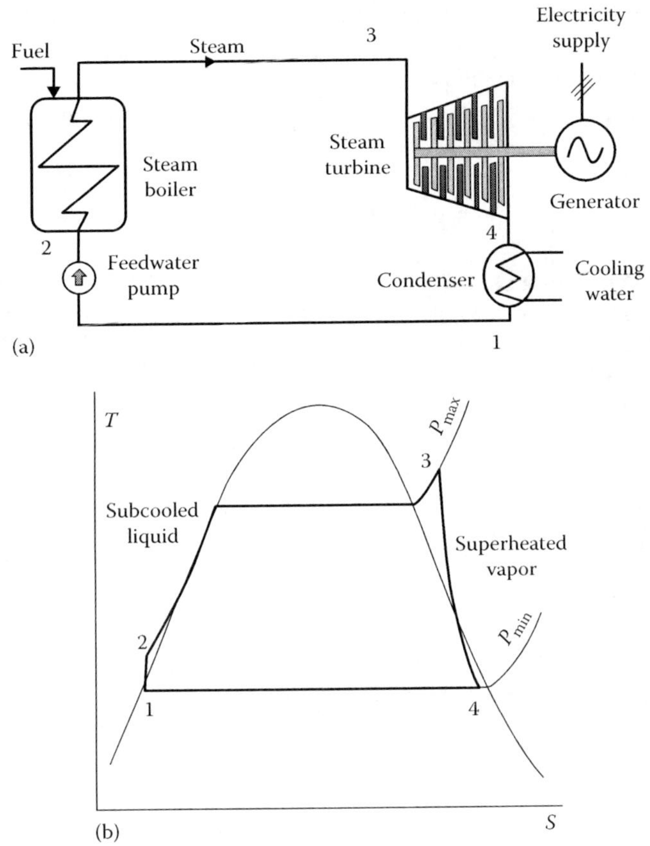

FIGURE 21.2 (a) Main components of a simple steam turbine cycle; (b) *T–S* diagram of a simple steam turbine ideal cycle.

and high-pressure superheated steam is then expanded adiabatically through a set of turbine blades that rotate the turbine–generator shaft, converting the energy in the steam to mechanical power. The expanded steam is finally drawn out of the last turbine stage and fed to a condenser where it is cooled to its liquid state again. In practice, it is not permissible to allow the quality of steam at the turbine exhaust to be lower than about 88%, as excess condensate in the steam flow can cause cavitation of the turbine blades (Leizerovich 2005). The cycle is completed by pumping the feedwater back to the boiler.

The thermal performance of a steam turbine cycle can be evaluated by its electrical efficiency, which is defined as the ratio of the net power output to the heat input to the working fluid in the boiler and can be expressed by the working fluid enthalpies as follows:

$$\eta_s = \frac{w_{net}}{q_{in}} = \frac{w_T - w_p}{q_{in}} = \frac{(h_3 - h_4) - (h_2 - h_1)}{h_3 - h_2} \quad (21.5)$$

where

- w_{net} is the cycle's specific net work output
- q_{in} is the specific heat input to the boiler
- w_T and w_p are the specific work output of the turbine and work input to the feedwater pump, respectively
- h_1, h_2, h_3, and h_4 are the specific enthalpy of the working fluid at marked state points on the *T–S* diagram

The advantage of a Rankine cycle is that the power consumption of the feedwater pump is usually very small (in the order of 2.5% of the turbine power output) and can be neglected. Hence, Equation 21.5 can be rewritten as follows:

$$\eta_s = \frac{w_T}{q_{in}} = \frac{h_3 - h_4}{h_3 - h_2} \tag{21.6}$$

For typical large steam turbine power plants, the efficiency of converting fuel energy into electricity is about 40%. This could be increased to about 45% with the modified Rankine cycle by expanding steam through a cascade of high- and low-pressure turbine stages with steam reheating processes in between. Figure 21.3a shows an example of a simple reheat Rankine cycle configuration. The reheat technique offers a way to limit the water condensate droplets at the turbine exit stages, reducing the effect of blade erosion and increasing energy conversion efficiency. Figure 21.3b shows that reheating of steam at the exit of the first turbine shifts the turbine expansion process away from the water vapor saturation line and toward the superheat region on the *T–S* diagram.

Generally, steam turbine cycles are characterized by the enormous quantities of thermal energy discharged into the heat sink as imposed by the principles governing the thermodynamic cycle. The high reliability and simplicity of steam turbine technology coupled with the development of electrical transmission and distribution systems has led to concentrating power generation in a few large power plants located in proximity of fuel energy sources serving many industrial and residential agglomerations over a wide geographic area.

Depending on required outlet steam pressure, steam turbines can be classified either as condensing or noncondensing (back-pressure) steam turbines (Perchers 2003).

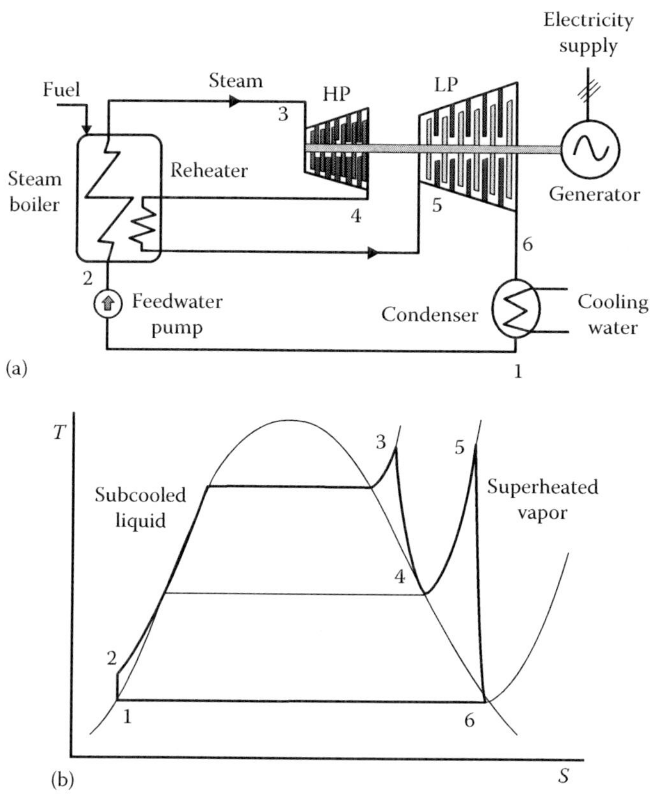

FIGURE 21.3 (a) Main components of a simple reheat steam turbine cycle; (b) *T–S* diagram of a simple reheat steam turbine ideal cycle.

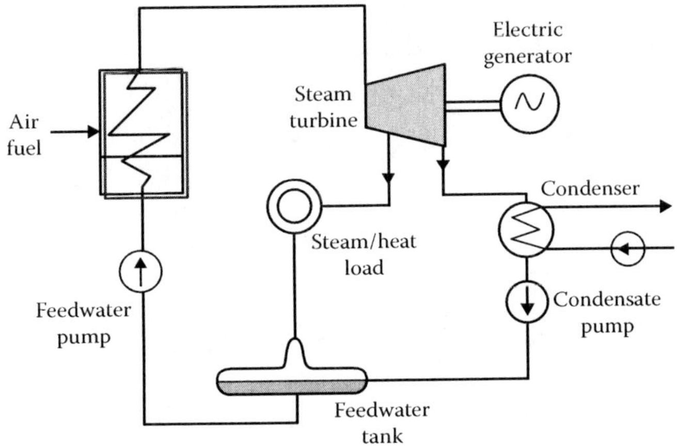

FIGURE 21.4 Extraction condensing steam turbine CHP.

Condensing steam turbine: In a condensing steam turbine, part or all of the inlet steam is discharged to a condenser where virtually a vacuum is maintained for maximum power output. In practice, condenser pressure of as low as 4 kPa could be achieved with the availability of a heat sink temperature of around 30°C. This in turn engenders huge quantities of attendant low-grade heat to be dissipated, requiring large cooling water flow rate from nearby sea, river, or erected cooling towers. It is noteworthy that in this case the discharged lukewarm cooling water has no real thermal value and represents a thermal pollution to the environment. In fact, most low-temperature heating applications such as space heating in buildings would require hot water temperatures at about 60°C–90°C, whereas medium- to high-temperature applications would require heat mostly in the form of steam. Steam can be bled off from carefully placed openings in the steam turbine casing to provide desired steam flow rates at the required process pressure. This type of steam turbines is known as extraction condensing steam turbines. A schematic of an extraction steam turbine CHP plant is shown in Figure 21.4.

Back-pressure steam turbine: The main application for noncondensing steam turbines is the heat supply in the form of steam where the electrical power generation is considered as a by-product. This is mainly due to discharging spent steam in the back-pressure steam turbine at a pressure often higher than atmospheric pressure, thus reducing its power output and electrical efficiency. Typical electrical efficiencies of a back-pressure steam turbine range from below 10% to over 30%. A back-pressure steam turbine has the advantage of simplicity and cost effectiveness compared to its condensing turbine counterpart. A schematic of a back-pressure steam turbine is shown in Figure 21.5.

21.2.2.2 Gas Turbines

The development of the concept of gas turbine for stationary applications to convert thermal energy to electrical energy originated as an offshoot from the aeroengines industry. Historically, the first successful implementation of gas turbine for power generation was back in 1937 in Neuchatel, Switzerland, when Brown Boveri Company introduced a 3 MW gas-turbine-driven electrical power system (Wetson 1992). However, the market adoption of this aeroderivative gas turbine technology was very slow, and for decades, it was mainly used by energy utilities for peaking and standby power generation duties. It was until the 1990s that the development of gas turbine as a competing technology to the conventional large steam turbines took off as a result of electric utilities deregulation in major industrialized countries, a relaxation of a ban on using natural gas for power generation in Europe, and the growing concern of environmental effect of coal-fired steam power stations.

Today, gas turbines' market share for power generation is growing at a fast pace. The technology is reliable and mature and is widely used by energy utilities and industrial power generators in

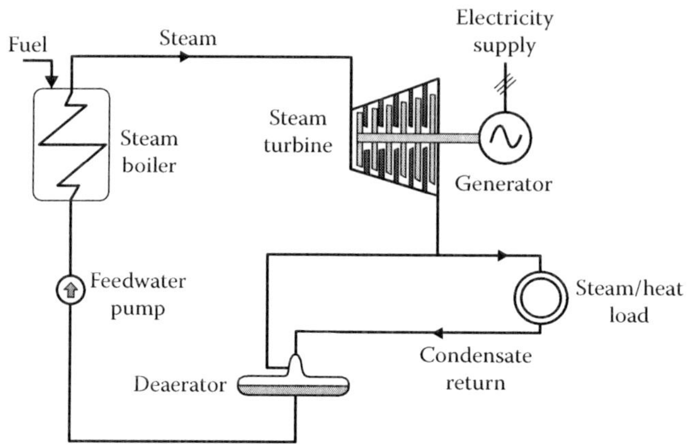

FIGURE 21.5 Back-pressure steam turbine CHP.

power outputs ranging from as low as 30 kW for newly developed microgas turbines to over 300 MW for large-scale gas turbines such as Siemens H-class gas turbines.

The basic principle of a gas turbine is that a high-energy-carrying fluid in the form of high-temperature and high-pressure combustion gases is expanded through a series of free-wheeled turbine blades mounted on a common shaft with the effect of generating mechanical energy. The high-pressure combustion air is continuously supplied to the combustion chamber by a compressor to provide required oxygen for fuel combustion. In a single-shaft gas turbine (as shown in Figure 21.6a), the compressor is mounted on a common shaft to the gas turbine, whereby consuming a large proportion of the gas turbine mechanical power. In practice, the gas turbines operate on an open thermodynamic referred to as the Brayton or Joule cycle.

According to the Brayton model, the working fluid undergoes four thermodynamic processes. First, the working fluid (air) is drawn in by the compressor at ambient conditions and compressed to high pressure. Modern gas turbine generating systems employ axial flow multistage compressors achieving compression ratios of up to 1.252:1 per stage at optimum efficiency (Boyce 2012). The air is then supplied to the combustion chamber where it is mixed with a gaseous fuel such as natural gas or atomized liquid fuel for combustion at constant pressure. The high-pressure and high-temperature combustion gases are then expanded in the gas turbine through a series of fixed nozzles that are placed in between rows of freely rotating blades mounted on a common shaft that converts fuel energy into mechanical power *shaft power*. The exhaust gases rejected at the turbine outlet are at high temperatures and contain a high proportion of high-grade thermal energy. As shown in the *T–S* diagram of Figure 21.6b, the compression and expansion of the working fluid in the compressor and turbine, respectively, are not isentropic processes and rather follow states 1–2–3–4. This contributes to entropy increase of the cycle and hence a loss in efficiency of the whole cycle.

The fuel energy conversion to mechanical power efficiency of a simple adiabatic gas turbine cycle can be expressed as

$$\eta_{i,g} = 1 - \frac{1}{r^{(k-1)/k}} \tag{21.7}$$

where

r is the compression ratio of the compressor
k is the specific heat capacity ratio of the working fluid

Equation 21.7 shows that the efficiency of an ideal Brayton cycle depends mainly on the pressure compression ratio of the cycle. However, in practice, the isentropic efficiency of the turbine, η_T, and compressor, η_c, is lower than unity, and the efficiency of the Brayton cycle is mainly affected by

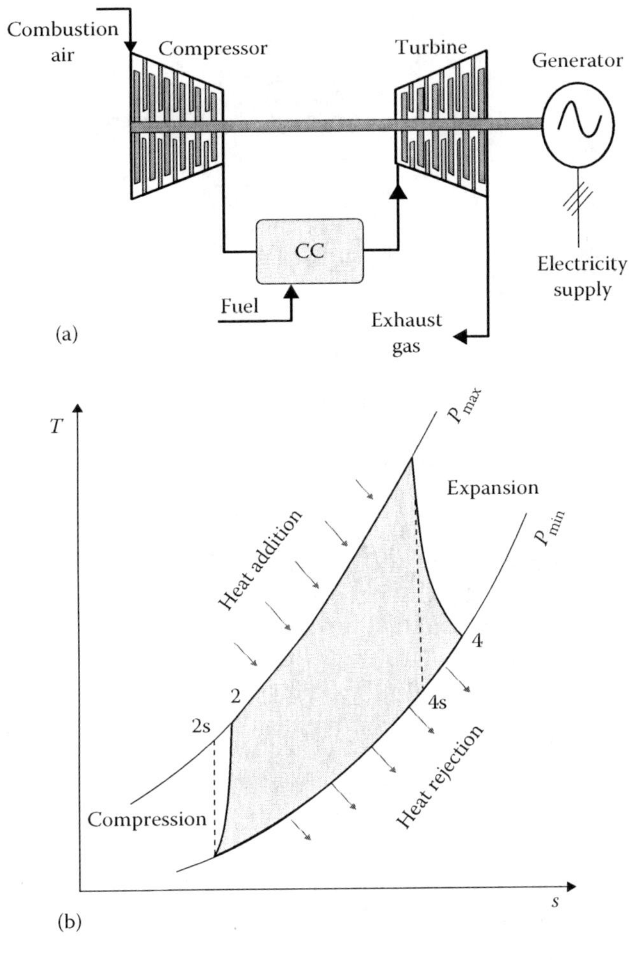

FIGURE 21.6 (a) Components of a single shaft gas turbine cycle; (b) *T–S* diagram.

the working fluid inlet temperature, T_1, firing temperature, T_3, and compression ratio, *r*. This can be expressed as follows (Boyce 2012):

$$\eta_{cyc,g} = \frac{\eta_T T_3 - (T_1/\eta_c) r^{(k-1)/k}}{T_3 - T_1\left(1 + (r^{(k-1)/k} - 1)/\eta_c\right)}\left(1 - \frac{1}{r^{(k-1)/k}}\right) \tag{21.8}$$

Modern power generation gas turbines operate at high compression ratios and employ various techniques such as using a recuperator, aftercoolers, and reheaters to increase the mechanical conversion efficiency to be as high as 42% (Poullikkas 2005). Most industrial gas turbines for CHP applications are designed to operate both on natural gas as the fuel of choice and oil for standby operations. High temperature of exhaust gases makes the gas turbine particularly a good candidate for high-grade thermal energy supply in the form of steam or hot water for industrial processes and district heating application.

21.2.2.3 Microgas Turbines

A microgas turbine operates on the same principle as the industrial-type gas turbine. Microgas turbines are available in power output ranging from as low as 25 to 500 kW (Pilavachi 2002). Advanced microgas turbines are of compact construction and use a single stage with the compressor–turbine assembly rotating at speeds in excess of 100,000 rpm. The recuperated-type microgas turbines have

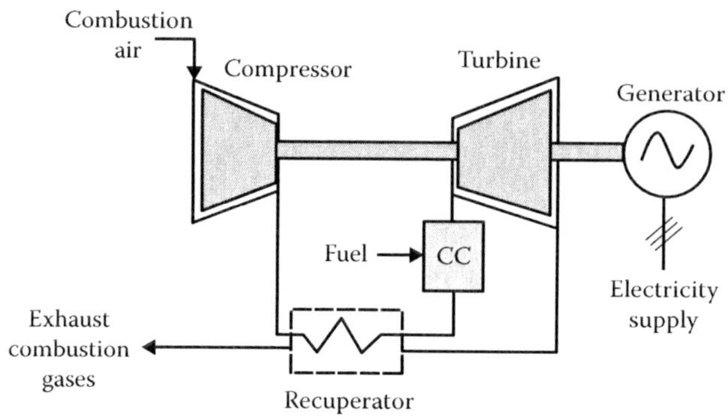

FIGURE 21.7 Microgas turbine components arrangement.

electrical efficiency in the range of 20%–30% (Chowdhury et al. 2009). A schematic of a microgas turbine is shown in Figure 21.7.

The exhaust gases leave the recuperator at relatively high temperature and carrying sufficient amount of energy that can be recovered in CHP packages for heating purposes, increasing overall fuel utilization efficiency to above 80%. Microgas turbines are found in small-scale CHP schemes such as commercial and institution buildings where there is adequate thermal base load that warrants the installation of a CHP system. Existing microgas turbine CHP packages are predominantly conceived to run on natural gas as a way to achieving low emissions. The introduction of new legislations in many countries banning flaring or releasing landfill gas to the atmosphere, because of its high global warming potential, makes microgas turbine a cost-effective candidate technology for power generation using landfill gas, while attendant heat could be used to accelerate decay rate of perishable products and enhance gas collection yield. Microgas turbines can also be found in applications for standby and peak shaving power generation. In addition, the recovered waste heat from microgas turbines can also be used to drive thermally activated cooling systems such as absorption chillers and desiccant dehumidifiers.

21.2.2.4 Combined Cycle Gas Turbines (CCGT)

A combined cycle gas turbine (CCG) plant consists of a high-temperature gas turbine cycle, referred to as the topping cycle, and a low-temperature steam turbine cycle known as the bottoming cycle. Heat rejected at high temperature (400°C–600°C) from the topping cycle is used to raise steam in an unfired or fully fired HRSG and then fed to the bottoming cycle. CCGT plants have undergone extensive development to become one of the preferred options for large-scale power generation. The electrical efficiencies of CCGTs have steadily increased from 40% for simple cycle aero derivative to over 60% for gas turbine cycles with intercooling and steam injection (Najjar 2001). However, efficiency higher than 60% is considered as the economic limit by the industry given the need for expensive high-temperature materials (Smock 1995). A simple CCGT arrangement with one pressure level at the HRSG is shown in Figure 21.8a. The simple HRSG is made up of an economizer where the feedwater is initially heated up close to its saturation temperature, an evaporator that further heats up the feedwater to steam at constant temperature and pressure with the saturated steam collected at the top of the drum, and a superheater where steam is superheated to the required temperature.

A *T–S* diagram of the working fluids processes is shown in Figure 21.8b. One of the important parameters defining the HRSG is the pinch-point temperature, which is the difference between the evaporator temperature on the water/steam side and the outlet temperature of the exhaust gases on the turbine side. Pinch-point temperatures typically range between 4°C and 12°C, depending on the specific design of the HRSG (Starr 2003). This temperature difference, which is unavoidable, contributes to the cycle's internal irreversibilities causing a decrease of the cycle efficiency.

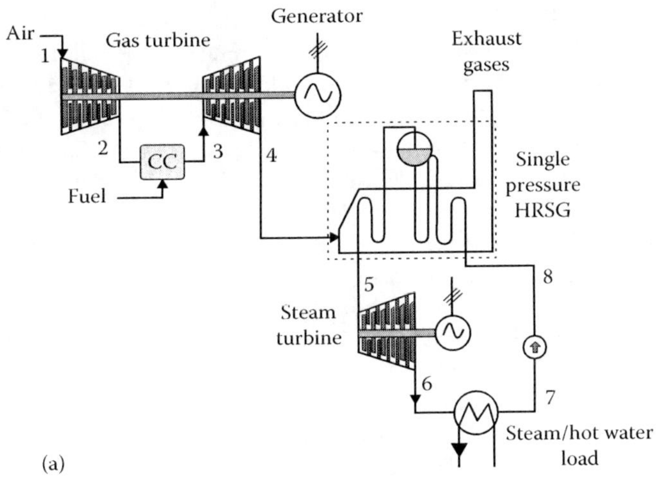

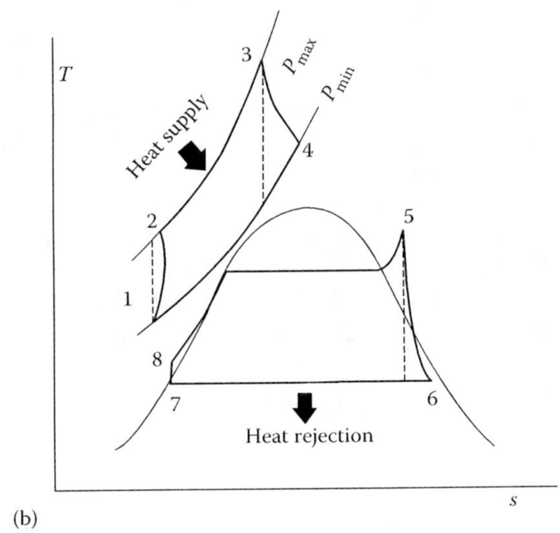

FIGURE 21.8 (a) A combined cycle of gas turbine and steam turbine; (b) *T–S* diagram of a gas turbine/steam turbine combined cycle.

The electrical efficiency of the CCGT cycle can be expressed as follows (Kehlhofer 2009):

$$\eta_{\text{CCGT}} = \eta_g + \eta_s(1-\eta_g)\varepsilon \tag{21.9}$$

where

η_g is the simple gas cycle efficiency
η_s is the simple steam cycle efficiency
ε is the effectiveness of the HRSG which is expressed as

$$\varepsilon = \frac{q_s}{q_g} \tag{21.10}$$

where

q_s is the heat recovered by HRSG
q_g is the maximum heat content of exhaust gases from the gas turbine

21.2.2.5 Internal Combustion Engines

IC engines are classified according to the way fuel is ignited in the engine cylinder, these being the spark ignition (Otto cycle) engine and the compression ignition (Diesel cycle) engine. Spark ignition engines use an electric spark plug to ignite a premixed air–fuel mixture in the cylinder of the engine to release fuel energy and convert it into shaft power. In compression ignition engines, fuel is injected into the cylinder after combustion air had been compressed and its temperature increased to the autoignition temperature of the fuel, sufficiently high to initiate the combustion process. These distinct design characteristics of the two types of IC engines have major effects on their thermal performance with compression ignition engines being able to reach compression ratios of fuel–air mixture in the cylinder ranging from 12 to 24 compared to that of spark ignition engines of 8–12 (van Basshuysen and Schafer 2002). This allows compression ignition engines to have superior thermal performance with efficiency of up to 45%, whereas that of its spark ignition counterpart is limited to about 35%. Spark ignition engines for stationary power generation applications are available in power output ratings of a few kW to over 200 kW for automotive derived engines and up to 5 MW for the industrial types. Compression ignition engines on the other hand are available at power output rating of as high as 20 MW with efficiency surpassing that of gas turbines including part load operation. In water-cooled IC engines, the fuel energy supply is approximately converted equally to useful shaft power, heat rejection through the cooling system and the exhaust gases. The heat removed from the engine's cylinder block to limit excessive temperature buildup is of low grade, providing low-temperature hot water suitable for space heating application in buildings. The high-temperature exhaust gases on the other hand could be used to supply low-pressure steam to the on-site processes or heat users. Figure 21.9 shows a simplified schematic of an IC engine where the rejected heat could be recovered for useful application as a part of a CHP installation.

Currently, IC engines are the market dominant choice for on-site power generation whether deployed as an emergency and standby generator, load peak shaving, and CHP or as the main power generator in an off-grid site. The main disadvantage of IC engines however is the high level of emission of pollutants (NO_x and CO_x) and particulates.

21.2.2.6 Other Prime Movers

Stirling engines: Stirling engine is an external combustion heat engine. It is built in many mechanical arrangements. Common to all basic Stirling engine embodiments is a hermetically and pressurized single- or multiple-cylinder–piston assembly that contains a fixed mass of working fluid such as air, nitrogen, helium, or hydrogen. Figure 21.10a shows a typical two-cylinder Stirling engine where two reciprocating pistons, a displacer and a power piston, are housed in each cylinder

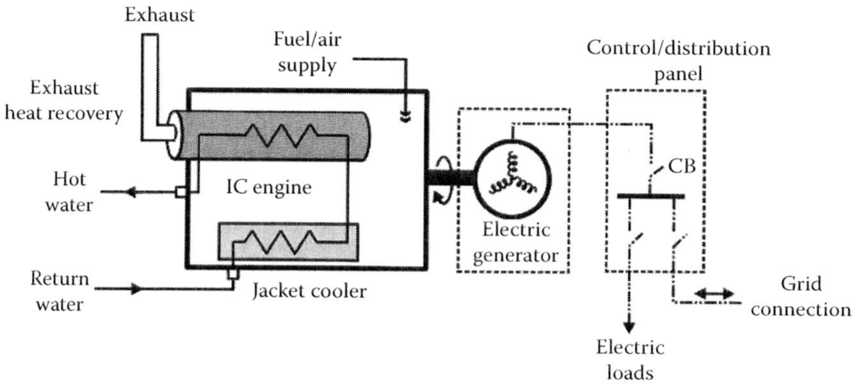

FIGURE 21.9 IC engine CHP.

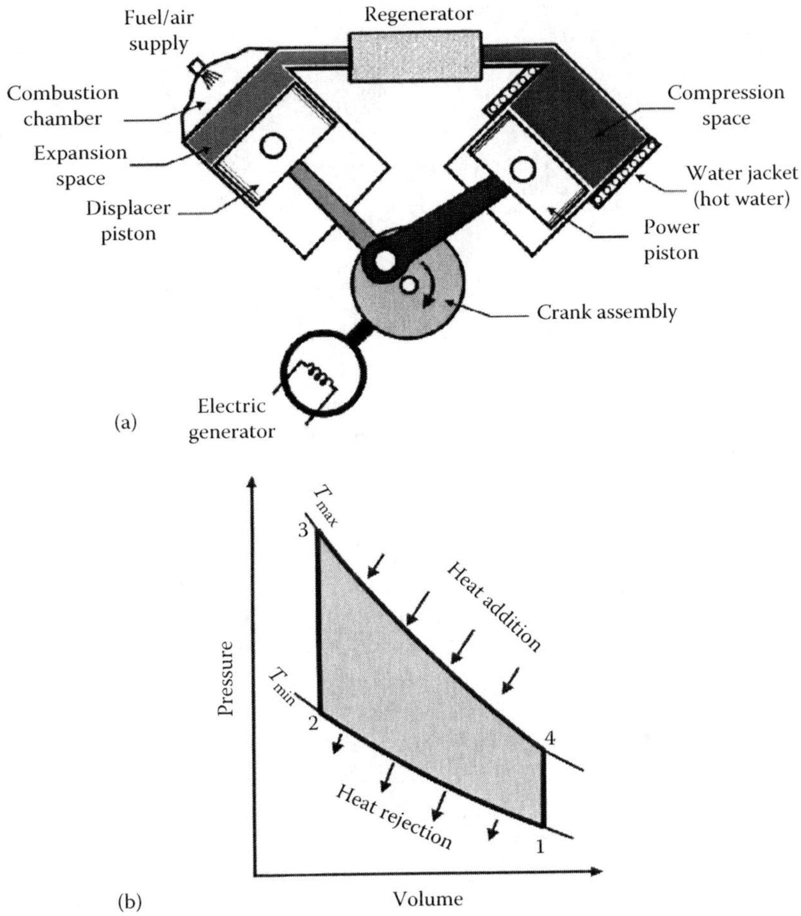

FIGURE 21.10 (a) Typical Stirling engine configuration and (b) *P–V* diagram.

(Walker et al. 1994). The main function of the displacer is to move the working fluid between the heat source and heat sink through a heat storage regenerator component, a role performed by a compressor in case of the Brayton cycle or a feedwater pump in a Rankine cycle. The Stirling cycle consists of four thermodynamic processes: isothermal heat addition and rejection and constant volume compression and expansion as shown in Figure 21.10b. Fuel energy is supplied to the engine at the hot end (expansion space) and low-grade heat is rejected at the cold end (compression space). The shuttling of the working fluid by the displacer piston between the heat source and heat sink causes it to expand and contract, which performs mechanical work on the power piston. The ideal Stirling cycle operating between two temperature limits has a theoretical thermal efficiency equals to that of the Carnot cycle. However, in practice, many irreversibilities that present in the engine limit the efficiency to a fraction of that of the ideal cycle.

There are two main types of Stirling engines: kinematic and free-piston Stirling engines (FPSE). Kinematic Stirling engines can be found in various mechanical designs with the displacer and power piston located in a single, split, or multiple cylinders and linked by rods and crankshaft to convert the reciprocating motion of the pistons to a rotary one. On the other hand, in FPSE, the reciprocating pistons are mounted on mechanical or pneumatic springs and move freely inside the cylinder in an oscillatory mode with the power piston used to drive an external load such as a linear electric generator.

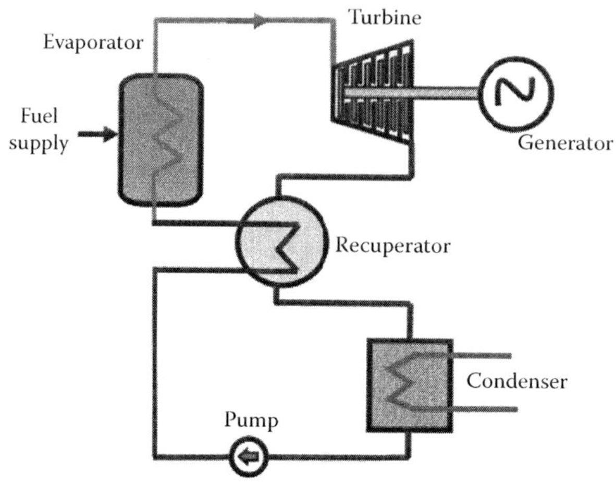

FIGURE 21.11 Schematic diagram of an ORC.

Currently, Stirling engine generators have power ratings ranging from a few watts to over 25 kW. The current market leader of Stirling engine generators is the WhisperGen micro-CHP unit with power and heat capacity rating of 1 and 8 kW, respectively, and achieving overall efficiency in excess of 90% (Harrison 2011). Furthermore, many advantages of Stirling engines including low noise and vibration, high efficiency, and multifuel capability have increased the interest for applications where IC engines are not compatible, for example, in using biomass fuels and solar energy.

ORC: ORC engines operate on the same thermodynamic principles that govern the conventional closed steam cycle. ORC has been designed primarily to convert low-grade heat energy from waste heat sources of industrial processes, renewable fuels, and geothermal energy into electricity. Unlike steam cycle where water is used as the working fluid, ORC uses a low vapor point organic working fluid such as hydrocarbons or refrigerants with favorable physical properties. When used as a CHP system, ORC derives further economic benefits by utilizing rejected heat for on-site heating applications. The core element of an ORC is the turbine in which the working fluid vapor expands to convert heat energy into mechanical power as shown in Figure 21.11. The working fluid is heated in an evaporator either directly or through a closed hot oil loop. The working fluid leaves the turbine as saturated vapor and passes through a recuperator to enhance the cycle efficiency by preheating the liquid organic fluid returning from the condenser.

Currently, customized ORC packaged units are available for power generation capacity ranging from 200 kW to 3 MW (Obernberger and Thek 2008, Vescovo 2009). Typically, the electrical efficiency of these packaged units varies between 8% and 18%, depending on the temperatures of available heat source and heat sink (Vescovo 2009).

21.3 BIOMASS-FUELLED CHP TECHNOLOGIES

Recent concerns over the ever-increasing cost of fossil fuels and their environmental impact have prompted many countries to shift their energy policies toward supporting energy-efficient technologies and the development of more sustainable forms of energy resources. Hence, biomass-fuelled CHP represents a commanding position in combining an efficient energy generation technology and a renewable/sustainable energy source with the potential to become a core element of the new energy strategies to decarbonize the power generation sector in the many parts of the world in the coming decades. In addition, biomass-fuelled CHP systems can be applied at various scales of the energy supply sector ranging from large industrial-scale CHP power plants to microscale CHP units suitable for domestic building (Liu 2011).

TABLE 21.1
Biomass-Fuelled CHP Technologies

Biomass Conversion Technology	Prime Mover
Combustion	Steam engine, steam turbine, Stirling engine, ORC turbine, hot air gas turbine
Gasification	Steam turbine, internal combustion engine, gas turbine/microturbine, fuel cell
Pyrolysis	Internal combustion engine
Biochemical/biological processes (fermentation and anaerobic digestion)	Internal combustion engine
Chemical/mechanical processes (biodiesel production)	Internal combustion engine

Source: Liu, H., Biomass fuels for small and micro combined heat and power (CHP) systems: Resources, conversion and applications, in: *Small and Micro Combined Heat and Power (CHP) Systems: Advanced Design, Performance, Materials and Applications*, ed. R. Beith, Woodhead Publishing Limited, Cambridge, U.K., 2011.

21.3.1 Introduction of Biomass-Fuelled CHP Technologies

Biomass-fuelled CHP technologies can be grouped by the type of prime mover technology, the size of the CHP system, or the biomass conversion technology. In this chapter, the latter classification has been adopted for the convenience of discussion. A summary of the various combinations of biomass conversion technologies and matched prime movers adopted by many biomass-fuelled CHP systems commercialized or under development is shown in Table 21.1. The main operational characteristics of the CHP prime mover listed in Table 21.1 have been described in Section 21.2, whereas most of the biomass conversion processes including gasification, pyrolysis, fermentation, anaerobic digestion, and biodiesel production have been discussed in detail in other chapters of this book.

21.3.2 Combustion of Solid Biomass Fuels

Generally, the combustion process of a solid biomass fuel undergoes three stages as shown in Figure 21.12. The first stage involves drying the fuel by the evaporation of its moisture content, the second stage is the devolatilization and volatile combustion process, and the last stage is the char combustion stage. The residue after the char combustion becomes clinker, slag, or bottom ash. Depending on the combustion conditions such as the combustion furnace, the biomass fuel itself, and the particle size, the three stages of combustion can be overlapping with each other.

Compared to coals, which have volatile contents of less than ca. 40 wt.%, most dry biomass materials contain about 80 wt.% of volatiles, indicating the greater importance of volatiles in terms of energy output and combustion process itself. The design of any biomass combustion facility

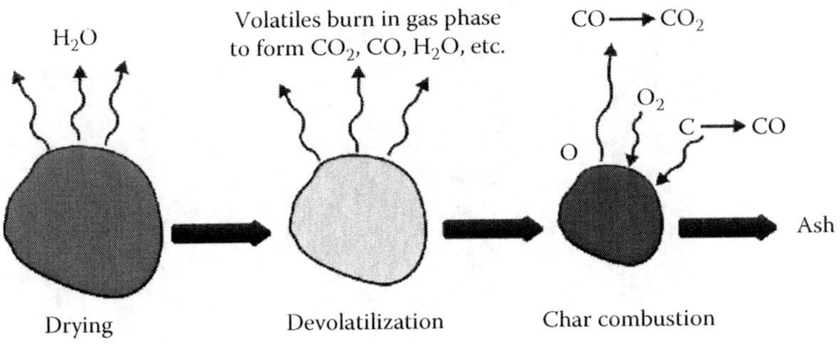

FIGURE 21.12 The main stages of solid biomass fuel combustion.

should ensure that the biomass volatiles are burnt to completion. The heterogeneous chemical reactions of biomass char with oxygen from air are equally important—the char still contains ca. 20% of the biomass energy but the combustion of char is the slowest stage of the three stages of biomass combustion. To achieve high overall combustion efficiency, sufficient combustion time should be provided for the char combustion stage. In addition, in practice, combustion is feasible only for biomass with moisture content of lower than ca. 50% (Goyal et al. 2008). Hence, for extremely wet biomass materials, biological conversion methods such as anaerobic digestion, rather than thermochemical conversion methods, may be the better route to extract maximum useful energy.

Today, direct combustion is the most common way of converting biomass to energy in the form of heat and electricity, providing over 90% of the energy generated from biomass worldwide (Van Loo and Koppejan 2010). Modern biomass combustion systems are proven technologies and are as varied as the solid biomass feedstocks. They are routinely applied at various scales ranging in size from small wood-burning stoves through domestic space and water heating systems to large-scale industrial boilers producing megawatts of heat or steam for power generation and district heating. Figure 21.13 shows an example of a logwood-fired heat-storing stove for space heating in domestic buildings.

Figure 21.14 shows an example of a 25 kW_{th} wood pellet boiler that can provide space heating and hot water in domestic buildings, installed at the Department of Architecture and Built Environment, University of Nottingham.

Biomass combustion systems can be easily integrated with existing large-scale infrastructure. Figure 21.15 shows an aerial photograph of Alholmens Kraft Biofuel-fired Power Plant in Pietarsaari, Finland, which was the largest CHP plant (when built in 2002) and operates on various types of biomass fuel including burning bark, sawdust, wood waste, commercial biofuel, and peat. The plant generates 240 MW of electricity, 100 MW_{th} of process steam, and 60 MW_{th} of district steam.

Similarly, Figure 21.16 shows Steven's Croft Biomass Power Station located in Lockerbie, Scotland, United Kingdom, which is the largest dedicated wood-fuelled power station in Britain with a power generation capacity of 44 MW.

Biomass fuels are different from coal in many other combustion-related properties. For example, biomass fuels have more free alkali in ash, which may aggravate the slagging and fouling problems than coal, and biomass chars have higher oxidation reactivity than coal chars (Yin et al. 2008).

FIGURE 21.13 Logwood-fired heat-storing stove.

FIGURE 21.14 A 25 kW wood pellet boiler.

FIGURE 21.15 Alholmens Kraft Biofuel-fired Power Plant, Pietarsaari, Finland. (From Pöyry, http://www.poyry.com/projects/alholmens-kraft-biofuel-fired-power-plant-pietarsaari-finland, accessed on December 23, 2012.)

Moreover, biomass fuels are generally bulky and expensive for long-distance transportation, which requires a large-scale biomass-fuelled CHP plant to be built at the proximity of the biomass fuel supplies. Provision of consistent good-quality biomass fuels with constant chemical and physical properties is also difficult. This requires the biomass conversion technology adopted by a biomass-fuelled CHP plant should be able to handle variable biomass fuel supplies.

FIGURE 21.16 Steven's Croft Biomass Power Station—Lockerbie, Scotland, United Kingdom. (Reproduced from James T M Towill and licensed for reuse under Creative Commons License.)

The two most common types of biomass combustion systems are those of grate-firing and fluidized bed combustion, both of which have good fuel flexibility and can burn 100% biomass or co-fire with coal (Yin et al. 2008). Both Alholmens Kraft Biofuel-fired Power Plant (Figure 21.15) and Steven's Croft Biomass Power Station (Figure 21.16) have adopted the concept of fluidized bed combustion with the former using circulation fluidized bed combustion and the latter using bubbling fluidized bed combustion. More than a dozen of examples of grate-firing biomass boilers have been listed by Yin et al. (2008), including a 33 MW_{fuel} CHP unit that produces 8.3 MWe and 20.8 MW_{th} and a 108 MW_{fuel} CHP unit that produces 35 MWe and 50 MW_{th}. Entrained flow or suspension combustion is usually adopted to co-fire milled biomass pellets or raw biomass with pulverized coal in large-scale power plants (Yin et al. 2008, Van Loo and Koppejan 2010). In the United Kingdom, fifteen large-scale coal-fired power plants co-fire biomass alongside coal (Fernando 2012). Of these, only two have achieved co-firing rates higher that about 6% (energy ratio of biomass to coal). Drax power plant, the largest coal-fired power plant in the United Kingdom, and the Fiddler's Ferry power plant have the capacity to co-fire at rates of up to 12.5% and 20%, respectively. More recently, the management of Drax power plant outlined plans to transform it into a predominantly biomass-fuelled generator with the initial stage to convert three out of the six generators to run on biomass and as early as the middle of 2013 (Drax 2012).

Like other fossil fuels counterparts, biomass combustion produces a range of pollutants including carbon monoxide, nitrogen oxides, particulates, and smoke and can have potential operational problems such as fouling and corrosion whether burning in dedicated combustion plants or co-firing with coal (Demirbas 2005, Yin et al. 2008, Khan et al. 2009, Van Loo and Koppejan 2010). However, biomass combustion can be CO_2 neutral if the cycle of biomass growth and harvest is sustained (Liu 2011). Further information is available from the handbook of biomass combustion and co-firing edited by Van Loo and Koppejan (2010), which provides a comprehensive review of biomass combustion basic principles and industrial applications.

21.3.3 Combustion-Based Biomass CHP Technologies

Direct combustion of biomass is the dominant method used in biomass CHP technologies. The heat released from biomass combustion is used to produce steam, hot air or organic working fluid vapor that can drive a steam turbine, a steam engine, a hot air gas turbine, or an ORC turbine to generate electricity. The following prime movers are the most popular for application with direct combustion biomass CHP: steam turbine/steam engine, hot air gas turbine, ORC, and Stirling engine.

21.3.3.1 Direct Biomass Combustion with Steam Turbine/Steam Engine CHP

Among the technologies listed in Table 21.1, the combination of *combustion* and *steam turbine* is most widely used, particularly for large-scale and medium-scale biomass-fuelled CHP systems. Indeed, the majority of current world biomass power plants are based on *combustion* and *conventional steam turbine Rankine cycle*. The Alholmens Kraft Power Plant (Figure 21.15) is a 240 MWe biomass CHP power plant based on *combustion* and *steam turbine* CHP technology (Pöyry 2012). It also produces about 100 MW_{th} of process steam and about 60 MW_{th} of district heating. The boiler is a circulating fluidized bed boiler with natural circulation and reheating and the steam turbine is a three-casing extraction condensing turbine with district heating and process steam supply. Steven's Croft Biomass Power Station in Scotland (Figure 21.16), which is owned by the power utility E.ON and started commercial operation in 2007, is a good UK example of a dedicated large biomass power plant (power generation capacity of 44 MWe) and based on the technologies of *combustion* and *steam turbine* (Power-Technology 2012). The Markinch Biomass CHP plant in Scotland (rated at 50 MWe) of RWE is another UK example of the application of the technologies, which is under construction and scheduled to enter the commissioning stage at the end of 2012 (RWE 2012).

The prohibitive cost of transporting dispersed and bulky biomass feedstock over long distance often limits the capacity of this type of biomass power/CHP plants to the order of 100 MWe. This results in the power generation efficiency to be in the region of 20%–30%, although the overall CHP efficiency can be kept as high as 80% or higher. If the same CHP technology, that is, the combination of the biomass conversion by combustion with the CHP prime mover of steam turbine, was applied to small- and microscale biomass CHP systems, it would lead to both unacceptable low power generation efficiency and high capital costs (Liu 2011). The minimum capacity of biomass-fired steam turbine CHP is likely to be around 2 MWe, whereas for biomass-fired steam engine CHP, it is likely to be around 200 kWe (BIOS 2010a,b).

21.3.3.2 Direct Biomass Combustion Hot Air Gas Turbine CHP

The combination of *combustion* and *hot air gas turbine* has been adopted by some research groups and developers of biomass CHP systems. Gaderer et al. (2010) reported the development of biomass-fired hot air gas turbine and a fluidized-bed wood combustor at the Institute for Energy Systems, Technical University of Munich, Germany. The system has an integrated high-temperature heat exchanger made out of structured steel tubes for indirect firing of a 100 kWe microturbines. Figure 21.17 shows the process flow diagram of the CHP system. Two key components of a biomass CHP with an externally fired air turbine are the high-temperature heat exchanger, which may require to be made of special materials, either metals or ceramics, and the turbine, which usually is of a standard or modified gas turbine (Kautz and Hanson 2007, Botala 2012, MacKay Group 2012).

Over the recent years, more indirectly fired gas turbines (IFGTs) have been specifically developed for the use of biomass-fuelled CHP. For example, Botala (2012) converted existing proven gas turbines to IFGT engines allowing them to run on biomass, waste products, and other fuels. MacKay Group (2012) has developed 250 kWe IFGT, which can be fuelled by biomass and waste products via external combustion.

21.3.3.3 Direct Biomass Combustion ORC CHP

The combination of *combustion* and *ORC turbine* is receiving increasing attention in the development of small- and microscale biomass CHP systems. Despite of low electrical efficiency of ORC, which is in the region of 6%–17% for small- and microscale power/CHP systems, new applications of ORC are consistently explored due to its possibility to utilize the low-grade waste heat from

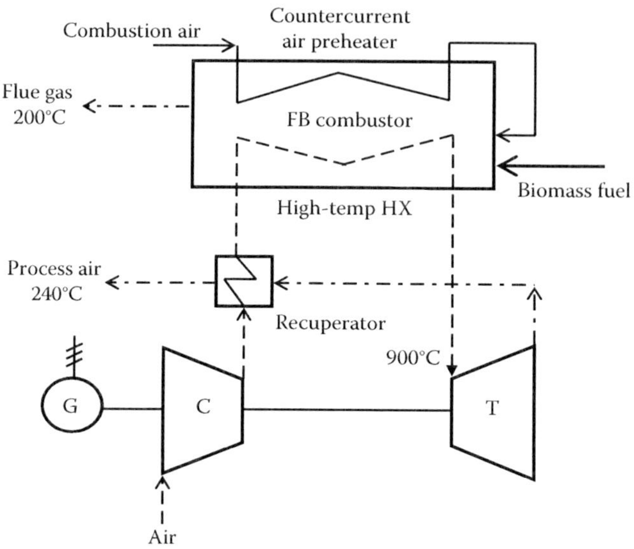

FIGURE 21.17 Process flow diagram of the biomass-fired hot air gas turbine with fluidized bed combustion. (From Gaderer, M. et al., *Appl. Therm. Eng.*, 30, 13, 2010.)

other processes (Schuster et al. 2009). The organic working fluid evaporates at lower temperatures than water and therefore ORC CHP systems can work at lower temperatures and pressures than the conventional steam Rankine cycle. For this reason, ORC is considered particularly suitable for small-scale and microscale biomass-fired CHP systems (Dong et al. 2009).

Biomass-fired ORC-based CHP plants with capacity higher than 200 kWe have been demonstrated and in operation across Europe for almost a decade (Dong et al. 2009). Two particular demonstration plants are worth mentioning: the 400 kWe Admont CHP plant and the 1000 kWe Lienz biomass CHP plant (Admont 2001, Obernberger et al. 2002, BIOS 2010c). For both ORC-based biomass CHP demonstration plants, the biomass combustion furnace/thermal oil boiler and the ORC process were coupled by a thermal oil cycle, whereas silicon oil was selected as the organic working fluid as shown in Figure 21.18. The condensation of the silicon oil takes place at a temperature level that allows the heat recovered to be utilized with the hot water feed temperature of about 80°C–90°C. The net electrical efficiency of the Admont CHP plant was reported to be 7.4% (Admont 2001). The net electric efficiency of the 1000 kWe Lienz biomass CHP plant amounted to 18% at nominal load and about 16.5% at 50% partial load at feedwater temperatures of 85°C (Obernberger et al. 2002). This underlines the excellent partial load behavior of the technologies, which is especially relevant for a heat-controlled operation of a CHP plant. Some manufacturers of ORC plants now offer commercial products of biomass-fired ORC-based CHP plants with capacity in the range of 200 kWe to 2000 kWe. For example, Turboden (2012) offers 6 different biomass-fired ORC CHP units with power capacities of 200, 700, 1000, 1300, 1800, and 2200 kWe, respectively. Turboden has more than 30 years of experience in ORC technology (Turboden 2012) and was the manufacturer and supplier of the ORC unit of the Lienz biomass CHP plant (Obernberger et al. 2002).

The development of microscale biomass ORC-based CHP systems has received more interests over the past few years, partly due to the increasing demand for the building applications of microscale CHP systems (Qiu et al. 2012). Qiu et al. (2012) investigated the feasibility of a 1 kWe biomass-fired ORC-based CHP. However, it was reported that the unit suffers from low power generation efficiency although an overall efficiency of over 80% was achieved. The low power generation efficiency was largely due to the combination of low turbine isentropic efficiency and

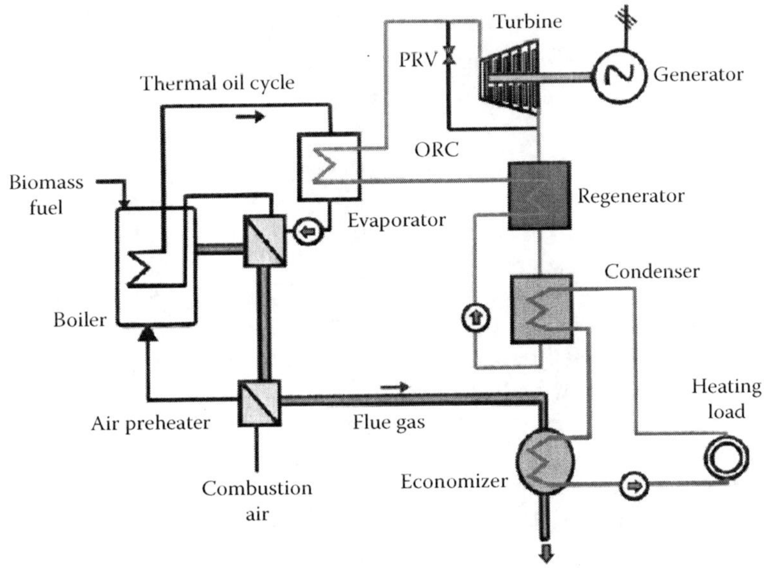

FIGURE 21.18 Schematic of the ORC-based biomass CHP plant of Lienz. (From Obernberger, I. et al., Description and evaluation of the new 1000 kWel organic Rankine cycle process integrated in the biomass CHP plant in Lienz, Austria. Euroheat & Power October 2002, Available at: http://www.turboden.eu/en/public/downloads/report_on_lienz_plant.pdf. Accessed on December 20, 2012, 2002; BIOS, BIOS BIOENERGIESYSTEME GmbH. http://www.bios-bioenergy.at/en/electricity-from-biomass/orc-process.html, Accessed on December 23, 2012, 2010c.)

low alternator efficiency (Qiu et al. 2012). As microscale turbines for organic fluids with capacity smaller than 10 kWe are not commercially available, many researchers have to use replacement turbines with their microscale ORC-based CHP systems. Scroll expanders is one of the popular choices for the replacement microscale ORC turbines (Peterson et al. 2008, Lemort et al. 2009, Wang et al. 2009, Liu et al. 2010, Quoilin et al. 2010), whereas modified air motors were used by Qiu et al. (2011, 2012). Further technical and economic advances on ORC turbines are needed before microscale biomass-fired ORC-based CHP technology can compete with other microscale biomass CHP technologies.

21.3.3.4 Direct Biomass Combustion Stirling Engine CHP

Stirling engine-based small- and microscale biomass CHP systems have been the focus of many researchers and developers. Biomass Stirling engine CHP systems rated 35 and 70 kWe had been developed and demonstrated by researchers of Austria and Denmark (Obenberger et al. 2003, Biedermann et al. 2004, BIOS 2010d). A 35 kWe plant was put into operation in 2002 and was successfully tested for more than 10,000 h, whereas a 70 kWe plant was in operation since autumn 2003 (BIOS 2010d). The overall electrical efficiency and the overall CHP efficiency of the 35 and 70 kWe plants were reported to be 11.7% and 88.3%, respectively (BIOS 2010d). Stirling (2012) currently offers biomass-fired Stirling engine CHP systems as single units of 35 kWe or in multiple units of up to 140 kWe. The biomass conversion technology of Stirling DK's CHP systems is updraft gasification followed by gas combustion (StirlingDK 2012). KWB (2013) is commercializing its Stirling engine-based biomass CHP systems—Sunmachine® *Pellet*, which is rated at 1.5–3.0 kWe. Recently, a SUNMACHINE® Pellet CHP system was tested in laboratories by Thiers et al. (2010). Although the performance of the CHP unit was not as good as the manufacturer had claimed, its electric power generation efficiency of 14.3% is far better than many of its competitors for such small biomass CHP systems.

21.3.4 Gasification-Based Biomass CHP Technologies

Gasification-based biomass CHP technologies are probably the second most popular category of technologies from large- to microscale biomass CHP systems.

21.3.4.1 Large-Scale Biomass Gasification Power/CHP Systems

For large-scale CHP systems, biomass gasification can be combined with an existing coal/gas-fired boiler and steam turbine to generate heat and power. After appropriate cleaning and conditioning, the combustible gas generated from the biomass gasification is combusted in the coal/gas-fired boiler, replacing part of the fossil fuel, and the steam produced in the boiler generates electricity and heat via the steam turbine cycle. This concept of biomass utilization is sometimes termed as *indirect co-firing* (Van Loo and Koppejan 2010) and has been successfully demonstrated with a 350 MW_{th} steam boiler fired with coal and natural gas at Kymijärvi Power Station, Lahti, Finland, since 1998 (Granatstein 2002, Palonen and Nieminen 2004). Lahti Energy's new power plant, Lahti Energy's Kymijärvi II, is now demonstrating a new version of the technology—the first gasification power plant in the world to use only solid recovered fuel (SRF)—the clean gas generated from the gasification process is directly fired in the gas boiler where high-temperature and high-pressure steam is generated for the production of electricity and district heat (Lahti Energy 2012). The SRF used in the plant consists of highly combustible waste that is not suitable for material recycling. The power plant produces 50 MW of electricity and 90 MW of district heat. Finnish-based Metso Oyj supplied the gasification reactors, gas purification systems, and the boiler of the power plant. The turbine and generator were manufactured by Siemens. Lahti Energy had the overall responsibility for the construction project (Lahti Energy 2012).

Biomass gasification can also be combined with gas turbines and steam turbines, the so-called biomass integrated gasification combined cycle (Bio-IGCC) as adopted by some global manufacturers of gas turbines (Brdar and Jones 2012, Siemens 2012a). Comparing with combustion-steam turbine cycle power generation, IGCC has significant advantages in terms of lower air emissions (particular matter, NOx, and SOx) and higher power generation efficiency. The gasifier and the gas turbine are the two key components of a biomass IGCC power/CHP plant. Although there are a number of well-known large-scale gasifier and gas turbine manufacturers such as Lurgi (2012), GE (2012), and Siemens (2012a,b), there are only limited number of successful IGCC power plants in operation around the world. The Willem-Alexander Power Plant of the Netherlands, which is a coal IGCC power plant, rated at 253 MWe, also co-fuelled by biomass, is an example of successful IGCC power plants (Nuon 2012). The power plant has been in successful commercial operation since 1993. However, the collapse of the UK ARBRE Biomass IGCC demonstration plant (Memorandum of RAE 2003) and the uncertainty related to the 6 MWe Varnamo Biomass IGCC demonstration plant (Rensfelt 2010) only serve as the reminder that there are many challenges for the future demonstration and commercialization of large-scale biomass IGCC CHP plants.

21.3.4.2 Small-/Microscale Gasification-Based CHP Systems

For small- and microscale biomass CHP systems, when gasification is adopted as the biomass conversion technology, there are three main choices for the prime mover: gas turbines, international combustion engines, and fuel cells (Liu 2011). The success of the small- and microscale gasification-based CHP systems will largely depend on the quality of the combustible gases produced by biomass gasification—there are specific requirements for each prime mover in terms of gas quality and cleanliness (Liu and Neubauer 2010). Gas cleaning and conditioning of the combustible gases produced by biomass gasification can be very costly and energy intensive for small- and microscale gasification-based CHP units (Liu and Neubauer 2010). Community Power Corporation (CPC 2012) has developed and commercialized small-scale biomass CHP systems, which are based on the technologies of gasification and IC engine, with fully automated modular systems in power blocks of 25 or 100 kWe and above (CPC 2012). SCHMITT ENERTEC GmbH of Germany (Schmitt Enertec

2012) commercially supplies biomass gasification and IC engine-based CHP systems with capacity between 250 and 1000 kWe. The specification of these CHP systems indicates that the electric power generation efficiency is ca. 26%, whereas the overall CHP efficiency is ca. 80% (Schmitt Enertec 2012). The development of biomass gasification-based microscale CHP systems is less advanced in comparison to the development of biomass combustion-based microscale CHP systems. This is partly due to developers' and general public's perceptions and concerns on the safety issues of biomass gasification. The technical and economic challenges of biomass gasification such as tar removal and gas cleaning have also contributed to this phenomenon, considering that biomass combustion is a mature and well-developed biomass conversion technology. There are some prototype microscale biomass gasification-based power generators available for field applications or R&D work such as the 10 and 20 kWe power pallet units of ALL Power Labs (All Power Labs 2012). Achieving long-term successful operation with a microscale biomass gasification-based power/CHP system requires not only a good technical design of the system itself but also sustainable fuel supplies integrated with viable business models (Buchholz et al. 2012).

21.3.5 Other Biomass CHP Technologies

There are various other technologies available for biomass CHP generation, such as using bioethanol/bio-oil/biodiesel and biogas to fuel IC engines and gas turbines/microturbines. Many IC engines can burn a blend of gasoline and bioethanol, a blend of petroleum diesel and biodiesel, and biogas, and hence IC engine-based CHP systems can be fuelled by liquid biofuels and biogas. The Holsworthy Biogas Plant is the United Kingdom's first and most productive, centralized anaerobic digestion CHP power station with a power generation capacity of 3.9 MWe and has been in operation since 2002 (Andigestion 2012). The biogas generated in the digestion plant is burned in the gas engines to produce power and heat. A UK-based company, Cantium Energy Ltd, offers biodiesel CHP systems within the size range of 36–280 kWe (Cantium Energy Ltd 2012). Another UK-based company, ENER-G Combined Power, commercially supplies biogas-fuelled CHP units within the size range of 100–1945 kWe (EnerG 2012). Over the past decade, many turbine/microturbine manufacturers have developed gas turbines and microgas turbines, which can successfully burn biogas such as the microturbines made by the Capstone Turbine Corporation of the United States (Capstone 2012). Capstone now commercially supplies biogas-fuelled microturbine power generating units within the size range of 30–1000 kWe.

21.4 CONCLUSION

Biomass is a renewable energy source that has the potential to replace a large portion of the fossil fuels used today and in the future. CHP is an efficient, clean, and reliable approach to generate power and thermal energy from a single fuel source. Biomass-fuelled CHP is a combination of a renewable energy source and an efficient energy generation technology that can help decarbonize the energy generations at various scales ranging from large-scale power plants to microscale CHP units suitable for domestic building applications.

Great advances have been made over the past decades with the development of biomass-fuelled CHP systems. Large-scale biomass-fuelled CHP systems based on biomass combustion and steam turbine have been built and in commercial operation for years, whereas the demonstrations of biomass IGCC power plants have experienced plenty of ups and downs. It is generally accepted that there are many potential long-term operational problems with the dedicated large-scale biomass-fuelled power/CHP plants such as slagging, fouling, and corrosion. Further R&D is needed to overcome/alleviate these potential problems that are associated with not only the plant design but also the biomass fuel properties.

There are various combinations of biomass conversion technologies and prime movers that have been used with the development of small-/microscale biomass-fuelled CHP systems. A range of combustion-based and gasification-based small-/microscale biomass CHP systems are emerging on the commercial market but their long-term economics and reliability are yet to be confirmed. The development and implementation of microscale biomass-fuelled CHP systems with capacity in the order of a few kWe, which are ideal for building applications, are slow and still hindered by technical and economic barriers. Significant R&D efforts are needed if the distributed microscale biomass-fuelled CHP systems are to be successfully demonstrated, commercialized, and widely deployed in the near future.

REFERENCES

Admont. 2001. Biomass fired CHP plant based on an ORC cycle—Project: ORC-STIA-Admont. http://ec.europa.eu/energy/res/sectors/doc/bioenergy/chp/bm_120_98.pdf. Accessed on December 20, 2012.

All Power Labs. 2012. The full solution: 10 kW and 20 kW power pallet. http://www.gekgasifier.com. Accessed on December 20, 2012.

Andigestion. 2012. The Holsworthy biogas plant. http://www.andigestion.co.uk/content/holsworthy. Accessed on December 27, 2012.

Biedermann, F., Carlsen, H., Obenberger, I., and Schöch, M. 2004. Small-scale CHP plant based on a 75 kWel Hermetic eight cylinder Stirling engine for biomass fuels—Development, technology and operational experiences. *2nd World Conference and Exhibition on Biomass for Energy, Industry and Climate Protection*, May 2004, Rome, Italy.

BIOS. 2010a. BIOS BIOENERGIESYSTEME GmbH. http://www.bios-bioenergy.at/en/electricity-from-biomass/steam-turbine.html. Accessed on December 23, 2012.

BIOS. 2010b. BIOS BIOENERGIESYSTEME GmbH. http://www.bios-bioenergy.at/en/electricity-from-biomass/screw-type-engine.html. Accessed on December 23, 2012.

BIOS. 2010c. BIOS BIOENERGIESYSTEME GmbH. http://www.bios-bioenergy.at/en/electricity-from-biomass/orc-process.html. Accessed on December 23, 2012.

BIOS. 2010d. BIOS BIOENERGIESYSTEME GmbH. http://www.bios-bioenergy.at/en/electricity-from-biomass/stirling-engine.html. Accessed on December 23, 2012.

Boukhanouf, R. 2011. Small combined heat and power (CHP) systems for commercial buildings and institutions. In: *Small and Micro Combined Heat and Power (CHP) Systems: Advanced Design, Performance, Materials and Applications*. R. Beith (ed.), Woodhead Publishing Limited, Cambridge, U.K.

Boyce, M.P. 2012. *Gas Turbine Engineering Handbook*, 4th edn., Elsevier, Amsterdam, the Netherlands.

Brdar, R.D. and Jones, R.M. 2012. GE IGCC technology and experience with advanced gas turbines-GE power systems GER-4207. Available at: http://www.netl.doe.gov/technologies/coalpower/turbines/refshelf/igcc-h2-sygas/IGCC%20GTs%20(GER4207).pdf. Accessed on December 22, 2012.

Btola. 2012. BTOLA indirectly fired gas turbine technology. http://www.btola.com/ifgt-technology.html. Accessed on December 20, 2012.

Buchholz, T., Da Silva, I.D., and Furtado, J. 2012. Power from wood gasifiers in Uganda: A 250 kW and 10 kW case study. *Proceedings of the Institution of Civil Engineers: Energy* 165: 181–196.

Bugge, J., Kjær, S., and Blum, R. 2006. High-efficiency coal-fired power plants development and perspectives. *Journal of Energy* 31: 1437–1445.

Cantium Energy Ltd. 2012. Renewable combined heat and power systems. http://www.cantiumenergy.co.uk/Products.php. Accessed on December 27, 2012.

Capstone. 2012. Capstone turbine corporation. http://www.capstoneturbine.com/prodsol/solutions/rrbiogas.asp. Accessed on December 23, 2012.

Chartered Institution of Building Services Engineers (CIBSE). 1999. Small scale CHP for buildings, CIBSE Application Manual: AM12, London, U.K.

Chowdhury, S., Chowdhury, S.P., and Crossley, P. 2009. *Microgrids and Active Distribution Networks*. The Institution of Engineering and Technology, Athenaeum Press Ltd., London, U.K.

CPC. 2012. CPC website. http://www.gocpc.com/. Accessed on December 21, 2012.

Demirbas, M.F. 2005. Potential applications of renewable energy sources, biomass combustion problems in boiler power systems and combustion related environmental issues. *Progress in Energy and Combustion Science* 31: 171–192.

Dong, L., Liu, H., and Riffat, R.B. 2009. Development of small-scale and micro-scale biomass-fuelled CHP systems—A literature review. *Applied Thermal Engineering* 29: 2119–2126.

Doty, S. and Wayne, C.T. 2009. *Energy Management Handbook*, 7th edn. The Fairmont Press Inc., Lilburn, GA.

Drax. 2012. About Drax's biomass plans. http://www.draxpower.com/biomass/cofiring_plans. Accessed on December 23, 2012.

EnerG. 2012. Combined heat and power biogas product range. http://www.energ-group.com/combined-heat-and-power/cogeneration/product-range. Accessed on December 24, 2012.

Energy Technology Support Unit (ETSU). 1999. Introduction to large-scale combined heat and power—Good practice guide 43. Prepared for the Department of the Environment, Transport and the Regions, London, U.K.

Fernando, R. 2012. *Cofiring High Ratios of Biomass with Coal*. IEA Clean Coal Centre, London, U.K. ISBN 978-92-9029-514-3.

Gaderer, M., Gallmetzer, G., and Spliethoff, H. 2010. Biomass fired hot air gas turbine with fluidized bed combustion. *Applied Thermal Engineering* 30: 13, 1594–1600.

GE. 2012. Gasification. http://www.ge-energy.com/products_and_services/products/gasification/index.jsp. Accessed on December 22, 2012.

Goyal, H.B., Seal, D., and Saxena, R.C. 2008. Bio-fuels from thermochemical conversion of renewable resources: A review. *Renewable and Sustainable Energy Reviews* 12: 504–517.

Granatstein, D.L. 2002. Case study on lahden lampovoima gasification project Kymijarvi Power Station, Lahti, Finland. IEA Bioenergy Agreement—Task 36. Available at http://www.ieabioenergytask36.org/Publications/2001-2003/Case_Studies/Case_Study_on_Lahden_Lampovoima_Gasification_Project.pdf. Accessed on December 27, 2012.

Harrison, J. 2011. Stirling engine systems for small and micro combined heat and power (CHP) applications. In *Small and Micro Combined Heat and Power (CHP) Systems: Advanced Design, Performance, Materials and Applications*, ed. R. Beith, Beith & Associates Ltd., Woodhead Publishing Ltd., Cambridge, U.K.

Kautz, M. and Hansen, U. 2007. The externally-fired gas-turbine (EFGT-Cycle) for decentralized use of biomass. *Applied Energy* 84: 795–805.

Kehlhofer, R. et al. 2009. *Combined-Cycle Gas and Steam Turbine Power Plants*, 3rd edn., PennWell Inc., Tulsa, OK.

Kelly, S. and Pollitt, M. 2010. An assessment of the present and future opportunities for combined heat and power with district heating (CHP-DH) in the United Kingdom. *Journal of Energy Policy* 38: 6936–6945.

Khan, A.A., De Jong, W., Jansens, P.J., and Spliethoff, H. 2009. Biomass combustion in fluidized bed boilers: Potential problems and remedies. *Fuel Processing Technology* 90: 21–50.

Lahti Energia. 2012. Power plant technology. http://www.lahtigasification.com/power-plant/power-plant-technology. Accessed on December 27, 2012.

Leizerovich, A.Sh. and Leĭzerovich, A.S. 2005. *Wet Steam Turbines*. PennWell Inc., Tulsa, OK.

Lemort, V., Quoilin, S., Cuevas, C., and Lebrun, J. 2009. Testing and modeling a scroll expander integrated into an Organic Rankine Cycle. *Applied Thermal Engineering* 29: 3094–3102.

Liu, G., Zhao, Y., Li, L., and Shu, P. 2010. Simulation and experiment research on wide ranging working process of scroll expander driven by compressed air. *Applied Thermal Engineering* 30: 2073–2079.

Liu, H. 2011. Biomass fuels for small and micro combined heat and power (CHP) systems: Resources, conversion and applications. In: *Small and Micro Combined Heat and Power (CHP) Systems: Advanced Design, Performance, Materials and Applications*. ed. R. Beith, Woodhead Publishing Limited, Cambridge, U.K.

Liu, H. and Neubauer, Y. 2010. Gasification. In: *High Temperature Processes in Chemical Engineering*. ed. M. Lackner, pp. 361–408. Verlag Processing Engineering GmbH. Vienna, Austria. ISBN: 978-3902655097.

Lurgi. 2012. Lurgi fixed bed dry bottom coal gasification technology. http://lurgi.com/website/Lurgi-Fixed-Bed-Dry-Bottom.887.0.html?&L=1. Accessed on December 22, 2012.

MacKay Group. 2012. Indirectly fired gas turbine (IFGT). http://www.themackaygroup.sk/mackay-technology/ifgt-turbine.html. Accessed on December 22, 2012.

Memorandum of RAE. 2003. Memorandum by the royal academy of engineering to the select committee on science and technology of the UK parliament. Available at: http://www.publications.parliament.uk/pa/ld200304/ldselect/ldsctech/126/126we40.htm. Accessed on December 22, 2012.

Najjar, Y.S.H. 2001. Efficient use of energy by utilizing gas turbine combined systems. *Applied Thermal Engineering* 21: 407–438.

Nuon. 2012. Willem-Alexander power plant. http://www.nuon.com/company/core-business/energy-generation/power-stations/buggenum. Accessed on December 26, 2012.

Obenberger, I., Carlsen, H., and Biedermann, F. 2003. State-of-art and future development regarding small-scale CHP systems with a special focus on ORC and Stirling engine technologies. *International Nordic Bioenergy 2003 Conference,* Jyväskylä, Finland. Available at: http://www.bios-bioenergy.at/uploads/media/Paper-Obernberger-SmallScaleCHP-NordicConference-2003-10-27.pdf. Accessed on December 21, 2012.

Obernberger, I. and Thek, G. 2008. Combustion and gasification of solid biomass for heat and power production in Europe—State of the Art and relevant future developments. *Proceeding of the 8th European Conference on Industrial Furnaces and Boilers*. Vilamoura, Portugal.

Obernberger, I., Thonhofer, P., and Reisenhofer, E. 2002. Description and evaluation of the new 1000 kWel organic Rankine cycle process integrated in the biomass CHP plant in Lienz, Austria. Euroheat & Power October 2002. Available at: http://www.turboden.eu/en/public/downloads/report_on_lienz_plant.pdf. Accessed on December 20, 2012.

Palonen, J. and Nieminen, J. 2004. Biomass CFB gasifier connected to a 350 MW_{th} steam boiler fired with coal and natural gas—THERMIE demonstration project at Kymijarvi CHP power station in Lahti, Finland. Foster Wheeler Technical Paper TP_GASFY_04_02. Available at http://fosterwheeler.com/publications/tech_papers/files/TP_GASFY_04_02.pdf. Accessed on December 27, 2012.

Perchers, N. 2003. *Combined Heating Cooling and Power Handbook: Technologies and Applications: An Integrated Approach to Energy Conservation/Resource Optimisation*. Fairmont Press Inc., Lilburn, GA.

Peterson, R.B., Wang, H., and Herron, T. 2008. Performance of a small-scale regenerative Rankine power cycle employing a scroll expander. *Proceedings of the Institution of Mechanical Engineers, Part A: Journal of Power and Energy* 222: 271–282.

Pilavachi, P.A. 2002. Mini- and micro-gas turbines for combined heat and power. *Applied Thermal Engineering* 22: 2003–2014.

Poullikkas, A. 2005. An overview of current and future sustainable gas turbine technologies. *Renewable and Sustainable Energy Reviews* 9: 409–443.

Power-Technology. 2012. Stevenscroft wood burning plant, UK. http://www.power-technology.com/projects/stevenscroftbiomass. Accessed on December 23, 2012.

Pöyry. 2012. Alholmens Kraft biofuel-fired power plant. http://www.poyry.com/projects/alholmens-kraft-biofuel-fired-power-plant-pietarsaari-finland. Accessed on December 23, 2012.

Qiu, G., Liu, H., and Riffat, S.B. 2011. Expanders for micro-CHP systems with organic Rankine cycle. *Applied Thermal Engineering* 31: 3301–3307.

Qiu, G., Shao, Y., Li, J., Liu, H., and Riffat, S.B. 2012. Experimental investigation of a biomass-fired ORC-based micro-CHP for domestic applications. *Fuel* 96: 374–382.

Quoilin, S., Lemort, V., and Lebrun, J. 2010. Experimental study and modeling of an Organic Rankine Cycle using scroll expander. *Applied Energy* 87: 1260–1268.

Rensfelt, E. 2010. VVBGC demonstration plant project. http://lnu.se/polopoly_fs/1.35436!vvbgc.pdf. Accessed on December 22, 2012.

RWE. 2012. Markinch biomass CHP. http://www.rwe.com/web/cms/en/429434/rwe-innogy/sites/power-from-biomass/united-kingdom/under-construction/markinch-biomass-chp-home. Accessed on December 23, 2012.

Schmitt Enertec. 2012. Woodgas CHP unit. http://www.schmitt-enertec.com/wood_gasification/woodgas_power.htm. Accessed on December 21, 2012.

Schuster, A., Karellas, S., Kakaras, E., and Spliethoff, H. 2009. Energetic and economic investigation of Organic Rankine Cycle applications. *Applied Thermal Engineering* 29: 1809–1817.

Siemens. 2012a. Integrated gasification combined cycle. http://www.energy.siemens.com/br/en/power-generation/power-plants/integrated-gasification-combined-cycle/integrated-gasification-combined-cycle.htm#content=Process%20. Accessed on December 22, 2012.

Siemens. 2012b. Siemens fuel gasification technology. http://www.energy.siemens.com/br/pool/hq/power-generation/fuel-gasifier/214_120449_WS_Fuel_Gasification_US.pdf. Accessed on December 22, 2012.

Simons G. 2009. Lessons learned from decades of CHP in California. Cogeneration and onsite power production. March–April 2009. PennWell Inc., Lilburn, GA. http://www.cospp.com/index.html.

Smock, R. 1995. Target: 60% thermal efficiency. Power Engineering International, March–April (4). http://www.powerengineeringint.com.

Starr, F. 2003. HRSG systems and implications for CCGT plant cycling. Power Plants: Operation Maintenances and Materials Issues (2) 1. ISSN 1740-5181. http://www.ommi.co.uk/main.asp.

StirlingDK. 2012. StirlingDK introduction. http://ida.dk/sites/prod.ida.dk/files/Stirling%20DK%20Commercial%20Update%20ver%204%20.pdf. Accessed on December, 2013.

KWB. 2013. Sunmachine: Pellet-BHKW mit Stirlingmotor. http://www.bhkw-prinz.de/sunmachine-pellet-bhkw-mit-stirlingmotor/141. Accessed on 3 Dec 2013.

Thiers, S., Aoun, B., and Peuportier, B. 2010. Experimental characterization, modeling and simulation of a wood pellet micro-combined heat and power unit used as a heat source for a residential building. *Energy and Buildings* 42: 896–903.

Turboden. 2012. Combined heat and power units. http://www.turboden.eu/en/products/products-chp.php. Accessed on 20 December 2012.

van Basshuysen, R. and Fred, S. 2002. *Basics, Components, Systems, and Perspectives*. SAE International, Warrendale, PA.

Van Loo, S. and Koppejan, J. 2010. *Biomass Combustion and Co-Firing*. Taylor & Francis, London, U.K.

Vescovo Recardo. 2009. ORC recovering industrial heat power generation from waste energy streams. Cogeneration and onsite power production. March–April 2009. PennWell Inc. http://www.cospp.com/index.html.

Walker, G. et al. 1994. *The Stirling Alternative*. Gordon and Breach Science Publishers, Yverdon, Switzerland.

Wang, H., Peterson, R.B., and Herron, T. 2009. Experimental performance of a compliant scroll expander for an organic Rankine cycle. *Proceedings of the Institution of Mechanical Engineers Part A: Journal of Power and Energy* 223: 863–872.

Weston, K. C. 1992. *Energy Conversion*. West Pub. Co., St. Paul, MN.

Yin, C., Rosendahl, L.A., and Kaer, S.K. 2008. Grate-firing of biomass for heat and power production. *Progress Energy Combustion Science* 34: 725–754.

Section IV

Sustainable Biorefinery Systems

22 Biorefinery Principles, Analysis, and Design

Zheng Liu and Mario Richard Eden

CONTENTS

22.1 INTRODUCTION

The increase of fossil fuel (crude oil) prices, depletion of natural resources, public awareness regarding energy security, as well as environmental pressures have led modern industries to embrace renewable energy sources (Tay et al. 2011). As a renewable energy source, biomass has an incredible potential to fulfill our future energy and chemical demands, as the biofuels and biochemicals derived from it are considered to have inherently lower carbon footprints than those derived from fossil sources (Sammons et al. 2008). For instance, the life cycle assessment shows that the potentials to reduce greenhouse gas (GHG) by corn ethanol and cellulosic ethanol are 52% and 86% over their gasoline counterpart, respectively (Bioenergy Knowledge Discovery Framework 2012). Additionally, the nation's reliance on foreign sources of oil has left the United States vulnerable to a shortage of petroleum-based fuel and products due to natural disasters, political disruptions, and price volatility. Therefore, complementing the nation's energy supply with renewable, sustainable, and domestically produced sources of energy, such as bioenergy, can help ensure our future energy security.

A biorefinery is defined by the National Renewable Energy Laboratory (NREL) as a "facility that integrates different biomass conversion processes and equipment to produce transportation biofuels, biochemicals, biomaterials and heat and power" (Fernando et al. 2006). An integrated biorefinery is designed to provide a sustainable supply of biofuels, as well as the production of bulk and fine biochemicals (e.g., methanol, syngas, glycerol, ether), with minimum waste generation. Biorefineries can be classified into three main categories based on the different conversion processes employed, namely, thermochemical conversion, chemical conversion, and biological conversion. Thermochemical conversion includes pyrolysis, gasification, combustion, and hydrothermal liquefaction. During thermochemical conversion, biomass is broken down to small gaseous or liquid molecules at elevated temperatures and/or pressures in the presence or absence of air or oxygen. Fermentation and anaerobic digestion are the two primary biological conversion processes used to convert biomass into alcohols (more commonly known as bioethanol) and biogas, respectively. The fermentation reactions are catalyzed by enzymes or acids, and the anaerobic digestion uses microorganisms to break down a biodegradable material in a low or zero oxygen environment. Transesterification is a common chemical process to produce biodiesel from vegetable oil and animal fats through the process of exchanging the organic group of an ester with the organic group of an alcohol. Transesterification reactions are often catalyzed by the addition of an acid or alkali catalyst. A summary of the conversion technologies in each category is given in Table 22.1 (E4Tech.com 2012). Selection of an appropriate conversion process is not trivial and depends on the availability and the characteristics of the biomass resource, the types of desired products, the process efficiency, economics, environmental requirements, and many other aspects related to the entire biorefinery supply chain.

A biorefinery supply chain typically has five stages, namely, feedstock production, feedstock logistics, biofuels and biochemicals production, bioproduct distribution, and bioproduct end use. Transport, storage, and handling are key issues throughout the supply chain and link different stages to each other. In addition, the following key issues are concerned within specific stages of the supply chain (Wisconsin Grasslands Bioenergy Network 2012). For feedstock producers, the key issues include crop selection, achieving sustainability, and improving farm profitability. For preprocessing, there are issues related to the scale of operation and whether to approach preprocessing as an opportunity for on-farm added value to biomass or to approach preprocessing as an opportunity for a new, separate industry. A major issue in conversion of biomass is scale-up of biofuel technologies. Among consumers, key issues are availability, affordability, and ease of use of the biofuel/bioenergy products. Note that a biorefinery supply chain can be a simple intraorganizational system consisting of different functional units of feedstock production, feedstock logistics, biofuels and biochemicals production, bioproduct distribution, and bioproduct end use that are all linked by the flows of goods, information, and finance or a more complex multisite intraorganizational system having several supplying, manufacturing, and distribution sites scattered in different locations with various material, information, and financial flows or even a most complex interorganizational one comprised of several intraorganizational supply chains in a large complex supply chain network.

Biorefinery supply chains must be designed and managed efficiently to reduce the cost of production and/or environmental impacts. Supply chain management (SCM) techniques can be used to analyze and integrate functional units along a supply chain and to manage and coordinate the flows of goods, services, information, and finance for the improvement of the competitiveness of the whole system (Yan et al. 2011). For supply chain design, which is also known as strategic supply chain network planning, the organizations should set strategy goals and development directions and then make the best investment decisions to satisfy the defined market requirements. Typical decisions involve the selection of bioproduct portfolio, the establishment or closure of manufacturing and distribution facilities, and the installation of major production lines. The objectives are most often financially oriented to be either profit maximization or cost minimization, subject to customer service and budget constrains (Goetschalckx 2002).

TABLE 22.1
Summary of the Conversion Technologies of Biorefinery

Category	Conversion Technology	Resource Type	Examples of Resources	Product	End Use	Technology Status
Thermochemical	Combustion	Mainly solid biomass	Wood logs, chips, and pellets Agricultural residues Chicken litter	Heat	Heat Electricity	Commercial
	Gasification	Mainly solid biomass	Wood chips and pellets Agricultural residues	Product gas	Heat Electricity Transport fuel	Demonstration/Early commercial
	Pyrolysis	Mainly solid biomass	Wood chips and pellets Agricultural residues	Pyrolysis oil By-products (product gas, char)	Heat Electricity	Demonstration
Chemical	Transesterification	Mainly solid and liquid biomass	Rape seeds Soybeans Oil or fat waste	Liquid fuels (biodiesel)	Heat Electricity Transport fuel	Demonstration/Early commercial
Biological	Fermentation	Mainly solid biomass	Wheat Maize Sugar beet Potatoes Colza Miscanthus	Liquid fuels (bioethanol)	Transport fuel	Commercial
	Anaerobic digestion	Wet biomass	Manure Sewage sludge	Biogas By-products	Heat Electricity Transport fuel	Commercial

Source: E4Tech.com., http://www.e-sources.com//biomass_converion.htm (accessed June 30, 2012).

The design and analysis of biorefinery systems is inherently challenging due to the complexity carried by the system under study, the optimality desired by the decision-maker, and the uncertainty associated with the data and knowledge. As stated, biorefinery conversion processes can transform available biomass materials in many different forms into different energy products including liquid fuels such as alcohols, biodiesel, and bio-oil; gaseous fuels such as biogas and syngas; and heat and power. This great diversity of biomass feedstocks, conversion technologies, and products could make any potential biorefinery system very complex. In addition, different criteria and concerns could be addressed at different supply chain stages of a biorefinery system, and an overall optimal design solution is always desired by the decision-maker, which results in more complexity in the design and analysis tasks. To further complicate matters, the available data and information about the biorefinery is frequently incomplete, imprecise, and uncertain. Various sources of uncertainty (e.g., future prices of energy, products, and feedstocks, climate policies, and process scale-up) often force decision-makers to make assumptions based on heuristics in filling this information gap when data are limited, which could greatly affect the quality of analysis results and planning decisions.

Many methodologies and tools have been developed recently to handle these challenges and help facilitate the tasks of design and analysis of biorefineries. In this chapter, a holistic overview of the research in this area will be given. The remainder of the chapter is organized as follows: first, we introduce the general principles of the sustainable biorefinery. Based on these principles, two specific research topics, namely, the analysis and the design of biorefineries, are described in detail. In each area, the research needs, challenges, existing methodologies and tools, and appropriate case studies are presented. Finally, we will conclude the significance of the chapter and discuss the types of uncertainties encountered in design and analysis of biorefineries and potential methods for handling them.

22.2 PRINCIPLES OF SUSTAINABLE BIOREFINERIES

As described in the International Energy Agency's (IEA) bioenergy task 42, biorefining is the sustainable processing of biomass into a spectrum of marketable products and energy (Cherubini et al. 2009). Although only a few general categories of biorefineries have been clearly identified and the number of different types of biorefineries is still growing, it is agreed that all kinds of biorefineries should be designed and operated in a sustainable way from the social, economic, and environmental (SEE) perspectives. In this regard, a biorefinery system should obey the following general sustainability principles:

1. Principle 1: A biorefinery should ensure the maximum economic benefits through minimizing the usage of raw material and energy resources as well as the cost involved in supply chain operations intrinsic to biorefining (Batsy et al. 2012).
2. Principle 2: A biorefinery should ensure minimum environmental impacts through minimizing the generation of hazardous wastes and the usage of nonrenewable energy resources.
3. Principle 3: A biorefinery should provide maximum social benefits through creating employment, supporting regional developments, improving public image, etc. (Cherubini et al. 2009).

   From the SCM point of view, principles 4–6 have been assigned to biorefinery systems by the IEA's bioenergy task 42 (Cherubini et al. 2009).
4. Principle 4: A biorefinery should meet the market demands on renewable energy products and the conformity of end-products to end-user requirements.
5. Principle 5: A biorefinery should not result in competition with food supply.
6. Principle 6: A biorefinery should have affordable biomass-to-products value chains.
7. Principle 7: A biorefinery system should have good conversion efficiency.

Note that those seven principles given earlier are overarching guidelines for the design and analysis of any biorefinery. More details are needed when they are applied to specific cases. For instance, principle 2 can be interpreted as a requirement on life cycle GHG emission reduction or no harm to soil, water, air, etc. In addition, other principles could also be adopted for biorefineries as necessary supplements, for instance, the twelve principles of green chemistry (Anastas and Warner 1998) and the principles of sustainable development (McKeown 2002).

22.3 SYSTEM ANALYSIS FOR BIOREFINERIES

Systems analysis is an explicit formal inquiry carried out to help decision-makers identify a better course of action and make a better decision than might otherwise have been made. Due to the inherent complexity and uncertainty carried by biorefinery systems, it is quite valuable to apply appropriate analysis to investigate the current status, the comparisons of different design and operation schemes, and the identification of improvement opportunities and recommendations. Among various types of analysis approaches, sustainability assessment and life cycle analysis (LCA) have been widely accepted and applied in industrial practice and academic research for the study of industrial problems including biorefineries. In this section, the general methodologies employed in these two approaches and their adoptions to biorefinery systems will be presented. Case studies will be given as well to demonstrate the methodological efficacy.

22.3.1 Sustainability Assessment of Biorefinery Systems

The most widely quoted definition of sustainability and sustainable development is given by the Brundtland Commission of the United Nations on March 20, 1987: "Sustainable development is a development that meets the needs of the present without compromising the ability of future generations to meet their own needs" (Brundtland Commission 1987). At the 2005 World Summit, it was decided that sustainability requires the reconciliation of environmental, social, and economic demands (United Nations General Assembly 2005), which are the so-called three pillars of sustainability. This view has later been expressed as an illustration using three overlapping ellipses indicating that the three pillars of sustainability are not mutually exclusive and can be mutually reinforcing (Forestry Commission of Great Britain 2009) (see Figure 22.1).

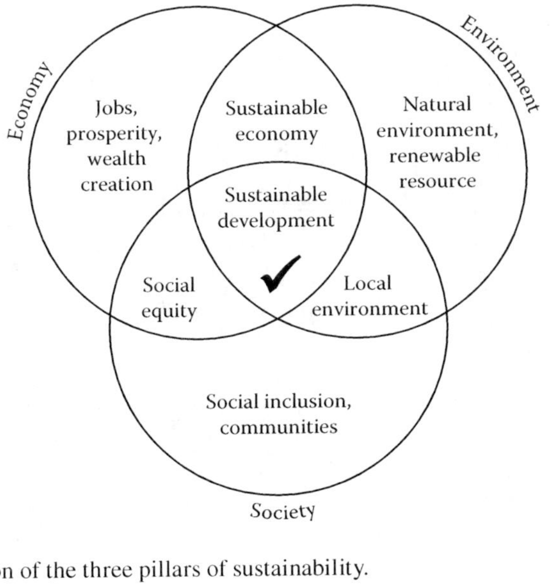

FIGURE 22.1 Illustration of the three pillars of sustainability.

As a broad subject, sustainability is studied and managed over many scales of time and space—from planet Earth to ecosystems, countries, economic sectors, individual lives, occupations, life styles, behavior patterns, and so on (Millennium Ecosystem Assessment 2003). Among those, a major branch is industrial sustainability, which focuses on how to pursue the short- to long-term sustainable development of industrial systems, such as a biorefinery, where material and energy efficiencies, waste reduction, safety, synergies among the subsystems, etc., are among the major concerns (Piluso et al. 2010).

22.3.1.1 Triple Bottom Lines of Industrial Sustainability

Among the three pillars of sustainability, economy is definitely the most critical one due to the intrinsic nature of industrial activities in creating wealth and reducing costs. Note that compared to conventional economics that historically demonstrated a close correlation between economic growth and environmental degradation, sustainable economics represents *a broad interpretation of ecological economics where environmental and ecological variables and issues are basic but part of a multidimensional perspective.* In fact, integrating economics with environmental and social concerns can provide opportunities for creating new benefits and business. For example, the economic benefits of sustainable waste reuse include savings from disposal costs, fewer environmental penalties, and reduced liability insurance. Moreover, it may lead to increased market share due to an improved social public image (Jackson and Clift 2008). As another example for energy systems, improvement of energy efficiency can also increase profits by reducing costs.

The environment must be protected during any types of industrial activities since all types of vital goods and services required by humans and other organisms are best provided by healthy ecosystems. However, human activities of industrial and energy systems most likely have negative impacts to the environment due to inherent resource depletion and waste generation. Therefore, environmental sustainability is required by modern industrial practices, which emphasize that the consumption of goods and services should be analyzed and managed at all scales through the chain of the product life cycle, where energy, materials, and water are key resources. Note that the design and analysis of environmental sustainability should be incorporated with the other two pillars of sustainability: social and economic sustainability.

Compared to the first two pillars, social sustainability is much more difficult to address and analyze. The reason is that sustainability is easily expressed in scientific, economic, and environmental terms, but social aspects are always related to nonscientific concerns such as national law, public image, local and individual life styles, and ethical consumerism (United Nations 1992). In general, social sustainability is the idea that future generations should have the same or greater access to social resources as the current generation. In this regard, the most fundamental principle of social sustainability is to meet human needs fairly and efficiently, which encompasses human rights, labor rights, and corporate governance. Therefore, the following criteria are commonly used to rate the social sustainability of industrial and energy systems, namely, community, diversity, employee relations, human rights, and process and product safety. Needless to say, those criteria are still quite difficult to be quantified exactly, which is why some soft-indicator-based approaches have been brought into practice for the assessment of social performance, that is, put a scaling system (for instance, from 0 to 10) on each social indicator to represent the relatively good or bad performance of industrial systems (Carvalho et al. 2008, Othman et al. 2010).

22.3.1.2 Sustainability Assessment Framework for Biorefineries

A general double-layered-multicriteria evaluation method for conducting sustainability assessment was introduced by Piluso et al. (2010), which can be used for biorefinery systems. There are five major tasks in the sustainability assessment: (1) to identify the sustainability metrics for the biorefinery system under study; (2) to select the system variables suitable for metrics evaluation and data collection; (3) to evaluate the economic, environmental, and social sustainability indices;

(4) to calculate the composite economic, environmental, and social sustainability; and (5) to calculate the overall sustainability.

Sustainability metrics selection: Since no specific sustainability metrics have been clearly claimed for biorefinery systems, decision-makers are suggested to carefully select their preferred metrics from various metric systems available for performing sustainability assessment, such as the IChemE (2002) and AIChE (Cobb et al. 2009) sustainability metrics that have been widely adopted by the chemical industries. These metrics are typically grouped for assessing economic, environmental, and social sustainability, and in each of these three categories, different numbers of indicators are assigned for the representation of various aspects. For instance, the IChemE sustainability metrics contain 24 environmental indicators related to the resource usage, emissions, effluents, waste, etc.; 14 economic indicators regarding the profit, value, tax, investments, etc.; and 11 social indicators focused on workplace and societal benefits. A set of sustainability metrics after selection is denoted as set S, which contains three subsets, each of which can have a number of specific indices:

$$S = \{E, V, L\} \quad (22.1)$$

where

$E = \{E_i | i = 1,2,\ldots,F\}$, the set of economic sustainability indices
$V = \{V_i | i = 1,2,\ldots,G\}$, the set of environmental sustainability indices $S = \{E, V, L\}$
$L = \{L_i | i = 1,2,\ldots,H\}$, the set of social sustainability indices

Top layer indicator calculation: The tasks of the top layer are the evaluation of the individual indicators, that is, E_i, V_i, and L_i, and the combination of them into the composite economic, environmental, and social sustainability. The data for the system variables required by each indicator should first be collected and then input into the quantification formula to obtain the corresponding scores. Note that those indicator values usually have different units, and as such, they must be normalized in order to be combined into a single composite sustainability value.

The dimensional data of each indicator can be normalized to a value in the range between 0 and 1, with *0* as the lowest sustainability and *1* as the highest sustainability. Specifically, if the engineering meaning of one indicator, $I \in \{E_i, V_i, L_i\}$, shows that a large value is more preferable from the sustainability point of view, then the normalized indicator, $\bar{I}$, can be derived using Equation 22.2:

$$\bar{I} = \frac{I - I_{min}}{I_{max} - I_{min}} \quad (22.2)$$

where I_{min} and I_{max} are the lower and upper bound values, respectively. On the contrary, if the engineering meaning of one indicator, $I \in \{E_i, V_i, L_i\}$, shows that a small value is more preferable, then Equation 22.3 should be used to derive the normalized indicator, $\bar{I}$:

$$\bar{I} = \frac{I_{max} - I}{I_{max} - I_{min}} \quad (22.3)$$

The last step of the top layer is to calculate the composite economic, environmental, and social sustainability for a biorefinery system P under study. This is done by combining the normalized indicators in the same sustainability category with a set of assigned weights, that is,

$$E(P) = \frac{\sum_{i=1}^{F} a_i \bar{E}_i(P)}{\sum_{i=1}^{F} a_i} \quad (22.4)$$

$$V(P) = \frac{\sum_{i=1}^{G} b_i \bar{V}_i(P)}{\sum_{i=1}^{G} b_i}, \tag{22.5}$$

$$L(P) = \frac{\sum_{i=1}^{H} c_i \bar{L}_i(P)}{\sum_{i=1}^{H} c_i}, \tag{22.6}$$

where a_i, b_i, and $c_i \in [1, 10]$ are the weighting factors associated with the corresponding indices, reflecting the relative importance of the individual indices in the overall assessment. The user is allowed to adjust these weighing factors. If all the factors are equally important, then they all have a default value of 1 (Liu and Huang 2012).

Bottom layer calculation: The task of this layer is to calculate the overall sustainability. To achieve that, the cube-based sustainability status representation proposed by Piluso et al. (2010) is applied. The proposed concept of a sustainability cube is shown in Figure 22.2, where the three coordinates represent the composite economic index, the composite environmental index, and the composite social index. Each composite index is set to have a value between 0 (meaning no sustainability) and 1 (meaning complete sustainability). With this representation, the corner coordinate of (0, 0, 0) represents the system's status of no sustainability, while the opposite corner having the coordinate (1, 1, 1) indicates complete sustainability. In the figure, the point, $S(P)$, represents the overall sustainability status of system P, which can be evaluated using the composite indices, $E(P)$, $V(P)$, and $L(P)$, with the weighting factors assigned again by the user, that is,

$$S(P) = \frac{\| (\alpha E(P), \beta V(P), \gamma L(P)) \|}{\| (\alpha, \beta, \gamma) \|} \tag{22.7}$$

where α, β, and γ each has a value of 1 (default) or greater. Naturally, $S(P)$ is still normalized.

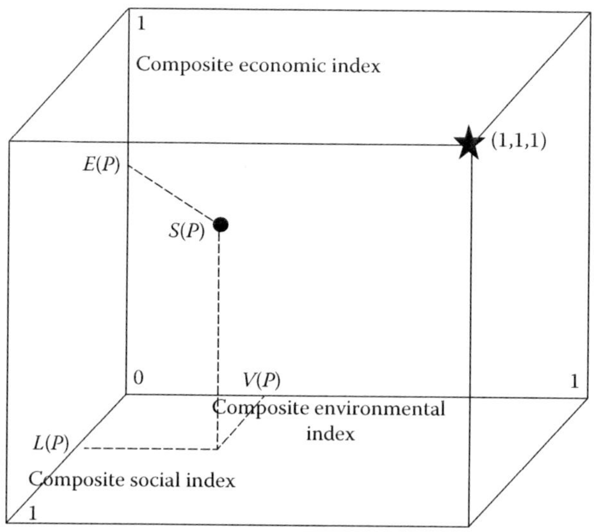

FIGURE 22.2 Cube-based sustainability evaluation: (●) the location of sustainability status of the system P, $S(P)$ and (★) the best possible sustainability status. (From Piluso, C. et al., *Ind. Eng. Chem. Res.*, 49(18), 8633, 2010.)

22.3.1.3 Case Studies of Sustainability Assessment

The sustainability assessment approach introduced earlier has been applied to several studies of biorefinery systems. Li et al. (2011) performed a comprehensive sustainability assessment on two biodiesel processes: the homogeneous catalyst process, which produces biodiesel using alkali-based catalyst, and the heterogeneous catalyst process operated in continuous mode, where the heterogeneous catalyst can be reused and regenerated. Note that in their work, the process efficiency was separated from the triple bottom lines of sustainability as the fourth evaluation category, and no overall sustainability was calculated. Therefore, the case study illustrated in the following was adapted from the work by Li et al. (2011) to be consistent with the sustainability assessment method presented in this chapter.

Three economic, five environmental, and one social indicator are selected for the sustainability assessment. The economic indicators are the following: (1) net present value (NPV), E_1, where a positive value means that the project is feasible and negative values indicate otherwise (Othman et al. 2010); (2) internal rate of return (IRR), E_2, which is designed to reflect the highest, after-tax interest or discount rate at which the project can just break even (Turton et al. 2003); and (3) the total profit, E_3. The environmental indicators contain four potential environmental impact (PEI) indicators and two indicators of process efficiency. PEI indicators can be calculated using the waste reduction (WAR) algorithm introduced by Young and Cabezas (1999), which focuses on the evaluation of the environmental impact only at the manufacturing stage within the overall life cycle of the chemical production and process. These PEI indicators are (1) PEI output rate, V_1; (2) PEI generation rate, V_2; (3) PEI output/mass of product, V_3; and (4) PEI generation/mass of product, V_4. The two indicators of process efficiency are (1) energy efficiency, V_5, and (2) exergy efficiency, V_6, which reflect the effective use of the portion of energy that can be converted to useful work. For social sustainability, the inherent safety index (L_1) modified from Heikkila (1999) was selected, which quantifies the safety of all the chemicals and equipment in a plant based on main principles of avoiding the use of hazardous materials and aiming at simpler processes.

Both the homogeneous catalyst process and the heterogeneous catalyst process for biodiesel manufacturing are simulated using Aspen Plus, where the design parameters, feed conditions, and reaction conditions are given in Table 22.2. Using the simulation data, the sustainability indicators are evaluated, and the results are given in Tables 22.3 and 22.4.

Comparisons show that the heterogeneous catalyst process is more sustainable than the homogeneous catalyst process, especially in the social and economic aspects.

TABLE 22.2
Feed Specifications and Reaction Conditions

	Homogeneous Catalyst Process (kg/h)	Heterogeneous Catalyst Process (kg/h)
	Feed Conditions	
Oil (triolein)	1050	1050
Methanol	121	114
NaOH	44.5	N/A
H_3PO_4	34.2	N/A
Water	50	N/A
	Reaction Conditions	
Temperature, °C	60	48
Pressure, bar	4	4

Source: Li, X. et al., *Ind. Eng. Chem. Res.*, 50, 2981, 2011.

TABLE 22.3
Sustainability Assessment Results of Homogeneous Catalyst Process

Indicators	Original Values	Normalized Values	Weighting Factors	Composite Sustainability
E_1: NPV, (M$)	1.8	0.014	0.333	0.123
E_2: IRR, (%)	7	0.140	0.333	
E_3: total profit, (M$)	0.54	0.216	0.333	
V_1: PEI output rate, (PEI/h)	146.7	0.022	0.125	0.591
V_2: PEI generation rate, (PEI/h)	82.25	0.452	0.125	
V_3: PEI output/mass of product, (PEI/kg)	0.139	0.073	0.125	
V_4: PEI generation/mass of product, (PEI/kg)	0.078	0.480	0.125	
V_5: Energy efficiency, (%)	0.879	0.879	0.250	
V_6: Energy efficiency, (%)	0.973	0.973	0.250	
L_1: Inherent safety index	41.6			0.168

TABLE 22.4
Sustainability Assessment Results of Heterogeneous Catalyst Process

Indicators	Original Values	Normalized Values	Weighting Factors	Composite Sustainability
E_1: NPV, (M$)	9.34	0.945	0.333	0.870
E_2: IRR, (%)	40	0.800	0.333	
E_3: total profit, (M$)	2.16	0.864	0.333	
V_1: PEI output rate, (PEI/h)	137.7	0.082	0.125	0.626
V_2: PEI generation rate, (PEI/h)	75.74	0.495	0.125	
V_3: PEI output/mass of product, (PEI/kg)	0.131	0.127	0.125	
V_4: PEI generation/mass of product, (PEI/kg)	0.072	0.520	0.125	
V_5: Energy efficiency, (%)	0.914	0.914	0.250	
V_6: Energy efficiency, (%)	0.978	0.978	0.250	
L_1: Inherent safety index	32			0.360

The overall sustainability is further calculated using Equation 22.7 for both biodiesel manufacturing processes, where the weighting factors, that is, α, β, and γ, each has a value of 1 (default). The results show that the heterogeneous catalyst process has a better overall sustainability (0.653) than the one using the homogeneous catalyst process (0.499), and the cube-based sustainability evaluation is plotted in Figure 22.3.

Other applications using the same sustainability assessment method can be found in Othman et al. (2010) and Liu and Huang (2012), where the first one compared the sustainability performance of two biodiesel manufacturing processes (alkali catalyst process vs. supercritical methanol process) and the second one solved a technology-based sustainability enhancement problem under uncertainty for a biodiesel plant. Recently, this assessment approach was implemented by Liu (2012) in a computational tool, which is named Industrial Sustainability Evaluation and Enhancement (ISEE). The tool is capable of conducting industrial sustainability assessment and providing decision-making, which can be used for sustainability status evaluation of biorefinery systems, comparisons between different biorefinery alternatives, identification of the best design for decision making, and potential biorefinery improvement recommendations.

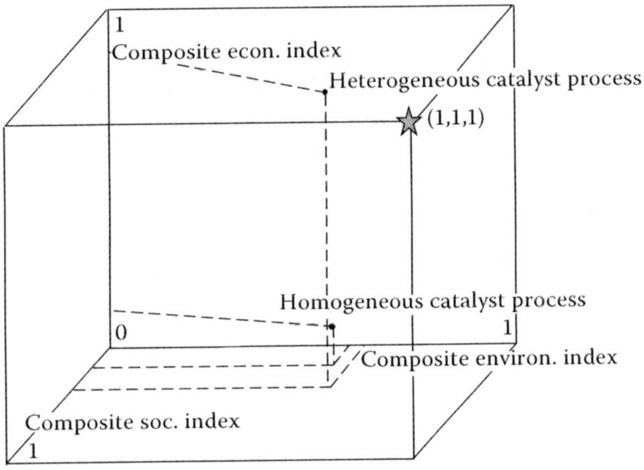

FIGURE 22.3 Cube-based sustainability evaluation of two biodiesel processes.

22.3.2 Life Cycle Analysis for Biorefinery Systems

With increased societal awareness of the issues related to natural resource depletion and environmental degradation, the chemical industry has to provide *greener* products and using *greener* processes. For the purpose of environmentally oriented decision making, a practical engineering tool, LCA, has been introduced for the estimation of the cumulative environmental impacts resulting from all stages in a product life cycle, often including impacts not considered in traditional analyses (e.g., raw material extraction, material transportation, ultimate product disposal). By including the impacts throughout the entire product life cycle, LCA provides a comprehensive view of the environmental aspects of the product or process and a more accurate picture of the true environmental trade-offs in product and process selection (Scientific Applications International Corporation 2006).

The term *life cycle* refers to the major activities in the product's life-span from its raw material acquisition, manufacturing, and use/reuse/maintenance to its final recycle/disposal. Cherubini and Jungmeier (2010) illustrated the possible life cycle stages that can be considered in the production chain of a biorefinery and compared them to the conventional fossil fuel-based reference system, as shown in Figure 22.4.

When conducting an LCA study on biorefinery systems, four technical phases should be completed, namely, goal definition and scoping, inventory analysis, impact assessment, and interpretation. A general illustration of those phases is adopted from Scientific Applications International Corporation (2006) and shown in Figure 22.5.

22.3.2.1 Goal Definition and Scoping

At the beginning of an LCA study, the goal and scope of the LCA project should be defined, which will determine the time and resources needed during the whole analysis process. The defined goal and scope will guide the entire process to ensure that the most meaningful results are obtained. Specifically, the following six basic decisions should be made according to the Scientific Applications International Corporation (2006):

1. Define the goal(s) of the project. The goals can be, for instance, supporting broader environmental assessments on biorefinery systems, ranking the relative contribution of individual biorefinery processes, etc.
2. Identify the questions that the decision-makers care about, and then define the type of information needed to inform the decision-makers.

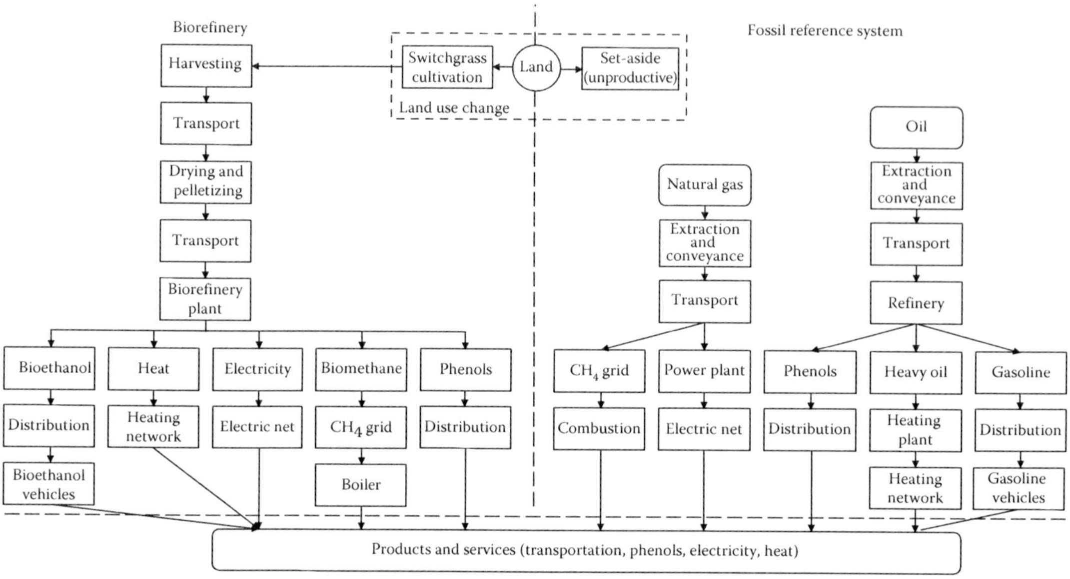

FIGURE 22.4 Comparison between the life cycle stages of biorefinery and fossil reference system. (From Cherubini, F. and Jungmeier, G., *Int. J. Life Cycle Assess.*, 15, 53, 2010.)

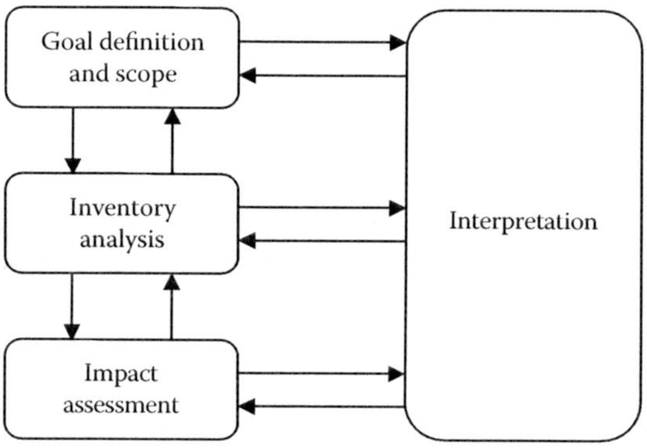

FIGURE 22.5 Phases of an LCA. (From Scientific Applications International Corporation, *Life Cycle Assessment: Principles and Practice*, Cincinnati, OH, EPA, 2006.)

3. Determine the required specificity, that is, to answer whether the data should be more general to represent common biorefinery practices or the data collected should be specific to one company or manufacturing facility.
4. Determine how the data should be organized and the results should be displayed. This step is essentially the identification of functional units that will be measured and displayed as the LCA results. Note that when two or more biorefinery products are compared by an LCA, an equivalent basis is always necessary for fair comparison.
5. Define the scope of the study, that is, to determine whether one or all of the stages should be included in the scope of the LCA.
6. Determine the ground rules for performing the work. This step is mainly for the LCA report preparation, where assumptions, quality assurance procedures, and reporting requirements should be carefully checked and recorded.

22.3.2.2 Life Cycle Inventory

After performing primarily linguistic tasks during the goal definition and scoping phase, this part starts to deal with numbers and calculations. A life cycle inventory is the process of quantifying energy and raw material requirements, emissions, solid wastes, and other releases from the entire life cycle of a biorefinery product or process (Scientific Applications International Corporation, 2006). A four-step framework of life cycle inventory is given in Environmental Protection Agency (EPA) 1993 and 1995 publications (Environmental Protection Agency 1993, EPA 1995), which are summarized in the following.

1. Develop a flow diagram of the biorefinery processes being evaluated. A flow diagram is a tool for mapping the inputs and outputs of a process or system. Inputs of a biorefinery system may contain materials and energy and the transportation of products. Outputs of a biorefinery system should include all kinds of products, coproducts, atmospheric emissions, solid wastes, and other possible releases. For a complex biorefinery, it is highly recommended to break it into several subsystems and develop flow diagram for each of them individually.
2. Develop a data collection plan. A data collection plan should be developed to ensure that the quality and accuracy of the data meet the expectations of the decision-makers. During this plan development, data sources, types, and quality should first be determined, and then a detailed data collection worksheet and checklist can be designed.
3. Collect data. Data can be collected through a variety of ways, on-site and/or off-site, direct and/or indirect, etc. Note that there are many commercial LCA software packages available, which may greatly reduce the efforts of biorefinery data collection.
4. Evaluate and report results. Life cycle inventory studies of biorefinery systems could generate a great deal of information of a disparate nature. Therefore, the analyst needs to select a presentation format and contents that are consistent with the purpose of the study and do not arbitrarily simplify the information solely for the sake of presenting it (Scientific Applications International Corporation 2006).

22.3.2.3 Life Cycle Impact Assessment

Life cycle impact assessment evaluates the potential human health and environmental impacts of the resources and the releases identified during the life cycle inventory. Scientific Applications International Corporation (2006) listed the following key steps for a life cycle impact assessment:

1. Impact category definition (e.g., global warming, acidification, terrestrial toxicity).
2. Linking life cycle inventory results to the impact categories (e.g., linking carbon dioxide emissions to global warming).
3. Modeling life cycle inventory impacts using science-based conversion factors (e.g., modeling the potential impact of carbon dioxide and methane on global warming).
4. Potential impact normalization and comparison (e.g., comparing the global warming impact of carbon dioxide and methane for two or more options).
5. Sorting or ranking the indicators (e.g., sorting the indicators by location: local, regional, and global).
6. Weighting the most important potential impacts.
7. Evaluating and reporting life cycle impact assessment results.

22.3.2.4 Life Cycle Interpretation

In the last phase of LCA, the task of life cycle interpretation is to identify, quantify, check, and evaluate information from the results of life cycle inventory and life cycle impact assessment and communicate them effectively. The International Organization for Standardization (ISO) has defined the following two objectives of life cycle interpretation (ISO 2006):

1. Analyze results, reach conclusions, explain limitations, and provide recommendations based on the findings of the preceding phases of LCA and report the results of the life cycle interpretation in a transparent manner.
2. Provide a readily understandable, complete, and consistent presentation of the results of LCA, in accordance with the goal and scope of the study.

Technically, a typical life cycle interpretation begins with a comprehensive review of the information obtained in the previous three phases. After that, the completeness, sensitivity, and consistency checks are performed in order to ensure that products/processes are fairly compared. Finally, conclusions and recommendations are drawn. Common conclusions and recommendations can include the contribution of different life cycle stages or groups of processes to the total biorefinery, the better biorefinery alternative in terms of certain life cycle impact, etc. Note that when drawing conclusions, it may not always be possible to state that one alternative is better than the others because of the uncertainty in the final results. However, the LCA will still provide decision-makers with a better understanding of the environmental and health impacts associated with each biorefinery alternative, where they occur (locally, regionally, or globally), and the relative magnitude of each type of impact (Scientific Applications International Corporation 2006).

In summary, the ability of LCA to track environmental impacts can help decision-makers fully characterize the entire environmental performance of biorefinery systems and manage trade-offs associated with biorefinery product or process alternatives. In addition, the methodological framework of performing LCA is not overly complicated and has been well designed and summarized in the open literature. A wide variety of LCA software tools are available nowadays, which can help facilitate the industrial practice and research on biorefinery design and decision making.

It should be emphasized that LCA has some limitations. For instance, conducting an LCA can be resource and time intensive, and for some biorefinery systems, gathering the data of LCA can be problematic, and the limited availability of data can greatly impact the accuracy of the final results. Also note that LCA provides only the environmental impact information, which has to be combined with other more comprehensive decision processes to assess the trade-offs between economic, social, and other concerns. Finally, value judgments are unavoidable in converting the impact results of LCA to a single-score result, which must be applied by the commissioner of the study or the modeler (Scientific Applications International Corporation 2006).

22.3.2.5 LCA Case Studies of Biorefineries

A recent example by Cherubini and Jungmeier (2010) has been adapted in this section for illustrating the application of LCA on biorefineries. The biorefinery system being studied incorporates a number of conversion technologies for the production of bioethanol, electricity, heat, and phenols from switchgrass, a lignocellulosic crop. The authors also provided a fossil reference system, which produces the same products/services from fossil resources as a comparison to the biorefinery system: gasoline for transportation, heat from natural gas (instead of biomethane), conventional phenols from an oil refinery, electricity from natural gas, and heat from heavy oil. The general life cycle stages of both the biorefinery system and the fossil reference system are given in Figure 22.4, while the details of the switchgrass conversion steps in the biorefinery are given in Figure 22.6.

Biomass production: In the life stage of biomass (switchgrass) production, it assumes that the switchgrass takes three years to reach its full production potential, that is, an average yield of 16 dry ton/ha per year (Smeets et al. 2009). At harvest, switchgrass has a water content of 20%, and the chemical and elemental compositions of it were obtained in energy efficiency and renewable energy (EERE) (2009). Baled switchgrass is harvested once a year and then transported by 40-ton trucks

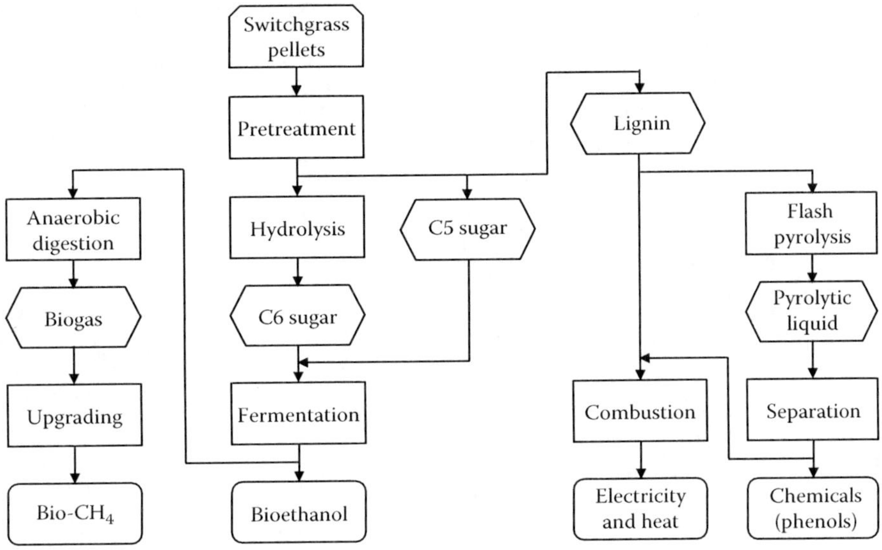

FIGURE 22.6 Simplified process scheme for switchgrass conversion in a biorefinery. (From Cherubini, F. and Jungmeier, G., *Int. J. Life Cycle Assess.*, 15, 53, 2010.)

to the pellet facility 20 km away from the field. In the pellet facility, switchgrass is dried using heat (from natural gas) to 10% water content, while pellets are produced using electricity from grid. The list of material and energy inputs required to produce switchgrass pellets (the feedstock for the biorefinery) is shown in Table 22.5. After that, the pellet feedstock is transported by 40-ton trucks for 100 km from the pellet facility to the biorefinery plant site. Note that during the switchgrass production, an emission of 0.042 g N_2O/g N fertilizer and 10 g CH_4/kg N is assumed for the agricultural land. It is also assumed that the initial soil carbon stock has a value of 77.90 tons C/ha, and that for the first 20 years the land has an average C sequestration rate of 0.6 tons C/(ha*a). Afterwards, a new equilibrium is reached, and the soil organic carbon (SOC) remains constant.

Biorefinery conversion: During the life stage of biorefinery conversion, the feedstock is first dried, which needs a heat input of 11.6 kWh_c/ton. Then, an uncatalyzed steam explosion occurs to hydrolyze the C5 sugars in the hemicellulose into xylose and arabinose with an efficiency of 85%. In the following enzymatic step, the cellulose is hydrolyzed to glucose with an efficiency of 90%; the remaining C6 polymers, galactan and mannan, are hydrolyzed to galactose and mannose with an efficiency of 82% and 89%, respectively, and a fraction of cellulose (9%) is set aside for bacteria and enzyme cultivation and production. All the sugar monomers are then sent to fermentation, while all the residues, together with the lignin, undergo thermochemical treatment (combustion or pyrolysis). The sugar fermentation is operated in a simultaneous saccharification and fermentation mode for the simultaneous fermentation of C5 and C6 sugars, where the ethanol conversion yields are 92.5% from C6 sugars and 85% from C5 sugars on a molar basis. To collect the final product, bioethanol is distilled to a purity of 98%. Residues from both the fermentation and bioethanol distillation are treated by anaerobic digestion to produce biogas at an average rate of 6 GJ/ton on a dry basis. Methane emissions to the atmosphere during biogas treatment are estimated as 3.47 mg/MJ, and the upgrading of biogas to biomethane needs another 5% of the energy content of the biogas product. The final digestate is discarded to a wastewater treatment plant.

In the thermochemical route, 20% of the lignin is first treated by flash pyrolysis, which results in 75% pyrolytic oil and 25% charcoal. Half of the phenol will be recovered from the pyrolytic oil, and the remaining pyrolytic oil will be combusted with charcoal, the remaining lignin fraction, and other residues (mainly unconverted sugar polymers) to generate electricity and heat with an efficiency of 25.5% and 44%, respectively. Ashes generated are discarded to a monitored landfill. The operation

TABLE 22.5
Material and Energy Inputs Required to Produce Switchgrass Pellets

Flow	Unit/Year	Value
Cultivation		
Seed	kg/ha	15
Fertilizers		
N	kg/ha	112
P	kg/ha	17
K	kg/ha	54
Lime	kg/ha	150
Herbicides	kg/ha	2.5
Emissions from land		
N_2O	g/g N	0.042
CH_4	g/kg N	10
Processing and transport		
Drying	MJ/t_{water}	2.26
Pelletizing	MJ_e/MJ	0.015
Transport	Mt?km	57.2

Source: Cherubini, F. and Jungmeier, G., *Int. J. Life Cycle Assess.*, 15, 53, 2010.

of the entire biorefinery system requires electricity in the following demands: 0.83 GJ/ton of dry feedstock, 0.03 GJ/GJ of pyrolytic oil produced in flash pyrolysis, and 0.54 GJ/ton of dry material in wastewater for biogas production and upgrading. The heat demand of the plant is based on 0.40 GJ/GJ bioethanol produced and 110 MJ/ton dry material in wastewaters. These energy needs can be fully supplied by heat and power internally produced through combustion of lignin and residues. Emissions from combustion of process residues for generation of electricity and heat are estimated using the information from Hamelinck et al. (2005).

Distribution and final use: Bioethanol is distributed to fuelling stations. It is assumed that the bioethanol will be consumed by vehicles at a specific consumption rate of 2.45 MJ/km, where the emission generation is estimated according to Halleux et al. (2008). Biomethane is fed to the national natural gas grid and then burnt in boilers where the emissions generated are estimated as well. It should be noted that the CO_2 released by the combustion of these biofuels is not counted for GHG due to their biological origin. Phenols are transported for 50 km and used for producing resins and finally disposed via incineration.

Life cycle inventory and impact assessment: The life cycle inventory is performed by using the LCA software tool SimaPro 7 (PRé Consultants 2010). For the impact assessment, the GHG emissions and cumulative primary energy demand (distinguished into fossil and renewable) are selected as the major LCA indicators, while other environmental impact categories (e.g., abiotic depletion, eutrophication) are assessed as well.

Table 22.6 shows the GHG balance of the biorefinery and the fossil reference system, and the GHG savings of the biorefinery system are also presented. It is clear that the biorefinery system has lower total GHG emissions than the fossil reference system. When examining the detailed categories, it can be seen that the CO_2 emissions of the biorefinery have a negative value during the first 20 years due to the sequestration of atmospheric carbon in the soil, while the amount after 20 years is still only 21.88% of the emissions by the fossil reference system. For CH_4 emissions, the amount

TABLE 22.6
GHG Balance of the Biorefinery and the Fossil Reference System

	Unit/a	Biorefinery 1–20 Year	Biorefinery >20 Years	Fossil Reference System
GHG emissions				
Total	kt $CO_{2\text{-eq}}$	60.5	126	281
CO_2	kt $CO_{2\text{-eq}}$	−8.6	58.2	266
N_2O	kt $CO_{2\text{-eq}}$	64.2	64.2	6.56
CH_4	kt $CO_{2\text{-eq}}$	4.86	4.86	8.91
GHG savings				
Per year	kt $CO_{2\text{-eq}}$	221	155	—
Per year	%	79%	55%	—
Per t_{dry} feedstock	t $CO_{2\text{-eq}}/t_{dry}$	0.46	0.33	—
Per hectare	kt $CO_{2\text{-eq}}$/ha	7.41	5.21	—

Source: Cherubini, F. and Jungmeier, G., *Int. J. Life Cycle Assess.*, 15, 53, 2010.

released by the biorefinery shows a 45.45% reduction compared to the fossil reference system before and after the first 20 years. However, the emissions of N_2O by the biorefinery are much higher than in the fossil reference system, which is due to the application of N fertilizers in agricultural soil.

Note that the CO_2 savings shown in Table 22.5 have different functional units, which are extremely important when assessing the effectiveness of policy decisions. For instance, improvements in one indicator may yield opposite results when other parameters are evaluated, and such trade-offs can only be adequately addressed if all possible parameters are taken into account. The contributions of the biorefinery to the total GHG emissions are shown in detail in Figure 22.7, which remains the same throughout the first 20 years and beyond. Among those categories, the switchgrass pellet

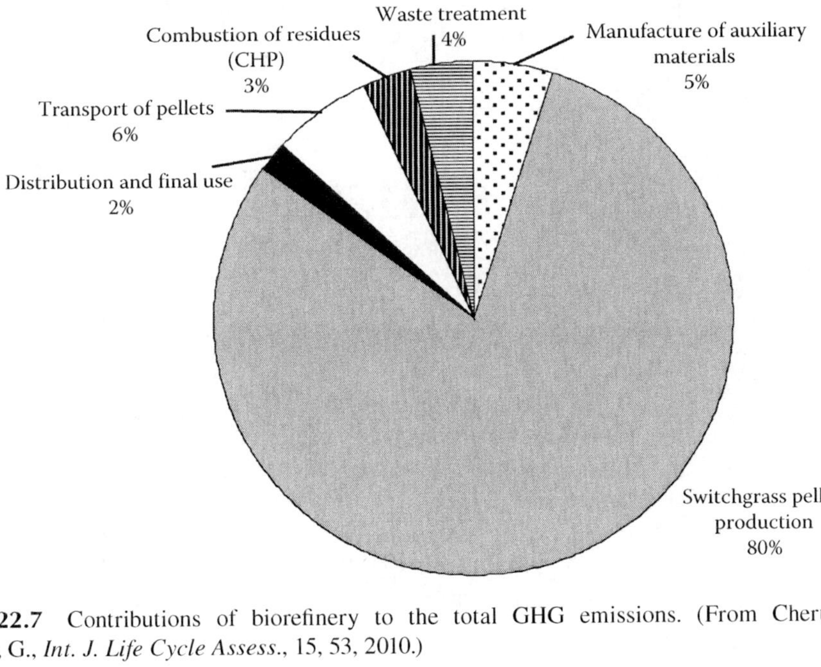

FIGURE 22.7 Contributions of biorefinery to the total GHG emissions. (From Cherubini, F. and Jungmeier, G., *Int. J. Life Cycle Assess.*, 15, 53, 2010.)

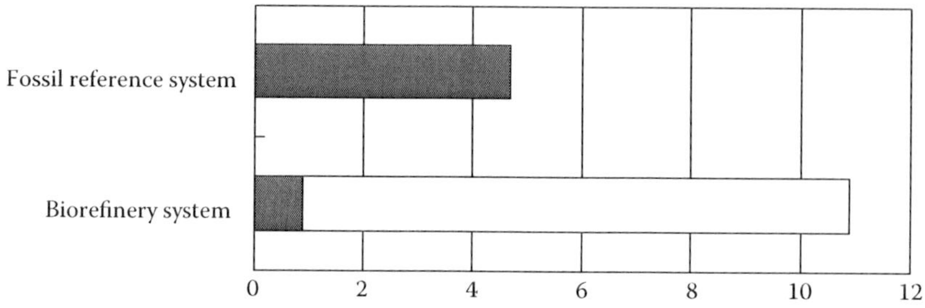

FIGURE 22.8 Primary energy demands of the biorefinery and fossil reference system. (From Cherubini, F. and Jungmeier, G., *Int. J. Life Cycle Assess.*, 15, 53, 2010.)

production accounts for 80% of the contribution, and the details of this major contribution are further broken down as follows:

- Emission of N_2O from agricultural land (40%)
- Processing and transport, including drying, pelletizing, and road transport (20%)
- Cultivation and harvest, including land preparation, seeds, herbicides, fertilizing, and harvesting (12%)
- Manufacture of fertilizers (28%)

For comparison, the total GHG emission of the fossil reference system is distributed as follows: the largest fraction of total GHG emissions originates from gasoline (86%), followed by electricity from natural gas (7%), heat from methane (5%), and then the others.

The primary energy demands of the biorefinery and fossil reference system are shown in Figure 22.8. Based on the LCA results, the biorefinery system requires a higher cumulative primary energy demand (10.9 PJ/a) than the fossil reference system (4.7 PJ/a). However, those energy demands of the biorefinery system consist mainly of renewable energy, which can be supplied by the manufacturing process itself. In this regard, the fossil energy demand of the biorefinery system is only 0.9 PJ/a, which saves about 80% of nonrenewable energy sources compared to the fossil reference system.

The PEIs of the biorefinery system were also quantified and compared with the fossil reference system, and the results are given in Table 22.7. Among all 10 impact categories, the biorefinery

TABLE 22.7
PEIs of Biorefinery System and Fossil Reference System

Impact Category	Unit/a	Biorefinery	Fossil Reference System
Abiotic depletion	kt Sb eq	0.42	1.94
Global warming (GWP100)	kt CO_2 eq	60.5	281
Ozone layer depletion (ODP)	kg CFC-11 eq	10.5	31.2
Human toxicity	kt 1,4-DB eq	34.2	187
Freshwater aquatic ecotoxicity	kt 1,4-DB eq	4.08	16.8
Marine aquatic ecotoxicity	Mt 1,4-DB eq	22.8	50.1
Terrestrial ecotoxicity	kt 1,4-DB eq	0.35	0.62
Photochemical oxidation	kt C_2H_4	0.06	0.28
Acidification	kt SO_2 eq	1.23	1.17
Eutrophication	kt PO_4 eq	2.82	0.17

Source: Cherubini, F. and Jungmeier, G., *Int. J. Life Cycle Assess.*, 15, 53, 2010.

system has lower impacts in eight of them. The two categories that the biorefinery has higher impacts are acidification and eutrophication. The reason is mostly due to the application of N fertilizers in switchgrass production.

Life cycle interpretation: The following conclusions are drawn by Cherubini and Jungmeier based on the life cycle inventory and impact assessment presented earlier. The LCA study performed in this work demonstrates a significant GHG and fossil energy savings when the biorefinery system is compared with a fossil reference system. However, investigation of other impact categories revealed that the biorefinery system has larger eutrophication and acidification potential impacts. In addition, since synthetic N fertilizers affect many impact categories, it is worth investigating alternative strategies for N fertilizer application. One option for reducing N_2O soil emissions is to use fertilizers more efficiently. For instance, the adoption of best fertilization practices could reduce agricultural N_2O emissions by 30%–40% (Council for Agricultural Science and Technology 2004). These practices include improved control of the amount, timing, and placement of fertilizers, as well as using manure and interseeding with nitrogen-fixing species (Sanderson et al. 2001, Samson et al. 2005).

LCA can be used as a scientific decision support technique to quantify the environmental implications of various biorefinery processes, feedstock, and integration options. The work by Cherubini and Jungmeier (2010) reviewed in this section demonstrated an effective application of LCA on one biorefinery system, which employs a combination of several conversion technologies for the production of bioethanol, electricity, heat, and phenols from switchgrass. There are many other good applications of LCA on biorefinery systems. For instance, DuPont used LCA to guide research and development on the most sustainable cellulosic ethanol biorefinery design in its integrated corn-based biorefinery (ICBR) project (U.S. Department of Energy 2006), which demonstrated the feasibility of a cellulosic ethanol biorefinery concept. Michigan State University used LCA in the estimation of the environmental performance of corn grain, corn stover, and the corn cob portion of the stover, grown under various farming practices for several corn-growing locations in the US Corn Belt (Kim et al. 2009). The value chain optimization (VCO) workshop on LCA and sustainability (2011) held in Vancouver, BC, provided an LCA analysis of the forest biorefinery, which contains 46 studies in the pulp and paper sector.

22.4 DESIGN AND DECISION MAKING FOR BIOREFINERIES

The integrated biorefinery has tremendous potential to play an important role in the development of a more sustainable economy, as it is capable of using renewable resources, generating fewer GHG emissions, accelerating regional economic development, maintaining intergenerational equity, etc. However, the increased use of biomass for energy and materials requires land and water and may compete with ecosystem services and other human demands. Therefore, the sustainable coexistence and positive contributions of biorefineries need to be guaranteed through design, planning, evaluation, and informed comparisons. The design and analysis of biorefinery systems is challenging due to the complexity of the system under study, the optimality desired by the decision-maker, and the uncertainty associated with the data and knowledge. Martin and Grossmann (2011) pointed out that process systems engineering (PSE) is capable of optimizing the design and operation of biorefineries. In this section, PSE-based design and decision-making methodologies for biorefineries are introduced. First, concepts and principles of product portfolio design will be defined. Then, a method for integrating portfolio design and multicriterion decision making (MCDM) is given. Finally, a detailed case study of cellulosic feedstock-based biorefinery design is described to show the efficacy of the introduced methods.

22.4.1 Concepts and Principles of Product Portfolio Design

A biorefinery platform design can be either brownfield, greenfield, or retrofit (Batsy et al. 2012). A project is described as brownfield when it uses infrastructure, equipment, or land abandoned by another manufacturer. Since this type of projects can create new jobs and higher tax revenues to the local community, revitalize business of neighbors, and solve any environmental issues left by the previous manufacturer, it is generally welcomed by the EPA and other federal partners. However, there are potential risks in handling hazardous substances, pollutants, and contaminants from its previous use (EPA 2008). A greenfield project is the one that lacks any constraints imposed by prior work, and a retrofit project is an integration of new technology within an existing operating plant structure. The methodology discussed in this section has been developed for greenfield projects, but can easily be extended to the other two types of design projects by incorporating the constraints describing the existing plants and environmental issues.

Regardless of whether the design is considered brownfield, greenfield, or retrofit, the design of a biorefinery system is essentially the development of a product portfolio, which is a set of multiple bio-based products related to each other through a common molecular architecture and/or manufacturing process. A product portfolio is usually designed based on a central platform commodity that can be easily converted to other commodities in its family (Batsy et al. 2012). Note that each product should be manufactured by certain processes within the scope of biorefinery systems. Therefore, the product design and process design should be integrated during the development of the whole biorefinery system. Due to the close dependency between various products and the processes used to manufacture them, the design of the whole biorefinery system is always complex.

In general, the principles introduced in Section 22.2 should be considered during the design, where the optimality for achieving one or some criteria or even a trade-off between several of them is typically set as the design objective. For instance, one biorefinery product portfolio design can be driven by the maximization of profit margins through advanced supply chain strategies.

22.4.2 Integrated Product and Process Design Framework

Batsy et al. (2012) pointed out that the choice of product portfolio is influenced by not only product functionalities but also the technological (process) constraints and product strategies. The generated scheme for identifying process–product platforms is presented in Figure 22.9. Sammons et al. (2008) proposed a direct technique for identification of the superstructure of processing routes by combining process and economic modeling, where the methodology framework is given in Figure 22.10. VCO workshop on LCA and sustainability (2011) proposed an overall biorefinery design framework where the design task is divided into three phases, namely, product design, process design, and MCDM (see Figure 22.11).

The first two works focus on product design and process design, respectively, and while the third one provides general ideas of integrating product and process design in a systematic way,

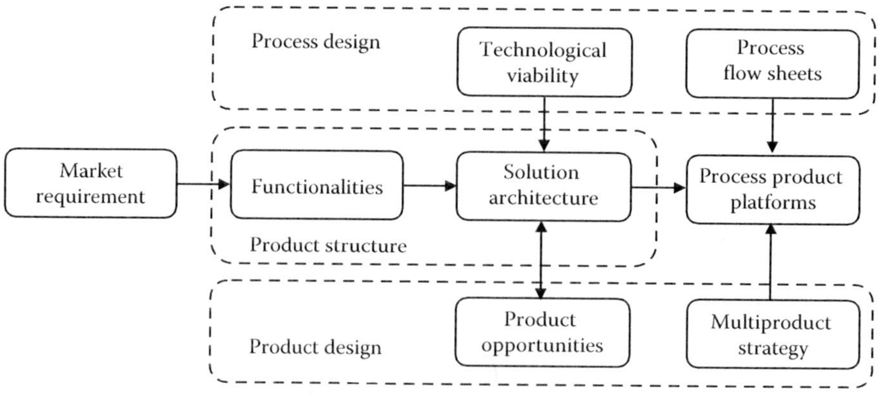

FIGURE 22.9 General process–product platform design by Batsy et al. (2012).

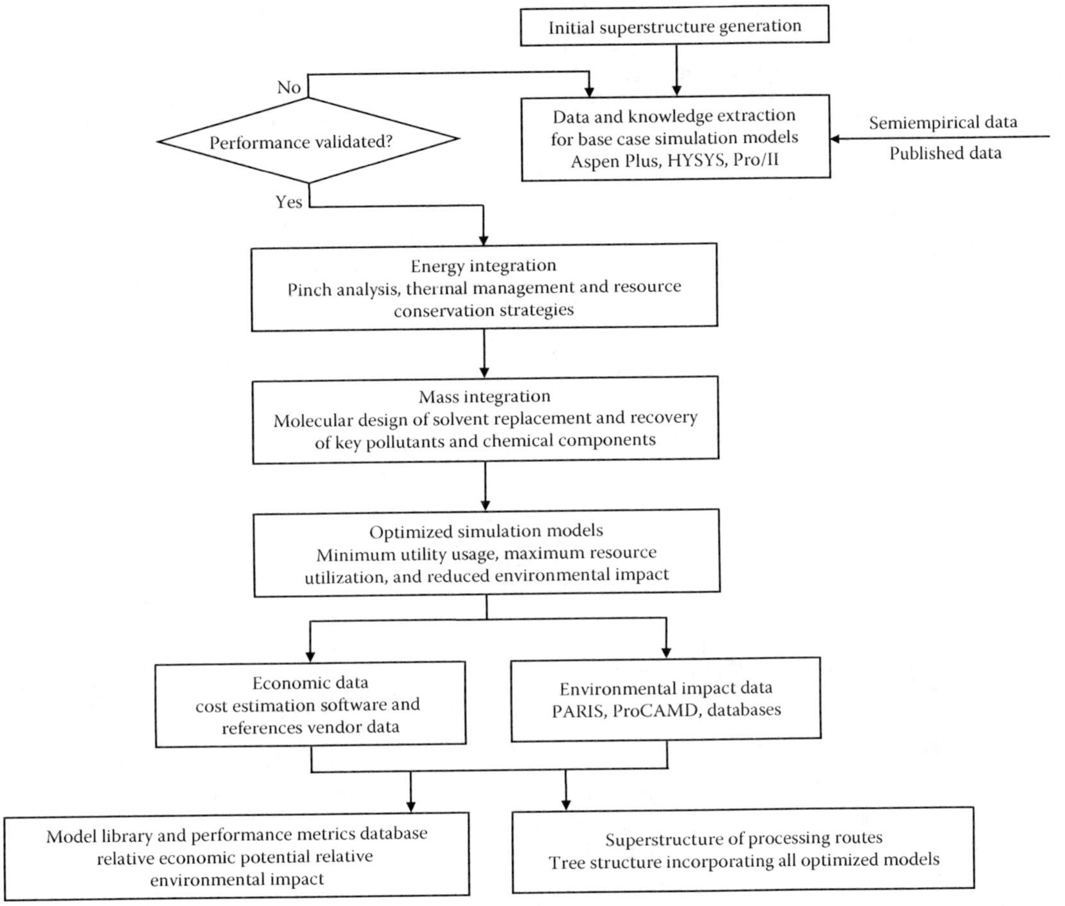

FIGURE 22.10 Strategy for identification of superstructure of processing routes by Sammons et al. (2008).

no detailed methodology was developed. In this chapter, those earlier works are combined into a complete biorefinery design framework, which is given in Figure 22.12. The key idea of this novel method is to obtain the superstructure of processing routes by sequential product and process design and then use MCDM to identify the optimal biorefinery system. The detailed methodology is illustrated in the following.

22.4.2.1 Initial Product Portfolio Selection

The introduced biorefinery design framework is composed of three functional modules: (1) initial product portfolio selection for the generation of promising biorefinery products and product constraints, (2) process design for the identification of the superstructure of process routes and process constraints, and (3) MCDM for finally obtaining the optimal biorefinery system.

The promising biorefinery products should be first selected by checking three types of information: the market requirements, product opportunities, and technology viability. The real selection is normally conducted in a heuristic way, where experience and expert judgment may be requested for obtaining the desired solutions. Note that in case no valid solutions can be obtained in the later MCDM, this initial selection step should be adjusted until a consistent solution is achieved for both products and processes. Product constraints should also be identified in this step, for instance, the specific composition requirements of some biorefinery products.

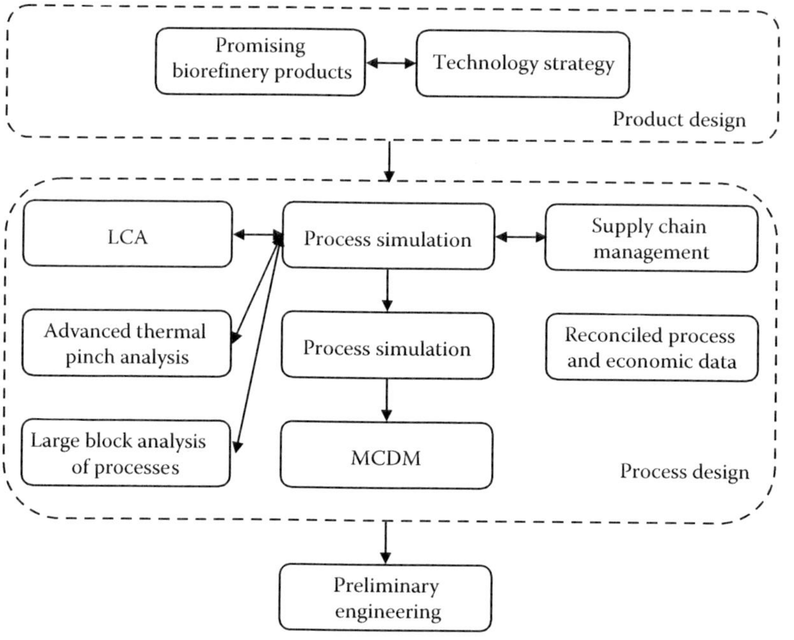

FIGURE 22.11 Overall biorefinery design framework by VCO workshop on LCA and sustainability (2011).

22.4.2.2 Process Design

Compared to the initial product portfolio selection, which is often handled heuristically, the process design can be performed in a much more algorithmic manner. One reason is that there are quite many tools available in the PSE area for process design and analysis, for instance, the sustainability assessment and LCA introduced in the previous sections; pinch analysis, thermodynamics-based analysis, molecular design of solvent replacement and recovery, etc., for mass and energy integration; and many other advanced methods. In addition, very detailed and reliable process models can be obtained by using powerful simulation engines (Aspen Plus, HYSYS, Pro/II, etc.), which greatly facilitate the process design work by providing flow sheets, operational parameters (e.g., pressure, temperature, reflux ratio), unit parameters (e.g., size, materials of construction), and many other useful pieces of information. Using these tools, a systematic process design framework can be described as follows. First, the desired biorefinery products and any identified constraints are sent to the process design module. Based on this information, the process alternatives that are able to manufacture the promising products are initially modeled using commercially available simulation software. After that, advanced process design tools (pinch analysis, thermodynamics-based analysis, molecular design of solvent replacement and recovery, etc.) will be applied to perform mass and energy integration on the possible combinations of process alternatives to obtain better overall process efficiency. When the processes are particularly complex, large-block analysis of processes, hierarchical analysis, SCM, etc., may provide great assistance. The optimized processes after mass and energy integration may need to be resimulated and then connected to the assessment tools to evaluate their performance (e.g., economic, environmental, and social performance; life cycle–based environmental impacts). The assessment results are used to screen out any low-performing process alternatives. Note that the process simulation, mass and energy integration, and performance assessment should be handled interactively, since a change or decision made in any of these tasks will cause changes in the other two. At the end, the superstructure of promising processing routes for manufacturing promising products as well as the related process/product constraints can be obtained.

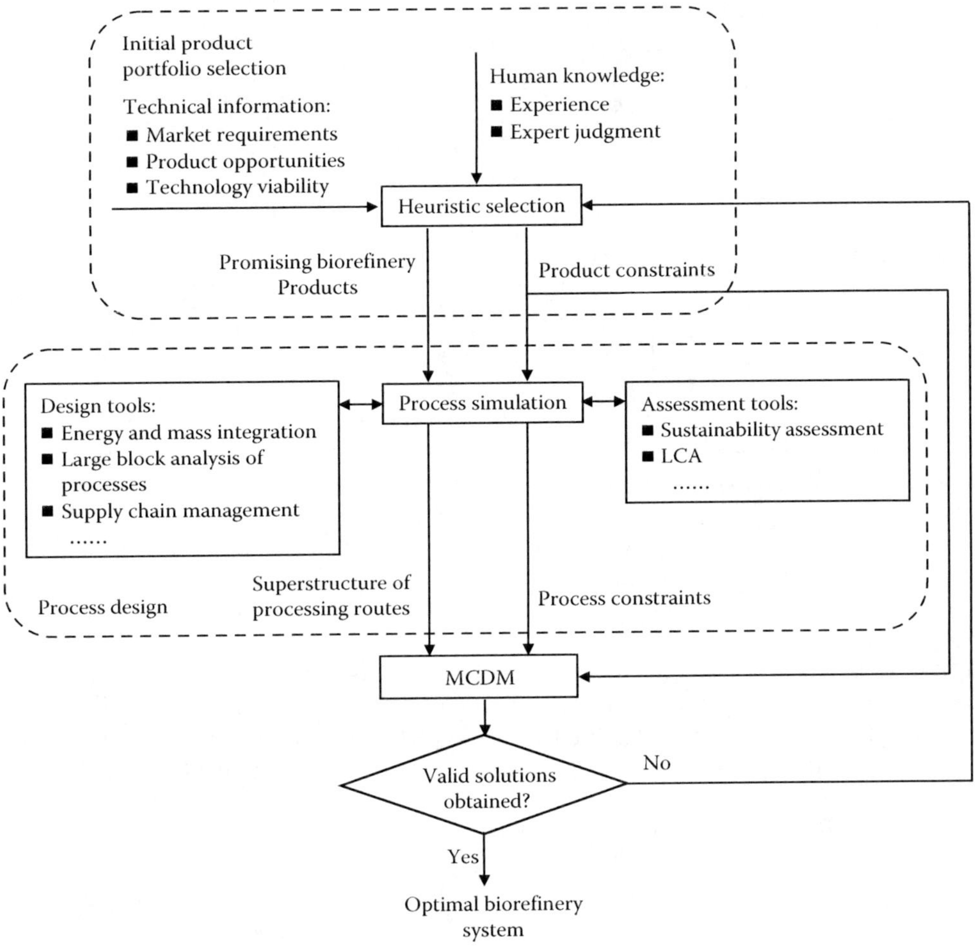

FIGURE 22.12 Biorefinery system design framework.

22.4.2.3 Multicriteria Decision Making

MCDM plays a key role in the overall biorefinery design scheme, as the optimal design solutions are identified in this functional module through the optimization of the superstructure of processing routes. As shown in Figure 22.12, three types of information are sent to the MCDM, namely, the superstructure of processing routes, the process constraints from the process design module, and the product constraints from the initial product portfolio module. Then, a mathematical programming problem adapted from Gani (2004) can be formulated as follows:

$$F_{OBJ} = \max f(\mathbf{x},\mathbf{y}) \tag{22.8}$$

$$h_1(\mathbf{x}) = 0 \tag{22.9}$$

$$h_2(\mathbf{x}) = 0 \tag{22.10}$$

$$h_3(\mathbf{x},\mathbf{y}) = 0 \tag{22.11}$$

$$l_1 \le g_1(\mathbf{x}) \le u_1 \tag{22.12}$$

$$l_2 \leq g_2(\mathbf{x},\mathbf{y}) \leq u_2 \tag{22.13}$$

$$l_3 \leq \mathbf{By} + \mathbf{Cx} \leq u_3 \tag{22.14}$$

where

- $\mathbf{x}$ is a vector of product and process variables
- $\mathbf{y}$ is a vector of binary integer variables representing the existence (or absence) of products or processing routes
- $h_1(\mathbf{x})$ is the set of equality constraints related to the process design parameters (e.g., pressure, reflux ratio)
- $h_2(\mathbf{x})$ is the set of equality constraints in terms of process models like mass and energy balances
- $h_3(\mathbf{x},\mathbf{y})$ is a set of equality constraints related to the product architecture
- $g_1(\mathbf{x})$ is a set of compositionally dependent inequality bounds on the process design specifications
- $g_2(\mathbf{x},\mathbf{y})$ is a set of composition and molecular architecture inequality bounds on the product Any other linear inequality constraints are described explicitly by Equation 22.14

Finally, the term $f(\mathbf{x},\mathbf{y})$ represents a vector of linear or nonlinear objective functions including profitability, environmental, and socioeconomic objectives (Batsy et al. 2012).

Note that specific criteria need to be assigned to each objective in Equation 22.8, which represents their relative importance to each other. A two-phase approach by Janssen (2009) demonstrates how an organization can handle the criterion decision. In the first phase, the decision structure, utility function, attributes, and decision criteria are established, while in the second phase, decision problems and a weighting procedure are introduced to a panel of experts and decision-makers. The members are made aware of the dependencies that may exist among criteria, and consequently, members are better equipped to address decisions under uncertainties. Once all criterion interpretations have been validated by panel members, the weighting method is used to assign a relative importance to each criterion.

Obviously, some of the objectives may be in conflict when analyzed simultaneously. Therefore, the best choice of strategies cannot be made by using simple common sense (Batsy et al. 2012). The mathematical programming problem in Equations 22.8 through 22.14 can be classified as either an mixed integer linear programming (MILP) or mixed integer nonlinear programming (MINLP) and then solved using numerical solvers. The optimal biorefinery system can be finally identified according to the valid optimized solutions derived from Equations 22.8 to 22.14. Note that in case no valid solution is obtained, the problem formulation and constraints are relaxed and the procedures repeated until a feasible (but not necessarily optimal) solution is found.

22.4.3 Case Study for Design of Biorefinery System

A case study presented by Sammons et al. (2008) and Odjo et al. (2011) has been adapted and is given in this section to demonstrate the efficacy of the biorefinery design framework proposed in Section 22.4.2. This simplified case study was performed on a potential biorefinery by using a chicken litter feedstock. In the initial product portfolio selection, technical information shows that syngas is a promising product of chicken litter through thermochemical conversion, which could be either sold on the market via a pipeline to a local customer or converted on-site into hydrogen or electricity. Therefore, a schematic of promising products and pathways from chicken litter is given in Figure 22.13. For this case study, no product constraints are involved.

In the process design module, the processes of biomass to syngas, syngas to hydrogen, and syngas to electricity are simulated and optimized using mass and energy integration. The resulting overall process (i.e., the superstructure of the biorefinery system) is simplified and given in Figure 22.14, where the detailed simulation results can be found in Sammons et al. (2008) and Larson et al. (2006). Next, the economic performance of the overall process in Figure 22.14 was evaluated, where

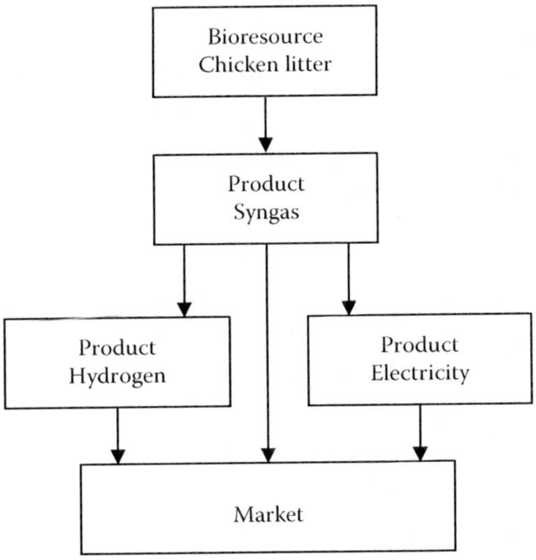

FIGURE 22.13 Promising products selected in the initial product portfolio selection. (From Sammons, N.E. et al., *Chem. Eng. Res. Des.*, 86, 800, 2008.)

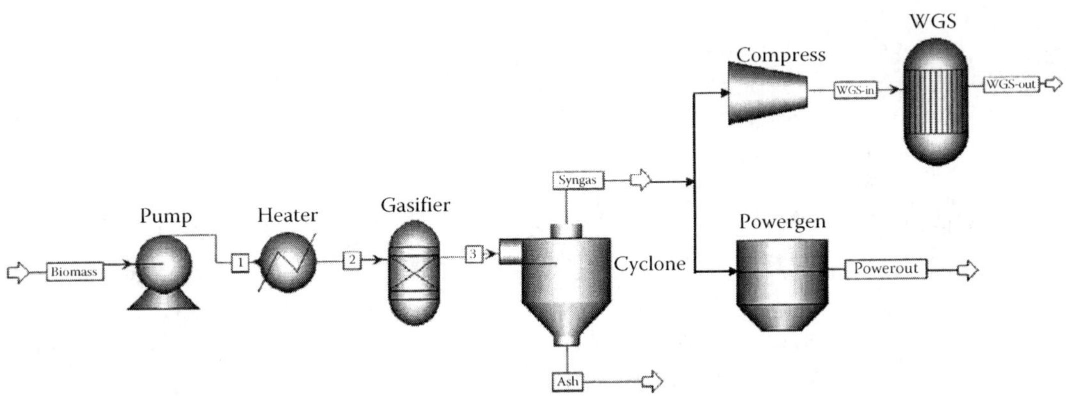

FIGURE 22.14 Process simulation of the overall biorefinery system.

the assessment results are given in Table 22.8. Finally, mass balances of the individual processes and the amount of feedstock availability (12.56 kg/s) are identified as process constraints.

With the superstructure of processing routes and the process constraints, a mathematical programming problem is formulated in the MCDM module, which has a single objective function given in Equation 22.15 for maximizing the overall profit of the system.

$$F_{OBJ} = \max\left(\sum (R_i - C_i)\right) \tag{22.15}$$

where

R_i and C_i are the revenue and cost of the *i*th main products, respectively
i is the index of syngas, hydrogen, and electricity

This objective function is subjected to constraints based on mass balances of the process superstructure in which a multiplicative conversion rate has been determined so that the conversion of feedstock to output of a given process is linear. The optimization program also contains constraints

TABLE 22.8
Economic Performance of the Overall Process

	Biomass to Syngas	Syngas to Electricity	Syngas to Hydrogen
Total fixed cost	\$112,302,000	\$100,091,000	\$461,527,000
Annualized fixed cost at 8% interest over 25 years	\$10,401,000	\$9,270,000	\$42,745,000
Total variable costs	\$13,618,000	\$15,301,000	\$202,114,000
Total annual product costs	\$24,019,000	\$24,571,000	\$244,859,000
Annual output	4.018×10^8 kg	1.065×10^6 MWe	8.957×10^{11} m^3
Cost per output	\$0.0598/kg	\$23.07/MWe	\$0.273/m^3

on the amount of feedstock available (12.56 kg/s) so that a predetermined feed basis will determine how much is produced, which in turn will be related to gross profit through market prices, processing cost, and feedstock cost (Sammons et al. 2008).

This optimization problem is finally an MILP, which can be solved by CPLEX. As the optimal solution, the optimal production pathway is to produce syngas from chicken litter and sell it directly to the market, which gives the maximum profit of \$1.922/s. The reason for exclusion of hydrogen and electricity production is mainly due to the high costs in converting the syngas to hydrogen or electricity. This simple case study can be expanded to include a much wider range of products, production pathways, and feedstock, which results in a crucial decision support study of the emerging field of biorefining.

22.5 DISCUSSION OF UNCERTAINTY IN DESIGN AND ANALYSIS OF BIOREFINERIES

The methods and tools discussed in the preceding sections are capable of handling the challenges in design and analysis of biorefinery systems due to the complexity carried by the system under study and the optimality desired by decision-makers. However, there is another challenge that has not yet been discussed in this chapter, which is the uncertainty associated with the data and knowledge. In fact, various sources of uncertainty (e.g., future prices of energy, products, and feedstocks; change of climate policies; and process scale-up) often force decision-makers to make assumptions based on heuristics to fill the information gap when data are limited, which could greatly affect the quality of the analysis results and planning decisions (Batsy et al. 2012). Uncertainties can be generally classified into two categories: the aleatory and the epistemic uncertainties (Parry 1966). The aleatory uncertainty refers to the variations associated with physical systems and/or the environment; it is objective and irreversible. By contrast, the epistemic uncertainty is carried due to the lack of knowledge and/or information; it is subjective and reducible (Liu and Huang 2012). The uncertainties encountered in the design and analysis of a biorefinery could be either aleatory or epistemic.

In most of the existing research on biorefinery design and analysis, the way for handling uncertainties is to use Monte Carlo simulation–based sensitivity analysis. Specifically, the design and analysis are first implemented without uncertainty to obtain the base case results. Then, a distribution function will be further introduced to each uncertain parameter. After that, a Monte Carlo–based simulation will be performed by randomly selecting values for each uncertain parameter from its distribution range and reconducting the design and analysis process as a new case. Since the values of uncertain parameters are different from the base case (also different from each new case), new results will be obtained for each new case. After running a large number of new cases, individual results of each new case are collected and aggregated, which generate a range and distribution on the design and analysis results. Those ranges and distribution provide the sensitive relationship between the base case results and the uncertain parameter distribution functions, which can help decision-makers review the risks of implementation on the base case solution under uncertainty. Sensitivity analysis has been used in many studies on biorefinery systems (Hytönen and Stuart 2009, Batsy et al. 2012, Svensson 2012). However, since the availability of distribution functions is scarce, a normal distribution was always assumed for uncertain parameters, which could be far from the real situation and may cause the analysis results and planning decisions to be unreliable.

Instead of using sensitivity analysis, another way to handle uncertainties is to introduce uncertainty on related parameters through unique approaches before conducting the design or analysis. Those unique approaches can be, for instance, stochastic-based approaches, fuzzy-logic-based approaches, and interval-based approaches, which all have attractive features for handling uncertainties. For instance, fuzzy-logic- and fuzzy programming-based approaches are attractive for formulating and manipulating epistemic uncertainties, where rigorous logic is used to deal with fuzzy information that is difficult to compute using conventional mathematical methods. Interval-parameter (IP)-based uncertainty handling does not require probability functions, but expresses parameter uncertainties by interval numbers, each of which have lower and upper bounds. Therefore, it is quite useful when known data are limited and uncertain, data ranges of parameters are known, but no data distribution information is available (Liu and Huang 2012). After introducing uncertainty on the parameters, the regular design and analysis methodological framework should be modified to be able to handle operations between different types of uncertain parameters (e.g., stochastic parameters, fuzzy numbers, interval-based parameters) and optimization based on specific types of uncertain parameters (e.g., stochastic-based optimization, fuzzy programming, IP-based optimization). Due to the unique benefits carried by those approaches, uncertainties can be effectively handled through the analysis and decision-making process, while the final solutions obtained provide answers that are suitable under a given uncertainty interpretation.

These types of uncertainty handling approaches have been successfully applied to the study of industrial sustainability problems, in which the biorefinery would be an interesting application example. However, it should also be noted that there are some disadvantages in using these methods. First, each unique uncertainty handling approach has some functional limits. For instance, when using fuzzy-logic-based approaches, the decision quality is largely affected by the definition of fuzzy sets and fuzzy numbers, where subjective judgments are used to a large extent because of the lack of sufficiently precise data. Not surprisingly, any poor judgment could be detrimental to decision quality (Liu and Huang 2012). Second, the modification of the regular design and analysis framework due to the introduction of uncertainty handling approaches could require great efforts, which makes the design and analysis process more complicated.

22.6 CONCLUSIONS

Biomass has incredible potential to fulfill future energy and chemical demands. Biofuels and biochemicals are considered to have inherent benefits on value addition, environmental protection, and energy security. A biorefinery is a facility that integrates different biomass conversion processes and equipment to produce transportation biofuels, biochemicals, biomaterials, and heat/power. Biorefineries must be designed and managed efficiently to reduce the cost of production and/or the environmental impacts. However, the design and analysis of biorefinery systems is always challenging due to the complexity carried by the system under study, the optimality desired by the decision-maker, and the uncertainty associated with the data and knowledge. Many methodologies and tools have been developed recently to handle those challenges and facilitate the tasks of design and analysis of biorefineries. In this chapter, a holistic overview of the research in this area was presented. The general principles of a sustainable biorefinery were described first. Based on those principles, two research topics, namely, the analysis and design of biorefinery systems, were further investigated by discussing the research needs, challenges, existing methodologies and tools, and case studies.

As demonstrated in this chapter, it is important to appropriately analyze biorefinery systems to investigate the status quo, to compare different design and operation schemes, and to identify improvement opportunities and recommendations. In addition, the integrated product and process design framework given in this chapter can improve the sustainable coexistence and positive contributions of the biorefinery. Finally, various types of uncertainties encountered in design and analysis of biorefinery systems can be handled by sensitivity analysis or by introducing advanced techniques into the regular design and analysis procedure.

REFERENCES

Anastas, P.T. and J.C. Warner. 1998. *Green Chemistry: Theory and Practice*. New York: Oxford University Press.

Batsy, D.R., C.C. Solvason, N.E. Sammons, V. Chambost, D. Bilhartz, M.R. Eden, M.M. El-Halwagi, and P.R. Stuart. 2012. Product portfolio selection and process design of the forest biorefinery, Chapter in *Integrated Biorefineries: Design, Analysis, and Optimization*, P.R. Stuart and M.M. El-Halwagi, Eds. Boca Raton, FL: CRC Press, pp. 3–36.

Bioenergy Knowledge Discovery Framework. 2012. Limiting greenhouse gas emissions to address global climate challenges. https://bioenergykdf.net//content//why-bioenergy (accessed June 30, 2012).

Brundtland Commission. 1987. Our Common Future. Report of the World Commission on Environment and Development, World Commission on Environment and Development. Oxford University Press. Oxford, UK.

Carvalho A., R. Gania, and H. Matos. 2008. Design of sustainable chemical processes: Systematic retrofit analysis generation and evaluation of alternatives. *Process Safety and Environment Protection*. 86: 328–346.

Cherubini, F. and G. Jungmeier. 2010. LCA of a biorefinery concept producing bioethanol, bioenergy, and chemicals from switchgrass. *Int J Life Cycle Assess* 15: 53–66.

Cherubini, F., G. Jungmeier, M. Wellisch, T. Willke, I. Skiadas, R.V. Ree, and E. Jong. 2009. Toward a common classification approach for biorefinery systems. *Biofuels Bioprod Biorefin*. DOI: 10.1002/bbb.

Cobb, C., D. Schuster, B. Beloff, and D. Tanzil. 2009. The AIChE sustainability index. *Chem Eng Prog*. 105: 60–63.

Council for Agricultural Science and Technology. 2004. *Climate Change and Greenhouse Gas Mitigation: Challenges and Opportunities for Agriculture*. Ames, IA: CAST, p. 120.

E4Tech.com. 2005. Biomass conversion processes. http://www.e-sources.com//biomass_converion.htm (accessed June 30, 2012).

Energy Efficiency and Renewable Energy. 2009. Biomass feedstock composition and property database. http://www1.eere.energy.gov//biomass//feedstock_databases.html (accessed June 07, 2009).

Environmental Protection Agency. 1993. Life cycle assessment: Inventory guidelines and principles, EPA/600/R-92/245. Cincinnati, OH: Office of Research and Development.

EPA. 1995. Guidelines for assessing the quality of Life Cycle Inventory Analysis, EPA/530/R-95/010. Washington, DC: Office of Solid Waste.

EPA. 2008. *Brownfields Tax Incentive Guidelines*. Washington, DC: U.S. Environmental Protection Agency.

Fernando, S., S. Adhikari, C. Chandrapal, and N. Murali. 2006. Biorefineries: Current status, challenges, and future direction. *Energy Fuels*. 20: 1727–1737.

Forestry Commission of Great Britain. 2009. Sustainability. Available online: UK.http://www.forestry.gov.uk/forestry/edik-59fmzf (accessed on January 5, 2014).

Gani, R. 2004. Chemical product design: Challenges and opportunities. *Comput Chem Eng*. 28: 2441–2457.

Goetschalckx, M. 2002. Strategic network planning. In *Supply Chain Management and Advanced Planning: Concepts, Models, Software and Case Studies*, H. Stadtler, and C. Kilger, Eds., pp. 105–121. Berlin, Germany: Springer.

Halleux, H., S. Lassaux, R. Renzoni, and A. Germain. 2008. Comparative life cycle assessment of two biofuels: Ethanol from sugar beet and rapeseed methyl ester. *Int J Life Cycle Assess*. 13(3): 184–190.

Hamelinck, N.C., G. Hooijdonk, and A.C. Faaij. 2005. Ethanol from lignocellulosic biomass: Techno-economic performance in short-, middle- and long-term. *Biomass Bioenerg*. 28: 384–410.

Heikkila, A.M. 1999. Inherent safety in process plant design: An index-based approach. PhD dissertation at Helsinki University of Technology, Espoo, Finland.

Hytönen, E. and P. Stuart. 2009. Biofuel production in an integrated forest biorefinery—Technology identification under uncertainty. *International Biorefinery Conference*, October 2009, Montreal, Quebec, Canada.

IChemE. 2002. The sustainability metrics. In *Sustainable Development Progress Metrics Recommended for Use in the Process Industries*, B. Tallis (ed.) Vol. 105, pp. 60–63. Rugby, U.K.: IChemE.

ISO (International Standards Organization). 2006. Environmental management—Life cycle assessment—Life cycle interpretation. ISO 14040. Available online: http://www.iso.org (accessed on January 5, 2014).

Jackson, T. and R. Clift. 2008. Where's the profit in industrial ecology? *Journal of Industrial Ecology*. 2(1): 3–5.

Janssen, M. 2009. Choice of a sustainable forest biorefinery product platform using an MCDM method. *Proceeding of Seventh International Conference on Foundations of Computer-Aided Process Design*. Breckenridge, CO.

Kim, S., B. Dale, and R. Jenkins. 2009. Life cycle assessment of corn grain and corn stover in the United States. *Int J Life Cycle Assess*. 14: 160–174.

Larson, E.D., S. Consonni, R.E. Katofsky, K. Iisa, and W.J. Frederick. 2006. A cost-benefit assessment of gasification-based biorefining in the Kraft pulp and paper industry, Vol. 1, Main Report. Princeton, NJ: Princeton University.

Li, X., A. Zanwar, A. Jayswal, H.H. Lou, and Y.L. Huang. 2011. Incorporating exergy analysis and inherent safety analysis for sustainability assessment of biofuels. *Ind Eng ChemRes*. 50: 2981–2993.

Liu, Z. 2012. Sustainable design of complex industrial and energy systems under uncertainty. PhD dissertation at Wayne State University, Detroit, MI.

Liu, Z. and Y.L. Huang. 2012. Technology evaluation and decision making for sustainability enhancement of industrial systems under uncertainty. *AIChE J*. 58(6): 1841–1852.

Martin, M. and I.E. Grossmann. 2011. Energy optimization of bioethanol production via gasification of switchgrass. *AIChE J*. 57: 3408–3428.

McKeown, R. 2002. *Education for Sustainable Development Toolkit*, Version 2. July 2002. Knoxville, TN: Energy, Environment and Resources Centre, University of Tennessee.

Millennium Ecosystem Assessment. 2003. Ecosystems and human well-being. Available online: http://www.unep.org/maweb/en/Framework.aspx (accessed on January 5, 2014).

Odjo, A.O., N.E. Sammons Jr., W. Yuan, A. Marcilla, M.R. Eden, and J.A. Caballero. 2011. A disjunctive-genetic programming approach to synthesis of process networks. *Ind Eng Chem Res*. 50(10): 6213–6228.

Othman, M.R., J.U. Repke, G. Wozny, and Y.L. Huang. 2010. A modular approach to sustainability assessment and decision support in chemical process design. *Ind Eng Chem Res*. 49: 7870–7881.

Parry, G.W. 1966. The characterization of uncertainty in probabilistic risk assessment of complex systems. *Reliab Eng Syst Saf*. 54: 119–126.

Piluso, C., J. Huang, Z. Liu, and Y.L. Huang. 2010. Sustainability assessment of industrial systems under uncertainty: A fuzzy-logic-based approach to short-to-mid-term predictions. *Ind Eng Chem Res*. 49(18): 8633–8643.

PRé Consultants. 2010. SimaPro LCA software. http://www.pre.nl//simapro//default.htm (accessed June 30, 2012).

Samson, R., S. Mani, R. Boddey, S. Sokhansanj, D. Quesada, S. Urquiaga, V. Reis, and C.H. Lem. 2005. The potential of C4 perennial grasses for developing a global BIO-HEAT industry. *Critical Rev Plant Sci*. 24: 461–495.

Sanderson, M.A., R.M. Jones, M.J. McFarland, J. Stroup, R.L. Reed, and J.P. Muir. 2001. Nutrient movement and removal in a switchgrass biomass-filter strip system treated with dairy manure. *J Environ Qual*. 30: 210–216.

Sammons, N.E., W. Yuan, M.R. Eden, B. Aksoy, and H.T. Cullinan. 2008. Optimal biorefinery product allocation by combining process and economic modeling. *Chem Eng Res Des*. 86: 800–808.

Scientific Applications International Corporation. 2006. Life cycle assessment: Principles and practice. Cincinnati, OH: EPA.

Smeets, E.W., I.M. Lewandowski, and A.C. Faij. 2009. The economical and environmental performance of *Miscanthus* and switchgrass production and supply chains in a European setting. *Renew Sust Energ Rev*. 13(6–7): 1230–1245.

Svensson, E. 2012. Optimization of investments for strategic process integration and pulp mill biorefinery projects under uncertainty. PhD dissertation of Chalmers University of Technology, Göteborg, Sweden.

Tay, D.H.S., D.K.S. Ng, N.E. Sammons, and M.R. Eden. 2011. Fuzzy optimization approach for the synthesis of a sustainable integrated biorefinery. *Ind Eng Chem Res*. 50: 1652–1665.

Turton, R., R. Bailie, W.B. Whiting, and J.A. Shaeiwi. 2003. *Analysis, Synthesis, and Design of Chemical Processes*, 2nd edn. Upper Saddle River, NJ: Prentice Hall.

United Nations. 1992. Report of the World Commission on Environment and Development. Rio de Janeiro, Brazil, June 3–14. Available online: sustainabledevelopment.un.org/content/documents/Agenda21.pdf (accessed on January 5, 2014).

United Nations General Assembly. 2005. 2005 world summit. September 14–16, New York, NY.

U.S. Department of Energy. 2006. Biomass program: Integrated corn-based bio-refinery. http://www1.eere.energy.gov//biomass//pdfs//corn-based_biorefinery.pdf (accessed June 30, 2012).

VCO Workshop on LCA and Sustainability. 2011. Life Cycle Analysis and Sustainability. Vancouver, BC: University of British Columbia. April 28, 2011.

Wisconsin Grasslands Bioenergy Network. http://www.wgbn.wisc.edu//biomass-supply-chain (accessed June 30, 2012).

Yan, F., A. Sophie, L. Luc, and N. Mustapha. 2011. Integrated bio-refinery and forest products supply chain network design using mathematical programming approach. *Expert Syst Appl*. 38: 13731–13748.

Young, D.M. and H. Cabezas. 1999. Designing sustainable processes with simulation: The waste reduction (WAR) algorithm. *Comput Chem Eng*. 23(10): 1477–1491.

23 Lignocellulosic Feedstock Biorefinery

Karin Arens and Shijie Liu

CONTENTS

23.1 INTRODUCTION

Lignocellulosic biomass (LB) provides an alternate source to fossil fuel for chemicals, materials, and energy. A biorefinery utilizing LB as raw material integrates multiple conversion processes to produce a range of products including liquid biofuels, biochemicals, steam/heat, and electricity. The challenge is to produce lignocellulosic products in an environmentally friendly system at

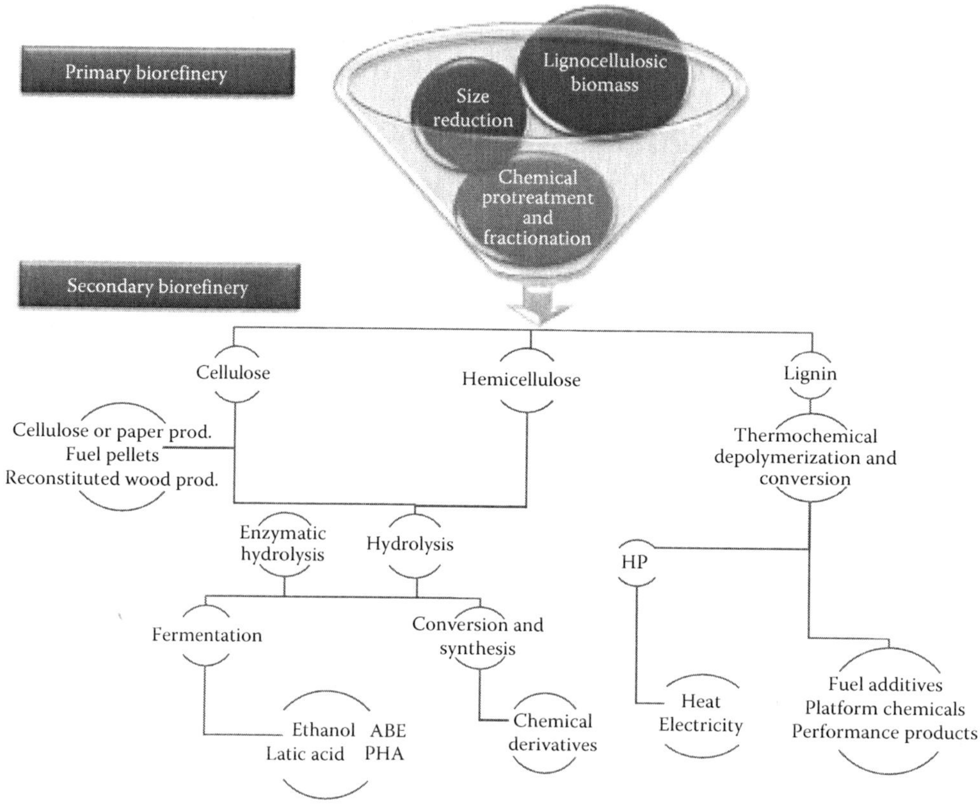

FIGURE 23.1 Schematic diagram of a sequential lignocellulose fractionation biorefinery.

low cost and on a large scale. The complex structure of plant cell walls makes LB more difficult to break down into sugars than starch. There are two main categorical processes used to convert LB into products: thermochemical and biochemical. Gasification to produce syngas and pyrolysis to produce bio-oil are two extreme thermochemical processes. They have the advantage of being relatively insensitive to the feedstock type but the drawback of producing outputs of low value.

A biochemical-based biorefinery normally integrates a series of systematical fractionation processes to treat the lignocellulosic material for optimal conversion of individual biomass components into desired products (Figure 23.1). In biochemical processes, the lignocellulosic material is pretreated to remove the physical barriers, after which cellulose and hemicellulose are hydrolyzed to produce pentoses and hexoses. These sugars are then fermented and converted to desired products such as ethanol, butanol, lactic acid, polyhydroxyalkanoids (PHA), and other biochemicals. Separation and/or physical fractionation is employed to isolate intermediates as well as final products. Biological conversions are selective processes and, hence, preferred over chemical conversions. However, one must consider that multiple biological conversion processes are required to treat LB due to the complexity of the raw material. Another consideration is the slower reaction in biological conversions, which requires lager facility footprints.

23.2 LIGNOCELLULOSIC BIOMASS

23.2.1 Chemical Composition

LB is plant biomass, such as wood, agricultural residues, energy crops, and municipal paper waste. The main components of trees and herbaceous materials are cellulose (40%–50%), hemicellulose

(25%–35%), and lignin (15%–20%), plus many traceable elements. Cellulose is a long linear chain of glucose with β 1-*O*-4 glycosidic bonds and a chemical formula of H-$(C_6H_{10}O_5)_n$-OH, where *n* represents the degree of polymerization (DP). In wood, the DP is approximately 10,000. Cellulose, being a structural material, does not break apart easily. The cellulose molecules aggregate to form microfibrils, which are organized in crystalline and amorphous regions. The crystalline regions are difficult to be penetrated by solvents or reagents, while the amorphous regions are easily penetrated and, therefore, more susceptible to hydrolysis reactions. In the crystalline region, the long cellulose chains are arranged side by side and stabilized by hydrogen bonds among the hydroxyl groups. Several different crystalline structures of cellulose are known, corresponding to the location of hydrogen bonds between and within strands. The hydrogen bonds provide the microfibrils with high tensile strength. In the plant cell wall, the microfibrils are aggregated into macrofibrils and in turn the macrofibrils are combined with noncellulosic material. This structure confers the rigidity to plant cells.

Hemicelluloses are a collection of heteropolysaccharide found mostly in the wood cell walls closely associated with cellulose and lignin. The common hemicelluloses sugars from LB are D-glucose, D-mannose, D-galactose, D-xylose, D-arabinose, and small amounts of L-rhamnose, from which D-xylose is the dominant sugar found in angiosperms. The hemicelluloses have a DP of 100–200 (Glaudemans and Timell 1958; Timell 1960; Goring and Timell 1962; Koshijima et al. 1965). Due to its low DP, hemicelluloses are relatively easy to be hydrolyzed to their monomeric sugars in addition to D-glucoronic acid, 4-*O*-methyl-D-glucoronic acid, and D-galacturonic acid. The structure and composition of hemicelluloses differ in a distinctive manner among softwoods and hardwoods, as illustrated in Table 23.1.

Aside from cellulose, lignin is one of the most abundant polymers on earth, utilizing one-third of nonfossil organic carbon, and constitutes 25%–30% of the dry mass of wood (Boerjan et al. 2003). An integral part of the secondary cell walls of plants, the lignin structure acts as a natural barrier to biological degradation to cellulose and hemicellulose in LB. It is relatively hydrophobic and aromatic in nature. Lignin is an amorphous heteropolymer with methoxylated phenylpropylene alcohol units that are interconnected by stable ether and ester linkages. The main lignin precursor in gymnosperms is coniferyl alcohol, while in angiosperms are *p*-coumaryl alcohol and sinapyl alcohol. Lignins are classified according to their structural elements into guaiacyl (G), *p*-hydroxyphenyl (H), and syringal (S) types, respectively. Many grasses mainly have G, while some palms mainly have S. All lignins contain small amounts of incomplete or modified monolignols, and other monomers are prominent in nonwoody plants (Sjöstrom 1993).

Extractives are compounds in the woody biomass that are soluble in neutral organic solvents or water at room temperature and atmospheric conditions. Depending on their solubility, they are grouped into lipophilic or hydrophilic types. The term resin is often used to refer to the lipophilic compounds soluble in nonpolar organic solvents. There are an extraordinary large number of extractive compounds. In the same tree, the amount and composition of extractives differ greatly from one part of the tree to another (roots, stem, bark, branches, needles, or leaves). Some of the extractives have similar chemical characteristics as degraded lignin. There are over 70 metal, earth elements, and inorganic compounds that are found in woody biomass. These components are the first ones that can be extracted from wood.

23.2.2 Lignocellulosic Biomass Size Reduction: Pretreatment

Biomass size reduction (BSR) is a prerequisite step in a biorefinery for further chemical or biochemical processing. Physical pretreatment increases the biomass surface area accessible to chemicals and enzymes during biochemical conversion to sugars and improves the flow through biorefinery processes. The energy demand for BSR varies according to the density of the biomass sources used. Herbaceous biomass does not need significant processing and energy to achieve desired BSR (Cadoche and Lopez 1989). On the other hand, lignocellulosic (wood) BSR is an energy-intensive step,

TABLE 23.1
Major Components of Hemicellulose and Extractives

	Type	Distribution, wt.%	
		Softwoods	Hardwoods
Hemicellulose		25–30	25–35
Galactoglucomannan (1:1:3)	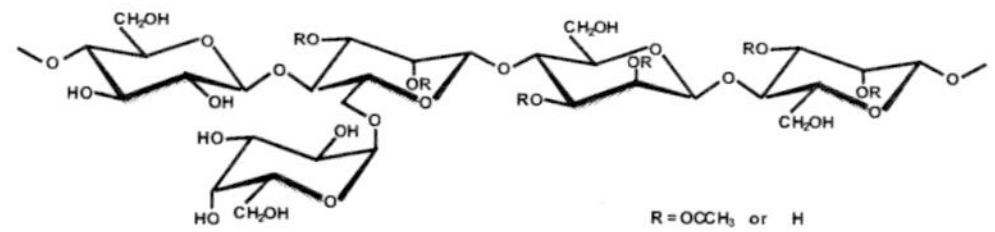	5–8	0
(Galacto)glucomannan (0.1:1:4)	The OH groups in the xylose units were partially substituted by OAc on the C-2 or C-3 positions, that is, R = CH_3CO (Ac) or H	10–15	0
Glucomannan (1:2-1:1)	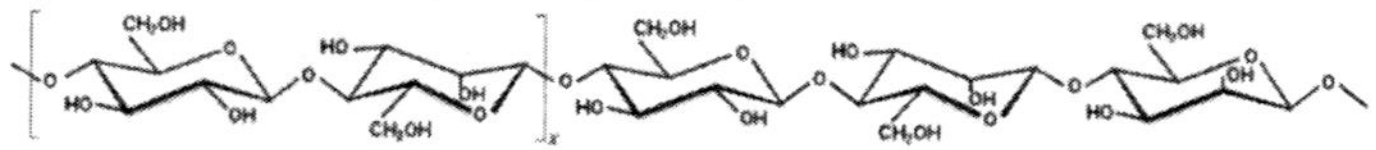	0	2–5
Arabinoglucuronoxylan	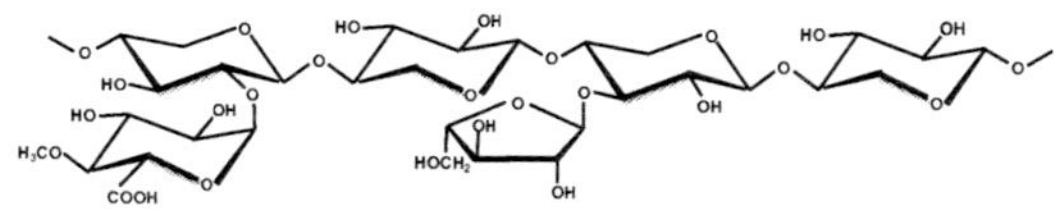	7–10	Trace
Glucuronoxylan	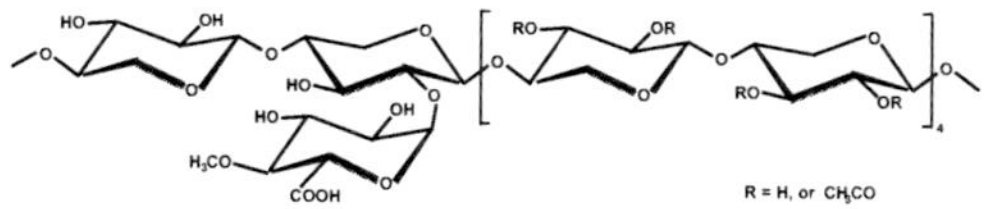 OH in the xylo units were partially substituted by OAc on the C-2 or C-3 positions (about 7 in 10 xylo units), that is, R = CH_3CO (Ac) or H	Trace	15–30
Extractives		5–8	2–4
Aliphatic and alicyclic	Terpenes, terpenoids, esters, fatty acids, alcohols, etc.		
Phenolics	Phenols, stilbenes, lignans, isoflavones, etc.		
Carbohydrates	Arabinose, galactose, glucose, xylose, raffinose, starch, pectic material		
Inorganics	Ca, K, Mg, Na, Fe, SO_4^{2-}, Cl^-, etc.		
Others	Cyclitols, tropolones, amino acids, protein, alkaloids, etc.		
Ash		0.2–0.5	0.2–0.8

Source: Compiled from Fengel, D. and Wegner, G., *Wood: Chemistry, Ultrastructure, Reactions*, Walter de Gruyter, Berlin, Germany, 1989; Gullichsen, J. and Paulapuro, H., *Pigment Coating and Surface Sizing of Paper*, Book 11, Finnish Paper Engineers' Association and TAPPI, 2000, pp. 2–61.

and therefore, well-defined process conditions are important in order to reduce process energy consumption. The quality of the biomass supply, such as the variable moisture content, soil particles, foreign matter, and initial chop length, may limit the degree of efficiency of subsequent processing. To increase efficiency, a design should simultaneously specify biomass feedstock supply systems, BSR equipment, and separation processes.

For wood, the first step in BSR is chipping. During harvesting, trees or logs and branches are reduced to wood chips with a size around 10–50 mm in two dimensions and 5–15 mm in the third dimension. The second step is fiber production or pulverization of wood chips to produce wood fibers/fiber bundles or flour. Of these BSR steps, wood chipping has relatively lower energy costs compared to fiberization or pulverization (Zhu and Pan 2010). Among the technologies available for BSR are hammer milling, knife milling, shredding, and disk or attrition milling. The two predominant technologies for large-scale production are hammer and disk milling. Hammer milling is mainly used for producing wood flours for composites and pellets. Disk milling is employed for wood fiber production at commercial scale of 1000 tons day^{-1}. In wood BSR, the disk milling conditions will affect the energy consumption and the substrate morphology and size (Tienvieri et al. 1999). The goal in a biorefinery is to optimize the milling process conditions to decrease energy consumption during wood BSR while maximizing the enzymatic cellulose saccharification efficiency.

BSR processes produce a wide range of fiber size distributions; therefore, biomass size characterization is necessary to maintain a control of the particles sizes used in a biorefinery. Techniques such as sieves, screens, and imaging analysis are used to determine the particle size distribution. Substrate surface area is most relevant to heat and mass transfer, chemical and/or enzyme accessibility, and energy consumption. Therefore, the substrate-specific surface should be used to characterize the size of a substrate. Specific surface area has been used to correlate energy consumption for comparing the efficiencies of several BSR processes (Holtzapple et al. 1989). When surface area is no longer a limiting factor, then chemical barrier has become the dominant factor. At this point, further surface area increases or substrate size reductions will not significantly affect substrate enzymatic digestibility.

23.2.3 Biomass Size Reduction, Post-Pretreatment

Pretreatments of LB can modify the chemical composition and physical structure of the biomass by partially removing and altering some cell wall components. Several studies confirm that milling pretreated wood chips significantly reduces the energy consumption (Figure 23.2) (Zhu et al. 2009; Zhu and Pan 2010). Chemical pretreatments prior to BSR lead to much higher liquid uptake in fibers and/or fiber bundles than wood chips fibers due to the porous and hydrophilic nature of wood fibers. A low liquid-to-biomass ratio in thermochemical pretreatments of wood chips can reduce the thermal energy consumption, produce a concentrated hemicellulose sugar stream favorable for fermentation, reduce downstream separation and distillation costs, and circumvent the energy-intensive operation of mixing high-consistency pulp with chemicals facilitating the liquid separation process after pretreatment.

23.3 LIGNOCELLULOSIC BIOMASS PRETREATMENT

Cellulosic biomass must be pretreated to realize high yields vital to commercial success in biological conversion (Mosier et al. 2005). Size reduction is a necessary step (Liu 2012) and has been discussed in the previous section. An effective pretreatment disrupts cell wall physical barriers as well as cellulose crystallinity and association with lignin so that hydrolytic enzymes can access the biomass macrostructure (McMillan 1994). Pretreatment is among the most costly steps and has a major influence on the cost of both prior (e.g., BSR) and subsequent (e.g., enzymatic hydrolysis and fermentation) operations. It influences downstream costs by determining fermentation toxicity, enzymatic hydrolysis rates, enzyme loadings, mixing power, product concentrations, product purification, waste treatment demands, power generation, and other process variables.

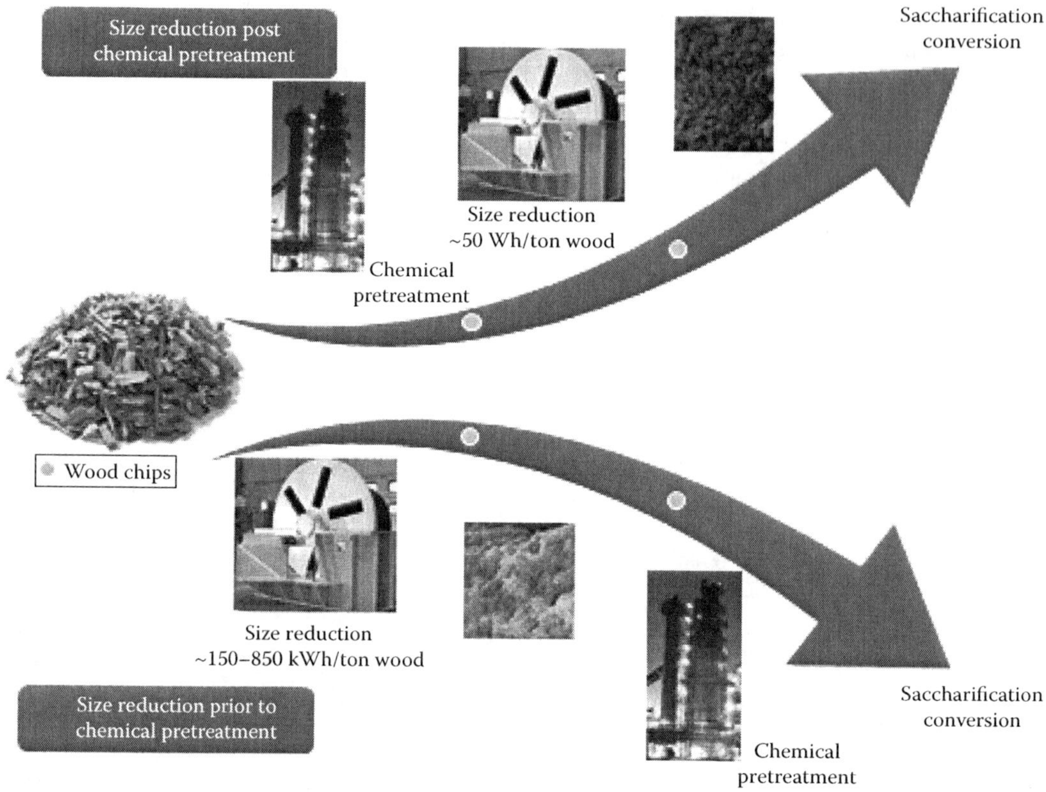

FIGURE 23.2 Schematic diagram comparing the energy use for mechanical wood size reduction operation prior chemical pretreatment and post chemical pretreatment. (Data from Zhu, J.Y. and Pan, X.J., *Bioresour. Technol.*, 101, 4992, 2010.)

23.3.1 Biological Pretreatment

The use of biological pretreatment is considered a safe, eco-friendly, and energy-saving process. It uses lignin-degrading microorganisms to treat the LB at low temperatures and in the absence of chemicals (Lee 1997; McMillan et al. 1999; Okano et al. 2005). Deterrents of this method are its slow reaction rate, microorganism consumption of carbohydrates, and microorganism poisoning by lignin derivatives (Hamelinck et al. 2003; Bohlmann 2006). Still, this method can be useful as a first step when followed by some of the other types of pretreatments.

23.3.2 Chemical Pretreatment

23.3.2.1 Acid Hydrolysis

Acid hydrolysis pretreatment of biomass is one of the most studied pretreatment methods. Dilute acid at moderate temperatures (140°C–190°C) effectively can remove and recover most of the hemicellulose as dissolved sugars, and glucose yields from cellulose increase with hemicellulose removal to almost 100% for complete hemicellulose hydrolysis. The data suggest that lignin is disrupted during this process, increasing cellulose susceptibility to enzymes (Yang and Wyman 2004; Wyman et al. 2005). A comparison study on aspen (hardwood), balsam (softwood), and switchgrass (energy crop) measured the total sugar yield (glucose and xylose; monomer plus oligomer) after enzymatic hydrolysis of the LB pretreated with dilute sulfuric acid (H_2SO_4) (Jensen et al. 2010). The results showed the high xylose yield was achieved at all acid concentrations (0.25–0.75 wt.%)

and temperatures (150°C–175°C) for aspen and balsam, as long as the reaction time was optimized. Switchgrass showed a relatively stronger dependence on acid concentration due to its higher neutralizing mineral content. Maximum total sugar yields from postenzymatic hydrolysis for aspen, balsam, and switchgrass were 88.3%, 21.2%, and 97.6%, respectively (Jensen et al. 2010).

The use of sulfur dioxide (SO_2) enhances yields in a way similar to dilute sulfuric acid, but dilute sulfuric acid is preferred because it is cheap, up to 90% hemicellulose yields are achieved, and enzymatic hydrolysis yields of glucose can be over 90% (Hsu 1996; Wyman et al. 2005). However, among the disadvantages of dilute acid pretreatment is the need for costly construction materials; high pressures; neutralization and conditioning of hydrolysate prior to biological steps; slow cellulose enzymatic digestion; production of potent inhibitors including furfural and hydroxymethylfurfural (HMF), which are by far regarded as the most toxic inhibitors present in lignocellulosic hydrolysate; and nonproductive enzymatic binding to lignin (Hsu 1996; Wyman et al. 2005).

23.3.2.2 Alkaline Hydrolysis

Alkali pretreatment processes utilize lower temperatures and pressures than other pretreatment technologies. When carried out at ambient conditions, the pretreatment time is on the order of hours or days rather than minutes or seconds, while at elevated temperatures, the contact time is reduced. Compared with acid hydrolysis, alkaline hydrolysis causes less sugar degradation, and many of the caustic salts can be recovered and/or regenerated. Sodium, potassium, calcium, and ammonium hydroxides are suitable alkaline agents for lignocellulosic pretreatment. Calcium hydroxide (slake lime) has been shown to be an effective pretreatment agent, is the least expensive per kilogram of hydroxide, and is also possible to be recycled by recovering calcium as insoluble calcium carbonate and regenerating calcium hydroxide by using established lime kiln technology. Lime pretreatment removes lignin and hemicellulose, which increases the crystallinity index. Slake lime pretreatment of corn stover increased the enzymatic hydrolysis by a factor of 9 compared to that of untreated corn stover (Karr and Holtzapple 2000).

The effect of alkaline pretreatment depends on the lignin content of the LB. Pretreatment with diluted sodium hydroxide (NaOH) was reported to decrease the lignin content from 24%–55% to 20% and increased the digestibility of NaOH-treated hardwood from 14% to 55%, while no effect was observed for softwoods with lignin content greater than 26% (Bjerre et al. 1996). Dilute NaOH treatment of LB has been found to cause swelling, leading to an increase in internal surface area, a decrease in the DP, a decrease in crystallinity, separation of structural linkages between lignin and carbohydrates, and disruption of the lignin structure (Fengel and Wegner 1989).

23.3.2.3 Ammonia Recycle Percolation

In the ammonia recycle percolation (ARP) process, aqueous ammonia (10–15 wt.%) passes through biomass at elevated temperatures (150°C–170°C) with a fluid velocity of 1 cm min^{-1} and a residence time of 14 min, after which the ammonia is recovered and recycled. Under these conditions, aqueous ammonia reacts primarily with lignin and causes depolymerization of lignin and cleavage of lignin–carbohydrate linkages. The ammonia pretreatment does not produce inhibitors for the downstream biological processes, so a water wash is not necessary (Kumar et al. 2009). This process is less efficient in pretreatment of softwood pulp (Mosier et al. 2005).

23.3.3 Physicochemical Pretreatment

23.3.3.1 Steam Explosion

Steam explosion pretreatment (SEP) is a technology used for breaking the crystalline lignocellulose structure. In this method, biomass is treated with high-pressure (0.69–4.83 MPa) saturated steam (160°C–260°C) for several seconds to a few minutes before the material is exposed to atmospheric pressure making the materials undergo an explosive decompression. The biomass/steam mixture is held for a period of time to promote hemicellulose hydrolysis. It is thought that hemicellulose undergoes autohydrolysis through acids that are derived from acetyl groups at high temperatures, while lignin is redistributed and to some extent removed from the material (Chen and Qiu 2010). The removal of hemicellulose exposes the cellulose surface and allows better access to cellulose microfibrils for an enzymatic attack (Alvira et al. 2010). Depending on the severity of the pretreatment, some degradation of the cellulose to glucose can take place. Grous et al. (1986) reported a 90% efficiency of enzymatic hydrolysis in 24 h for poplar chips pretreated by SEP, compared to only 15% hydrolysis of untreated chips. SEP is more effective on agricultural residues than wood due to lower acetic acid concentration in hemicellulose portion.

The factors that affect SEP are residence time, temperature, chip size, and moisture content. Optimal hemicellulose solubility and hydrolysis can be achieved by either high temperature and short residence time (270°C, 1 min) or lower temperature and longer residence time (190°C, 10 min). At high temperatures, water acts as an acid, but adding H_2SO_4 (or SO_2) or CO_2 [typically 0.3%–3% (w w^{-1})] in SEP can decrease time and temperature, effectively improve hydrolysis, decrease the production of inhibitory compounds, and lead to complete removal of hemicellulose (Kumar et al. 2009). For the pretreatment of softwoods, the addition of an acid catalyst is a prerequisite to make the substrate accessible to enzymes. The advantages of SEP include the low energy requirement compared to mechanical comminution and no recycling or environmental costs. The conventional mechanical methods require 70% more energy than SEP to achieve the same particle size reduction. SEP with the addition of a catalyst is the technology that has been claimed to be closest to commercialization. The pretreatment has been tested extensively for a large number of different LB feedstock (Hu et al. 2008). SEP is recognized as one of the most cost-effective pretreatment for hardwoods and agricultural residues, but it is less effective for softwoods. Some limitations are the partial xylan destruction, incomplete lignin–carbohydrate matrix disruption, limited lignin removal, and lignin redistribution on the cellulose surfaces (Chen and Qiu 2010).

23.3.3.2 Ammonia Fiber Explosion

Ammonia fiber explosion (AFEX) pretreatment is similar to steam explosion method. The biomass is treated with liquid ammonia under pressure at high temperature, and then the pressure is suddenly reduced, making the materials undergo an explosive decompression. The hemicellulose is degraded to oligomeric sugars and deacetylated, while the liquid ammonia swells the cellulose causing a phase change in the crystal structure from cellulose I to cellulose III, thus allowing a higher rate of cellulose hydrolysis. In a typical AFEX process, the dosage of liquid ammonia is 1–2 kg of ammonia per kg of dry biomass, the temperature is 90°C, and the residence time is 30 min. The AFEX pretreatment can significantly improve the fermentation rate of various herbaceous crops and grasses but is not a very efficient technology for LB with relatively high lignin (25%) content such as newspaper, woods, and nut shells (Teymouri et al. 2005). The moderate temperatures (<90°C) and pH values (<12.0) of AFEX minimize formation of sugar degradation products while giving high yields (Mosier et al. 2005).A drawback for this method is the cost of ammonia, especially of ammonia recovery, which drives the cost of the AFEX pretreatment (Kumar et al. 2009).

23.3.3.3 Hot-Water Pretreatment

Water pretreatments use elevated pressures to maintain the liquid form of water at high temperatures (van Walsum et al. 1996). Water under pressure penetrates the biomass cell structure, hydrates cellulose, and removes hemicellulose. At high temperature, the pKa of pure water is affected resulting in pH near 5.0 at 200°C. The high dielectric constant of water enables ionic substances to dissociate and dissolve from most LB. Hot water cleaves hemiacetal linkages and liberates acids that catalyze breakage of ether linkages in biomass (Wyman et al. 2005). Pretreatment using water at a controlled pH minimizes the hydrolysis of the oligosaccharides while causing hydration of the structure by liquid water at a pressure above the saturation vapor pressure of water at the selected temperature. The goal is to minimize degradation by avoiding monosaccharides formation that can be degraded to aldehydes during high-temperature pretreatment, and enzymes may be added at lower temperatures for the hydrolysis of the resulting cellulose and hemicellulose oligomers (Mosier et al. 2005; Wyman et al. 2005). This pretreatment applied to corn fiber from a corn wet milling operation, with subsequent enzymatic hydrolysis results in nearly complete conversion of cellulose and hemicellulose, as well as starch.

There are three types of liquid hot-water reactor configurations: cocurrent, countercurrent, and flow through (Figure 23.3). In cocurrent pretreatments, a slurry of LB and water is heated to the desired temperature and held at the pretreatment conditions for a controlled residence time before being cooled. Countercurrent pretreatment is designed to move water and LB in opposite directions through the pretreatment reactor. In a flow-through reactor, hot water passed over a stationary bed of LB, hydrolyzes, dissolves LB components, and carries them out of the reactor (Mosier et al. 2005).

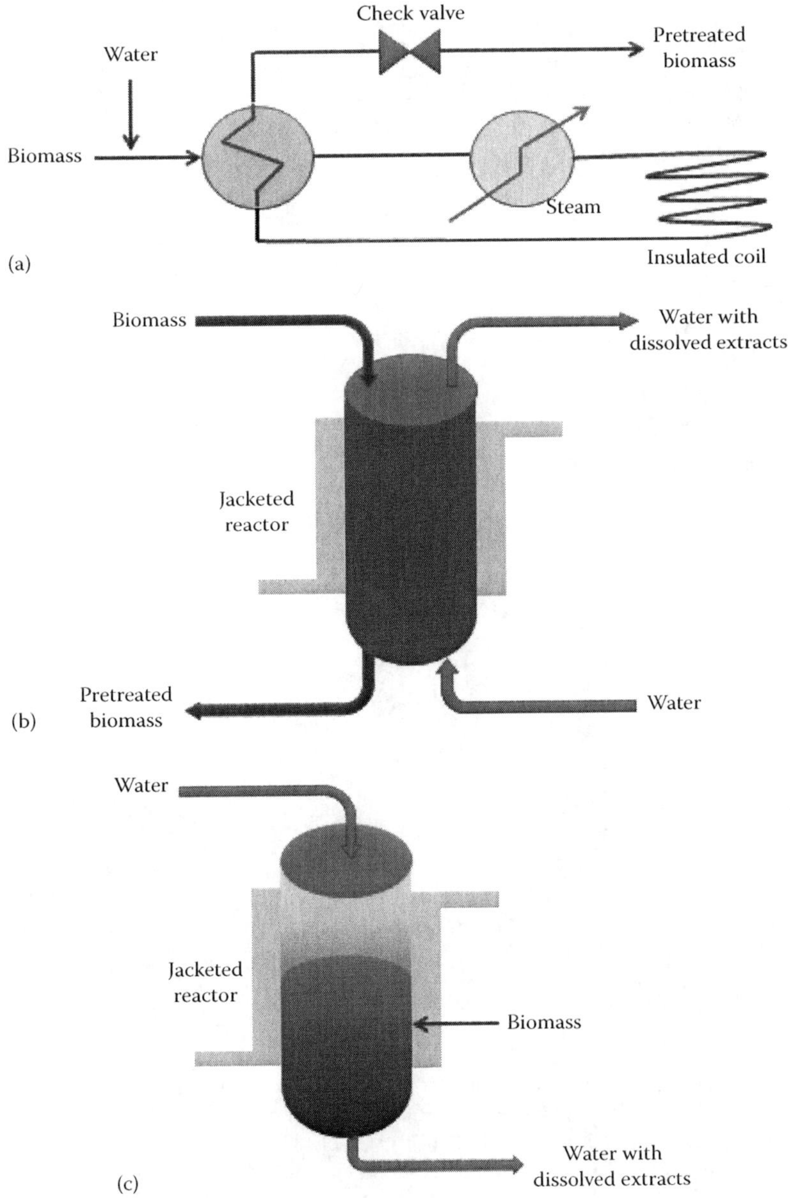

FIGURE 23.3 Schematic illustrations of cocurrent, countercurrent, and flow-through pretreatment methods: (a) cocurrent liquid hot-water pretreatment, (b) countercurrent reactor, (c) flow-through reactor.

23.4 SEQUENTIAL LIGNOCELLULOSE FRACTIONATION BIOREFINERY SYSTEM

Simultaneous isolation of cellulose, hemicelluloses, lignin, and extractives in a single process is not possible. The fractionation and conversion of LB to products requires a variety of integrated processes. Subjecting LB to a series of systematic fractionation processes optimizes conversion into its individual components. Of the major components of LB, the easiest to degrade is hemicellulose, while cellulose is the most difficult. An approach to sequential lignocellulose fractionation biorefinery is to initially treat LB with a hemicellulose-degrading step (Amidon and Liu 2009; Liu 2008). Figure 23.4 shows a schematic of a biorefinery that utilizes lignocellulosic feedstock. There are five main components in the biorefinery based on sequential fractionation of LB: (1) hot-water extraction (Liu 2008); (2) hydrolysis of hot-water extracts (Hu et al. 2010); (3) separation of xylan, sugars, and acetic acid (Liu et al. 2008); (4) fermentation of sugars to ethanol, butanol, or bioplastics (Liu 2010); and (5a) pulping of hot-water extracted woodchips followed by bleaching and papermaking (Liu et al. 2009), (5b) use of extracted woodchips/woody biomass for reconstituted wood products such as particle board, or (5c) further separation/hydrolysis of the extracted LB to platform chemicals: aromatics and lignocellulosic sugars. These components are designed such that the process can be readily adapted into existing industrial operations.

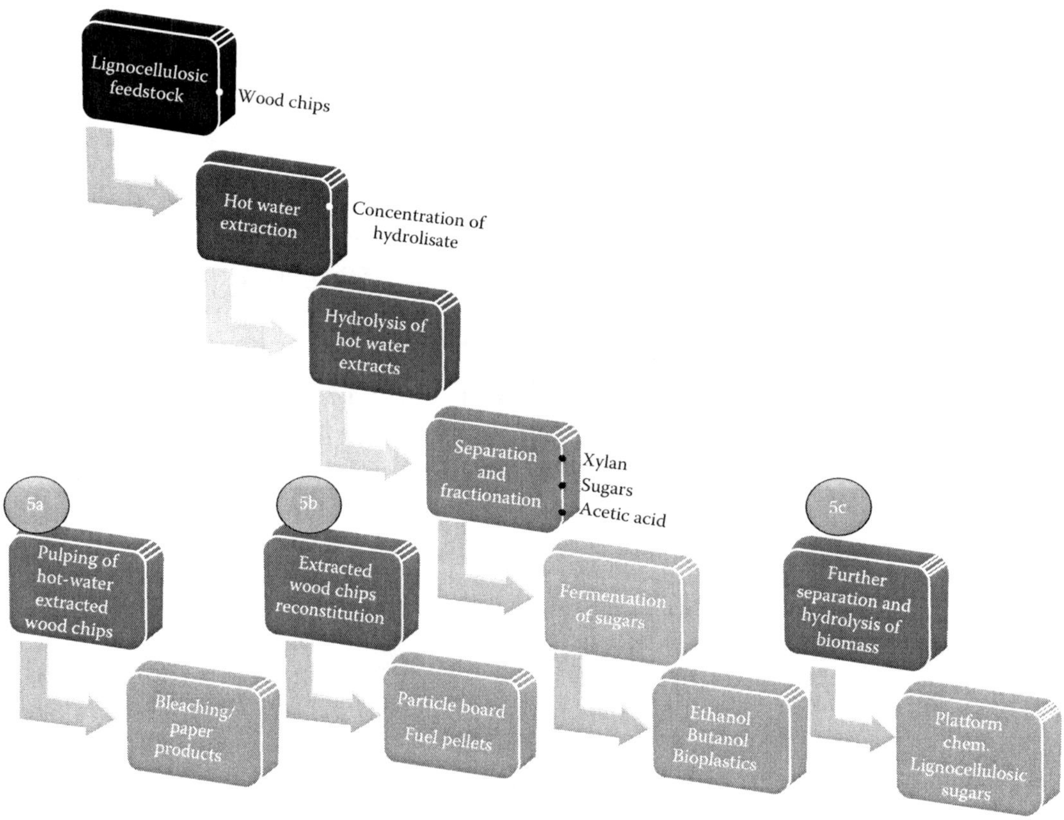

FIGURE 23.4 Schematic diagram of a sequential lignocellulose fractionation biorefinery. There are five main components to a biorefinery based on sequential fractionation: (1) hot-water extraction (Liu 2008); (2) hydrolysis of hot-water extracts (Hu et al. 2010); (3) separation of xylan, sugars, and acetic acid (Liu et al. 2008); (4) fermentation of sugars to ethanol, butanol, or bioplastics (Liu 2010); and (5a) pulping of hot-water extracted woodchips followed by bleaching and papermaking (Liu et al. 2009), (5b) use of extracted woodchips/woody biomass for reconstituted wood products such as particle board, or (5c) further separation/hydrolysis of the extracted LB to platform chemicals: aromatics and lignocellulosic sugars.

23.4.1 Hot-Water Extraction

Hot-water extraction is the first step in the sequential lignocellulose fractionation. This process is a complex dynamic system consisting of (1) acetic acid formation, (2) hemicelluloses cleavage and oligomers dissolution, (3) oligomeric sugars degradation to monomeric sugars, (4) monomeric sugars decomposition to HMF and furfural, and (5) further reactions related to HMF and furfural. Hydrogen ions present in the extraction mixture catalyze the hemicellulose depolymerization (Overend and Chornet 1989; Heitz et al. 1991; Aoyama et al. 1995; Saska and Ozer 1995; Weil et al. 1997; Garrote et al. 2001a,b; Amidon et al. 2008). The initial hydrogen ion concentration during hot-water extraction is very low. As the extraction progresses, the acetic, uronic, and phenolic acids present in lignin and hemicellulose are hydrated causing acidification of the liquor and thus formation of hydrogen ions, which produces an autocatalytic reaction. This autocatalytic extraction process has the advantage of limiting corrosion problems (Lamptey et al. 1985), absence of sludge production, reduced capital and operation costs (Overend and Chornet 1989), and minimal cellulose degradation under normal operating conditions (Amidon et al. 2008). In the presence of protons, the monomeric sugars can be further dehydrated to products, such as furfural, humic acid, and levulinic acid.

23.4.1.1 Extraction Kinetics

The overall extraction and hydrolysis kinetics of hemicellulose and extractives is described in a kinetic model presented by Liu (2008). In this model, the extraction and hydrolysis reaction rates are functions of proton concentration. It allows us to calculate the distribution of oligosaccharides, the polydispersivity, the DP, and the concentration of bound and free acetate, as functions of time, solid-to-water ratio, and temperature. The total dissolution rate (r_E) is obtained by adding up the rates of all the individual species leaving the solid phase:

$$r_E = \sum_{m=1}^{N} m(-r_{0_m}) = \sum_{m=1}^{N} mk_E\theta_m - \sum_{m=1}^{N} m \sum_{n=m+1}^{N} \frac{k_E}{n}\theta_n \tag{23.1}$$

where

θ_m is the surface concentration of component m on the LB surface

k_E is effective extraction rate constant

Equation 23.1 is reduced to

$$r_E = \frac{1}{2}k_E\left(\sum_{m=1}^{N} m\theta_m + \sum_{m=1}^{N} \theta_m\right) = \frac{1}{2}k_E P\left(1 + \frac{M_1}{M_n}\right) \tag{23.2}$$

where

P is the total extractable substance concentration in terms of the monosaccharide units in the LB

M_1 is the monosaccharide molecular weight

M_n is the number averaged molecular weight of the extractable molecules in the LB

Equation 23.2 can also be written in terms of the DP:

$$r_E = \frac{1}{2}k_E\left(1 + \frac{1}{\text{DP}}\right)P \tag{23.3}$$

The overall extraction process is approximated to be a pseudo-first-order process for the overall mass concentration in the LB. This reaction proceeds slower at the beginning, where DP is high and the proton concentration is low, and accelerates toward the extraction completion, where proton concentration reached its maximum and the DP is low. The overall concentrations based on mass or the monomer units are convenient measures when detailed distribution is not available.

23.4.1.2 Effect of Operation Conditions

The extent of extraction and amount of extracted solids are essential for extraction performance analysis and reactor design. The dissolved solids in the extraction liquor are the extracted substances that are not volatile at a temperature of 105°C and have undergone minimum degradation. The amount of mass removed from LB during hot-water extraction is used as a measure of the extent of extraction for short residence times. Residence time, temperature, and wood-to-water ratio are the main variables that affect the amount of mass removed and need to be controlled in order to produce the targeted final products. Studies on hot-water extraction of hardwood chips exhibit the increase of acetic acid formation and higher oligomeric sugars conversion to monomeric sugars with increasing time and temperature (Figures 23.5 and 23.6) (Liu et al. 2008; Li et al. 2010). At temperatures higher than 190°C, the extraction reaction proceeds quickly due to thermal degradation of LB. Acetic acid can accelerate the hemicellulose extraction and conversion.

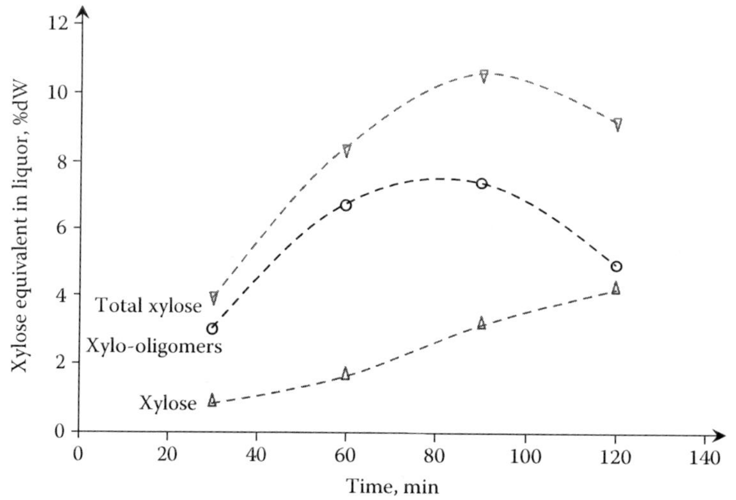

FIGURE 23.5 Xylo units in the hot-water extraction liquor of an Eastern Canada hardwood chips mixture obtained at 170°C. (Data from Li, H. et al., *J. Wood Chem. Technol.*, 30, 48, 2010.)

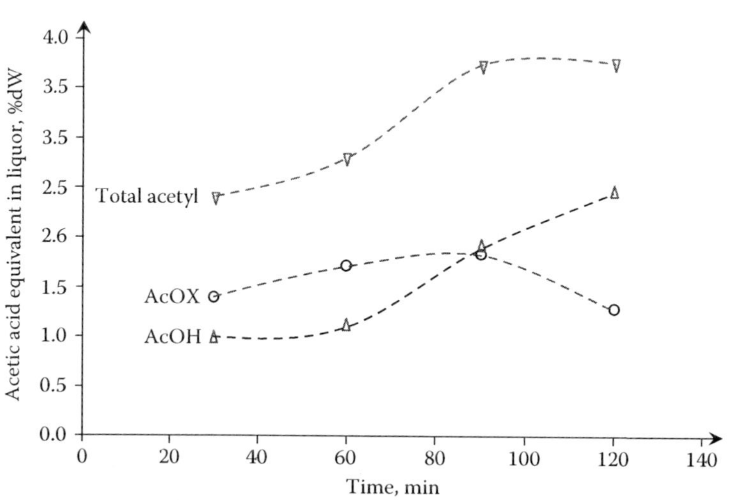

FIGURE 23.6 Acetyl solubilized in the HWWE of an Eastern Canada hardwood chips mixture obtained at 170°C. (Data from Li, H. et al., *J. Wood Chem. Technol.*, 30, 48, 2010.)

The hemicellulose extraction time can be shortened and the throughput of the process increased by adding acetic acid and/or recycling the extraction liquor (Liu et al. 2012).

23.4.2 Membrane Separation and Concentration

The hot-water wood extract (HWWE) consists of polysaccharides, monosaccharides, acetic acid, formic acid, methanol, furfural, HMF, aromatic compounds, and other low molecular weight extractable substances. In order to increase the process efficiency and the further use of the hydrolysate, the HWWE needs to be concentrated, and molecules such as acetic acid, formic acid, furfural, HMF, and other by-products should be removed. Volatile gas stripping units can be utilized at the end of the extraction to remove the low molecular weight volatile compounds.

An efficient and cost-effective method for the separation and fractionation of HWWE is essential to optimize production. There are various methods available for liquid mixture separation, including distillation, centrifugation, chemical separation, pervaporation, and filtration. All separation methods present advantages and disadvantages (see Table 23.2). Using the membrane separation method to concentrate and fractionate the HWWE has the advantages of preserving the values of the separated streams, being simple to implement, and medium-priced, while its disadvantage is membrane clogging

TABLE 23.2
Advantages and Disadvantages of Various Methods Available for Liquid Mixture Separation

Process	Based on	Advantages	Disadvantages
Distillation	Vapor–liquid equilibrium. Uses volatility differences of components in a liquid mixture.	Wide applications: Refineries Petrochemicals Chemical plants Natural gas processing plants Quality control	High equipment cost High operating cost High-energy input
Centrifugation	Density differences Uses g-force to separate partially miscible particles in a medium.	Wide applications: Liquid clarification and purification Solid demoistening Particle fractionation and sorting Solid washing Liquid–liquid separation Solids or liquids extraction Cyclonic separation	High equipment cost High operating cost High-energy input
Chemical separation	Difference in chemical properties. Chemical reactions bond certain compounds to other molecules, allowing them to be more easily removed from the medium.		Does not readily yield the desirable product stream Chemical consumption in other for separation to occur
Pervaporation	Difference of partial pressures. Uses differences in transport rates through a membrane to separate individual components present in a medium.	Selective separation Mild process Simple to be implemented In-line nature	Potential fouling of membrane surface or internal pores Temperature sensitive Relatively low-energy efficiency in ethanol–water separation

TABLE 23.3
Classes of Filtration

Class	Pore Size, Å	Applications
RO	1–10	Desalination Sugar concentration
NF	8–50	Sugar concentration Water softening Pesticide removal Sulfate removal
Ultrafiltration (UF)	10–200	Electrocoat paint concentration Food processes Cold sterilization Waste recovery: latex and milk production
Microfiltration (MF)	200–1000	Sterilization Particle removal
Electrodialysis (ED)	NA: Charge-oriented	Salt recovery/desalination Cheese whey purification Acid recovery

Source: Eykamp, W., Microfiltration and ultrafiltration. In *Membrane Separations Technology: Principles and Applications*, eds. R.D. Noble and S.A. Stern, Elsevier Science B.V., New York, 1995.

or fouling. Table 23.3 shows the five main classes of filtration (Eykamp 1995). The physicochemical properties of the solution are important to the selection of membranes, since the size, shape, and chemical affinity of molecular species present in a solution can vary considerably. When the molecular size is greater than the membrane pore size, the molecules are not able to permeate through. Therefore, the molecular weight that corresponds to the pore size is termed molecular weight cutoff. The molecular weights of hexoses, pentoses, furfural, HMF, and acetic acid present in the HWWE are 180.16, 150.13, 96.09, 110.11, and 60.05 g mol^{-1}, respectively. Considering the molecular weight cutoffs, the nanofiltration (NF) and reverse osmosis (RO) membranes are suitable to be used for the sugar purification and concentration, acetic acid recovery, and elimination of furfural and HMF. In a sequential filtration process, the first step is to filter the high molecular weight molecules such as aromatics, which are unfavorable to the fermentation process, from the HWWE. The lignin fragments or aromatics can be used as a chemical feedstock or direct energy recovery, through burning or gasification. The following steps are designed to concentrate and purify the sugars and isolate acetic acid (Liu et al. 2008).

A schematic representation of a membrane separation unit is shown in Figure 23.7. A pump feeds the HWWE or hydrolysates in one end of the unit, while at the other end the raffinate stream is recycled and the permeate stream is collected in one tank. Small molecules are preferentially removed to the permeate stream. The permeate flow rate increases with increasing pressure drop and decreases with increasing extract concentration. The pore size distributions in an industrial membrane is nonuniform; therefore, molecules larger than the average molecular weight cutoff may still pass through a commercial membrane but at lower rates.

23.5 HYDROLYSIS

23.5.1 Hydrolysis of Concentrated Hot-Water Wood Extracts

Hydrolysis of HWWE converts polysaccharides into monosaccharides, which are more readily used by microorganisms during fermentation. Hydrolysis of HWWE can be obtained by extending the

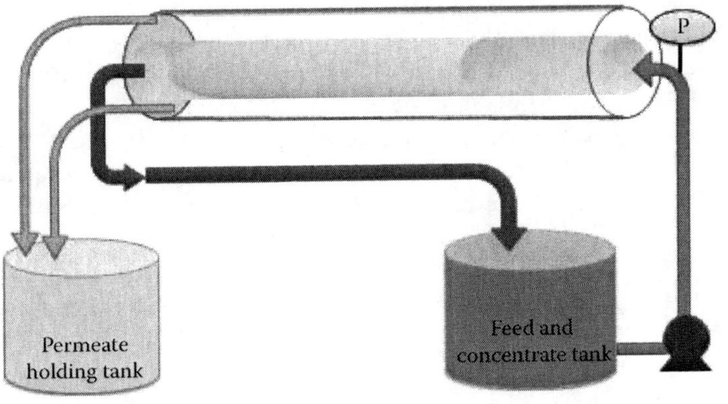

FIGURE 23.7 Schematic diagram of the membrane separation unit.

residence time during hot-water extraction. Unfortunately, monosaccharides degrade into other non-sugar molecules when subjected for extended times to relatively high temperatures and acidic conditions. In order to increase the overall process efficiency, the HWWE needs to be concentrated prior to hydrolysis. The hydrolysis reaction rate can be accelerated with either a chemical or an enzymatic catalyst. Dilute acid hydrolysis with sulfuric acid has been shown to catalyze the cleavage of β(1–4) glycosidic linkages present in wood extracts (Hu et al. 2010; Liu et al. 2012). Studies showed that the hydrolysis carried out at relatively higher acid concentration and temperature produced more xylose monomers in a comparatively shorter period of reaction time. In the hydrolysate from hardwood extract, xylose accounts for over 80% on a weight basis; in addition, comparatively small amounts of arabinose, glucose, rhamnose, mannose, and galactose are present. The maximum xylose concentration obtained after hydrolysis of sugar maple hot-water extract was achieved with the addition of 6.2% sulfuric acid, at 105°C temperature and 20 min reaction time at atmospheric pressure. Arabinose is released faster at higher acid concentration and temperatures and degrades quicker than xylose; therefore, the final amount of arabinose obtained from the acid hydrolysis is very low (Åman 1985; Albrecht et al. 1987; Jung et al. 1998; Gunnarsson et al. 2008). As the average concentration of glucose (7.45 g L^{-1}) and mannose (8.27 g L^{-1}) in all of the hydrolysates were similar, they may come from glucomannan and galactoglucomannan. The xylosidic bonds adjacent to the nonreducing terminal group react almost 1.8 times faster than interior bonds under low temperature and dilute acid concentration (Kamiyama and Sakai 1979). Therefore, the acetyl groups bonded to the hemicelluloses xylose backbone are hydrolyzed very quickly, resulting in nearly unchanged concentrations with time and minimal concentration differences among reaction temperatures. There also exist polymers and some degradation products of lignin or more precisely aromatic materials. The initial solid deposit is due to lignin precipitation at low pH. As wood extracts or hydrolysate remained at a high temperature for longer time periods and at higher acid concentrations, more solid deposits are formed, which can be attributed to furfural formation and subsequent dehydration polymerization (Liu et al. 2008).

23.5.2 Enzymatic Hydrolysis

Enzymatic hydrolysis is a highly selective reaction that converts specific polysaccharides into its individual building blocks at an accelerated rate of reaction. Cellulose is the major component remaining after pretreatments of LB that can be hydrolyzed by cellulases to produce glucose. There are five general types of cellulases, which are classified by the reaction they catalyze: endocellulases, exocellulases, cellobiases, oxidative cellulases, and cellulose phosphorylases. These complex enzyme structures are found in many bacteria to be organized in supramolecular complexes, called cellulosomes (Bayer et al. 1998). At least three classes of enzymes act synergistically to hydrolyze cellulose: (1) endocellulase (EC 3.2.1.4), which randomly breaks internal bonds at amorphous sites that create new chain ends; (2) exocellulases (EC 3.2.1.91), which cleaves two to four units from the ends of the exposed chains produced by endocellulase, resulting in tetrasaccharides or disaccharides; and (3) cellobiase (EC 3.2.1.21) or β-glucosidase, which hydrolyses the exocellulase product into individual monosaccharides. The enzyme activity is affected by temperature, pH, substrate concentration, product concentration, activators, inhibitors, and cellulose structure (Coughlan and Ljungdahl 1988). Cellobiose is an end-product inhibitor for endo- and exocellulases; therefore, the β-glucosidase activity is crucial for the efficiency of the hydrolysis process (Galbe and Zacchi 2002; Rabinovich et al. 2002; Zhang et al. 2006). Studies indicate that most cellulose enzymes show an optimum activity at temperatures in the range of 45°C–55°C and at pH values between 4 and 5 (Galbe and Zacchi 2002). The cellulose features known to affect the rate of hydrolysis include (1) molecular structure of cellulose, (2) crystallinity of cellulose, (3) surface area of cellulose fiber, (4) degree of swelling of cellulose fiber, (5) DP, and (6) associated lignin or other materials (Detroy and Julian 1982). A more pure and refined cellulose facilitates the cellulase enzymatic activity, thus lowering the enzyme load and cost for the enzymatic hydrolysis process.

23.6 FERMENTATION

Sugars produced via hydrolysis and purified (by membrane separation) can be used as feedstock to microorganisms for the production of (1) biofuels such as ethanol, propanol, and butanol; (2) biochemicals; and (3) biopolymers such as polylactic acid (PLA) and polyhydroxyalkanoates (PHAs). The presence of phenolics and other high molecular weight nonsugar compounds in the sugar stream could hinder the conversion efficiency. In order to avoid a detoxification step, it is preferable to adapt microorganism to medium (Liu 2010). Recovery of coproducts should be implemented when convenient in process. The traits of microorganism's fermentation ability are evaluated by their performance parameters, such as temperature range, pH range, growth rate, productivity, substrate utilization, specificity, yield, alcohol tolerance, osmotic tolerance, inhibitor tolerance, and genetic stability (Demirbas 2005). In addition, the microorganism's compatibility with existing products, processes and equipment is also of importance. The characteristics required for an industrially suitable microorganism are summarized in Table 23.4 (Dien et al. 2003).

23.6.1 Bioethanol

For biofuel production utilizing LB hydrolysate as feedstock, the microorganisms used need to be able to ferment not only hexoses but also pentoses, more specifically D-xylose and L-arabinose. Some ethanologenic microorganisms with their advantages and disadvantages for industrial production are shown in Table 23.5. Among microorganisms known to produce ethanol from hexoses and pentoses are the bacterium *Escherichia coli* and the yeast *Pichia stipitis* (Claassen et al. 1999). *E. coli*

TABLE 23.4
Important Traits for Bioethanol Fermentations

Trait	Requirement
Bioethanol yield	Greater than 90% of theoretical yield
Bioethanol tolerance	Greater than 40 g L^{-1}
Bioethanol productivity	Greater than 1 g L^{-1} h^{-1}
Robust grower and simple growth requirements	Inexpensive medium formulation
Able to grow in undiluted hydrolysate	Resistance to inhibitors
Culture growth conditions retard contaminants	Acidic pH or higher temperature

TABLE 23.5
Advantages and Disadvantages of Various Natural Microorganisms Regarding Industrial Ethanol Production

	Natural Sugar Utilization Pathways					Major Products		Tolerance				
Organism	**Glu**	**Man**	**Gal**	**Xyl**	**Ara**	**EtOH**	**Other**	**Alcohols**	**Acids**	**Hydrolysate**	**O_2 Needed**	**pH**
Anaerobic bacteria	+	+	+	+	+	+	+	–	–	–	–	Neutral
E. coli	+	+	+	+	+	–	+	–	–	–	–	Neutral
Zymomonas mobilis	+	–	–	–	–	+	–	+	–	–	–	Neutral
S. cerevisiae	+	+	+	–	–	+	–	++	++	++	–	Acidic
P. stipitis	+	+	+	+	+	+	–	–	–	–	+	Acidic
Filamentous fungi	+	+	+	+	+	+	–	++	++	++	–	Acidic

Source: Hahn-Hagerdal, B. et al., *Appl. Microbiol. Biotechnol.*, 74, 937, 2007.

ethanologenic strains can ferment all lignocellulose-derived sugars and do not require complex growth factors prior to industrial use. Some disadvantages with the use of *E. coli* are the narrow and neutral pH range (6–8) making bacterial fermentation susceptible to contaminations, low tolerance to lignocellulose-derived inhibitors, low ethanol tolerance, and mixed-product formation (ethanol, acetic acid, lactic acid, and others). The yeast *P. stipitis* is able to produce hemicellulolytic enzymes efficiently and ferment xylose under oxygen-limited conditions (Hahn-Hagerdal et al. 2007). A study using *P. stipitis* to ferment a sugar maple hydrolysate diluted with distilled water in a 1:1 ratio showed that controlled aeration helps increase ethanol production and the sugar fermentation rate (Shupe and Liu 2012). Limitations of *P. stipitis* are low tolerance to pH, ethanol, and hydrolysate inhibitors.

23.6.2 Biobutanol

In nature, the production of biobutanol is performed exclusively by bacteria strains of the genus *Clostridia* via acetone–butanol–ethanol (ABE) fermentation (Keis et al. 1995). ABE fermentation and microorganisms that can do ABE fermentation have been widely studied in an effort to understand the regulation and genetics of complex and primary metabolism. Butanol is used to produce various important derivatives such as butyl acrylate and methacrylate esters (used in latex surface coatings, enamels, and lacquers), butyl glycol ether, butyl acetate, plasticizers, amino resins, and butylamines (Kirschner 2006). It is also used as a diluent for brake fluid formulations, base for perfumes, and a solvent for the manufacturing of antibiotics, vitamins, and hormones (Lee et al. 2008). Moreover, butanol can be used as fuel in unmodified gasoline engines and therefore is considered to be part of the next-generation biofuel. Additionally, butanol as an additive can be blended with diesel to reduce soot emissions. The comparison of butanol with other fuels is shown in Table 23.5.

23.6.3 Lactic Acid

Lactic acid (2-hydroxypropanoic acid) is a chiral carboxylic acid with two optically active configurations: D- and L-enantiomers (Garlotta 2001; Tuominen 2003; Naveena et al. 2005). A naturally occurring nontoxic hygroscopic molecule miscible with water or ethanol, lactic acid, has numerous commercial applications (see Table 23.6) (Narayanan et al. 2004; Jin et al. 2005). This renewable raw material can be produced by chemical synthesis or microbial fermentation. The chemical synthesis yields a racemic mixture of L(+) and D(−) isomers, while microbial fermentation is a selective process where L(+)-, D(−)-, and DL isomers are formed depending on the genus and species used (Garlotta 2001; Liu 2003; Rao et al. 2008). The bulk of the commercially produced lactic acid worldwide is done via microbial fermentation by homolactic organisms, such as lactic acid bacteria (LAB) strains from the genus *Lactobacilli* (Jin et al. 2005). LAB can ferment several carbohydrates including hemicellulosic sugars from wood and agricultural residues (Garde et al. 2002; Saha 2003; Bustos et al. 2005). Two main challenges in the utilization of wood hydrolysate in fermentation to produce lactic acid are the detoxification of inhibitory compounds and the fermentation of pentoses present in the hemicellulose. Some methods studied to remedy the toxicity problem are neutralization, overliming, evaporation, the use of ion-exchange resins, and activated charcoal adsorption (Liu 2010). To address the fermentation of xylose, the most abundant pentose in hemicellulose, research has been focused on (1) isolating naturally occurring LAB strains, which can ferment both xylose and hexoses, and (2) genetic engineering to develop microbial strains with the desired characteristics (Picataggio et al. 1998; Dien et al. 2002). Some of the microorganisms studied are *Lactobacillus xylosus*, *L. pentosus*, *L. lactis* IO-1, *L. coprophilus*, *L. hilgardii*, and *L. casei* subsp. *rhamnosus* (Iyer et al. 2000). A batch process is commonly used for commercial fermentation of lactic acid. This process takes about 3–6 days to completely use sugar at concentrations of 5%–10% and yield about 2 g of acid per liter of broth per hour. High concentrations of lactic acid in the broth lead to toxicity and growth inhibition. Therefore, either neutralization or extraction methods need to be applied in order to maintain cell growth and production and to obtain high efficiency (Garlotta 2001).

TABLE 23.6
Lactic Acid Commercial Applications

Use	Lactic Acid/ Derivative	Commercial Application
Polymer precursor	Lactide	Biodegradable polyester
	PLA	Biodegradable plastic
		(Valuable medical properties)
		(Strong mechanical properties)
Chemical industry		Metal etching
		Acrylic acid production
		Acidulant in leather tanning
		Descaling agent
		Solvent
		Cleaning agent
		Low-acid releasing agent
		Humectant (environmentally safe)
Pharmaceutical and cosmetic		Water-soluble lactates
		Moisturizer
		Skin lightener and rejuvenator
		Antibacterial agent
		Adjust acidity in topical preparations.
		Disinfectant (keratolytic properties)
Food industry	Calcium lactate	Coagulates casein
		Acid flavoring agent
		Buffering agent
		Lower pH and increase the body of beer
		In winemaking: reduces flavor sharpness
		Food preservative
		Bacterial inhibitor
		Dough conditioner
	Sodium lactate	Dough conditioner
		Emulsifier

Sources: John et al. 2007; Jin, B. et al., *J. Ind. Microbiol. Biotechnol.*, 32, 678, 2005; Narayanan, N. et al., *Electron. J. Biotechnol.*, 7, 167, 2004; Monteagudo et al. 1997; Schmidt and Padukone 1997.

23.6.4 Polyhydroxyalkanoates (PHA)

PHAs are linear polyesters synthesized by bacterial fermentation of sugars or lipids as an intracellular carbon and energy storage compound. The biosynthesis of PHA is caused by certain deficiency conditions as the lack of macroelements, such as phosphorus, nitrogen, trace elements, or oxygen, and the excess supply of carbon sources. There are over 150 different monomers that can be combined to produce a range of materials with very different properties. The mechanical properties and biocompatibility of PHA can be altered by modifying the surface or combining PHA with other polymers, inorganic materials, and enzymes. PHAs are used in the production of biodegradable plastics and biomedical materials. Currently, the major limiting problem inhibiting the development and wide use of PHAs is their high cost compared to petroleum-derived plastics. One of the main contributing factors to the high production cost is the high cost of substrate, which may account for as much as 50% of the cost. The utilization of LB as feedstock represents an excellent substrate for the production of PHAs (Mai et al. 2004). Extensive studies on glucose and xylose fermentation for the production of PHAs have been done throughout the years (Bertrand et al. 1990; Young et al. 1994; Ramsay et al. 1995; Jang and Rogers 1996; Valentin and Dennis 1997; Lee 1998; Shang et al. 2004; Sun et al. 2009). These studies demonstrate that the microorganism growth rate, product yield, and polymer composition vary according to the substrate or combination of substrates used.

23.6.5 Simultaneous Saccharification and Fermentation

The cost of cellulase enzymes is a key barrier to economical production of bioethanol from lignocellulosic material, since it accounts for 30%–50% of the total production cost versus 8% when using starch or sugar crops. The simultaneous saccharification and fermentation (SSF) bioprocess integrates the cellulase production, cellulose hydrolysis, and product fermentation in a single unit operation. As a result, the cost of using cellulase is reduced, the volumetric productivity is increased, and there is reduction in capital investment.

23.7 CONCLUSION

A biorefinery using lignocellulosic feedstock normally integrates a series of systematical fractionation processes to treat the lignocellulosic material in order to achieve optimal conversion of individual biomass components into desired products. The LB is pretreated to remove the physical barriers, after which cellulose and hemicellulose are hydrolyzed to produce pentoses and hexoses. These sugars are fermented and converted to desired products such as liquid biofuels, biochemicals, steam/heat, and electricity. In a sequential lignocellulose fractionation biorefinery, the initial step is to treat lignocellulose biomass with a hemicellulose-degrading step. A suggested design with five main components to a biorefinery based on sequential lignocellulose fractionation includes (1) hot-water extraction; (2) hydrolysis of hot-water extracts; (3) separation of xylan, sugars, and acetic acid; (4) fermentation of sugars to ethanol, butanol, or bioplastics; and (5a) pulping of hot-water extracted woodchips followed by bleaching and papermaking, (5b) use of extracted woodchips/woody biomass for reconstituted wood products such as particle board, or (5c) further separation/hydrolysis of the extracted LB to platform chemicals: aromatics and lignocellulosic sugars.

Studies demonstrate that hot-water extraction rate increases with the extent of extraction and higher temperature. Membrane filtration can be employed to concentrate and fractionate HWWE to platform chemicals. Hydrolysis of concentrated HWWE produces more xylose monomers at higher acid concentration and higher temperature in a shorter time. Sugars produced via hydrolysis and purified by NF membrane separation can be used as feedstock to microorganisms for the production of (1) biofuels such as ethanol, propanol, and butanol; (2) biochemicals; and (3) biopolymers such as PLA and PHA.

REFERENCES

Albrecht, K.A., W.F. Wedin, and D.R. Buxton. 1987. Cell-wall composition and digestibility of alfalfa stems and leaves. *Crop Science*. 27: 735–741.

Alvira, P., E. Tomas-Pejo, M. Ballesteros, and M.J. Negro. 2010. Pretreatment technologies for an efficient bioethanol production process based on enzymatic hydrolysis. *Bioresource Technology*. 101: 4851–4861.

Åman, P. 1985. Chemical composition and in vitro degradability of major chemical constituents in botanical fractions of red clover harvested at different stages of maturity. *Journal of the Science of Food and Agriculture*. 36(9): 775–780.

Amidon, T.E. and S.J. Liu. 2009. Water-based woody biorefinery. *Biotechnology Advances*. 27: 542–550.

Amidon,T.E., C.D. Wood, A.M. Shupe, Y. Wang, M. Graves, and S.J. Liu. 2008. Biorefinery: Conversion of woody biomass to chemicals, energy and materials. *Journal of Biobased Materials and Bioenergy*. 2: 100–120.

Aoyama, M., K. Seki, and N. Saito. 1995. Solubilization of bamboo grass xylan by steaming treatment. *Holzforschung*. 49: 193–196.

Bayer, E.A., H. Chanzy, R. Lamed, and Y. Shoham. 1998. Cellulose, cellulases and cellulosomes. *Current Opinion Structural Biology*. 8: 548–557.

Bertrand, J.L., B.A. Ramsay, J.A. Ramsay, and C. Chavarie. 1990. Biosynthesis of poly-p-hydroxyalkanoates from pentoses by pseudomonas pseudoflava. *Applied Environmental Microbiology*. 56: 3133–3138.

Bjerre, A.B., A.B. Olesen, T. Fernqvist, A. Ploger, and A.S. Schmidt. 1996. Pre-treatment of wheat straw using combined wet oxidation and alkaline hydrolysis resulting in convertible cellulose and hemicellulose. *Biotechnology and Bioengineering*. 49: 568–577.

Boerjan, W., J. Ralph, and M. Baucher. 2003. Lignin biosynthesis. *Annual Review of Plant Biology* 54: 519–546.

Bohlmann, G.M. 2006. Process economic considerations for production of ethanol from biomass feedstocks. *Industrial Biotechnology*. 2: 14–20.

Bustos, G., A.B. Moldes, J.M. Cruz, and J.M. Domínguez. 2005. Production of lactic acid from vine-trimming wastes and viticulture lees using a simultaneous saccharification fermentation method. *Journal of the Science of Food and Agriculture*. 85: 466–472.

Cadoche, L. and G.D. Lopez. 1989. Assessment of size reduction as a preliminary step in the production of ethanol from lignocellulosic wastes. *Biological Wastes*. 30(2): 153–157.

Chen, H.Z. and W.H. Qiu. 2010. Key technologies for bioethanol production from lignocellulose. *Journal of Biotechnology Advances*. 28: 556–562.

Claassen, P.A.M., A.M. Lopez-Contreras, L. Sijtsma, R.A. Weusthuis, J.B. van Lier, E.W.J. van Niel, A.J.M. Stams, and S.S. De Vries. 1999. Utilization of biomass for the supply of energy carriers. *Applied Microbiology and Biotechnology*. 52: 741–755.

Coughlan, M.P. and L.G. Ljungdahl. 1988. Comparative biochemistry of fungal and bacterial cellulolytic enzyme system. In *Biochemistry and Genetics of Cellulose Degradation*, eds. J.P. Aubert, P. Beguin, and J. Millet, pp. 11–30, Academic Press, London, U.K.

Demirbas, A. 2005. Bioethanol from cellulosic materials: A renewable motor fuel from biomass. *Energy Sources*. 21: 327–337.

Detroy, R.W. and G.S. Julian. 1982. Biomass conversion: Fermentation chemicals and fuels. *Critical Reviews in Microbiology*. 10: 203–228.

Dien, B.S., M.A. Cotta, and T.W. Jeffries. 2003. Bacteria engineered for fuel ethanol production: Current status. *Applied Microbiology and Biotechnology*. 63: 258–266.

Dien, B.S., N.N. Nichols, and R.J. Bothast. 2002. Fermentation of sugar mixtures using *Escherichia coli* catabolite repression mutants engineered for production of L-lactic acid. *Journal of Industrial Microbiology and Biotechnology*. 29: 221–227.

Eykamp, W. 1995. Microfiltration and ultrafiltration. In *Membrane Separations Technology: Principles and Applications*, eds. R.D. Noble and S.A. Stern, Elsevier Science B.V., New York.

Fengel, D. and G. Wegner. 1989. *Wood: Chemistry, Ultrastructure, Reactions*, Walter de Gruyter, Berlin, Germany.

Galbe, M. and G. Zacchi. 2002. A review of the production of ethanol from softwood. *Applied Microbiology and Biotechnology*. 59: 618–628.

Garde, A., G. Jonsson, A.S. Schmidt, and B.K. Ahring. 2002. Lactic acid production from wheat straw hemicellulose hydrolysate by Lactobacillus pentosus and Lactobacillus brevis. *Bioresource Technology*. 81: 217–223.

Garlotta, D. 2001. A literature review of polylactic acid. *Journal of Polymers and the Environment*. 9: 63–84.

Garrote, G., H. Dominguez, and J.C. Parajo. 2001a. Generation of xylose solutions from Eucalyptus globulus wood by auto hydrolysis-posthydrolysis processes: Posthydrolysis kinetics. *Bioresource Technology*. 79: 155–164.

Garrote, G., H. Dominguez, and J.C. Parajo. 2001b. Kinetic modeling of corncob autohydrolysis. *Process Biochemistry*. 36: 571–578.

Glaudemans, C.P.J. and T.E. Timell. 1958. The polysaccharides of white birch (*Betula papyrifera*). *Svensk Papperstidn*. 61(1): 1–9.

Goring, D.A.I. and T.E Timell. 1962. Molecular weight of native celluloses. *Tappi Journal*. 45: 454–460.

Grous, W.R., A.O. Converse, and H.E. Grethlein. 1986. Effect of steam explosion pre-treatment on pore size and enzymatic hydrolysis of poplar. *Enzyme and Microbial Technology*. 8: 274–280.

Gullichsen, J. and H. Paulapuro. 2000. *Pigment Coating and Surface Sizing of Paper*, Book 11, Finnish Paper Engineers' Association and TAPPI, pp. 2–61.

Gunnarsson, S., H. Marstorp, A.S. Dahlin, and E. Witter. 2008. Influence of non-cellulose structural carbohydrate composition on plant material decomposition in soil. *Biology and Fertility of Soils*. 45(1): 27–36.

Hahn-Hagerdal, B., K. Karhumaa, C. Fonseca, I. Spencer-Martins, and M.F. Gorwa-Grauslund. 2007. Towards industrial pentose-fermenting yeast strains. *Applied Microbiology and Biotechnology*. 74: 937–953.

Hamelinck, C.N., G. van Hooijdonk, and A.P.C. Faaij. 2003. Prospects for ethanol from lignocellulosic biomass: Techno-economic performance as development progresses. In Report NWS-E-2003-55, 35. Utrecht University, Copernicus Institute, Science Technology and Society, Utrecht, the Netherlands.

Heitz, M., E. Capek-Menard, P.G. Koeberle, J. Garne, E. Chornet, R.P. Overend, J.D. Taylor and E. Yu. 1991. Fractionation of populous-tremuloides at the pilot plant scale: Optimization of steam pretreatment conditions using the STAKE II technology. *Bioresource Technology*. 35: 23–32.

Holtzapple, M.T., A.E. Humphrey, and J.D. Taylor. 1989. Energy requirements for the size reduction of poplar and aspen wood. *Biotechnology and Bioengineering*. 33: 207–210.

Hsu, T.A. 1996. Pretreatment of biomass. In *Handbook on Bioethanol, Production and Utilization*, ed. Wyman, C.E., Taylor & Francis, Washington, DC.

Hu, G., J.A. Heitmann, and O.J. Rojas. 2008. Feedstock pretreatment strategies for producing ethanol from wood, bark and forest residues. *Bioresources*. 3(1): 270–294.

Hu, R.F., L. Lin, T. Liu, and S. Liu. 2010. Dilute sulfuric acid hydrolysis of sugar maple wood extract at atmospheric pressure. *Bioresource Technology*. 101: 3586–3594.

Iyer, P.V., S. Thomas, and Y.Y. Lee. 2000. High-yield fermentation of pentoses into lactic acid. *Applied Biochemistry and Biotechnology: Enzyme Engineering and Biotechnology*. 84–86: 665–677.

Jang, J.H. and P.L. Rogers. 1996. Effect of levulinic acid on cell growth and poly-β-hydroxyalkanoate production by Alcaligenes sp. SH-69. *Biotechnology Letters*. 18: 219–224.

Jensen, J.R., J.E. Morinelly, K.R. Gossen, M.J. Brodeur-Campbell, and D.R. Shonnard. 2010. Effects of dilute acid pretreatment conditions on enzymatic hydrolysis monomer and oligomer sugar yields for aspen, balsam, and switchgrass. *Bioresource Technology*. 101(7): 2317–2325.

Jin, B., P. Yin, Y. Ma, and L. Zhao. 2005. Production of lactic acid and fungal biomass by Rhizopus fungi from food processing waste streams. *Journal of Industrial Microbiology and Biotechnology*. 32: 678–686.

John, R.P., K.M. Nampoothiri, and A. Pandey. 2007. Fermentative production of lactic acid from biomass: An overview on process developments and future perspectives. *Applied Microbiology and Biotechnology Journal*. 74: 524–534.

Jung, H.G., T.A. Morrison, and D.R. Buxton. 1998. Degradability of cell-wall polysaccharides in maize internodes during stalk development. *Crop Science*. 38: 1047–1051.

Kamiyama, Y. and Y. Sakai. 1979. Rate of hydrolysis of xylo-oligosaccharides in dilute sulfuric acid. *Carbohydrate Research*. 73(1): 151–158.

Karr, W.E. and T. Holtzapple. 2000. Using lime pretreatment to facilitate the enzymatic hydrolysis of corn stover. *Biomass and Bioenergy*. 18: 189–199.

Keis, S., C.F. Bennett, V.K. Ward, and D.T. Jones. 1995. Taxonomy and phylogeny of industrial solvent-producing Clostridia. *International Journal of Systematic Bacteriology*. 45: 693–705.

Kirschner, M. 2006. N-Butanol. *Chemical Market Reporter*. 269(4): 42.

Koshijima, T., T.E. Timell, and M. Zinbo. 1965. The number-average molecular weight of native hardwood xylans. *Journal of Polymer Science, Part C Polymer Symposium*. 11: 265–279.

Kumar, P., D.M. Barrett, M.J. Delwiche, and P. Stroeve. 2009. Methods for pretreatment of lignocellulosic biomass for efficient hydrolysis and biofuel production. *Industrial and Engineering Chemical Research*. 48: 3713–3729.

Lamptey, J., C.W. Robinson, and M. Moo-Young. 1985. Enhanced enzymatic hydrolysis of lignocellulosic biomass pretreated by low pressure steam autohydrolysis. *Biotechnology Letters*. 7(7): 531–536.

Lee, J. 1997. Biological conversion of lignocellulosic biomass to ethanol. *Journal of Biotechnology*. 56: 1–24.

Lee, S.Y. 1998. Poly(3-hydroxybutyrate) production from xylose by recombinant *Escherichia coli*. *Bioprocess and Biosystems Engineering*. 18: 397–399.

Lee, S.Y., J.H. Park, S.H. Jang, L.K. Nielsen, J. Kim, and K.S. Jung. 2008. Fermentative butanol production by Clostridia. *Biotechnology and Bioengineering*. 101: 209–228.

Li, H., A. Saeed, M.S. Jahan, Y. Ni, and A. van Heiningen. 2010. Hemicellulose removal from hardwood chips in the extraction step of the Kraft-based dissolving pulp production process. *Journal of Wood Chemistry and Technology*. 30(1): 48–60.

Liu, S.J. 2008. A kinetic model on autocatalytic reactions in woody biomass hydrolysis. *Journal of Biobased Materials and Bioenergy*. 2: 135–147.

Liu, S.J. 2010. Woody biomass: Niche position as a source of sustainable renewable chemicals and energy and kinetics of hot-water extraction/hydrolysis. *Biotechnology Advances*. 28: 563–582.

Liu, S.J. 2012. Woody biomass conversion: Sustainability and water-based processes. *Journal of Bioprocess Engineering and Biorefinery* 1(1): 6–32.

Liu, S.J., T.E. Amidon, and C.D. Wood. 2008. Membrane filtration: Concentration and purification of hydrolyzates from biomass. *Journal of Biobased Materials and Bioenergy*. 2(2): 121–134.

Liu, S.J., H. Lu, R. Hu, A. Shupe, L. Lin, and B. Liang. 2012. A sustainable woody biomass biorefinery. Biotechnology Advances. http://dx.doi.org//10.1016//j.biotechadv.2012.01.013.

Liu, S.J., G. Mishra, T.E. Amidon, and K. Gratien. 2009. Effect of hot-water extraction of woodchips on the Kraft pulping of *Eucalyptus* woodchips. *Journal of Biobased Materials and Bioenergy*. 3(3): 363–372.

Liu, S.Q. 2003. Practical implications of lactate and pyruvate metabolism by lactic acid bacteria in food and beverage fermentations. *International Journal of Food Microbiology*. 83: 115–131.

Mai, C., U. Kues, and H. Militz. 2004. Biotechnology in the wood industry. *Applied Microbiology and Biotechnology*. 63: 477–494.

McMillan, J.D. 1994. Pretreatment of lignocellulosic biomass. In *Enzymatic Conversion of Biomass for Fuels Production*, eds. Himmel, M.E., J.O. Baker, and R.P. Overend, pp. 292–324, American Chemical Society, Washington, DC.

McMillan, J.D., M. Newman, D. Templeton, and A. Mohagheghi. 1999. Simultaneous saccharification and cofermentation of dilute-acid pretreated yellow poplar hardwood to ethanol using xylose-fermenting *Zymomonas mobilis*. *Applied Biochemistry and Biotechnology*. 79: 649–665.

Monteagudo, J.M., L. Rodriguez, J. Rincon, and J. Fuertes. 1997. Kinetics of lactic acid fermentation by *Lactobacillus delbrueckii* grown on beet molasses. *Journal of Chemical Technology and Biotechnology*. 68: 271–276.

Mosier, N., C. Wyman, B. Dale, R. Elander, Y.Y. Lee, M. Holtzapple, and M. Ladisch. 2005. Features of promising technologies for pre-treatment of lignocellulosic biomass. *Bioresource Technology*. 96: 673–686.

Narayanan, N., P.K. Roychoudhury, and A. Srivastava. 2004. L (+) lactic acid fermentation and its product polymerization. *Electronic Journal of Biotechnology*. 7: 167–179.

Naveena, B.J., M. Altaf, K. Bhadriah, and G. Reddy. 2005. Selection of medium components by Plackett-Burman design for production of L(+) lactic acid by Lactobacillus amylophilus GV6 in SSF using wheat bran. *Bioresource Technology*. 96: 485–490.

Okano, K., M. Kitagaw, Y. Sasaki, and T. Watanabe. 2005. Conversion of Japanese red cedar (*Cryptomeria japonica*) into a feed for ruminants by white-rot basidiomycetes. *Animal Feed Science and Technology*. 120: 235–243.

Overend, R. and E. Chornet. 1989. Steam and aqueous pretreatments: Are they prehydrolyses? In *Wood Processing and Utilization*, eds. Kennedy, J.F., G.O. Phillips and P.A. Williams, pp. 395–400, Elsevier, New York.

Picataggio, S.K., M. Zhang, M.A. Franden, J.D. McMillan, and M. Finkelstein. 1998. Patent no. US00579823A, United States Patent.

Rabinovich, M.L., M.S. Melnik, and A.V. Boloboba. 2002. Microbial cellulases (review). *Applied Biochemistry and Microbiology*. 38: 305–321.

Ramsay, J.A., M.C. Alyhassan, and B.A. Ramsay. 1995. Hemicellulose as a potential substrate for production of poly(β-hydroxyalkanoates). *Canadian Journal of Microbiology*. 41: 262–266.

Rao, C.S., R.S. Prakasham, A.B. Rao, and J.S. Yadav. 2008. Production of L (+) lactic acid by Lactobacillus delbrueckii immobilized in functionalized alginate matrices. *World Journal of Microbiology and Biotechnology*. 24: 1411–1415.

Saha, B.C. 2003. Hemicellulose bioconversion. *Journal of Industrial Microbiology and Biotechnology*. 30: 279–291.

Saska, M. and E. Ozer. 1995. Aqueous extraction of sugarcane bagasse hemicellulose and production of xylose syrup. *Biotechnology and Bioengineering*. 45: 517–523.

Schmidt, S. and N. Padukone. 1997. Production of lactic acid from wastepaper as a cellulosic feedstock. *Journal of Industrial Microbiology and Biotechnology*. 18: 10–14.

Shang, L., S.C. Yim, H.G. Park, and H.N. Chang. 2004. Sequential feeding of glucose and valerate in a fed-batch culture of *Ralstonia eutropha* for production of poly(hydroxybutyrate-co-hydroxyvalerate) with high 3-hydroxyvalerate fraction. *Biotechnology Progress*. 20: 140–144.

Shupe, A.M. and S.J. Liu. 2012. Ethanol fermentation from hydrolyzed hot-water wood extracts by pentose fermenting yeasts. *Biomass and Bioenergy*. 39: 31–38.

Sjöstrom, E. 1993. *Wood Chemistry: Fundamentals and Applications*, Academic Press, New York.

Sun, N., M. Rahman, Y. Qin, M.L. Maxim, H. Rodriguez, and R.D. Roger. 2009. Complete dissolution and partial delignification of wood in the ionic liquid 1-ethyl-3-methylimidazolium acetate. *Green Chemistry*. 11: 646–655.

Teymouri, F., L. Laureano-Perez, H. Alizadeh, and B.E. Dale. 2005. Optimization of the ammonia fiber explosion (AFEX) treatment parameters for enzymatic hydrolysis of corn stover. *Bioresource Technology*. 96(18): 2014–2018.

Tienvieri, T., E. Huusari, J. Sundholm, P. Vuorio, J. Kortelainen, H. Ystedt, H., and A. Artamo. 1999. Thermomechanical pulping. In *Mechanical Pulping*. Book 5. *Papermaking Science and Technology*, eds. J. Sundholm, J. Gullichsen, and H. Paulapuro, pp. 157–221, Fapet, OyJyvaskyla, Helsinki, Finland.

Timell, T.E. 1960. Determination of the degree of polymerization of reducing pentose and hexose oligosaccharides. *Svensk Papperstidn*. 63: 668–671.

Tuominen, J. 2003. Chain linked lactic acid polymers: Polymerization and biodegradation studies. PhD dissertation, Helsinki University of Technology, Espoo, Finland.

Valentin, H.E. and D. Dennis. 1997. Production of poly(3-hydroxybutyrate-co-4-hydroxybutyrate) in recombinant *Escherichia coli* grown on glucose. *Journal of Biotechnology*. 58: 33–38.

van Walsum, G.P., S.G. Allen, M.J. Spencer, M.S. Laser, M.J. Antal, and L.R. Lynd. 1996. Conversion of lignocellulosics pretreated with liquid hot water to ethanol. *Applied Biochemistry and Biotechnology*. 57–58: 157–170.

Weil, J., A. Sarikaja, S.I. Rau, J. Goez, C.M. Ladisch, M. Brewer, R. Hendrickson, and M.R. Ladisch. 1997. Pretreatment of yellow poplar sawdust by pressure cooking in water. *Applied Biochemistry and Biotechnology*. 68: 21–40.

Wyman, C.E., B.E. Dale, R. Elandert, M. Holtzapple, M.R. Ladisch, and Y.Y. Lee. 2005. Coordinated development of leading biomass pretreatment technologies. *Bioresource technology*. 96(18): 1959–1966.

Yang, B. and C.E. Wyman. 2004. Effect of xylan and lignin removal by batch and flow through pretreatment on the enzymatic digestibility of corn stover cellulose. *Biotechnology and Bioengineering*. 86 (1): 88–95.

Young, F.K., J.R. Kastner, and W.M. Sheldon. 1994. Microbial production of poly-β-hydroxybutyric acid from D-xylose and lactose by Pseudomonas cepacia. *Applied Environmental Microbiology*. 60: 4195–4198.

Zhang, Y.H.P., M.E. Himmel, and J.R. Mielenz. 2006. Outlook for cellulase improvement: Screening and selection strategies. *Biotechnology Advances*. 24: 452–481.

Zhu, J.Y. and X.J. Pan. 2010. Woody biomass pretreatment for cellulosic ethanol production: Technology and energy consumption evaluation. *Bioresource Technology* 101: 4992–5002.

Zhu, J.Y., G.S. Wang, X.J. Pan, and R. Gleisner. 2009. Specific surface to evaluate the efficiencies of milling and pretreatment of wood for enzymatic saccharification. *Chemical Engineering Science*. 64: 474–485.

24 Whole-Crop Biorefinery*

Apostolis A. Koutinas, Carol Lin, Chenyu Du, and Colin Webb

CONTENTS

24.1 INTRODUCTION

Nowadays, it is widely accepted that the development of viable and sustainable biorefineries is highly dependent upon the utilization of all biomass components and, if possible, the production of multiple products that could satisfy different markets. Integration of opportunities in current industrial sectors through the utilization of waste and by-product streams should be also explored in order to reduce capital investment and operating costs, promote re-utilization of the same feedstock/product in a cascade mode, and improve the environmental impact of current industrial processes. Furthermore, biomass-based biorefineries should be developed in an analogous manner that petroleum refineries evolved since the 1860s from relatively simple processes using atmospheric distillation to produce kerosene as a substitute for whale oil in oil lamps to complex and integrated distillation and conversion processes that fractionate and convert petroleum into several products. The development of petroleum refineries for the production of fuels, chemicals, plastics,

* This chapter is dedicated to the memory of Dr. Ruohang Wang, our collaborator and beloved friend, who passed away in July 2010.

and materials went through a series of improvements and restructuring based upon numerous innovations such as the introduction of vacuum distillation, thermal cracking and reforming, alkylation, and catalytic polymerization, cracking, and reforming. The low production cost of most products derived nowadays from petroleum refineries is highly dependent on the advantages of economies of scale, high energy density, continuous mode of operation, and ease of transportation provided by petroleum.

Contrary to petroleum, conventional fermentation technologies using agricultural crops as raw materials usually operate in a batch mode and generally do not take advantage of the entire potential of all components in crops. However, a successful whole-crop biorefinery is dependent on the exploitation of all components and the manufacture of products that can satisfy different market outlets. Corn wet milling processes are nowadays considered as mature industrial processes. The success of the corn milling technology is based on the fact that corn is refined into highly marketable products, such as oil, corn gluten meal, corn gluten feed, and starch that can be used for the production of various products including glucose syrups, high-fructose corn syrups, adhesives, cross-linked waxy maize starches, cationic starches, hydroxyethyl starches, and ethanol (Peckham 2001). The economic viability of whole-crop biorefineries is also dependent on the combined production of both specialty chemicals and bulk commodities.

This chapter aims to present whole-crop biorefining concepts for the production of a spectrum of products including food, fuels, chemicals, biopolymers, biomaterials, and energy. Special focus is given on wheat-based biorefineries including analysis of minor components and presentation of extraction or bioconversion technologies that lead to production of both specialty and commodity products.

24.2 WHEAT-BASED BIOREFINERIES

Since the 1990s, wheat-based biorefineries appeared as generic processing concepts exploiting wet or dry fractionation combined with bioconversion or green chemical conversion for the production of chemicals and materials, apart from food and biofuels. Professor Colin Webb as the director of the Satake Centre for Grain Process Engineering (SCGPE) was among the first researchers that introduced the concept of whole-wheat biorefining (Webb 1994; Webb and Wang 1997; Koutinas et al. 2004, 2006, 2007a; Webb et al. 2004). This section presents wheat composition and structure, patented or published extraction technologies of value-added components contained in wheat kernel and potential whole-wheat biorefining concepts.

24.2.1 Wheat Structure and Composition

The structure of the wheat grain, botanically known as caryopsis, has certain features that influence significantly wheat processing and utilization. The outmost layers of this caryopsis, namely, outer and inner pericarp, constitute approximately 3.3%–5.6% of whole grain weight on a dry basis. The seed coat and the nucellar tissues (together approximately constitute 2.2% of the wheat grain weight) are located between the inner pericarp and the aleurone layer providing a complete coverage of the endosperm and embryo. The aleurone layer is a one-cell thick layer, constituting approximately 6.7%–7% of the wheat grain weight on a dry basis, located between the nucellar tissues and the endosperm and part of the embryo. The endosperm predominantly contains starch and gluten. The embryo or germ (2.5%–3.6% of the wheat grain weight) is mainly composed of the embryonic axis and the scutellum and is located at the lower dorsal side of the wheat kernel. Evers and Bechtel (1988), Pomeranz (1987, 1988), MacMasters et al. (1971), and Toepfer et al. (1972) provide thorough description of the wheat grain structure and chemical composition. Table 24.1 presents the chemical composition of wheat kernel as was reported by Toepfer et al. (1972).

TABLE 24.1
Chemical Constituents of Wheat Grain (dw)

Component	Content	Component	Content
Major Components		Nonstarch Polysaccharides	
Starch (%)	61.2–66.2	Arabinoxylans (%)	7.27–7.62
Protein (N×5.83, %)	11.7–15.6	Hemicellulose (%)	6.24–6.45
Fat (%)	2.86–3.56	Nonreducing sugar (%)	2.92–3.25
Fatty acid (%)	2.03–2.78	Cellulose (%)	2.30–2.71
Reducing sugar (%)	0.55–0.59	Crude fiber (%)	2.47–2.77
Ash (%)	1.73–2.03	Lignin (%)	0.54–0.67
Minerals		Amino acids	
K (μg/g)	4140–4940	Glutamic acid	3440–5313
P (μg/g)	3400–3700	Proline	1168–1643
Mg (μg/g)	1589–1864	Leucine	812–1137
Ca (μg/g)	340–390	Serine	652–860
Fe (μg/g)	37–44	Aspartic acid	660–879
Mn (μg/g)	32–38	Phenylalanine	567–812
Na (μg/g)	30–50	Arginine	612–760
Zn (μg/g)	5–30	Valine	581–748
Sn (μg/g)	5.6–7.9	Glycine	504–648
Cu (μg/g)	4.5–5.1	Isoleucine	464–672
Pb (μg/g)	0.4–1.0	Alanine	458–586
Ni (μg/g)	0.3–0.5	Ammonia	393–584
Se (μg/g)	0.5	Threonine	393–841
Cr (μg/g)	0.3–0.4	Tyrosine	344–476
Cd (μg/g)	0.1	Lysine	360–433
		Histidine	271–406
Sugars		Cystine	223–306
Glucofructosans (%)	0.94–1.14	Tryptophan	225–266
Raffinose (%)	0.19–0.68	Methionine	206–274
Glucodifructose (%)	0.26–0.41		
Maltose (%)	0.04–0.18	Vitamins	
Sucrose (%)	0.54–1.55	Niacin (μg/g)	72–111
Glucose (%)	0.03–0.09	Tocopherols	54–58
Fructose (%)	0.06–0.08	Thiamine (μg/g)	5.4–6.7
Galactose (%)	0.02	Vitamin B_6 (μg/g)	3.3–4.3
		Riboflavin (μg/g)	1.1–1.2

Source: Toepfer et al., 1972.

24.2.2 Wheat Processing

Wheat is a complex biological entity, and each layer of the wheat kernel has a different composition that could be used for the production of different end products. The distribution of these chemical components in the structures of wheat grain is the basis of wheat processing. Conventional wheat processes focused on the production of white flour are based on sequential milling and sieving operations to separate wheat endosperm that constitutes more than 80% of the whole-wheat kernel from wheat bran and germ fractions. Prior to the application of milling and sieving unit operations, wheat grains are cleaned, conditioned with water or steam, and subsequently allowed to rest in temper bins for 4–20 h, a process causing bran and endosperm layers to become tougher and softer, respectively, in order to facilitate their separation in the subsequent stages. Complete separation of wheat fractions is not possible by the conventional dry milling process. For this reason, wheat milling by-products contain mainly the bran fraction but also a starch fraction (approximately 20% on a dry basis). Many value-added components in the wheat grain are contained in the bran and germ layers. The separation of these components in the conventional wheat dry milling operations should be carried out through processing of the entire bran-rich by-product stream that includes all bran layers and the aleurone layer.

Debranning or pearling of wheat grains has been developed as an advanced fractionation process of wheat providing the flexibility to remove sequentially the outer bran layers reducing the contamination of milling products with high fiber content (Dexter and Wood 1996). As in the case of conventional milling, conditioning with water is an important step prior to debranning that facilitates the separation of individual bran fractions by controlling the quantity of water added and the duration that wheat grains are conditioned with water. The quantity of water added is usually 1%–3% (w/w) depending on the moisture content and the hardness of the wheat variety used (e.g., hard or soft wheat varieties). Contrary to conventional dry milling operations, wheat conditioning prior to debranning should be applied at shorter residence times in order to allow limited penetration of water and loosen only the outer bran layers leading to high separation efficiency of these layers. Pearling is achieved via sequential friction and abrasion stages that remove progressively most of the bran layers and germ without affecting the endosperm. There are various pearling devices that have been presented in previous publications (Tkac 1992; Wellman 1993; Satake et al. 1995; Dexter and Wood 1996; Koutinas et al. 2006). The level of bran removal is dependent upon operating conditions (e.g., pressure of the abrasion rolls, grain flow rate) (Dexter and Wood 1996). Pearling could also lead to the production of fractions nutritionally enriched in aleurone cells. Debranning operations usually remove 6%–10% of the wheat kernel weight representing 65%–75% of the bran content.

Debranning can be applied at successive time intervals for the fractionation of individual bran layers to produce fractions enriched in specific value-added components, such as phytic acid, lipids, vitamins, arabinoxylans, oligosaccharides, and β-glucan (Figure 24.1). Individual fractions may have varying therapeutic, physicochemical, and nutritional properties significantly different than those of conventional wheat milling fractions (Dexter and Wood 1996; Cui et al. 1999). For instance, Singh et al. (1998) applied debranning for 30, 60, 90, and 120 s to different wheat varieties to produce fractions with ash contents that progressively decreased with the increasing pearling time. Generally, pearling efficiency will be influenced by pearling duration of sequential operations, temperature, and wheat moisture content and hardness.

24.2.3 Value-Added Products from Wheat Components

24.2.3.1 Dietary Fiber and Associated Components

Partially due to the fact that many fruits, vegetables, and grains considered by consumers to be high in fiber actually contain low to moderate levels, the average fiber intake of consumers is only 14–15 g/day. This is much lower than the recommended 20–35 g by nutritionists. In recent years, there is a steadily increasing positive view on the effect of dietary fiber on human health (Mehta 2005, 2009). Detailed reports are available in the literature about different types of fiber and analytical methodologies for quantitative analysis of total dietary fiber as well as specific components (Nelson 2001; De Vries and Rader 2005; Mehta 2009). Dietary fiber includes many components (Table 24.2) and is generally defined as the part of plants the majority of which cannot be digested by the enzymes of the human digestive tract. One of the most established definitions for dietary fiber is given by the American Association of Cereal Chemists (AACC): "Dietary fibre is the edible parts of plants or analogous carbohydrates that are resistant to digestion and absorption in the

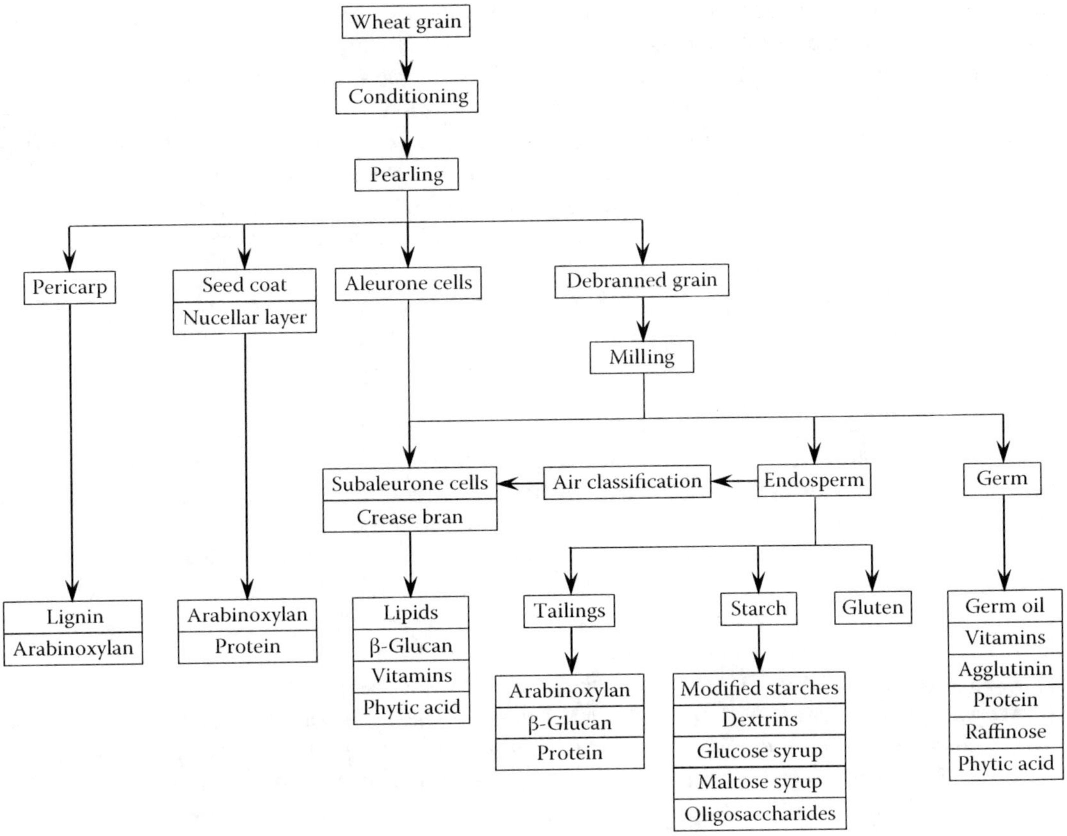

FIGURE 24.1 Relationship between wheat processing and value-added products.

TABLE 24.2
Constituents of Dietary Fiber

Nonstarch polysaccharides and resistant oligosaccharides: cellulose, hemicellulose (arabinoxylans, arabinogalactans), polyfructoses (inulin, oligofructans), galactooligosaccharides, β-glucans, mucilages, pectins

Analogous carbohydrates: indigestible dextrins (resistant maltodextrins, resistant potato dextrins), synthesized carbohydrate compounds (polydextrose, methyl cellulose, hydroxypropylmethyl cellulose), indigestible (*resistant*) starches

Lignin

Substances associated with the nonstarch polysaccharide and lignin complex in plants: waxes, phytate, cutin, saponins, suberin, tannins

Source: De Vries, J.W., *Cereal Foods World*, 46, 112, 2001.

human small intestine with complete or partial fermentation in the large intestine. Dietary fibre includes polysaccharides, oligosaccharides, lignin, and associated plants substances. Dietary fibres promote beneficial physiological effects including laxation, and/or blood cholesterol attenuation, and/or blood glucose attenuation." Isolation of dietary fiber from wheat will provide marketable fiber supplement and increase wheat utilization and market value.

As included in the definition of dietary fiber, it is nowadays regarded that its consumption could lead to weight loss, prevention of obesity (Mahta 2009), and reduction of the risk for cancer and heart disease (Fernandez 2001; McCann et al. 2001). Kahlon and Chow (2000) reported that binding of bile acid by wheat bran, among other cereal brans including oat, rice, and corn, contributes to cancer prevention and other healthful properties (Kahlon and Chow 2000). Furthermore, dietary fiber promotes gastrointestinal health and relief of constipation.

Dietary fiber can be divided into water-soluble and water-insoluble fractions that contain different components and provide different health benefits. Water-soluble fiber contains mainly nonstarch polysaccharides, such as β-glucan and arabinoxylan, as well as inulin, fructo-oligosaccharides, polydextrose, and resistant starches and maltodextrins. It can increase viscosity in the intestine reducing the rate of intestinal transit, delaying gastric emptying, and reducing glucose and sterol absorption. On the other hand, water-insoluble fiber contains lignin as well as nonstarch polysaccharides including water-unextractable arabinoxylan. Lignin is a lipophilic polymer that can absorb bile acids. The structural components of lignin are phenolic compounds that exert a variety of physiological effects.

The unbranched polysaccharides of β-glucan, consisted of (1→4)- and (1→3)-linked β-D-glucopyranosyl units in varying proportions, are important components of dietary fiber. It is recognized that β-glucan contributes health benefits such as blood serum cholesterol lowering capabilities, hypoglycemic effects assisting in the metabolic control of diabetes, and reduction of the glycemic index (Wood 1993; Colleoni-Sirghie et al. 2003; Maki et al. 2003). High molecular weight β-glucans are viscous due to labile cooperative associations, whereas lower molecular weight β-glucans can form soft gels as the chains are easier to rearrange to maximize linkages. Wheat bran contains low quantities of β-glucans as compared to other cereals grains such as oats.

Arabinoxylan can be isolated in both water-extractable and water-unextractable forms. It can be found in cereal tissues and particularly, wheat bran contains considerable quantities of arabinoxylan. Arabinoxylans are nonstarch polysaccharides with a linear backbone of xylose units linked with β-(1→4) bonds that are highly substituted with arabinose units linked with primarily single α-(1→3)- and/or α-(1→2) bonds as well as arabino-oligosaccharides, D-galactose, D-4-*O*-methylglucuronic acid, and ferulic acid residues. They can be found in both water-extractable and water-unextractable forms in cereal grains and tissues, and they are nowadays considered as an important dietary fiber.

Among all cereal grains, wheat contains moderate amount of dietary fiber (Nelson 2001), which is mainly distributed in the bran (Table 24.3). From outer pericarp to nucellus, crude fiber, cellulose, and arabinoxylan occur in decreasing amounts, except arabinoxylan in nucellus. The procedures for the isolation and purification of dietary fiber (Burley and Blundell 1995) and the techniques involved in their quantitative and structural analysis (Asp et al. 1992) were developed for the

TABLE 24.3
Composition of Soluble and Insoluble Dietary Fiber in Wheat and Conventional Wheat Milling Products

	Soluble			Insoluble				
Material	Neutral Sugars	Uronic Acids	Total	Neutral Sugars	Uronic Acids	Klason Lignin	Total	Total Fiber
Whole-wheat flour[a]	2.7	0.1	2.8	7.9	0.2	N/A	8.1	10.9
White flour[a]	1.0	0.1	1.1	2.0	0	0.2	2.2	3.3
Bran[a]	2.9	0.3	3.2	37.5	0.9	N/A	38.4	41.6
Germ[b]	1.0	0.1	1.1	11.5	0.8	1.3	13.6	14.7

[a] Cummings and Englyst (1987).
[b] Marlett (1992).

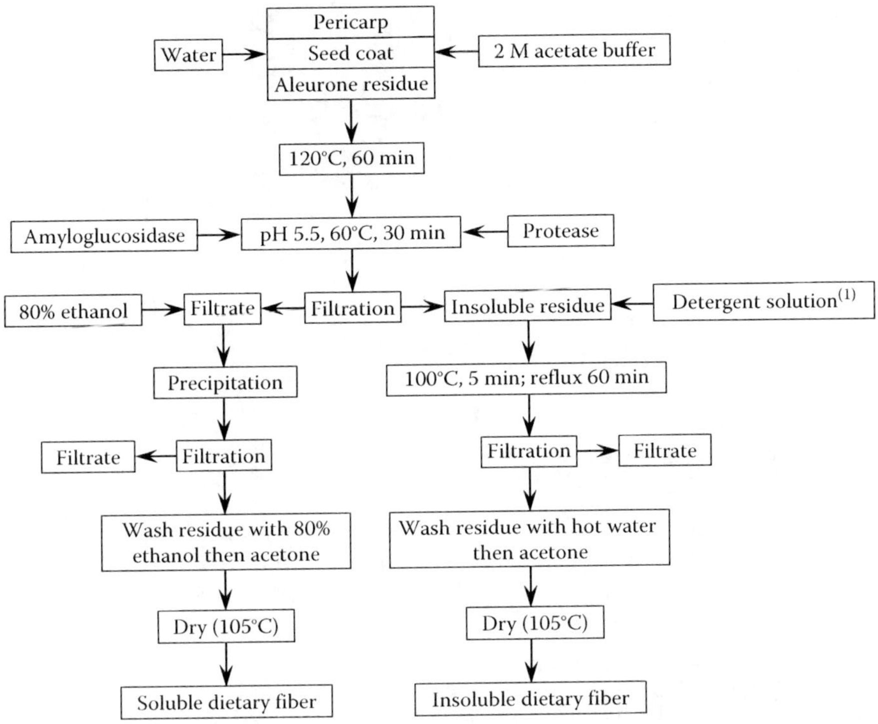

FIGURE 24.2 Flow diagram for the isolation of dietary fiber from wheat bran constituents. (1) Neutral detergent solution: Dissolve 148.5 g sodium lauryl sulfate in 3 L water. Mix 92.12 g disodium ethylenediaminetetraacetic acid and 33.70 g sodium borate decahydrate with 1 L water, heat until dissolved, and add to first solution. Dissolve 22.57 g anhydrous disodium hydrogen phosphate in 1 L water and add to first solution. Final pH of the solution: 6.9–7.1.

isolation of dietary fiber from conventional milling products. The combination of the technique of wheat debranning and the simplified milling process will produce processing streams of more specified botanical constituents such as the outer pericarp, the inner pericarp, the seed coat, the aleurone cells, the embryo, and the starchy endosperm. Targeting at particular dietary fiber in each of these constituents according to fiber distribution in wheat grain, these isolation procedures would be simplified and their products more purified.

The enzymatic gravimetric procedure presented in Figure 24.2 that was originally developed by Asp et al. (1983) could be modified for the isolation of total dietary fiber from wheat bran. The replacement of ethanol with dioxane may modify this procedure to extract dioxane-extractable lignin from pericarp tissues. Treatment of other wheat bran tissues with alkaline solutions could lead to lignin extraction (Schwarz et al. 1989). The procedure presented in Figure 24.3 could be employed for the isolation of water-soluble and water-insoluble pure arabinoxylans from wheat flour (Yeh et al. 1980; Faurot et al. 1995). Water-insoluble arabinoxylans could be produced by sequential treatment of the insoluble residue with proteolytic enzymes and α-amylase. Soluble proteins could be removed via precipitation with $Ba(OH)_2$ or $ZnSO_4$ before starch hydrolysis with α-amylase (Nandini and Salimath 2001).

Research on the extraction and analysis of β-glucan from wheat is not extended as in the case of β-glucan contained in barley and oats. Wheat contains comparatively lower quantities of β-glucan (0.5%–2% w/w, on a dry basis) compared to oats and barley (2%–7% w/w, on a dry basis) (Miller et al. 1993; Wood 1994; Havrlentová and Kraic 2006; Tiwari and Cummins 2009). There is varying level of difficulty on the ability to extract β-glucan from different cereal grains (Wood 2010). In general, β-glucan extraction from cereal grains or individual tissues can be accomplished with various

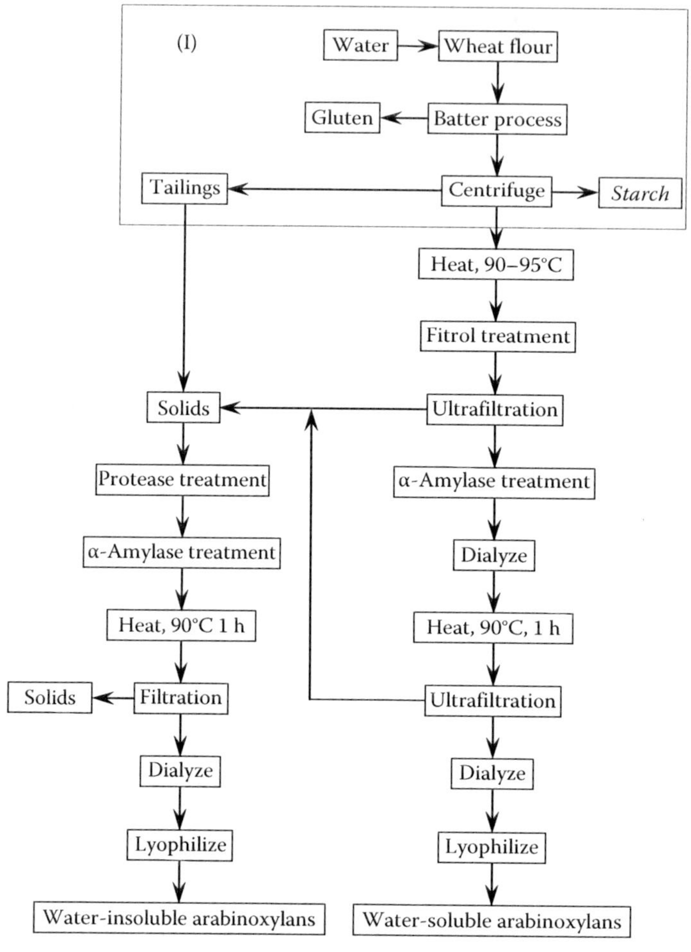

FIGURE 24.3 Flow diagram for the isolation of arabinoxylans from wheat flour and starch tailing.

methodologies including treatment with Na_2CO_3, NaOH, hot water, or enzymes. Hot water can be employed to extract β-glucan from oat and barley in high yield and purity (Ragaee et al. 2008). Li et al. (2006) reported that fractionation of high purity wheat β-D-glucan (91.58%) could be accomplished from white wheat bran after alkali extraction and multiprecipitation with ammonium sulfate.

24.2.3.2 Phytic Acid

Phytic acid (inositol hexaphosphate) is a 6-phosphate ester of inositol (IP6). Phytic acid stores 60%–90% of phosphorus contained in plants in the form of phytin. Wheat bran is the main location of phytic acid in the wheat kernel. However, its content in wheat brans varies considerably. Garcia-Estepa et al. (1999) reported that the phytic acid content in wheat brans was in the range of 25–58 mg/g. Furthermore, phytic acid content in hard and soft wheat brans was 42–54 and 46–67 mg/g, respectively (Camire and Clydesdale 1982; Knuckles et al. 1982). Phytic acid cannot be digested by nonruminants, and the utilization of bran-rich wheat milling by-products as animal feed results in high phytic acid content in animal waste. The removal of phytic acid from wheat bran will result in the production of a value-added coproduct and the reduction of phosphorus pollution when wheat bran is used as a generic animal feed.

Phytic acid contained in wheat bran contributes favorably to the reduction of early biomarkers of colon carcinogenesis (Jenab and Thompson 2000). Fox and Eberl (2002) stressed that there is strong evidence based on studies carried out on animals that phytic acid may have a role in both the prevention and treatment of many forms of cancer including breast, colon, liver, prostate, rhabdomyosarcoma, and skin. Numerous mechanisms of action have been proposed including the increase in natural killer of cell activity, alteration in signal transduction, stimulation of genes toward greater cell differentiation, and antioxidant activity (Fox and Eberl 2002). Phytic acid is well known for chelating multivalent metal ions (e.g., magnesium, calcium, zinc, iron) causing possible antinutritional effects because the complexes formed are highly insoluble resulting in reduced absorption of these metals in the gastrointestinal tract (Garcia-Estepa et al. 1999). However, this is a controversial issue because other studies (Ullah and Shamsuddin 1990) have shown that long-term administration of phytic acid does not lead to demonstrable removal of metals via chelation. The antioxidant properties of phytic acid render it a useful preservative in the food industry, such as fish paste products and oils (Graf 1983). Phytic acid stabilizes both natural and artificial coloring agents including the prevention of discoloration of caviar and fresh vegetables (Graf 1983). Phytic acid can be used to improve protein analysis by capillary electrophoresis (Veraart et al. 1997) and coagulate soybean protein improving the texture and flavor of tofu (Graf 1983). It has been used to reduce the duration of fermentation processes. Furthermore, phytic acid has been extensively applied in the surface protection of metals due to its chelating ability with many metals (Yang et al. 2005). Phytic acid provides good resistance to corrosion, and phytic acid pretreatment is used as a substitute for toxic chromate pretreatment to improve the corrosion resistance of organic coating on metal and alloy surfaces (Cui et al. 2012). Phytic acid could also be used in oral care and in the cosmetic industry.

The examples given previously indicate the potential utilization of phytic acid in various industrial fields including food, chemical, medical applications, and cosmetics. Thus, phytic acid isolation from wheat bran could lead to the generation of a value-added product with many market outlets. Phytic acid purification from various plant sources can be carried out by the treatment with either dilute organic acids (e.g., formic, trichloroacetic, lactic, oxalic, citric acids) or dilute mineral acids (e.g., hydrochloric and nitric acids). Isolation of a crude precipitate enriched in phytic acid can be accomplished via neutralization and subsequent treatment with various alkaline solutions. Further purification of phytic acid from the crude precipitate is possible. Deactivation of potential phytase before phytic acid extraction may be necessary via heat treatment to prevent hydrolysis of phytic acid. These processes could be also adapted to phytic acid separation from wheat bran.

24.2.3.3 Ferulic Acid

Ferulic acid is contained in wheat bran in higher quantities (up to 0.34 mg/g total steryl ferulates) when compared to whole-wheat flour (approximately 0.063 mg/g) (Hakala et al. 2002). The purification of ferulic acid from wheat bran is mainly dependent on the breakage of ester bonds to other cell wall components, including mainly arabinoxylans, lignin, and cellulose. Wheat bran could be used for the extraction of ferulic acid using ferulic acid esterase and xylanase (Barberousse et al. 2009). The antioxidant capacity of a purified fraction enriched in ferulic acid produced from wheat bran by the combined action of ferulic acid esterase and xylanase resulted in a better DPPH scavenging efficiency than the ferulic acid standard (Barberousse et al. 2009). Extrusion followed by treatment with plant cell wall hydrolyzing enzymes led to the separation of ferulic acid and arabinoxylan from cereal brans (Hwang et al. 2001). Ferulic acid and its derivatives, steryl ferulates, are compounds with antioxidant properties that can be in various food and nonfood applications including skin lotions, sunscreens, and wound dressings (Sancho et al. 2001). Ferulic acid from wheat bran has been also evaluated for the microbial production of the value-added fine chemical vanillin (Di Gioia et al. 2009).

24.2.3.4 Wheat Germ Oil and Associated Components

The whole-wheat grain contains 2%–4% lipids that are mainly contained in wheat germ (8%–14% on a weight basis). Dissected germ, free of endosperm and bran, contains 25%–30% acyl lipids, as determined by acid hydrolysis (Pomeranz 1988). Dunford (2005) estimated that approximately 10 million tonnes of wheat germ can be produced worldwide from conventional wheat milling operations considering that wheat germ is approximately 2% of the whole-wheat grain. Wheat germ oil contains approximately 80% of unsaturated fatty acids, mostly linoleic (18:2) and linolenic (18:3) acids (Wang and Johnson 2001), that are considered of significant importance in human metabolism.

Wheat germ oil can be used as a high specialty food supplement due to the high content of bioactive molecules (e.g., tocopherols, octacosanols, phytosterols) and a wide spectrum of cosmetics and personal care products. It is rich in α and β tocopherols, and in particular, it is considered as one of the richest natural sources of D-α-tocopherols (vitamin E) that has much higher biological activity than the commercially available synthetic vitamin E product (Dunford 2005). Furthermore, wheat germ oil also contains octacosanol, the chemical formula of which is $CH_3(CH_2)_{26}CH_2OH$. There are indications that octacosanol could be used as supplement to improve athletic performance and reduce blood cholesterol among other potential benefits (Taylor et al. 2003). Phytosterols is a major fraction (approximately 35%) of the unsaponifiable content in wheat germ oil (Dunford 2005). Phytosterols may reduce serum or plasma total cholesterol and low density lipoprotein cholesterol levels (Ling and Jones 1995). In addition, phytosterols may offer a wide spectrum of other therapeutic effects including protection from the most common cancers in Western societies, such as colon, breast, and prostate cancer (Awad and Fink 2000).

Oil extraction from wheat germ has been carried out commercially via mechanical pressing and solvent extraction, while extraction using supercritical fluids is also possible, but this technology has not been applied commercially yet (Dunford 2005). Mechanical pressing extracts only 50% of the total oil content and is usually employed when wheat germ of high purity and low bran content is used. Solvent extraction of oil from wheat germ, mainly with hexane, is the most widely used method. Refining of the extracted oil, targeting the removal of phospholipids via degumming, free fatty acids, and odor compounds, could be applied if the quality and stability of the oil should be improved. Supercritical fluid extraction can be used as an alternative to solvent extraction to either extract or fractionate wheat germ oil for the production of high-quality products (Dunford 2005). Gomez and De la Ossa (2000) reported that extraction of wheat germ oil via supercritical CO_2 (150 bar pressure, 40°C temperature and 1.5 L/min CO_2 flow rate) could be achieved at similar yields as the oil extracted with hexane, but at higher quality resulting in a cost-competitive process. Under more precisely controlled conditions, further fractionation of tocopherols or other high-value components could be isolated from wheat germ or crude wheat germ oil as it has been demonstrated in other cases, such as olive pomace (Ibanez et al. 2000).

The pearling technology could be employed to increase the purity of germ-rich fractions generated during wheat milling. Conventional milling operations with no special germ-separation equipment achieves the recovery yields of food-grade germ of only 0.5%, while with appropriate equipment the yield increases to 1.5%–2% (Dunford 2005).

24.2.3.5 Oligosaccharides

Oligosaccharides have attracted significant attention as health-promoting substrates that could be used as additives in the food industry due to their functional properties including prebiotic properties, low caloric content, alleviation of constipation, enhancement of mineral absorption, reduced risk of infection and diarrhea, and improved immune response (Boler and Fahey 2012). As oligosaccharides are defined as carbohydrates with low number of monosaccharides, either 2–20 or no more than 10 depending on the authority that issued the specific definition (IUB-IUPAC 1982; Food and Drug Administration 1993). The oligosaccharides that cannot be digested or absorbed in the upper gastrointestinal tract and may be fermented by resident microbial population in the large intestine are called *nondigestible* oligosaccharides. Specific nondigestible oligosaccharides are called prebiotics when they can be fermented in the colon by beneficial indigenous microorganisms to enhance the health of the host. To date, the only nondigestible polysaccharides that are proved to be potent prebiotics are fructans, galactooligosaccharides, and lactulose (Boler and Fahey 2012). Oligosaccharides with prebiotic properties are effective in promoting the growth and activity of bifidobacteria and lactic acid bacteria in the large intestine that results in growth inhibition of pathogenic bacteria due to production of short-chain fatty acids, reduction of colonic pH, production of antimicrobial compounds, and/or competition for growth substrates and adhesion sites (Manderson et al. 2005). Oligosaccharides are used as bifidogenic substances in some infant products in the hope that these compounds may provide some of the benefits attributed to oligosaccharides in human milk (Rivero-Urgell and Santamaria-Orleans 2001). The ESPGHAN Committee on Nutrition recommends further evaluation of oligosaccharide supplementation in infancy as a prophylactic or therapeutic measure (Agostoni et al. 2004).

Wheat contains variable quantities of oligosaccharides that could be divided in two groups: (1) galactosyl derivatives of sucrose (raffinose series) and (2) fructosyl derivatives of sucrose (fructan series) (Henry and Saini 1989). The exact distribution of oligosaccharides in the wheat kernel is not fully identified. However, literature-cited reports suggest that oligosaccharides are distributed in all milling products (MacMasters et al. 1971; Yamada et al. 1993). In particular, wheat germ contains significant quantities of raffinose family oligosaccharides (7.2% on a dry basis).

Xylo-oligosaccharides (XOS) have been suggested as potential prebiotics and food ingredients promoting gastrointestinal health and growth of various lactic acid bacteria and bifidobacteria (Madhukumar and Muralikrishna 2012). XOS consist of xylose units linked with β-1,4 bonds and can be produced using various natural sources of hemicelluloses via enzymatic hydrolysis (e.g., with xylanase), chemical treatment (e.g., hydrothermal or dilute mineral acid treatment), or a combinations of these methods (Vasquez et al. 2000). Wang and Lu (2013) applied repeated microwave-assisted enzymatic hydrolysis combined with activated carbon adsorption to produce 3.2 g of purified dry XOS from 50 g dry wheat bran powder. Depending on the type of hemicelluloses, branched XOS could be produced linked with arabinose units, at varying ratios with xylose, or other residues such as uronic acids (Vasquez et al. 2000). Arabinoxylan oligosaccharides (AXOS) produced from wheat bran have been evaluated as potential prebiotics selectively increasing fecal bifidobacteria (Maki et al. 2012). Francois et al. (2012) reported that intake of 10 g wheat bran extracts containing arabinoxylan oligosaccharides on a daily basis exerts beneficial effects on gut health parameters.

Research is also focused on the development of cost-effective processes for the production of oligosaccharide formulations with prebiotic properties (Manderson et al. 2005). The utilization of by-products from current food processes or future biorefineries is a promising option. However, the production of specific oligosaccharide formulations and their association with other macromolecules that should be removed increases the difficulty for the development of cost-competitive processes. In the case of wheat processing, water-soluble oligosaccharides could be produced from wheat flour by exploring their water solubility during, for instance, the water washing process commonly applied for gluten separation and the formation of a starch slurry. After the removal of starch by centrifugation, the residual paste could be used for the production of oligosaccharides. In addition, water-soluble and water-insoluble arabinoxylans produced by the process presented in Figure 24.3 could be employed for the production of XOS or AXOS via enzymatic treatment. Furthermore, separation of oligosaccharides from both the wheat germ residue after oil extraction and the aleurone layer harvested from debranning might be developed based on the solubility of oligosaccharides in 80% ethanol solution (Henry and Saini 1989). Purification of oligosaccharide concentrates could be achieved by various methods, such as high-performance liquid affinity chromatography (Zopf et al. 1989).

24.2.3.6 Starch

Starch contains two types of polymers, amyose consisting of linear α-(1→4)-linked glucose units and amylopectin that is a branched polymer consisting of glucose units linked by α-(1→4) and α-(1→6)-glucosidic linkages. Wheat starch contains approximately 25% amylase and 75% amylopectin.

Conventional dry milling operations have been developed mainly aiming to isolate the starch-rich endosperm from bran and germ in the form of flour that contains more than 80% starch on a dry basis.

The industrial separation of starch and gluten from wheat flour was initially carried out by the Martin process that commenced in the 1830s (Kerkkonen et al. 1976). The Martin process generally begins with the formation of a dough by mixing wheat flour (10 parts) with water (6–7 parts) and then separates starch from the gluten matrix in the resulted mixture through washing the dough with water. In the Martin process, the quantity of water usage is usually 6–10 times of the quantity of flour used, while in the same time up to 20% of the dry content of flour is lost in the diluted wastewater (Kerkkonen et al. 1976). Waste effluents can be processed via anaerobic digestion or concentrated via evaporation and used in fermentation processes. Other processes, such as the modified Fesca process and the *batter process*, were developed in the twentieth century with a common target to reduce water requirements through the formation of a free-flowing smooth slurry or batter at the beginning of both processes (Anderson et al. 1958; Johnston and Fellers 1971). Traditional processes are based on the formation of a dough or batter with a developed gluten matrix where starch separation from gluten is achieved via kneading and/or screening unit operations (Sayaslan et al. 2010). Several modified versions of these processing schemes are nowadays available, including the Posner, Alfa-Laval/Raisio, hydrocyclone, Pillsbury Hydromilling, Far-Mar-Co., and high-pressure disintegration (Jansma et al. 1985; Czuchajowska and Pomeranz 1995; Sayaslan 2004; Peighambardoust et al. 2009; Sayaslan et al. 2010). Modern processes are mainly based on the initial formation of a sheared flour-water or dough-water dispersion with partially developed gluten, and starch is separated from gluten via hydrocyclone or centrifugation (Sayaslan et al. 2010).

Starch has many conventional applications in various industrial sectors, predominantly the food and chemical industries (Table 24.4). Starch hydrolysis expands the industrial uses of this versatile raw material even further. Starch hydrolysis products are classified according to their dextrose equivalent (DE) number that describes the degree of starch hydrolysis as it represents

TABLE 24.4
Industrial Application of Starch

Industries	Applications
Food	Mayonnaise, baby food, bread, buns, meat sausages, meat rolls and loaves, ketchup, soups, snacks, pizza sauces, sauces, low-fat foods, noodles
Beverage	Soft drinks, beer, alcohol, coffee
Confectionery	Jelly gums, high-boiled sweets, jellies, marshmallows, marmalade, jam, ice cream, dairy cream fruit fillings
Animal feed	Pellets, by-products
Agriculture	Seed coating, fertilizer
Fermentation	Vinegar, enzymes
Plastic	Biodegradable plastic
Textile	Warp, fabrics, yarn
Nonwoven	Hygienic diapers, baby diapers, sanitary napkins
Pharmacy	Tablets, dusting powder
Paper	Corrugated board, cardboard, paper, printing paper, packaging material
Building	Mineral fiber tiles, gypsum board, concrete, gypsum plaster
Various	Foundries, water treatment, coal, detergent, oil drilling stain remover, glue, foamed starch

Source: Evers, A.D. and Bechtel, D.B., Microscopic structure of the wheat grain, in: *Wheat: Chemistry and Technology*, ed. Y. Pomeranz, pp. 47–95, Vol. I., St. Paul, MN: AACC, 1988.

the percentage of the quantity of reducing sugars present in a starch hydrolysate as related to the reducing power of pure dextrose (D-glucose). In this respect, glucose has a DE value of 100, while nonhydrolyzed starch has a DE value of zero. Enzymatic hydrolysis of starch has nowadays almost entirely replaced acid hydrolysis in the industrial sector leading to products with varying DE values, for example, dextrins (DE lower than 10), maltodextrins (DE 10–20), various glucose syrups (DE 20–80), starch hydrolysates (DE higher than 80), pure dextrose, and high-fructose syrup.

Industrial hydrolysis of starch into glucose commences with liquefaction using α-amylase produced from either *Bacilli* or *Aspergilli* that cleaves only α-1,4-D- and not 1,6-α-D-glucosidic bonds. The utilization of α-amylase produced by *Bacillus licheniformis* enabled simultaneous starch liquefaction and gelatinization as this enzyme can be used at temperatures up to 110°C. Starch liquefaction usually leads to the production of approximately 8–12 DE values. In the subsequent saccharification stage, glucose is produced from dextrins using mainly amyloglucosidase coupled with pullulanase that cleaves 1,6-α-D-glucosidic linkages. In a biorefinery concept, starch hydrolysates will be used in fermentation processes for the production of platform chemicals, biofuels, and biodegradable polymers.

24.2.3.7 Wheat Gluten

Wheat gluten can be separated from wheat flour by the wet milling processes presented in the previous section. Commercial wheat gluten contains mainly protein (77.5% w/w, on a dry basis), moisture (6.4%), ash (5.7%), ash (0.7%), fiber, and starch (Maningat et al. 1994). The unique physical properties of gluten, including thermosetting properties, film-forming ability, and viscoelasticity, have resulted in the application of gluten in many food industries. Wheat gluten can be employed in the production of edible films and coatings that can serve as moisture and oxygen barriers and extend the shelf life of foods (Maningat et al. 1994; Mojumdar et al. 2011). In wheat-based biorefineries, gluten will be produced in high quantities and novel applications in nonfood industrial sectors should be developed. Maningat et al. (1994) reported that wheat gluten and its derivatives could be used for the production of adhesive tapes, cigarette filters, and pharmaceutical tablets. Nordqvist et al. (2012) reported that wheat gluten constitutes a potential renewable resource for the production of sustainable wood adhesives. Woerdeman et al. (2004) reported that incorporation of thiol-containing molecules into wheat gluten could lead to the production of a modified gluten biopolymer with modified properties that could be used for the production of biodegradable high-performance engineering plastics and composites. Derivatives of wheat gluten (e.g., succinylated wheat gluten) could be applied in the production of paper and cardboard coatings (Kersting et al. 1994). Wheat gluten is also a promising renewable resource for the production of thermosetting materials for the production of high-performance products for the industry (Raquez et al. 2010). Low-cost complex media derived from wheat, including gluten hydrolysates, could be used as alternative fermentation media to conventional LB media for the production of EAK16 peptide as fusions with SUMO (Satakarni et al. 2009).

24.2.4 Whole-Wheat Biorefinery Concepts

An optimized wheat-based biorefinery should produce a range of products to ensure sustainability and economic viability. Therefore, biorefinery concepts based on wheat should utilize mainly the starch fraction as carbon source and part of the protein fraction as nitrogen source in fermentation processes, whereas remaining protein and other components (e.g., arabinoxylans, germ oil, and ferulic acid) should be used as valuable coproducts. In the University of Manchester (United Kingdom), Professor Colin Webb's research group at the SCGPE has developed various wheat-based biorefinery schemes (Figure 24.4). Wheat bran could be utilized for the production of various products reported in previous sections. Wheat starch is hydrolyzed by enzymes, produced during fungal fermentations of *Aspergillus awamori*, to generate a glucose-rich

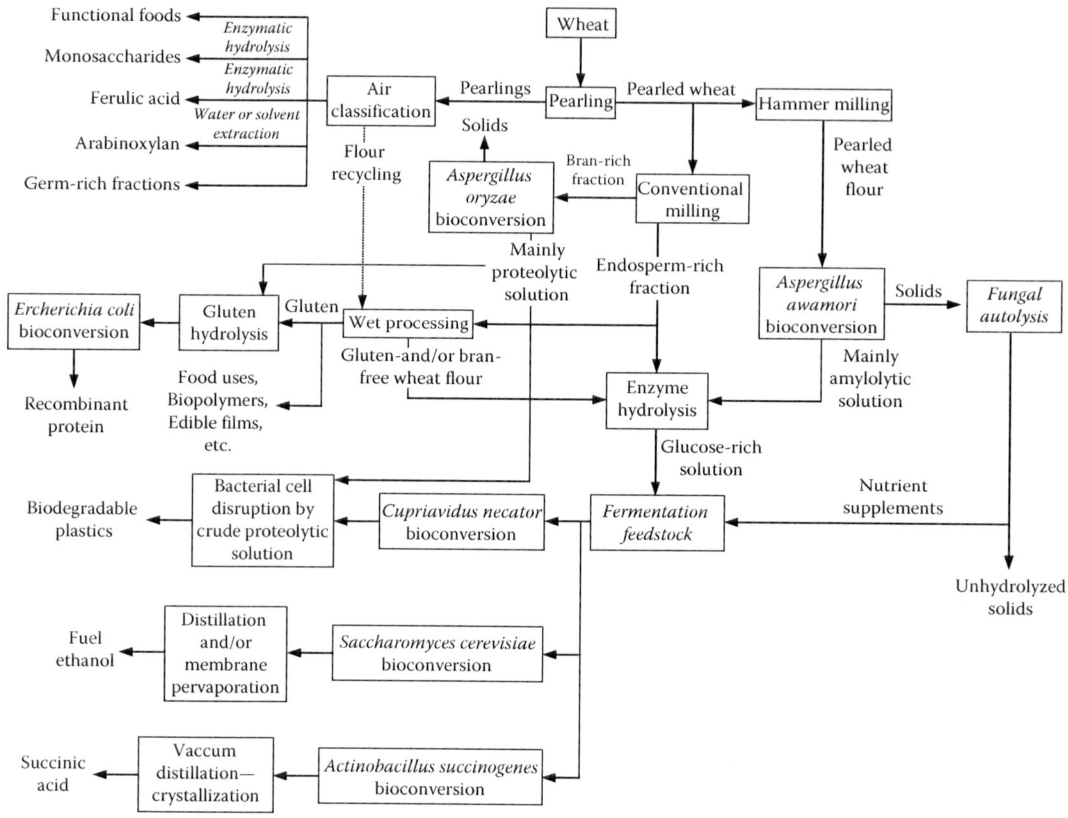

FIGURE 24.4 Wheat-based biorefinery approaches developed in SCGPE. (From Koutinas, A.A. et al., *Biofuels Bioprod. Biorefin.*, 1, 24, 2007a.)

fermentation medium (Koutinas et al. 2004, 2007b). Fungal cultivation on whole-wheat flour, pearled wheat flour, or wheat milling by-products can be carried out both in submerged and solid-state fermentations (Du et al. 2007, 2008). Submerged fermentations of *A. awamori* produced an enzymatic consortium containing glucoamylase, protease, and phosphatase (Koutinas et al. 2007b). The use of crude broth filtrates, for the production of wheat flour hydrolysates, resulted in the simultaneous hydrolysis of wheat protein and phytic acid, increasing free amino nitrogen (FAN) and phosphorous concentrations (Du et al. 2007). A solution rich in peptides, phosphorous, amino acids, and vitamins can also be obtained by a fungal autolysis process (Koutinas et al. 2005). Glucose- and nitrogen-rich media, combined in appropriate proportions, provide a generic microbial feedstock ideal for the growth of many microorganisms that can produce a wide spectrum of metabolites including bioethanol, polyhydroxybutyrate (PHB), and succinic acid (Koutinas et al. 2007a,b). In the following sections, various wheat-based biorefineries will be presented.

24.2.4.1 Wheat Grain as Raw Material

Bioethanol production in EU countries was increased 11-fold in the period from 2000 to 2009 (Figure 24.5), and the main raw material used was wheat (Figure 24.6). In the conventional bioethanol production process from wheat, gluten and bran are not separated from starch, and this is the main reason for the protein and fiber content in the dried distillers grains with solubles (DDGS) that constitutes the main by-product of the conventional process. DDGS is currently used as animal feed, but increasing quantities of wheat utilization for bioethanol production will oversaturate this market, which creates

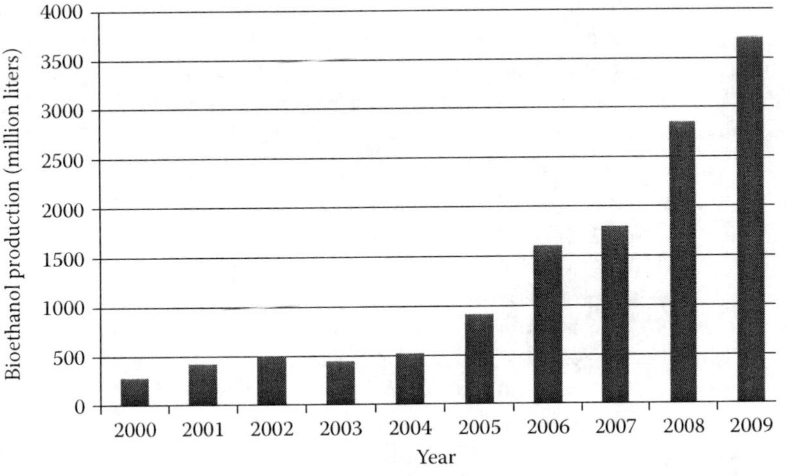

FIGURE 24.5 EU bioethanol production from 2000 to 2009. (Courtesy of Renewable Fuels Association, Washington, DC, http://ethanolrfa.org, visited on 07-01-2013.)

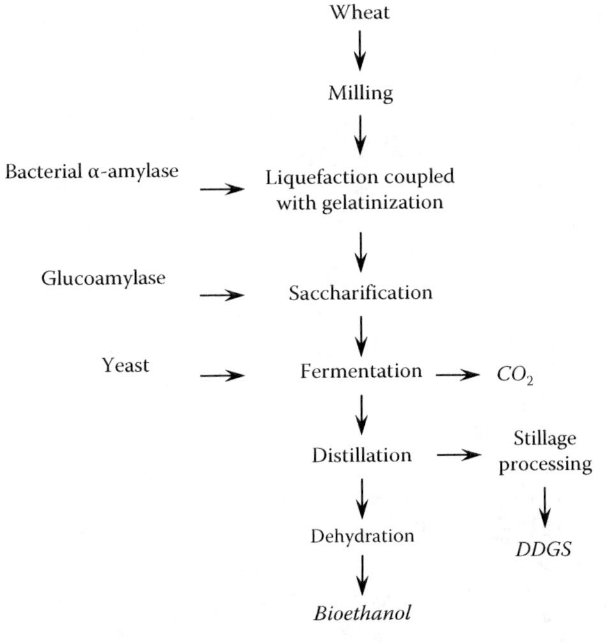

FIGURE 24.6 Schematic diagram of a typical first-generation wheat-based bioethanol production process.

the need for novel uses for DDGS or development of novel wheat-based biorefineries. DDGS could be incorporated into a poly-(butylene adipate-*co*-terephthalate) matrix to create green composites with enhanced biodegradability (Muniyasamy et al. 2013). DDGS could be used for the production of XOS that may exhibit beneficial health properties as functional food ingredients (Samala et al. 2012). Oil extracted from DDGS has been evaluated for the production of biodiesel (Moser and Vaughn 2012). The CO_2 could also be considered as an additional coproduct capable to generate additional income since it could be supplied to greenhouses or utilized in fermentation processes where CO_2 fixation increases product formation as in the case of succinic acid production (Nghiem et al. 2010). Conventional bioethanol production processes from wheat or other cereal grains will not be described in this chapter as they have been reviewed in other publications (McAloon et al. 2000; Smith et al. 2006).

A novel wheat-based upstream processing concept has been developed in SCGPE (Figure 24.4) to generate fermentation media for bioethanol production and valuable coproducts in order to improve process economics (Arifeen et al. 2007a). This process was developed on continuous mode and included on-site enzyme production, integration of starch hydrolysis and fungal autolysis, and production of gluten and bran as value-added coproducts. This upstream process could produce 120 m^3/h wheat-based fermentation media at a unitary cost of \$0.126 per kg glucose (Arifeen et al. 2007a). This fermentation medium contained glucose and FAN concentrations of 250 and 0.85 g/L, respectively. However, this cost estimation study was based on a wheat cost of \$0.16 per kg, which is nowadays considered low as the cereal grain prices have increased. For instance, in December 2012, the wheat price in the United Kingdom was increased to £0.227 per kg. The increase of wheat price significantly affected the wheat-based biorefining industry, especially the bioethanol plants in the EU.

Arifeen et al. (2007b) utilized this upstream process, in combination with yeast fermentation with recycling, ethanol concentration (up to 40 mol%) via a pervaporation membrane system, and pressure swing distillation (PSD) for the production of fuel-grade ethanol, to optimize the proposed wheat-based biorefinery system using the software General Algebraic Modelling System (GAMS). At the best scenario discussed by Arifeen et al. (2007b), the ethanol production cost could be as low as \$0.96–0.50 per gal ethanol (\$0.25–0.13 per L ethanol). In the current high cereal grain price market, the cost of wheat-derived bioethanol would be inevitably higher than the cost estimated in this study.

Arifeen et al. (2007b) demonstrated that if suitable by-products could be generated, then the bioethanol production cost could be decreased significantly. Arifeen et al. (2009) reported that higher sugar to ethanol conversion yields, glucose consumption rate, and ethanol production rate are achieved when gluten is extracted from wheat flour before the production of fermentation media.

From 2003 to 2005, the effect of grain protein content on bioethanol production was evaluated to demonstrate that alcohol yields increased by approximately 7 L/dry tonne for every 1% decrease in grain protein related inversely to grain protein content (Anonymous 2007). This observation was consistent with direct replacement of protein by starch (Anonymous 2007). Therefore, gluten extraction is beneficial not only because it improves the cost-competitiveness of the process, but also because it improves bioethanol production efficiency. As mentioned earlier, Satakarni et al. (2009) reported that gluten hydrolysates could be employed as alternative media to commercial media formulations (i.e., LB media) for the production of recombinant proteins (i.e., intracellular SUMO-EAK16 fusion protein).

Apart from gluten, arabinoxylan is another valuable coproduct that could be extracted from wheat bran to improve the overall economics of bioethanol production (Du et al. 2009; Misailidis et al. 2009). The extraction of arabinoxylan in a bioethanol production process is benefited also by the fact that its extraction could be achieved with ethanol. Arabinoxylan extraction from wheat bran has been evaluated by Du et al. (2009), and the cost of the AX-rich fraction produced (80% purity) was approximately £3.7–4.5 per kg that could reduce the ethanol selling price (Misailidis et al. 2009).

Researchers at the SCGPE have also developed wheat-based biorefineries for the production of succinic acid and PHB (Du et al. 2008; Dorado et al. 2009; Xu et al. 2010). Fermentations carried out in wheat-derived media with the microbial strain *Actinobacillus succinogenes* resulted in the production of 64 g/L succinic acid when $MgCO_3$ was used (Du et al. 2008). Lin et al. (2011) utilized seawater and wheat-derived media in fermentations with *A. succinogenes* to produce 49 g/L succinic acid with a yield of 0.94 g/g and a productivity of 1.12 g/(L·h). Xu et al. (2010) reported that fed-batch fermentations with only wheat-derived media led to the production of 162.8 g/L PHB at a productivity of 0.89 g/L/h.

24.2.4.2 Wheat Straw Biorefinery

Wheat straw is an abundant agricultural residue that can be used as feedstock for the production of fuels, energy, and chemicals through the development of second-generation biorefineries.

Deswarte et al. (2008) reported that approximately 4 out of 10 million tonnes of wheat straw produced in the United Kingdom on an annual basis have no commercial market. The main components of wheat straw are cellulose (31.5%–48.6%), hemicelluloses (22.6%–38.8%), lignin (5.3%–19%), crude protein (1.9%–5.7%), and ash (1.4%–10.2%) (Lee et al. 2007). Various sources of lignocellulosic biomass (including wheat straw) contain macromolecules (such as cellulose, hemicelluloses, and lignin) in high quantities and high-value molecules (such as waxes, sterols, and terpenoids) in low quantities. In a second-generation biorefinery, high-value molecules should be extracted from lignocellulosic biomass via solvent or supercritical fluid extraction prior to fractionation of macromolecules. For instance, supercritical carbon dioxide extraction has been applied for the extraction of waxes, polycosanol, and sterol in an integrated straw-based biorefinery (Clark et al. 2006; Dewarte et al. 2007).

Budarin et al. (2011) combined CO_2 extraction and low-temperature microwave pyrolysis to develop an integrated close to zero-waste wheat straw biorefinery that can produce a variety of products, including energy and CO_2, which can be internally recycled to sustain the processes. Secondary metabolites (e.g., fatty acids, wax esters, and fatty alcohols) can be extracted with CO_2. Higher-quality oils and chars can be produced with less energy by low-temperature microwave pyrolysis (<200°C) than conventional pyrolysis. The economic viability of a wheat straw biorefinery could be improved through integration of the technologies developed by Budarin et al. (2011).

After the extraction of high-value and low-volume compounds, major components of wheat straw could be fractionated or even isolated and utilized for the production of fuels, chemicals, materials, and energy. The isolation and characterization of macromolecules in wheat straw have been presented in previous publications (Sun et al. 1996, 1997, 2004). Various chemical pretreatment methods could be employed including alkali, dilute acid, steam explosion, organosolv process, and thermochemical pretreatment (Talebnia et al. 2010). Curreli et al. (1997) developed a pretreatment protocol based on a two-step mild alkaline/oxidative pretreatment on wheat straw, including treatment with 1% sodium hydroxide solution for 24 h to solubilize hemicelluloses in the first step and treatment with a solution of 1% sodium hydroxide and 0.3% hydrogen peroxide for 24 h in the second step, which achieved 81% lignin degradation and 90% cellulose recovery. Zeng et al. (2011) employed solid-state fermentation of *Phanerochaete chrysosporium* on wheat straw that resulted in the degradation of 25% of total lignin within 7 days, coupled with an approximately 250% higher efficiency on the release of sugars from wheat straw via enzymatic hydrolysis. Salvachua et al. (2013) optimized process parameters, including inoculum preparation, wheat straw particle size, moisture content, organic and inorganic supplementations, and mild alkali washing during solid-state fermentation, in biological pretreatment of wheat straw with the white-rot fungus *Irpex lacteus* to achieve glucose yields from wheat straw of 68% after 21 days as compared to yields of 62% and 33% for cultures grown without supplementation or on untreated raw material, respectively.

Cellulose and hemicelluloses fractions could be converted into C6 and C5 sugars for fermentation processes to produce biofuels, platform chemicals, and biopolymers using various enzymatic consortia. One of the challenges in second-generation biorefineries is to identify microbial strains that can consume efficiently both C5 and C6 sugars for fermentative production of value-added fuels, chemicals, and biopolymers. The lignin constituent is generally used for energy generation, but utilization as raw material for chemical and biopolymer production is currently a challenging research objective. Potential platform for chemicals and biopolymers have been reviewed in recent publications (Bozell and Petersen 2010; Rehm 2010; Jang et al. 2012).

Panagiotopoulos et al. (2013) investigated hydrogen production with the extreme thermophilic bacterium *Caldicellulosiruptor saccharolyticus* by mixing wheat straw with wheat grain hydrolysates. This approach could be beneficial for fermentative hydrogen production in a biorefinery. Fermentations carried out with mixtures of the two hydrolysates at sugar concentrations up to 20 g/L resulted in a hydrogen production of 82–97% compared to the hydrogen produced from the fermentations with pure sugars.

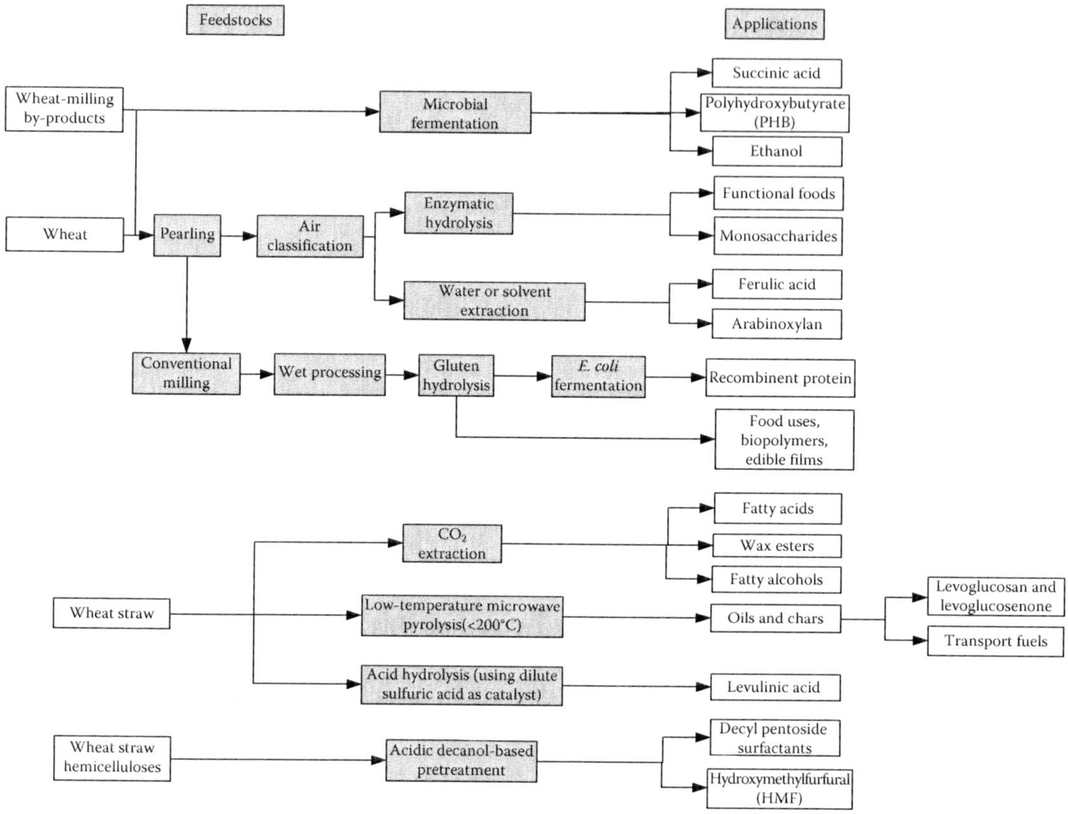

FIGURE 24.7 Whole-wheat biorefinery applications for chemicals, materials, and fuel production.

Chang et al. (2006, 2009) employed dilute sulfuric acid as a catalyst for the production of levulinic acid from wheat straw. Marinkovic et al. (2012) investigated the acidic decanol-based pretreatment of wheat straw, mainly targeting the conversion of hemicelluloses, through a single-step operation, into decyl pentoside surfactants. The remaining material could be subsequently used as potential feedstocks for fuel or platform chemical production (e.g., hydroxymethylfurfural). The wheat-based surfactants produced from this process exhibited superior surface properties compared to fossil-based polyethoxylates decyl alcohol or alkyl oligoglucosides. Han et al. (2010) evaluated wheat straw pretreatment with steam explosion in order to modify wheat straw characteristics, regarding morphology, acidity, wettability, and ash and silicon contents, for potential application in panel manufacturing.

Based on the results presented previously, Figure 24.7 presented potential biorefining concepts that integrate wheat grain and straw utilization.

24.3 OILSEED-BASED BIOREFINERIES

In 2010, the worldwide biodiesel production was higher than 16 million tonnes (Lamers 2012). The expected increase of biodiesel production will inevitably lead to proportional generation of by-products. The biodiesel industry using oilseeds as raw material will have to develop sustainable technologies to valorize by-product streams, namely, oilseed meals and crude glycerol generated during transesterification of vegetable oils as a 10% (w/w) by-product stream. Several recent reviews have been devoted to the utilization of crude glycerol for chemical production via fermentation or chemical conversion (Koutinas et al. 2007a; Chatzifragkou and Papanikolaou 2012; Yang et al. 2012).

24.3.1 Utilization of Rapeseed and Sunflower Meals

In 2012, 18.8 million tonnes of rapeseed (*Brassica carinata*), 6.95 million tonnes of sunflower (*Helianthus annuus*), and 0.95 million tonnes of soybean (*Glycine max*) were produced in EU-27 countries (Anonymous 2013). In 2012, the worldwide production of soybean, rapeseed, and sunflower meals (SFM) were 269.5, 59.3, and 36.37 million tonnes, respectively (Anonymous 2013). If oilseed imports are also taken into consideration, then the total quantity of oilseeds that were processed for oil extraction in 2011 in EU-27 countries was more than 41 million tonnes (Krautgartner et al. 2011). Rapeseed (about 6.4 million tonnes) and sunflower (about 1 million tonnes) are the main oilseeds used for biodiesel production in EU-27 countries (Krautgartner et al. 2011). Together with imports, the total quantity of oilseed meals utilized in EU-27 countries as feed and waste was more than 52 million tonnes (Krautgartner et al. 2011).

Rapeseed meals (RSM) and SFM contain the protein, carbohydrate, and mineral content that remain after oil extraction (Table 24.5). According to Lomascolo et al. (2012), the mean crude protein content of RSM and SFM are approximately 35.0% and 32.0%, respectively. Both meals contain significant protein content and hence they are good sources of amino acids. Ramachandran et al. (2007) reported that the main amino acids in SFM are arginine, leucine, valine, glycine, and phenylalanine. RSM contains a protein of good quality with adequate content of lysine as well as leucine and arginine as the major amino acids (Lomascolo et al. 2012). As shown in Figure 24.5, SFM contains approximately 40% of lignocellulosics. The main carbohydrate content in RSM is pectins, pentosans, and cellulose (Lomascolo et al. 2012). Chlorogenic and caffeic acids are the main phenolic constituents (up to 70%) in SFM among others, such as *p*-hydroxy-benzoic, *p*-coumaric, cinnamic, vanillic and sinapic acids (Dominguez et al. 1995). SFM and RSM are also rich sources of minerals. For instance, RSM is a source of calcium, iron, and phosphorus. Despite the antinutritional compounds (e.g., glucosinolates) present in RSM, it is a protein-rich feedstock, and significant quantities are used as animal feed. Figure 24.8 presents potential applications of RSM and SFM.

Edible oilseed meals were initially studied and used as feed or a source of protein, which was either extracted or hydrolyzed to produce protein hydrolysates, while nonedible oilseed meals were used as biocontrol agents, fertilizers, or energy generation via combustion (Ramachandran et al. 2007). Kolesarova et al. (2011) investigated the potential of an integrated

TABLE 24.5
Mean Composition of Rapeseed and Sunflower Meals

Component	Rapeseed Meal (%)	Sunflower Meal (%)
Moisture	8.7	9.0
Dry matter	91.3	91.0
Crude proteins	35.0	32.0
Crude lipids	2.7	1.5
Ash	6.3	6.5
Total phenolics	5310–6940 μg of sinapic acid equivalents/g	3.5
Crude fiber	11.6	41.0
Lignin	N/A	6.8–8.4
Hemicelluloses	N/A	13.0
Cellulose	N/A	23.0

Source: Lomascolo, A. et al., *Appl. Microbiol. Biotechnol.*, 95, 1105, 2012.

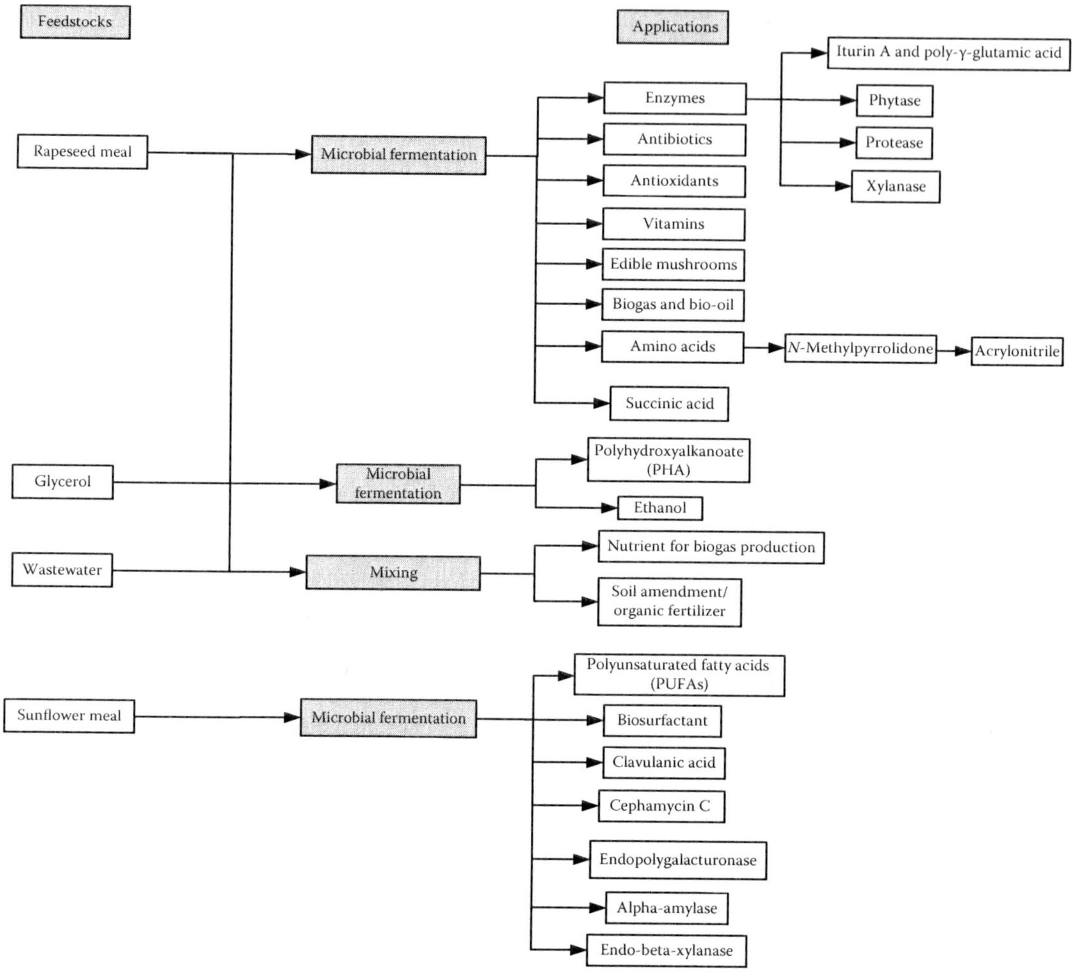

FIGURE 24.8 Utilization of RSM and SFM as fermentation raw materials for chemical, fuel, energy, and biopolymer production.

solution for biogas production from the waste streams of biodiesel plants involving the crude glycerol stream, the RSM stream, and the wastewater stream from the purification unit of fatty acid methyl esters (FAME). Furthermore, RSM could be utilized as an organic fertilizer for the cultivation of energy crops (Bonnardeaux 2007).

Lomascolo et al. (2012) presented a comprehensive review regarding the potential of RSM and SFM in biotechnological applications including the production of enzymes, antibiotics, antioxidants, and other bioproducts. In recent years, novel alternative uses of RSM and SFM were proposed, which exploit the structural and physicochemical properties of oilseed meals. This became feasible since biodiesel production increased to levels that resulted in the generation of relatively cheap and surplus quantities of oilseed meals. First-generation biodiesel plants could be converted into second-generation processes through the exploitation of oilseed cakes that could be used for the production of various products including enzymes, antibiotics, antioxidants, vitamins, edible mushrooms, biogas, and bio-oil (Ramachandran et al. 2007).

RSM and SFM have been recently utilized for the production of fermentation nutrient supplements involving a two-stage bioprocess. Wang et al. (2010) utilized RSM for the production of fermentation nutrient supplements by exploiting the crude enzyme consortia produced via solid-state fermentation of *Aspergillus oryzae* cultivated on RSM. Fermented solids were subsequently mixed with unprocessed RSM to produce fermentation supplements that have been evaluated for the growth of *Saccharomyces cerevisiae* in aerobic cultures (Wang et al. 2010), the production of poly(3-hydroxybutyrate-*co*-3-hydroxyvalerate) P(3HB-*co*-3HV) using *Cupriavidus necator* DSM 545 (Garcia et al. 2013) and microbial oil using the oleaginous yeast *Rhodosporidium toruloides* Y4 (Kiran et al. 2012). RSM contained essential nutrients for microbial growth and product formation in all cases. Garcia et al. (2013) utilized RSM hydrolysates and crude glycerol as carbon source in fed-batch fermentations carried out in shake flasks to produce 10.9 g/L of P(3HB-*co*-3HV) with a content of 55.6% (w/w). Kiran et al. (2012) utilized RSM hydrolysates in fed-batch fermentation in a bioreactor for the production of 82.3% g/L of yeast biomass (*R. toruloides*) that contained a lipid content of 48% (w/w). The dry cell weight obtained was higher than that obtained by using conventional yeast extract as a nitrogen source.

Chen et al. (2011) demonstrated that a range of fermentable sugars, including glucose, fructose, sucrose, and arabinose, could be released from RSM via sequential treatment with dilute sulfuric acid pretreatment followed by hydrolysis with commercial enzymes, such as pectinase, celluclast, or viscozyme. Fed-batch simultaneous saccharification and fermentation of dilute-acid-pretreated RSM resulted in the production of 23.4 g/L succinic acid with the bacterial strain *A. succinogenes*.

Kachirmanidou et al. (2013) have used SFM for the production of a nutrient supplement for the production of P(3HB-*co*-3HV) using *C. necator* DSM 545. Shake flask fermentations carried out in fed-batch mode resulted in the production of 9.9 g/L of P(3HB-*co*-3HV) that contained 97% 3HB and 3% 3HV, respectively.

Lammens et al. (2012) described different potential sources for amino acids that could be used for the production of bulk chemicals in a biorefinery using agricultural by-product streams, including RSM, SFM, soybean meal, Jatropha, animal slaughterhouse waste, and algae. It was stressed that there is sufficient availability of raw materials for the production of biobased chemicals, such as *N*-methylpyrrolidone, from amino acids. Bulk chemicals, such as acrylonitrile, can partly be replaced in the future by their biobased equivalent, depending on the amounts of biofuels that will be produced in the future. However, cost-effective and highly efficient technologies for the isolation of amino acids from protein-rich by-product streams should be developed.

24.3.2 Utilization of Jatropha Seed Cakes

Jatropha curcas (JC) is a tropical plant that is cultivated in Latin America, Asia, and Africa. The dried fruit of JC is mainly constituted of the shell (35%–40%) and the seed (60%–65%) (Vyas and Singh 2007). The JC fruit is initially deshelled and the remaining seed contains 40%–42% husk and 58%–60% kernel, which contains approximately 50% oil (Singh et al. 2008). The kernel can be subsequently processed via mechanical and/or solvent extraction to extract the oil that can be used for the production of biodiesel. The properties of JC-derived biodiesel are similar to petrodiesel according to the European standards (Jingura et al. 2010). Apart from biodiesel production, JC oil could be also used as carbon source in microbial fermentations. Ng et al. (2010) utilized JC oil to produce PHB via fermentation with *C. necator* H16. Their results demonstrated that JC oil could substitute food-grade oils for large-scale production of PHB. The remaining by-products after oil extraction are shells, husk, and meal (Singh et al. 2008). Figure 24.9 describes the process of JC utilization for oil extraction, and Table 24.6 presents the composition of JC husks, oil, and meal (Martin et al. 2010).

JC seeds have been used in medicinal, soap, and cosmetics applications in various tropical countries (Martínez-Herrera et al. 2006). Although the JC meal that remains after oil extraction is rich in protein, it is not suitable for the direct use as food and feed because it contains significant amounts of toxic compounds, such as trypsin inhibitor, lectin activities, and phorbol esters. The meal was

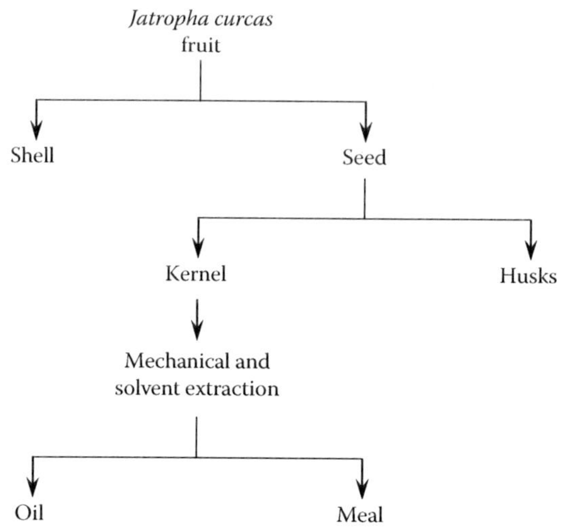

FIGURE 24.9 Schematic diagram of the production of shells, husks, oils and meal from the JC fruit.

TABLE 24.6
Compositions of JC Oil, Meal, and Husks

Component	Jatropha Meal	Jatropha Husks
Protein (%)	59.1	7.5
Hemicelluloses (%)	2.3	N/A
Cellulose (%)	6.1	16.9
Lignin (%)	1.7	45.5
Ash (%)	11.7	4.7
Xylan (%)	N/A	16.2
Arabinan (%)	N/A	1.7
Acetyl groups (%)	N/A	2.0
Unidentified (%)	N/A	5.5
Major fatty acids in Jatropha oil (%, w/w)		
Palmitic	16.9	
Palmitoleic	0.8	
Stearic	6.7	
Oleic	35.2	
10-Octadecenoic	1.4	
Linoleic	39.0	

Source: Martin, C. et al., *Biomass Bioenergy*, 34, 533, 2010.

identified as nontoxic to rats after the removal of the toxic and heat-stable compounds with 92% methanol (Martínez-Herrera et al. 2006). *Jatropha* press cake proteins could be used as emulsifier or paper adhesive as they presented very good adhesive and emulsifying properties (Lestari et al. 2011). The de-oiled JC meal could be utilized in solid-state fermentations for the production of industrial enzymes such as protease and lipase (Mahanta et al. 2008). Rao et al. (2007) has employed JC seed meal for the production of itaconic acid at a concentration of 24.45 g/L after 120 h fermentation with the fungal strain *Aspergillus terreus*.

Sharma et al. (2009) employed a lignocellulolytic fungal consortium for the bioconversion of JC hulls in order to alleviate the unfavorable physicochemical characteristics (e.g., high pH and phenolic content) and produce a compost that can be applied to acidic soil as a remedial organic manure. The high content of cellulose in the compost showed that the hull fermented by the fungal consortium could be further used for ethanol production. Wever et al. (2012) reported that the high lignin content in JC shells indicates that they could be used for the production of particle boards, while a water extract of the JC shells results in a fraction being rich in alkali metals and phosphorous that could be used as fertilizer. Namasivayam et al. (2007) indicated that JC husk could be used for the production of an activated carbon that could be used as an adsorbent for the removal of toxic pollutants from water. JC meal can be used as feedstock in a two-stage anaerobic process without chemical addition for pH adjustment that can lead to high methane yield (Sinbuathong et al. 2012).

JC is suitable for worldwide cultivation in degraded wasteland (Pandey et al. 2012). JC cultivation could provide many benefits as it can be used as effective phytoremediator, carbon sequester, degraded land developer, and soil erosion controller (Pandey et al. 2012). All parts of JC (e.g., seeds, leaves, and bark) are used in traditional medicine and active components are being investigated. Figure 24.10 presents potential applications for JC.

24.4 BIOREFINERIES BASED ON SUGAR CROPS

Sugar crops together with cereal grains and oilseeds are the main agricultural products that could lead to the development of whole-crop biorefineries. Since biorefineries based on sugar crops have been presented in other studies (Filho et al. 2011; Cavalett et al. 2012; Canilha et al. 2012; Chong and O'Shea 2012), this section will provide only recent developments and future trends regarding this type of whole-crop biorefineries based on sugarcane (Figure 24.11).

Sugarcane is one of the most important sugar crops in tropical regions such as South America (e.g., Brazil, Colombia) and India (Moncada et al. 2012). The National Alcohol Program (PROALCOOL) that was started in Brazil in the 1970s has gradually led to the creation of a mature bioethanol market that can nowadays evolve into the future sugarcane biorefinery era. The conventional sugarcane industry in Brazil that produces bioethanol, sweeteners, and energy in an efficient process could be restructured into an advanced biorefinery concept for the production of biodegradable polymers and platform chemicals (Filho et al. 2011). This will be developed through the effective utilization of the whole cane crop and the exploitation of the alcohol chemistry and fermentative routes (Filho et al. 2011). The main feedstocks derived from traditional sugarcane industry that can be used for the production of fuels, chemicals, and biopolymers are cane juice, molasses, and bagasse. Cane juice and molasses are the products of first-generation sugarcane mills, while bagasse will be integrated as feedstock in the second-generation sugarcane biorefineries. Bagasse is the lignocellulosic pulp residue that remains after juice extraction, which is dried and utilized as fuel in traditional sugarcane mills. In the second-generation biorefineries, bagasse could be fractionated into cellulose, hemicellulose, and lignin, each of which could be utilized for chemical, material, fuel, and energy production. Uppal and Kaur (2011) utilized sugarcane bagasse for the production of furfural via treatment with various concentrations of different acids (e.g., acetic acid, formic acid, phosphoric acid, and nitric acid). Furfural can be used in various applications including petroleum refining and pharmaceuticals, plastics, and agrochemical production. Sugarcane bagasse is a competitive raw material for furfural production due to its high carbohydrate to lignin content, the constant supply that is independent on seasonal growth and the on-site availability in conventional sugarcane mills. Levulinic acid is another important platform chemical that could be produced via chemical treatment of sugarcane bagasse (Yan et al. 2008).

Pye et al. (2012) patented a bagasse biorefining process that could be integrated in a traditional sugarcane mill to concurrently produce ethanol and coproducts. Rabelo et al. (2011) investigated the combined production of fuel ethanol, biogas, and heat from sugarcane bagasse that could be incorporated in a conventional sugarcane mill. In the case of Colombia, conventional sugarcane plants

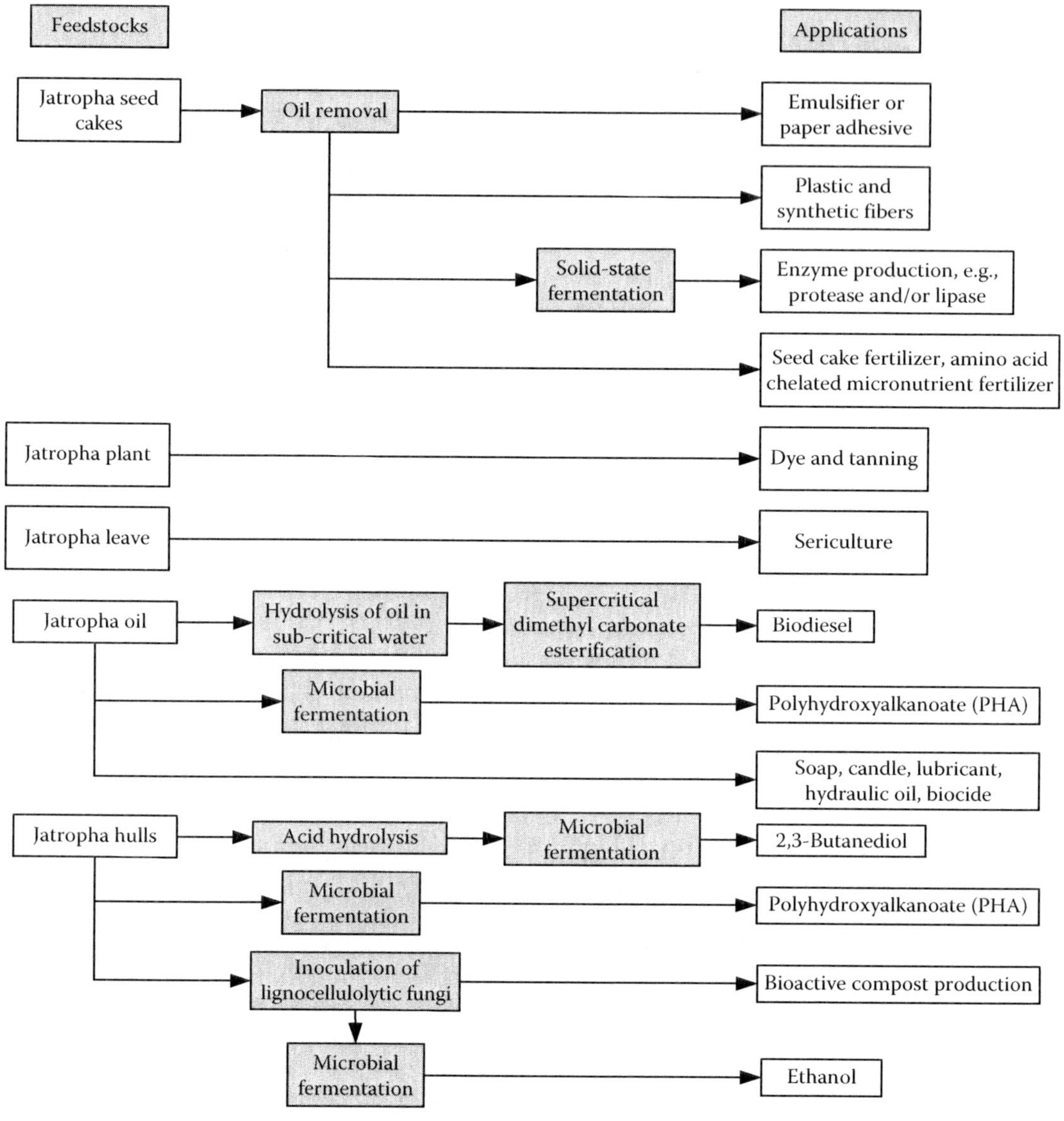

FIGURE 24.10 *Jatropha*-based feedstocks and applications for chemicals, materials, and fuel production.

produce sugar for food purposes from cane juice, fuel ethanol from molasses, and electricity from cane bagasse (Moncada et al. 2012). The development of future biorefinery plants could be based on the integrated utilization of fresh sugarcane and the respective sugarcane bagasse generated at the processing stage as feedstocks for the combined production of sugar, fuel ethanol, PHB, anthocyanins, and electricity (Moncada et al. 2012).

In the last stage of sugar production, the clarified sugarcane juice is treated via evaporation and crystallization in order to produce sugar crystals, while the remaining liquor is molasses. Molasses contain various value-added constituents, such as organic acids, that could be isolated and converted into other products that could contribute to the transformation of conventional sugarcane mills into novel biorefinery schemes. Piang-Siong et al. (2012) evaluated the production of value-added esters from aconitic acid in sugarcane-derived molasses. The polyfunctional esters that were produced via the esterification of aconitic acid with various alcohols of natural origin could be considered as bio-based molecules with various applications to replace petroleum-derived products, which are banned

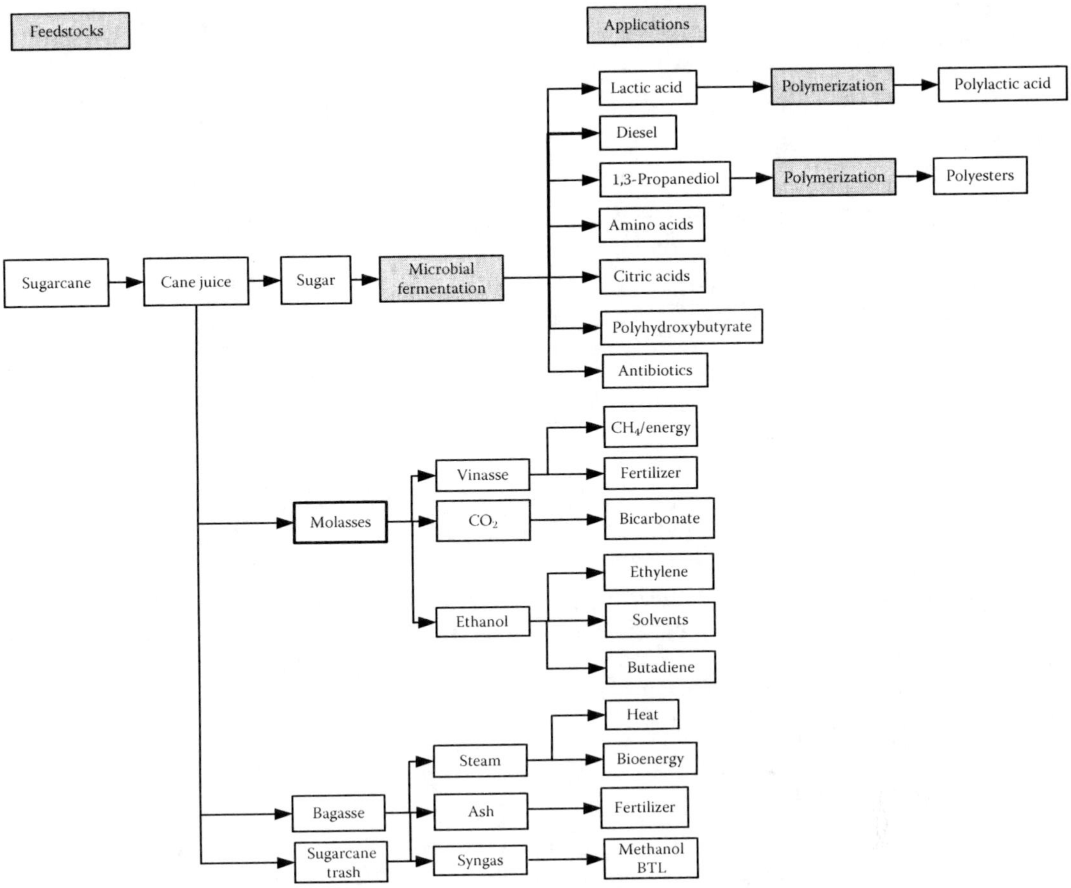

FIGURE 24.11 Potential sugar crop biorefinery scenarios for the production of chemicals, materials, fuels, and energy.

by regulations (Piang-Siong et al. 2012). Minor constituents present in sugarcane molasses could be utilized in a similar way as the minor constituents present in wheat bran and germ. Furthermore, ethanol produced nowadays as fuel could be utilized as a platform chemical for the production of various chemicals such as ethylene and butadiene.

24.5 FUTURE TRENDS: UTILIZATION OF CEREAL-BASED FOOD WASTE

From the viewpoint of food waste management, Hang (2004) suggested the use of large quantities of liquid and solid wastes (food processing residues) from the food industry for production of valuable bioproducts. For example, fermentative production of lactic acid from a mixture of restaurant food waste, bakery by-product, barley, and wheat bran was demonstrated (Figure 24.12). Also, a membrane-integrated process for lactic acid production from sugarcane juice was economically evaluated using pilot plant scale data (Sikder et al. 2012). Production of biopolymers from various food industry wastes and agricultural crops is shown to be techno-economically feasible for replacing petroleum-derived plastics (García et al. 2013). Similarly, succinic acid could be produced from wheat-based and cereal-based renewable feedstocks (Du et al. 2008; Dorado et al. 2009; Lin et al. 2012).

Succinic acid (SA) can be produced by the fermentation of bakery waste, and the overall yield (1.16 g SA/g total sugar) is the highest among other food waste-derived media (Leung et al. 2012;

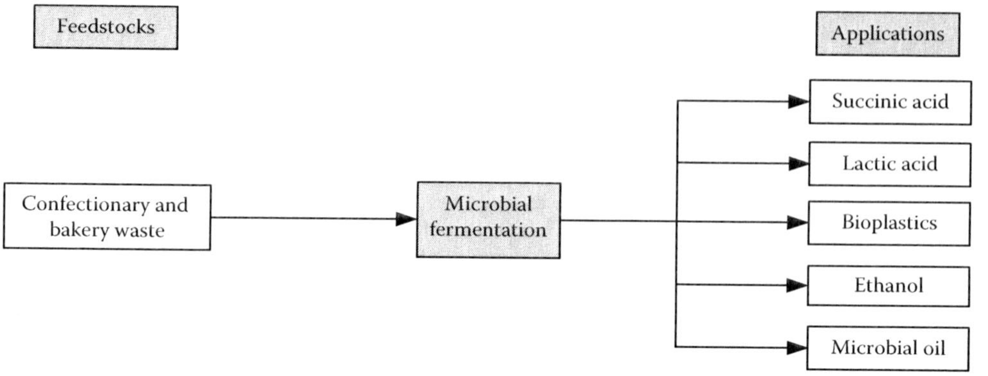

FIGURE 24.12 Cereal-based food waste feedstocks and applications of chemicals, materials, and fuel production by different technologies.

Zhang et al. 2013). Lam et al. (2013) evaluated the economic feasibility of the whole production process of SA including downstream separation and purification using computer simulation. They assessed the profitability of a pilot-scale facility to be built in Hong Kong for treating bakery waste. The total capital investment for the plant and the total production cost were US$1,118,243 and US$230,474 per year, respectively. Overall revenue generated from the process was US$374,041 per year. The return on investment, payback period, and internal rate of return of the project were 12.8%, 7.2 years, and 15.3%, respectively. The results indicate that it is economically feasible to ferment bakery waste into succinic acid.

24.6 CONCLUSIONS

This chapter provides an overview of whole-crop biorefineries with particular emphasis on wheat, specific oilseeds, and sugarcane biorefineries. The successful evolution of the first-generation biorefinery concepts into advanced generation processes is highly dependent on the exploitation of waste and by-product streams for the production of fuels, chemicals, materials, and energy.

ACKNOWLEDGMENTS

We would like to express our sincere gratitude to the late Dr. Ruohang Wang, who stood beside us like a guiding star. Memories of him will always be cherished by all members of Prof. Colin Webb's research group. Dr. Carol Sze Ki LIN acknowledges the Industrial Technology Funding from the Innovation and Technology Commission (ITS/323/11) in Hong Kong, the donation from the Coffee Concept (Hong Kong) Ltd. for the *Care for Our Planet* campaign, as well as a grant from the City University of Hong Kong (Project No. 7200248). Dr. Apostolis Koutinas acknowledges the financial support from *BIOREF* (09SYN-81-715) research project implemented within the National Strategic Reference Framework (NSRF) 2007–2013 and cofinanced by national (Greek Ministry—General Secretariat of Research and Technology) and community funds (EU—European Social Fund).

REFERENCES

Agostoni, C., I. Axelsson, O. Goulet et al. 2004. Prebiotic oligosaccharides in dietetic products for infants: A commentary by the ESPGHAN Committee on Nutrition. *Journal of Pediatric Gastroenterology and Nutrition* 39: 465–473.

Anderson, R.A., V.F. Pfeifer, and E.B. Lancaster. 1958. Continuous batter process for separating gluten from wheat flour. *Cereal Chemistry* 35: 449–457.

Anonymous. 2007. Growing wheat for alcohol/biofuel production. Project Progress 14. Home-Grown Cereals Authority (HGCA), http://publications.hgca.com/publications/documents/cropresearch/PP14.pdf

Anonymous. 2013. United States Department of Agriculture. Foreign Agricultural Service. Production, Supply and Distribution Online. Accessed February 10, 2013. http://www.fas.usda.gov/psdonline/psdquery.aspx.

Arifeen, N., I.K. Kookos, R. Wang, A.A. Koutinas, and C. Webb. 2009. Development of novel wheat biorefining: Effect of gluten extraction from wheat on bioethanol production. *Biochemical Engineering Journal* 43: 113–121.

Arifeen, N., R. Wang, I. Kookos, C. Webb, and A.A. Koutinas. 2007a. Optimization and cost estimation of novel wheat biorefining for continuous production of fermentation feedstock. *Biotechnology Progress* 23: 872–880.

Arifeen, N., R. Wang, I.K. Kookos, C. Webb, and A.A. Koutinas. 2007b. Process design and optimization of novel wheat-based continuous bioethanol production system. *Biotechnology Progress* 23: 1394–1403.

Asp, N.G., C.G. Johansson, H. Hallmer, and M. Siljeström. 1983. Rapid enzymatic assay of insoluble and soluble dietary fibre. *Journal of Agricultural and Food Chemistry* 31: 476–482.

Asp, N.G., T.F. Schweizer, D.A.T. Southgate, and O. Theander. 1992. Dietary fibre analysis. In: *Dietary Fibre—A Component of Food, Nutritional Function in Health and Disease*, eds. T.F. Schweizer and C.A. Edwards, pp. 57–101, London, U.K.: Springer-Verlag.

Awad, A.B. and C.S. Fink. 2000. Phytosterols as anticancer dietary components: Evidence and mechanism of action. *Journal of Nutrition* 130: 2127–2130.

Barberousse, H., A. Kamoun, M. Chaabouni, J.-M. Giet, O. Roiseux, M. Paquot, C. Deroanne, and C. Blecker. 2009. Optimization of enzymatic extraction of ferulic acid from wheat bran, using response surface methodology, and characterization of the resulting fractions. *Journal of the Science of Food and Agriculture* 89: 1634–1641.

Boler, B.M.V. and G.C. Fahey Jr. 2012. Prebiotics of plant and microbial origin. In: *Direct-Fed Microbials and Prebiotics for Animals: Science and Mechanisms of Action*, eds. R. Callaway and S.C. Ricke, Chapter 2, pp. 13–26, New York: Springer Science+Business Media.

Bonnardeaux, J. 2007. *Uses for Canola Meal.* Western Australia: Department of Agriculture and Food, Western Australia.

Bozell, J.J. and G.R. Petersen. 2010. Technology development for the production of biobased products from biorefinery carbohydrates—The US Department of Energy's "Top 10" revisited. *Green Chemistry* 12: 525–728.

Budarin, V.L., P.S. Shuttleworth, J.R. Dodson, A.J. Hunt, B. Lanigan, R. Marriott, K.J. Milkowski et al. 2011. Use of green chemical technologies in an integrated biorefinery. *Energy and Environmental Science* 4: 471–479.

Burley, V.J. and J.E. Blundell. 1995. Dietary fiber intake and the pattern of energy intake. In: *Dietary Fiber in Health and Disease*, eds. D. Kritchevsky and C. Bonfield, pp. 243–256. Saint Paul, MN, Eagan Press.

Camire, A.L. and F.M. Clydesdale. 1982. Analysis of phytic acid in foods by HPLC. *Journal of Food Science* 47: 575–578.

Canilha, L., A.K. Chandel, T.S. dos Santos Milessi, F.A.F. Antunes, W.L. da Costa Freitas, M. das Graças Almeida Felipe, and S.S. da Silva. 2012. Bioconversion of sugarcane biomass into ethanol: An overview about composition, pretreatment methods, detoxification of hydrolysates, enzymatic saccharification, and ethanol fermentation. *Journal of Biomedicine and Biotechnology* 2012: Article ID 989572.

Cavalett, O., T.L. Junqueira, M.O.S. Dias, C.D.F. Jesus, P.E. Mantelatto, M.P. Cunha, H.C.J. Franco et al. 2012. Environmental and economic assessment of sugarcane first generation biorefineries in Brazil. *Clean Technologies and Environmental Policy* 14: 399–410.

Chang, C., X. Ma, and P. Cen. 2006. Technology for production of levulinic acid as new platform chemical from wheat straw. *Nongye Gongcheng Xuebao/Transactions of the Chinese Society of Agricultural Engineering* 22: 161–164.

Chang, C., X. Ma, and P. Cen. 2009. Kinetic studies on wheat straw hydrolysis to levulinic acid. *Chinese Journal of Chemical Engineering* 17: 835–839.

Chatzifragkou, A. and S. Papanikolaou. 2012. Effect of impurities in biodiesel-derived waste glycerol on the performance and feasibility of biotechnological processes. *Applied Microbiology and Biotechnology* 95: 13–27.

Chen, K., H. Zhang, Y. Miao, P. Wei, and J. Chen. 2011. Simultaneous saccharification and fermentation of acid-pretreated rapeseed meal for succinic acid production using *Actinobacillus* succinogenes. *Enzyme and Microbial Technology* 48: 339–344.

Chong, B.F. and M.G. O'Shea. 2012. Developing sugarcane lignocellulosic biorefineries: Opportunities and challenges. *Biofuels* 3: 307–319.

Clark, J.H., V. Buldarni, and F.I.E. Deswarte. 2006. Green chemistry and the biorefinery: A partnership for a sustainable future. *Green Chemistry* 8: 853–860.

Colleoni-Sirghie, M., I.V. Kovalenko, J.L. Briggs, B. Fulton, and P.J. White. 2003. Rheological and molecular properties of water soluble (1,3) (1,4)-β-D-glucans from high-β-glucan and traditional oat lines. *Carbohydrate Polymers* 52: 439–447.

Cui, X.F., G. Jin, E.B. Liu et al. 2012. Corrosion resistance of organic coating pretreated by phytic acid. *Surface Engineering* 28: 657–661.

Cui, W., P.J. Wood, J. Weisz, and M.U. Beer. 1999. Nonstarch polysaccharides from preprocessed wheat bran: Carbohydrate analysis and novel rheological properties. *Cereal Chemistry* 76: 129–133.

Cummings, J.H. and H.N. Englyst. 1987. The development of methods for the measurement of dietary fibre in food. In: *Cereals in a European Context. First European Congress of Food Science and Technology*, ed. I.D. Morton, pp. 188–220, New York: VCH Publishers.

Curreli, N., B.M. Fadda, A. Rescigno, A.C. Rinaldi, G. Soddu, F. Sollai, S. Vaccargiu, E. Sanjust, and A. Rinaldi. 1997. Mild alkaline/oxidative pretreatment of wheat straw. *Process Biochemistry* 32: 665–670.

Czuchajowska, Z. and Y. Pomeranz. 1995. Process for fractionating wheat flours to obtain protein concentrates and prime starch. U.S. patent number 5,439,526.

Deswarte, F.E.I., J.H. Clark, J.J.E. Hardy, and P.M. Rose. 2008. The fractionation of valuable wax products from wheat straw using CO_2. *Green Chemistry* 8: 39–42.

De Vries, J.W. 2001. The definition of dietary fibre. *Cereal Foods World* 46: 112–129.

De Vries, J.W. and J.I. Rader. 2005. Historical perspective as a guide for identifying and developing applicable methods for dietary fiber. *Journal of AOAC International* 88(5): 1349–1366.

Dewarte, F.E.I., J.H. Clark, A.J. Wilson, J.J.E. Hardy, R. Marriott, S.P. Chahal, C. Jackson, and G. Heslop. 2007. Toward an integrated straw-based biorefinery. *Biofuels, Bioproducts and Biorefineries* 1: 245–254.

Dexter, J.E. and P.J. Wood. 1996. Recent applications of debranning of wheat before milling. *Trends in Food Science and Technology* 7: 35–41.

Di Gioia, D., L. Sciubba, M. Ruzzi, L. Setti, and F. Fava. 2009. Production of vanillin from wheat bran hydrolyzates via microbial bioconversion. *Journal of Chemical Technology and Biotechnology* 84: 1441–1448.

Dominguez, H., M.J. Núñez, and J.M. Lema. 1995. Aqueous processing of sunflower kernels with enzymatic technology. *Food Chemistry* 53: 427–434.

Dorado, M.P., S.K.C. Lin, A. Koutinas, C. Du, R. Wang, and C. Webb. 2009. Cereal-based biorefinery development: Utilisation of wheat milling by-products for the production of succinic acid. *Journal of Biotechnology* 143: 51–59.

Du, C., G.M. Campbell, N. Misailidis, F. Mateos-Salvador, J. Sadhukhan, M. Mustafa, and R.M. Weightman. 2009. Evaluating the feasibility of commercial arabinoxylan production in the context of a wheat biorefinery principally producing ethanol. Part 1. Experimental studies of arabinoxylan extraction from wheat bran. *Chemical Engineering Research and Design* 87: 1232–1238.

Du, C., S.K.C. Lin, A. Koutinas, R. Wang, P. Dorado, and C. Webb. 2008. A wheat biorefining strategy based on solid-state fermentation for fermentative production of succinic acid. *Bioresource Technology* 99: 8310–8315.

Du, C., S.K.C. Lin, A. Koutinas, R. Wang, and C. Webb. 2007. Succinic acid production from wheat using a biorefining strategy. *Applied Microbiology and Biotechnology* 76: 1263–1270.

Dunford, N.T. 2005. Germ oils from different sources. In: *Bailey's Industrial Oil and Fat Products*, ed. F. Shahidi, 6th edn., pp. 195–231, New York: John Wiley & Sons.

Evers, A.D. and D.B. Bechtel. 1988. Microscopic structure of the wheat grain. In: *Wheat: Chemistry and Technology*, ed. Y. Pomeranz, pp. 47–95, Vol. I., St. Paul, MN: AACC.

Faurot, A.-L., L. Saulnier, S. Berot et al. 1995. Large-scale isolation of water-soluble and water-insoluble pentosans from wheat flour. *Food Science and Technology* 28: 436–441.

Fernandez, M.-L. 2001. Soluble fibre and non-digestible carbohydrate effects on plasma lipids and cardiovascular risk. *Current Opinion in Lipidology* 12: 35–40.

Filho, V.M., C. Araujo, A. Bonfá, and W. Porto. 2011. Chemistry based on renewable raw materials: Perspectives for a sugar cane-based biorefinery. *Enzyme Research* 2011, Article ID 654596.

Food and Drug Administration. 1993. Food labeling: Mandatory status of nutrition labeling and nutrient content revision, format for nutrition label. *Federal Register* 58: 2079.

Fox, C.H. and M. Eberl. 2002. Phytic acid (IP6), novel broad spectrum anti-neoplastic agent: A systematic review. *Complementary Therapies in Medicine* 10: 229–234.

Francois, I.E., O. Lescroart, W.S. Veraverbeke et al. 2012. Effects of a wheat bran extract containing arabinoxylan oligosaccharides on gastrointestinal health parameters in healthy adult human volunteers: A double-blind, randomised, placebo-controlled, cross-over trial. *British Journal of Nutrition* 108: 2229–2242.

García, I.L., J.A. López, M.P. Dorado, N. Kopsahelis, M. Alexandri, S. Papanikolaou, M.A. Villar, and A.A. Koutinas. 2013. Evaluation of by-products from the biodiesel industry as fermentation feedstock for poly(3-hydroxybutyrate-co-3-hydroxyvalerate) production by Cupriavidus necator. *Bioresource Technology* 130: 16–22.

Garcia-Estepa, R.M., E. Guerra-Hernandez, and B. Garcia-Villanova. 1999. Phytic acid content in milled cereal products and breads. *Food Research International* 32: 217–221.

Gomez, A.M. and E.M. De la Ossa. 2000. Quality of wheat germ oil extracted by liquid and supercritical carbon dioxide. *Journal of the American Oil Chemists' Society* 77: 969–974.

Graf, E. 1983. Applications of phytic acid. *Journal of the American Oil Chemists' Society* 60: 1861–1867.

Hakala, P., A.-M. Lampi, V. Ollilainen, U. Werner, M. Murkovic, K. Wahala, S. Karkola, and V. Piironen. 2002. Steryl phenolic acid esters in cereals and their milling fractions. *Journal of Agriculture and Food Chemistry* 50: 5300–5307.

Han, G., J. Deng, S. Zhang, P. Bicho, and Q. Wu. 2010. Effect of steam explosion treatment on characteristics of wheat straw. *Industrial Crops and Products* 31: 28–33.

Hang, Y.D. 2004. Management and utilization of food processing wastes. *Journal of Food Science* 69: CRH104–CRH107.

Havrlentová, M. and J. Kraic. 2006. Content of β-D-glucan in cereal grains. *Journal of Food and Nutrition Research* 45: 97–103.

Henry, R.J. and H.S. Saini. 1989. Characterization of cereal sugars and oligosaccharides. *Cereal Chemistry* 66: 362–365.

Hwang, J., B. Park, and J. Yun. 2001. Biologically active materials such as ferulic acid and arabinoxylan obtained from cereal bran by extrusion and enzyme treatment. PCT International Applications WO 01/67891 A1.

Ibanez, E., J. Palacios, F.J. Senorans, G. Santa-Maria, J. Tabera, and G. Reglero. 2000. Isolation and separation of tocopherols from olive by-products with supercritical fluids. *Journal of the American Oil Chemists' Society* 77: 187–190.

IUB-IUPAC. 1982. Joint commission on biochemical nomenclature and (JCBN): Abbreviated terminology of oligosaccharide chains: Recommendations 1980. *Journal of Biological Chemistry* 257: 3347–3351.

Jang, Y.-S., B. Kim, J.H. Shin, Y.J. Choi, S. Choi, C.W. Song, J. Lee, H.G. Park, and S.Y. Lee. 2012. Bio-based production of C2–C6 platform chemicals. *Biotechnology and Bioengineering* 109: 2437–2459.

Jansma, W., J. Mars, P.G. Stoutjesdijk, and H.J. Vegter. 1985. Separation of gluten and starch from wheat flour. U.S. patent number 4,494,530.

Jenab, M. and L.U. Thompson. 2000. Phytic acid in wheat bran affects colon morphology, cell differentiation and apoptosis. *Carcinogenesis* 21: 1547–1552.

Jingura, R.M., D. Musademba, and R. Matengaifa. 2010. An evaluation of utility of *Jatropha curcas* L. as a source of multiple energy carriers. *International Journal of Engineering, Science and Technology* 2: 115–122.

Johnston, P.H. and D.A. Fellers. 1971. Process for protein-starch separation in wheat flour: 2. Experiments with a continuous decanter-type centrifuge. *Journal of Food Science* 36: 649–652.

Kachrimanidou, V., N. Kopsahelis, A. Chatzifragkou, S. Papanikolaou, S. Yanniotis, I. Kookos, and A.A. Koutinas. 2013. Utilisation of by-products from sunflower-based biodiesel producton processes for the production of fermentation feedstock. *Waste Biomass Valor* 4: 529–537.

Kahlon, T.S. and F.I. Chow. 2000. In vitro binding of bile acids by rice bran, oat bran, wheat bran, and corn bran. *Cereal Chemistry* 77: 518–521.

Kerkkonen, H.K., K.M.J. Laine, M.A. Alanen, and V.H. Renner. 1976. Method of separating gluten from wheat flour. U.S. patent number 3,951,938.

Kersting, H.-J., M.G. Lindhauer, and W. Bergthaller. 1994. Application of wheat gluten in non-food industry—Wheat gluten as a natural cobinder in paper coating. *Industrial Crops and Products* 3: 121–128.

Kiran, E.U., A. Salakkam, A.P. Trzcinski, U. Bakir, and C. Webb. 2012. Enhancing the value of nitrogen from rapeseed meal for microbial oil production. *Enzyme and Microbial Technology* 50: 337–342.

Knuckles, B.E., D.D. Kuzmicky, and A.A. Betschart. 1982. HPLC analysis of phytic acid in selected foods and biological samples. *Journal of Food Science* 47: 1257–1262.

Kolesarova, N., M. Hutnan, I. Bodik, and V. Špalkova. 2011. Utilization of biodiesel by-products for biogas production. *Journal of Biomedicine and Biotechnology* 2011, Article ID 126798.

Koutinas, A.A., N. Arifeen, R. Wang, and C. Webb. 2007b. Cereal-based biorefinery development: Integrated enzyme production for cereal flour hydrolysis. *Biotechnology and Bioengineering* 97: 61–72.

Koutinas, A.A., R. Wang, G. Campbell, and C. Webb. 2006. A whole crop biorefinery system—A closed system for the manufacture of non-food products from cereals. In: *Biorefineries—Industrial Processes and Products*, eds. B. Kamm, P.R. Gruber, and M. Kamm, pp. 165–191, Weinheim, Germany: Wiley-VCH.

Koutinas, A.A., R.-H. Wang, and C. Webb. 2004. Restructuring upstream bioprocessing: Technological and economical aspects for the production of a generic microbial feedstock from wheat. *Biotechnology and Bioengineering* 85: 524–538.

Koutinas, A.A., R.-H. Wang, and C. Webb. 2005. Development of a process for the production of nutrient supplements for fermentations based on fungal autolysis. *Enzyme and Microbial Technology* 36: 629–638.

Koutinas, A.A., R.-H. Wang, and C. Webb. 2007a. The biochemurgist—Bioconversion of agricultural raw materials for chemical production. *Biofuels, Bioproducts and Biorefining* 1: 24–38.

Krautgartner, R., M.-C. Henard, S. Lieberz, M. Boshnakova, B. Flach, J. Wilson, A. Wideback, O. Bettini, M. Guerrero, and K. Bendz. 2011. Oilseed and products annual: Modest rebound in EU-27 oilseed production. Gain report number E60016. Global Agricultural Information Network (GAIN).

Lam, K.F., C.C.J. Leung, M.L. Ho, and C.S.K. Lin. 2013. Economic feasibility of a pilot-scale fermentative succinic acid production from bakery wastes. *Food and Bioproduct Processing*. DOI:10.1016/j.fbp.2013.09.001.

Lamers, P. 2012. International biodiesel markets. Developments in production and trade. Published http://www.ecofys.com//en//publication//international-biodiesel-markets/. Accessed on February 2013.

Lammens, T.M., M.C.R. Franssen, E.L. Scott, and J.P.M. Sanders. 2012. Availability of protein-derived amino acids as feedstock for the production of bio-based chemicals. *Biomass and Bioenergy* 44: 168–181.

Lee, D., V.N. Owens, A. Boe, and P. Jeranyama. 2007. Composition of herbaceous biomass feedstocks. Report (SGINC1-07) sponsored by Sun Grant Initiative, North Central Center, South Dakota State University, Brookings, SD. http://ncsungrant.sdstate.org//uploads//publications//SGINC1-07.pdf.

Lestari, D., W.J. Mulder, and J.P.M. Sanders. 2011. Jatropha seed protein functional properties for technical applications. *Biochemical Engineering Journal* 53: 297–304.

Leung, C.C.J., A.S.Y. Cheung, A.Y.-Z. Zhang, K.F. Lam, and C.S.K. Lin. 2012. Utilisation of waste bread for fermentative succinic acid production. *Biochemical Engineering Journal* 65: 10–15.

Li, W., S.W. Cui, and Y. Kakuda. 2006. Extraction, fractionation, structural and physical characterization of wheat β-D-glucans. *Carbohydrate Polymers* 63: 408–416.

Lin, C.S.K., R. Luque, J. Clark, C. Webb, and C. Du. 2011. A seawater-based biorefining strategy for fermentative production and chemical transformations of succinic acid. *Energy and Environmental Science* 4: 1471–1479.

Lin, C.S.K., R. Luque, J.H. Clark, C. Webb, and C. Du. 2012. Wheat-based biorefining strategy for fermentative production and chemical transformations of succinic acid. *Biofuels, Bioproducts and Biorefining* 6: 88–104.

Ling, W.H. and P.J.H. Jones. 1995. Dietary phytosterols: A review of metabolism, benefits and side effects. *Life Sciences* 57: 195–206.

Lomascolo, A., E. Uzan-Boukhris, J.C. Sigoillot, and F. Fine. 2012. Rapeseed and sunflower meal: A review on biotechnology status and challenges. *Applied Microbiology and Biotechnology* 95: 1105–1114.

MacMasters, M.M., J.J.C. Hinton, and D. Bradbury. 1971. Microscopic structure and composition of the wheat kernel. In: *Wheat: Chemistry and Technology*, ed. Y. Pomeranz, 2nd edn., pp. 51–113, St. Paul, MN: AACC.

Madhukumar, M.S. and G. Muralikrishna. 2012. Fermentation of xylo-oligosaccharides obtained from wheat bran and Bengal gram husk by lactic acid bacteria and bifidobacteria. *Journal of Food Science and Technology* 49: 745–752.

Mahanta, N., A. Gupta, and S.K. Khare. 2008. Production of protease and lipase by solvent tolerant *Pseudomonas aeruginosa* PseA in solid-state fermentation using *Jatropha curcas* seed cake as substrate. *Bioresource Technology* 99: 1729–1735.

Maki, K.C., G.R. Gibson, R.S. Dickmann et al. 2012. Digestive and physiologic effects of a wheat bran extract, arabino-xylan-oligosaccharide, in breakfast cereal. *Nutrition* 28: 1115–1121.

Maki, K.C., F.S. Shinnick, A. Seeley, P.E. Marlyn et al. 2003. Food products containing free tall oil-based phytosterols and oat β-glucan lower serum total and LDL cholesterol in hypercholesterolemic adults. *Journal of Nutrition* 133: 808–813.

Manderson, K., M. Pinart, K.M. Tuohy et al. 2005. In vitro determination of prebiotic properties of oligosaccharides derived from an orange juice manufacturing by-product stream. *Applied and Environmental Microbiology* 71: 8383–8389.

Maningat, C.C., S. Bassi, and Hesser, J.M. 1994. Wheat gluten in food and non-food systems. *AIB Research Department Technical Bulletin* XVI: 1–8.

Marinkovic, S., J. Le Bras, V. Nardello-Rataj, M. Agach, and B. Estrine. 2012. Acidic pretreatment of wheat straw in decanol for the production of surfactant, lignin and glucose. *International Journal of Molecular Sciences* 13: 348–357.

Marlett, J.A. 1992. The content and composition of dietary fibre in 117 frequently consumed foods. *Journal of the American Dietetic Association* 92: 175–186.

Martín, C., A. Moure, G. Martın, E. Carrillo, H. Domınguez, and J.C. Parajo. 2010. Fractional characterisation of jatropha, neem, moringa, trisperma, castor and candlenut seeds as potential feedstocks for biodiesel production in Cuba. *Biomass and Bioenergy* 34: 533–538.

Martínez-Herrera, J., P. Siddhuraju, G. Francis, G. Dávila-Ortíz, and K. Becker. 2006. Chemical composition, toxic/antimetabolic constituents, and effects of different treatments on their levels, in four provenances of *Jatropha curcas* L. from Mexico. *Food Chemistry* 96: 80–89.

McAloon, A., F. Taylor, W. Yee, K. Ibsen, and R. Wooley. 2000. Determining the cost of producing ethanol from corn starch and lignocellulosic feedstocks. A Joint Study Sponsored by: U.S. Department of Agriculture and U.S. Department of Energy, NREL/TP-580-28893.

McCann, S.E., K.B. Moysich, and C. Mettlin. 2001. Intakes of selected nutrients and food groups and risk of ovarian cancer. *Nutrition and Cancer* 39: 19–28.

Mehta, R. 2009 (January/February). Dietary fiber I. *AIB International* XXXI(1): 1–7.

Mehta, R.S. 2005. Dietary fiber benefits. *Cereal Foods World* 50(2): 66–71.

Miller, S.S., P.J. Wood, L.N. Pietrzak, and R.G. Fulcher. 1993. Mixed linkage β-glucan, protein content, and kernel weight in Avena species. *Cereal Chemistry* 70: 231–233.

Misailidis, N., G.M. Campbell, C. Du, J. Sadhukhan, M. Mustafa, F. Mateos-Salvador, and R.M. Weightman. 2009. Evaluating the feasibility of commercial arabinoxylan production in the context of a wheat biorefinery principally producing ethanol: Part 2. Process simulation and economic analysis. *Chemical Engineering Research and Design* 87: 1239–1250.

Mojumdar, S.C., C. Moresoli, L.C. Simon, and R.L. Legge. 2011. Edible wheat gluten (WG) protein films. *Journal of Thermal Analysis and Calorimetry* 104: 929–936.

Moncada, J., M.M. El-Halwagi, and C.A. Cardona. 2012. Techno-economic analysis for a sugarcane biorefinery: Colombian case. *Bioresource Technology*. http://dx.doi.org//10.1016//j.biortech.2012.08.137.

Moser, B.R. and S.F. Vaughn. 2012. Biodiesel from corn distillers dried grains with solubles: Preparation, evaluation, and properties. *Bioenergy Research* 5: 439–449.

Muniyasamy, S., M.M. Reddy, M. Misra, and A. Mohanty. 2013. Biodegradable green composites from bioethanol co-product and poly(butylene adipate-co-terephthalate). *Industrial Crops and Products* 43: 812–819.

Namasivayam, C., D. Sangeetha, and R. Gunasekaran. 2007. Removal of anions, heavy metals, organics and dyes from water by adsorption onto a new activated carbon from Jatropha husk, an agro-industrial solid waste. *Process Safety and Environmental Protection* 85: 181–184.

Nandini, C.D. and P.V. Salimath. 2001. Carbohydrate composition of wheat, wheat bran, sorghum and bajra with good chapati/roti (Indian flat bread) making quality. *Food Chemistry* 73: 197–203.

Nelson, A.L. 2001. *High-Fibre Ingredients*. St. Paul, MN: Eagan Press.

Ng, K.S., W.Y. Ooi, L.K. Goh, R. Shenbagarathai, and K. Sudesh. 2010. Evaluation of jatropha oil to produce poly(3-hydroxybutyrate) by Cupriavidus necator H16. *Polymer Degradation and Stability* 95: 1365–1369.

Nghiem, N.P., K.B. Hicks, and D.B. Johnston. 2010. Integration of succinic acid and ethanol production with potential application in a corn or barley biorefinery. *Applied Biochemistry and Biotechnology* 162: 1915–1928.

Nordqvist, P., D. Thedjil, S. Khosravi, M. Lawther, E. Malmströn, and F. Khabbaz. 2012. Wheat gluten fractions as wood adhesives—Glutenins versus gliadins. *Journal of Applied Polymer Science* 123: 1530–1538.

Panagiotopoulos, I.A., R.R. Bakker, T. de Vrije, P.A.M. Claassen, and E.G. Koukios. 2013. Integration of first and second generation biofuels: Fermentative hydrogen production from wheat grain and straw. *Bioresource Technology* 128: 345–350.

Pandey, V.C., K. Singh, J.S. Singh, A. Kumar, B. Singh, and R.P. Singh. 2012. *Jatropha curcas*: A potential biofuel plant for sustainable environmental development. *Renewable and Sustainable Energy Reviews* 16: 2870–2883.

Peckham, B.W. 2001. The first hundred years of corn refining in the United States. *Starch/Stärke* 53: 257–260.

Peighambardoust, S.H., A.J. Van Der Goot, R.J. Hamer, and R.M. Boom. 2009. Process for the separation of gluten and starch. Patent publication number U.S. 2009/0202689 A1.

Piang-Siong, W., P. De Caro, C. Lacaze-Dufaure, A. Shum Cheong Sing, and W. Hoareau. 2012. Effect of catalytic conditions on the synthesis of new aconitate esters. *Industrial Crops and Products* 35: 203–210.

Pomeranz, Y. 1987. *Modern Cereal Science and Technology*. New York: VCH Publishers.

Pomeranz, Y. 1988. Chemical composition of kernel structures. In: *Wheat: Chemistry and Technology*, ed. Y. Pomeranz, pp. 97–158, Vol. I., St. Paul, MN: AACC.

Pye, E.K., M. Rushton, and J.R. Maclachlan. 2012. Organosolv biorefining of whole sugar cane. Patent application number 20120094348.

Rabelo, S.C., H. Carrere, R. Maciel Filho, and A.C. Costa. 2011. Production of bioethanol, methane and heat from sugarcane bagasse in a biorefinery concept. *Bioresource Technology* 102: 7887–7895.

Ragaee, S.M., P.J. Wood, Q. Wang, S. Tosh, and Y. Brummer, 2008. Extractability, structure and molecular weight of β-glucan from Canadian rye (Secale cereale L.) whole meal. *Cereal Chemistry* 85: 283–288.

Ramachandran, S., S.K. Singh, C. Larroche, C.R. Soccol, and A. Pandey. 2007. Oil cakes and their biotechnological applications—A review. *Bioresource Technology* 98: 2000–2009.

Rao, M.D., J.S.M.D. Hussain, P.V. Rangadu, K. Subramanyam, G.S. Krishna, and A.V.N. Swamy. 2007. Fermentative production of itaconic acid by *Aspergillus terreus* using Jatropha seed cake. *African Journal of Biotechnology* 6: 2140–2142.

Raquez, J.-M., M. Deléglise, M.-F. Lacrampe, and P. Krawczak. 2010. Thermosetting (bio)materials derived from renewable resources: A critical review. *Progress in Polymer Science* 35: 487–509.

Rehm, B.H.A. 2010 Bacterial polymers: Biosynthesis, modifications and applications. *Nature Reviews Microbiology* 8: 578–592.

Rivero-Urgell, M. and A. Santamaria-Orleans. 2001. Oligosaccharides: Application in infant food. *Early Human Development* 65: S43–S52.

Salvachua, D., A. Prieto, M.E. Vaquero, A.T. Martínez, and M.J. Martínez. 2013. Sugar recoveries from wheat straw following treatments with the fungus *Irpex lacteus. Bioresource Technology* 131: 218–225.

Samala, A., R. Srinivasan, M.P. Yadav, T.J. Kim, and L. Prewitt. 2012. Xylo-oligosaccharides production by autohydrolysis of corn fiber separated from ddgs. *BioResources* 7: 3038–3050.

Sancho, A.I., B. Bartolome, C. Gomez-Cordoves, G. Williamson, and C.B. Faulds. 2001. Release of ferulic acid from cereal residues by barley enzymatic extracts. *Journal of Cereal Science* 34: 173–179.

Satakarni, M., A.A. Koutinas, C. Webb, and R. Curtis. 2009. Enrichment of fermentation media and optimization of expression conditions for the production of EAK16 peptide as fusions with SUMO. *Biotechnology and Bioengineering* 102: 725–735.

Satake, S., T. Ishii, and Y. Tokui. 1995. Vertical pearling machines and apparatus for preliminary treatment prior to flour milling using such pearling machines. U.S. Patent number 5,390,589.

Sayaslan, A. 2004. Wet-milling of wheat flour: Industrial processes and small-scale test methods. *Lebensmittel-Wissenschaft und-Technologie* 37: 499–515.

Sayaslan, A., P.A. Seib, and O.K. Chung. 2010. Properties of starch and vital gluten isolated from wheat flour by three different wet-milling methods. *Journal of Applied Biological Sciences* 4: 57–62.

Schwarz, P.B., V.L. Youngs, and D.R. Shelton. 1989. Isolation and characterization of lignin from hard red spring wheat bran. *Cereal Chemistry* 66: 289–295.

Sharma, D.K., A.K. Pandey, and Lata. 2009. Use of *Jatropha curcas* hull biomass for bioactive compost production. *Biomass and Bioenergy* 33: 159–162.

Sikder, J., M. Roy, P. Dey, and P. Pal. 2012. Techno-economic analysis of a membrane-integrated bioreactor system for production of lactic acid from sugarcane juice. *Biochemical Engineering Journal* 63: 81–87.

Sinbuathong, N., P. Sirirote, B. Sillapacharoenkul, J. Munakata-Marr, and S. Chulalaksananukul. 2012. Biogas production from two-stage anaerobic digestion of Jatropha curcas seed cake. *Energy Sources, Part A: Recovery, Utilization and Environmental Effects* 34: 2048–2056.

Singh, N., H., Singh, and M.S. Bakshi. 1998. Determining the distribution of ash in wheat using debranning and conductivity. *Food Chemistry* 62: 169–172.

Singh, R.N., D.K. Vyas, N.S.L. Srivastava, and M. Narra. 2008. SPRERI experience on holistic approach to utilize all parts of *Jatropha curcas* fruit for energy. *Renewable Energy* 33: 1868–1873.

Smith, T.C., D.R. Kindred, J.M. Brosnan, R.M. Weightman, M. Shepherd, and R. Sylvester-Bradley. 2006. Wheat as a feedstock for alcohol production. HGCA Research Review No. 61.

Sun, R., J.M. Lawther, and W.B. Banks. 1996. Fractional and structural characterization of wheat straw hemicelluloses. *Carbohydrate Polymers* 29: 325–331.

Sun, R., J.M. Lawther, and W.B. Banks. 1997. A tentative chemical structure of wheat straw lignin. *Industrial Crops and Products* 6: 1–8.

Sun, X.F., R.C. Sun, P. Fowler, and M.S. Baird. 2004. Isolation and characterisation of cellulose obtained by a two-stage treatment with organosolv and cyanamide activated hydrogen peroxide from wheat straw. *Carbohydrate Polymers* 55: 379–391.

Talebnia, F., D. Karakashev, and I. Angelidaki. 2010. Production of bioethanol from wheat straw: An overview on pretreatment, hydrolysis and fermentation. *Bioresource Technology* 101: 4744–4753.

Taylor, J.C., L. Rapport, and G.B. Lockwood. 2003. Octacosanol in human health. *Nutrition* 19: 192–195.

Tiwari, U. and E. Cummins. 2009. Factors influencing β-glucan levels and molecular weight in cereal-based products. *Cereal Chemistry* 86: 290–301.

Tkac, J.J. 1992. Process for removing bran layers from wheat kernels. U.S. Patent number 5,082,680.

Toepfer, E.W., M.M. Polansky, J.F. Eheart, H.T. Slover, E.R. Morris, F.N. Hepburn, and F.W. Quackenbush. 1972. Nutrient composition of selected wheats and wheat products. XI. Summary. *Cereal Chemistry* 49: 173–186.

Ullah, A. and A.M. Shamsuddin. 1990. Dose-dependent inhibition of large intestinal cancer by inositol hexaphosphate in F344 rats. *Carcinogenesis* 11: 2219–2222.

Uppal, S.K. and R. Kaur. 2011. Hemicellulosic furfural production from sugarcane bagasse using different acids. *Sugar Tech* 13: 166–169.

Vasquez, M., J. Alonso, H. Dominguez, and J.C. Parajo. 2000. Xylo-oligosaccharides: Manufacture and applications. *Trends in Food Science and Technology* 11: 387–393.

Veraart, J.R., Y. Schouten, C. Gooijer, and H. Lingeman. 1997. Evaluation of phytic acid as a buffer additive for the separation of proteins in capillary electrophoresis. *Journal of Chromatography A* 768: 307–313.

Vyas, D.K. and R.N. Singh. 2007. Feasibility study of Jatropha seed husk as an open core gasifier feedstock. *Renewable Energy* 32: 512–517.

Wang, R., S.M. Shaarani, L.C. Godoy, M. Melikoglu, C.S. Vergara, A. Koutinas, and C. Webb. 2010. Bioconversion of rapeseed meal for the production of a generic microbial feedstock. *Enzyme and Microbial Technology* 47: 77–83.

Wang, T. and L. Johnson. 2001. Refining high-free fatty acid wheat germ oil. *Journal of the American Oil Chemists' Society* 78: 71–76.

Wang, T.-H. and S. Lu. 2013. Production of xylooligosaccharide from wheat bran by microwave assisted enzymatic hydrolysis. *Food Chemistry* 138: 1531–1535.

Webb, C. 1994. Grain processing: A challenge for chemical engineers. *Chemical Technology Europe* I(5): 21–23.

Webb, C., A. Koutinas, and R. Wang. 2004. Developing a sustainable bioprocessing strategy based on a generic feedstock. *Advances in Biochemical Engineering/Biotechnology* 86: 195–268.

Webb, C. and R. Wang. 1997. Development of a generic fermentation feedstock from whole wheat flour. In: *Cereals—Novel Uses and Processes*. eds. Cambpell, G.M., C. Webb and S.L. McKee, pp. 205–218, Springer Science + Business Media, New York.

Wellman, W. 1993. Process for milling cereal grains. U.S. Patent number 5,186,968.

Wever, D.A.Z., H.J. Heeres, and A.A. Broekhuis. 2012. Characterization of physic nut (*Jatropha curcas* L.) shells. *Biomass and Bioenergy* 37: 177–187.

Woerdeman, D.L., W.S. Veraverbeke, R.S. Parnas, D. Johnson, J.A. Delcour, and I. Verpoest. 2004. Designing new materials from wheat protein. *Biomacromolecules* 5: 1262–1269.

Wood, P.J. 1993. Physicochemical characteristics and physiological properties of oat (1–3), (1–4)-beta-D-glucan. In: *Oat Bran*, ed. Wood, P.J. pp. 83–112, St. Paul, MN: AACC.

Wood, P.J. 1994. Evaluation of oat bran as a soluble fibre source. Characterization of oat β-glucan and its effects on glycaemic response. *Carbohydrate Polymers* 25: 331–336.

Wood, P.J. 2010. Review oat and rye β-glucan: Properties and function. *Cereal Chemistry* 87: 315–330.

Xu, Y., R.-H. Wang, A.A. Koutinas, and C. Webb. 2010. Microbial biodegradable plastic production from a wheat-based biorefining strategy. *Process Biochemistry* 45: 153–163.

Yamada H., K. Itoh, Y. Morishita, and H. Taniguchi. 1993. Structure and properties of oligosaccharides from wheat bran. *Cereal Foods World* 38: 490–492.

Yan, L., N. Yang, H. Pang, and B. Liao. 2008. Production of levulinic acid from bagasse and paddy straw by liquefaction in the presence of hydrochloride acid. *Clean—Soil, Air, Water* 36: 158–163.

Yang, F., M.A. Hanna, and R. Sun. 2012. Value-added uses for crude glycerol—A byproduct of biodiesel production. *Biotechnology for Biofuels* 5: 13.

Yang, H.F., Y. Yang, Y.H. Yang et al. 2005. Formation of inositol hexaphosphate monolayers at the copper surface from a Na-salt of phytic acid solution studied by in-situ surface enhanced Raman scattering spectroscopy, Raman mapping and polarization measurement. *Analytica Chimica Acta* 548: 159–165.

Yeh, Y.F., R.C. Hoseney, and D.R. Lineback. 1980. Changes in wheat flour pentosans as a result of dough mixing and oxidation. *Cereal Chemistry* 57: 144–148.

Zeng, J., D. Singh, and S. Chen. 2011. Biological pretreatment of wheat straw by *Phanerochaete chrysosporium* supplemented with inorganic salts. *Bioresource Technology* 102: 3206–3214.

Zhang, A.Y.-Z., Z. Sun, C.C.J. Leung, W. Han, K.Y. Lau, M. Li, and C.S.K. Lin. 2013. Valorisation of bakery waste for succinic acid production. *Green Chemistry* 15: 690–695.

Zopf, D., S. Ohlson, J. Dakour, W. Wang, and A. Lundblad. 1989. Analysis and purification of oligosaccharides by high-performance liquid affinity chromatography. *Methods in Enzymology* 179: 55–64.

25 Green Biorefinery

H.S. Shekhar Sharma and Michael Mandl

CONTENTS

25.1 INTRODUCTION

This chapter comprises a review of green biorefinery systems, including processing steps, new technologies being developed, and future prospects for bioproducts. Biorefining is defined as "a processing and conversion facility that efficiently separates its biomass raw materials into individual components and converts these components into marketplace products, including biofuels, biopower, and conventional and new bioproducts" (Anon, 2002). During the past decade, European research groups have been investigating the biorefining of green biomass including ryegrass, clover, and lucerne (Andersen and Kiel, 2000; Grass, 2004; Kromus et al., 2004; Mandl, 2010; Venus, 2006, 2009; Lyko et al., 2009). These initiatives have assisted in the successful emergence of green biorefineries in Austria, Denmark, Germany, the Netherlands, Switzerland, and the United Kingdom. This sector will be influenced by factors, such as the availability of green or ensiled biomass, product range, conversion process employed, efficient utilization of the energy content of biomass, and production capacity (Brown, 2007).

25.1.1 Historical

The pioneering research carried out by Pirie (1942) led to the development of protocols for the extraction of chlorophyll, carotene, and protein from green crop plants, for example, lucerne, maize, immature wheat, and ryegrass. The objective of the investigation was to develop protein concentrates suitable for animal and human nutrition. The procedure involved pulping of the green leafy material and separation of the protein from juice by coagulation, followed by drying of the protein fraction. However commercial exploitation of the technology could not be implemented at that time, due to difficulties associated with the handling of fibrous crops and a lack of suitable processing machinery. The residual liquid fraction remaining after the coagulation of protein is rich in nutrients (Chiesa and Gnansounou, 2011) and may be utilized as organic fertilizer for crop production.

25.1.2 Generation I, II, and III Biorefineries

The first-generation biorefineries utilizing traditional food crops have focused on an edible part of a plant using technology based on the processing of maize, oilseed, and sugarcane. Utilization of all by-products was not considered, although residual products can be used as animal feed. This type of biorefinery uses a single feedstock to produce one major product, indicating the lack of processing capabilities for multiple products. The input material, that is, corn or oilseed, is usually processed after dry milling, or screw pressed, producing bioethanol and biodiesel, respectively. Typical examples are the early-generation biorefining plants operating in Europe and the United States using edible feedstocks (Kamm et al., 2006).

The first-generation biorefineries can only process one type of feedstock (e.g. oilseed or corn/wheat) to produce a single product, compared to second-generation plants which are capable of utilizing multiple feedstocks and produce various end products, including bioethanol, biodiesel, chemicals, and biomaterials, depending on the product demand. Corn biorefinery plants in the United States can produce a range of materials, including starch, high-fructose corn syrup, ethanol, corn oil, and gluten-free animal feed (Kamm et al., 2006; Brown, 2007). The benefits of using lignocellulose as a feedstock are lower cost and absence of competition with food production, with additional reductions in the environmental impact, energy usage, and production cost. Consequently, the conversion of lignocellulosic biomass into bioenergy and other specific value-added products could revolutionize this sector. However, key bottlenecks remain in the development of low-cost and high-yielding enzymes to enhance economic feasibility. Ultimately, all of the first-generation biorefineries, for example, biodiesel units, can be converted into the second-generation plants by fractionating additional products, such as glycerine, to other chemical products (Demirbas, 2009). In the early 1960s, the oil industry adopted process chemistry and engineering for an enhanced product range of high-value feedstock chemicals, in addition to their conventional refining of fuels from crude oil (Realff and Abbas, 2004). The green biorefining sector needs to adopt a similar initiative.

The third-generation biorefineries are still at a design and planning stage. These will be capable of processing a mixture of biomass feedstocks to produce multiple products, using a number of discrete technologies, including the production of protein, organic acids, bioethanol, nanofibrillated cellulose (NFC), and dyes. It is anticipated that the input materials could be green biomass and lignocellulosic materials from agriculture and/or forestry (Clark and Deswarte, 2008). Future green biorefinery initiatives must consider a holistic life cycle assessment (LCA), which includes all inputs from farm to factory to provide the full interdependencies of crop production, transportation, and green biorefining (Brown, 2007; Demirbas, 2009; Luguel, 2011a,b).

25.1.3 Future Land Use Scenario in Europe

Recent reports (Angus et al., 2009; Mandl, 2010) on livestock energy balances reveal that the proportion of pasture land required for animal production varies widely across the European countries depending on animal density, the provision of feed crops, and pasture yields (Smit et al., 2008). The majority of pasture areas in Belgium, Ireland, and the Netherlands are required for animal production, while in the remaining countries, including Denmark, France, Germany, and the United Kingdom, only 50%–70% of the available pasture land is utilized for animal production (Kleijn et al., 2009). In eastern European countries, livestock energy balances showed 30% use of pasture land for animal production. The surplus land can supply additional biomass, as a source for biorefining to generate bioproducts and bioenergy in integrated processing plants.

25.2 GREEN BIOMASS

Crop plants are cultivated for an edible part, with the remainder left on the field to be plowed in the soil for maintaining soil fertility or used as animal fodder. The desirable portion of the crop, that is, seed head or root, is harvested and further processed or supplied as fresh food. In contrast, unwanted leaves, stems, and roots of different crop plants are treated as agricultural waste. Crops in general can be used for both food and nonfood applications to generate both high- and low-value products.

25.2.1 Fodder Pellet Production

Researchers in Denmark initiated the whole crop green biorefinery concept with the objective of utilizing solid fraction as animal feed pellet and the juice as raw material for microbial fermentation to produce high-value products (Carlsson and Hanczakowski, 1989; Andersen and Kiel, 2000). They have also considered the potential threat of the environmental pollution of waterways by juice runoff from fields when used to fertilize grassland. In order to minimize energy consumption in the drying of perennial crop plants (e.g., ryegrass, clover/grass, and lucerne), the moisture content of the harvested biomass needs to be reduced by mechanical treatment, resulting in the production of green juice before drying the biomass as feed pellets.

Alternatively, the green crops can be heated at 80°C using steam before pressing, resulting in the coagulation of a large proportion of the protein. The liquid residue after the separation of protein is known as brown juice. Both the green and brown fractions have a dry matter (DM) content of approximately 6%. Green juice contained 13% carbohydrate and 35% protein and brown juice contained 33% carbohydrate and 26% protein (Andersen and Kiel, 2000).

Crops are usually harvested wet and transported directly to another site for drying and processing. Since biorefining plants are not always based in the vicinity of crop production sites, the transportation of dry biomass is a better option to reduce cost and energy requirements for downstream processing. Therefore, an on-site or mobile processing facility, including wet fractionation, will increase the density of transported biomass. The amount of fossil fuel saved by the reduction of transportation cost of biomass could compensate for the additional inputs in energy required for on-site processing with a mobile screw press (Wachendorf et al., 2009). The feasibility of mobile processing facilities should be evaluated to enhance the process efficiency of a regional biorefining plant, which can cater for raw materials from a surrounding 20–40 km^2.

25.2.2 Composition of Biomass

The first processing step is wet fractionation, employing a screw press to separate the fibrous material and juice. The liquid fraction is a nutrient-rich juice containing protein, minerals, and lactic and amino acids (AAs). Compositional differences between biomass fractions of ryegrass, clover, and lucerne are presented in Table 25.1 (Andrighetto et al., 1995; Oleskowicz-Popiel et al., 2011; Sharma et al., 2012a,b). A more rational utilization of the feedstock, capable of processing different fractions present in crops (listed in Table 25.1) other than fermentable sugars, could represent significant progress. Products of special interest could originate not only from sugars but also from the minor components, for example, fructose, AAs, and lignin. This approach could provide a product range similar to the petroleum industry. Improving the process efficiency of green biorefining is a crucial factor for boosting industrial development.

The composition of ryegrass fractions (i.e., juice and fiber fractions) is dependent on the time of harvest, type of feedstock (i.e., fresh or ensiled), DM content, and equipment employed to separate the fiber and juice fractions. In general, fresh plant juices have a lower DM content compared to silage juices, mainly because of the higher moisture content of fresh plant material. Table 25.2 shows the composition of polysaccharides, fructans, and organic acids in fresh juice, thereby

TABLE 25.1
Composition (g kg^{-1} DM) of Ryegrass (Green and Silage), Clover (Green and Silage), and Lucerne

Crop	Cellulose	Hemicellulose	Fructose	Lignin	Protein	Extractives	DM	Ash
Lucerne								
Green (Andrighetto et al., 1995)	575.8	81.7	—	71.9	216	—	917	91.9
Clover								
Green (Oleskowicz-Popiel et al., 2011)	232	146	193	135		275	184	18
Silage (Oleskowicz-Popiel et al., 2011)	210	14.2	0.0	53		418	174	20
Ryegrass								
Green (Sharma et al., 2012b)	226	196	58–120	31	175	—	229	69
Silage (Sharma et al., 2012b)	225	50	—	29	129	—	161	94

Note: Ryegrass fructose data from Table 25.4.

TABLE 25.2
Composition ($g\ kg^{-1}$ DM) of Green Italian Ryegrass Juice (GIRJ), Green Clover Juice (GCJ), and Green Lucerne Juice (GLJ); Brown Grass Juice (BGJ); and Ryegrass Silage Juice (RSJ)

Fractions ($g\ kg^{-1}$ DM)	GIRJ (Andersen and Kiel, 2000)[a]	GCJ (Andersen and Kiel, 2000)[a]	GLJ (Andersen and Kiel, 2000)[a]	BGJ (Andersen and Kiel, 2000)[a]	GLJ (Vodnar et al., 2010)	GLJ (Vodnar et al., 2010)	GLJ (Venus, 2006)	RSJ (Ecker et al., 2012b)[b]
Water-soluble carbohydrate	24.2	19.6	8.2	15.6	—	—	—	—
Free carbohydrate	15.2	13.0	8.2	12.0	12.5	8.3	8.8	1.1
Fructan	8.9	6.6	0.0	3.6	—	—	—	—
Succinic acid	0.8	0.3	0.2	0.2	—	—	—	—
Malonic acid	1.0	0.3	3.2	0.2	—	—	—	—
Citric acid	0.5	0.9	0.5	0.6	—	—	—	—
Malic acid	2.3	2.2	2.0	0.8	—	—	—	—
Acetic acid	0.0	0.0	0.0	0.2	—	4.1	—	5.9
Lactic acid	0.2	0.0	0.0	2.1	—	6.6	—	31.2
Formic acid	0.2	0.0	0.0	0.0	—	—	—	—
Total organic acid	5.0	3.7	5.9	4.1	—	0.0	—	—
Protein (N×6.25)	9.4	15.7	21.0	7.3	23.1	26.3	21.6	—
Amino acids	—	—	—	—	—	—	—	18.2
DM (%)	5.4	5.9	6.0	3.4	5.4	8.0	5.5	11.2

[a] Derived from ($g\ kg^{-1}$) into ($g\ kg^{-1}$ DM).
[b] Derived from ($g\ L^{-1}$) into ($g\ kg^{-1}$ DM) using density of 1.0 ($gm\ L^{-1}$).

predisposing the juice to fermentation. The latter is an important constraint for processing fresh juice, as rapid stabilization with an acid treatment is necessary to stop unwanted microbial contamination (Andersen and Kiel, 2000). In contrast, silage juice has a pH range of 3.5–4.5 and is stable (Danner et al., 2000).

25.2.3 Integration of Green Biorefinery

A third-generation green biorefinery must be optimized to process available biomass feedstocks and produce commercially viable products. A number of technical and processing hurdles must be cleared to achieve a high degree of process optimization, including pretreatment, improved batch and continuous flow fermentation, efficient biocatalysts, and the use of thermo/chemical/mechanical protocols. The process steps require the integration of different conversion technologies into an integrated third-generation biorefinery.

Economically viable pretreatment step(s) will enhance conversion of biomass and open up a large and diverse array of feedstocks, such as ryegrass, lucerne, agricultural crop residues, and forest brash. Several studies on the evaluation of sustainable green biorefineries have shown that this type of enterprise can be viable, if the process components are linked to high- and low-value products, along with the conventional use of ryegrass as fodder or silage for feeding animals (Kromus et al., 2004; Sharma et al., 2012a,b). O'Keefe et al. (2012) have shown that farmers in Ireland could be offered more than the production cost to maintain biomass availability, within the catchment area supplying the green biorefining plant. They also concluded that the choice of product range, that is, low and high value, was of significance for the green biorefinery design. Products such as NFC, AAs, lactic acid (LA), bioethanol, biogas, and fiber for insulation can be manufactured using fresh or ensiled ryegrass.

25.3 PRETREATMENT OF BIOMASS

Many methods for the pretreatment of biomass (Taherzadeh and Karimi, 2008; Alvira et al., 2010; Agbor et al., 2011; Galbe and Zacchi, 2012) have been investigated and are still in development. The evaluation of pretreatment technologies is complex, due to their contributions in the upstream and downstream processing steps, chemical recycling, and waste treatment systems (Aden et al., 2002; Ingram et al., 2011). However, in an integrated green biorefinery system, mass balance analysis can be used to determine the benefits of a pretreatment process. A thorough economic analysis is necessary to determine the optimal method suitable for a particular green feedstock (Aden et al., 2002; Sharma et al., 2012b). Galbe and Zacchi (2012) reported chemical treatment methods grouped by process pH, such as acidic, alkaline, and neutral pretreatments. A broader review classified processes into physical, chemical, biological, and combinatorial pretreatments (Agbor et al., 2011). In the latter process, temperature, pressure, and biological steps are combined with chemical treatments, designated as physicochemical methods (Sun and Cheng, 2002; Itoh et al., 2003; Girio et al., 2010; Ingram et al., 2011). The strategies adopted in the physicochemical treatment protocols can enhance biomass accessibility, thereby increasing recovery of target constituents.

Figure 25.1 gives the flowchart showing the process steps for pressed juice and fiber fractions from fresh grass/silage. Mechanical fiber preparation is followed by the evaluation of quality by the visible and near-infrared spectroscopy (Vis–NIRS)-based decision support system to grade fiber for low- and high-value process steps yielding roof insulation, biocomposites, intermediate platform chemicals, bioethanol, and NFC. Pressed fresh grass juice is separated for coagulated protein and plant biostimulant and the use of sugars/fructose for bioethanol. Silage juice is supplied to the anaerobic digester for biogas production to generate combined heat and power (CHP) and/or separated by an ion-exchange (IE) or hybrid process to fractionate AAs, LA, and other products (see Figures 25.3 and 25.4 later in the chapter; Ecker et al., 2012a,b; Sharma et al., 2012a,b). Figure 25.2 shows the processing of fresh ryegrass for the production of NFC.

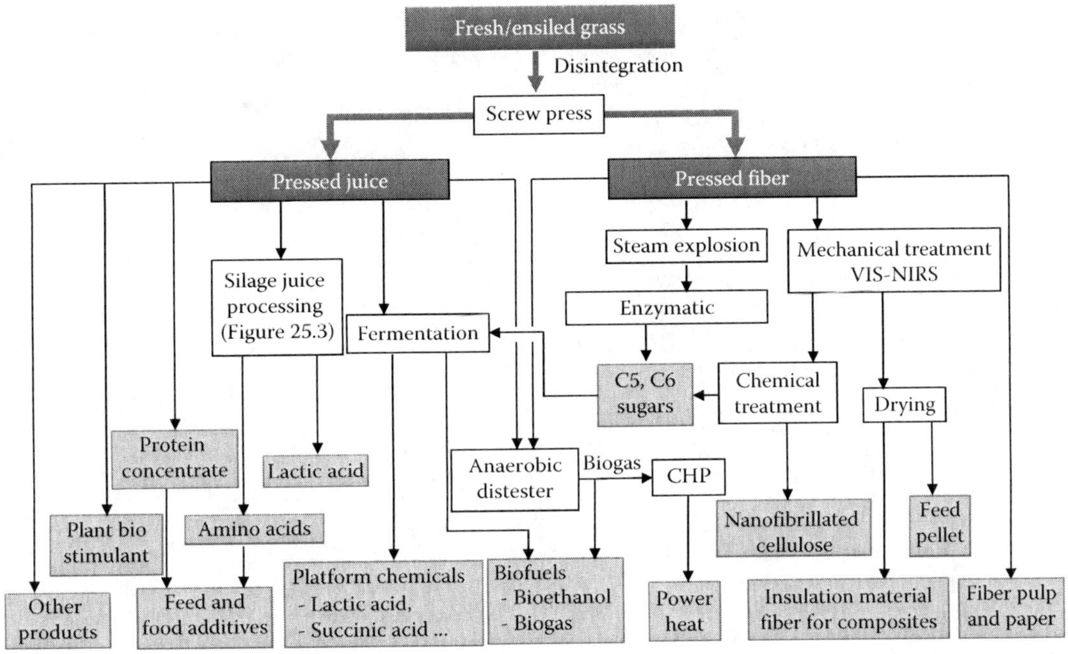

FIGURE 25.1 Flowchart showing process steps for pressed juice and fiber fractions from fresh grass/silage.

25.3.1 Hydrothermal Treatment

The green biomass is cut to a particle size of 4–8 cm lengths to facilitate the dewatering step. The process involves a mechanical pressing stage to separate juice from the fibrous materials, with or without hydrothermal conditioning steps to enhance the removal of soluble constituents (Wachendorf et al., 2009). In addition, the hydrothermal treatment assisted the flow of biomass during screw press operation, which results in the abrasion of biomass under high pressure (Wachendorf et al., 2009; Sharma et al., 2012a,b). Wachendorf et al. (2009) also described a protocol for hydrothermal conditioning and mechanical dehydration of silage, in which water and silage were mixed (4:1 ratio) and heated to 60°C or 80°C under continuous stirring for 15 min. The treatment macerated plant cells in the conical screw press, which allows the release of proteins, carbohydrates, and minerals, such as K, Mg, P, and Cl, in the juice fraction. The resulting solid fiber cake is suitable for thermal combustion after drying to 85% DM content, and the liquid fraction can be utilized as a nutrient source for biogas production. Different types of pressure-assisted dewatering protocols, for example, centrifuge, filter, and belt/roller press, can be employed to reduce the moisture content of green biomass. The use of a less effective hydraulic press has also been described by King et al. (2012).

Recently, a French research group (Arlabosse et al., 2011; Mahmoud et al., 2011) published reports on the use of a thermally assisted mechanical dewatering (TAMD) process to separate lucerne and spinach into solid and liquid fractions, without the initial addition of water in wide temperature (21°C–90°C) and pressure (300–3000 kPa) ranges. TAMD is an energy-efficient liquid–solid separation process that enhances the separation efficiency by supplying heat, during the mechanical dewatering step. It has been shown that the TAMD can be used to remove up to 83% of the inherent water from lucerne under moderate processing conditions (T_{piston} = 80°C) compared to 43% in a conventional mechanical dewatering process (Arlabossea et al., 2011; Mahmoud et al., 2011). Steam pretreatment is a commonly used physicochemical method for biomass treatment. Hemicellulose is the predominant fraction hydrolyzed during the treatment at high temperature (160°C–240°C) and pressure (0.7 and 4.8 MPa) for 0.3–1 min (Saddler et al., 1993; McMillan, 1994;

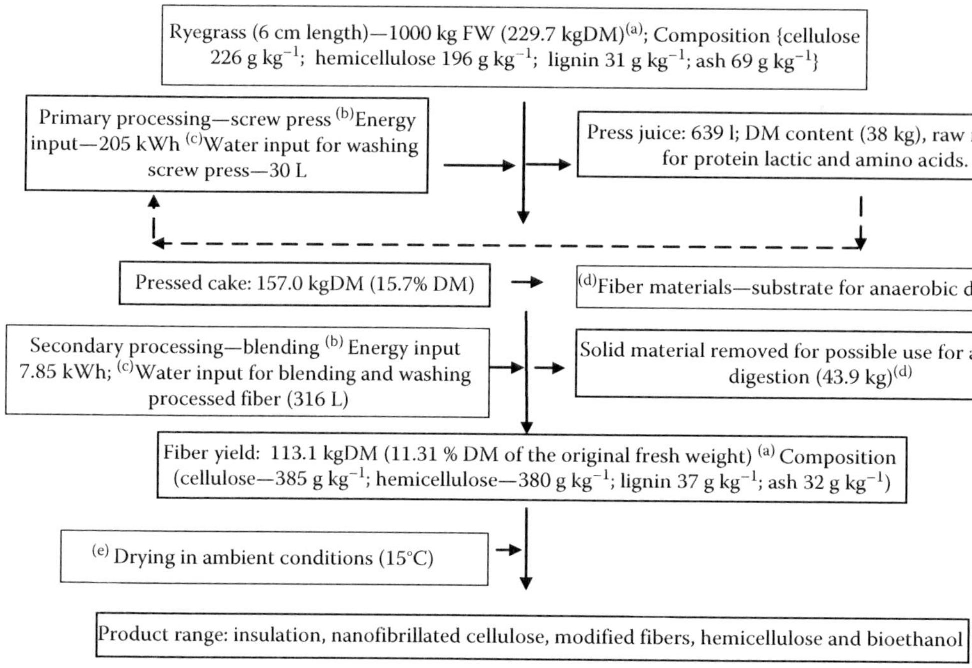

(*a*) *Mean composition of fiber fractions at the start and end of processing (FW, fresh weight; DM, dry matter).*

(*b*) *Energy inputs at each stage of processing measured by using a power monitoring module; note power requirement differences (primary processing of fresh grass with conical Pondorf screw press and secondary processing with* 5 L *commercial blender—note power input differences; subsequent loss of fiber materials during the second processing step).*

(*c*) *Part of the water used for washing processed samples can be recycled after filtration.*

(*d*) *Anaerobic digester for biogas generation using fiber and dilute wash water as input materials. Loss of DM between primary and secondary processing steps (27.07% DM).*

(*e*) *Drying of the processed fiber samples carried out in a plastic house to minimize dry-back (hydrogen bonding between residual hemicellulose, cellulose, and other components).*

FIGURE 25.2 Processing of fresh ryegrass for the production of NFC: mass, water, and energy balances of ryegrass processing showing changes in major fractions and potential products and mass, water, chemical, and energy balances of NFC production using four physical and chemical process steps, nanocellulose yield (kg t^{-1} DM); solid lines denote flow of material and broken lines denote flow of recycled water. (From Sharma, H.S.S. et al., *RSC Adv.*, 2, 6424, 2012b.)

Chandra et al., 2007). The residual lignin is transformed as a result of the high-temperature treatment and the cellulosic fraction in biomass becomes more accessible, resulting in a less severe downstream processing requirement.

25.3.2 Physical Treatment

Milling of fiber cake to a particle size of 1.0–0.5 mm is necessary to enhance degradability of the lignocellulosic fraction. The milling of biomass increases available specific surface area and reduces both the degree of polymerization during subsequent process steps (Sun and Cheng 2002). Vibratory ball milling is more effective than ordinary ball milling in reducing the degree of polymerization of plant fiber. Disk milling of fiber is more efficient in enhancing cellulose hydrolysis than hammer milling, due to a larger surface area of the finer fiber bundles (Zhu et al., 2009). Delgenes et al. (2002) have shown that the milling of input raw material increases biogas and bioethanol yields of biomass. However, the high energy requirement of milling biomass on an industrial scale would not be economically viable (Hendricks and Zeeman, 2009). Alternatively, the milling process can be performed after the chemical pretreatment, resulting in a reduced process energy requirement (Zhu et al., 2009; Zhu et al., 2010).

25.3.3 Chemical Treatment

Chemical treatments employing acid, alkali, organic solvent, and ionic liquid have been reported to indicate a significant effect on the structure of lignocellulosic biomass (Fengel and Wegener, 1984; Swatloski et al., 2002). Alkaline pretreatment of biomass disrupts the lignin structure by breaking the lignin–carbohydrate bonds and permitting carbohydrate fractions in the heterogeneous matrix to be more accessible. Furthermore, the accessibility and reactivity of the polysaccharide complex increases, as the lignin is selectively removed. Acetyl and other uronic acid substitutions on hemicellulose can reduce the accessibility of enzymes to the cellulose surface, but are removed by alkaline pretreatment (Mosier et al., 2005; Chandra et al., 2007; Galbe and Zacchi, 2012). Alkaline pretreatment is the most effective with low lignin content of biomass, such as ryegrass fiber cake (Sharma et al., 2012b), but becomes less effective with high lignin-containing biomass, such as willow and miscanthus. Dilute solutions (1%–4%) of sulfuric, hydrochloric, and phosphoric acid have been used to hydrolyze ryegrass fiber cake (Torget et al., 1992; Nguyen et al., 2000), resulting in the breakdown of hemicellulose to its monomeric units and also rendering the cellulose more accessible. The residence time ranges from a few seconds to minutes, depending on the temperature of pretreatment (140°C–215°C, Nguyen et al., 2000).

Organosolv pretreatments are conducted at a temperature range of 100°C–250°C using organic solvents, such as methanol and ethanol (Thring et al., 1990). Ethanol pretreatment removes lignin extensively, resulting in hemicellulose solubilization by disruption of the internal lignin bonds, as well as the ether and 4-*O*-methylglucuronic acid ester bonds between lignin and hemicellulose, and breaking down the glycosidic bonds in hemicellulose and cellulose depending on the process conditions (Zhao et al., 2009).

Ionic liquids are a class of solvents with low melting points (≤100°C), high polarities, high thermal stabilities, and negligible vapor pressure, consisting entirely of ions (cations and anions) (Zavrel et al., 2009). Most ionic liquids used in biomass fractionation are imidazonium salts (1-allyl-3-methylimidazonium chloride [AMIMCl] and 1-butyl-3-methylimidazonium chloride [BMIMCl]), which can be used effectively as a nonderivatizing solvent for the dissolution of cellulose at temperatures below 100°C (Zhu et al., 2006; Zavrel et al., 2009). Dissolution is achieved by the ionic liquid competing with lignocellulosic components for hydrogen bonding, thus disrupting the 3D network (Moulthrop et al., 2005). The cellulose fraction can be recovered by precipitating in either water/ethanol or acetone. The solvent can be recovered by employing various methods including reverse osmosis, salting out, and IE.

25.3.4 Biological Treatment

Fungi (e.g., basidiomycetes) can selectively degrade lignin and hemicellulose with their hydrolytic enzymes, liberating cellulose (Sun and Cheng, 2002; Talebnia et al., 2010). Brown rot fungi primarily attack cellulose with cellulase enzymes, whereas white and soft rot fungi attack both cellulose and lignin, via the production of lignin-degrading enzymes, such as lignin peroxidases, polyphenol oxidases, manganese-dependent peroxidases, and laccases (Agbor et al., 2011). Researchers have developed a cellulase-less fungal mutant for the selective degradation of lignin and to prevent the loss of cellulose, but in a majority of applications, the rate of hydrolysis is very slow for industrial application (Hatakka, 1994). However, biological pretreatment is energy efficient, due to a reduced need for subsequent mechanical treatments (Sun and Cheng, 2002; Talebnia et al., 2010). A recent study has shown that brown rot fungi can initiate hydrolysis of hemicelluloses and cellulose (Prasad et al., 2007), and such pretreatment improved the subsequent enzymatic hydrolysis of cellulose.

Additional biological pretreatment can be performed by applying hydrolytic or laccase enzymes to the fiber cake. Itoh et al. (2003) reported the impact of biological pretreatment of beech wood with white rot fungi, resulting in energy savings of 15% during the downstream ethanolysis of beech wood undergoing bioorganosolv pretreatment.

Ensiling of green fodder crops is practiced throughout the world for conserving nutrients in the biomass. LA bacteria dominate the ensiling process at low pH caused by the hydrolysis of sugars, hemicellulose, and proteins, thereby preserving the feedstock from further degradation. The whole crop silage (e.g., immature wheat and ryegrass) consists of a valuable lignocellulosic fraction, for example, cellulose, hemicellulose, lignin, and starch. Oleskowicz-Popiel et al. (2011) investigated the microbial fermentation of ensiled maize, rye, and mixture of clover/ryegrass to test efficiency of enzymatic hydrolysis for assessing glucose yield, sugar recovery, and further conversion of sugar to ethanol by *Saccharomyces cerevisiae* and *Kluyveromyces maxianus*. The yields significantly increased after hydrothermal pretreatment ranging from 72.8% to 80.7% of the theoretical ethanol yield for C6 sugar content of the ensiled plant materials (Oleskowicz-Popiel et al., 2011).

25.4 SEPARATION OF JUICE AND FIBER

The processing of ryegrass (e.g., ensiled or fresh) is carried out in two stages: the primary processing to separate the juice and fiber and the secondary processing of fiber cake to further refine and remove residual juice. The primary processing of ryegrass requires a high-volume screw press (0.5–1 tonnes h^{-1}). The advantages of employing a mobile screw press have been discussed by Wachendorf et al. (2009). In a large-scale trial carried out in Northern Ireland, the primary processing employed a screw press turning at 11–14 revolutions per minute with the pitch set at 1:7.5 (Sharma et al., 2012a,b). The resulting fiber cake is processed in a mechanical blender with additional water to remove residual juice and air-dried for 48–72 h under ambient conditions (12°C–30°C) to minimize dryback. The mass balance of the input material is presented in Figure 25.2 to show differences in the yield of extruded fiber at primary and secondary processing steps. Mandl et al. (2006) reported results of processing trials of clover/grass and lucerne silages, showing that 90% of LA and 55%–65% of crude protein (CP) can be recovered from the juice by repeated (i.e., double) screw pressing of the biomass.

25.4.1 Bioethanol from Fresh Juice

Fresh ryegrass juice represents a potential feedstock that could be used for the production of biofuels or other commercially valuable feedstock intermediates. The microbial fermentation of ryegrass *Lolium perenne* (cv. Aberdart) fructans to bioethanol has been demonstrated by Martel et al. (2010). In this study, fructans in grass juice are first hydrolyzed to bioavailable fructose monomers, using a novel core domain of bacterial (*Lactobacillus paracasei*) fructosidase, then fermented to ethanol using *S. cerevisiae*. Bioethanol yields are higher with enzyme-pretreated (36.5 mg mL^{-1}) than enzyme-untreated grass juice (28.2 mg mL^{-1}) feedstocks (Martel et al., 2010). The advantages of microbial species that can simultaneously saccharify and ferment complex lignocellulosic substrate to biofuels are widely known. However, identifying such species suitable for bioethanol production is challenging, due to the poor efficiency of enzymes involved in the hydrolysis of plant materials (Alper and Stephanopoulos, 2009). For example, *Bacillus subtilis* produces both cellulase and inulinase enzymes, but under fermentative conditions, *B. subtilis* produces lactate, acetate, and butanediol, with only trace levels of ethanol (Cruz-Ramos et al., 2000). Ethanol yields can similarly be very high with *S. cerevisiae*, but enzymes required to degrade lignocellulose are not produced. Consequently, Martel et al. (2011) initiated the use of genetically engineered yeast for the bioconversion of juice, as an alternative to the enzyme prehydrolysis of substrate. Bioethanol yields were 32.7 and 36.8 mg mL^{-1} of juice for recombinant yeast (with construct AH22:psacC; Martel et al., 2011) and enzyme-prehydrolyzed grass juice (Martel et al., 2010), respectively. This represents a significant breakthrough in the production of bioethanol from grass juice.

25.4.2 Plant Biostimulant from Fresh Juice

Biostimulants can reduce fertilizer use, improve yields, increase resistance to water and temperature stresses, and positively affect plant growth and physiology (Richardson et al., 2004). A recent plant bioassay study has shown that the liquid fraction of ryegrass can be used as a biostimulant/elicitor for application on field and horticultural crops (Selby, C., AFBI, UK, unpublished data, 2013). After the separation of all the high-value components, including fiber cake, protein, AAs, and minor fractions, a clear solution containing minerals and residual organic compounds is left in the residual fraction. The heat treatment of extracts from four cultivars (i.e., Abermagic, Dromara, Spelga, and Tyrella) of ryegrass by autoclaving the supernatant neither increased nor decreased the elicitor activity. The elicitor(s) fractionated from ryegrass is water soluble at ambient temperature and heat stable (Selby, C., AFBI, UK, unpublished data, 2013). This is a promising area of research with an extensive market potential for biostimulant, organic, and mineral products (Mora et al., 2010; Paradikovic et al., 2011; Sharma et al., 2013).

25.4.3 Production of LA from Fresh Juice

Thomsen et al. (2004) have developed a LA fermentation process using an inoculum of *Lactobacillus salivarius* subspecies *Salivarius*, and most of carbohydrates and organic acids in the juice are converted to LA by the *Lactobacillus* sp. increasing yield of LA. Supplementation of the juice with an additional carbohydrate source has been suggested by Andersen and Kiel (2000). Potential sources include molasses or carbohydrate products from sugar beet or wheat processing mills as additives (Thomsen, 2005). The results of LA fermentation of fresh lucerne juice enriched with glucose (55 g L^{-1}) and yeast extract (15 g L^{-1}) have been reported showing lactate yields of 91% (Venus, 2006; Vodnar et al., 2010). Fresh juice is currently used as an additive for LA fermentation to reduce the application of expensive ingredients, such as peptone and yeast extract in their processing plant (Section 25.6.4).

One of the major challenges of LA production is the development of efficient process steps, during acid recovery and subsequent purification, which could amount to almost 50% of the final product cost (Evangelsita et al., 1994). Additional separation strategies, such as chromatography in combination with membrane technology, need to be adopted in order to meet the required technical and yield target for enhancing process efficiency. Polylactic acid is a biodegradable polymer manufactured by the polymerization of LA, which is showing great potential, as a biopolymer for many products including food packaging. However, the production cost of polylactic acid must be reduced to compete with conventional packaging materials, suggesting a requirement for cheaper sources of LA.

25.4.4 Protein from Fresh Juice

The protein fraction in juice can be recovered by heat treatment following acidification (Pirie, 1942; Chiesa and Gnansounou, 2011). Further degradation of the protein with protease enzyme is necessary to recover AA building blocks. Dutch researchers have advanced technologies linked to separation, isolation, and purification of the AA mixtures to fractionate food and feed grade protein concentrates (Hulst et al., 2000).

25.4.5 Processing of Silage Juice

Danner et al. (2000) have evaluated the composition of silage extensively. They reported that the DM content of silage juice can range between 257 and 428 g kg^{-1} DM (optimum range 280–350 g kg^{-1})

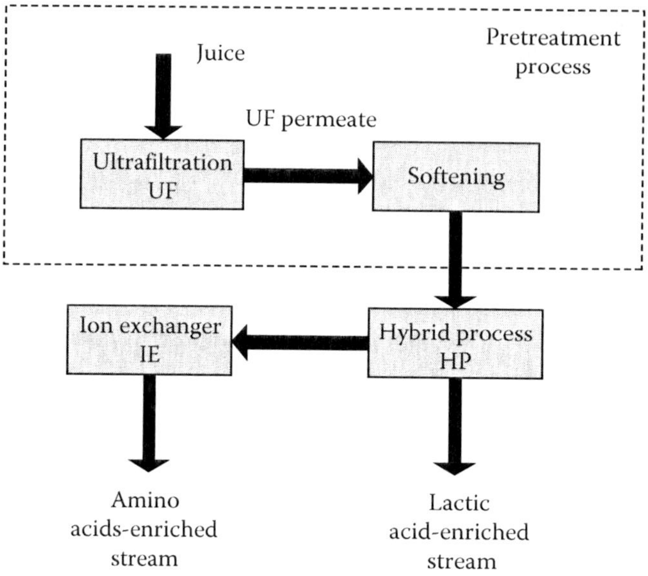

FIGURE 25.3 Process flowchart for biorefining silage juice, showing major process steps to separate amino and LA by employing UF and a hybrid process of NF and ED. (From Ecker, J. et al., *Sep. Purif. Technol.*, 96, 237, 2012b.)

with an LA content of 85–177 g kg^{-1} DM in the pressed silage juice, but the yield of LA in the extract ranged between 31 and 96 g kg^{-1} DM of silage. The variability in LA content of silage samples can be attributed to the heterogeneous nature of silage (Danner et al., 2000; Thang and Novalin, 2008). Furthermore, the LA content of silage juice is related to the quality of ryegrass at harvest, and subsequent ensiling conditions. More advanced processing protocols are presented in Figure 25.3, showing pretreatment of silage juice followed by filtration steps (Ecker et al., 2012a,b). They have demonstrated a pilot-scale process to fractionate lactic and AAs from silage juice, following a primary treatment step to remove solids and dispersed organic particles by sedimentation. The supernatant is processed by ultrafiltration (UF) (cut off at 300 kDa) to remove impurities, such as macromolecules and plant fibers, followed by filtering through a column containing a strong cation exchanger to reduce cations (Ca^{2+} and Mg^{2+}). Subsequently, the prefiltered and softened silage juice is processed through a two-stage nanofiltration (NF) (cut off at 1 kDa) process. In the NF hybrid process (Figure 25.3), the AA is recovered in the NF retentate (Figure 25.4a) and the LA is mainly passed into the NF permeates (Figure 25.4b; Ecker et al., 2012b). The NF permeates are desalted by electrodialysis (EDI) (Figure 25.4a) and the purified fraction is collected in a vessel, followed by reverse osmosis to increase concentration. A second ED device (EDII) (Figure 25.4a) is employed to separate LA from the concentrated stream, yielding 54.6% (Figure 25.4a). In the second product stream, the AA-enriched NF2 retentate solution is processed by IE to recover AA mixtures of 77.3% concentration (Figure 25.4b). Both product streams require further processing to increase the concentration of lactic and AAs. For example, the AA fraction can be spray-dried to produce a dry and stable product of 80% free AA content. Further purification of the AA-enriched fraction can deliver a specified grade of AA purity (Ecker et al., 2012b).

25.4.6 Feed for Anaerobic Digester

All farm wastes, including grass or forage juice (Demirbas, 2009), can be used to produce biogas, which is a renewable resource consisting mainly of methane and carbon dioxide, produced during anaerobic digestion. However, high-quality silage is required to maximize methane yield

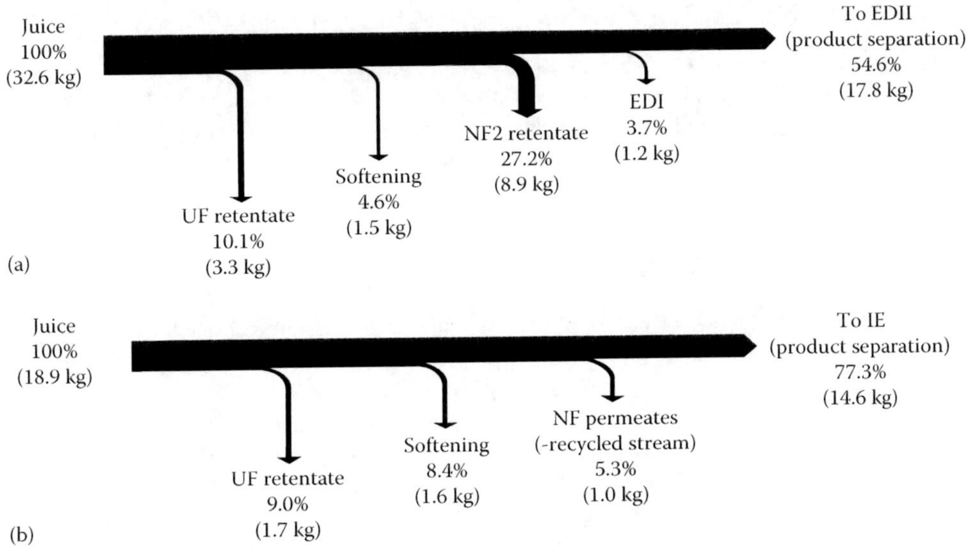

FIGURE 25.4 Product partition steps to separate (a) lactic (32.6 g kg^{-1}) and (b) AA (18.9 g kg^{-1}) using UF, NF, ED, and IE separation technologies. (From Ecker, J. et al., *Sep. Purif. Technol.*, 96, 237, 2012b.)

(Prochnow et al., 2009). Currently, silage fiber cake is anaerobically digested and the resulting biogas is widely used for power generation in Europe; a total of 2429 anaerobic digesters (AD) is in operation (AD-Nett accessed on July 12, 2012, http://www.adnett.org/) with many more at the commissioning stage. Alternatively, after extraction of valuable solid fractions from the silage, the residual liquid fraction can be fermented to generate biogas in an AD plant. Ryegrasses (highly digestible: 73%–80% D value; Sharma et al., 2012c) can yield 200–250 m^3 CH_4/kg_{VDS} of biogas (Murphy and Power, 2009).

Anaerobic digestion is an important technology for implementing green biorefining, since it is capable of utilizing residual liquid or solid organic residues after biomass processing. In addition, the integration of green biomass processing with an anaerobic digester unit can deliver heat from the CHP for drying products or hydrothermal pretreatment of feedstock. Grass silage is not widely used in Europe as feedstock for anaerobic digesters due to its fibrous consistency, lower density compared to maize silage, and the requirement for preprocessing before anaerobic digestion. If the proportion of input silage is high (40%), the parasitic energy demand to operate the paddle in the digestion tank would increase, thereby reducing energy output, and the biogas yield of the digester can also be inhibited by high nitrogen content of silage. Therefore, a mechanical pretreatment step to homogenize a mixture of silage and slurry is employed before feeding into the fermenter tank. In Northern Ireland, seven commercial anaerobic digester plants are currently operating using grass/maize silage, root vegetables, and waste effluents from dairy, pig, and poultry farms to generate biomethane.

25.4.7 Other Products

Lee and Chang (1994) reported that supplementation with a complex nitrogen source, such as tryptone, could enhance 3-polyhydroxybutyrate (PHB) production by recombinant *Escherichia coli* in a defined medium containing glucose as sole carbon source. Koller et al. (2005) studied the production of 3-PHB with *Wautersia eutropha* using juice extracted from green grass and silage and concluded that silage juice can be used as a sustainable substrate for the production of microbial polyhydroxyalkanoates. Pakarinen et al. (2009) investigated H_2 and CH_4 production, with silage pretreated with NaOH, and concluded that the fiber cake can produce thermophilic H_2 and mesophilic CH_4 in a two-stage process, while the silage juice can be utilized as substrate for CH_4 production.

25.5 PROCESSING OF FIBER CAKE

The fiber cake resulting from the primary processing stage (i.e., screw press) requires a secondary treatment to remove the residual plant tissue and juice from the fiber. After the fiber cake is cut to fiber lengths of 8–10 cm, it is mixed with water at a ratio of 1:10 (mass/volume) and processed in a commercial blender. The resulting fibrous material requires further treatment with water to wash off residual juice.

25.5.1 Fiber Quality

Vis–NIRS has been used widely for the quality assessment of agricultural produce, such as forages, feeds, and silages. Recently, Perbandt et al. (2010) evaluated Vis–NIRS technology to estimate nitrogen, ash, and mineral contents of pressed silage cakes of various plant materials, including grasses from seminatural grassland sites in Germany. Their objective was to develop rapid protocols for the quality assessment of biomass for bioenergy applications. Fibers from ryegrass and other forage crops have been characterized using modified textile protocols by European researchers (Sfilgoj-Smole et al., 2004, 2005). Recently, Sharma et al. (2012a) have developed Vis–NIRS calibration equations for evaluating fiber quality, indicated by cleanliness and fineness of fiber from ryegrass (green/silage). The calibrations developed from this study can be used for monitoring quality of processed fiber, determining downstream process routes for high- or low-value products in a green biorefining plant. The quality can be measured rapidly and accurately by a Vis–NIR spectrometer, with a mean error of 1.26 dtex. This degree of precision is sufficient for biorefining plants to determine specific end use application of coarse fiber for insulation products, pots and mulch boards, and fine/clean fiber for bioethanol, NFC, and other high-value products. Other major factors determining fiber quality are the amounts of residual hemicellulose and lignin on the fiber.

The important products listed in this section are bioethanol, NFC, and intermediate feedstocks, such as succinic and acetic acid. The factors for selecting the products are based on a strong promotion of bioethanol as a transport fuel in Europe (DT, 2005), high commercial potential of NFC for fabricating nanoproducts (Sharma et al., 2011b), and excellent feedstock potential of succinic and acetic acid (Bechthold et al., 2008).

25.5.2 Nanofibrillated Cellulose Production

A recent study has demonstrated the process steps required for the biorefining of ryegrass to produce fiber cakes, as raw material for NFC (Sharma et al., 2012b). Four protocols, consisting of a combination of mechanical and chemical steps, were compared for yield of NFC from ryegrass. NFC prepared by employing NaOH pretreatment followed by mechanical and high-pressure microfluidization treatments has been identified as a promising material due to the enhanced thermal and mechanical performance of the NFC–polyvinyl alcohol composite (5 wt% NFC content), which also showed enhanced uniformity and filler dispersion compared with pure polyvinyl alcohol polymer sheets and other composite treatments (Sharma et al., 2012b). Composites containing NFC could be used for manufacturing high- or low-value products, such as biomedical packaging, drain pipes, gutters, and flower pots. Large-scale production of NFC could be carried out using an environmentally sustainable biorefinery protocol. Estimated NFC yields extracted from ryegrass ranged between 36.1 and 43.8 kg tonne^{-1} (Sharma et al., 2012b).

25.5.3 Bioethanol Production

Neureiter et al. (2004) were one of the firsts to demonstrate that the pressed fiber cake can be hydrolyzed and used as feedstock for bioethanol production. They concluded that the cold water pretreatment of the materials removed residual LA and other fractions, which can reduce efficiency during microbial processing. Further treatment of the materials with sulfuric acid (2.4% [w/w] for 20 min at 120°C) resulted in monomeric sugar yields of 1.64 g kg^{-1} DM, which is approximately 25% of available sugars and 60% of the available hemicelluloses (Neureiter et al., 2004). Martin et al. (2008) reported preliminary experiments on bioethanol fermentation by the hydrolysis of clover and ryegrass mixtures using *Mucor indicus*. They evaluated wet oxidation pretreatment and enzymatic hydrolysis followed by saccharification and fermentation of clover/ryegrass mixtures to enhance bioethanol yield. Assuming a DM yield of 10–12 tonnes ha^{-1} of high-sugar grass, such as Abermagic, one hectare of grassland in Wales and England could potentially produce 3000–4500 L of bioethanol, indicating suitability of ryegrass, as a renewable material for complying with road transport fuel obligation targets (IBERS, 2009).

25.5.4 Intermediate Feedstocks

The use of biobased feedstocks within the petrochemical industry can lead to a sizeable reduction of fossil fuel consumption (Brehmer et al., 2009). A typical example is succinic acid, which is a common intermediate in the metabolic pathway of several anaerobic and facultative microorganisms (Song and Lee, 2006; Bechthold et al., 2008). The succinic acid-producing bacteria from the fermentation produce succinate, as well as ethanol and lactic, acetic, and formic acid (Luo et al., 2010). The yields of succinic and acetic acids from glucose are reported to be 1.4 and 0.29 mol mol^{-1}, respectively (Okino et al., 2008). Currently, succinic acid is mainly produced by chemical processes from *n*-butane/butadiene via maleic anhydride (Bechthold et al., 2008). Major applications of succinic acid are in the manufacture of surfactant/detergent, iron chelator used during electroplating to prevent corrosion and pitting of metals, an acidulent/pH modifier, flavoring, and antimicrobial agent and the production of health-promoting products (Zeikus et al., 1999). The world market for succinic acid was estimated at 30,000 tonnes in 2009–2010 (Luo et al., 2010), but this sector is growing by 10% per year.

25.6 EUROPEAN GREEN BIOREFINERY

Currently, seven green biorefineries are operating commercially or as pilot plants in Europe, and brief descriptions of the process and products are as follows:

25.6.1 AgroFerm

The Danish research group (Andersen and Kiel, 1999, 2000) developed an acidification process to stabilize green and brown juices, which constitute by-products of animal feed pellet production for the livestock sector. The treatment allows storage of the juice fraction before processing. AgroFerm was established as a spin-off company in 2002, to manufacture L-lysine by fermentation of fresh and stabilized brown juice using a patented protocol (Andersen and Kiel, 1999). Lysine is an essential AA, used as a feed additive for pig and poultry to ensure optimal feedstuff utilization, better growth rates, improved meat quality, and reduced environmental impact due to the lower nitrate content in the slurry. The production plant is based in Esbjerg, Denmark, and currently operates under a new brand name VitaLys®. The production capacity of the factory is 12,500 tonnes of L-lysine per year using refined sugar with proprietary additives. The substrate is treated in a continuous sterilizer and transferred into an aerated stirred tank reactor. The fermenters are inoculated with a culture of *Corynebacterium glutamicum* and processed by batch fermentation. Website was accessed on July 12, 2012 (http://www.agroferm.dk/Default.asp?ID=35).

25.6.2 Biowert

In 2005, an industrial-scale grass biorefinery plant (Biowert Industrie, GmbH) was established in Germany. Meadow grass from farms in Odenwald is used as raw material for the production of high-value products. The company patented a protocol in which moist biomass is fractionated into a fiber cake and juice, without the application of chemicals or organic solvents (Gass, 2011). The core business of Biowert is the production of high-quality cellulosic fibers from grass for a range of applications. The fiber processing unit is integrated with an anaerobic digester by utilizing residual liquid fraction for biogas generation. The latter delivers power and heat for the processing of fiber, and the residual digestate is supplied back to the primary agricultural production cycle as organic fertilizer. The key product of Biowert is high-quality cellulose fiber (production capacity 2500 tonnes $year^{-1}$), which can be used for *blow-in* insulation material as well as AgriPlast®, a biocomposite material suitable for injection molding or extrusion applications. Website was accessed on July 12, 2012 (http://www.biowert.de/company/biowert).

25.6.3 Green Biorefinery Upper Austria

A pilot-scale green biorefinery (Green Biorefinery Upper Austria F&E GmbH) utilizing grass silage as feedstock was established in Utzenaich, Austria, in 2008. The pilot facility is integrated with an anaerobic digester plant to generate biogas from the solid and liquid residues. The silage grass is mechanically treated to open up the compressed layers of ensiled grass, followed by mechanical pressing to separate the juice and fiber fractions. Currently, the fiber cake is digested for the generation of biogas, linked to power generation units of 500 kW capacity, and the silage juice is processed through a cascade of purification steps. After prefiltration, a hybrid process is employed for the separation and purification of lactic and AAs. Further downstream processing steps include separation technologies, including UF and NF. LA is separated by ED from the nanofiltered fraction (Ecker et al., 2012a,b). In addition, a reversed osmosis process is employed to increase the concentration of LA in the final product. Different AA mixtures are recovered from nanofiltered retentate using IE chromatography. The demonstration plant has an operating capacity of 4000 L day^{-1} silage juice, to separate 55 kg of AA and 65 kg of LA per day. Further scaling up of the production facility is planned to increase capacity and enhance the economic viability of the enterprise.

25.6.4 Brandenburg Pilot Plant

A pilot plant to process green crops and cereals (e.g., rye and wheat) has been set up in Brandenburg in Germany. The biorefining plant is focused on demonstrating the LA production from a variety of feedstocks, including green biomass, starch crops, and sugar beet. The processing of fresh juice has been achieved by fermentation with *Lactobacillus* sp. (Venus and Richter 2006). The protocols employed in the plant have been developed from a collaborative research program participated by Austrian, Danish, and German groups (Venus 2006). The pilot plant can operate as either a batch or a continuous fermenter (up to 400 L vessel) with a capacity for 10 tonnes of LA per annum (website: http://www.atb-potsdam.de/hauptseite-deutsch/Institut/Abteilungen/Abt1/mitarbeiter/jvenus/Pilot_2006.pdf).

25.6.5 British Sugar

The company is one of the largest processors of sugar beet to manufacture sugar, bioethanol, animal feed, betaine, and sugar beet lime in Wissington, United Kingdom. After the extraction of sugar from sugar beet, the pulp is pressed to generate over 500,000 tonnes of high-quality animal feed supplied as either moist pulp or dried pellets. The animal feed is produced in accordance with stringent feed assurance standards and provides energy food for a wide range of stock including cattle, sheep, pigs, goats, and horses. It contains a high level of digestible fiber, which helps promote stable rumen conditions, and is supplied to the livestock industry by Trident Feeds, established as a subsidiary of British Sugar in 1985. The company is able to demonstrate full traceability of the whole process from crop production to fermentation and distribution, in such a way that environmental benefits are delivered. Bioethanol is produced by the fermentation of sugars and followed by distillation. The plant utilizes a CHP generation system to produce both steam and electricity. Producing up to 70 million liters of bioethanol per annum, the refinery uses around 110,000 tonnes of sugar, equivalent to 650,000 tonnes of sugar beet consumption (http://www.britishsugar.co.uk/Bioethanol.aspx).

25.6.6 Grassa

During the past decade, Dutch research groups and industrial partners have developed green biorefining protocols to process fresh ryegrass. A pilot plant has been set up in Oenkerk with a processing capacity of 500 kg of biomass per hour. The plant is designed to process fresh grass for the production of two main products: (1) a protein fraction suitable for animal feed and food supplementation and (2) a fiber fraction for the pulp and paper sector for manufacturing packaging films. The company is also focused on developing and implementing a mobile processing unit to reduce transportation cost of high-moisture feedstock. The processing technology developed for ryegrass could also be used for processing green leaves from sugar beet, currently left behind on the field after harvesting the roots. The enterprise has great potential for increasing the added value of products in the feed, pulp, and paper sectors. The company is aiming to reduce European Union (EU) dependency on imported protein, such as soya bean from regions outside of Europe by developing new opportunities (website: http://www.grassanederland.nl).

25.6.7 Gramitech

A green biorefinery plant has been established in Switzerland to process and market fiber products. The brand Gramitherm® comprises a variety of flexible insulation products of different densities and widths. The product complies with Swiss and EU standards, has a thermal conductivity of 0.040 Wm^{-1} K^{-1}, and is suitable for interior use in office and residential buildings due to high vapor permeability. Additional positive product features are excellent acoustic properties and negative CO_2 emission. The total embodied energy required for manufacturing ryegrass fiber insulation products is 20 kWh m^{-2}, which is only 10% of the energy required for the production of conventional insulation products, such as polystyrene. Embodied energy is defined as the amount of energy consumed in the manufacture, transportation, installation, and disposal of a material or a product. The juice from the processing plant is integrated with an anaerobic digester to generate biogas and power. The annual yield of 1 ha grassland can deliver about 200 m^3 insulation product.

25.6.8 Impact and Commercial Value of Bioproducts

The technologies and bioproducts highlighted (Tables 25.3 and 25.4) show an array of products, which can be manufactured by employing suitable process steps. However, large-scale green biorefineries are still in the planning and development stage due to a lack of pilot-scale trials of complex processing and separation technologies. The potential economic impact on feed, biogas, NFC, bioethanol/biodiesel, pulp, and paper is most promising due to the potential size of the market sectors (Table 25.4). Therefore, an integrated production of high-volume/low-revenue products (i.e., animal feed/biogas) in conjunction with small-volume/high-revenue products (e.g., protein/amino/LAs/NFC) and in the required scale of production should provide an ideal product range for economic viability. The most challenging task in green biorefining is achieving economic viability, by careful selection of target products linked to specific process technologies (Table 25.3).

Apart from the fermentation of ryegrass juice to bioethanol, most of the technologies listed earlier have been proven by various groups. However, the protocols have not been implemented in the recent second-generation biorefining plants possibly due to various factors, including economic

TABLE 25.3
Comparison of Product Matrix to Identify Viable Options of High- and Low-Value Products

Options	Product Processing Steps for Juice (Green/Silage)	Product Type	Product Processing Steps for Fiber Cake (Silage/Green)	By-Products
1	Protein/amino/lactic acid	A	Biogas	Organic fertilizer
2	Fresh juice to bioethanol	B	Fibers for pulp and paper/NFC	Lignin
3	Silage juice to biogas	C	Bioethanol from lignocellulose	Lignin

The specified product matrix can consequently lead to a combination of products and processes.

1+A	1+B	1+C
2+A	2+B	2+C
	3+B	3+C

viability of the processing steps. Future business opportunities for green biorefining are potentially excellent, and process development efforts should focus on enhancing the added value of the product range. In the future, high-value products (e.g., AAs for body care or nutritional supplement and functionalized NFC) are likely to be the focus of green biorefining rather than the production of bulk products.

25.7 POLICY, ENVIRONMENTAL, AND STRATEGIC PLAN

25.7.1 Policy Drivers

Although bioproducts are largely ignored in the current *biofuel versus food* debate, the impact of green biorefining on the availability of food and animal feed should be accounted for, and policy initiatives should consider the economic benefits of high-value bioproducts. Ajanovic (2011) investigated the relationship between an increase of biofuel production and rise in agricultural commodity prices. The major conclusions are that no significant impact of biofuel production on feedstock prices can be observed, and the adoption of second-generation biofuel will increase the bioproduct range, thereby reducing potential conflict with feedstock demand. Sorda et al. (2010) also reviewed biofuel policies, including the *food versus fuel* debate, across the developing and developed economies of the world, to provide key factors responsible for the expansion of biofuel economy. They concluded that recent policies are strongly distortionary, due to the use of food crops, for example, sugar beet, sugarcane, corn, and oilseed, as feedstock for bioenergy production. Consequently, this sector has been at least partially blamed for the increase in food prices between 2003 and 2008 (Mercer–Blackman et al., 2008; Mitchell, 2008). However, the principal food crops in most countries are rice and wheat rather than corn and soybean, which are mainly used for livestock feed, processed foods, and biofuels. However, the effects of the biofuel component of the grain commodity price increase of 2008 on food price in developing countries must have been significant. The same interpretation can be used for the rise in the food price index in 2011 (FAO, 2011). Similarly, estimated indices for 2012 (FAO, 2012) for cereal grains, oilseed, and sugar prices are, respectively, 124.4%, 142.5%, and 235.8% higher than the 2002–2004 average values. Oil price increases per barrel have had a much larger effect on food price index, as indices are interlinked (Canadian Renewable Fuels, http://www.greenfuels.org/en.aspx). Current political and scientific initiatives in Europe on biorefinery and industrial biotechnology are largely uncoordinated, and in addition, a high degree of policy uncertainty exists considering the time frames required for establishing an integrated green biorefinery operation (Clark and Deswarte, 2008; Luguel, 2011a).

TABLE 25.4
Evaluation of Green/Silage Ryegrass (1 DM Tonne Calculated Using 70% Moisture Content) for Product Range, Yield (g^{-1} kg^{-1} DM) of Refined Product, Cost of Product after Biorefining, and Current Market Value

Technology Pathway	Maturity of Technology (1)	Fraction Used	Utilized Component	End Product	Yield [g kg^{-1} DM Feedstock] (2)	Current Market Value [€kg^{-1} Product] (3)	Revenue Potential [€t^{-1} DM Feedstock] (4)
Membranes, Ion exchange	3	Silage juice	Amino acid	Amino acids mixtures (a)	70–90 (a1) 45–58 (a2)	1.5–8.0 (b)	105–720 67.5–464
Membranes, electrodialysis	3	Silage juice	Racemic lactic acid of ensilage	Racemic lactic acid	49–63 (c1) 19–28 (c2)	1.3–1.60 (c)	63.7–78.4 24.7–44.8
Anaerobe digestion	4	Silage juice	Silage juice, 12% DM	Methane	96–125 (e) 1_{nCH4} kg^{-1} DM in press	0.7–1.44 (d) €m^{-3}	67.2–180
Agglomeration, ultrafiltration, extraction	3	Fresh juice	Protein	Protein feed grade (f)	41–62 (g) 84–126 (h)	0.56–0.94 (i)	23.0–58.3 47.0–118.4
Ultrafiltration, extraction	2	Fresh juice	Protein	Protein food grade (j)	84–126 (k)	0.95–2.50 (l)	79.8–315
Specific fermentation	2	Fresh juice	Fructose	First generation bioethanol	58–75 (p)	0.625–1.04 (q)	36.3–78
Specific fermentation, specific DSP	4	Silage juice, grass juice	Lactic acid sugars, nutrients	Lysine	n.a.	1.7–1.75	0–0
Specific fermentation, fresh juice	2	Grass juice	Sugars, nutrients	Lactic acid	45–75 (r)	1.00–1.60 (v)	45–120
Pressing or refining, special fiber preparation	4	Grass press cake	Bulk press cake	Insulation material, fleece blow-in insulation fiber	700 (o)	80 €m^3 (o) 0.28 (o1)	339.4 196
Pressing or refining, special fiber preparation	3	Grass press cake	Bulk press cake	Fiber fraction for pulp and paper	500–700 (m)	0.035–0.12 (n)	17.5–84
Anaerobe digestion	4	Grass press cake	Bulk press cake	Methane	150–195 (e) 1_{nCH4} kg^{-1} DM in press	0.7–1.44 (d) €m^{-3}	105–280.8
Pressing, drying, pelleting	5	Grass, luceme	Fresh press	Feed pellets	850–950	0.25–0.30	212.5–285
Pressing or refining, special fiber	4	Grass press cake	cellulose	Fiber for composites	300–700 (s)	0.45–0.60	135–420

(continued)

TABLE 25.4 (continued)

Evaluation of Green/Silage Ryegrass (1 DM Tonne Calculated Using 70% Moisture Content) for Product Range, Yield (g^{-1} kg^{-1} DM) of Refined Product, Cost of Product after Biorefining, and Current Market Value

Technology Pathway	Maturity of Technology (1)	Fraction Used	Utilized Component	End Product	Yield [g kg^{-1} DM Feedstock] (2)	Current Market Value [€kg^{-1} Product] (3)	Revenue Potential [€t^{-1} DM Feedstock] (4)
Pretreatment, enzymatic hydrolysis, fermentation, specific DSP	1	Grass press cake	Cellulose	Second generation bioethanol	95–177 (t)	0.625–1.04 (q)	59.4–184.1
Alkaline pretreatment post green biorefinery pressing	2	Grass press cake	Cellulose	Nanofibrillated cellulose	173–409 (z1)	1.0–2.5 (z2)	173–1023
Pretreatment, enzymatic hydrolysis, fermentation, specific DSP	1	Grass press cake	Cellulose	Succinic acid	187–349 (z)	1.95–2.80	364.7–977.2
Pretreatment, enzymatic hydrolysis, fermentation, specific DSP	1, 2	Grass press cake	Cellulose + hemicellulose	L or D lactic acid	367–424 (u)	1.00–1.60 (v)	367–678.4
Separation after enzymatic	1, 2	Grass press cake	Lignin	Lignin pellets	50–150 (x)	0.22–0.26 (y)	11–39

(1) Maturity level is classified: 1 *discussion*, theoretical; 2 *laboratory or bench scale experiments* in scientific papers; 3 *in piloting stage* technology upscale; 4 in demonstration, first production, market uptake phase; 5 technology implemented and widely used.

(2) Yield [g kg-DM in]: specific yield of product per kg DM (green or ensiled) feedstock; yield includes losses (e.g., in downstream processing).

(3) Current market value [€kg^{-1} prod.]: revenue per kg of product in €; dependent on quality of product and end use (e.g., feed, food, pharmaceuticals).

(a) AA product, 80% free AA; (a1) yields based on the following: high protein quality feedstock CP 200 g kg^{-1} DM; CP recovery rate in juice 50% (Mandl et al., 2006); lower value—AA recovery rate 70% from juice, higher-value AA recovery rate 90% from juice. (a2) Yields based on the following: medium protein quality feedstock CP 130 g kg^{-1} DM; CP recovery rate in juice 50% (Mandl et al., 2006); lower value—AA recovery rate 70% from juice, higher-value AA recovery rate 90% from juice.

(b) Strongly dependent of application sector (feed, food, nutrition additive, and body care).

(c) Food grade quality (minimum LA content 80%): (c1) yields based on the following: high-quality silage feedstock LA 100 g kg^{-1} DM; LA recovery rate in juice 70%, first pressing (Mandl et al., 2006); lower value—LA recovery rate 60% from juice, higher-value LA recovery rate 90% from juice. (c2) Yields based on the following: high medium silage feedstock LA 45 g kg^{-1} DM; LA recovery rate in juice 70%, first pressing (Mandl et al., 2006); lower value—LA recovery rate 60% from juice, higher-value LA recovery rate 90% from juice.

(d) Value dependent on incentive or national pricing system in Europe.

(e) Unpublished data (Mandl, M.G.; Berghold H).

(f) Protein concentrate, CP content 40% comparable to soya bean meal.
(g) Calculated yield based on the following: 150–225 g kg^{-1} DM CP in feed; transfer rate in to liquid 55%, recovery rate from the juice 50%.
(h) Optimized scenario based on the following: 150–225 g kg^{-1} DM CP in feed; transfer rate in to liquid 70%, recovery rate from the juice 80%.
(i) Protein market value is assessed in relation to soya protein price; in general quite volatile market value.
(j) Food grade protein with functional properties 150–225 g kg^{-1} DM CP in feed; transfer rate in to liquid 55%, food grade proportion 25%.
(k) Calculated yield.
(l) Estimate.
(m) Yield is dependent on specifications of fine/coarseness.
(n) Market value of grass fibers for pulping is strongly dependent on functional properties: lower value for *filler*, higher-value fibers comparable to recycled paper quality.
(o) Source Stefan Grass, CEO Gramitherm, Switzerland.
(p) Yield calculated on the following: pressing: 1000 kg fresh grass (22% DM) deliver 450 kg juice at 5% DM; → 1 t DM into press delivers 2045 L juice; specific ethanol yield 28.2–36.8 mL per mL juice (Martel et al., 2010).
(q) Wholesale market value under EU conditions.
(r) Yield based on a fructan content of 45–75 g kg^{-1} DM in fresh juice as fermentable sugar.
(s) Estimate.
(t) Yield based on the following: cellulose content 230–430 g kg^{-1} DM; efficiency of pretreatment and enzymatic hydrolysis, 72% of theoretical C6 sugar yield; conversion rate of C6 sugars to ethanol in fermentation at 0.51 g g^{-1} (source EUBIO); DSP efficiency 95% for ethanol recovery.
(u) Yield (cellulose and hemicelluloses) based on the following: 650–750 g kg^{-1} DM; efficiency of pretreatment and enzymatic hydrolysis, 72% of theoretical C6, C5 sugar yield; conversion rate of LA fermentation, 90%; recover rate of total downstream processing, 70%.
(v) Dependent on quality: lower value, food grade; higher value, quality for PLA production. Dependent on maturity and type of feedstock.
(x) Market value stated represents use as solid fuel for combustion.
(y) Market value stated represents use as solid fuel for combustion; caloric value at 21 MJ kg^{-1} (approx 1.2 times of wood pellets); market value stated based on wood pellets multiplied factor 1.2; benchmark for wood pellets is 180–215 € t^{-1}.
(z) Estimated yield based on the following: cellulose content 230–430 g kg^{-1} DM; efficiency of pretreatment and enzymatic hydrolysis, 72% of theoretical C6 sugar yield; fermentation, C6 sugar conversion rate to succinic acid, 1,1 g succinic acid per g C6 sugar (M Jansen 2012); DSP efficiency 85% for product recovery.
(zl) Yield based on the following: cellulose content 230–430 g kg^{-1} DM; lower value/higher value, efficiency on NFC recovery 75%/95%.
(z2) Estimate.
(4) Revenue potential range derived from specific yields (2) times market values.

25.7.2 Environmental Impact

More efficient use of grassland, including low-input set-aside land, can significantly minimize the environmental impact of green biorefining (Smit et al., 2008; Thomas et al., 2009). For example, Tilman et al. (2006) have proven that carbon-negative biofuels can be produced from low-input high-diversity grassland biomass. Tscharntke et al. (2012) reviewed previous studies including groundbreaking reports (Phalan et al. 2011a,b) on global food security, biodiversity, and the intensification of agriculture. They concluded that the linking of agricultural intensification with biodiversity conservation and hunger reduction requires coordinated regional and targeted solutions, which the current debate on food versus fuel has failed to consider.

Some initiatives support the development of plant biorefinery, while other regulations pose barriers to the introduction of bioproducts. Current policies do not take advantage of key factors, including the stimulus provided by renewable energy targets and the synergies with existing biomass-based industries (e.g., fuel pellets, wood chip, and paper), or the evaluation of feedstocks necessary to support biorefinery initiatives. Current European policy, such as the biomass action plan (EU, 2005), encourages the use of biomass for production of energy and biofuels but does not address the need to integrate both energy and high-/low-value bioproducts (i.e., NFC, fibers, lactic and AAs, protein, and insulation boards) from ryegrass silage. The major deficiency appears to be the lack of regular communication between scientists and government policy groups on the rapid advances being made in the field of biotechnology and material research. Consequently, national legislations do not take into account the favorable characteristics of biobased materials and therefore create implementation problems for recycling the bioproducts (i.e., composites). Wiloso et al. (2012) reviewed LCA studies carried out in the past 6 years and identified key issues necessary for resolving LCA practice of first- and second-generation bioethanol production with lignocellulosic biomass. They concluded that a classification of input biomass to define boundaries is required for comparison of LCA results, based on system components for attributing environmental burden of different biomass feedstocks. Furthermore, the second-generation bioethanol performs better than fossil fuel at least for the two impact categories, that is, net energy output and global warming (Wiloso et al., 2012).

25.7.3 Strategic Plan

The implementation of green biorefinery will require advances in the following areas: the availability of better biomass feedstock, the preprocessing or compaction of biomass for more efficient transport, and the more efficient processing of green/lignocellulosic biomass (Werpy and Petersen, 2004; Luguel, 2011a,b). Anaerobic fermentation of biomass fractions should provide novel bioconversion processes for intermediate bulk chemicals. The focus of such a system should be on the utilization of substrates that are more reduced or oxidized than sugars, the use of cosubstrates, and the production of mixtures of high added-value products (Weusthuis et al., 2011). An efficient utilization of grassland will be needed to supply demand for both animal feed and biomass for green biorefining. Since significant expansion of the land given over to crops for nonfood products or applications is unlikely, the increased demand for green biomass will have to be met by increasing productivity of ryegrass or the use of set-aside land (Mandl, 2010). In addition, the full value chain needs to be considered from farm to end use application (Sharma et al., 2011). The initiatives will need to focus on resource efficiency using a cost competitive intermediate chemical from biorefining crops, such as ryegrass and lucerne.

Recent reports (Lugel, 2011a,b) have estimated that intermediate feedstocks derived from biorefining will make up 30% of the chemical sector value by 2030. To achieve this target, the manufacturing sector requires a coordinated strategy from industry, policymakers, and research groups (Koops et al., 2010; Scott et al., 2010; Weusthuis et al., 2011) to develop the following:

Technical perspective

- To develop environment-friendly technologies able to deal with a range of intermediate feedstocks, and enhancing the efficiency of conversion to high-value products
- To reduce energy required to process biomass and also to reduce environmental footprint of high- and low-value bioproducts
- To reduce costs of processing by integrating process steps to minimize use of energy, water, and raw material
- To exploit available intermediate products from genetically modified crops expressing useful intermediate feedstock chemicals (i.e., lysine and itaconic acid)
- To develop novel anaerobic fermentation processes using low-cost sugars from hydrolyzed lignocelluloses for the production of intermediate feedstock chemicals

Cooperation perspective

- To establish research momentum by forming alliances of national research groups
- To initiate regular communication between researchers and industrial end user groups and provide feedback at an early stage to speed up innovation
- Underpin collaboration with an exchange of ideas and results
- Create incentives for innovative companies to exploit technology

25.7.4 Future Prospects

Agriculture will be core to the biobased economy, providing raw materials for commodity items such as AAs, lipids, organic acids, and vitamins (Brown, 2007; USDA, 2008; Demirbas, 2009; OECD, 2009). Biobased products are already widely used in the pharmaceutical, food and feed, detergent, and cosmetic sectors. In addition, intermediate platform chemicals, such as 1,3-propandiol and fumaric, lactic, malic, succinic, and itaconic acids, are used in the manufacture of polymers or as food acidulants (Sanders et al., 2007; USDA, 2008; Koops et al., 2010; Scott et al., 2010). The Organisation of Economic Co-operation and Development (OECD) report (2009) on bioeconomy has highlighted many uses of traditional materials, such as wood and cotton, and emphasized the potential of new biobased products from packaging, textile, electronics, and car component sectors. While there are some specialized niche materials, the most important biomaterials to date are biocomposites and bioplastics manufactured from natural fibers/NFC and LA. Although these sectors are in their infancy, the industry has been growing rapidly driven by new synergies, collaborations, and research in Europe, China, and North America. The biobased resource is defined as commercially important monomers and intermediates, as feedstock for established polymer manufacturing processes. Global production capacity for biobased polymers was estimated to be 360 million tonnes in 2007, with an annual market growth rate of 48% in Europe and 38% globally (USDA, 2008; Luguel, 2011a). Recent developments in processes combining biotechnology and chemical synthesis are leading to the emergence of materials, such as ethylene derived from bioethanol, but their higher production cost currently limits commercial viability. Green biorefining technologies cannot be developed on an industrial scale in isolation, but need to be established as part of a value chain to integrate process schemes from feedstock to end products (Luguel 2011a,b). However, the scale of investment needed and the perceived risks including innovative technology, feedstock quality, product value, and legislation are major obstacles for large-scale commercialization. Many key chemical intermediates, such as lactic and succinic acid, can be produced from plant biomass by employing a combination of chemical and biological process steps. The intermediate feedstock can be used in a range of downstream processes, such as the production of green solvent (e.g., ethyl lactate) and polylactic acid. During the next decade, the chemical sector will use a growing proportion of renewable materials derived from green/lignocellulosic resources (Table 25.5).

TABLE 25.5
Projected Value (US $ bn) of World Chemical Production 2010 and 2025

	2010			2025		
Products	Total Value	Biobased Value	Biobased Share	Total Value	Biobased Value	Biobased Share
Commodity	550	5–11	0.9–2.0	875	50–86	50–86
Speciality	435	87–110	20–25.3	679	300–340	44.2–40.1
Fine	125	25–32	20–25.6	195	88–98	45.1–50.3
Polymer	290	15–30	5.2–10.3	452	45–90	10.0–19.9
All chemicals	1400	132–183	9.4–13.1	2183	483–614	22.1–28.1

Source: USDA, U.S. biobased products market potential and projections through 2025, February 2008, http://www.usda.gov/oce/reports/energy/BiobasedReport2008.pdf (accessed on July 12, 2012).

Recent reports on future markets of feedstock chemicals derived from bioproducts have suggested that the sector will reach 22%–28% in 2025, increasing subsequently to 30% by 2030 (USDA, 2008; OECD, 2009). The reports (USDA, 2008; Luguel, 2011b) predicted that most high-performance polymers will be manufactured from biomass and the polymers will be competitive with the petrochemical route especially at the high-value applications. Currently, most thermopolymer/thermoset products are made from basic monomers, such as ethylene, due to their availability at a low unit cost. The polymer industry is not likely to utilize feedstock chemicals from biomass for a long time due to high unit cost of production. Therefore, it appears that the potential evolution in the polymer sector will have a smaller impact on mass market products and commodities manufactured using polyolefins, such as polyethylene and polypropylene. Feedstock chemicals for the polymer sector may be largely fossil fuel derived or be imported from countries where cost of biomass is low (Luguel, 2011a,b). The sugar used as a raw material for biotechnology processes will be derived from the hydrolysis of biomass. In the lignocellulosic-bioethanol biorefining system, the resulting lignin by-product will form the basis of process streams for commercial production of biopolymers and aromatic compounds. A number of new biocatalysts and advanced fermentation processes will be developed to support existing chemical processing plants, and these will need to be more efficient, less expensive, and more versatile than comparable chemical catalysts (Luguel, 2011a,b). Many processes will rely on specialty enzymes tailored to specific production processes and environmental conditions. The biobased chemical processing technology will become an economic reality in the next decade. Processes using LA, carbohydrates, proteins, and vegetable oils will have become more efficient and will enjoy larger market share. A significant switchover to biobased products may occur first in specialty products and niche markets, where the returns are higher, but as technology evolves and production can be reduced, the sector may expand into bulk markets (USDA, 2008).

New developments in green biorefining should focus on maximizing the unique properties of plant biomass and should underpin a future European biobased economy. A combination of sustained implementation of EU directives with public and private cofunding and a development of the necessary science and technology will be needed, to exploit the full potential benefit within the next 5–10 years. Currently, with the exception of the pharmaceutical and some specialized chemical sectors, plant biomass is not routinely used as a chemical feedstock. Bioenergy (i.e., bioethanol, biodiesel, and biocompressed natural gas) products are the most realistic options for reducing the fossil fuel component in the energy sector, and this sector will form the core business in many future green biorefinery developments.

ACKNOWLEDGMENTS

We wish to acknowledge the Strategic Investment Fund and the Chancellor's Innovation Fund for financial support and the Department of Agriculture and Rural Development for supporting the research program. The research leading to these results has received funding from the European Community's Seventh Framework Programme (FP7/2007-2013) under grant agreement n° 312117 (BIOFECTOR).

REFERENCES

Aden, A., Ruth, M., Ibsen, K., Neeves, J., Sheehan, J., Wallace, J., Montague, L., Slayton, A., and Lukas, J. (2002). Lignocellulose biomass to ethanol process design and economics utilizing concurrent dilute acid prehydrolysis and enzymatic hydrolysis of corn stover. NREL.2002:TP-510-32438. Colarado. Accessed on 9 October, 2012.

Agbor, V.B., Cicek, N., Sparling, R., Berlin, A., and Levin, D.B. (2011). Biomass pretreatment: Fundamentals toward application. *Biotechnology Advances* 29, 675–685.

Ajanovic, A. (2011). Biofuels versus food production: Does biofuels production increase food prices? *Energy* 36, 2070–2076.

Alper, H. and Stephanopoulos, G. (2009). Engineering for biofuels: Exploiting innate microbial capacity or importing biosynthetic potential? *Nature Review of Microbiology* 7, 715–723.

Alvira, P., Tomas-Pejo, E., Ballesteros, M., and Negro, M.J. (2010). Pre-treatment technologies for an efficient bioethanol production process based on enzymatic hydrolysis: a review. *Bioresource Technology* 101, 4851–4861.

Andersen, M. and Kiel, P. (1999). Method for treating organic waste materials. European Patent Application WO 00/56912.

Andersen, M. and Kiel, P. (2000). Integrated utilization of green biomass in the green biorefinery. *Industrial Crops and Products* 11, 129–137.

Andrighetto, I., Cozzi, G., Magni, G., Hartman, B., Hinds, M., and Sapienza, D. (1995). Comparison of in situ degradation kinetics of Lucerne germplasm by Anova of non-linear models. *Animal Feed Science and Technology* 54, 287–299.

Angus, A., Burgess, P.J., Morris, J., and Lingard, J. (2009). Agriculture and land use: demand for and supply of agricultural commodities, characteristics of the farming and food industries, and implications for land use in the UK. *Land Use Policy* 26, 230–242.

Anon. (2002). Roadmap for biomass technologies in the United States. Biomass Research and Development Technical Advisory Committee, United States Department of Energy and United States Department of Agriculture. USDE, Colarado.

Arlabossea, P., Blanca, M., Kerfaïc, S., and Fernandez, A. (2011). Production of green juice with an intensive thermo-mechanical fractionation process. Part I: Effects of processing conditions on the dewatering kinetics. *Chemical Engineering Journal* 168, 586–592.

Bechthold, I., Bretz, K., Kabasci, S., Kopitzky, R., and Springer, A. (2008). Succinic acid: a new platform chemical for biobased polymers from renewable resources. *Chemical Engineering Technology* 31, 647–654.

Brehmer, B., Boom, R.M., and Sanders, J. (2009). Maximum fossil fuel feedstock replacement potential of petrochemicals via biorefineries. *Chemical Engineering Research and Design* 87, 1103–1119.

Brown, R.C. (2007). The future of biorefining for fuels and chemicals. In: *Production Issues in New Crops and New Uses*, eds. Janick, J. and Whipkey, A., ASHS Press, Alexandria, VA.

Carlsson, R. and Hanczakowski, P. (1989). Waste green parts of plants as raw material for leaf protein concentrate production. *Biological Waste* 28, 83–90.

Chandra, R.P., Bura, R., Mabee, W.E., Berlin, A., Pan, X., and Saddler, J.N. (2007). Substrate pretreatment: The key to effective enzymatic hydrolysis of lignocellulosics? *Advances in Biochemical Engineering and Biotechnology* 108, 67–93.

Chiesa, S. and Gnansounou, E. (2011). Protein extraction from biomass in a bioethanol refinery—Possible dietary applications: Use as animal feed and potential extension to human consumption. *Bioresource Technology* 102, 427–436.

Clark, J.H. and Deswarte, F.E.I. (2008). The biorefinery concept—An integrated approach. In: *Introduction to Chemicals from Biomass*, eds. Clark, J. and Deswarte, F., John Wiley & Sons Ltd., Chichester, U.K.

Cruz-Ramos, H., Hoffmann, T., Marino, M., Nedjari, H., Presecan-Siedel, E., Dreesen, O., Glaser, P., and Jahn, D. (2000). Fermentative metabolism of *Bacillus subtilis*: Physiology and regulation of gene expression. *Journal of Bacteriology* 182, 3072–3080.

Danner, H., Madzingaidzo, L., Holzer, M., Mayrhuber, L., and Braun, R. (2000). Extraction and purification of lactic acid from silage. *Bioresource Technology* 75, 181–187.

Delgenes, J.P., Penaud, V., and Moletta R. (2002). Pretreatment for the enhancement of anaerobic digestion of solid water chapter 8. In *Biomethanization of the Organic Fraction of Municipal Solid Waste*. IWA Publishing, London, U.K., pp. 201–228.

Demirbas, A. (2009). Biorefineries: Current activities and future developments. *Energy Conversion Management* 50, 2782–2801.

DT. (2005). Feasibility study on certification for a Renewable Transport Fuel Obligation. http://www.dft.gov.uk/pgr/roads/environment/rtfo/secrtfoprogdocs/annex2feasibilitystudy.pdf (accessed online June 12, 2012).

Ecker, J., Raabb, T., and Haraseka, M. (2012a). Nanofiltration as key technology for the separation of LA and AA. *Journal of Membrane Science* 389, 389–398.

Ecker, J., Schaffenberger, M., Koschuh, W., Mandl, M., Böchzelt, H.G., Schnitzer, H., Harasek, M., and Steinmüller, H. (2012b). Green biorefinery upper Austria—Pilot plant operation. *Separation and Purification Technology* 96, 237–247.

EU. (2005). Communication from the Commission of 7 December 2005—Biomass Action Plan COM(2005) 628final—Official Journal C 49 of February 28, 2005.

Evangelista, R.L., Mangold, A.J., and Nikolov, Z.L. (1994). Recovery of lactic acid by sorption, resin evaluation. *Applied Biochemistry Biotechnology* 45/46, 131–145.

FAO. (2011). FAO Statistics. http://faostat.fao.org/site/339/default.aspx. Accessed on September 12, 2012.

FAO. (2012). *Food and Agricultural Organisation Statistical Yearbook 2012*. FAO, Rome, Italy, pp. 369.

Fengel, D. and Wegener, G. (1984). *Wood Chemistry, Ultrastructure, Reactions*. De Gruyter, Berlin, New York.

Galbe, M. and Zacchi, G. (2012). Pre-treatment: The key to efficient utilization of lignocellulosic materials. *Biomass and Bioenergy* 70, 70–78.

Gass, M. (2011). Plastic material having increased fire resistance, method for producing same and plastic product made of such a plastic material. Publication No. WO/2011/047804.

Girio, F.M., Fonseca, C., Varvalheiro, F., Duarte, L.C., Marques, S., and Bogel-Lukasik, R., (2010). Hemicellulose for fuel ethanol: A review. *Bioresource Technology* 101, 4775–4800.

Grass, S. (September 11, 2004). *Utilisation of Grass for Production of Fibres, Protein and Energy*. OECD Publication Service, Paris, France.

Hatakka, A. (1994). Lignin-modifying enzymes from selected white-rot fungi: Production and role from in lignin degradation. *FEMS Microbiol Review* 13, 125–135.

Hendricks, A.T. and Zeeman, G. (2009). Pretreatments to enhance the digestibility of lignocellulosic biomass. *Bioresource Technology* 100, 10–18.

Hulst, A.C., Ketelaars, J.J., and Sanders, J.P. (2000). Separating and recovering components from plants. Patent WO 00/40788; PCT/NL99/00805.

IBERS. (2009). *Ryegrass: A New Option for Biofuel*, Vol. 60. Institute of Biological, Environmental and Rural Sciences, Aberystwyth, Wales.

Ingram, T., Wörmeyer, K., Lima, J.C.I., Bockemühl, V., Antranikian, G., Brunner, G., and Smirnova, I. (2011). Comparison of different pretreatment methods for lignocellulosic materials. Part I: Conversion of rye straw to valuable products. *Bioresource Technology* 102, 5221–5228.

Itoh, H., Wada, M., Honda, Y., and Kuwahara, M. (2003). Bioorganosolve pretreatments for simultaneous saccharification and fermentation of beech wood by ethanolysis and white-rot fungi. *Journal of Biotechnology* 103, 273–280.

Kamm, B., Kamm, M., Gruber, P., and Kromus, S. (2006). Biorefinery systems—An overview. In: *Biorefineries—Industrial Processes and Products. Status Quo and Future Directions*, Vol. 1., eds. Kamm, B., Kamm, M., Gruber, P., Wiley-VCH, Weinheim, Germany.

King, C., McEniry, J., O'Kiely, P., and Richardson, M. (2012). The effects of hydrothermal conditioning, detergent and mechanical pressing on the isolation of the fibre-rich press-cake fraction from a range of grass silages. *Biomass and Bioenergy* 42, 179–188.

Kleijn, D., Kohler, F., Baldi, A., Batary, P., Concepcione, E.D., Clough, Y., Díaz, M. et al. (2009). On the relationship between farmland biodiversity and land-use intensity in Europe. *Proceedings of the Royal Society B* 276, 903–909.

Koller, M., Bona, R., Hermann, C., Horvat, P., Martinz, J., Neto, J., Pereira, L., Varila, P., and Braunegg, G. (2005). Biotechnological production of poly (3-hydroxybutyrate) with Wautersia eutropha by application of green grass juice and silage juice as additional complex substrates. *Biocatalysis and Biotransformation* 23, 329–337.

Koops, A.J., Brumbley, S., Poirier, Y., Laat, A. de, Scott, E.L., Sanders, J.P.M., and van der Meer, I.M. (2010). Plant production of chemical building blocks. In: *The Biobased Economy: Biofuels, Materials and Chemicals in the Post-Oil Era*, eds. Langeveld, J.W.A. and Sanders, J.P.M., Earthscan, London, U.K..

Kromus, S., Wachter, B., Koschuh, W., Mandl, M., Krotscheck, C., and Narodoslawsky, M. (2004). The green biorefinery Austria—Development of an integrated system for green biomass utilization. *Chemical Biochemical Engineering Quarterly* 18, 7–12.

Lee, S.Y. and Chang, H.N. (1994). Effect of complex nitrogen source on the synthesis and accumulation of poly(3-hydroxybutyric acid) by recombinant *Escherichia coli* in flask and fed-batch cultures. *Journal of Environmental Polymer Degradation* 2, 169–176.

Luguel, C. (2011a). *European Biorefinery Joint Strategic Research Roadmap of 2020*. Star-Colibiri, EU Publications, Brussels.

Luguel, C. (2011b). *Joint Biorefinery Vision for 2030, Star-Colibiri, Strategic Targets for 2020—Collaboration Initiative on Biorefineries*. EU Publications, Brussels.

Luo, L., van der Voet, E., and Huppes, G. (2010). Biorefining of lignocellulosic feedstock—Technical, economic and environmental considerations. *Bioresource Technology* 101, 5023–5032.

Lyko, H., Deerberg, G., and Weidner, E. (2009). Coupled production in biorefineries—Combined use of biomass as a source of energy, fuels and materials. *Journal of Biotechnology* 142, 78–86.

Mahmoud A., Arlabosse, P., and Fernandez, A. (2011). Application of a thermally assisted mechanical dewatering process to biomass. *Biomass and Bioenergy* 35, 288–297.

Mandl, M. (2010). Status of green biorefinery in Europe. *Biofuels Bioproducts and Biorefining* 4, 268–274.

Mandl, M., Graf, N., Thaller, A., Böchezelt, H., and Schnitzer, H. (2006). Grüne Bioraffinerie- Aufbereitung und Verwertung der Grasfaserfraktion. Berichte aus Energie- und Umweltforschung 67/2006; accessible at: www.fabrikderzukunft.at/nw_pdf/0667_grasfaserfraktion2.pdf. Accessed on September 12, 2012.

Martel, C.M., Parker, J.E., Jackson, C.J., Warrilow, A.G.S., Rolley, N., Greig, C., Morris, S.M., Donnison, I.S., Kelly, D.E., and Kelly, S.L. (2011). Expression of bacterial levanase in yeast enables simultaneous saccharification and fermentation of grass juice to bioethanol. *Bioresource Technology* 102, 1503–1508.

Martel, C.M., Warrilow, A.G.S., Jackson, C.J., Mullins, J.G.L., Togawa, R.C., Parker, J.E., Morris, M.S., Donnison, I.S., Kelly, D.E., and Kelly, S.L. (2010). Expression, purification and use of the soluble domain of Lactobacillus paracasei b fructosidase to optimise production of bioethanol from grass fructans. *Bioresource Technology* 101, 4395–4402.

Martín, C., Thomsen, M.H., Hauggaard, H., and Thomsem, A.B. (2008). Wet oxidation pretreatment, enzymatic hydrolysis and simultaneous saccharification and fermentation of clover–ryegrass mixtures. *Bioresource Technology* 99, 8777–8782.

McMillan, J.D. (1994). Pretreatment of lignocellulosic Biomass. In: *Enzymatic Conversion of Biomass for Fuels Production*, eds. Himmel, M.E., Baker, J.O., and Overend, R.P., American Chemical Society, Washington, DC, symposium series 566, pp. 292–324.

Mercer-Blackman, V., Samiei, H., and Cheng, K. (2008). Biofuel demand push up food prices. *International Monetary Fund Survey Magazine: IMF Research*, October 17, 2008.

Mitchell, D. (2008). A note on rising food prices. Policy Research Working Paper 4682, The World Bank, Washington, DC.

Mora, V., Bacaicoa, E., Zamarreno E-M., Aguirre, E., Garnica, M., Fuentes, M., and Garcia-Mina, J.M. (2010). Action of humic acid on promotion of cucumber shoot growth involves nitrate related changes associated with the root-to-shoot distribution of cytokinins, polyamines and mineral nutrients. *Journal of Plant Physiology* 167, 633–642.

Mosier, N., Wyman, C.E., Dale, B.E., Elander, R, Lee Y.Y., Holtzapple, M.T., and Ladisch, M. (2005). Features of promising technologies for pretreatment of lignocellulosic biomass. *Bioresource Technology* 96, 673–686.

Moulthrop, J.S., Swatloski, R.P., Moyna, G., and Rogers R.D. (2005). High resolution 13C-NMR studies of cellulose and cellulose oligomers in ionic liquid solutions. *Chemical Communication* 12, 1557–1559.

Murphy, J.D. and Power, N.M. (2009). An argument for using biomethane generated from grass as a biofuel in Ireland. *Biomass and Bioenergy* 33, 504–512.

Neureiter, M., Danner, H., Fruhauf, S., Kromus, S., Thomasser, C., Braun, R., and Narodoslawsky, M. (2004). Dilute acid hydrolysis of press cakes from silage and grass to recover hemicellulose-derived sugars. *Bioresource Technology* 92, 21–29.

Nguyen, Q.A., Tucker, M.P., Keller, F.A., and Eddy F.P. (2000). Two-stage dilute-acid pretreatment of softwoods. *Applied Biochemical and Biotechnology* 84–86, 561–575.

O'Keeffe, S., Schulte, R.P.O., Sanders, J.P.M., and Struik, P.C. (2012). I. Technical assessment for first generation green biorefinery (GBR) using mass and energy balances: Scenarios for an Irish GBR blueprint. *Biomass and Bioenergy* 35, 4712–4723.

OECD. (2009). OECD report The Bioeconomy to 2030: designing a policy agenda. http://www.oecd.org/document/48/0,3343,en_2649_36831301_42864368_1_1_1_1,00.html. Accessed on September 12, 2012.

Okino, S., Noburyu, R., Suda, M., Jojima, T., Inui, M., and Yukawa, H. (2008). An efficient succinic acid production process in a metabolically engineered *Corynebacterium glutamicum* strain. *Applied Microbiology and Biotechnology* 81, 459–464.

Oleskowicz-Popiel, P., Thomsen, A.B., and Schmidt, J.E. (2011). Ensiling e Wet-storage method for lignocellulosic biomass for bioethanol production. *Biomass and Bioenergy* 35, 2087–2092.

Pakarinen, O.M., Tahti, H.P., and Rintala, J.A. (2009). One-stage H_2 and CH^4 and two-stage H_2+CH_4 production from grass silage and from solid and liquid fractions of NaOH pre-treated grass silage. *Biomass and Bioenergy* 33, 1419–1427.

Paradikovic, N., Vinkovic, T., Vrcek, I.V., Zuntar, I., Bojic, M., and Medic-Saric, M. (2011). Effect of natural biostimulants on yield and nutritional quality: An example of sweet yellow pepper (*Capsicum annuum* L.) plants. *Journal of the Science of Food and Agriculture* 91, 146–2152.

Perbandt, D., Reulein J., Richter, F., Stulpnagel, R., and Wachendorf, M. (2010). Assessment of mass flows and fuel quality during mechanical dehydration of silages using near infrared reflectance spectroscopy. *Bioenergy and Research* 3, 194–203.

Phalan, B., Balmford, A., Green, R.E., and Scharlemann, J.P.W. (2011a). Minimising harm to biodiversity of producing more food globally. *Food Policy* 36, 62–71.

Phalan, B., Onial, M., Balmford, A., and Green, R.E. (2011b). Reconciling food production and biodiversity conservation: land sharing and land sparing compared. *Science* 333, 1289–1291.

Pirie, N.W. (1942). Green leaves as a source of proteins and other nutrients. *Nature* 149, 251.

Prasad, S., Singh, A., and Joshi, H.C. (2007). Ethanol as an alternative fuel from agricultural, industrial and urban residues. *Resources, Conservation and Recycling* 50, 1–39.

Prochnow, A., Heiermann, M., Plöchl, M., Linke, B., Idler, C., Amon, T., and Hobbs, P.J. (2009). Bioenergy from permanent grassland—A review: 1. Biogas. *Bioresource Technology* 100, 4931–4944.

Realff, M.J. and Abbas, C. (2004). Industrial symbiosis—Refining the biorefinery. *Journal of Industrial Ecology* 7, 5–9.

Richardson, A.D., Aikens, M., Berlyn, G.P., and Marsh, P. (2004). Drought stress and paper birch (*Betula papyrifera*) seedlings: Effects of an organic bio-stimulant on plant health and stress tolerance, and detection of stress effects with instrument-based, non-invasive methods. *Arboricultural Journal* 30, 52–60.

Saddler, J.N., Ramos, L.P., and Breuil, C. (1993). Steam pretreatment of lignocellulosic residues. In: *Bioconversion of Forest and Agricultural Plant Residues*, ed. Saddler, J.N., Oxford, U.K., CAB International, Chapter 3, pp. 73–92.

Sanders, J., Scott, E., Weusthuis, R., and Mooibroek, H. (2007). Bio-refinery as the bio-inspired process to bulk chemicals. *Macromolecular Bioscience* 7, 105–117.

Scott, E.L., Haveren, J. van, and Sanders, J.P.M., 2010. The production of chemicals in a biobased economy. In: *The Biobased Economy: Biofuels, Materials and Chemicals in the Post-Oil Era*, eds. Langeveld, J.W.A. and Sanders, J.P.M., Earthscan, London, U.K.

Sfilgoj-Smole, M., Kleinschek, K.S., Kreze, T., Strnad, S., Mandl, M., and Wachter, B. (2004). Physical properties of grass fibres. *Chemical and Biochemical Engineering Q.* 18, 47–53.

Sfilgoj-Smole, M., Kreze, T., Strnad, S., Stana Kleinschek, K., and Hribernik, S. (2005). Character*isation of grass* fibres. *Journal of Material Science* 40, 5349–5353.

Sharma, H.S.S., Carmichael, E., Muhamad, M., McCall, D., Andrews, F., Lyons, G., McRoberts, C., and Hornsby, P.R. (2012b). Biorefining of perennial ryegrass for the production of nanofibrillated cellulose. *RSC Advances* 2, 6424–6437.

Sharma, H.S.S., Fleming, C., Selby, C., Rao, J.R., and Martin, T. (2013). Plant biostimulants: A review on the processing of macroalgae and use of extracts for crop management to reduce abiotic and biotic stresses. *Journal of Applied Phycology.* Online published: August 14. DOI: 10.1007/s10811-013-0101-9.

Sharma, H.S.S., Lyons, G., and McRoberts, C. (2011). Biorefining of perennial grasses: A potential sustainable option for Northern Ireland grassland production. *Chemical Engineering Research and Design* 89, 2309–2321.

Sharma, H.S.S., McCall, D., Andrews, F., Carmichael, E., Lyons, G., and McRoberts, C. (2012a). Biorefining of perennial grasses: Development of visible and near infrared tools for assessing fineness of extracted fibre. *Industrial Crops and Products* 36, 476–484.

Sharma, H.S.S., Mellon, R., Givens, D.I., Park, R.S., Archer, J., Lyons, G., and Fletcher, H. (2012c). Evaluation of perennial ryegrasses (*Lolium perenne* L.) for dry organic matter digestibility using thermogravimetry. *Journal of Animal Feed Science and Technology* 177, 30–39.

Smit, H.J., Metzger, M.J., and Ewert, F. (2008). Spatial distribution of grassland productivity and land use in Europe. *Agricultural System* 98, 208–219.

Song, H. and Lee, S.Y. (2006). Production of succinic acid by bacterial fermentation. *Enzyme Microbial Technology* 29, 352–361

Sorda, G., Banse, M., and Kemfert, C. (2010). An overview of biofuel policies across the world. *Energy Policy* 38, 6977–6888.

Sun, Y. and Cheng, J. (2002). Hydrolysis of lignocellulosic materials for ethanol production: A review. *Bioresource Technology* 83, 1–11.

Swatloski, R.P., Spear, S.K., Holbrey, J.D., and Rogers, R.D. (2002). Dissolution of cellulose with ionic liquids. *Journal of American Chemical Society* 124, 4974–4975.

Taherzadeh, M.J. and Karimi, K. (2008). Pretreatment of lignocellulosic wastes to improve ethanol and biogas production: A review. *International Journal of Molecular Science* 9, 1621–1651.

Talebnia, F., Karakashev, D., and Angelidaki I. (2010). Production of bioethanol from wheat straw: an overview on pretreatment, hydrolysis and fermentation. *Bioresource Technology* 101, 4744–4753.

Thang, V.H. and Novalin, S. (2008). Green biorefinery: Separation of lactic acid from grass silage juice by chromatography using neutral polymeric resin. *Bioresource Technology* 99, 4368–4379.

Thomas, V.M., Choi, D.G., Luo, D., Okwo, A., and Wang, J.H. (2009). Relation of biofuel to bioelectricity and agriculture: Food security, fuel security, and reducing greenhouse emissions. *Chemical Engineering Research and Design* 87, 1140–1146.

Thomsen, M.H. (2005). Complex media from processing of agricultural crops for microbial fermentation. *Applied Microbiology Biotechnology* 68, 598–606.

Thomsen, M.H., Bech, D., and Kiel, P. (2004). Manufacturing of stabilized brown juice for L-lysine production—From university lab scale over pilot scale to industrial production. *Chemical and Biochemical Engineering Quarterly* 18, 37–46.

Thring, R.W., Chornet, E., and Overend, R. (1990). Recovery of a solvolytic lignin: Effects of spent liquor/acid volume ratio, acid concentrated and temperature. *Biomass* 23, 289–305.

Tilman, D., Hill, J., and Lehman, C. (2006). Carbon negative biofuels from low-input high-diversity grassland biomass. *Science* 314, 1598–1600.

Torget, R.W., Himmel, M., and Grohmann K. (1992). Dilute acid pretreatment of two short-rotation herbaceous crops. *Applied Biochemistry and Biotechnology* 34/35, 115–123.

Tscharntke, T., Clough, Y., Wanger, T.C., Jackson, L., Motzke, I., Perfecto, I., Vandermeer, J., and Whitbread, A. (2012). Global food security, biodiversity conservation and the future of agricultural intensification. *Biological Conservation* 151, 53–59.

USDA. (2008). U.S. biobased products market potential and projections through 2025, February 2008. http://www.usda.gov/oce/reports/energy/BiobasedReport2008.pdf (accessed on July 12, 2012).

Venus, J. (2006). Processing and bioconversion of renewable resources into lactic acid. *BioTopics–Journal of Biotechnology* 28, 14–15.

Venus, J. (2009). Utilization of renewables for lactic acid fermentation. *Biotechnology Journal* 1, 1428–1432.

Venus, J. and Richter, K. (2006). Production of lactic acid from barley: Strain selection, phenotypic and medium optimization. *Engineer Life Science*, 6(5), 492–500.

Vodnar, D.C., Venus, J., Schneider, R., and Socaciu, C. (2010). Lactic acid production by *Lactobacillus paracasei* in discontinuous fermentation using lucerne green juice as nutrient substitute. *Chemical Engineering Technology* 33, 468–474.

Wachendorf, M., Richter, F., Fricke, T., Graß, R., and Neff, R. (2009). Utilisation of semi-natural grassland through an integrated generation of solid fuel and biogas from biomass I: Effects of hydrothermic conditioning and mechanical dehydration on mass flows of organic and mineral plant compounds, and nutrient balances. *Grass Forage Science* 64, 132–143.

Werpy, T. and Petersen, G. (2004) Top value added chemicals from biomass [U.S. Department of Energy, Office of scientific and technical information, No.: DOE/GO-102004-1992, www.osti.gov/bridge. Accessed on September 12, 2012.

Weusthuis, R.A., Lamot, I., van der Oost, J., and Sanders, J.P.M. (2011). Microbial production of bulk chemicals: Development of anaerobic processes. *Trends in Biotechnology* 29, 153–158.

Wiloso, E.I., Heijungs, R., and de Snoo, G.R. (2012). LCA of second generation bioethanol: A review and some issues to be resolved for good LCA practice. *Renewable and Sustainable Energy Reviews* 16, 5295–5308.

Zavrel, M., Bross, D., Funke, M., Buchs, J., and Spiess, A.C. (2009). High-throughput screening for ionic liquids dissolving ligno-cellulose. *Bioresource Technology* 100, 2580–2587.
Zeikus, J.G., Jain, M.K., and Elankovan, P. (1999). Biotechnology of succinic acid production and markets for derived industrial products. *Applied Microbiology Biotechnology* 51, 545–552.
Zhao, X., Cheng, K., and Liu, D.H. (2009). Organosolv pretreatment of lignocellulosic biomass for enzymatic hydrolysis. *Applied Microbiology Biotechnology* 82, 815–827.
Zhu, J.Y., Pan, X., Ronald, S., and Zalesny, J. (2010). Pretreatment of woody biomass for biofuel production: energy efficiency, technologies, and recalcitrance. *Applied Microbiology Biotechnology* 87, 847–857.
Zhu, J.Y., Wang, G.S., Pan, X.J., and Gleisner, R. (2009). Specific surface to evaluate the efficiencies of milling and pretreatment of wood for enzymatic saccharification. *Chemical Engineering Science* 64, 474–485.
Zhu, S.D., Wu, Y.X., Chen, Q.M., Yu, N., Wang, C.W., Jin SW, Ding, Y., and Wu, G. (2006). Dissolution of cellulose with ionic liquids and its application: A mini-review. *Green Chemistry* 8, 325–327.

Index

W